Trace Elements in Soils

Trace Elements in Soils

Edited by

PETER S. HOODA

*School of Geography, Geology and the Environment,
Kingston University London, UK*

A John Wiley and Sons, Ltd., Publication

This edition first published 2010
© 2010 Blackwell Publishing Ltd

Registered office
John Wiley & Sons Ltd, The Atrium, Southern Gate, Chichester, West Sussex, PO19 8SQ, United Kingdom

For details of our global editorial offices, for customer services and for information about how to apply for permission to reuse the copyright material in this book please see our website at www.wiley.com.

The right of the author to be identified as the author of this work has been asserted in accordance with the Copyright, Designs and Patents Act 1988.

All rights reserved. No part of this publication may be reproduced, stored in a retrieval system, or transmitted, in any form or by any means, electronic, mechanical, photocopying, recording or otherwise, except as permitted by the UK Copyright, Designs and Patents Act 1988, without the prior permission of the publisher.

Wiley also publishes its books in a variety of electronic formats. Some content that appears in print may not be available in electronic books.

Designations used by companies to distinguish their products are often claimed as trademarks. All brand names and product names used in this book are trade names, service marks, trademarks or registered trademarks of their respective owners. The publisher is not associated with any product or vendor mentioned in this book. This publication is designed to provide accurate and authoritative information in regard to the subject matter covered. It is sold on the understanding that the publisher is not engaged in rendering professional services. If professional advice or other expert assistance is required, the services of a competent professional should be sought.

The publisher and the author make no representations or warranties with respect to the accuracy or completeness of the contents of this work and specifically disclaim all warranties, including without limitation any implied warranties of fitness for a particular purpose. This work is sold with the understanding that the publisher is not engaged in rendering professional services. The advice and strategies contained herein may not be suitable for every situation. In view of ongoing research, equipment modifications, changes in governmental regulations, and the constant flow of information relating to the use of experimental reagents, equipment, and devices, the reader is urged to review and evaluate the information provided in the package insert or instructions for each chemical, piece of equipment, reagent, or device for, among other things, any changes in the instructions or indication of usage and for added warnings and precautions. The fact that an organization or Website is referred to in this work as a citation and/or a potential source of further information does not mean that the author or the publisher endorses the information the organization or Website may provide or recommendations it may make. Further, readers should be aware that Internet Websites listed in this work may have changed or disappeared between when this work was written and when it is read. No warranty may be created or extended by any promotional statements for this work. Neither the publisher nor the author shall be liable for any damages arising herefrom.

Copyright Acknowledgments
A number of chapters in *Trace Elements in Soils* have been written by a government employee in the United States of America. Please contact the publisher for information on the copyright status of such works, if required.
Works written by US government employees and classified as US Government Works are in the public domain in the United States of America.

Library of Congress Cataloging-in-Publication Data

Trace elements in soils / edited by Peter S. Hooda.
  p. cm.
  Includes bibliographical references and index.
  ISBN 978-1-4051-6037-7 (cloth)
  1. Soils—Trace element content. I. Hooda, Peter S.
  S592.6.T7T7267 2010
  631.4′1—dc22
                                    2010003983

A catalogue record for this book is available from the British Library.

ISBN: 978-1-405-16037-7 (H/B)

Set in 10/12pt Times by Integra Software Services Pvt. Ltd., Pondicherry, India
Printed and bound in the United Kingdom by CPI Antony Rowe, Chippenham, Wiltshire.

Contents

Preface xv
List of Contributors xvii

SECTION 1 BASIC PRINCIPLES, PROCESSES, SAMPLING AND ANALYTICAL ASPECTS 1

1 Introduction 3
Peter S. Hooda
References 7

2 Trace Elements: General Soil Chemistry, Principles and Processes 9
Filip M. G. Tack

    2.1 Introduction 9
    2.2 Distribution of Trace Elements in the Soil 10
    2.3 Chemical Species 11
    2.4 Sorption and Desorption 13
        2.4.1 Sorption Mechanisms 13
        2.4.2 Sorption Isotherms 16
    2.5 Precipitation and Dissolution 18
    2.6 Mobilization of Trace Elements 19
        2.6.1 pH and Redox Potential 19
        2.6.2 Influence of Soil Constituents 23
    2.7 Transport 25
    2.8 Plant Uptake 28
    2.9 Concluding Remarks 31
    References 32

3 Soil Sampling and Sample Preparation 39
Anthony C. Edwards

    3.1 Introduction 39
    3.2 Soil Sampling 40

	3.3	Errors Associated with Soil Sampling and Preparation	41
	3.4	Overview of the Current Situation	46
	3.5	Scale and Variability	48
	3.6	Conclusions	49
		References	49

4 Analysis and Fractionation of Trace Elements in Soils — 53
Gijs Du Laing

4.1	Introduction	53
4.2	Total Analysis	54
	4.2.1 Matrix Dissolution	54
	4.2.2 Instrumental Analysis Techniques	55
	4.2.3 Nondestructive Methods	59
4.3	Fractionation of Trace Elements	61
	4.3.1 Single Extractions	61
	4.3.2 Sequential Extraction Procedures	65
	4.3.3 Fractionation Based on Particle Size	71
4.4	Species-Retaining and Species-Selective Leaching Techniques	71
4.5	Equipment for Direct Speciation of Trace Elements in Soil	73
4.6	Conclusions	74
	References	74

5 Fractionation and Speciation of Trace Elements in Soil Solution — 81
Gijs Du Laing

5.1	Introduction	81
5.2	Soil Solution Sampling, Storage and Filtration	82
5.3	Particle Size Fractionation	83
5.4	Liquid–Liquid Extraction	86
5.5	Ion-Exchange Resins and Solid-Phase Extraction	86
5.6	Derivatization Techniques to Create Volatile Species	87
5.7	Chromatographic Separation of Trace Element Species	88
5.8	Capillary Electrophoresis	90
5.9	Diffusive Gradients in Thin Films	91
5.10	Ion-Selective Electrodes	93
5.11	Donnan Membrane Technique	94
5.12	Voltammetric Techniques	96
5.13	Microelectrodes and Microsensors	98
5.14	Models for Predicting Metal Speciation in Soil Solution	100
5.15	Conclusions	102
	References	103

SECTION 2 LONG-TERM ISSUES, IMPACTS AND PREDICTIVE MODELLING — 111

6 Trace Elements in Biosolids-Amended Soils — 113
*Weiping Chen, Andrew C. Chang, Laosheng Wu,
Albert L. Page and Bonjun Koo*

- 6.1 Introduction — 113
- 6.2 Biosolids-Borne Trace Elements in Soils — 115
  - 6.2.1 Land Application and Trace Element Loading — 115
  - 6.2.2 Trace Element Availability in Biosolids-Amended Soils – A Time Bomb? — 116
  - 6.2.3 Plant Response to Trace Elements in Biosolids-Amended Soils – Is There a Plateau? — 118
- 6.3 Assessing Availability of Trace Elements in Biosolids-Amended Soils — 120
  - 6.3.1 Source Assessment — 120
  - 6.3.2 End Measurement — 121
- 6.4 Long-Term Availability Pool Assessment through a Root Exudates-Based Model — 122
  - 6.4.1 Rationale for Root Exudate-Based Trace Element Phytoavailability — 122
  - 6.4.2 Case Studies — 125
- 6.5 Conclusions — 128
- References — 129

7 Fertilizer-Borne Trace Element Contaminants in Soils — 135
Samuel P. Stacey, Mike J. McLaughlin and Ganga M. Hettiarachchi

- 7.1 Introduction — 135
- 7.2 Phosphatic Fertilizers — 136
- 7.3 Micronutrient Fertilizers — 139
- 7.4 Long-Term Accumulation of Fertilizer-Borne Trace Element Contaminants — 139
- 7.5 Trace Element Contaminant Transfer to Crops and Grazing Animals — 141
  - 7.5.1 Arsenic — 142
  - 7.5.2 Cadmium — 143
  - 7.5.3 Fluorine — 145
  - 7.5.4 Lead — 145
  - 7.5.5 Uranium — 147
- 7.6 Conclusions — 147
- References — 148

8 Trace Metal Exposure and Effects on Soil-Dwelling Species and Their Communities — 155
David J. Spurgeon

- 8.1 Introduction — 155
- 8.2 Hazards and Consequences of Trace Metal Exposure — 156

		8.2.1	Effects on Individuals, Risk Assessment and the Prediction of Population Effects	156
		8.2.2	Populations and Communities	158
	8.3	Routes of Exposure, Uptake and Detoxification		162
		8.3.1	Uptake Routes and Speciation Models	162
		8.3.2	Toxicokinetics and Compartment Models	163
		8.3.3	Molecular Mechanisms of Detoxification and Effect	166
	8.4	Conclusions		167
	References			168

9 Trace Element-Deficient Soils — **175**
Rainer Schulin, Annette Johnson and Emmanuel Frossard

	9.1	Introduction		175
	9.2	The Concept of Trace Element-Deficient Soils		176
		9.2.1	The Role of Trace Elements as Essential Micronutrients	176
		9.2.2	Soil Trace Element Concentrations and Micronutrient Deficiencies	177
		9.2.3	Trace Element Deficiency as a Disturbance-Related Concept	178
	9.3	Methods to Identify and Map Soil Trace Element Deficiencies		179
		9.3.1	Detection and Diagnosis of Trace Element Deficiency	179
		9.3.2	Mapping of Trace Element Deficiencies	181
	9.4	Soil Factors Associated with Trace Element Deficiencies		182
		9.4.1	General Relationships between Soil Factors and Micronutrient Deficiencies	182
		9.4.2	Boron	183
		9.4.3	Cobalt	186
		9.4.4	Copper	187
		9.4.5	Iron	188
		9.4.6	Manganese	188
		9.4.7	Molybdenum	189
		9.4.8	Selenium	190
		9.4.9	Zinc	191
		9.4.10	Other Micronutrients	192
	9.5	Treatment of Soils Deficient in Trace Elements		192
	References			194

10 Application of Chemical Speciation Modelling to Studies on Toxic Element Behaviour in Soils — **199**
Les J. Evans, Sarah J. Barabash, David G. Lumsdon and Xueyuan Gu

	10.1	Introduction	199
	10.2	The Structure of Chemical Speciation Models	201
	10.3	The Species/Component Matrix	203

	10.4	Aqueous Speciation Modelling	204
		10.4.1 Calculating the Concentration of Soluble Species of Toxic Elements	205
	10.5	Modelling of Surface Complexation to Mineral Surfaces	207
		10.5.1 Proton and Toxic Element Binding to Oxide Minerals	207
		10.5.2 Proton and Toxic Element Binding to Clay Minerals	213
	10.6	Modelling of Surface Complexation to Soil Organic Matter	217
	10.7	Discussion	219
	References		222

SECTION 3   BIOAVAILABILITY, RISK ASSESSMENT AND REMEDIATION — 227

11 Assessing Bioavailability of Soil Trace Elements — 229
Peter S. Hooda

	11.1	Introduction	229
	11.2	Speciation, Bioavailability and Bioaccumulation: Definitions and Concepts	230
	11.3	Bioavailability Assessment Approaches	234
		11.3.1 Single Chemical Extraction Procedures	234
		11.3.2 Sequential Extraction Procedures	238
		11.3.3 Soil Solution Concentration and Speciation	240
		11.3.4 Other Approaches	244
		11.3.5 Bioaccessibility	249
		11.3.6 Bioassays, Biosensors and Bioavailability	251
	11.4	Discussion and Conclusions	253
	References		255

12 Bioavailability: Exposure, Dose and Risk Assessment — 267
Rupert L. Hough

	12.1	Introduction	267
		12.1.1 The 'Classical' Risk Assessment Model	268
	12.2	Hazard Identification	270
		12.2.1 Approaches, Uncertainties, Issues for Discussion	270
	12.3	Exposure Assessment	272
		12.3.1 Approaches to Estimating Exposure from Trace Elements in Soils	272
		12.3.2 Environmental Measurements and Influence of Bioavailability	276
	12.4	Dose–Response	280
		12.4.1 High- to Low-Dose Extrapolation	280
	12.5	Risk Characterization	284
	12.6	Assessment of Mixtures and Disparate Risks	287
	12.7	Conclusions	288
	References		288

13 Regulatory Limits for Trace Elements in Soils 293
Graham Merrington, Sohel Saikat and Albania Grosso

    13.1   Introduction 293
    13.2   Derivation of Regulatory Limits for Trace Elements 296
            13.2.1 Environmental Protection Limit Values for Soils 296
            13.2.2 Human Health Protection Limit Values for Soils 299
    13.3   National and International Initiatives in Setting Limit Values 301
    13.4   Forward Look 303
    13.5   Conclusions 304
    References 305

14 Phytoremediation of Soil Trace Elements 311
Rufus L. Chaney, C. Leigh Broadhurst and Tiziana Centofanti

    14.1   Introduction 311
    14.2   The Nature of Soil Contamination where Phytoextraction may be Applied 315
    14.3   Need for Metal-Tolerant Hyperaccumulators for Practical Phytoextraction 316
    14.4   Phytoremediation Strategies: Applications and Limitations 317
            14.4.1 Phytomining Soil Nickel 317
            14.4.2 Soil Cadmium Contamination Requiring Remediation to Protect Food Chains 321
            14.4.3 Phytoextraction or Phytovolatilization of Soil Selenium 325
            14.4.4 Phytoextraction of Soil Cobalt 327
            14.4.5 Phytoextraction of Soil Boron 327
            14.4.6 Phytovolatilization of Soil Mercury 328
            14.4.7 Induced Phytoextraction of Soil Gold 329
            14.4.8 Induced Phytoextraction of Soil Lead 329
            14.4.9 Phytoextraction of Soil Arsenic 331
            14.4.10 Phytoextraction of Other Soil Elements 333
    14.5   Phytostabilization of Zinc-Lead, Copper, or Nickel Mine Waste or Smelter-Contaminated Soils 334
    14.6   Recovery of Elements from Phytoextraction Biomass 336
    14.7   Risks to Wildlife during Phytoextraction Operations 336
    14.8   Conclusions 337
    References 339

15 Trace Element Immobilization in Soil Using Amendments 353
Jurate Kumpiene

    15.1   Introduction 353
    15.2   Soil Amendments for Trace Element Immobilization 354
            15.2.1 Metal Oxides 354
            15.2.2 Natural and Synthetic Aluminosilicates 358

		15.2.3	Ashes	361
		15.2.4	Phosphates	364
		15.2.5	Organic Amendments	365
		15.2.6	Liming Compounds	367
		15.2.7	Gypsum	368
	15.3	Method Acceptance		369
	15.4	Concluding Remarks		370
	References			371

SECTION 4 CHARACTERISTICS AND BEHAVIOUR OF INDIVIDUAL ELEMENTS — 381

16 Arsenic and Antimony — 383
Yuji Arai

	16.1	Introduction		383
	16.2	Geogenic Occurrence		385
		16.2.1	Arsenic	385
		16.2.2	Antimony	385
	16.3	Sources of Soil Contamination		386
	16.4	Chemical Behavior in Soils		387
		16.4.1	Arsenic Speciation and Solubility	387
		16.4.2	Arsenic Retention in Soils	388
		16.4.3	Arsenic Desorption in Soils	392
		16.4.4	Antimony Speciation and Solubility	393
		16.4.5	Antimony Adsorption and Desorption in Soils	394
	16.5	Risks from Arsenic and Antimony in Soils		396
	16.6	Conclusions and Future Research Needs		400
	References			400

17 Cadmium and Zinc — 409
Rufus L. Chaney

	17.1	Introduction	409
	17.2	Geogenic Occurrence and Sources of Soil Contamination	409
	17.3	Chemical Behavior in Soils	415
	17.4	Plant Accumulation of Soil Cadmium and Zinc	416
	17.5	Risk Implications for Cadmium in Soil Amendments	419
	17.6	Plant Uptake of Cadmium and Zinc in Relation to Food-Chain Cadmium Risk	422
	17.7	Food-Chain Zinc Issues	427
	References		429

18 Copper and Lead — 441
Rupert L. Hough

	18.1	Introduction	441
	18.2	Copper	443

	18.2.1	Sources and Content of Copper in Soils	443
	18.2.2	Chemical Behaviour in Soils	445
18.3	Lead		446
	18.3.1	Sources and Content of Lead in Soils	446
	18.3.2	Chemical Behaviour in Soils	448
18.4	Risks from Copper and Lead		449
	18.4.1	Essentiality and Metabolism	449
	18.4.2	Exposure and Toxicology	450
18.5	Concluding Remarks		452
References			453

19 Chromium, Nickel and Cobalt — 461
Yibing Ma and Peter S. Hooda

19.1	Introduction		461
19.2	Geogenic Occurrences		463
19.3	Sources of Soil Contamination		464
19.4	Chemical Behaviour in Soils		465
	19.4.1	Chromium	465
	19.4.2	Nickel	467
	19.4.3	Cobalt	468
19.5	Environmental and Human Health Risks		470
	19.5.1	Chromium	470
	19.5.2	Nickel	472
	19.5.3	Cobalt	474
19.6	Concluding Remarks		474
References			475

20 Manganese and Selenium — 481
Zhenli L. He, Jiali Shentu and Xiao E. Yang

20.1	Introduction		481
20.2	Concentrations and Sources of Manganese and Selenium in Soils		482
	20.2.1	Manganese	482
	20.2.2	Selenium	483
20.3	Chemical Behaviour of Manganese and Selenium in Soils		484
	20.3.1	Solution and Solid Forms	484
	20.3.2	Ion-Exchange and Sorption–Desorption Reactions	485
	20.3.3	Precipitation–Dissolution and Oxidation–Reduction Reactions	487
	20.3.4	Availability of Manganese and Selenium in Soils	489
20.4	Effects on Plant, Animal and Human Health		490
References			493

21	**Tin and Mercury**	**497**
	Martin J. Clifford, Gavin M. Hilson and Mark E. Hodson	
	21.1 Introduction	497
	21.2 Geogenic Occurrence	500
	21.2.1 Tin	500
	21.2.2 Mercury	501
	21.3 Sources of Soil Contamination	502
	21.3.1 Tin	502
	21.3.2 Mercury	503
	21.4 Chemical Behaviour in Soils	505
	21.4.1 Tin	505
	21.4.2 Mercury	506
	21.5 Risks from Tin and Mercury in Soils	506
	21.5.1 Tin	506
	21.5.2 Mercury	507
	References	509
22	**Molybdenum, Silver, Thallium and Vanadium**	**515**
	Les J. Evans and Sarah J. Barabash	
	22.1 Introduction	515
	22.2 Molybdenum	517
	22.2.1 Geochemical Occurrences and Soil Concentrations	517
	22.2.2 Sources of Soil Contamination	518
	22.2.3 Chemical Behavior in Soils	518
	22.3 Silver	523
	22.3.1 Geochemical Occurrences and Soil Concentrations	523
	22.3.2 Sources of Soil Contamination	523
	22.3.3 Chemical Behavior in Soils	524
	22.4 Thallium	528
	22.4.1 Geochemical Occurrences and Soil Concentrations	528
	22.4.2 Sources of Contamination	529
	22.4.3 Chemical Behavior in Soils	529
	22.5 Vanadium	534
	22.5.1 Geochemical Occurrences and Soil Concentrations	534
	22.5.2 Sources of Contamination	534
	22.5.3 Chemical Behavior in Soils	535
	22.6 Environmental and Human Health Risks	540
	22.6.1 Molybdenum	540
	22.6.2 Silver	541
	22.6.3 Thallium	542
	22.6.4 Vanadium	542
	References	543

23 Gold and Uranium — 551
Ian D. Pulford

    23.1 Introduction — 551
    23.2 Geogenic Occurrence — 553
        23.2.1 Gold — 553
        23.2.2 Uranium — 554
    23.3 Soil Contamination — 555
        23.3.1 Gold — 555
        23.3.2 Uranium — 556
    23.4 Chemical Behaviour in Soils — 557
        23.4.1 Gold — 557
        23.4.2 Uranium — 559
    23.5 Risks from Gold and Uranium in Soils — 560
        23.5.1 Gold — 560
        23.5.2 Uranium — 561
    23.6 Concluding Comments — 562
    References — 562

24 Platinum Group Elements — 567
F. Zereini and C.L.S. Wiseman

    24.1 Introduction — 567
    24.2 Sources of PGE in Soils — 568
        24.2.1 Geogenic Sources — 568
        24.2.2 Anthropogenic Sources — 569
    24.3 Emissions, Depositional Behavior, and Concentrations in Soils — 570
    24.4 Geochemical Behavior in Soils — 573
    24.5 Bioavailability — 573
    24.6 Conclusions — 574
    References — 575

Index — 579

Preface

Trace elements occur naturally in soils, some are essential micronutrients for plants and animals and are thus important for human health and food production. At elevated levels, all trace elements (TEs), however, become potentially toxic. Anthropogenic input of TEs into the natural environment thus poses a range of ecological and health problems. Because of the growing awareness of these problems, TEs in soils have received widespread scientific and legislative attention during the last 40 years. Consequently, significant progress has been made in a number of areas, such as the important role soil properties can play in determining trace element sorption–desorption processes and how they subsequently influence plant uptake, leaching to groundwater or ecotoxicity. The significance of appropriate experimental design has become readily apparent. For example, initial research which involved using soils spiked with soluble trace element salts cannot be usefully extrapolated to situations where TEs inputs arise through sewage-sludge disposal or other less readily soluble sources.

The number of publications covering TEs in soils continues to grow, but they are dispersed across many different journals and books, often covering specific areas of TEs in soils (general chemistry, sampling and analysis, speciation, bioavailability, ecotoxicity, risk assessment, modelling, and remediation). This situation can make it difficult to grasp the direction and progress being made on the whole subject. This book brings together an up-to-date, balanced and comprehensive review of key aspects relating to TEs in soils.

The book comprises four sections:

- Basic principles, processes, sampling and analytical aspects
- Long-term issues, impacts and predictive modelling
- Bioavailability, risk assessment and remediation
- Characteristics and behaviour of individual elements

Written as an authoritative guide for scientists in soil science, geochemistry, environmental science and analytical chemistry areas at postgraduate level and beyond, this book is intended to serve as a synthesis of much of the current knowledge on trace elements in soils. The book is also intended for professionals in regulatory, land management and environmental planning and protection organizations.

Peter S. Hooda
Editor
Kingston University London, UK

List of Contributors

Yuji Arai Clemson University, Department of Entomology, Soils and Plant Sciences, Clemson, SC, USA.

Sarah J. Barabash University of Guelph, School of Environmental Sciences, Ontario, Canada.

C. Leigh Broadhurst University of Maryland, Department of Civil and Environmental Engineering, College Park, MD, USA.

Tiziana Centofanti University of Maryland, Department of Civil and Environmental Engineering, College Park, MD, USA.

Rufus L. Chaney United States Department of Agriculture, Agricultural Research Service, Environmental Management and Byproduct Utilization Laboratory, Beltsville, MD, USA.

Andrew C. Chang University of California, Department of Environmental Sciences, Riverside, CA, USA.

Weiping Chen University of California, Department of Environmental Sciences, Riverside, CA, USA, and State Key Laboratory of Urban and Regional Ecology, Research Center For Eco-Environmental Sciences, Chinese Academy of Science, Beijing, China.

Martin J. Clifford University of Reading, School of Agriculture, Policy and Development, Reading, UK.

Gijs Du Laing Ghent University, Department of Applied Analytical and Physical Chemistry, Ghent, Belgium.

Anthony C. Edwards Nether Backhill, Ardallie, Aberdeenshire, UK.

Les J. Evans University of Guelph, School of Environmental Sciences, Ontario, Canada.

Emmanuel Frossard Institute of Plant Sciences, ETH Zurich, Switzerland.

Albania Grosso WCA Environment Ltd, Faringdon Oxfordshire, UK.

Xueyuan Gu Nanjing University, School of the Environment, China.

Zhenli L. He University of Florida, Institute of Food and Agricultural Sciences, Indian River Research and Education Center, Fort Pierce, FL, USA.

Ganga M. Hettiarachchi Kansas State University, Department of Agronomy, USA.

Gavin M. Hilson University of Reading, School of Agriculture, Policy and Development, Reading, UK.

Mark E. Hodson University of Reading, Department of Soil Science, Reading, UK.

Peter S. Hooda Kingston University London, School of Geography, Geology and the Environment, Kingston upon Thames, UK.

Rupert L. Hough The Macaulay Land Use Research Institute, Aberdeen, UK.

Annette Johnson Eawag, Dübendorf, Switzerland.

Bonjun Koo California Baptist University, Department of Natural and Mathematical Sciences, Riverside, USA.

Jurate Kumpiene Luleå University of Technology, Division of Waste Science and Technology, Luleå, Sweden.

David G. Lumsdon The Macaulay Land Use Research Institute, Aberdeen, UK.

Yibing Ma Institute of Agricultural Resources and Regional Planning, Chinese Academy of Agricultural Sciences, Beijing, China.

Mike J. McLaughlin University of Adelaide, School of Earth and Environmental Sciences, and CSIRO Land and Water, Glen Osmond, Australia.

Graham Merrington WCA Environment Ltd, Faringdon Oxfordshire, UK.

Albert L. Page University of California, Department of Environmental Sciences, Riverside, USA.

Ian D. Pulford University of Glasgow, Environmental Chemistry, Chemistry Department, Glasgow, UK.

Sohel Saikat Health Protection Agency, London, UK.

Rainer Schulin Institute of Terrestrial Ecosystems, ETH Zurich, Switzerland.

Jiali Shentu Zhejiang University, College of Environmental and Resource Sciences, Hangzhou, China.

David J. Spurgeon Centre for Ecology and Hydrology, Monks Wood, Abbots Ripton, Huntingdon, Cambridgeshire, UK.

Samuel P. Stacey University of Adelaide, School of Earth and Environmental Sciences, Glen Osmond, Australia.

Filip M.G. Tack Ghent University, Department of Applied Analytical and Physical Chemistry, Ghent Belgium.

C.L.S. Wiseman University of Toronto, Centre for Environment, Toronto, Ontario, Canada.

Laosheng Wu University of California, Department of Environmental Sciences, Riverside, USA.

Xiao E. Yang Zhejiang University, College of Environmental and Resources Sciences, Hangzhou, China.

F. Zereini Institute For Atmospheric and Environmental Sciences, J. W. Goethe University, Frankfurt, Germany.

Section 1

Basic Principles, Processes, Sampling and Analytical Aspects

1

Introduction

Peter S. Hooda

*School of Geography, Geology and the Environment, Kingston University London,
Kingston upon Thames, Surrey, UK*

Chemical elements in soil are referred to as trace elements (TEs) because of their occurrence at concentrations less than 100 mg kg^{-1}. As a matter of fact, many of these elements are present at concentrations much lower than this. Most of the trace elements of environmental and human/animal health significance are metals, for example cadmium, chromium, cobalt, copper, gold, lead, manganese, mercury, molybdenum, nickel, palladium, platinum, rhodium, silver, thallium, tin, vanadium and zinc. Other important TEs belong to the metalloid (for example boron, arsenic, and antimony), nonmetal (for example selenium), actinoid (for example uranium) and halogen (for example iodine and fluorine) groups of elements.

Trace elements have also been termed 'toxic metals', 'trace metals' or 'heavy metals', although none of the terms is entirely satisfactory from a chemical viewpoint. 'Heavy metals' is the most popularly used and widely recognized term for a large groups of elements with density greater than 6 g cm^{-1} but not all TEs are metals. Likewise, the term 'toxic metals' is not appropriate as TEs become toxic to living organisms only when they are exposed to excess levels. For this reason, TEs are also often referred as potentially toxic trace elements (PTEs); this term is more inclusive and appropriate than toxic or heavy metals.

The term 'trace element' is useful as it embraces metals, metalloids, nonmetals and other elements in the soil–plant–animal system, but it is somewhat vague because it can include any element regardless of its function. Seven elements, chlorine (Cl) manganese (Mn), iron (Fe), zinc (Zn), boron (B), copper (Cu) and molybdenum (Mo) are essential nutrients required in trace amounts for plant growth as well as human and animal health [1], although Cl and Fe within soils and plants are not TEs because their average concentration is generally greater than 100 mg kg^{-1}. These elements are necessary for maintaining the life processes in plants and/or animals including humans and therefore are essentially micronutrients (see Chapter 9). Cobalt (Co), chromium (Cr), fluorine (F), iodine (I), nickel (Ni) and selenium (Se) found in plants are not essential nutrients as such, but animals have developed a dependency on these elements for use in their metabolic processes [2]. Cobalt (Co) is also required by microorganisms for atmospheric-nitrogen fixation and by ruminants for their rumen bacteria. Although the biological role of these elements is not fully understood, they are considered as important beneficial TEs. There is little evidence to suggest that arsenic (As), cadmium (Cd), lead (Pb) and mercury (Hg) play a nutritive role in higher plants and animals.

Trace elements occur naturally in soils. However, production-oriented policies in the twentieth century, which exploited land for mineral extraction, manufacturing industry and waste disposal have resulted in the input and accumulation of large quantities of TEs in the soils. There are a variety of both natural and anthropogenic input sources of trace elements into soils. The major natural sources include weathering (including erosion and deposition of wind-blown particles), volcanic eruptions, forest fires and biogenic sources [3,4]. While inputs via natural sources constitute a significant burden of TEs in the soil, the contribution from anthropogenic sources for many elements is several times that from natural sources [4,5], raising obvious environmental and health concerns. The major anthropogenic sources of trace elements input to soils are:

- Atmospheric deposition, arising from coal and gasoline combustion, nonferrous and ferrous metal mining, smelting, and manufacturing, waste incineration, production of phosphate fertilizers and cement, and wood combustion;
- Land application of sewage sludge, animal manure and other organic wastes and co-products from agriculture and food industries;
- Land disposal of industrial co-products and waste, including paper industry sludge, coal fly ash, bottom fly ash and wood ash;
- Fertilizers, lime and agrochemicals (pesticides) use in agriculture.

Globally in the 1980s, atmospheric deposition was the single largest source of trace elements, in most cases being responsible for more than 50–80 % of their inputs to soils [4]. With the introduction of efficient flue gas desulfurization installations in smelters and power plants, and the banning of lead additives to gasoline, major reductions in the atmospheric emission of trace elements have been achieved in Europe and the USA [6–9] as well as in other developed countries. For instance, emissions of Hg, Cd and Pb to the atmosphere have decreased by more than 4, 5 and 10 times, respectively compared with their peaks in the mid 1960s and the early 1970s [7,9]. Despite these major reductions in the atmospheric fallout in recent years, this still appears to remain the main source of trace elements inputs to soils, as estimated in a UK study [10] in which this source accounted for 25–85 % of their (Zn, Cu, Ni, Pb, Cr, As and Hg) total input. While the direct inputs via industrial and municipal waste disposal tend to elevate contaminant content to a greater level, albeit limited to the affected soils when compared with the diffused-atmospheric fallout, the latter is expected to remain a significant source of TEs input to a larger landmass. Studies of trace elements distribution in ecosystems show that soils near industrial and mining sites, major road networks or those that have received heavy applications of sewage sludge contain much greater levels of these elements compared with their local background concentrations.

The soil is the primary source of trace elements for plants, animals and humans. Elevated levels of TEs in the soil as a consequence of human activities therefore pose a range of environmental and health risks. Trace elements, unlike organic contaminants, are retained in soils essentially indefinitely because they are not degradable. Consequently, soils contaminated with TEs pose a long-term risk of increased plant uptake and leaching [for example 11–16], with potentially adverse implications for the wider environment, including human health. Arsenic, Cd, Hg, Pb and Se are the most important in terms of the food-chain contamination and ecotoxicity viewpoints [17]. Excess intake of As is mostly likely to occur from elevated concentrations of As in drinking water, which has been associated with increased risk of skin cancer. Exposure to Hg is primarily via food, fish being a major source of methyl mercury [18]. Selenium toxicity and other health problems generally seem to arise in areas where the soils are naturally rich in Se [17,19]. Cadmium in contaminated soils is of the greatest concern in terms of its entry into the food chain as it can be taken up by food plants in large amounts relative to its concentration in the soil. Furthermore, Cd accumulates over a lifetime in the body, and recent data suggest that adverse health effects of this element, mainly in the form of renal dysfunction but possibly also effects on bone and fractures, may occur at lower exposure levels than previously anticipated [18]. The evidence of human health effects due to low-level Cd contamination is contentious, however, as manifestations such as renal dysfunction have occurred only in situations of gross soil contamination, combined with significant exposure pathways modified by other interacting factors. For example, in rice-based economies, the diet is often deficient in Fe, Zn and Ca. This appears to enhance Cd absorption into duodenal cells and steepen the dose–response relationship between Cd intake and adverse health outcomes [17,20].

Low soil concentrations of essential TEs (micronutrients) can result in their inadequate supply to plants, affecting plant growth and development, which ultimately can cause deficiency disorders further up the food chain (see Chapter 9). Deficiencies of Zn, Cu and Mn are not uncommon in many parts of the world, limiting the growth and healthy development of many field crops. Likewise deficiency disorders associated with Co, Cu, I, Fe, Mn, Se and Zn, following low accumulation in the feed/fodder plants, are a common occurrence in grazing livestock. Such deficiencies in agricultural soils can be rectified by fertilization with appropriate micronutrients, although this has been limited, largely to Zn and Cu [21].

Trace elements have received widespread scientific and legislative attention because of their ubiquity, persistence and potential toxicity/deficiency. As a result, the general behaviour of TEs, such as the role of soil properties in their retention–release processes and how they subsequently influence plant uptake or leaching is now generally well characterized. This has initiated a broad understanding of the concentrations at which (crop) plants are protected from 'excessive accumulation' of potentially toxic elements. These 'safe' concentrations of TEs are the basis of current regulations (for example EU sludge Directive limits 86/278/EEC) for their inputs to soils, especially through the application of sewage sludge in European, North American and other countries. The regulatory limits for maximum permissible concentrations (MPCs) of TEs in soils differ between European Union Member States.

Initially the UK opted for the maximum concentrations permissible in the EU Sludge Directive, that is 300, 200 and 110 mg kg^{-1} of Zn, Cu and Ni, respectively for soil pH >7; lower limits apply for soil pH <7. For Cd and Pb the maximum possible concentrations in the Sludge Directive, that is 3 and 300 mg kg^{-1} Cd and Pb, respectively, were allowed. However, it soon became clear that 300 mg kg^{-1} sludge-borne Zn could have adverse effects on soil microbial biomass and on nitrogen fixation by cyanobacteria and by the nitrogen-fixing bacterium *Rhizobium* [22,23], with the key microbial processes of litter breakdown and nutrient cycling at stake. Following these concerns and the evidence obtained from long-term field trails, the UK government has accepted the recommendation of a reduction in the national standard for Zn from 300 to 200 mg kg^{-1}.

Recent field experiments indicate that the current UK soil total Cd limit of 3 mg kg^{-1} may not be adequately protective against producing wheat grain with concentrations of Cd above the EU grain Cd MPC of 0.235 mg kg^{-1}, unless the soil pH is maintained above 6.8 [24]. This is likely to bring down the UK limit of soil total Cd concentration, as suggested [24]. Clearly, the continued refinement of a regulatory framework that protects the wider environment in the long term requires future research studies to focus on the low levels of TEs permitted in the current regulations.

Assessing the bioavailability and speciation of trace elements is crucial to determining the environmental impact of contaminated soils, as it is widely recognized that the total TE content of soils is not a good indicator/predictor of their plant uptake or ecotoxicity. Numerous bioavailability and speciation assessment soil tests have been used and proposed [25] (see Chapter 11), but there is no general consensus on how metal speciation and bioavailability can assist the risk-assessment process. This has hindered the development of a framework for quantitative risk assessment of TEs in soils that incorporates their bioavailability.

Since the early 1990s, remediation of soils contaminated with TEs has been the focus of much research. This has involved either the use of hyperaccumulator plants (for example *Thlaspi caerulescens*) with exceptional metal-accumulating capacities (phytoremediation) or the use of soil amendments to reduce metal mobility. Phytoremediation in principle is a cost-effective and environment-friendly cleanup technology for TE-contaminated soils. Its progress has been constrained, however, by the generally low biomass production of metal hyperaccumulator plant species, prompting the use of some high biomass-producing as well as metal-accumulating field crops (for example Indian mustard). The use of such field crops, has not been effective in cleaning up contaminated soils, because of the inability of the plants to mobilize non-plant-available forms of TEs, which are often a significant fraction of their total soil content. This has led to the so-called 'assisted phytoextraction' approaches, which involve the use of acidifying or chelating agents to mobilize TEs for plant uptake – this, however, has become a contentious issue, as it increases the risk of spreading the contaminants into the wider environment at least in the short-term (see Chapter 14).

The use of soil amendments (for example natural and synthetic aluminosilicates, rock phosphate, various sources of Fe, Mn and Al oxides, liming compounds, ashes, organic waste) for stabilization/immobilization of TEs can be useful for the 'remediation' of soils that are only marginally contaminated or when the purpose is to contain the contaminants, such as reclamation of mining-affected sites. The long-term effectiveness of such stabilization/immobilization techniques is yet to be fully established (see Chapter 15).

This book brings together comprehensive coverage of all aspects of TEs in soils, presented in 24 chapters arranged in four sections. After a brief introduction, Section 1 presents general TE soil chemistry (Chapter 2), issues in soil sampling and sample preparation (Chapter 3) and reviews of TEs analytical techniques (Chapters 4 and 5), including speciation. Section 2 includes reviews on issues and/or sources of trace element inputs in soil that are likely to remain important, including sewage sludge- and fertilizer-borne TEs, their impact on soil ecology, application of chemical speciation modelling for understanding their behaviour in soils, and deficiency of TEs in soils (Chapters 6–10). Section 3 deals with soil tests for TE bioavailability (Chapter 11) and bioavailability and risk assessment (Chapter 12), and presents rationale underpinning regulatory limits for TEs (Chapter 13), before considering remediation and management techniques (Chapters 14 and 15). Section 4, the final section, provides information on the behaviour of individual trace elements (arsenic antimony, cadmium, chromium, cobalt, gold, lead, manganese, mercury, molybdenum, nickel, palladium, platinum, rhodium, selenium, silver, thallium, tin, uranium, vanadium and zinc). These elements are presented in nine chapters (Chapter 16–24), each chapter covering a group of elements, largely according to the authors' research interests.

References

[1] Bennett, W.F. (ed.), Nutrient Deficiencies & Toxicities in Crop Plants; APS Press, St. Paul, MN, 1993.
[2] Johnston, A.E., Trace elements in soil: status and management; in Essential Trace Elements for Plants, Animals and Humans; NJF Seminar No. 370, Reykjavík, Iceland, 15–17 August 2005, pp. 71–74.
[3] Adriano D.C., Trace Elements in the Terrestrial Environment; Springer, New York, 1986.
[4] Nriagu, J.O.; Pacyna, J.M., Quantitative assessment of worldwide contamination of air, water and soils by trace metals; Nature 1988, 333, 1341–139.
[5] Nriagu, J.O., A global assessment of natural sources of atmospheric trace metals; Nature 1989, 338, 47–49.
[6] Pirrone, N.; Keeler, N.J.; Nriagu, J.O., Regional differences in worldwide emissions of mercury to the atmosphere; Atmos. Environ., 1996, 30, 2981–2987.
[7] Callender, E., Heavy metals in the environment – historical trends; in H. D. Holland and K.K. Truekian (eds), Treatise on Geochemistry; Elsevier, 2007, Chapter 9.03, Vol. 9, pp. 67–105.
[8] Pacyna, E.G.; Pacyna, J.M.; Fudala, J.; Strzelecka-Jastrzab, E.; Hlawiczka S.; Panasiuk, D.; Nitter, S.; Pregger, T.; Pfeiffer, H.; Friedrich, R., Current and future emissions of selected heavy metals to the atmosphere from anthropogenic sources in Europe; Atmos. Environ., 2007, 41, 8557–8566.
[9] Pacyna, J.M.; Pacyna, E.G.; Wenche, A., Changes of emissions and atmospheric deposition of mercury, lead, and cadmium; Atmos. Environ., 2009, 43, 117–127.
[10] Nicholson, F.A.; Smith, S.R.; Alloway, B.J.; Carlton-Smith, C., Chambers, B.J., An inventory of heavy metals inputs to agricultural soils in England and Wales. Sci. Total Environ. 2003, 311, 205–219.
[11] Alloway, B.J.; Jackson, A.P., The behaviour of heavy-metals in sewage sludge-amended soils; Sci. Total Environ., 1991, 100, 151–176.
[12] Alloway B.J., Heavy Metals in Soils, 2nd edn; Blackie Academic and Professional, London, 1995.

[13] Hooda, P.S.; McNulty, D.; Alloway, B.J.; Aitken, M.N., Plant availability of heavy metals in soils previously amended with heavy applications of sewage sludge; J. Sci. Food Agric. 1997, 73, 446–454.

[14] Xue, H.; Nhat, P.H.; Gachter, R.; Hooda, P.S, The transport of Cu and Zn from agricultural soils to surface water in a small catchment; Adv. Environ. Res. 2003, 8, 69–76.

[15] Alvarez-Ayuso, E.; Garcia-Sanchez, A.; Querol, X.; Moyano, A., Trace element mobility in soils seven years after the Aznalcollar mine spill; Chemosphere 2008, 73, 1240–1246.

[16] Zhao, L.Y.L.; Schulin, R.; Nowack, B., Copper and Zn mobilization in soil columns percolated by different irrigation solutions; Environ. Pollut. 2009, 157, 823–833.

[17] McLaughlin, M.J.; Parker, D.R.; Clarke, J.M., Metals and micronutrients – food safety issues; Field Crop Res, 1999, 60, 143–163.

[18] Jarup, L., Hazards of heavy metal contamination; Br. Med. Bull. 2003, 68, 167–182.

[19] Tinggi, U., Selenium toxicity and its adverse health effects; Rev. Food Nutr. Toxic. 2005, 4, 29–55.

[20] Chaney, R.L.; Reeves, P.G.; Ryan, J.A.; Simmons, R.W.; Welch, R.M.; Angle, J.S., An improved understanding of soil Cd risk to humans and low cost methods to phytoextract Cd from contaminated soils to prevent soil Cd risks; BioMetals 2004, 17, 549–553.

[21] Bell, R.W.; Dell, B., Micronutrients for Sustainable Food, Feed, Fibre and Bioenergy Production; IFA, Paris, France, 2008.

[22] Giller, K.E.; McGrath, S.P.; Hirsch, P.R., Absence of nitrogen-fixation in clover growing on soil subject to long-term contamination with heavy metals is due to survival of only effective *Rhizobium*; Soil Biol. Biochem. 1989, 21, 841–848.

[23] Chaudri, A.M.; McGrath, S.P.; Giller, K.E.; Rietz, E.; Hirsch, P.R., Enumeration of indigenous *Rhizobium leguminosarum* biovar *Trifolii* in soils previously treated with metal-contaminated sewage sludge; Soil Biol. Biochem. 1993, 25, 301–309.

[24] Chaudri, A.; McGrath, S.; Gibbs, P.; Chambers, B.; Carlton-Smith, C.; Godley, A.; Bacon, J.; Campbell, C.; Aitken, M., Cadmium availability to wheat grain in soils treated with sewage sludge or metal salts; Chemosphere 2007, 66, 1415–1423.

[25] McLaughlin, M.J.; Zarcinas, B.A.; Stevens, D.P.; Cook, N., Soil testing for heavy metals; Commun. Soil Sci. Plant Anal., 2000, 31, 1661–1700.

2

Trace Elements: General Soil Chemistry, Principles and Processes

Filip M. G. Tack

Department of Applied Analytical and Physical Chemistry, Ghent University, Gent, Belgium

2.1 Introduction

Several trace elements occurring naturally in soil are required in low concentrations to sustain biological processes and therefore living organisms (see Chapter 1). Trace elements can also be introduced into soil through a variety of anthropogenic sources, including aerial deposition, commercial fertilizers, liming materials, sewage sludge, manure, pesticides, coal combustion residues, metal-smelting industries and transport emissions [1]. Although natural processes such as volcanic emissions also contribute to aerial transport of trace elements, air pollution due to anthropogenic emissions is the most important source of trace element contamination in industrialized countries [2].

Regardless of their biological essentiality, trace elements become toxic when taken up in excessive amounts. Some severe historical incidences of trace element toxicity problems have prompted widespread public and scientific debate on trace element contamination issues. Widespread methylmercury poisoning in the region around Minamata Bay, Japan, manifested during the 1950s into a variety of health problems. Collectively termed as 'Minamata disease', it occurred mainly among fishermen and their families, and originated from the consumption of fish and shellfish contaminated with methylmercury discharged from a local chemical plant [3]. Also in the 1950s, a major incidence of severe cadmium toxicity, causing the *itae-itae* disease, occurred in Toyama, Japan [4].

A general increase in the risk of leaching and plant uptake of trace elements is related to their total content in soils, although actual mobility is strongly related to their physicochemical forms of occurrence and the local environmental conditions [5,6]. The significance of site conditions is evident from a simple comparison between the total Cd contents of Toyama rice fields (5 mg kg^{-1}) and those of Shipham in the UK or Stolberg in Germany, where Cd levels were as high as 50–150 mg kg^{-1}. In contrast to the Toyama case, few or no effects of Cd were observed in the latter two cases [7–9]. The difference in soil chemistry was a main factor responsible for this paradoxical outcome. In Toyama (Japan), the low soil pH favoured Cd mobilization and uptake by rice, whereas neutral pH of the Shipham and Stolberg soils restricted Cd uptake by vegetables. The solubility of trace elements in soils is a key factor in determining the likelihood and extent of their environmental effects. Trace elements in the soil solution constitute the fraction that is readily available for migration to other soil compartments, leaching through the soil profile or uptake by biota. This chapter reviews the key soil processes and underlying principles with respect to the fate of trace elements in soils.

2.2 Distribution of Trace Elements in the Soil

In soils, trace elements are distributed over the different soil compartments where they exist in a variety of chemical forms, strongly differing in reactivity (Figure 2.1 [10]). Typical concentrations in the soil solution of uncontaminated soils are in the order of

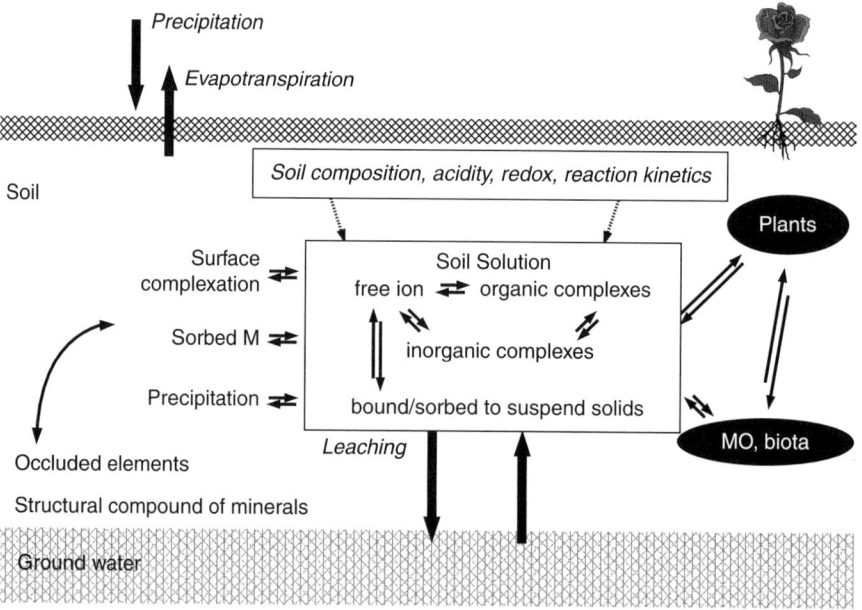

Figure 2.1 *Schematic representation of various pools of trace elements in the soil system. The soil solution is the central gateway through which various forms of elements interact with the soil solid phase and soil biological activity (modified after [10] and [17])*

micrograms per litre [11]. Trace elements in soil solution may exist in their most simple form as hydrated cations or oxyanions, but might also exist as inorganic or organic complexes. In addition to the truly dissolved form of elements, soil solution also contains colloids and fine suspended particulate material that might carry a significant fraction of the trace elements in the soil solution. In ground water polluted by landfill leachate, 24–56 % of Cd, Ni and Zn, and 86–99 % of Cu and Pb were associated with the colloidal fraction [12]. The significance of colloidal and suspended material is much overlooked in environmental studies.

A major fraction of the total trace element content is usually associated with the solid phase. For example, 0.5 mg kg^{-1} dry soil represents a typical baseline concentration for Cd [13]. The soil solution concentration of Cd in uncontaminated soils generally ranges from about 0.01 up to 5 μg Cd l^{-1} [11]. Assuming 1 μg Cd l^{-1} as the concentration in soil solution and standard soil properties (volumetric moisture content 20 %; soil bulk density 1500 kg m^{-3}), the dissolved Cd content represents less than 0.00003 % of the total Cd content in this example. Yet this tiny soluble fraction might have important environmental significance.

Forms of trace elements present in the soil solid phase include adsorbed, complexed or occluded pools (Figure 2.1). Trace elements that are superficially adsorbed or complexed with solid-phase components such as clay minerals, hydrated oxides of iron and manganese, or organic matter are more or less exchangeable with the soil solution phase. When occluded within soil solids, they are at least temporary unavailable for exchange with the soil solution, unless the phase concerned is dissolved or destroyed. Trace elements structurally incorporated in soil minerals are not likely to become available. They may be released over large timescales upon the gradual weathering of the minerals.

Soil physical, chemical and biological processes will determine the speciation, redistribution, mobility and ultimately the bioavailability of trace elements. The timescale by which the different processes occur in soils can vary between nanoseconds and centuries or more [14]. Because of this range in reaction kinetics and since soil is continuously subjected to varying conditions such as changes in temperature and moisture content, a soil is never in true chemical equilibrium.

2.3 Chemical Species

It is only since the 1980s that focus started to shift from determining total elemental contents to identifying the different chemical forms of occurrence of the elements in the environment and particularly, in soils [15]. A chemical species is a specific form of an element defined on the basis of its isotopic composition, electronic or oxidation state, and/or complex or molecular structure [16]. The distribution of trace elements between the various chemical species in soil solid or solution phase is referred to as speciation [16]. Broadly, categories of chemical species may be defined as follows (adapted from [17]):

- Soil solution-phase species:
  - free ion;
  - inorganic complex;

- organic complex;
- bound to suspended colloids (clay, organic matter, sesquioxides) as in the solid phase.

- Soil solid phase species:
  - exchangeably bound to charged surfaces;
  - complexed with or occluded within organic matter;
  - adsorbed or occluded in hydrated oxides of iron and manganese;
  - adsorbed or occluded in carbonates;
  - as precipitate (carbonates, phosphates, sulfides);
  - as structural component in minerals.

In the soil environment, many trace elements exist only in a single dominant oxidation state. A few, such as, As, Cr and Se can occur in different oxidation states. They can undergo oxidation or reduction when they interact with mineral or organic compounds that act as oxidants or reducing agents [18]. Many trace metals in soil solution exist in their most simple form as charged hydrated cations. The free trace metal ions (for example Zn^{2+}, Cu^{2+}, Ni^{2+}) tend to be surrounded by six water molecules arranged in an octahedron. This can be written, for example, as $Zn(H_2O)_6^{2+}$, but commonly a simplified representation is used, for example Zn^{2+}, where the hydrated water is omitted [19]. Other elements exist in their most simple form as oxyanions. An example is Mo, which in soil solution occurs as the molybdate anion (MoO_4^{2-}) [19]. The highly toxic, hexavalent form of Cr in the soil solution is the anion CrO_4^{2-} [18,20,21]. This toxic form is typically introduced in the environment due to human activities. It is rather unstable in surface soils, but may nevertheless be formed under conditions prevalent in many field soils [20,21]. Also inorganic As in solution occurs in oxyanion species. In the normal pH range of soils (pH 4–8), arsenate in soil solution will be present as $H_2AsO_4^-$ or $HAsO_4^-$. Arsenous acid (H_3AsO_3) is usually present as the undissociated acid. It is converted to $H_2AsO_3^-$ only above pH 8 [22]. Also the nonmetal element B exists as a neutral species, H_3BO_3, in solution [23]. For selenium, both selenate (SeO_4^{2-}) and selenite (SeO_3^{2-}) may be stable in soils for long periods. Between the different selenate forms, only SeO_4^{2-} exists in the common pH range of soils. The dominant form of selenite below soil pH 8.5 is $HSeO_3^-$ [19].

Some elements can exist as organometallic compounds, where the element is chemically bound to a carbon atom of an organic group. Examples of organoarsenicals include monomethylarsonate [$CH_3AsO_3^{2-}$] and dimethylarsonate [$(CH_3)_2AsO_2^-$] [22,24]. Organic forms of Se are analogous to those of S and include seleno amino acids (for example selenocysteine and selenomethionine), methylselenides [for example $(CH_3)_2Se$] and methylselenones [for example $(CH_3)_2SeOO$] [24]. The most common organic mercury compound in the environment is methylmercury – also known as monomethylmercury (CH_3Hg^+). It is particularly prone to accumulating in fish and marine mammals to levels that are many times greater than its levels in the surrounding water [25]. Dimethylmercury [$(CH_3)_2Hg$] has less tendency to accumulate because it is a volatile organomercury compound that can escape to the atmosphere [26]. Owing to the use of tetraethyl lead [$(CH_3CH_2)_4Pb$] and tetramethyl lead [$(CH_3)_4Pb$] as anti-knock additives in gasoline, alkyl lead compounds are now commonly present in the environment [27]. Environmental contamination is also the most important source of organotin compounds in the environment, although methyltin can originate from natural biomethylation processes [28].

Complexes consist of one or several metal cations bound to one or several ligands. The metal behaves as a Lewis acid, a substance capable of accepting an electron pair to form a bond. Ligands are Lewis bases and as such can provide a free electron pair to form a chemical bond. The ligand can be an ion or a neutral molecule. It might be a mineral substance such as chloride or phosphate, or an organic substance, in which case it could be a simple organic acid such as citric acid, or a high molecular weight compound such as a fulvic acid molecule. A complex is termed a chelate when an internal ring structure is created through complexation. A chelating agent is typically an organic agent that has two or more groups capable of complex formation.

Many metallic cations tend to form complexes with common inorganic anions such as chloride, sulfate, nitrate, etc., to a varying degree. The OH^- anion is the most important inorganic ligand. Most transition elements hydrolyse in water to form hydroxo-complexes such as $ZnOH^+$ [29]. Carbonate complexes are of importance for Cu^{2+} and UO_2^{2+}. The Cl^- anion is an important ligand for metals such as Ag^+, Hg^{2+} and Cd^{2+}. Although these complexes usually are weak and labile, they might significantly influence the behaviour of trace elements. For example, significant increases in Cd uptake by potato were attributed to the effects of chloride on the availability of Cd [30].

Complexes of elements with organic substances are usually much more stable and of greater importance. While simple organic acids, such as acetic acid, are found in the soil solution phase, a large part of the soluble organic matter consists of compounds of a higher molecular weight. Organic matter will therefore significantly influence the behaviour of many elements, in particular Fe, Cu and Pb (see Section 2.6.2).

The vast majority of trace elements contained in soils are associated with the solid soil fraction [13], where they exist in a variety of physicochemical forms and associations. Trace elements might be bound at the surface of the solid phase by sorption. Because of their low concentrations, they rarely exist as discrete precipitates in soils. Elements might also be included in different solid phases existing in the soil system, such as organic matter, hydrated oxides of iron and manganese, carbonates, or, in reduced soils, as sulfides. Such chemically different phases do not necessarily occur as discrete phases in the soil [17]. There is an intensive interaction between different compounds such as organic matter, clay minerals and hydrated oxides, causing the formation of aggregates that are mixtures of the various compounds. Such entities may be very stable, so that subsequent reactions with ions may take place with either the organic or the inorganic part [31]. The most immobile fraction of trace elements is the fraction included in mineral structures. They will be released only upon weathering of the minerals, which usually is a very slow process.

2.4 Sorption and Desorption

2.4.1 Sorption Mechanisms

'Sorption' is a general term that denotes the loss of a solute from aqueous solution [32]. It includes both adsorption, the process by which a solute clings to a solid surface, and absorption, the process by which the solute diffuses into a porous solid and clings to interior surfaces [33]. There are various mechanisms that cause elements to be removed

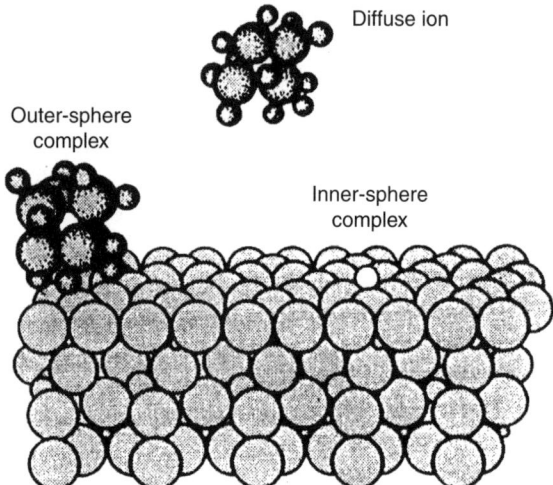

Figure 2.2 Different mechanisms of sorption (Reproduced from G. Sposito, P.W. Schindler, Reactions at the soil colloid-soil solution interface, in L. Landner, Speciation of Metals in Water, Sediment and Soil Systems, *Springer Verlag, Berlin, 1986, pp 683–699. Reproduced with kind permission from Springer Science + Business Media [32].)*

from the solution. In a more specific definition, sorption occurs when matter accumulates at a surface without building up a three dimensional structure [32]. Precipitation, in contrast, will cause the development of a solid phase whose molecular ordering is three-dimensional.

Sorption of an ion may occur according to three mechanisms: (1) inner-sphere-complexation, (2) outer-sphere-complexation, and (3) diffuse ion swarm [32,34] (Figure 2.2). When no water molecule is present between a functional group on a solid surface it is termed inner-sphere. Upon inner-sphere complexation, an ionic or covalent chemical bond is formed. The term 'chemisorption' is often used for this type of binding to a surface. The stability of these bonds depends strongly on the specific properties of the cation such as ionic size and electronic structure, and on steric factors involved in the binding. This mechanism is the basis of specific or selective adsorption [32].

Various factors influence the selective sorption of trace elements [32]:

- properties of the substance (ionic radius, polarity, hydrated radius, equivalent conductivity, hydration enthalpy and entropy;
- availability of pH-dependent sorption sites;
- steric factors;
- formation of hydroxy complexes;
- affinity of the ions for organo-mineral complex formation and their stability;
- interaction with amorphous hydroxides.

In an outer-sphere complex, the ion is involved in the binding as a hydrated species. One or two water molecules are in between the surface and the ion. Because the ion is masked by the hydrated water, the properties of the ion itself will be far less important in determining the binding. Hence, this mechanism will contribute to nonspecific sorption.

The third category of sorption is where a hydrated ion is merely attracted to a surface by electrostatic forces arising from a charged surface. The counter-ions are influenced by two equal but opposing forces. Ions of opposite charge of that of the surface are attracted, while ions with the same charge are repelled. The diffusive or thermal forces, the forces responsible for Brownian motion, cause ions to move from regions with higher concentration to regions with lower concentration. The balance of these two forces gives rise to a distribution of ions in water adjacent to the clay surface, which follows the Boltzmann distribution law [35]:

$$c = c_0 \exp\left(\frac{-neU}{kT}\right) \quad (2.1)$$

where c and c_0 are the concentrations of the ion at the considered distance and at infinite distance (the bulk solution), respectively, n is the valence of the ion, e is the elemental charge, U represents the difference in potential between the charged surface and at distance x, k is the Boltzmann constant and T is the absolute temperature. A resulting distribution is illustrated in Figure 2.3 for cations sorbing on a negative surface. This distribution, described as a diffuse electrical double layer or simply diffuse double layer, is made up of the negatively charged clay surface and the spread-out (diffuse) distribution of the counter-ions. Depending on the ionic strength of the solution, the thickness of this layer can be between 1 and 10 nm [36]. There are different electrostatic models available, which can be distinguished by the way the double layer at the solid–solution interface is described. The three models that are used most are the constant capacitance model, the diffuse layer model and the triple-layer model, which describe the double layer by two, three and four potential adsorption planes [34]. Elements retained by electrostatic forces are easily exchangeable with other ions of similar charge in the soil solution and are thus readily available for plant uptake. Hence, also this mechanism contributes to nonspecific sorption.

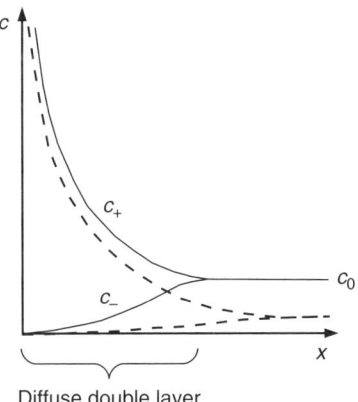

Figure 2.3 Concentration of cations (c_+) and anions (c_-) as a function of the distance x from a negatively charged surface. Dotted lines represent a case where the concentration in the bulk solution (c_0) is lower

In most soils, cation exchange is dominant over anion exchange [37]. The colloidal fraction in soils mainly consists of clay particles (fraction > 2 μm), amorphous hydrated oxides of iron and manganese, and organic colloids. In the normal pH range of soils (pH 4–8), this fraction is negatively charged and behaves as a cation exchanger. The cation exchange capacity (CEC) of a soil is a measure of the quantity of surface sites that can retain positively charged ions (cations) by electrostatic forces. It is expressed in moles of charge per kg weight of soil ($cmol_+ \ kg^{-1}$ or $cmol_c \ kg^{-1}$), although the old, but numerically equivalent unit of meq $(100 \ g)^{-1}$ (milliequivalents of charge per 100 g of dry soil) might still be encountered. Anion exchange capacity in soils is generally small, and is only of practical significance in more acidic soils. In more acidic soils, sesquioxides of iron and aluminium, allophane and even kaolinite may acquire positive charges and are then capable of attracting anions in an electrical double layer [37].

Specific sorption is not limited to the surface of minerals, but may also occur into the inner surfaces of mineral structures such as interlayers in silicate clays. Slow diffusion of metals into goethite, oxides of manganese and illite and smectite clays has been demonstrated [17]. The velocity of the diffusion decreased in the order Ni < Zn < Cd. This agrees with an increase in ionic radius of these elements (0.069, 0.074 and 0.097 nm, respectively) [17]. It is hypothesized that diffusion occurs in micro pores or through discontinuities caused by faults in the crystal structure. This slow diffusion is one of the mechanisms explaining the decrease in biological activity of metal contamination with time, a process referred to as ageing or natural attenuation [38–40].

2.4.2 Sorption Isotherms

Empirical sorption isotherms describe the equilibrium between the amount of an element sorbed and its concentration in solution at a constant temperature. They are useful to empirically describe sorption of trace elements and other chemicals, including nutrients, to soils [34]. To determine soil sorption characteristics, a series of batch experiments is performed, where solutions containing known concentrations of the dissolved substance are in contact with the sorbent (the soil). After a certain period of time, after which it is assumed that equilibrium has been established, the concentration in solution is measured. The amount sorbed is calculated from the decrease in solution concentration.

In its most simple form, sorption behaviour is linear, that is, there is a linear relation between the concentration of an element/chemical in solution (C_w) and that in the solid phase (C_s):

$$C_s = K_d \cdot C_w \quad (2.2)$$

where K_d is the soil/solution partition coefficient. The value of K_d depends on the properties of the sorbing substance, the temperature and the soil.

For low solution concentrations, sorption of trace elements usually can adequately be described by a linear model. At higher concentration levels, sorption becomes nonlinear. Several empirical models derived from other areas of application have been used to describe nonlinear sorption to soils. The Freundlich and the Langmuir models are most widely used and are usually adequate to empirically describe and summarize sorption behaviour over a wider concentration range [41–44]. The *Freundlich isotherm* is an exponential relation, such that

$$C_s = K_d \cdot C_w^n \qquad (2.3)$$

where n is between 0 and 1. It can be linearized by taking the logarithm:

$$\log C_s = \log K_d + n \log C_w \qquad (2.4)$$

To determine the sorption parameters K_d and n for a particular soil, one experimentally determines C_w and C_s over a range of values, that is, using a series of increasing initial concentrations in solution. The parameters are estimated by plotting log C_s against log C_w. The best fitting line is calculated using linear regression. The intercept yields log K_d and the slope is equal to n.

The Langmuir equation assumes a hyperbolic relation between C_s and C_w.

$$C_s = \frac{K_d C_{max} C_w}{1 + K_d C_w} \qquad (2.5)$$

K_d and C_{max} are essentially empirical constants, but C_{max} may be interpreted as a maximum amount that can be sorbed. The Langmuir equation can be linearized through inversion, yielding

$$\frac{1}{C_s} = \frac{1}{C_{max}} + \frac{1}{K_d C_{max}} \cdot \frac{1}{C_w} \qquad (2.6)$$

A plot of $1/C_s$ versus $1/C_w$ yields an intercept of $1/C_{max}$ and a slope of $1/(K_d C_{max})$. An alternative linear form is

$$\frac{C_w}{C_s} = \frac{1}{K_d C_{max}} + \frac{C_w}{C_{max}} \qquad (2.7)$$

If the sorption data set follows the Langmuir behaviour, the transformed data will plot on a straight line.

Deviations between experimental datasets and calculated behaviour have been explained by the competition of different sorbates for the sorption sites on the surface. A well-known situation for competitive behaviour is the influence of pH on sorption. To accommodate for these effects, modified equations have been used [34]. For example, the Freundlich model has been used in a modified form to include pH and organic carbon content [45,46]:

$$C_s = K_d \cdot C_{org}^o \cdot (H^+)^q \cdot C_w^n \qquad (2.8)$$

where C_{org} is the organic carbon content, (H^+) is the proton activity, and K_d, o, q, n are empirical constants. A model that also included the Ca^{2+}-activity in solution was used successfully to describe Cd sorption in sandy soils [47,48]. The competitive Langmuir equation takes the form [34]:

$$C_s = \frac{K_1 C_{max} C_1}{1 + K_1 C_1 + K_2 C_2} \qquad (2.9)$$

where K_2 and C_2 are the empirical parameter and concentration, respectively, of another compound that competes for the sorption sites.

2.5 Precipitation and Dissolution

Precipitation involves the formation of a solid compound. Such precipitates include oxides, oxyhydroxides, hydroxides, carbonates, phosphates and silicates [49]. In anaerobic conditions, where biological reduction of SO_4^{2-} occurs, sulfide precipitates may be formed. The extent of dissolution and precipitation of a mineral can be described by its solubility product. For example, Cu^{2+} might precipitate according to [49]

$$CuSO_4(\text{chalcocyanite}) \leftrightarrow Cu^{2+} + SO_4^{2-}, \qquad \log K_s = 3.72 \qquad (2.10)$$

The solubility product is given by the equilibrium constant of the dissolution reaction:

$$K_s = (Cu^{2+})(SO_4^{2-}) = 10^{3.72} \qquad (2.11)$$

When the product of the activities of Cu^{2+} and SO_4^{2-} exceeds the solubility product, the solution is said to be oversaturated with respect to the solid compound, and the compound is expected to form. When the ionic product is lower than the solubility product, the solution is undersaturated. The precipitate will tend to dissolve until the solubility product is satisfied or until all of the precipitate has dissolved. Reaction kinetics of solution/dissolution might be very slow. It is therefore not uncommon in the environment and particularly in soils to find solutions that are under- or oversaturated with respect to solids that are present in the soil.

Several metals such as Fe, Al, Cu, Fe, Zn and Cd can be precipitated as hydroxides under neutral to alkaline pH conditions [29]. For example, the dissolution of α-Zn(OH)$_2$ is described by [49,50]:

$$\alpha - Zn(OH)_2(s) + 2H^+ \leftrightarrow Zn^{2+} + 2H_2O \qquad \log K = 12.48 \qquad (2.12)$$

In logarithmic form, the equilibrium constant can be written as

$$\log(Zn^{2+}) = 12.48 - 2\,\text{pH} \qquad (2.13)$$

This equation shows that the activity of Zn^{2+} in solution controlled by α-Zn(OH)$_2$(s) is expected to decrease by a factor of 100 with each increase in pH of one unit. At pH 7, the activity of Zn^{2+} will be 0.03 mol l^{-1}, which corresponds to about 2 g l^{-1}. At pH 8, the activity of the free ion would be 100 times less, at about 20 μg l^{-1}. The solubility of the total metal, however, might be higher than that indicated by the free ion activity because of the formation of soluble metal complexes. Figure 2.4 shows the activity of the free Zn^{2+} and various Zn hydroxide complexes in equilibrium with α-Zn(OH)$_2$(s) as a function of pH. The formation of higher hydroxy complexes causes an increase in the activity of total Zn above pH 11.

Distinct mineral phases of trace elements have been reported rarely in soils [29]. Many mineral phases in soils consist of mixed solids in which the trace element constitutes only a small proportion of the precipitate. Homogeneous nucleation and precipitation practically occur only when the solubility product has been exceeded to a certain extent [51]. The presence of other mineral surfaces reduces the extent of supersaturation needed for precipitation, which explains why heterogeneous nucleation is probably the most important process of crystal formation in soil systems. The mineral reduces the energy barrier for the nuclei of the new crystals to form from solution by providing a sterically similar,

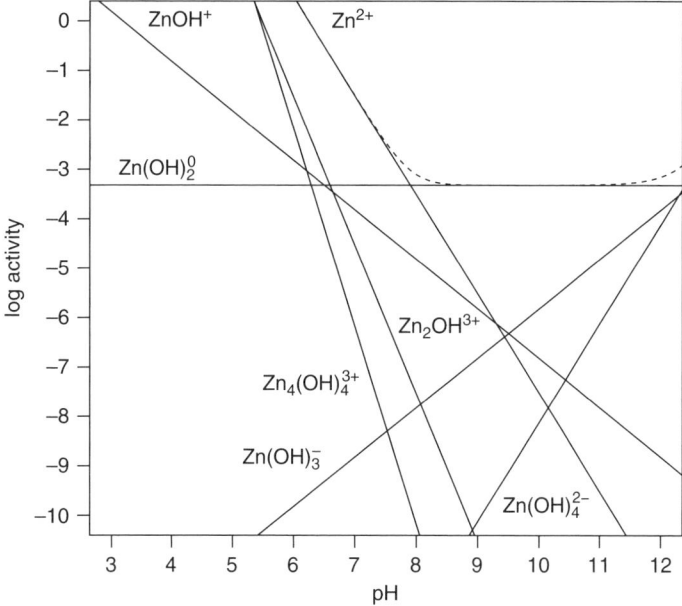

Figure 2.4 Activity of free Zn and various zinc hydroxide complexes in equilibrium with solid α-Zn(OH)$_2$ as a function of pH. Constants are taken from [50]. The dotted line represents the total of the activities of the different species

though chemically foreign, surface for nucleation. The similarity of lattice dimensions is an important factor here. The energy barrier arises from the fact that the small crystallites that initially form in the crystallization process are more soluble than large crystals because of surface effects [51]. Heterogeneous nucleation can result in the formation of a coating on the surface of a solid phase. Precipitation may also involve the formation of a solid mixture either by inclusion or by co-precipitation.

2.6 Mobilization of Trace Elements

2.6.1 pH and Redox Potential

The soil solution pH and redox potential (E_h) directly and indirectly influence all chemical processes, and consequently also determine the behaviour of trace elements in soil. The combined effects of pH and E_h on the mobility of trace elements are complex and highly element specific. Broad trends for different groups of elements are summarized in Figure 2.5.

The solubility of the trace elements that can occur as free hydrated cations generally increases with decreasing pH. Various factors explain this behaviour: (i) competition for sorption; (ii) decreasing pH-dependent negative charge of the sorption complex; and

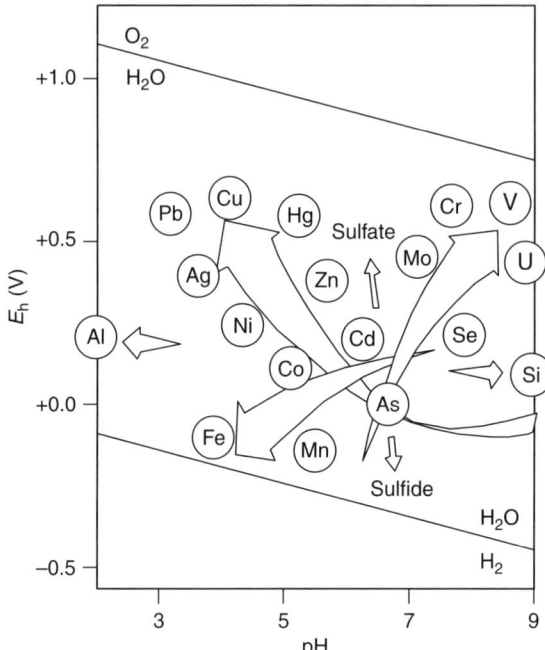

Figure 2.5 Schematic representation of major trends for increasing element mobilities in soils (broadening arrows) as a function of redox potential and pH (Reprinted from U. Förstner, Metal speciation – general concepts and applications, *International Journal of Environmental and Analytical Chemistry*, 1993, 51, 5–23. Reprinted by permission of Taylor & Francis Ltd [52].)

(iii) dissolution of soil components. (i) With a decrease in pH of the soil solution, there is an increase in the activity of H^+, Fe^{3+}, Al^{3+} and their positively charged hydroxide in the soil solution. These cations will compete with the trace elements for negative sorption sites. (ii) The total amount of negative sorption sites decreases with decreasing pH. The pH-dependent negative charges on the solid phase, caused by the dissociation of surface OH groups on minerals or functional groups on organic colloids, are neutralized by protonation. In addition, positive charges are created by covalent binding of H^+ on hydrated oxides of iron and manganese, or organic functional groups. The overall negative charge on the sorption complex therefore decreases. Below the point of zero charge, the soil colloids acquire a net positive charge. (iii) Several soil components become unstable with decreasing pH. While free calcium carbonate is stable only in soils with pH 7.5 and higher, hydroxides of aluminium will significantly dissolve at pH values below 5.5, and those of Fe when pH is lower than 3.5 [49]. When relative mobility of metals is expressed as the percentage of the total content dissolved as the pH is decreased below 6, the mobility of metals decreases in the order Cd > Zn > Ni > Mn > Cu > Pb > Hg [10].

Elements that exist as anions, for example As, Mo, Se, Cr(VI), are more mobile in alkaline conditions (Figure 2.5). Anions are increasingly sorbed with decreasing pH because soil colloids increasingly acquire additional positive charge. This presumes that the element continues to exist as a negatively charged species at lower pH. The dominant ionic form of Mo remains MoO_4^{2-} down to a pH as low as 4.5 [19]. In contrast, inorganic trivalent As in solution predominantly exists as the uncharged species $H_3AsO_3^0$ in solution within the normal pH range of soils (pH 4–8). It is therefore less subjected to electrostatic sorption than the oxidized species, arsenate, that in soils is present as a negatively charged oxyanion, $H_2AsO_4^-$ between pH 4 and 7, and $HAsO_4^{2-}$ at higher pH values [53]. Arsenate therefore strongly sorbs onto iron oxide surfaces in acidic to near-neutral conditions [54].

Trace elements tend to be less mobile in reducing conditions than in oxidizing conditions (Figure 2.5). Insoluble large molecular humic material and sulfides are likely to control the behaviour of several elements in reducing conditions [55]. Because of the extremely low solubility of many metal sulfides, even small activities of sulfide will cause these metals to precipitate. For example, Billon *et al.* [56] found that the pore water of sediments was strongly oversaturated with respect to various metal sulfides, even in the top layer with acid-volatile sulfide levels in the order of only 0.07 g kg^{-1}.

Within a few days microbial activity induces reducing conditions in soils when diffusion of atmospheric oxygen into the soil is limited or hindered, for example during flooding [57]. In order to use carbon sources for their growth, microorganisms utilize different oxidation–reduction reactions in the order of decreasing energy that is yielded from the reaction. Oxygen disappears at redox potentials of about 300 mV. Nitrate is reduced between 200 and 300 mV, followed by the reduction of Mn(IV) and Fe(III), and SO_4^{2-}. When there is a sufficient amount of sulfur in the system, FeS is less soluble and will precipitate rather than $FeCO_3$. At redox potentials below –300 mV, NH_4^+ becomes the dominant nitrogen form, and methanogenesis proceeds [58].

Upon flooding of an oxidized soil, the solubility of various metals strongly increases as the soil becomes reduced [57]. This is caused by the reduction of Fe(III) and Mn(III–IV), which in oxidizing conditions are present as largely insoluble hydrated oxides. The reduced species Fe^{2+} and Mn^{2+} exhibit a much higher solubility, as evidenced from the increasing pore water concentrations with time in an initially oxic soil that becomes permanently flooded (Figure 2.6a). The dissolution of the hydrated oxides of Fe and Mn results in the release of adsorbed and occluded elements, explaining an initial increase in pore water concentrations, which is illustrated for Cd in Figure 2.6b. After this initial increase, Cd soil concentrations decreased to very low levels in the reduced treatment, possibly owing to the reduction of SO_4^{2-} and consequential formation of CdS.

A reverse change, from reducing to oxidizing conditions, will also involve periodically high levels of trace elements in the soil pore water. Changes from reducing to oxidizing conditions involve transformations of sulfides and a shift to more acidic conditions [59,60]. Upon oxidation of a reduced soil or sediment, the important components that buffer against a decrease in pH are carbonates, exchangeable cations, clays and Al hydroxides [59]. The oxidation of metal sulfides, mainly pyrite, is a major cause of acidification in soils that are drained and aerated. This is particularly exemplified in acid sulfate soils. Pyrite soils originate in conditions where seawater or water high in sulfates is flooding iron oxide-bearing sediments and in the presence of organic matter [61]. When these soils are exposed to oxygen, for example after drainage, acid sulfate soils develop. The oxidation of pyrite (FeS_2) results in the release of sulfuric acid:

$$4\,FeS_2(s) + 14\,H_2O + 15\,O_2(g) \leftrightarrow 4\,Fe(OH)_3(s) + 8\,H_2SO_4(aq) \qquad (2.14)$$

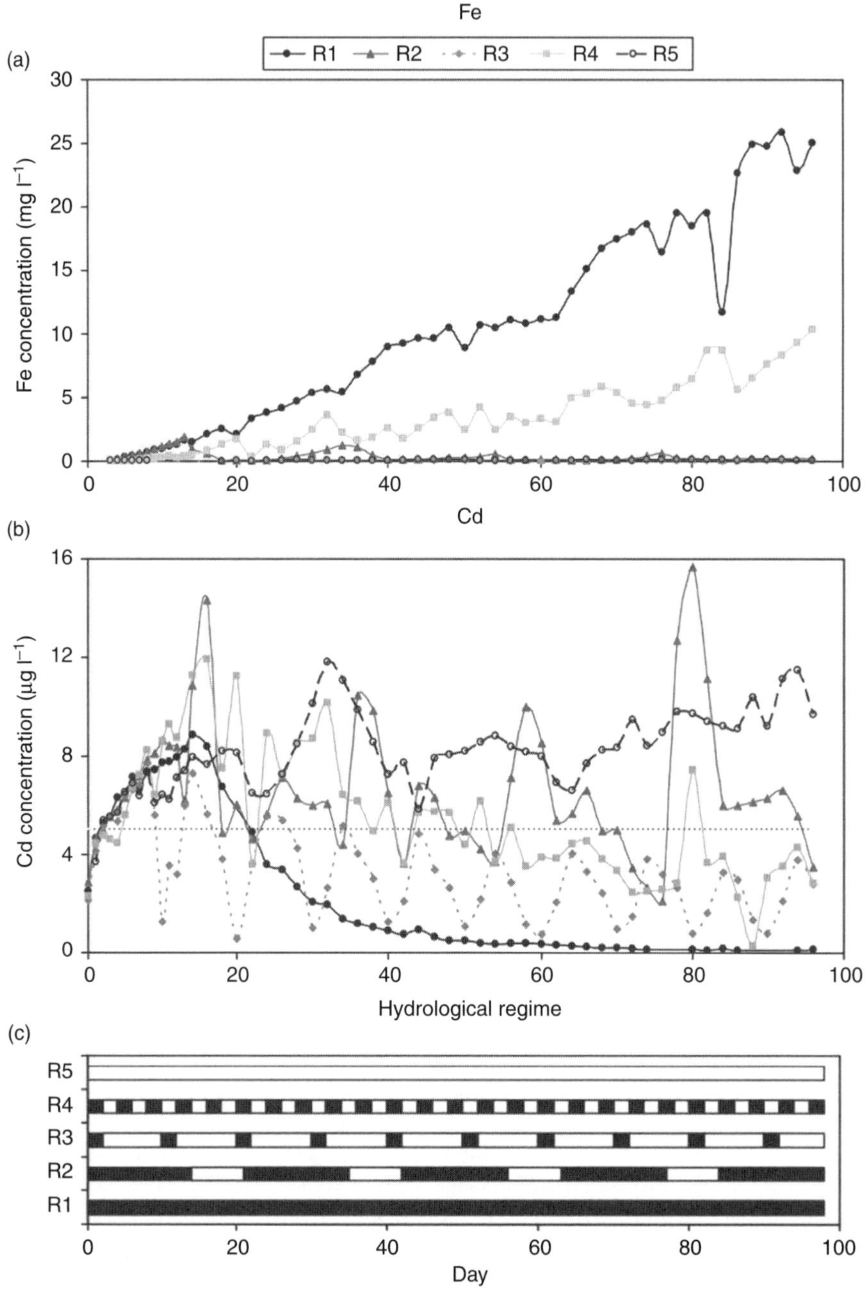

Figure 2.6 Concentrations of Fe and Cd in pore water of a soil, subjected to different hydrological regimes: R1, permanently flooded; R2, 3 and 4, intermittently flooded and dry; and R5, permanently at field capacity (Reprinted from Environmental Pollution, 147, G. Du Laing, D. Vanthuyne, B. Vandecasteele, F. Tack and M. Verloo, Influence of hydrological regime on pore water metal concentrations in a contaminated sediment-derived soil, 615–625. Copyright 2007, with permission from Elsevier [57].)

The decrease in pH to values as low as 2–3 causes a considerable release of Fe, Al and toxic trace metals in such soil.

Oxidizing, aerobic conditions also favour the mineralization of organic matter and thus the release of elements associated with organic matter. This will increase the mobility of elements such as Hg, Zn, Pb, Cu, and Cd [52]. These released metals will then be mostly immobilized by other processes, though somehow less effectively than in a reducing environment [62].

While changes in the redox state of an environment causes strong changes in the behaviour and mobility of trace elements, many of these such as Pb, Cd, Zn, do not change their valence state in the natural environment. In specific circumstances, Cu(II) might be reduced to the monovalent form, subsequently leading to Cu_2S precipitation [63]. Elements with more complex redox chemistry include Hg, Cr, Se, As among others.

2.6.2 Influence of Soil Constituents

The soil texture is an important factor in trace element retention or release. In general, coarse-grained soils exhibit a lower tendency for trace element sorption than fine-grained soils. The fine-grained soil fraction contains soil particles with large surface reactivities and large surface areas such as clay minerals, iron and manganese oxyhydroxides, humic acids, and others [34]. Clay minerals are a significant source of negative surface charges in soil and are a major contributor to their cation exchange capacity, particularly in mineral soils. They are therefore an important solid phase for retaining positively charged ions through electrostatic sorption. Besides the amount, the type of clay mineral is of great importance. For example, montmorillonite is characterized by a very high CEC, between 80 and 100 $cmol_+\ kg^{-1}$, illite has an average CEC, between 15 and 40 $cmol_+\ kg^{-1}$ while kaolinite has a relatively low CEC between 3 and 15 $cmol_+\ kg^{-1}$ [37]. Generally, clay particles are negatively charged silicate minerals and therefore preferentially sorb positively charged ions. However, it has been reported that sorption of As oxyanions from soil solution occurs by chemisorption or ligand exchange on clay surfaces, mainly by replacing or competing with phosphate [22]. Boron is effectively sorbed by clay minerals by means of ligand exchange, and is very little subjected to anion exchange [64].

Hydrous oxides occur generally as partial coatings on the silicate minerals rather than as discrete, well-crystallized minerals. This allows the oxides to exert chemical activity which is far out of proportion to their concentration [65]. Hydrous Fe oxides have a zero charge at pH ranging from 7 to 10 [66,67]. Soils are generally below pH 8.5 and Fe oxide surfaces are expected to be mostly positively charged. Despite their net positive charge, they retain a high affinity and sorption capacity for cationic trace elements. The mechanisms for sorption include isomorphic substitution of divalent or trivalent cations for Fe and Mn ions, and cation exchange. The range of pH for 50 % sorption on amorphous $Fe_2O_3\ H_2O$ was reported to be 5.9–7.3 for Cd, 6.1–6.7 for Zn, 5.2–5.8 for Cu, and 4.1–5.0 for Pb [68]. Kinniburgh et al. [66] reported pH values for 50 % sorption on freshly precipitated Fe oxide gels as 7.8 for Mg, 7.4 for Sr, 6.0 for Co, 5.8 for Cd, 6, 5.6 for Ni, 5.4 for Zn, 4.4 for Cu, and 3.1 for Pb. For Al oxide gels, this was 9.2 for Sr, 8.1 for Mg, 6.6 for Cd, 6.5 for Co, 6.3 for Ni, 5.6 for Zn, 5.2 for Pb, and 4.8 for Cu. The net positive charge also favours sorption of anions, and this becomes more prominent with decreasing pH.

By ligand exchange with surface hydroxyl groups anions such as arsenate [22], borate [64], molybdate [69] and chromate [70] are specifically adsorbed on oxide minerals. Although B exists predominantly as undissociated boric acid in solution, direct evidence has been provided for the sorption of both trigonal B (undissociated acid) and tetrahedral B (borate anion) on the surface of amorphous Al hydroxide [71]. As pH becomes alkaline, anions start desorbing due to the change in the iron-oxide net surface charge from positive to negative. Fresh, amorphous oxide gels are most effective in sorbing trace metals. In moist upland soils, temporary anoxic conditions lead to some reductive dissolution and reprecipitation, which keeps the oxides in a highly reactive state with respect to trace metal and anion sorption [65,72].

Although soil organic matter might constitute only 2–10 % of the soil composition, it helps maintaining a good soil structure and plays a key role in the various physical, biological and chemical processes in soils, including retention of trace elements. Organic matter is a rich source of negative charges and therefore can make a significant contribution to the CEC of a soil. The CEC of soil organic matter ranges in the order of 100–300 $cmol_+$ kg^{-1}, depending upon its nature and composition [37]. It plays a very significant role in retaining essential nutrients and trace elements that are essential for plant growth. During decay, it constitutes a slow but continuous source of nutrient elements, including trace metals for plant growth [73].

Soil organic matter includes undecayed plant and animal tissues, their partial decomposition products, and the soil biomass. Thus, it includes (i) identifiable, high-molecular-weight organic materials such as polysaccharides and proteins, (ii) simpler substances such as sugars, amino acids and other small molecules, and (iii) humic substances [74]. Traditionally, humic substances have been classified according to their solubility in fulvic acids, humic acids and humins. Fulvic acids are soluble in acidic and alkaline medium. They include the lighter coloured, low-molecular-weight (order 2000–10 000 dalton) fraction. Humic acids are only soluble in alkaline medium and constitute a darker coloured fraction of higher molecular weight on the order of 100 000–200 000. Fulvic acids have higher oxygen but lower carbon contents than humic acids. They contain more acidic functional groups, particularly COOH. Humins, the third major group of humic substances, are insoluble in water and are the most resistant to decomposition [73].

The retention mechanisms of trace elements by organic matter involve not only the formation of inner-sphere complexes but also ion exchange and precipitation reactions. According to Stevenson and Ardakani [75] the stability constants (log K) of metal–fulvic acid complexes at pH 3.5 decrease according to the sequence (with log K in parentheses) Cu (5.8) > Fe (5.1) > Ni (3.5) > Pb (3.1) > Co (2.2) > Ca (2.0) > Zn (1.7) > Mn (1.5) > Mg (1.2). At pH 5, the sequence is Cu (8.7) > Pb (6.1) > Fe (5.8) > Ni (4.1) > Mn (3.8) > Co (3.7) > Ca (2.9) > Zn (2.3) > Mg (2.1). Thus, the stability of metal–organic complexes increases with pH. The stability constants reflect the great affinity of Cu^{2+}, Pb^{2+} and Fe^{3+} for organic complex formation. Above pH 6–7, most metals in solution exist as organic complexes. Organic complexes of Cu and Pb remain stable until pH 4, while complexes of Cd and Zn are less stable and dissociate when the pH is below 6 [73].

Fulvic acids are fairly soluble in both acidic and alkaline environments and therefore tend to contribute to an enhanced mobility of trace elements. Humic acids exhibit a more complex interaction and solubility. They are insoluble at low pH, but become more soluble as pH increases. They behave as colloids that are flocculated in the presence of sufficient dissolved salts, and particularly divalent cations. When the electrolyte concentration is low, humic acids become deflocculated as colloidal suspension and move up and down the soil profile depending on the ground-water currents [31].

At a pH below 6, metal–organic complexes, which are mostly negatively charged, may sorb onto iron oxides, which acquire a net positive charge in these conditions. Moreover, the solubility of humic acids decreases with decreasing pH. Both factors explain why organic complexes of metals might still be present in acidic soils (pH < 4). Significant fractions of Pb and Cu can be sorbed by organic matter at pH 3, and of Cd between pH 3 and 4 [17,76]. Solid-phase complexes of Zn with organic matter are not stable below pH 5. Soils high in organic matter therefore can bind metals efficiently even in acidic soil conditions. At pH > 6 or 7, the concentration of metals in solution can increase due to the formation of soluble metal–organic complexes.

Overall, soil organic matter exhibits a low solubility. Water-extractable organic matter typically constitutes less than a few per cent of the total soil organic matter [77]. It nevertheless plays a dominant role in the mobilization and transport of trace elements, in particular Cu and Pb, as soluble organic complexes [78–80]. It also considerably influences the toxicity of elements. Metals are known to be less toxic to aquatic organisms when they exist as a complex with dissolved organic matter [81].

The presence of free $CaCO_3$ generally reduces the solubility of trace elements, as $CaCO_3$ raises the soil pH. Furthermore, the accompanying carbonate/bicarbonate ions will form metal carbonates that may be precipitated or are only sparingly soluble. Free $CaCO_3$ in soils therefore controls the solubility of the trace elements via its influence on pH and the formation of metal carbonates [65]. Sorption on carbonate phases to some extent accounts for the low solubility of trace metals, and precipitation of Cd and Cu have been observed in some cases [82,83]. Also, carbonates constitute a metal solubility-controlling phase in reduced environments [55]. Boron and arsenate are significantly sorbed to carbonate mineral surfaces by ligand exchange or chemisorption at pH below 10 [22,64]. Sorption decreases with higher pH, owing to the carbonates acquiring a negative charge above that pH [22].

2.7 Transport

Trace elements in mobile forms, that is in true solution and those associated with colloidal and suspended material, can migrate downward. The transport of solutes in the liquid phase is governed by the processes of advection and hydrodynamic dispersion [84,85]. Advection is the movement of dissolved or suspended particles along with the solution. Hydrodynamic dispersion is the combined effect of molecular diffusion and mechanical dispersion. Molecular diffusion is caused by Brownian motion of the particles. Because of this random motion, particles tend to migrate from high-concentration zones towards places in the solution where the concentration is lower. Mechanical dispersion occurs as a result of the irregular shape of the soil particles, which causes individual particles to follow different pathways in the porous structure (Figure 2.7). As a result, the true, microscopic velocity of the particles is different from the average macroscopic velocity.

26 *Basic Principles, Processes, Sampling and Analytical Aspects*

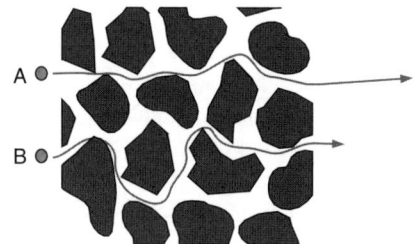

Figure 2.7 *Illustration of the effect of mechanical dispersion. Owing to the irregularity of the porous medium, characterized by the tortuosity factor, pollutant B will be retarded compared to pollutant A*

The advection is largely determined by the hydraulic conductivity of soils, which in turn depends on the soil texture. Well-sorted sand or gravel is pervious, with hydraulic conductivities between 10^2 and 10^{-1} cm s^{-1}. The hydraulic conductivity of very fine sands and silt is in the order of 10^{-2} to 10^{-5} cm s^{-1}, while clays have hydraulic conductivities below 10^{-6} cm s^{-1}, and therefore are practically impervious [86]. Accordingly, the risk of downward migration or leaching of metals to ground water is much greater in coarser- than finer-textured soils.

Sorption to the solid phase during percolation considerably slows down the migration of particles compared with the flow of the solution. A simulated migration of Zn leached from a waste material with time is presented in Figure 2.8 [87]. Leaching tests were used to estimate the change of concentrations in the leachate from the waste with time. The initial leachate Zn concentration was 120 mg Zn l^{-1} and decreased to about 5 mg l^{-1} at the end of the leaching test. The liquid to solid ratios of the leaching test can be related to a time scale that indicates the time needed to reach these liquid to solid ratios in the field situation under consideration [88]. In a scenario where annual precipitation was 780 mm and the height of the waste material was 1 m, the laboratory-observed leaching behaviour would correspond with a decrease in leachate Zn concentration from an initial level of 120 mg l^{-1} that is, after disposal of the waste material, to 10 mg l^{-1} after 30 years, and to about 5 mg l^{-1} after more than 100 years [87]. These leachate concentrations were used as input for a simple one-dimensional diffusion dispersion model. Soil sorption characteristics were estimated using batch experiments for determining sorption isotherms (see Section 2.4.2). The resulting migration of Zn with time is presented in Figure 2.8 for three soils with different Zn sorption behaviour.

The influence of the sorption properties of different soils is highly important and determines the extent to which the contaminants migrate to the deeper soil layers and eventually into the ground water (Figure 2.8). In the light sandy soil, metals quickly migrate over a larger depth. This results in a dilution effect. The accumulation in the soil is modest, but extends over a larger depth. In contrast, metals are very strongly retained in the clay soil (Figure 2.8). There is a high accumulation, restricted in this example to at most 40–50 cm in the long term.

With these model calculations it was estimated that the velocity at which metals move downwards was in the order of 1–5 cm year^{-1} for Zn, 0.1–1 cm year^{-1} for Cd,

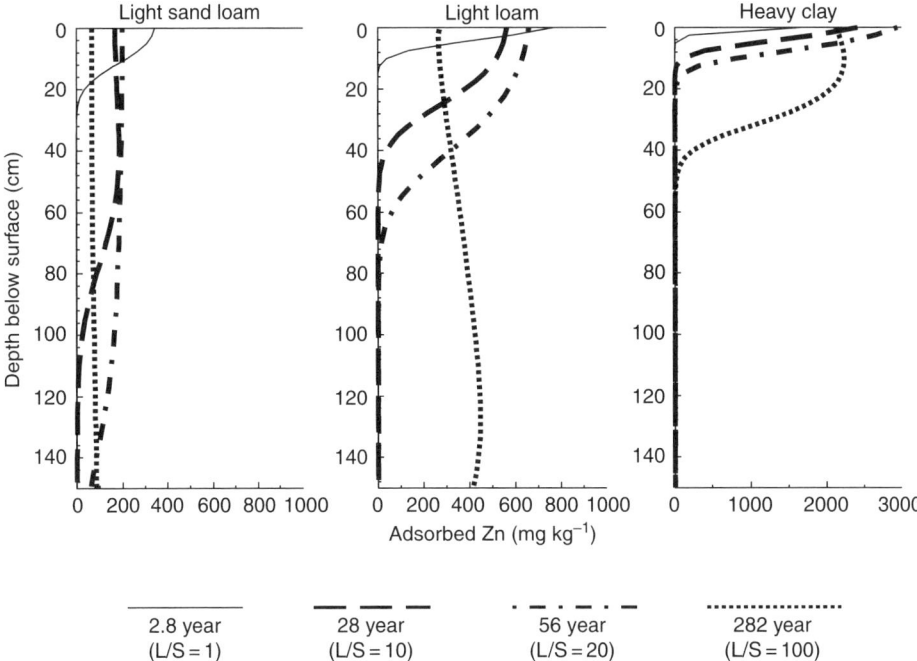

Figure 2.8 *Calculated migration profiles of Zn as a function of time for a light sand loam, a light loam and a heavy clay soil (Adapted from F.M.G. Tack et al, Leaching behaviour of granulated non-ferrous metal slags, in J.P. Vernet, Environmental Contamination, Elsevier, Amsterdam, 1993, 103–117 [87].)*

and 0.01–0.8 cm year^{-1} for Cu and Pb. The macroscopic pore water velocity was 1.1 cm day^{-1} in the modelled example. Thus, because of sorption, the relative velocity of metal movement was between 0.3 % and 0.003 % of the velocity at which water moved through the soil. Gerritse *et al.* [89] reported values between 0.1 % and 0.01 %.

Two factors cause metal leaching under field conditions to be greater than that observed in column metal leaching studies performed on homogeneous soils [90]. First, the speciation of the dissolved elements might alter the sorption behaviour. Dissolved organic matter in particular might complex metals. This formation of soluble organometallic complexes accounts for a much larger mobility of metals, particularly Cu, than what is expected from sorption characteristics of Cu^{2+} in soils determined in column sorption studies [78,90]. The second factor is preferential flow, that is the rapid transport of water and solutes through cracks and macropores in the soil that bypasses a large part of the soil matrix [79,91]. Within a period of 20 years, an estimated 43 % of Cu, 23 % of Cd and 38 % of Zn were lost from the topsoil of a heavily loaded sludge application site [79]. Examination of the bulk subsoil did not indicate a statistically significant increase in metal concentrations. Losses were accounted to preferential flow and metal complexation with soluble organics.

2.8 Plant Uptake

Plants acquire nutrients, including trace elements, from the soil for their growth. Also trace elements that are not needed for development and growth are taken up. A relative measure of the transfer of trace elements from the soil to the plant is expressed by the transfer factor (TF), also referred to as accumulation index or bioconcentration factor [92,93]:

$$\text{TF} = \frac{\text{concentration in the plant}}{\text{concentration in the soil}} \qquad (2.15)$$

The TF can be used as a measure of element accumulation efficiency [94]. It reflects the relative mobility of elements in the soil–plant system, which is element- and plant-specific.

The tolerance index (TI) represents the ratio of biomass for plants grown in soils with elevated levels of elements compared to plants grown in control soils with baseline elemental contents [92,93].

$$\text{TI} = \frac{\text{biomass}_{\text{contaminated soil}}}{\text{biomass}_{\text{reference soil}}} \qquad (2.16)$$

TI values lower than 1 indicate a net decrease in biomass and suggests that the plants are stressed, whereas TI values equal to 1 indicate no difference relative to non-contaminated control conditions [94].

Trace element contents in plants are only poorly related to soil total element contents [94]. Plants mostly take up trace elements from the soil solution. The availability of an element for plant uptake is related to the concentration in the soil solution at a particular moment in time (the intensity factor, I). Equally important is the capacity of the soil to maintain a certain concentration in the soil solution (the capacity factor, Q). Additionally, the kinetic aspect, that is the speed at which elements can be released from the solid phase of the soil to a form available for plant uptake, is of great influence [17].

Although a direct exchange between plant roots and soil particles is possible, it is mostly the concentration in the soil solution that determines the availability of trace elements to plants [17]. Plant roots absorb trace elements from the solution, eventually depleting their concentrations in the rhizosphere. This can lead to resupply via the dissociation of metal complexes present in the solution phase as well through the release of labile elements associated with the solid phase. This resupply will be determined by the concentration of total labile micronutrient or contaminant in solution, its diffusional supply, its labile concentration in the solid phase, and the rate at which it is released from solid phase to solution [95]. The effective availability of trace elements to plants generally is limited because of their low solubility in the soil solution, particularly in noncontaminated environments. Any factor that affects the solubility of trace elements in soils will also tend to affect uptake of elements by plants. However, plant-specific mechanisms of plant metal tolerance and homeostasis mechanisms will play a major role in determining the effective uptake of elements [96].

Physicochemical factors that influence the solubilization and mobilization of trace elements have been discussed in the previous sections. In order to maintain the concentration of essential trace elements within physiological limits and to minimize the detrimental effects of nonessential metals, plants, like all other organisms, have evolved a complex network of homeostatic mechanisms that serve to control the uptake, accumulation, trafficking and detoxification of metals [96]. These plant specific physiological responses, which are highly species- and even clone-specific [97], can greatly weaken the relations between mobile element concentrations as determined by soil solution concentrations and extractable contents, and metal concentrations in plant tissues.

Trace element accumulation rates in plants are dependent consecutively on mobilization and uptake from the soil, compartmentalization and sequestration within the root, efficiency of xylem loading and transport, distribution between metal sinks in the aerial parts, sequestration and storage in leaf cells [98]. To increase the uptake of nutrients, plants actively change the soil environment in the rhizosphere by the exudation of substances that are involved in the uptake mechanism. The processes responsible for changes in the rhizospheric pH involve the evolution of CO_2, the release of root exudates, the excretion or uptake of H^+ and HCO_3^-, and microbial production of organic acids [99]. Plant roots are known to release Fe by either reduction to Fe(II) or complexation with Fe(III)-chelating phytosiderophores [98].

Following mobilization, an element has to be captured by root cells. Elements are first bound by the cell wall, which acts as an ion exchanger of comparatively low affinity and low selectivity [98]. Transport systems and intracellular high-affinity binding sites then mediate and drive uptake across the plasma membrane. Specific metal transporter proteins are involved in the active uptake of elements. In recent years the molecular understanding of the entry of metal ions into plant cells has increased tremendously [96].

Many examples can be found in literature where relations between soil physicochemical properties and concentrations of trace elements in the plant are evidenced. An accurate general prediction of plant metal contents from soil properties nevertheless remains difficult. Several models have been developed to predict the phytoavailability of trace elements, especially of Cd, Zn, Cu and Pb, but they are rather limited to a given plant and specific growth conditions [11]. At most, ranges of expected metal contents can be delineated, which for some elements might differ depending on soil characteristics.

Factors that limit the solubility and mobility of elements in the soil solution will also generally tend to limit their uptake by plants. The soil pH is one of the most significant factors influencing trace elements in the solution, and hence is expected to influence plant uptake to a great extent. In a greenhouse experiment, Cd in oat (*Avena sativa* L.), ryegrass (*Lolium multiflorum* L.), carrot (*Daucus carota* L.) and spinach (*Spinacea oleracea* L.) in most cases decreased with increasing soil pH, controlled by liming. There was a strong negative correlation between soil pH and the log transfer factor for Cd at pH 4.5–7.2 for experimental data from past soil-crop surveys for Cd [100].

The influence of organic matter on the uptake of trace elements by plants is more complex. Plant uptake might increase due to an increasing activity of elements in the soil solution through organic complexation [101]. However, organic matter also has immobilizing effects (see Section 2.6.2), and the net effect is element specific and is dependent on the specific soil environment [102].

Uptake of metal cations is significantly influenced by their presence as complexes in the soil solution. Increases in Cd availability in the field due to the presence of chlorides have repeatedly been observed [30,103,104]. According to the free ion hypothesis for uptake of metals by plants, the uptake of a metal is governed by the activity of the free ion in solution [105]. This explains why Cd uptake decreased with increasing Cl activity in a nutrient solution where the Cd^{2+} activity was not buffered [106]. Subsequent work has shown that the free ion hypothesis holds only to a limited extent [106–110]. This may be caused by differences in buffering of the free metal ion activity. In soil solutions, the free ion activity is to an extent buffered by the soil solid phase, which will release free metal cations when the activity in the soil solution decreases, for example as a result of uptake by plants. Strong complexing agents in solution also can buffer the free metal activity. In hydroponic experiments, where different chelator-buffered systems were employed, there was no single relationship between the activity of the free metal ion in solution and metal uptake by plants. At any given free metal ion activity, plant metal uptake depended on the type of ligand in solution [107]. The reasons for differences in plant element concentrations between different chelator-buffered systems might be that either intact metal–ligand complexes are taken up, or ligands interact in decreasing diffusional limitations to free metal uptake in the zone adjacent to the plant root and in the apoplast [107].

The uptake of trace elements by plants is also affected by nutrient interactions. These are generally measured in terms of growth response and change in concentration of nutrients [111]. The interaction is synergistic when the increase in growth response is more than when adding each nutrient separately. An antagonistic interaction occurs when the response to two nutrients together is lower than the responses to each nutrient individually. There is no interaction when the effects are additive. A nutrient may interact simultaneously with more than one other nutrient. This may induce deficiencies or toxicities, or may modify growth responses, and/or nutrient composition of the plant tissue [111]. The interactions can be complex, as illustrated for example by McKenna *et al.* [112]. Increasing solution Cd increased Zn concentrations in young leaves of lettuce but not of spinach, regardless of Zn levels. Cadmium concentrations in young leaves of both crops decreased exponentially with increasing solution Zn at low (3.5 $\mu g\ l^{-1}$) but not at high (35 $\mu g\ l^{-1}$) solution Cd [112].

Uptake of trace elements and distribution into the plants is highly species specific, and even clone specific. For example, 19 inbred lines of maize (*Zea mays* L.) when grown in identical conditions showed shoot Cd concentrations ranging from 0.9 to 9.9 mg kg^{-1} dry weight [113].

Field soils are characterized by a degree of heterogeneity that is also responsible for variations in the uptake of elements by plants. Spatial heterogeneity of metal concentrations in soils is an important source of uncertainty that must be considered in risk assessment [114]. In a field survey, metal contents in stinging nettle (*Urtica dioica* L) had no clear relationships with soil properties or soil metal contents for most metals [115]. Only for Zn, was there an effect (Figure 2.9). The plant Zn content ranged between 50 and 500 mg kg^{-1} for soils low in clay or organic matter content. For soils with clay content higher than 10 %, and/or organic matter content higher than 3 %, Zn varied within a more limited range, between 50 and 100 mg kg^{-1}.

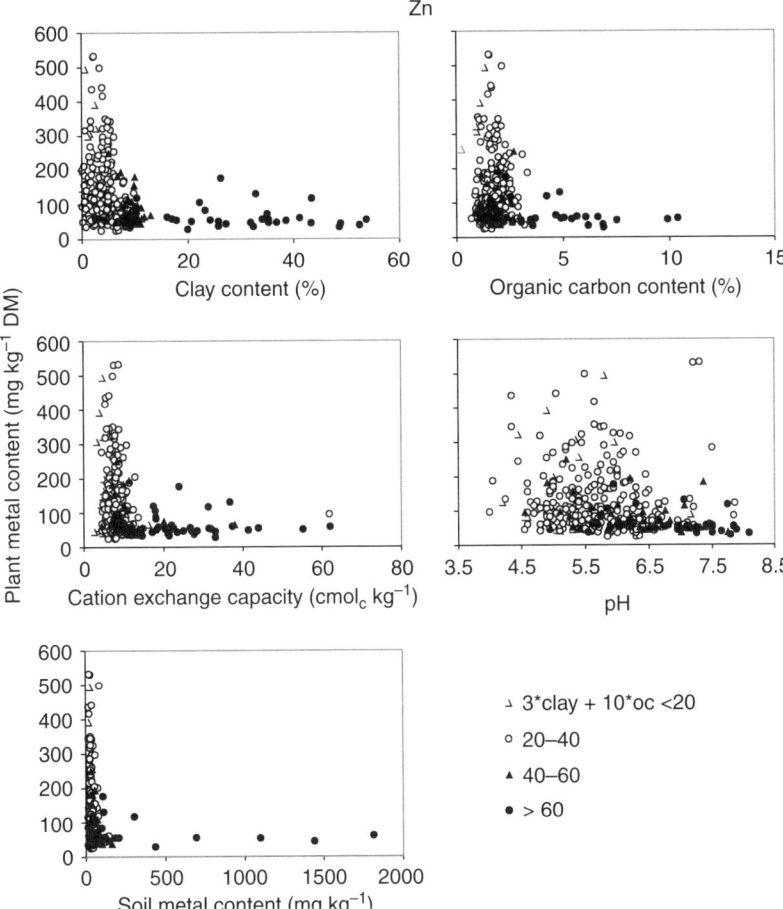

Figure 2.9 *Relations between Zn content in stinging nettle* (Urtica dioica L.) *and soil properties (Reproduced from Sci. Total Environ., 192, F.M. Tack and M.G. Verloo, Metal contents in stinging nettle* (Urtica dioica L.) *as affected by soil characteristics, 31–39. Copyright 1996, with permission from Elsevier [115].)*

2.9 Concluding Remarks

Some trace elements are absolutely essential in trace amounts for biological life. Any trace element becomes toxic when taken up in larger amounts. The soil is the most important reservoir of trace elements in terrestrial systems. Trace elements may be taken up from the soil by plants or biota and cycle in biological tissues before ultimately returning to the soil through decaying biological remains. They might be transported with soil water and be removed by leaching to the underground, or by runoff to surface water. The extent by which this cycling is initiated is strongly determined by the different processes to which trace elements in soils are subjected.

Trace elements can exist in various chemical forms of association with different reactivities, which are highly element specific. These forms and their subsequent transformations are determined by the dynamic characteristics of a specific soil environment, which is defined by the composition of the soil, biological activity, water flows, and temperature. Different processes occurring simultaneously in soils might have opposite effects. The most dominant processes will ultimately determine the overall changes in reactivity, mobility and plant availability of trace elements.

The soil environment in general tends to have an immobilizing effect on most trace elements. Mobile amounts that move between the different environmental compartments are usually a tiny fraction of the total amount present in the soil system. In order to have significant short-term hazardous effects of trace elements, conditions already must be quite extreme in terms of trace element input combined with unfavourable soil conditions for the retention of elements, for example a low pH. The soil generally protects the ecosystems and ultimately humans against the potential hazardous effects of trace elements that have entered the soil environment through anthropogenic sources. However, environmental contamination can go unnoticed for a long time before adverse effects reach a stage where they can no longer be ignored. Trace elements as chemical elements in fact cannot be degraded. They can only be converted between different physical and chemical forms. Because of their generally low mobility in the soil system, they will tend to accumulate in an ecosystem. Environmental hazards can suddenly manifest themselves within a relative short time span in response to slow alterations in a chemical environment over time. For example, the depletion of free $CaCO_3$ in a contaminated environment might cause a decrease in pH of perhaps one unit or more. This would be accompanied by a sudden and strong increase in trace element levels in the soil solution to which the ecosystem is not adapted. Such possibility for sudden changes in the behaviour of contaminants has been referred to as a 'chemical time bomb' [116]. A precautionary principle must therefore be adopted when manipulating trace elements in industrial, agricultural and domestic applications.

Once significant hazards start to occur, it becomes extremely difficult to reverse the trends. To manage trace elements in soils, a good understanding of the various processes and mechanisms that govern their behaviour is a first prerequisite. Yet, quantitative prediction of metal behaviour in a specific environment will remain a difficult task because of the inherit complexity and diversity of soil systems.

References

[1] Adriano, D.C. Trace elements in the terrestrial environment; Springer Verlag, New York, 1986.
[2] Kloke, A.; Sauerbeck, D.R.; Vetter, H., The contamination of plants and soils with heavy metals and the transport of metals in terrestrial food chains; in Nriagu, J.O. (ed.), Changing Metal Cycles and Human Health; Springer, Berlin, 1984, pp. 113–141.
[3] Sakamoto, M.; Nakano, A.; Akagi, H., Declining Minamata male birth ratio associated with increased male fetal death due to heavy methylmercury pollution; Environ. Res. 2001, 87, 92–98.
[4] Nogawa, K., Itai-Itai disease and follow-up Studies; in Nriagu, J. (ed.); Cadmium in the Environment, Part 11, Health Effects; John Wiley & Sons, Inc., New York, 1981, pp. 1–37.

[5] Nelson, A.; Donkin, P., Processes of bioaccumulation: the importance of chemical speciation; Mar. Pollut. Bull. 1984, 16, 164–169.
[6] Lund, W., Speciation analysis – why and how; Fresenius' J. Anal. Chem. 1990, 337, 557–564.
[7] Inskip, H.; Beral, V.; McDowall, M., Mortality of Shipham residents: 40-year follow-up; Lancet, 1982, 1, 896–899.
[8] Ewers, U.; Brockhaus, A.; Dolgner, R.; Freier, I. et al., Environmental exposure to cadmium and renal function of elderly women living in cadmium-polluted areas of the Federal Republic of Germany; Int. Arch. Occup. Environ. Health 1985, 55, 217–239.
[9] Alloway, B.; Thornton, I.; Smart, G.; Sherlock, J.; Quinn, M., The Shipham report. An investigation into cadmium contamination and its implications for human health. Metal availability; Sci. Total Environ. 1988, 75, 41–69.
[10] Cottenie, A.; Verloo, M., Analytical diagnosis of soil pollution with heavy metals; Fresenius J. Anal. Chem. 1984, 317, 389–393.
[11] Kabata-Pendias, A., Soil-plant transfer of trace elements—an environmental issue; Geoderma, 2004, 122, 143–149.
[12] Jensen, D.L.; Ledin, A.; Christensen, T.H., Speciation of heavy metals in landfill-leachate polluted groundwater; Water Res. 1999, 33, 2642–2650.
[13] Kabata-Pendias, A.; Pendias, H., Trace Elements in Soils and Plants; CRC Press, Boca Raton, FL, 1984.
[14] Stumm, W.; Morgan, J.J., Aquatic Chemistry: An introduction emphasizing chemical equilibria in natural waters; John Wiley & Sons, New York, 1981.
[15] Bernhard, M.; Brinckman, F.E.; Sadler, P.J., The Importance of Chemical 'Speciation' in Environmental Processes; Springer Verlag, Berlin, 1986.
[16] Templeton, D.M.; Ariese, F.; Cornelis, R.; Danielsson, L.G.; Muntau, H.; Van Leeuwen, H.P.L.R., Guidelines for terms related to chemical speciation and fractionation of elements. Definitions, structural aspects, and methodological approaches (IUPAC Recommendations 2000); Pure Appl. Chem. 2000, 72, 1453–1470.
[17] Brümmer, G.W., Heavy metal species, mobility and availability in soils; in Bernhard, M.; Brinckman, F.E.; Sadler, P.J. eds; The Importance of Chemical 'Speciation' in Environmental Processes; Springer Verlag, Berlin, 1986, pp. 169–192.
[18] Brown, G.E.; Foster, A.L.; Ostergren, J.D., Mineral Surfaces and Bioavailability of Heavy Metals: a Molecular-Scale Perspective; Proc. Natl. Acad. Sci. U.S.A., 1999, 96, 3388–3395.
[19] Barrow, N.J., The four laws of soil chemistry: the Leeper lecture 1998; Aust. J. Soil Res. 1999, 37, 787–829.
[20] Bartlett, R.; James, B., Behavior of chromium in soils: III. Oxidation; J. Environ. Qual. 1979, 8, 31–35.
[21] Masscheleyn, P.H.; Pardue, J.H.; DeLaune, R.D.; Patrick, W.H.J., Chromium redox chemistry in a lower Mississippi Valley bottomland hardwood wetland; Environ. Sci. Technol. 1992, 26, 1217–1226.
[22] Sadiq, M., Arsenic chemistry in soils: An overview of thermodynamic predictions and field observations; Water Air Soil Pollut. 1997, 93, 117–136.
[23] Goldberg, S.; Lesch, S.M.; Suarez, D.L., Predicting boron adsorption by soils using soil chemical parameters in the constant capacitance model; Soil Sci. Soc. Am. J. 2000, 64, 1356–1363.
[24] Masscheleyn, P.H.; Delaune, R.D.; Patrick, W.H.J., Arsenic and selenium chemistry as affected by sediment redox potential and pH; J. Environ. Qual. 1991, 20, 522–527.
[25] ATSDR Toxicological Profile for Mercury; ATSDR/U.S. Public Health Service, Atlanta, GA, 1999.
[26] Manahan, S.E., Environmental Chemistry, 7th edn; CRC Press, Boca Raton, FL, 2000.
[27] Teeling, H.; Cypionka, H., Microbial degradation of tetraethyl lead in soil monitored by microcalorimetry; Appl. Microbiol. Biotechnol. 1997, 48, 275–279.
[28] Hoch, M., Organotin compounds in the environment – an overview; Appl. Geochem. 2001, 16, 719–743.
[29] Evans, L.J., Chemistry of metal retention by soils; Environ. Sci. Technol. 1989, 23, 1046–1056.

[30] McLaughlin, M.J.; Palmer, L.T.; Tiller, K.G.; Beech, T.A.; Smart, M.K., Increased soil salinity causes elevated cadmium concentrations in field-grown potato tubers; J. Environ. Qual. 1994, 23, 1013–1018.
[31] Livens, F.R., Chemical reactions of metals with humic material; Environ. Pollut. 1991, 70, 183–208.
[32] Sposito, G.; Schindler, P.W., Reactions at the soil colloid-soil solution interface.; in Landner, L. (ed.), Speciation of Metals in Water, Sediment and Soil Systems; Springer Verlag, Berlin, 1986, pp. 683–699.
[33] Fetter, C., Contaminant Hydrogeology; Macmillan, New York, 1993.
[34] Bradl, H.B., Adsorption of heavy metal ions on soils and soils constituents; J. Colloid Interface Sci. 2004, 277, 1–18.
[35] Bolt, G.H.; Bruggenwert, M.G.M., Soil Chemistry. A. Basic Elements; Elsevier, Amsterdam, 1976.
[36] Bolt, G.H., Soil Chemistry. B. Physico-chemical models; Elsevier, Amsterdam, 1979.
[37] Brady, N.C.; Weil, R.R., The Nature and Properties of Soils, 10th edn; Macmillan, New York; 1990.
[38] Tagami, K.; Uchida, S., Aging effect on bioavailability of Mn, Co, Zn and Tc in Japanese agricultural soils under waterlogged conditions; Geoderma 1998, 84, 3–13.
[39] Seuntjens, P.; Tirez, K.; Simunek, J.; van Genuchten, M.; Cornelis, C.; Geuzens, P., Aging effects on cadmium transport in undisturbed contaminated sandy soil columns; J. Environ. Qual. 2001, 30, 1050.
[40] Lock, K.; Janssen, C.R., The effect of ageing on the toxicity of zinc for the potworm *Enchytraeus Albidus*; Environ. Pollut. 2002, 116, 289–292.
[41] Sterritt, R.M.; Lester, J.N., The value of sewage sludge to agriculture and effects of the agricultural use of sludges contaminated with toxic elements: a review; Sci. Total Environ. 1980, 16, 55–90.
[42] Hooda, P.S.; Alloway, B.J., Cadmium and lead sorption behaviour of selected English and Indian soils; Geoderma 1998, 84, 121–134.
[43] van Gestel, C.; Koolhaas, J., Water-extractability, free ion activity, and pH explain cadmium sorption and toxicity to *Folsomia candida* (Collembola) in seven soil-pH combinations; Environ. Toxicol. Chem. 2004, 23, 1822–1833.
[44] Serrano, S.; Garrido, F.; Campbell, C.G.; Garcia-Gonzalez, M.T., Competitive sorption of cadmium and lead in acid soils of Central Spain; Geoderma 2005, 124, 91–104.
[45] Boekhold, S.; van der Zee, S.E.A.T.M., Significance of soil chemical heterogeneity for spatial-behavior of cadmium in field soils; Soil Sci. Soc. Am. J. 1992, 56, 747–754.
[46] Tichý, R.; Nýdl, V.; Kuzel, S.; Kolár, R., Increased cadmium availability to crops on a sewage-sludge amended soil; Water Air Soil Pollut. 1997, 94, 361–372.
[47] Temminghoff, E.J.M.; Van Der Zee, S.E.A.T.M.; De Haan, F.A.M., Speciation and calcium competition effects on cadmium sorption by sandy soil at various pHs; Eur. J. Soil Sci. 1995, 46, 649–655.
[48] Seuntjens, P., Field-scale cadmium transport in a heterogeneous layered soil; Water Air Soil Pollut. 2002, 140, 401–423.
[49] Lindsay, L.W. Chemical Equilibria in Soils; John Wiley & Sons, Inc., New York, 1979.
[50] Smith, R.M.; Martell, A.E.; Motekaitis, R.J., NIST critically selected stability constants of metal complexes database Version 3.0; U.S. Department of Commerce, Gaithersburg, MD, 1997.
[51] McBride, M., Processes of heavy metal and transition metal sorption by soil minerals; in Bolt, G.; De Boodt, M.; Hayes, M.; McBride, M.; De Strooper, E., eds; Interactions at the Soil Colloid–Soil Solution Interface; Kluwer Academic Publishers, Dordrecht, The Netherlands, 1991, pp. 149–176.
[52] Förstner, U., Metal speciation – general concepts and applications; Int. J. Environ. Anal. Chem. 1993, 51, 5–23.
[53] Masscheleyn, P.H.; Delaune, R.D.; Patrick, W.H.J., Effect of redox potential and pH on arsenic speciation and solubility in a contaminated soil; Environ. Sci. Technol. 1991, 25, 1414–1419.

[54] Fuller, C.C.; Dadis, J.A.; Waychunas, G.A., Surface chemistry of ferrihydrite: Part 2. Kinetics of arsenate adsorption and coprecipitation; Geochim. Cosmochim Acta 1993, 57, 2271–2282.
[55] Guo, T.; DeLaune, R.D.; Patrick, Jr, W.H., The influence of sediment redox chemistry on chemically active forms of arsenic, cadmium, chromium, and zinc in estuarine sediment; Environ. Int. 1997, 23, 305–316.
[56] Billon, G.; Ouddanea, B.; Laureyns, J.; Boughriet, A., Chemistry of metal sulfides in anoxic sediments; PCCP 2001, 3, 3586–3592.
[57] Du Laing, G.; Vanthuyne, D.; Vandecasteele, B.; Tack, F.; Verloo, M., Influence of hydrological regime on pore water metal concentrations in a contaminated sediment-derived soil; Environ. Pollut. 2007, 147, 615–625.
[58] Reddy, K.R.; Patrick, W.H.J., Effects of aeration on reactivity and mobility of soil constituents; in Ellis, R. (ed.), Chemical Mobility and Reactivity in Soil Systems; Soil Science Society of America, Madison, WI, 1983, pp. 11–33.
[59] Satawathananont, S.; Patrick, W.H.J.; Moore, P.A.J., Effect of controlled redox conditions on metal solubility in acid sulfate soils; Plant Soil 1991, 133, 281–290.
[60] Tack, F.M.; Callewaert, O.W.J.J.; Verloo, M.G., Metal solubility as a function of pH in contaminated, dredged sediment affected by oxidation; Environ. Pollut. 1996, 91, 199–208.
[61] Berner, R., Sedimentary pyrite formation: An update; Geochim. Cosmochim. Acta 1984, 48, 605–615.
[62] Gambrell, R.P., Trace and toxic metals in wetlands - a review; J. Environ. Qual. 1994, 23, 883–891.
[63] Simpson, S.; Rosner, J.; Ellis, J., Competitive displacement reactions of cadmium, copper, and zinc added to a polluted, sulfidic estuarine sediment; Environ. Toxicol. Chem. 2000, 19, 1992–1999.
[64] Goldberg, S., Reactions of boron with soils; Plant Soil 1997, 193, 35–48.
[65] Jenne, E.A., Controls on Mn, Fe, Co, Ni, Cu, and Zn concentrations in soils and water: the significant role of hydrous Mn and Fe oxides; Adv. Chem. Ser. 1968, 73, 337–387.
[66] Kinniburgh, D.G.; Jackson, M.L.; Syers, J.K., Adsorption of alkaline earth, transition, and heavy metal cations by hydrous oxide gels of iron and aluminum; Soil Sci. Soc. Am. J. 1976, 40, 796–799.
[67] Tripathy, S.S.; Kanungo, S.B., Adsorption of Co^{2+}, Ni^{2+}, Cu^{2+} and Zn^{2+} from 0.5 M NaCl and major ion sea water on a mixture of delta-MnO_2 and amorphous FeOOH; J. Colloid Interface Sci. 2005, 284, 30–38.
[68] Benjamin, M.M.; Leckie, J.O., Multiple-site adsorption of Cd, Cu, Zn, and Pb on amorphous iron oxyhydroxide; J. Colloid Interface Sci. 1981, 79, 209–221.
[69] Goldberg, S.; Forster, H.S.; Godfrey, C.L., Molybdenum adsorption on oxides, clay minerals, and soils; Soil Sci. Soc. Am. J. 1996, 60, 425–432.
[70] Becquer, T.; Quantin, C.; Sicot, M.; Boudot, J.P., Chromium availability in ultramafic soils from New Caledonia; Sci. Total Environ. 2003, 301, 251–261.
[71] Su, C.; Suarez, D., Coordination of adsorbed boron: a FTIR spectroscopic study; Environ. Sci. Technol. 1995, 29, 302–311.
[72] Cavallaro, N.; McBride, M., Effect of selective dissolution on charge and surface properties of an acid soil clay; Clays clay miner. 1984, 32, 283–290.
[73] Stevenson, F.J., Humus Chemistry: Genesis, Composition, Reactions; Wiley, New York, 1982.
[74] Saar, R.A.; Weber, J.H., Fulvic acid: modifier of metal-ion chemistry; Environ. Sci. Technol. 1982, 16, 510–517.
[75] Stevenson, F.J.; Ardakani, M.S., Organic matter reactions involving micronutrients in soils; in Mortvedt, J.J.; Giordano, P.M.; Lindsay, W.L. (eds), Micronutrients in Agriculture.; Soil Science Society of America, Madison, WI, 1972, pp. 79–114.
[76] Brümmer, G.; Herms, U., Influence of soil reaction and organic matter on the solubility of heavy metals in soils; in: Ulrich, B.; Pankrath, J. (eds), Effects of Accumulation of Air Pollutants in Forest Ecosystems; Reidel, Dordrecht, 1983, pp. 233–243.
[77] Kaiser, M.; Ellerbrock, R., Functional characterization of soil organic matter fractions different in solubility originating from a long-term field experiment; Geoderma 2005, 127, 196–206.

[78] Domergue, F.L.; Védy, J.C., Mobility of heavy metals in soil profiles; Int. J. Environ. Anal. Chem. 1992, 46, 13–23.

[79] Richards, B.K.; Steenhuis, T.S.; Peverly, J.H.; McBride, M.B., Metal mobility at an old, heavily loaded sludge application site; Environ. Pollut. 1997, 99, 365–377.

[80] Hoffmann, C.; Marschner, B.; Renger, M., Influence of DOM-quality, DOM-quantity and water regime on the transport of selected heavy metals; Phys. Chem. Earth 1998, 23, 205–209.

[81] Paquin, P.R.; Santoreb, R.C.; Wua, K.B.; Kavvadasa, C.D.; Di Tor, D.M., The biotic ligand model: a model of the acute toxicity of metals to aquatic life; Environ. Sci. Policy 2000, 3, 175–182.

[82] Cavallaro, N.; McBride, M., Copper and cadmium adsorption characteristics of selected acid and calcareous soils; Soil Sci. Soc. Am. J. 1978, 42, 550–556.

[83] Martin, H.W.; Kaplan, D.I., Temporal changes in cadmium, thallium, and vanadium mobility in soil and phytoavailability under field conditions; Water Air Soil Pollut. 1998, 101, 399–410.

[84] van Genuchten, M.; Wierenga, P., Mass transfer studies in sorbing porous media I. Analytical solutions; Soil Sci. Soc. Am. J. 1976, 40, 473–480.

[85] De Smedt, F., Simulation of ion transport in porous media; in Rondia, D. (ed.), Belgian Research on Metal Cycling in the Environment; Scope Committee, Brussels, Belgium, 1986, pp. 253–266.

[86] Bear, J., Dynamics of Fluids in Porous Media; Dover, NY, 1972.

[87] Tack, F.M.G.; Masscheleyn, P.H.; Verloo, M.G., Leaching behaviour of granulated non-ferrous metal slags; in Vernet, J.P. ed.; Environmental Contamination; Elsevier, Amsterdam, 1993, pp. 103–117.

[88] van der Sloot, H.A.; Comans, R.N.J.; Hjelmar, O., Similarities in the leaching behaviour of trace contaminants from waste, stabilized waste, construction materials and soils; Sci. Total Environ. 1996, 178, 111–126.

[89] Gerritse, R.; Vriesema, J.; Dalenberg, J.; De Roos, H., Effect of sewage sludge on trace element mobility in soils; J. Environ. Qual. 1982, 11, 359–364.

[90] Camobreco, V.J.; Richards, B.K.; Steenhuis, T.S.; Peverly, J.H.; McBride, M.B., Movement of heavy metals through undisturbed and homogenized soil columns; Soil Sci. 1996, 161, 740–750.

[91] Bundt, M.; Zimmermann, S.; Blaser, P.; Hagedorn, F., Sorption and transport of metals in preferential flow paths and soil matrix after the addition of wood ash; Eur. J. Soil Sci. 2001, 52, 423–431.

[92] Cottenie, A.; Camerlynck, R.; Verloo, M.; Velghe, G.; Kiekens, L.; Dhaese, A., Essential and non essential trace elements in the system soil-water-plant; Laboratorium voor Analytische en Agrochemie, Rijksuniversiteit Gent, Gent, 1979.

[93] Kiekens, L.; Camerlynck, R., Transfer characteristics for uptake of heavy metals by plants; Landwirtschaftliche Forschung 1982, 39, 255–261.

[94] Audet, P.; Charest, C., Heavy metal phytoremediation from a meta-analytical perspective; Environ. Pollut. 2007, 147, 231–237.

[95] Lehto, N.J.; Davison, W.; Zhang, H.; Tych, W., Analysis of micro-nutrient behaviour in the rhizosphere using a DGT parameterised dynamic plant uptake model; Plant Soil 2006, 282, 227–238.

[96] Clemens, S., Molecular mechanisms of plant metal tolerance and homeostasis; Planta 2001, 212, 475.486.

[97] Greger, M.; Landberg, T., Use of willow in phytoextraction; Int. J. Phytorem. 1999, 1, 115–123.

[98] Clemens, S.; Palmgren, M.G.; Kramer, U., A long way ahead: understanding and engineering plant metal accumulation; Trends Plant Sci. 2002, 7, 309–315.

[99] Tao, S.; Liu, W.X.; Chen, Y.J. *et al.*, Evaluation of factors influencing root-induced changes of copper fractionation in rhizosphere of a calcareous soil; Environ. Pollut. 2004, 129, 5–12.

[100] del Castilho, P.; Chardon, W.J., Uptake of soil cadmium by three field crops and its prediction by a pH-dependent Freundlich sorption model; Plant Soil 1995, 171, 263–266.

[101] Antoniadis, V.; Alloway, B.J., The role of dissolved organic carbon in the mobility of Cd, Ni and Zn in sewage sludge-amended soils; Environ. Pollut. 2002, 117, 515–521.
[102] Alloway, B.J.; Jackson, A.P., The behaviour of heavy metals in sewage sludge-amended soils; Sci. Total Environ. 1991, 100, 151–176.
[103] Norvell, W.; Wu, J.; Hopkins, D.; Welch, R., Association of cadmium in durum wheat grain with soil chloride and chelate-extractable soil cadmium; Soil Sci. Soc. Am. J. 2000, 64, 2162–2168.
[104] Weggler-Beaton, K.; McLaughlin, M.J.; Graham, R.D., Salinity increases cadmium uptake by wheat and Swiss chard from soil amended with biosolids; Aust. J. Soil Res. 2000, 38, 37–46.
[105] Parker, D.; Chaney, R.; Norvell, W., Chemical equilibrium models: Applications to plant nutrition research; in: Loeppert, R.; Schwab, A.; Goldberg, S. (eds), Chemical Equilibrium and Reaction Models; Special publication 42; Soil Science Society of America, Madison, WI, 1995, pp. 163–200.
[106] Smolders, E.; McLaughlin, M.J., Effect of Cl on Cd uptake by Swiss Chard in nutrient solutions; Plant Soil 1996, 179, 57–64.
[107] McLaughlin, M.; Smolders, E.; Merckx, R.; Maes, A., Plant uptake of Cd and Zn in chelator-buffered nutrient solution depends on ligand type; in Ando, T.; Fujita, K.; Mae, T.; Matsum, H.; Mori, S. and Sekiya, J. eds; Plant Nutrition for Sustainable Food Production and Environment; Springer, 1997, pp. 113–118.
[108] Parker, D.R.; Pedler, J.F.; Ahnstrom, Z.A.S., Resketo, M., Reevaluating the free-ion activity model of trace metal toxicity toward higher plants: experimental evidence with copper and zinc; Environ. Toxicol. Chem. 2001, 20, 899–906.
[109] Hough, R.L.; Tye, A.M.; Crout, N.M.J.; McGrath, S.P.; Zhang, H.; Young, S.D., Evaluating a 'free ion activity model' applied to metal uptake by *Lolium perenne* L. grown in contaminated soils; Plant Soil 2005, 270, 1–12.
[110] Degryse, F.; Smolders, E.; Merckx, R., Labile Cd complexes increase Cd availability to plants; Environ. Sci. Technol. 2006, 40, 830–836.
[111] Fageria, V.D., Nutrient interactions in crop plants; J. Plant Nutr. 2001, 24, 1269–1290.
[112] McKenna, I.M.; Chaney, R.L.; Williams, F.M., The effects of cadmium and zinc interactions on the accumulation and tissue distribution of zinc and cadmium in lettuce and spinach; Environ. Pollut. 1993, 79, 113–120.
[113] Florijn, P.J.; Van Beusichem, M.L., Uptake and distribution of cadmium in maize inbred lines; Plant Soil 1993, 150, 25–32.
[114] Millis, P.R.; Ramsey, M.H.; John, E.A., Heterogeneity of cadmium concentration in soil as a source of uncertainty in plant uptake and its implications for human health risk assessment; Sci. Total Environ. 2004, 326, 49–53.
[115] Tack, F.M., Verloo, M.G., Metal contents in stinging nettle (*Urtica dioica* L.) as affected by soil characteristics; Sci. Total Environ. 1996, 192, 31–39.
[116] Stigliani, W.M.; Doelman, P.; Salomons, W.; Schulin, R.; ter Meulen-Smidt, G.R.B.; van der Zee, S.E.A.T.M., Chemical time bombs – predicting the unpredictable; Environment 1991, 33, 4–9.

3
Soil Sampling and Sample Preparation

Anthony C. Edwards

Nether Backhill, Ardallie, By Peterhead, Aberdeenshire, Scotland, UK

3.1 Introduction

Soils are sampled and analysed in order to characterize a range of attributes and provide a value (or range of values) that is indicative of a specified area. For trace elements, sampling may be undertaken for the purpose of providing their soil concentration, which can then be used to assess the degree of site contamination. Trace elements are introduced in soils through a variety of sources which differ widely in their composition. Typical delivery is to the soil surface through a combination of passive (for example, atmospheric deposition) or active (for example, spreading of contaminated materials) mechanisms and supplement background 'geochemical' sources. The change in total content (dG/dt) over a time period (t) of a soil contaminant (G) depends on the input rate at the soil surface (plough) layer (A), the leaching rate at the lower boundary of the system (L) and the removal by harvesting plants (U) [1]:

$$\frac{dG}{dt} = A - L - U \quad (3.1)$$

The source plus delivery mechanism influence not only the absolute metal concentrations likely to be present in soil, but also a combination of the spatial distribution and chemical form/species present. Some historical knowledge of the contributing sources and site management can be of considerable help to the design of a sampling plan, sample collection and preparation stages. An indication of the degree of accumulation or loss of any target element can be obtained through a comparison of its concentration with some reference element [2].

Obtaining a representative sample is therefore a central objective of soil sampling [3]. One key soil attribute is its spatial variability in composition and this is responsible for the continuing uncertainty that surrounds describing soil properties. Soils develop a natural variability in their properties, which can result in a strong horizontal and vertical (profile) spatial dependence. Heterogeneity in analytical terms means that the composition of any one small proportion will not correspond to the average composition of the whole sample. Enrichment maybe extremely localized and the term 'hot spot' is frequently used in this context and may be regarded as [4]:

- an area of contamination within an otherwise uncontaminated site;
- an area of greater contamination within a site that is generally contaminated;
- an area of contamination above a guideline 'trigger' concentration;
- contaminant concentrations say, 2 standard deviations (2SD) above the 'background'; or
- contaminant concentration above some specific arbitrary values.

Soil chemical composition may vary considerably over small distances, both horizontally and vertically. Obtaining a representative field sample presents specific problems which, if ignored, may render all subsequent analytical work worthless [5]. The analytical significance that can be associated with spatial heterogeneity in metal distribution is large as recently demonstrated for forest ecosystems [6]. Analytical uncertainty generally develops from inadequate soil sampling strategies. Two potential sources of variability have been defined: (a) heterogeneity of the parent material from which soil is formed, and (b) heterogeneity as a result of soil-forming processes [7]. These are supplemented by a range of land management operations, such as spreading of livestock or industrial wastes and cultivation.

3.2 Soil Sampling

Four basic stages are commonly defined (for example, [8]) which help determine the representativeness and reliability of any results: Stage 1, presampling assessment and plan; Stage 2, soil sampling; Stage 3, soil preanalysis treatment (which includes soil preparation and storage); and Stage 4, soil analysis. The emphasis in this chapter is placed upon the first three stages, with aspects of the analysis stage described in Chapter 4. Various terms are used to describe the practice of soil sampling [9]. Here we consider sampling (Figure 3.1), which in the case of trace elements often starts from the selection of the site to be investigated. Samples are then collected at the different sampling points and accurately located. Most soil samples are collected using a specific corer or auger, and these primary samples (Figure 3.1) are subsequently either combined (composite/aggregated) or kept and analysed separately. The logistics, costs, information obtained and interpretation that is likely to be gained from either of these two approaches is very different. The number and the relative position of the sampling points depend on the scope of sampling and thus on the particular sampling strategy chosen, which can be selected on a statistical basis. Using the terminology proposed by De Zorzi *et al.* [10] and Figure 3.1, cores are taken from each

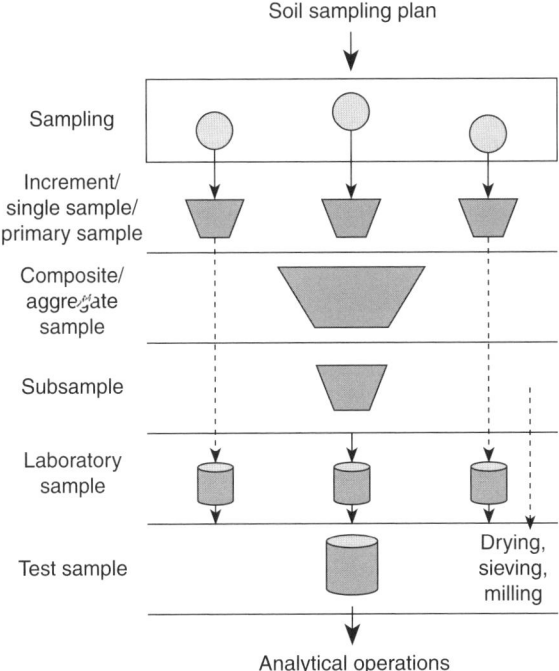

Figure 3.1 Sampling operations (Adapted from P. De Zorzi, S. Barbizzi, M. Belli et al., Terminology in soil sampling (IUPAC Recommendations 2005); Pure Appl. Chem. 2005, 77, 827–884 [10].)

sampling point to produce a primary sample. When these primary samples are mixed together, a composite/aggregate sample is obtained. A sample ready for the laboratory can be obtained either directly from the primary sample or from the composite sample. During this phase some form of representative sampling method may be employed such as coning or quartering, riffling and/or grinding [10].

3.3 Errors Associated with Soil Sampling and Preparation

While ideally the same person (analyst) should be involved in each stage, this is seldom the actual case. While instrumental detection limits, multielemental capabilities and sample throughput are continually increasing, the accuracy of any analytical information gained still relies heavily upon the rigour of initial soil sampling and preparation stages. Detection limits of the $\mu g\ kg^{-1}$ level or lower are readily achievable and these can create problems related to sampling and storage together with contamination from handling and reagents [11]. The increased focus on quality and the accreditation of laboratory systems has contributed to increased knowledge about the uncertainty in the digestion procedures and the chemical analysis. The emphasis of data uncertainty assessments in environmental analysis have typically focused on the final analysis steps (Stage 4), while sampling and preanalytical sample treatment are largely ignored [12].

$$\sigma^2_{total} = \sigma^2_{Stage\ 1} + \sigma^2_{Stage\ 2} + \sigma^2_{Stage\ 3} + \sigma^2_{Stage\ 4} \qquad (3.2)$$

The total error, deviation, or uncertainty associated with an analysis could be defined as the sum of errors (σ^2) incurred in each stage (Equation 3.2). Separating the relative contributions that arise from individual stages is difficult and involves some degree of independence. Referenced soil material is widely available and provides an opportunity to test the analytical aspects of Stage 4 but it has always been difficult to test the earlier stages. For example, while the $\sigma^2_{Stage\ 1} + \sigma^2_{Stage\ 2}$ will be location specific, much of the error associated with later laboratory and analysis stages, $\sigma^2_{Stage\ 3} + \sigma^2_{Stage\ 4}$, will be fixed common to a particular method and analytical technique. The error which arises from Stage 1 + 2 is typically larger than that linked to the preparation, manipulation and analyses [13]. One estimate of the likely error associated with each individual stage during instrumental multielement analysis has been suggested to be up to 1000 % for sampling, 100–300 % for sample preparation, and 2–20 % for instrument measurement, while error in data evaluation is estimated to be up to 50 % [14]. Using the same sampling protocol individual samplers from nine organizations produced uncertainty of ~ 50 % in the estimated mean lead concentration from a single contaminated site. Differences introduced by the use of individual samplers were small and the uncertainty was due entirely to the sampling of a heterogeneous site, rather than the analytical measurement as all the analyses were carried out in random sequence within a single batch [15].

Individual Stages

Sample Plan (Stage 1)

It is useful to develop a basic sampling plan which should have sufficient flexibility to enable it to be capable of modification to best fit individual situations. There are various aspects of sampling that may be standardized, although some flexibility is required, taking into account factors such as objectives, site characteristics (for example, area), logistics, sampling equipment. There are various reasons why soil is required to be sampled, and these include (a) determining some regional average element concentration, perhaps to be used in the making of element budget calculations or to quantify risk; (b) identifying the level of contamination and some associated ecological impact; (c) identifying the presence of hot spots for remediation purposes; and (d) understanding soil processes. A clear understanding of the purpose and use of any information gathered help to decide upon sampling plan. Soil immediately under or adjacent to features such as power cables and galvanized fences should be avoided (unless they are the specific topic of research) as they represent sources of copper and zinc, respectively. Sampling schemes should always take into account the project aims. For example, where contamination of surface soil is likely and indirect ingestion of soil represents the main transfer pathway the suggested focus should be on sampling the upper, contactable layer of topsoil to assess exposure risk [16].

Table 3.1 Factors that can be used to stratify soils

Stratum	Factor
Soil type	Common mineralogy
Texture	For example, sandy loam vs clay loam
Topography	Influences natural drainage properties (streams bottoms, valley slopes and ridge tops are appropriate strata)
Land uses	Cropland, forest areas, pasture, industrial/domestic areas
Practices	Reduced tillage vs ploughed land
Horizons	A horizon, B horizon, and C horizon (surface (A) usually has more organic matter in the soil)

Source: Modified from B.J. Mason, Preparation of soil sampling protocols: sampling techniques and strategies; EPA/600/R-92/128; EPA, 1992 [18].

An initial sampling plan should incorporate decisions based on the site information [17]:

- choice of sampling units;
- choice of sampling depth;
- choice of sampling density;
- selection of individual sampling position;
- choice of sampling tools and their use.

The sampling plan should take account of any prior knowledge of the area to be sampled. Stratification is an approach that uses common site attributes that link environmental processes and certain properties. Examples of site attributes that can be used to select individual layers or stratum are listed in Table 3.1. The variance within the strata should be smaller than the variance between strata. Soil types are frequently used as a means of stratification, especially if they are quite different in physical and chemical properties. Sampling of pedagogical soil horizons is also a commonly used form of stratification.

Applying some form of stratification increases the precision of the estimates and helps control the sources of variation in the data, making the units within each stratum more homogeneous than the total population. Stratification must remove some of the variation from the sampling error or else there is no additional benefit to be gained from the effort, other than perhaps a better geographic spread of sample points [18]. A limit will exist where any improvement in sampling error will not be worth the effort of undertaking further stratification. The various aspects of random and grid sampling schemes have recently been discussed for Brownfield sites [16] and the schemes have a wider more general relevance. Targeted sampling uses existing knowledge, which may involve some form of stratification or relevant historical site data, in order to use judgment to produce site-specific, spatially resolved information. Fixed patters of sampling points tend to be used in those situations which lack reliable information; examples include regular grid patterns, that is systematic. A choice of distance between sampling points and total number of samples to be collected must be made with direct logistical and cost implications. It may be appropriate to employ two or more different sampling schemes for a particular site characterization. In this situation an exploratory investigation might be followed by the main sampling programme and a final targeted sampling to provide more specific information. A systematic sampling scheme should reduce the error compared to those associated with simple random sampling. The extent to which precision is improved by using a systematic sampling scheme will depend upon site factors such as the spatial heterogeneity in metal distribution and will therefore differ between elements measured [19]; up to 10-fold improvement in precision has been reported [20]. As the level of uncertainty increases, the likelihood increases that localized areas of contamination (hot spots) will become progressively less reliable. Traditional sampling schemes such as following W or X patterns tend to introduce bias, with the latter having a central bias of the area and leaving large areas relatively unstapled [21]. Regular grid patterns overcome this problem, but the sampling points within each subunit should be totally random and without bias [17].

Sampling (Stage 2)

Depending upon the level of prior site information the sampling plan might require some late modification in the field. Unforeseen factors include the problems raised by natural variability within the soil profile that might effect sample collection, such as horizon depth and continuity, or physical aspects such as stone content or water-table depth. The twin aspects of sampling depth and soil bulk density are fundamentally important to developing a successful soil sampling campaign. Usually the final number of samples collected represents a compromise between a statistically acceptable number and logistical and economic constraints. There is often a choice to be made between analysing a single composite sample or a number of individual samples. A well-homogenized sample made up from two or more samples collected from the block of soil will normally exhibit a smaller variance [22]. Compositing of samples also reduces the final number of analyses and therefore costs, but also leads to a loss of information on spatial variability. A composite sample is prepared from numerous, approximately equal, subsamples taken from a defined area of land, for example a field, from roughly separated points on a predetermined pattern or a smaller area. Typically soil samples of ~500 g are taken and the mass also affects the extent of sampling uncertainty, which is predicted [23] to be inversely proportional to the square root of sample mass:

$$s^2_{samp} \propto \frac{1}{mass} \qquad (3.3)$$

where s^2_{samp} is sampling variance. Increasing the sample mass by a factor of 2 is predicted to reduce the sampling standard deviation by a factor of $\sqrt{2}$, that is by a factor of 1.4 [16]. Even where a 500 g sample is taken to be representative of a 1 ha area, this represents a sample of <0.0001 % of the total soil present and usually only a test sample of 1 g is actually used for any subsequent analysis.

A unique, common sampling strategy cannot be followed for different land uses, which for certain situations may have a greater influence than soil type (Table 3.1). Selecting an appropriate sampling depth can be especially important and should be kept flexible depending upon the element(s) involved and project aims. For example, the effect of contrasting land management between arable vs. grassland soils, where the former has a disturbed physically mixed A_{pe} (plough) horizon compared to an undisturbed grassland profile has been highlighted [24]. The effect of depth within the surface horizon should be of less significance for cultivated arable soils compared with grasslands, where a strong concentration gradient can develop for surface applied elements. Therefore, while soil sampled from the 0–15 cm layer of an arable field was representative of the A_{pe} horizon, a similar depth in a grassland soil resulted in a dilution effect in the concentration of several radionuclide's (especially ^{137}Cs), since it accumulated in the 5 cm top layer. In addition to the loss of spatial compositional information, a further aspect of sampling depth relates to the unnecessary dilution of contaminants which could have potential consequences for analytical detection limits.

Most analytical data are quoted as concentrations and relate to guidelines or regulatory limits. Additional information that might be required include an estimate of soil bulk density to enable the calculation of elemental budgets where results expressed on a mass per unit area (mg m^{-2}) or mass per unit volume (mg m^{-3}) basis are required. Cross-contamination problems can be eliminated or minimized through the use of dedicated sampling equipment [25] and thorough cleaning between samples.

Preanalysis (Stage 3)

Drying allows for reliable subsampling of soils, which is difficult for many soil in their field-moist condition. Drying at 40 °C is recommended [26] although the temperature used varies widely. While drying soils at higher temperatures may have few consequences for determining total elemental concentrations, some effects on selective fractionation and distribution of individual elements between chemical forms could be expected. Sieving is often initially <2 mm screening followed by something finer; 100 mesh may require grinding. Grinding of both organic and mineral soil samples improved (2–4 times) the uncertainty for Zn determination over samples that had only been sieved [27]. Different sieving techniques altered the mean value of Zn while having little influence on uncertainty. Reanalysis of the samples after 15 months' storage, but without mixing, resulted in a decrease of between 7 % and 27 % for Zn [27], although other studies showed an increase in Zn [28].

Subsampling may be done by careful cone and quartering; scooping directly from a storage bottle/bag can introduce error due to a partial fractionation according to particle size. Volumetric sampling can lead to major errors that can result from the degree of aggregation, which is variable and related to differences in texture and organic matter among samples [29]. Soil test samples can be taken gravimetrically or volumetrically, the latter – because of speed – often being the chosen method in routine test laboratories. A comparison of 54 samples taken using a 4.25 cm^3 Urbana laboratories standard soil scoop from a single air-dried, roller crushed and 2-mm-sieved sample demonstrated the level of variability that could be expected [29]. While the averaged weight was 4.4 g per scoop it ranged from 3.4 to 6.7 g per scoop and bulk densities averaged 0.96 g cm^{-3} and ranged between 0.87 and 1.13 g cm^{-3}. Pulverizing the sample and increasing the scoop weight can reduce the error. Increasing the weight of sample taken for analysis from 1 g to 10 g greatly reduced the size of any variability between aliquots [30].

Soil consists of a heterogeneous mixture of inorganic and organic material having a range of physical and chemical properties and long-term storage can result in the physical separation due to differences in particle size and density. Once collected and prepared, soil samples can be stored prior to analysis, and storage time should have few implications where a total analysis is being undertaken. The possibility of physical (size of density) separation occurring prior to sampling must be minimized through ensuring the homogeneity of the stored sample by using suitable mixing and/or subsampling procedures.

The effect of various storage conditions, air-dry at either 22 °C, 3 °C or −21 °C for up to 1942 days, on extractable Zn concentrations were compared for forest soil (organic and mineral) horizons [28]. Under these conditions extractable Zn increased in soil stored at 22 °C but showed no change at −21°C and the authors suggest storage should not exceed 50 days at room temperature but recommend −21°C in airtight containers. The changes were suggested to be due to the effects of biological and chemical activity. Some guidelines on long-term storage of samples are provided in ISO 18512 [31].

Recently it was concluded [32] that the operations of

- fragment removal,
- drying,
- disaggregating before sieving: rolling, crushing or grinding,
- sieving techniques and tools, and
- sample reduction methods (splitting, subsampling)

require further investigation with respect to the uncertainty and variability caused by segregation of components, contamination and alterations in chemical, physical and biological properties during the pretreatment of soil samples in the laboratory.

3.4 Overview of the Current Situation

Comparison of a few recent studies chosen at random indicates that a range of sampling protocols is employed (Table 3.2). Despite the very considerable differences that are evident, it is difficult to assess their significance in terms of the final values and any intercomparisons.

A recent attempt to address some of the basic issues related to soil sampling has been undertaken as part of the European Community Commission Standard, Measurement & Testing Programme. A Comparative Evaluation of European Methods for Sampling and Sample Preparation (Project PL95-3090) CEEM-Soil programme was organized. Some of the project aims included a review of the existing European soil sampling guidelines, an assessment of the comparability between approaches and methodologies together with implications for data quality; and an exploration of sources of uncertainty in order to asses the need for harmonized sampling guidelines with suggestions for standard operating procedures (SOP).

Information from this review has been summarized in a series of papers in *Science of the Total Environment* (volume 264, see [32]) and the main points are discussed below. A survey of the soil sampling strategies for 15 European countries has been summarized [8]. These ranged from national to local strategies and the complexity and details provided varied widely. For example, 5 EU countries had a sampling plan, 12 lacked preanalysis detail, the depth sampled varied widely and 7 sampled organic horizon separately, and the majority composited samples, using 2–20 cores. The sampling pattern varied from point to various 'W' and 'X' shaped patterns. With respect to technical aspects of sampling, tools used ranged through corer, auger or spade; the majority did not mention size or composition of the sampling tool. Perhaps surprisingly, many sampling strategies did not mention the size of sieving or drying conditions. This produced the following useful summary points adapted

Table 3.2 Comparison of information provided in relation to preanalysis stages for a few randomly selected international publications

Factor	Jones et al. [33]	Tack et al. [34]	Horckmansa et al. [35]	Senwo and Tazisong. [36]	Loska et al. [2]
Soil depth	Three depths (15 cm increments)	0–20 cm		Horizons	0–20 cm
Sample type	Drilled cores (180) 10 replicates	Hand borer			Composite of 30–40 subsamples
Sample pattern	Zigzag	Four subsamples collected within a 20 m radius			From 10 000 m^2
Preparation	Pestle mortar	Ground	Aggregates crushed	Crushed and ground	
Storage		Polyethylene bags	Glass bottles		
Dried		Air-dried	40 °C	Air-dried	Air-dried (75 °C)
Sieve size	<2 mm	<2 mm	Quartered <1 mm	<2 mm	Coning and quartering <2 mm sub-samples < 0.01 mm
Land use	Agricultural	Mixed		Forested	
Aim	Metal concentrations after 20 years cropping	Baseline trace elements contents		Background concentrations of metals	
Country	USA	Belgium	Luxembourg	USA	Poland

from [8]: 'Soil sampling guidelines differ as to whether they are applied by law, or used throughout the country. In some countries these are ISO/DIS related or based [37–38], or are produced by a scientific society or a standardization body.'

- Not all sampling guidelines (strategy) clearly describe the sampling scale, the specifications for contamination risk precautions, the sampling plan and protocol structure and the preanalysis treatment of the soil samples.
- The purpose for sampling, in descending order of frequency, is soil pollution, soil fertilization, general soil monitoring, background risk assessment, or else it is not specified.
- The majority of countries do not sample the top organic matter separately. Sampling depth is either related to the morphogenetic horizon or to ad hoc sampling depth, which is not specified in all cases.

- The stratification criterion for area, site, unit, subunit, and point selection are mainly based on pedology and land use, following the history and prescreening information or geology, or is site related.
- The sampling pattern is mainly grid sampling, grid and random sampling, or not mentioned.
- Sampling density inside the sampling unit either varies greatly or it is not mentioned, while the size of the sampling unit varies widely.
- Most guidelines require the collection of composite instead of simple samples, while some prefer sampling soil profiles.

While this summarizes the current European situation it is not unreasonable to suggest that they apply in some general way to the global situation. Some indication of the potential range in preanalysis stages used in a randomly selected group of international, peer-reviewed papers, is provided in Table 3.2. The likely consequences that any individual or combined effects of these differences might have in terms of specific contributions to error generated from the pre-analysis stage cannot currently be quantified. However, the following points were regarded as being particularly critical for the comparability of the results of soil contamination studies from the CEEM Soil project [32]:

- sampling depth and depth sections;
- definition of soil horizons;
- treatment of the organic layers of natural soil profiles;
- subdivision of the investigation site;
- selection and allocation of sampling points;
- sampling techniques and tools.

This should provide at least an initial way forward in terms of prioritizing future international research emphasis within the general area of minimizing error associated with preanalysis stages of soil sampling.

3.5 Scale and Variability

The choice of sampling density and scale at which analysis is undertaken is becoming more confused with the development of analytical techniques capable of spot/point analysis (for example, electron microprobe and synchrotron x-ray microfluorescence spectroscopy). While the primary aim of most sampling protocols is to produce a homogeneous sample that is representative of some given area, the significance of finer scale variation for soil processes is increasingly evident. The significance of short-scale variability in concentrations of metal contaminants on bacterial activity community structure has been demonstrated [39]. Metal concentration and metabolic potential varied by as much as 30-fold and 10 000-fold respectively in soil samples taken <1 cm apart. The response of microorganisms to metal contamination of soils varies significantly from one investigation to another. One explanation is that metals are heterogeneously distributed at spatial scales relevant to microbes and that microorganisms are able to avoid zones of intense contamination. The microscale (< millimetre) distribution of copper (Cu) in a vineyard soil indicate a strongly heterogeneous pattern to the distribution of Cu. Entire regions of the thin sections were virtually devoid of Cu, whereas highly localized 'hotspots' have Cu signal intensities thousands of times higher than background [40]. Although something of an oversimplification, it is useful to define two broad scales at which sampling and analysis is undertaken. At the larger geographical scale multiple soil cores are taken and used to describe a composited type average concentration, or if analysed separately provide some indication of spatial variability. At a microscale, relatively undisturbed soil samples can be sampled and analysed directly. The information gained from these two approaches is very different and currently it is difficult to reconcile these differences.

3.6 Conclusions

The error associated with each of the four stages of soil sampling and analysis has been shown to be considerable. While greatest emphasis of most laboratory accreditation schemes has tended to focus on the final analysis stage it remains difficult to determine error associated with the initial sampling and preparation stages. Where these errors have been determined they are considerable and larger than those associated with the final analytical stage. Variability in soil properties and the spatial distribution of target elements represent the main cause of sampling errors. The error can be reduced through the development of site-related sampling plans that incorporate adequate sample numbers and a stratified system. National and organizational sampling strategies vary widely in terms of many of their basic steps and because they do not lend themselves to standardization the guidelines contain only broad recommendations. The need for greater emphasis to be placed on this general topic has been raised by various authors, and is particularly important with the greater emphasis placed upon international soil quality standards.

References

[1] Boekhold, A.E.; van der Zee, S.E.A.T.M., Long term effects of soil heterogeneity on cadmium behaviour in soil; J. Contam. Hydrol. 1991, 7, 371–390.
[2] Loska, K.; Wiechuła, D.; Pelczar, J., Application of enrichment factor to assessment of zinc enrichment/depletion in farming soils; Commun. Soil Sci. Plant Anal. 2005, 36, 1117–1128.
[3] Jackson, K.W.; Eastwood, I.W.; Wild, M.S., Stratified sampling protocol for monitoring trace metal concentrations in Soil; Soil Sci. 1987, 143, 436–443.
[4] Nortcliff, S., Sampling and pre-treatment – some observations from the United Kingdom; Sci. Total Environ. 2001, 264, 163–168.
[5] Marr, I.L.; Cresser, M.S, Environmental Chemical Analysis; Blackie & Sons Ltd, London, 1983, p. 258.
[6] MacDonald, J.D.; Hendershot, W.H., Spatial variability of trace metals in podzols of northern forest ecosystems: sampling implications; Can. J. Soil Sci. 2003, 83, 581–587.
[7] Houba, V.J.G.; Novazamsky, I.; van deer Lee, J.J., Quality aspects in laboratories for soil and plant analysis; Commun. Soil Sci. Plant Anal. 1996, 27, 327–348.
[8] Theocharopoulosa, S.P.; Wagnerb, U,G.; Sprengartb, J. *et al*, European soil sampling guidelines for soil pollution studies; Sci. Total Environ. 2001, 264, 51–62.

[9] Soil Science Society of America; Internet Glossary of Soil Science Terms; Soil Science Society of America, Madison, WI; 2001, https://www.soils.org/publications/soils-glossary (accessed 26 November 2009).

[10] De Zorzi, P.; Barbizzi, S.; Belli, M. et al., Terminology in soil sampling (IUPAC Recommendations 2005); Pure Appl. Chem. 2005, 77, 827–884.

[11] Sturgeon, R.E., Current practice and recent developments in analytical methodology for trace element analysis of soils, plants, and water; Commun. Soil Sci. Plant Anal. 2000, 31, 1479–1512.

[12] Muntau, H.; Rehnert, A.; Desaules, A. et al., Analytical aspects of the CEEM soil project; Sci. Total Environ. 2001, 264, 27–49.

[13] Fortunati, G. U.; Pasturenzi, M., Quality in soil sampling; Quimica Analitica 1994, 13, S5–S29.

[14] Markert B., Quality assurance of plant sampling and storage; in Quevauviller, P. (ed.), Quality Assurance in Environmental Sampling and Sample Pre-analysis. Wiley-VCH Verlag GmbH, Weinheim; 1995, pp. 215–254.

[15] Ramsey, R.H.; Argyraki, A., Estimation of measurement uncertainty from field sampling: implications for the classification of contaminated land; Sci. Total Environ. 1997, 198, 243–257.

[16] Taylor, P.D.; Ramsey, M.H., Sampling strategies for contaminated brownfield sites; Soil Use Manag. 2006, 21, 440–449.

[17] Bacon, J.R.; Hudson, G., A flexible methodology for the characterisation of soils: a case study of the heavy metal status of a site at Dornach; Sci. Total Environ. 2001, 264, 153–162.

[18] Mason, B.J., Preparation of soil sampling protocols: sampling techniques and strategies; EPA/600/R-92/128; EPA, 1992.

[19] McBratney, A.B.; Webster, R., How many observations are needed for regional estimation of soil properties; Soil Sci. 1983, 135, 177–183.

[20] Webster, R., Quantitative and Numerical Methods in Soil Classification and Survey; Oxford University Press, Oxford, 1977.

[21] Berrow, M.L., Sampling of soils and plants for trace element analysis; Anal Proc. 1988, 25, 116–118.

[22] Barth, D.S; Mason, B.J., Soil sampling quality assurance and the importance of an exploratory study; ACS Symposium Series 1984, 267, 97–104.

[23] Gy, P., Sampling of Particulate Materials – Theory and Practice; Elsevier, Amsterdam, 1979.

[24] Sastre, J.; Vidal, M.; Rauret, G. et al., A soil sampling strategy for mapping trace element concentrations in a test area; Sci. Total Environ. 2001, 264, 141–152.

[25] US EPA, Environmental Response Team; Standard Operating Procedures – Soil Sampling 2012; EPA, 2000.

[26] ISO 11464, Soil quality – pretreatment of samples for physico-chemical analysis; 2006, http://www.iso.org/iso/iso_catalogue/catalogue_ics/catalogue_detail_ics.htm?csnumber=37718 (accessed 11 November 2009).

[27] Wickstrøm, T.; Ogner, G.; Remedios, G., Effects of different pretreatments (sieving, milling, and grinding) on quality of determination of Kjeldahl nitrogen, pH, and extractable elements in forest soils; Commun. Soil Sci. Plant Anal. 2004, 35, 369–384.

[28] Ogner, G.; Randem, G.; Remedios, G.; Wickstrøm, T., Increase of soil acidity and concentrations of extractable elements by 1 M ammonium nitrate after storage of dry soil for up to 5 years at 22 °C; Commun. Soil Sci. Plant Anal. 2001, 32, 675–684.

[29] Glenn, R.C., Reliability of volumetric sampling as compared to weighed samples in quantitative soil test interpretation; Commun. Soil Sci. Plant Anal. 1983, 14, 199–207.

[30] Gilbert, R.O.; Pamela, G.D., Determining the number and size of soil aliquots for assessing particulate contaminant concentrations; J. Environ. Qual. 1985, 14, 286–292.

[31] ISO 18512, Soil quality – Guidance on long and short term storage of soil samples; International Standards Organization, 2007, http://www.iso.org/iso/iso_catalogue/catalogue_tc/catalogue_detail.htm?csnumber=38721 (accessed 11 November 2009).

[32] Wagner, G.; Desaules, A.; Muntau, H., Harmonisation and quality assurance in pre-analytical steps of soil contamination studies – conclusions and recommendations of the CEEM Soil project; Sci.Total Environ. 2001, 264, 103–117.

[33] Jones, C.A.; Jacobsen, J.; Lorbeer, S., Metal Concentrations in three Montana soils following 20 years of fertilization and cropping; Commun. Soil Sci. Plant Anal. 2002, 33, 1401–1414.
[34] Tack, F.M.G.; Verloo, M.G.; Vanmechelen, L. *et al.*, Baseline concentration levels of trace elements as a function of clay and organic carbon contents in soils in Flanders (Belgium); Sci. Total Environ. 1997, 201, 113–123.
[35] Horckmansa, L.; Swennena, R.; Deckers, J. *et al.*, Local background concentrations of trace elements in soils: a case study in the Grand Duchy of Luxembourg; Catena 2005, 59, 279–304.
[36] Senwo, Z.N.; Tazisong, I.A., Metal contents in soils of Alabama; Commun. Soil Sci. Plant Anal. 2004, 35, 2837–2848.
[37] ISO 10381-1, Part 1: Guidance on the design of sampling programs; International Standards Organization, 2002.
[38] ISO 10381-2, Part 2: Guidance on sampling techniques; International Standards Organization, 2002.
[39] Becker, J.M.; Parkin, T.; Nakatsu, C.H. *et al.*, Bacterial activity, community structure, and centimeter-scale spatial heterogeneity in contaminated soil; Microbial Ecol. 2006, 51, 220–231.
[40] Jacobson, A.R.; Dousset, S.; Andreux, F. *et al.*, Electron microprobe and synchrotron X-ray fluorescence mapping of the heterogeneous distribution of copper in high-copper vineyard soils; Environ. Sci. Technol. 2007, 41, 6343–6349.

4

Analysis and Fractionation of Trace Elements in Soils

Gijs Du Laing

*Ghent University, Laboratory of Analytical Chemistry and Applied Ecochemistry,
Faculty of Bioscience Engineering, Ghent, Belgium*

4.1 Introduction

Most standards and sanitation thresholds for trace elements in soils are based on their total contents. Several techniques have been developed for the determination of total trace element contents in soils. It is generally recognized, however, that the behaviour of trace elements in the environment is determined by their specific physicochemical forms rather than by their total concentration. This has led to the development and application of a wide range of chemical speciation and fractionation methods.

According to The International Union for Pure and Applied Chemistry, IUPAC [1], a chemical species is the 'specific form of an element defined as to its isotopic composition, electronic or oxidation state, and/or complex or molecular structure'. Speciation analysis is defined as the 'analytical activities of identifying and/or measuring the quantities of one or more individual chemical species in a sample. To be considered in speciation analysis are nuclear (isotopic) composition, electronic or oxidation state, inorganic compounds and complexes, organometallic compounds, and organic and macromolecular complexes. Strictly speaking, whenever an element is present in different states, these states must be regarded as different species. In practice, however, the relevance of the different species and the level of understanding required will determine whether different species should be grouped or measured separately. The term fractionation, often used for speciation is defined as the 'process of classification of an analyte or a group of analytes from a certain sample according to physical (for example, size, solubility) or chemical (for example, bonding, reactivity) properties' [1]. Fractionation basically defines the soil phases that trace metals are associated with, regardless of their speciation (for example their isotopic composition, electronic state or oxidation state), by measuring the metal contents which are operationally released during the decomposition of particular soil phases. As some metal species are preferentially associated with specific soil fractions, fractionation is often seen as a form of 'operational speciation' [2]. Fractionation by size is carried out by the separation of samples into different particle size fractions usually during sampling, whereas the concept of chemical leaching is based on the idea that a particular chemical solvent is either selective for a particular phase or selective in its dissolution process [3]. Unfortunately, speciation analysis made directly on the solid soil matrix is generally limited since few of the techniques available have sufficient sensitivity for trace element studies [4]. Chemical species are also often not stable enough to be determined individually [3], whereas special procedures for species-retaining sampling of soil have not yet been developed and validated for most species [5]. Moreover, in many cases it is technically impossible to determine the large numbers of individual species [3]. Therefore, so far mainly fractionation procedures have been developed and used in soil science, with the exception of some speciation procedures which are directly applicable to the soil solid phase. Further, some recent studies have focused on developing species-retaining and species-selective extraction procedures. In this chapter, the most commonly used techniques for the determination of total trace element contents in soils are reviewed, as well as fractionation and speciation procedures for the soil solid phase.

4.2 Total Analysis

Some techniques for the determination of trace elements in soil are nondestructive. However, most commonly used techniques rely upon an initial dissolution stage. In the dissolution-based procedures, trace elements are first released from the soil solid phase, often using strong acids to decompose the matrix. This is followed by their determination, which is performed by instrumental analytical techniques.

4.2.1 Matrix Dissolution

Most decomposition techniques involve the use of strong acids and an external heat source to decompose the sample matrix, often followed by filtration to remove the residue. An individual acid or a combination of acids is chosen. This choice depends upon the characteristics of the acid and the nature of the matrix to be decomposed (Table 4.1). Incomplete dissolution means that most of the procedures are considered as pseudo-total analysis techniques. Hydrofluoric acid (HF) is the only acid that can dissolve silicates. As the use of HF is rather dangerous, aqua regia (nitric–hydrochloric acid: 1:3 vol/vol mixture) is often used as an alternative. The latter liberates the trace elements associated

Table 4.1 Some acids commonly used for wet decomposition of soil and biological samples

Acid	Characteristics
Hydrochloric acid (HCl)	Weakly reducing, generally not used to dissolve organic matter, but useful for metals bound to phosphates, carbonates, some oxides and sulfides
Sulfuric acid (H_2SO_4)	Good oxidizing properties for ores and (hydr)oxides; useful for releasing volatile products. Often used in combination with HNO_3. Cannot be used in PTFE vessels
Nitric acid (HNO_3)	Oxidizing attack on phases which are not dissolved by HCl; liberates the trace elements as soluble nitrate salt
Perchloric acid ($HClO_4$)	A strong oxidizing agent for organic matter. Explosive reactions may occur, so samples are normally pre-treated with HNO_3 prior to addition of $HClO_4$
Hydrofluoric acid (HF)	Dissolves silica-based materials; forms SiF_6^{2-} in acid solution; only plastic vessels are used as it reacts with glass
Aqua regia (HNO_3–HCl 1:3 vol/vol mixture)	Mixture of acids which is more oxidizing compared to the individual acids. Forms a reactive intermediate, NOCl. Dissolves sulfides and other ores

with most soil fractions although silicate minerals are not attacked. This is perfectly acceptable for environmental analysis as it is unlikely that the silicate-bound trace elements will leach from soil and/or become available in the short or medium term. Most commonly, the acid digestion of soil samples is carried out in open glass vessels (beakers, flasks or tubes), which are heated using a hot-plate or a multiple-sample digester. Decompositions carried out in open vessels may, however, be subject to losses of analyte by volatilization or may be contaminated by dust. These losses and contamination errors are minimized by decomposition in closed systems at elevated temperature and pressure, using PTFE (polytetrafluoroethylene) autoclave or microwave oven systems [6]. Fusion is another alternative, which involves the addition of an excess of fusion reagent (for example sodium carbonate, lithium meta- or tetraborate or potassium pyrosulfate, 10- to 20-fold the amount of sample) to a finely ground sample. This mixture is placed in a platinum crucible and then heated in a furnace (300–1000 °C), which results in the formation of a clear 'melt'. After cooling, this melt can be dissolved in a mineral acid. By applying this technique, some substances, for example silicates and oxides, which are normally not destroyed by the action of mineral acids, can be dissolved. The addition of fusion reagent, however, leads to a high risk of contamination, and the high salt content in the final solution may also lead to problems during the subsequent analysis. For determination of some elements which are hard to extract, such as Cr, Y and Zr, it is convenient to combine acid decomposition with final fusion of insoluble residues [6].

4.2.2 Instrumental Analysis Techniques

After dissolution, the trace element contents are determined in the liquid phase by a suitable instrumental analysis technique. The most important techniques are flame atomic absorption spectrometry (F-AAS), graphite furnace or thermoelectric atomic absorption spectrometry (GF-AAS, ET-AAS), inductively coupled plasma-atomic emission spectrometry (ICP-AES), hydride generation-AAS (HG-AAS), cold vapour-AAS (CV-AAS), inductively coupled plasma-mass spectrometry (ICP-MS), anodic stripping voltammetry (ASV) and neutron activation analysis (NAA). These techniques differ mainly in their purchase and operating costs, sensitivity, potential interferences, and ease-of-use. The principles of most techniques are briefly mentioned below. Anodic stripping voltammetry is discussed in Chapter 5, as it is also considered as a technique for speciation analysis.

4.2.2.1 Atomic Absorption Spectrometry (AAS) Techniques

In AAS, a light beam from a light source passes through a medium in which the elements from the sample are brought into an atomic state, before being focused on the entrance slit of a monochromator which selects a very narrow wavelength interval to be analysed in the detector. The concentration can be deduced from the measurement of light absorption by the atoms which were brought to their atomic state. The light source is mostly a hollow cathode lamp (HCL), a discharge lamp which contains a fill gas (argon or neon). The emission spectrum depends upon the element which constitutes the cathode. In principle, a hollow cathode lamp emits light only at wavelengths which are specifically absorbed by the element being measured. Hence, to measure an element such as lead, the cathode must contain lead. An alternative is the use of electrode-less discharge lamps (EDL) whose light intensity is about 10–100 times greater, although they are not as stable as the hollow cathode lamps. These lamps contain a salt of the element of interest along with an inert gas. The element of interest is excited and ionized due to the application of a radiofrequency (RF) field and emits light at wavelengths which are specifically absorbed by the element being measured. The electrode-less discharge lamps are in general used for elements such as As, Hg, Sb, Bi and P. The need for individual lamps for each element constitutes a major drawback of atomic absorption spectrometry. The medium in which the elements from the sample are brought to an atomic state can be a flame (in F-AAS) or a graphite furnace (in GF-AAS or ET-AAS). When a flame is chosen, the atomic aerosol is provided by a combination of a nebulizer and a burner. As the chemical reactivity of the flame depends on the composition of the combustible gaseous mixture, this composition should be optimized depending on the element to be analysed. When a graphite furnace is used, a tube of graphite with a small cavity holds a precise amount of sample, often only a few microlitres. The tube behaves as an ohmic resistor that can attain 3000 K. In comparison with F-AAS, GF-AAS produces a very high atom density and a longer confinement period, which results in an overall improvement in the sensitivity by a factor of 1000. Some elements, like As, Bi, Sn and Se are difficult to atomize in a flame when they are in higher oxidation states. Therefore, the sample is treated with a reducing agent (sodium borohydride) prior to analysis, which for example, results in the reaction

$$3\,BH_4^- + 3\,H^+ + 4\,H_3AsO_3 \rightarrow 3\,H_3BO_3 + 4\,AsH_3 + H_2O$$

The volatile hydride which is formed during this pretreatment is swept up by a make-up gas into a quartz cell which is placed in the flame of the burner (HG-AAS). The hydrides are easily thermolysed and release the element in its atomic state (for example $2\,AsH_3 \rightarrow 2\,As^0 + 3\,H_2$). Mercury is not transformed to a hydride upon chemical reduction with $SnCl_2$ but rather remains in the metallic state ($Sn^{2+} + Hg^{2+} \rightarrow Sn^{4+} + Hg^0$). Consequently, a special cell which does not require to be put in the flame is used in this 'cold vapour' AAS technique (CV-AAS) used for Hg analysis [7].

4.2.2.2 Inductively Coupled Plasma (ICP)-based Techniques

Plasma can be defined as the coexistence in a confined space of positive ions, electrons and neutral species of an inert gas, typically argon or helium. The plasma which is most widely used for trace element analysis is the inductively coupled plasma (ICP), although other types of plasma are also available, such as the direct-current plasma (DCP), microwave-induced plasma (MIP) and glow discharge. The ICP is formed within the confines of three concentric glass tubes of a plasma torch. Located around the outer glass tube is a coil of copper tubing through which power input is achieved, typically in the range 0.5–1.5 kW at a frequency of 27 or 40 MHz. An oscillating magnetic field is induced, which causes transfer of elements of the plasma gas and the sample to their excited states, as well as ionization of these elements. The inert gas which is used to maintain the plasma is mainly introduced through the outer two tubes of the torch, whereas the sample is injected in the plasma through the inner tube. Prior to injection, the sample is turned into an aerosol by a nebulizer and sent through a spray chamber to decrease turbulence and reduce the aerosol particle size. The use of a nebulizer and spray chamber results in only a small portion of the sample reaching the ICP, with the transport efficiency often being lower than 2 %. Different types of nebulizers and spray chambers can be used. The nebulizers differ mainly in their transport efficiency, their ability to desolvate the aerosol and to process only small sample amounts or liquids with relatively high dissolved solids contents. The spray chambers differ mainly in their transport efficiency, the time needed to remove all existing sample prior to the introduction of the next sample (that is the 'wash-out' time) and their ability to reduce the turbulence and produce a stable signal.

Each atom, once in the excited state in the plasma can lose its excess energy by emission of photons whose energies can have many different values. In ICP-AES, also often called inductively coupled plasma–optical emission spectrometry or ICP-OES, the intensity of light emission at a characteristic wavelength is seen as a measure of the concentration of an element. However, interfering emission by the matrix also occurs, which is minimized by measuring the light emission at multiple wavelengths. From a single run, a multielement analysis can be obtained easily when an echelle grating with prism focusing device is used as dispersive system (Figure 4.1). In contrast to the sequential AAS-based techniques, more than 75 elements can be determined by ICP-AES in as little as one minute, although only elements for which a calibration has been carefully undertaken can really be measured [7]. Such fast multielement analysis, however, also requires a high performance instrument capable of locating weak intensity lines in close proximity to other much more intense.

In ICP-MS, ions are directly sampled from the plasma and analysed by using a mass spectrometer based on their mass/charge ratio. A lot of interfering substances can be formed prior to detection, for example isobaric interferences of polyatomic Ar_2 and $CaCl$ occur when analysing Se on mass 78 or 80 and As on mass 75, respectively. In recent years, different types of reaction and collision cells (and a combination of both) have been developed, in which most interferences are removed by reaction and/or collision with a gas (for example CH_4, NH_3 or O_2 for reaction and He for collision) prior to detection. This makes ICP-MS much more interesting as a technique for routine trace element analysis.

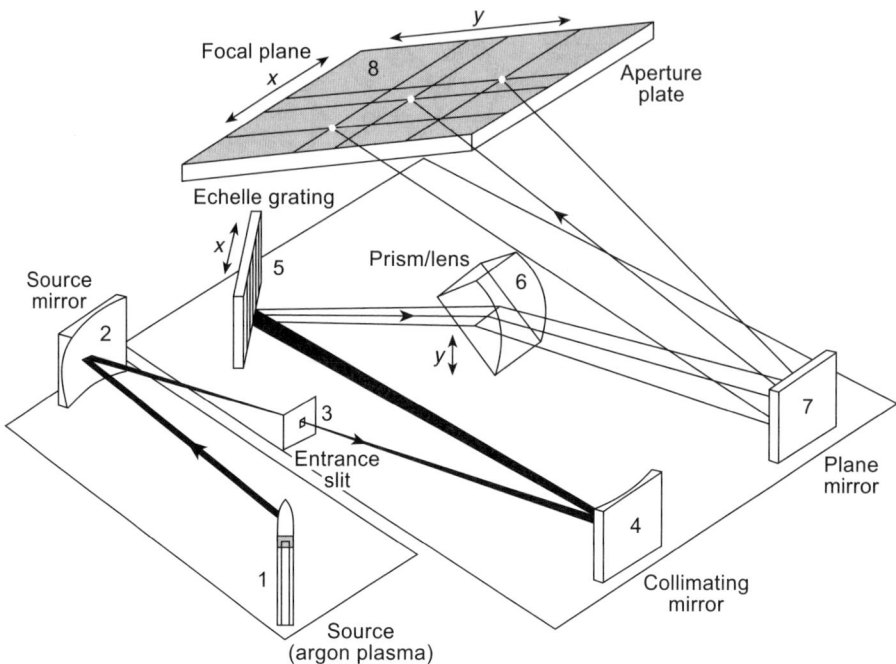

Figure 4.1 Optical scheme for an ICP-AES spectrometer with an echelle grating. For clarity only the central section of the beam issuing from source 1 (plasma) is represented (this beam should cover the whole mirror 2). The echelle grating 5 separates the radiation arriving from the source in the horizontal plane (in x). The prism then deviates the radiation in the vertical plane (in y). The path of three different spectral lines is shown. The images of the entrance aperture 2 are in the focal plane 8. In the past, to detect these radiations, photomultiplier tubes (PMTs) of reduced size were installed in specific planes, but now charge transfer devices (charge-coupled or charge injection devices, CCD/CID) are used, as an electronic equivalent of photographic plates. This allows a continuous spectral coverage from 190 to 800 nm with excellent resolution. Sensors of 500–2000 pixels (each 12–12 µm) are now used (Reproduced with permission from F. Rouessac and A. Rouessac, Chemical Analysis. Modern Instrumentation Methods and Techniques, 2nd edition. Copyright 2007, John Wiley & Sons [7].)

Mass spectrometers differ with respect to their sensitivity and spectral resolution. Most ICP-MS equipment used for routine analyses in environmental laboratories use a quadrupole mass spectrometer. In a quadrupole, elements are subjected to oscillatory paths by applying a direct current (DC) and radiofrequency (RF) voltage to two opposite pairs of rods. With the selection of appropriate RF and DC voltages, only ions of a given mass/charge ratio are able to traverse the full length of the rods and enter into the detector. A high-resolution mass analyser can even eliminate isobaric interferences as it focuses not only on mass/charge ratio, but also on kinetic energy. As the ions are directly introduced and collected in the detector and no optical system is required, the ICP-MS technique is much more sensitive than ICP-AES and F-AAS, but also more expensive. An ICP-MS can easily be directly coupled to chromatographic techniques, for example size exclusion, ion or reversed-phase chromatography, which allows its use as a sensitive detector in speciation analysis (see Chapter 5). Moreover, isotopes can be measured using ICP-MS. The analysis of the different isotopes of an element can be considered as speciation technique itself. Likewise, when using ICP-MS, enriched stable isotopes which are commercially available can be used in spiking experiments to study trace element behaviour in the environment. Isotope dilution analysis can also be used as a better calibration alternative for external calibration with standard solutions or internal calibration by standard addition [8].

4.2.2.3 Radiochemical Neutron Activation Analysis

In neutron activation analysis (NAA) the element of interest is converted from a stable isotope to a radioisotope by neutron bombardment, for example in the case of sodium, stable sodium (^{23}Na) is converted to radioactive sodium (^{24}Na). This new radioactive nucleus decays by emission of an electron and one or more characteristic gamma-rays into a more stable configuration at a rate according to the half-life of this nucleus. The gamma-ray emission spectrum is characteristic of each radionuclide as the gamma-rays possess unique energies that are characteristic of the radioisotope undergoing decay. Gamma-rays detected at a particular energy are usually indicative of the presence of a radioisotope, and thus also of the stable isotope it originates from. Several types of neutron sources can be used, but megawatt nuclear reactors offer the highest sensitivity for most elements [7]. Radiochemical neutron activation analysis (RNAA) is used when the radioactivity of the radionuclides prepared from the trace elements of interest could be masked by a high matrix radioactivity. In this technique, the resulting radioactive sample is chemically decomposed and the radionuclides are isolated by chemical separations into a single fraction or several fractions free of interfering radioactivities, each with one radionuclide or relatively few radionuclides. RNAA has a high sensitivity for many rare earth elements. Its accuracy and precision are comparable to those of ICP-MS. However, the technique is rather time consuming and the radiochemical procedures are often very complicated [9].

4.2.3 Nondestructive Methods

Nondestructive techniques have also been used for trace element analysis in soils, with measurements being made directly on the solid sample. These techniques do not require complicated handling, so all problems related to the decomposition of soil samples and separation (dissolution, contamination and losses) are avoided. These techniques, however, are less sensitive for some environmentally important trace elements, such as Cd, Cu and Zn, compared with other techniques (for example ICP-MS), so they are less common in trace element analysis.

4.2.3.1 Instrumental Neutron Activation Analysis

Instrumental neutron activation analysis (INAA) is also a type of NAA, but the decomposition of the radioactive sample is not necessary as in RNAA. In this technique, the radionuclides are determined on the basis of the differences in decay rates via measurements at different decay intervals using high-resolution detectors. The technique is multi-elemental and suitable for sample sizes from micrograms to kilograms; with relatively low detection limits up to 50 elements can be analysed. For some environmentally important elements, for example Cd, Cu, Ni and Pb, the technique is not sensitive at low concentrations. Moreover, it is necessary to have access to a nuclear reactor and the technique is rather time consuming. It takes about 2 to 4 weeks to obtain information on the concentration of elements involving radionuclides with long half-lives. The development of non-nuclear techniques, such as ICP-MS, has caused a decline in the use of NAA. Nevertheless, there are still some fields where NAA is irreplaceable, for example the analysis of solid materials which are difficult to dissolve or specific applications where the risk of contamination during handling and decomposition would be very high [10].

4.2.3.2 Particle-Induced X-ray Emission and X-ray Fluorescence Spectrometry

In particle induced X-ray emission (PIXE) and X-ray fluorescence (XRF) spectrometry an electron is ejected from an inner shell of an atom and an electron from a higher shell fills the gap in this lower shell. This results in the emission of an X-ray photon equal in energy to the energy difference between the two shells. The X-ray spectrum is made up of radiations with wavelengths and intensities that are characteristic to the atoms present in the sample and their concentrations. The spectrum depends only slightly on the chemical combination or state of the elements in the sample. The difference between the two techniques is the mechanism by which the inner-shell electron is emitted. In the PIXE technique, protons or sometimes helium ions are accelerated towards the sample and this ejects inner-shell electrons. In the XRF technique, high-energy X-ray photons are directed at the sample and eject the inner-shell electrons.

The PIXE technique is capable of analysing very small sample quantities down to 10^{-4}g for solids and around 1 μl for liquids. It allows the simultaneous detection of all elements heavier than Na, by a nondestructive and rapid analysis, with part per million (ppm) elemental sensitivity and submillimetre spatial resolution [11]. However, the technique also has some limitations which hamper its use for routine analysis, mainly the need for a very expensive particle accelerator and the occurrence of some interferences and background problems [12]. Moreover, the ultimate sensitivity of PIXE is not as good as for XRF. Alamin and Spyrou [13] compared the feasibility of using INAA and PIXE for trace element analysis in coal, sawdust, fly ash and landfill materials. Using both techniques, it was possible to determine 34 elements. Some could be determined only by INAA, such as Th, Ta, Sm, La, Eu, Hf, Zb, Ho and Hg, whereas others were determined only by PIXE analysis such as Cu, Ni, Cl, Sr and Cd, while elements such as K, Ca, Ti, V, Cr, Mn, Fe, Co, Zn and As could be determined by both techniques. INAA displayed better detection limits for the elements V, Cr, Zn and As, while elements such as K, Ca, Ti, Mn and Fe showed better detection limits using the PIXE technique.

Although semiquantitative analysis using XRF does not present any major difficulty, quantitative analysis is not quite simple, because a significant part of the emitted radiation from the outer surface of the sample is reabsorbed before being able to escape from the sample [7]. Moreover, when the surface of the sample is heterogeneous and/or not representative for the rest of the sample, the measured concentration might not represent the total sample concentration if the sample is not prepared by, for example, fusion or pellet formation prior to the analysis. X-ray fluorescence exists in various forms and is now considered as a powerful, well established and mature technique for environmental analysis, especially as the technique has been adapted for true field portable use. Among all X-ray spectrometric techniques, energy dispersive X-ray fluorescence (EDXRF) is the most popular one. It has increasingly been applied in the last 20 years for the analysis of aerosols, waters, sediments, soils, solid waste and other environmental samples with sufficient sensitivity for the determination of many trace elements [14]. Field portable XRF (FPXRF) technology has gained widespread acceptance in the environmental community as a viable analytical approach for field application due to the availability of efficient radioisotope source excitation combined with highly sensitive portable detectors and their associated electronics. FPXRF methodology now provides a viable, cost- and time-effective approach for on-site analysis of a variety of environmental samples [15]. However, FPXRF analysers are generally less sensitive than laboratory methods (AAS, ICP-OES or ICP-MS). Moreover, FPXRF results are typically surface measurements. Therefore, they are mainly used as field screening tools to guide on-site decision making for mapping strategies and for real-time identification of contaminated sites [14].

4.3 Fractionation of Trace Elements

Fractionation procedures for the soil solid phase usually consist of a sequence of 'selective' chemical extraction techniques, which include the (successive) removal of soil phases and their associated trace elements. The binding form of a trace element in the soil solid phase determines the intensity of the release to the soil solution and hence the likelihood of mobilization and uptake by organisms [2]. While the release of trace elements is important from nutrition or ecotoxicity viewpoints, their retention is important in designing specific physicochemical treatments for binding them into immobile fractions. To assess the mobility and solubility of trace element fractions in the soil solid fraction, a range of extraction reagents is used, with the aim of releasing trace elements from specific soil phases. The improving detection limits of the analytical techniques for trace element determination, together with the increasing evidence that exchangeable elements correlate better with plant uptake, has led extraction methods to evolve towards the use of less and less aggressive solutions [16]. These solutions are sometimes called 'soft', 'mild' or partial extractants and are based on nonbuffered salt solutions although diluted acids and complexing agents are also included in the group [17]. As the specificity and selectivity of most extraction procedures is, however, often limited due to the complexity of possible reactions and often unknown reaction kinetics, they are generally considered as 'operationally defined'. Chemical fractionation procedures include single and sequential extractions. In addition to fractionation based on chemical extraction procedures, fractionation based on particle size can also be distinguished. In the latter, particles from a soil suspension are separated into different size fractions and the trace element contents are measured in each of these fractions.

4.3.1 Single Extractions

The use of single extractions to ascertain chemically distinguishable pools of trace elements within the soil matrix has been investigated since the early 1970s. Since then, the number of extraction procedures has grown exponentially. As a result, a wide range of extraction procedures have been employed to assess the mobility of trace elements in soils. These methods vary with respect to the extraction conditions: chemical nature and concentration of leaching solutions, solution/soil ratio, operational pH, and extraction time [18]. Single extraction procedures have mainly been developed for the extraction of cationic trace metals. Table 4.2 illustrates that a wide range of extractants can be used to assess trace elements associated with specific soil fractions.

4.3.1.1 Unbuffered and Buffered Salt Solutions

A number of buffered and unbuffered solutions (Table 4.2) are used to extract exchangeable trace elements in soils. The exchangeable fraction includes weakly adsorbed elements retained on the solid surface by relatively weak electrostatic interactions, elements that can be released by ion-exchange processes and elements that can be coprecipitated with carbonates [4]. Changes in the ionic composition, affecting adsorption–desorption reactions, or lowering of pH could cause remobilization of elements from this fraction [28]. Exchangeable ions are a measure of those trace elements which are released most readily into the environment [29]. The exchangeability of positively charged ions has previously been assessed by a wide variety of extraction reagents, mainly unbuffered salt solutions. Furthermore, the concentration of the extraction reagent, the liquid to solid ratio (L:S), equilibration/extraction time and temperature are important parameters [3]. The use of 0.01 M $CaCl_2$ as an extraction reagent was proposed by Novozamsky et al. [19] to estimate

Table 4.2 Most common single extraction reagents for trace metals according to Rauret [17]

Group	Chemical and solution strength	Reference
Acid extraction	HNO_3 0.43–2 M	[19]
	Aqua regia	[20]
	HCl 0.1–1 M	[19]
	CH_3COOH 0.1 M	[21]
	Melich 1:	[22]
	HCl 0.05 M + H_2SO_4 0.0125 M	
Chelating agents	EDTA 0.01–0.05 M, variable pH	[19]
	DTPA 0.005 M+ TEA 0.1 M	[23]
	Melich 3:	[24]
	CH_3COOH 0.02 M + NH_4F 0.015 M + HNO_3 0.013 M + EDTA 0.001 M	
Buffered salt solution	NH_4 acetate, acetic acid buffer pH 7; 1 M	[25]
	NH_4 acetate, acetic acid buffer pH 4.8; 1 M	[19]
Unbuffered salt solution	$CaCl_2$ 0.1 M	[19]
	$CaCl_2$ 0.05 M	[19]
	$CaCl_2$ 0.01 M	[19]
	$NaNO_3$ 0.1 M	[16]
	NH_4NO_3 1 M	[19]
	$AlCl_3$ 0.3 M	[26]
	$BaCl_2$ 0.1 M	[27]

Reprinted from Talanta, 46, G. Rauret, Extraction procedures for the determination of heavy metals in contaminated soil and sediment, 449–455. Copyright 1998, with permission from Elsevier [17].

the bioavailability of metals and nutrients in air-dry soil samples. Advantages of this particular protocol include [30,31] the following:

- Ca^{2+} is also the predominant cation, with a similar ionic strength in typical soils.
- Since the extraction solution is not buffered at a fixed pH, the reagent does not modify soil pH during extraction.
- The divalent cation also assures a good coagulation of the colloidal material in the suspension. As a result, higher reagent concentrations as required when using monovalent cation-based (Na^+, K^+, or NH_4^+) extractants are not necessary.

This extraction protocol has, for example, been used for the assessment of availability of metals to plants [32] or toxicity and metal uptake by soil organisms [33,34]. Alternative procedures which have been reported in monitoring programmes include higher-strength $CaCl_2$ solutions, or 1 M NH_4NO_3, 1 M $MgCl_2$, 0.1 M $NaNO_3$, etc. [19,35–38]. The relative exchangeability of cationic trace elements is determined by the affinity of the exchanging cation for the solid phase. This affinity increases with increasing valency and decreasing radius of the hydrated cation. Although heterogeneous soil systems may deviate from this ideal behaviour, the selectivity of soils for cations was frequently observed to increase according to the order Na < K < Mg < Ca. Consequently, under comparable conditions (for example concentration, extraction time, soil/solution ratio), the efficiency of cations to exchange trace metals increased as Li < Na < K < Mg < Ca < Ba. Therefore, salts of Ca and Ba were regarded as most effective and selective agents in extracting exchangeable trace metals. Unfortunately, both cations may cause serious background problems (interferences) during determination. Usually, this can be resolved by at least a 10-fold dilution of the extracts prior to measurement, decreasing the detection limit by the same ratio. For that reason, the use of easily volatilizable salts during analysis (that is $Mg(NO_3)_2$ or NH_4NO_3) has been proposed [3]. Moreover, the use of nitrate as counterion does not cause additional complexation and mobilization, compared to the use of chloride-containing extraction reagents. On the other hand, at a high pH, NH_3 may also complex metals in the NH_4NO_3 extraction protocol and induce additional release [37]. An extraction reagent containing 1 M NH_4OAc (ammonium acetate) has also frequently been used for the assessment of exchangeable elements. Acetate has a higher complexing ability compared to chloride and nitrate [3] and can thus effectively prevent metal readsorption during extraction. The NH_4OAc-solution can be buffered at pH 7 when carbonate dissolution is to be avoided [39,40], or it can be buffered at a lower pH perhaps to mimic that of the rhizosphere [41,42].

4.3.1.2 Dilute Acid Solutions

Nitric acid has been used to assess the geochemically active trace element fraction in the soil [43]. The term 'active' implies susceptibility to chemical interactions with the soil solids that control solution concentrations. However, this proposed terminology may be somewhat vague, and the fraction may therefore best be referred to as dilute acid-extractable metals. Diluted acids partially dissolve trace elements associated with different fractions such as exchangeable, carbonates, iron and manganese oxides and organic matter [17]. Nitric acid as an extraction reagent has a particular benefit that no interference from the counterion in mobilization of metals is expected, unlike the case for extraction reagents such as acetic acid or hydrochloric acid [38]. Extractions using 0.43–2 M HNO_3 [19,43] have also been used. In relation to this, the 0.1 M HCl-extractable metals at a liquid to solid (L:S) ratio of 25:1 have been defined as those operationally released by moderate acid attack [44,45]. However, there is no agreement on the acid concentration to be used [38]. In this context, the recently developed continuous leach inductively coupled plasma mass spectrometry (CL-ICP-MS) can be seen as a promising technique [46]. It addresses some of the shortcomings of bulk leaching techniques while overcoming the uncertainties associated with interpreting selective extraction data. It provides information on the specific geochemical sites and mineral phases from which elements are being released using real-time data generated by continually analysing progressively reactive solutions from water to 30 % nitric acid as they are pumped through the sample directly into an ICP-MS set-up. Mineral breakdown reactions can be monitored from the major elements released, thereby eliminating uncertainties related to host phase/trace element associations. Results from single mineral phases, mixtures of mineral phases, and natural ore samples indicated that the release of elements from specific minerals is not obscured in more complex samples and that reprecipitation and back reactions are not a concern with this method [46]. Low-molecular-weight organic acids (LMWOA) typical of those in plant exudates have been used to assess phytoavailability within the rhizosphere of many plants. In this regard, mainly acetic acid has been used (for example [21]), but some authors (for example [47]) have proposed the use of LMWOA mixture solution, consisting of a combination of acetic, formic, citric and malic acids at a ratio of 2:2:1:1, which mimics the soil solution composition in the rhizosphere.

4.3.1.3 Fraction Extractable by Chelating Agents

Chelating agents dissolve not only the exchangeable element fraction but also the elements forming complexes with organic matter, elements that are fixed on the soil iron and manganese hydroxides [17] and elements bound to carbonates [48]. Quevauviller ([48], [49], and references cited therein) reported on the standardization of some extraction procedures using chelating agents and acetic acid, and the production of certified reference materials for these procedures. Ethylenediamine tetraacetic acid (EDTA) 0.05 M (pH 7) was used in the certification of two sewage-sludge amended soils (CRM 483 and 484), a calcareous soil (CRM 600) and an organic soil (CRM 700). This solution is assumed to extract both carbonate-bound and organic-matter bound fractions of metals and was hence also considered to be suitable for calcareous soils. Mixed acid ammonium acetate / EDTA reagent was previously also introduced, but seems less suitable as there is some evidence that EDTA at pH 5.5 can precipitate Cr, Pb and Zn. Diethylenetriamine pentaacetic acid (DTPA) 0.005 M (pH 7.3) was thought to be a possibility for the determination of extractable Cd, Cu, Fe and Mn, but would be less suitable for Cr and Ni. In addition, this procedure is more complicated than that with EDTA, and the DTPA extracts less than the EDTA (at the specified concentrations), which might lead to sensitivity problems. As a result, the EDTA was found to be the method of preference. It is also worth mentioning that shaker type and speed were found to be responsible for major differences in extraction results between laboratories ([48], [49], and references cited therein). It should be noted that chelating extraction solutions have especially been developed to assess the bioavailability of trace elements [19,23]. As a result, a major decision factor during the standardization of extraction procedures using chelating agents was their potential for assessing bioavailability, which is discussed in Chapter 11.

4.3.1.4 Water-Soluble Fraction

The determination of the water-soluble fraction has grown in popularity during recent years, mainly due to the analytical equipment becoming more sensitive and less expensive. The water-soluble phase contains free ions and ions complexed with soluble organic matter and inorganic ligands. Soil solution can be directly sampled *in situ* at the actual soil water content, in combination with a size fractionation, using *in situ* filtration, dialysis, and so on (see Chapter 5). However, when the soil is too dry and/or pore water cannot be sampled, the water-soluble elements can be assessed by extracting them from the soil with deionized water at a certain soil–water ratio, followed by centrifugation or filtration. As the ionic strength in the extracts is very low and adsorption of trace elements to the filters or percolation of colloidal material can occur, filtration nevertheless should be conducted carefully. Moreover, the extraction time and L:S ratio should be chosen carefully as an equilibrium is not reached rapidly [50], as shown by Sinaj *et al.* [51] where it took 72 h before steady state was reached when extracting Zn with ultrapure water from two unpolluted soils.

4.3.2 Sequential Extraction Procedures

Sequential extraction schemes involve a number of extractions (mostly 3 to 8), which are sequentially applied to a solid sample aliquot. In all schemes, extractants are applied in order of increasing reactivity so that the successive fractions obtained correspond to trace elements associated with lower mobility. The extractants more commonly used in sequential extraction schemes fall generally within the following groups: unbuffered salts, weak acids, reducing agents, oxidizing agents and strong acids [17]. Sequential extraction schemes provide detailed information about the origin, mode of occurrence, mobility, biological and physicochemical availability of the trace elements in solid samples and have been shown to provide a convenient means to determine the trace elements associated with principal accumulative phases in soils and sediments.

4.3.2.1 Fractions in Sequential Extraction Procedures

There is a general agreement that the partitioning of trace elements obtained by sequential extraction procedures is operationally defined; that is, extractants suffer from a lack of selectivity and therefore leaching of specific species from the target mineralogical phase is troublesome. Although some authors classify sequential extraction schemes as speciation procedures, the schemes should be more correctly referred to as fractionation methods as a consequence of their more operational character. Most procedures involve the chemical separation of trace elements into the following broad fractions: (a) soluble in water, (b) exchangeable and/or acid soluble, (c) reducible or those associated with Fe and Mn oxides, and (d) oxidizable or those associated with organic matter and sulfides, often supplemented by (e) residual or those associated with silicates. Extraction reagents used in some representative sequential extraction schemes applied between 1973 and 2000 are given in Table 4.3.

Water-soluble, Exchangeable and Acid-soluble Fractions
Extraction procedures to assess the water-soluble and exchangeable trace element fractions have already been discussed above (see Section 4.3.1). In most extraction schemes,

Table 4.3 Extraction reagents used in some representative sequential extraction schemes applied between 1973 and 2000

Scheme	Extraction step						
	1	2	3	4	5	6	7
MacLaren and Crawford [52]	$CaCl_2$	HOAc	$K_4P_2O_7$	$NH_4Ox/$ HOx	DCB		
Gibbs [53]	$MgCl_2$	DCB	NaOCl/ DCB				
Engler et al. [54]	NH_4OAc	$NH_2OHs \cdot$ HCl	$H_2O_2/$ NH_4OAc	DCB			
Tessier et al. [55]	$MgCl_2$	NaOAc	$NH_2OHs \cdot$ HCl / HOAc	$H_2O_2/$ NH_4OAc			
Meguellati et al. [56]	$BaCl_2$	$H_2O_2/$ NH_4OAc	NaOAc	$NH_2OH \cdot$ HCl			
Shuman [57]	$Mg(NO_3)_2$	NaOCl	$NH_2OH \cdot$ HCl	$NH_4Ox/$ HOx			
Salomons and Förstner [58]	NH_4OAc	NaOAc	$NH_2OH \cdot$ HCl	$NH_4Ox/$ HOx	$H_2O_2/$ NH_4OAc		
Miller et al. [59]	$Ca(NO_3)_2/$ $Pb(NO_3)_2$	HOAc / $Ca(NO_3)_2$	$NH_2OH \cdot$ HCl	$K_4P_2O_7$	$NH_4Ox/$ HOx	$NH_4Ox/$ HOx	
Elliot et al. [60]	$MgCl_2$	NaOAc	$NH_4Ox/$ HOx	$Na_4P_2O_7$			
Ure et al. (BCR)[a] [21]	HOAc	$NH_2OH \cdot HCl$	$H_2O_2/$ NH_4OAc				
Krishnamurti et al. [61]	$Mg(NO_3)_2$	NaOAc	$Na_4P_2O_7$	$NH_2OH \cdot$ HCl	$H_2O_2/$ $Mg(NO_3)_2$	NH_4Ox	$NH_4Ox/$ AA
Campanella et al. [41]	NH_4OAc	$NH_2OH \cdot HCl$ /HOAc	HCl	NaOH	HNO_3		
Sahuquillo et al. [62] (Modified BCR)[b]	HOAc	$NH_2OH \cdot$ HCl^b	$H_2O_2/$ NH_4OAc				

[a]BCR, Bureau Communautaire de Reference.
[b]This scheme differs from that of Ure et al. [21], mainly in the hydroxylamine hydrochloride concentration (0.5 instead of 0.1 mol l^{-1}) and the pH of this extractant (1.5 instead of 2).
Adapted from A.V. Filgueiras, I. Lavilla, C. Bendicho, Chemical sequential extraction for metal partitioning in environmental solid samples, J. Environ. Monit. 2002, 4, 823–857. With permission of The Royal Society of Chemistry [4].

water-soluble trace elements are extracted as part of the exchangeable fraction. In contrast to the single extractions used to assess an acid-extractable fraction, the acid-soluble fraction of the sequential extractions schemes is primarily intended to contain mainly trace elements associated with carbonates, and not those associated with Fe/Mn oxides and organic matter. Therefore, 1 M sodium acetate–acetic acid buffered to pH 5 is often used [55,62,63], although a significant amount of trace metals sorbed to other phases may also be solubilized at this pH and metal complexation must be assumed [64]. Other extractants for carbonates can be HCl and acetic acid (pH 3–3.5) and the complexing agent Na$_2$EDTA buffered to pH 4.6, but these are less selective [3]. In some procedures, assessment of the acid-soluble fraction is not preceded by an extraction of the exchangeable fraction, which implies that exchangeable trace elements are also included in the acid-soluble fraction.

Fraction Associated with Fe and Mn oxides or Reducible Fraction

Manganese and iron oxides are strong scavengers for trace elements in soils [61,65]. Scavenging of trace elements by secondary oxides, present as coatings on mineral surfaces or as fine discrete particles, can occur by any or a combination of the following mechanisms: coprecipitation, adsorption, surface complex formation, ion exchange, and penetration into the lattice [66]. The oxides contribute to a 'dynamic' trace element fraction in the soil. Their reduction in a reducing environment (for example a flooded soil) can result in the release of associated trace elements. Upon subsequent oxidation of the soil, reprecipitation of amorphous oxides with coprecipitation of associated trace elements can occur [67]. In an oxidizing, dry environment, they can further evolve to more crystalline and stable structures, which can further immobilize trace elements. Effects of the pH in this relation are caused by the competitive exchange of protons with the trace elements sorbed on hydrous oxides [68] but also by pH-dependent precipitation/dissolution and oxidation/reduction of the hydrated oxides [69]. Manganese oxides differ from iron oxides in terms of their solubility in the natural environment, and their dependence on the oxidation–reduction conditions and pH [61,70]. Moreover, 'new' oxides (amorphous forms) are easily reducible, but 'aged' oxides with a more crystalline character (crystalline oxides) are more resistant to changes in the oxidation–reduction conditions and soil pH [71]. On the basis of these principles, a range of extraction procedures has been developed to discriminate between trace elements associated with amorphous and crystalline oxides, often as part of a sequential extraction scheme [55,57,71–73].

Practically, the reducible fraction can split into three subfractions: an easily reducible fraction (Mn oxides), a moderately reducible fraction (amorphous Fe oxides) and a poorly reducible fraction (crystalline Fe oxides). To assess the easily reducible fraction, hydroxylamine hydrochloride in a nitric acid medium is often used, which, however, also releases substantial amounts of trace elements bound to organic matter [64]. The moderately reducible fraction, that is trace elements associated with amorphous Fe oxides, can be assessed with 1 M hydroxylamine hydrochloride in an acetic acid medium or with an oxalic acid/ammonium oxalate buffered to pH 3 in the dark. The first reagent is capable of breaking the bonds between trace elements and amorphous and poorly crystallized Fe oxides without attacking either the silicates or the organic matter fraction [74], whereas trace elements associated with organic matter might also leach out due to the complexing capacity of oxalate in the latter case [75]. Extraction of trace elements associated with the organic matter by sodium hypochlorite prior to extraction of trace elements associated with the Fe–Mn oxides is recommended [57]. A sodium citrate–sodium dithionite buffer (DCB) is adequate for dissolving the crystalline Fe oxides, and is implemented in some schemes [26]. This reagent, however, easily attacks silicates and is difficult to purify [55]. In this context, an ascorbic acid–ammonium oxalate mixture under illumination offers a better alternative [57]. It has been included in some schemes, for example Krishnamurti *et al.* [61]. Although the terms 'amorphous' and 'crystalline' are used in the fractionation schemes, it should again be emphasized that these are essentially operationally defined [69].

Fraction Associated with Organic Matter and Sulfides or Oxidizable Fraction

Trace elements may be associated with various forms of organic material such as living organisms, detritus or coatings on mineral particles through complexation or bioaccumulation processes. They are assumed to be retained in the soil for longer periods but may be mobilized by decomposition processes [4,76]. As trace elements bound to organic matter can be easily released under oxidizing conditions, an oxidation process is usually applied to leach them, which also results in the release of metals precipitated as sulfides or bound to sulfides. The most common oxidants are hydrogen peroxide in an acidic medium [55,61], NaOCl at pH 9.5 [57,61], $Na_4P_2O_7$ at pH 9.5 [77] and $K_4P_2O_7$ [78]. A hydrogen peroxide–ammonium acetate (H_2O_2–NH_4OAc) mixture, where the addition of NH_4OAc prevents readsorption of extracted trace elements onto the oxidized substrate, has been adopted in most schemes for leaching trace elements associated with organic matter and sulfides. However, the oxidation can be incomplete in the presence of high levels of recalcitrant organic substances and sulfide minerals. Sodium hypochlorite (NaOCl) at pH 9.5 dissolves neither Fe/Mn oxides nor carbonate, and was found superior to H_2O_2 in removing organic carbon, whereas potassium pyrophosphate ($K_4P_2O_7$) is selective for the easily soluble organic fraction (that is metals associated with humic and fulvic acids). Unlike sodium hypochlorite and sodium pyrophosphate, H_2O_2 can attack the moderately reducible and crystalline Fe–Mn oxides; consequently, this reagent must be applied after dissolution of the complete reducible phase. Other oxidizing mixtures, such as H_2O_2–ascorbic acid or HNO_3–HCl can dissolve sulfides with enhanced selectivity, but silicates are also attacked to some extent [4], whereas NaOH is less selective and might induce the precipitation of the trace elements as hydroxides [3]. According to Hlavay and Polyák [3], the extraction of organically bound trace metals by competing synthetic chelates, for example EDTA or DTPA, should be considered after removal of the most labile oxide fraction, as an alternative to the destruction of the organic ligands.

Residual Fraction

This fraction mainly contains silicate/crystalline bound trace elements and is most commonly dissolved with concentrated acids, as previously described in Section 4.2.1.

After applying each of the above-mentioned single or sequential extractions, the trace element contents should be determined in the individual liquid phases separated from the solid residue by centrifugation or filtration using analysis techniques described earlier (see Section 4.2.2).

4.3.2.2 Standardization of Sequential Extraction Procedures

Uniformity on which reagents should be employed in sequential extraction procedures as well as their application order has been lacking in the past. The release of trace elements associated with soil organic matter has been the most controversial. The variations in extraction parameters for the oxidizable fraction with concentrations between 0.1 and 1 M, solution/soil ratios between 10:1 and 100:1 and extraction times from 1 h to 24 h indicate that results obtained by different procedures are hardly comparable and are likely to extract nonorganically bound trace elements to varying extents [3]. Moreover, some workers recommended a prior decomposition of the organic matter so that the release in the subsequent fractions is facilitated [57,61]. In contrast, in most sequential extraction procedures the reducible fraction and the fraction associated with carbonates are isolated prior to the fraction associated with organic matter since the acidic conditions involved in the leaching of the latter fraction can cause premature leaching of both fractions [21,55,63,79]. Alternatively, extraction of the fraction associated with organic matter can be conducted before the extraction of the moderately reducible fraction, mainly when the latter fraction is extracted with oxalate [4,59].

The various sequential extraction procedures that have been developed, aim at defining the total trace element pool into fractions that have relevance when assessing their mobility or bioavailability. Most procedures address a wide range of trace elements, but some extraction schemes were developed for specific elements or groups of elements, taking into account their specific chemical properties [80,81]. The procedure proposed by Tessier *et al.* [55] has been widely used and several subsequent techniques are modifications of this procedure. Some comprehensive reviews have been written on the widespread use of sequential extraction schemes, their application, harmonization, validation, etc. (for example [4,82,83]). Recent research attention has emphasized harmonization and development of a standardized three-step BCR sequential extraction procedure [3,4,62].

4.3.2.3 *Sample Storage and Pretreatment*

Changes in the speciation and the fractionation of trace elements often occur as a result of drying, oxidation or reduction during sample storage and preparation of the soil sample. Wet storage of oxidized soils at ambient temperature is inadequate because of the rapid microbial induced shift from oxidizing to reducing conditions. Drying, however, seems to accelerate the crystallization of solids such as Fe–Mn oxides and promotes Fe, Mn and S oxidation, causing an increase in the amount of trace elements bound to Fe and Mn oxyhydroxides to the detriment of more labile phases made up of the exchangeable and carbonate fractions [84]. Moreover, aeration causes the oxidation of sulfide to sulfate, hence releasing metals and protons. Although fractionation procedures should thus preferably be conducted on fresh soil samples [85], problems concerning sample handling, homogenization and representative subsampling may arise when the soils are not dried and sieved prior to the analysis. When freeze-drying, air-drying and oven-drying at 105 °C were compared, it was found that Cd, Cu and Pb were the metals most sensitive to pretreatments, whereas freeze-drying and air-drying were the least disturbing procedures [86]. Davidson *et al.* [87], in a study with industrially contaminated soil, found that the reproducibility was higher for air-dried samples than for field-moist soils, but also larger amounts of the metals were extracted. The latter again indicates the occurrence of alterations in the metal distribution during drying. Another aspect to be considered is the need for remixing before extraction. The metal extractability from a soil sample was found to be affected by grinding, so rehomogenization merely by manual shaking has been recommended [88]. Cryogrinding, that is continuous cooling with liquid nitrogen during grinding after shock-freezing, is found to be well suited for speciation analysis, but not for fractionation between groups of species which are differently immobilized at naturally occurring surfaces in the sample of interest [5].

4.3.2.4 *Disadvantages of the Current Procedures and Recent Developments*

By conducting one or a combination of sequential extraction procedures, one should be able to assess which part of the total trace element pool will be released in the long

70 Basic Principles, Processes, Sampling and Analytical Aspects

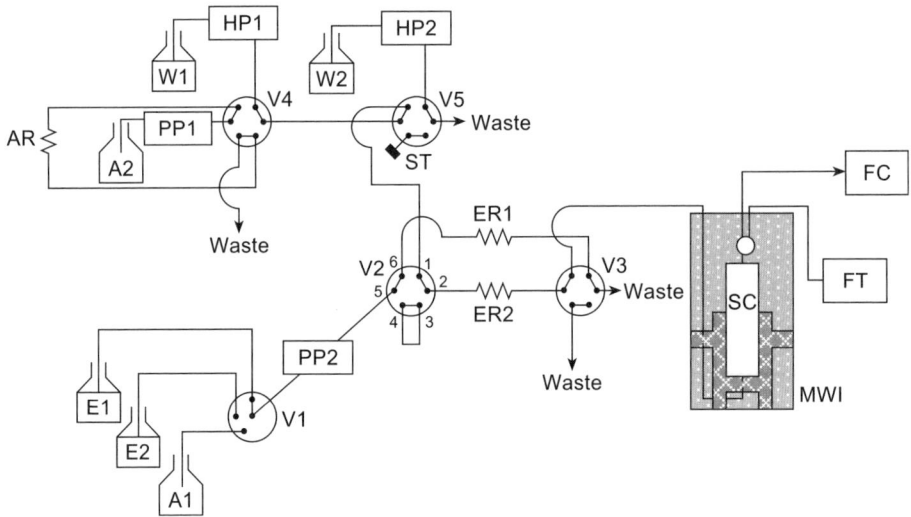

Figure 4.2 *Scheme of a microwave-assisted continuous-flow sequential extraction system according to Nakazato et al. [100]. SC, column packed with a soil sample; MWI, microwave irradiator; FT, fluoroptic thermometer; FC, fraction collector; V1, 6-position switching valve with 6-ports; V2–V5, 2-position switching valves with 6-ports; 1–6, connection ports of the two-position valve; ST, stopper; E1 and E2, extractants for steps 1 and 2; A1 and A2, air; W1 and W2, pure water; AR, reservoir tube for air; ER1 and ER2, reservoir tubes for E1 and E2; PP1 and PP2, peristaltic pumps; HP1 and HP2, high-pressure plunger pumps (Reproduced from T. Nakazato, M. Akasaka and H. Tao, A rapid fractionation method for heavy metals in soil by continuous-flow sequential extraction assisted by focused microwaves, Anal. Bioanal. Chem. 2006, 386, 1515–1523. Reproduced with kind permission from Springer Science + Business Media [100].)*

term upon the creation of, for example, a high ionic strength or an oxidizing, reducing or acidic environment in a soil. However, most extraction reagents are not entirely selective. These procedures also have other potential disadvantages: they are time-consuming; the experimental conditions cannot be extrapolated to natural conditions; and readsorption and redistribution of trace element can occur during the various sample preparation and analysis steps. Moreover, precautions should be taken against changes in speciation during the different steps of the extraction procedure itself (for example [63,87,89–93]).

To make sequential extraction procedures less time-consuming and labour-intensive, ultrasound-assisted [94] or microwave-assisted procedures [95], as well as dynamic, flow-through approaches [96–101] have recently been developed. When using a continuous-flow sequential extraction procedure (Figure 4.2), not only fractional distribution data for evaluation of the mobility or potential environmental impact of the trace elements are generated, but also extraction profiles (extractograms) which can be used to study elemental association and to demonstrate the degree of anthropogenic contamination [102].

Bacon and Davidson [83] considered sequential extraction to have a healthy future in the 21st century, but they stated that its continued usefulness, in particular for environmental monitoring, requires researchers to be aware of the limitations. Studies based on sequential extractions are more likely to be successful if certain conditions are met [83]:

1. A 'standard' sequential extraction scheme (for example the revised BCR procedure) should be used whenever possible because of the availability of certified reference materials for quality control and the possibility of direct comparison between studies.
2. When using such a 'standard' scheme it is critical that the specified procedure is adhered to strictly or, at the very least, any variations should be reported.
3. There should if possible be some degree of comparison within a study – that is, comparing the difference in 'before and after' situations, spatial or temporal variability, the effect of some form of treatment or other changes within a system.
4. Data should be presented in terms of absolute concentrations instead of or in addition to percentage values.
5. Data should be interpreted according to the 'operational speciation' definition of Ure [103]. For example, potentially toxic elements recovered in Step 2 of the BCR procedure should be described as associated with the 'reducible fraction', rather than as being 'bound to iron/manganese oxyhydroxides'.
6. It is important that sequential extractions are not applied uncritically and users should take cognizance of the scope and limitations of the approach.
7. Care should be taken when drawing conclusions concerning bioavailability based on sequential extraction results.

4.3.3 Fractionation Based on Particle Size

Fractionation of trace elements in soils can also be conducted on the basis of particle size. Soil suspensions can be separated into different size fractions and the trace elements associated with each of these particle sizes measured individually. Fractionation using centrifugation and filtration with particle size thresholds of, for example, <0.2 and <0.45 μm, combined with subsequent analysis of trace elements in the collected leachates (or those retained on the filter) can easily be conducted in most laboratories. These size fractions coincide with the distinction between dissolved and particulate loads in many studies and represent the two most common operational fractions isolated by traditional membrane filtration [104]. However, it is well known that in natural waters many trace elements are associated with colloidal (that is, between 1 nm and 1 μm) rather than true dissolved forms. Methods to establish a further fractionation below 0.2 μm have therefore been developed and include ultrafiltration (UF), field-flow fractionation (FFF) or size exclusion chromatography (SEC), often directly coupled to an ICP-MS. As they can also be used to fractionate trace elements in soil solution, these techniques are discussed in Chapter 5.

4.4 Species-Retaining and Species-Selective Leaching Techniques

Most of the previously mentioned extraction reagents and procedures can also identify soil fractions the trace elements are associated with. However, changes in trace element species during the extraction itself are often observed. If these species transformations do not occur during the extraction, speciation analysis can be conducted on the leachate as described for the soil solution in Chapter 5 and the results can be used to assess the trace element speciation in the soil solid phase. Moreover, if different trace element species can be leached from the soil solid phase selectively, leaching techniques themselves can be used as part of a speciation procedure. Such species-retaining and species-selective extraction procedures have especially been developed for organometallic compounds, as these are apparently not involved in mineralogical processes in soils, but can bind onto the soil surface [105].

The basic approach to releasing organometallic compounds from soil or sediment samples involves acid leaching (HCl, HBr or HOAc) into an aqueous or methanolic medium by sonication, stirring, shaking or Soxhlet extraction with an organic solvent. In order to increase the extraction yield, the addition of a complexing agent (for example diethyldithiocarbamate, DDTC or tropolone) is mandatory. The majority of the procedures reported have not only been extremely time-consuming (from 1 hour to 2 days), but have also been inefficient in terms of analyte recovery and are, in general, unreliable. In general, the more polar the species to be extracted (for example monobutyltin $BuSn^{3+}$ compared to tributyltin Bu_3Sn^+), the lower is its recovery and the longer the leaching procedure necessary [105]. In this context, several modern extraction techniques, such as microwave-assisted procedures, supercritical fluid extraction (SFE) or accelerated solvent extraction (ASE), are very promising. They offer improved efficiency and selectivity over classical extractions. Microwave-assisted leaching has been shown to reduce the time necessary for leaching of organometallic compounds (for example methylmercury or organotin species) from soil samples, but also to increase the recovery of compounds that are most difficult to extract. Of course, the preservation of the organometallic moiety is the prerequisite of a successful leaching procedure prior to speciation analysis. It can be achieved by a careful optimization of the conditions of the microwave attack. The use of a focused low-power microwave field is preferred to keep the carbon–metal bonds intact [105].

Supercritical fluid extraction is an interesting technique which can obtain extraction recoveries close to 100 %. The properties of supercritical fluids that are attractive from an extraction point of view include: (a) considerably great diffusion coefficients, leading to efficient and rapid extractions; (b) low viscosity and absence of surface tension, which facilitate pumping in the extraction process; (c) density close to that of liquids, enabling the greater interactions on a molecular level necessary for the solubilization. Temperature and pressure changes, near the supercritical point, can affect the solubility by a factor of as much as 100, or even 1000. Moreover, the use of fluids with low critical temperature values (for example CO_2 and N_2O) allows extractions under thermally mild conditions, protecting labile compounds [106]. Supercritical fluid extraction was mainly developed to extract several compounds from biological materials. By *in situ* derivatization prior to SFE and use of modifiers (HCO_2H, AcOH, or methanolic HCl) or complexing agents, an efficient extraction of different trace element species from soils and sediments can also be established [107]. The efficiency of metal extraction using the *in situ* chelation-SFE method depends on a number of factors including (a) stability and solubility of ligand, (b) solubility of metal chelate, (c) water and pH, (d) temperature and pressure, (e) chemical form of metal species, and (f) matrix. Several different types of chelating agents, including dithiocarbamates, beta-diketones, organophosphorus reagents and macrocyclic compounds, have been tested for their ability to extract heavy metals from different sample matrices, for example soils and sediments. With a proper selection of the ligands, separation of different organic and inorganic metal species (for example Cr, Sn, Pb and Hg) has been achieved using the SFE technique [107–111].

Another new technique is accelerated solvent extraction (ASE), which uses liquid solvents at high pressures and temperatures, above 200 bar ($2 \times 10^{7-}$ Pa) and 200°C respectively. Although ASE has been used successfully for the isolation of butyltin species from sediments [112], its use for trace element speciation in soil or sediment samples has rarely been studied.

4.5 Equipment for Direct Speciation of Trace Elements in Soil

Many of the known analytical methods for elemental speciation analysis are restricted to fluid samples, for instance the whole range of chromatographic methods (see Chapter 5). Consequently, these methods are applicable to solid samples only if it is possible to dissolve the analyte without destruction of the species information, which is often a tedious and error-prone operation [113], as discussed previously. An analytical method circumventing this problem by enabling direct species determination in solid materials is thus highly desirable. Such a technique is provided by X-ray absorption fine structure (XAFS) spectroscopy, which has been reviewed elsewhere [113–116]. A positive property of XAFS spectroscopy is that it yields the species information without being significantly disturbed by the matrix due to its high degree of element selectivity. This contrasts with other methods that enable the determination of the chemical speciation only in a pure or at least highly concentrated solid sample, such as X-ray diffraction (XRD) or infrared (IR) spectroscopy. As in other absorption spectroscopic techniques, the dependence of the absorption on the wavelength of the incoming beam, here consisting of X-rays, is measured in X-ray absorption spectroscopy (XAS). Unlike the absorption for example in the UV/VIS region, the X-ray absorption spectrum shows no peaks but edges, at which the absorption coefficient increases abruptly. This is attributed to the electrons being excited to the continuum in XAS, whereas the transitions in optical UV spectroscopy take place between energetically well defined orbitals. The edge position corresponds to the energy needed to lift a core electron to the continuum, which is specific for every element. Small perturbations of the core ground states due to the redox state lead to a 'chemical shift' of the edge position depending on the oxidation state of the absorbing atom, whereas oscillations can also be found above the edge. The latter contain much more information about the sample than simply detecting the oxidation state of the absorber atom, for example information about the nature, number and distance of the next neighbouring atoms. Analysis of these oscillations is used in XAFS. The first 50–100 eV above the edge is explored in X-ray absorption near-edge structure (XANES) spectroscopy, whereas the region above the XANES region is explored in extended X-ray absorption fine structure (EXAFS) spectroscopy. These techniques provide insight into the mineralogy, molecular level speciation and atomic coordination, the oxidation state, bond distances and degree of complexation of the trace element of interest. Some general limitations are the need for homogeneous samples, the limited availability of synchrotron radiation source beamlines for XAFS measurements, and the problem of choosing suitable reference substance spectra. Nowadays, XAFS spectroscopy can be performed with a spatial resolution in the micrometer range when using focused X-ray optics in μ-XANES or μ-EXAFS. It can generally be said that XAFS spectroscopy is not a technique that is suited for trace analysis, although the actual detection limit is highly dependent on the matrix properties. In general, the detection limit is lower in samples that contain a heavy element as analyte in a light element matrix. Moreover, the use of fluorescence detectors decreases the detection limit by a factor between 10 and 1000 [113]. As a result, XANES and EXAFS have been widely used for studies of trace element speciation in solid soil samples (for example [117–122]).

4.6 Conclusions

A wide range of techniques to assess the (pseudo-)total trace element contents in soils are currently available. Most of them are dissolution based, although some nondestructive techniques are also available and being further developed. The present state of knowledge on speciation and fractionation of trace elements in the soil solid phase is still, however, somewhat unsatisfactory. Techniques for direct speciation in the soil solid phase are not sensitive enough for trace elements, and they require major capital investment to set up and hence are not easily available to most environmental laboratories. Moreover, chemical species are often not stable enough to be determined individually, and special procedures for species-retaining sampling and extraction of soils are still lacking, though they are currently being developed. As a result, operationally defined fractionation schemes, consisting of one or more single and/or sequential extractions, are considered as a valuable tool to distinguish between trace metal fractions which are empirically related to mobility classes, despite their limitations.

References

[1] IUPAC, Guidelines for terms related to chemical speciation and fractionation of elements. Definitions, structural aspects, and methodological approaches; Pure Appl. Chem. 2000, 72, 1453–1470.
[2] Tack, F.M.G.; Verloo, M.G., Chemical speciation and fractionation in soil and sediment heavy-metal analysis - a review; Int. J. Environ. Anal. Chem. 1995, 59, 225–238.
[3] Hlavay, J.; Polyák, K., Sample preparation – fractionation (sediments, soils, aerosols, and fly ashes); in: R. Cornelis, H.; Crews, J.; Caruso, K.; Heumann (eds), Handbook of Elemental Speciation: Techniques and Methodology; John Wiley & Sons, Ltd, Chichester, 2003, pp. 119–146.
[4] Filgueiras, A.V.; Lavilla, I.; Bendicho, C., Chemical sequential extraction for metal partitioning in environmental solid samples; J. Environ. Monit. 2002, 4, 823–857.
[5] Emons, H., Sampling: collection, processing and storage of environmental samples; in: R. Cornelis, H.; Crews, J.; Caruso, K.; Heumann (eds), Handbook of Elemental Speciation: Techniques and Methodology; John Wiley & Sons, Ltd, Chichester, 2003, pp. 7–22.
[6] Medved, J.; Stresko, V.; Kubova, J.; Polakovicova, J., Efficiency of decomposition procedures for the determination of some elements in soils by atomic spectroscopic methods; Fresen. J. Anal. Chem. 1998, 360, 219–224.

[7] Rouessac, F.; Rouessac, A., Chemical Analysis. Modern Instrumentation Methods and Techniques, 2nd edn; John Wiley & Sons, Chichester, 2007.
[8] Heumann, K.G., Isotope-dilution ICP-MS for trace element determination and speciation: from a reference method to a routine method?; Anal. Bioanal. Chem. 2004, 378, 318–329.
[9] Shinotsuka, K.; Ebihara, M., Precise determination of rare earth elements, thorium and uranium in chondritic meteorites by inductively coupled plasma mass spectrometry – a comparative study with radiochemical neutron activation analysis; Anal. Chim. Acta 1997, 338, 237–246.
[10] Witkowska, E.; Szczepaniak, K.; Biziuk, M., Some applications of neutron activation analysis: a review; J. Radioan. Nucl. Ch. 2005, 265, 141–150.
[11] Cruvinel, P.E.; Crestana, S.; Artaxo, P.; Martins, J.V.; Armelin, M.J.A., Studying the spatial variability of Cr in agricultural field using both particle induced X-ray emission (PIXE) and instrumental neutron activation analysis (INAA) technique; Nucl. Instrum. Meth. B 1996, 109, 247–251.
[12] Pillay, A.E., A review of accelerator-based techniques in analytical studies; J. Radioan. Nucl. Ch. 2000, 243, 191–197.
[13] Alamin, M.B.; Spyrou, N.M., Elemental characterization of different matrices including coal, sawdust, fly-ash and landfill waste samples using INAA and PIXE analyses; J. Radioan. Nucl. Ch. 1997, 216, 41–45.
[14] Melquiades, F.L.; Appoloni, C.R., Application of XRF and field portable XRF for environmental analysis; J. Radioan. Nucl. Ch. 2004, 262, 533–541.
[15] Kalnicky, D.J.; Singhvi, R., Field portable XRF analysis of environmental samples; J. Hazard. Mater. 2001, 83, 93–122.
[16] Gupta, S.K.; Aten, C., Comparison and evaluation of extraction media and their suitability in a simple-model to predict the biological relevance of heavy-metal concentrations in contaminated soils; Int. J. Environ. Anal. Chem. 1993, 51, 25–46.
[17] Rauret, G., Extraction procedures for the determination of heavy metals in contaminated soil and sediment; Talanta 1998, 46, 449–455.
[18] Beckett, P.H.T., The use of extractants in studies on trace metals in soils, sewage sludges and sludge-treated soils; Adv. Soil Sci. 1989, 9, 143–179.
[19] Novozamsky, I.; Lexmond, T.M.; Houba, V.J.G., A single extraction procedure of soil for evaluation of uptake of some metals by plants; Int. J. Environ. Anal. Chem. 1993, 51, 47–58.
[20] Colinet, E.; Gonska, H.; Griepink, B.; Muntau, H., The certification of the contents of cadmium, copper, mercury, lead and zinc in a calcareous loam soil: BCR No. 141. EUR Report 8833 EN, Commission of the European Communities, Luxembourg, 1983, p 57.
[21] Ure, A.M.; Quevauviller, P.; Muntau, H.; Griepink, B., Speciation of heavy-metals in soils and sediments - an account of the improvement and harmonization of extraction techniques undertaken under the auspices of the BCR of the commission of the European communities; Int. J. Environ. Anal. Chem. 1993, 51, 135–151.
[22] Mulchi, C.L.; Adamu, C.A.; Bell, P.F.; Chaney, R.L., Residual heavy-metal concentrations in sludge amended coastal-plain soils. 2. Predicting metal concentrations in tobacco from soil test information; Comm. Soil Sci. Plan. 1992, 23, 1053–1069.
[23] Lindsay, W.L. and Norvell, W.A., Development of a DTPA soil test for zinc, iron, manganese, and copper; Soil Sci. Soc. Am. J. 1978, 42, 421–428.
[24] Melich, M., Melich-3 soiltest extractant: a modification of the Melich-2 extractant; Comm. Soil Sci. Plan. 1984, 15, 1409–1416.
[25] Ure, M.; Thomas, R.; Littlejohn, D., Ammonium acetate extracts and their analysis for the speciation of metal-ions in soils and sediments; Int. J. Environ. Anal. Chem. 1993, 51, 65–84.
[26] Hughes, J.C.; Noble, A.D., Extraction of chromium, nickel and iron and the availability of chromium and nickel to plants from some serpentinite-derived soils from the eastern Transvaal as revealed by various single and sequential extraction techniques; Comm. Soil Sci. Plan. 1991, 22, 1753–1766.
[27] Juste, C.; Solda, P., Changes in the cadmium, manganese, nickel and zinc bioavailability of a sewage sludge-treated sandy soil as a result of ammonium-sulfate, acid peat, lime or iron compound addition; Agronomie 1988, 8, 897–904.

[28] Sutherland, R.A.; Tack, F.M.G.; Tolosa, C.A.; Verloo, M.G., Operationally defined metal fractions in road deposited sediment, Honolulu, Hawaii; J. Environ. Qual. 2000, 29, 1431–1439.
[29] Kabata-Pendias, A., Behavioural properties of trace metals in soils; Appl. Ecochem. 1993, 2, 3–9.
[30] Houba, V.J.G.; Lexmond, T.M.; Novozamsky, I.; Van der Lee, J.J., State of the art and future developments in soil analysis for bioavailability assessment; Sci. Total Environ. 1996, 178, 21–28.
[31] Houba, V.J.G.; Temminghoff, E.J.M.; Gaikhorst, G.A.; Van Vark, W., Soil analysis procedures using 0.01 M calcium chloride as extraction reagent; Comm. Soil Sci. Plan. 2000, 31, 1299–1396.
[32] Brun, L.A.; Maillet, J.; Hinsinger, P.; Pepin, M., Evaluation of copper availability to plants in copper-contaminated vineyard soils; Environ. Pollut. 2001, 111, 293–302.
[33] Weitje, L., Mixture toxicity and tissue interactions of Cd, Cu, Pb and Zn in earthworms (oligochaeta) in laboratory and field soils: a critical evaluation of data; Chemosphere 1998, 36, 2643–2660.
[34] Van Peijnenburg, W.J.G.M.; Posthuma, L.; Zweers, P.G.P.C.; Baerselman, R.; de Groot, A.C.; Van Veen, R.P.M. et al., Prediction of metal bioavailability in Dutch field soils for the oligochaete *Enchytraeus crypticus*; Ecotox. Environ. Saf. 1999, 43, 170–186.
[35] Belotti, E., Assessment of a soil quality criterion by means of a field survey; Appl. Soil Ecol. 1998, 10, 51–63.
[36] Maddocks, G.; Reichelt-Brushett, A.J.; Vangronsveld, J., An assessment of bioaccumulation in *Eisenia fetida* after exposure to metal loaded Bauxsol^{TM}; Environ. Toxicol. Chem. 2005, 24, 554–565.
[37] Pueyo, M.; Lopez-Sanchez, J.F.; Rauret, G., Assessment of $CaCl_2$, $NaNO_3$ and NH_4NO_3 extraction procedures for the study of Cd, Cu, Pb and Zn extractability in contaminated soils; Anal. Chim. Acta 2004, 504, 217–226.
[38] Meers, E.; Du Laing, G.; Unamuno, V.; Ruttens, A.; Vangronsveld, J.; Tack, F.M.G. et al., Comparison of cadmium extractability from soils by commonly used single extraction protocols; Geoderma 2007, 141, 247–259.
[39] Baker, A.J.M.; Reeves, R.D.; Hajar, A.S.M., Heavy metal accumulation and tolerance in British populations of the metallophyte *Thlaspi caerulescens* J. & C. Presl.; New Phytol. 1994, 127, 61–68.
[40] Gommy, C.; Perdrix, E.; Galloo, J.-C.; Guillermo, R., Metal speciation in soil: extraction of exchangeable cations from a calcareous soil with a magnesium nitrate solution; Int. J. Environ. Anal. Chem. 1998, 72, 27–45.
[41] Campanella, L.; D'Orazio, D.; Petronio, B.M.; Pietrantonio, E., Proposal for a metal speciation study in sediments; Anal. Chim. Acta 1995, 309, 387–393.
[42] Yu, S.; He, Z.L.; Huang, C.Y.; Chen, G.C.; Calvert, D.V., Copper fractionation and extractability in two contaminated variable soils; Geoderma 2004, 123, 163–175.
[43] Tipping, E.; Rieuwerts, J.; Pan, G.; Ashmore, M.R.; Lofts, S.; Hill, M.T.R. et al., The solid-solution partitioning of heavy metals (Cu, Zn, Cd, Pb) in upland soils of England and Wales; Environ. Pollut. 2003, 125, 213–225.
[44] Barreto, S.R.G.; Nozaki, J.; De Oliveira, E.; Do Nascimento Filho, V.F.; Aragão, P.H.A.; Scarminio, I.S. et al., Comparison of metal analysis in sediments using EDXRF and ICP-OES with the HCl and Tessier extraction methods; Talanta 2004, 64, 345–354.
[45] Fiszman, M.; Pfeiffer, W.C.; De Lacerda, L.D., Comparison of methods used for extraction and geochemical distribution of heavy metals in bottom sediments from Sepetiba Bay, R.J.; Environ. Technol. 1984, 5, 567–575.
[46] MacFarlane, W.R.; Kyser, T.K.; Chipley, D.; Beauchemin, D.; Oates, C., Continuous leach inductively coupled plasma mass spectrometry: applications for exploration and environmental geochemistry; Geochem.-Explor. Env. A. 2005, 5, 123–134.
[47] Wang, W.-S.; Shan, W.-Q.; Wen, B.; Zhang, S.-Z., Relationship between the extractable metals from soils and metals taken up by maize roots and shoots; Chemosphere 2003, 53, 523–530.
[48] Quevauviller, Ph., Operationally defined extraction procedures for soil and sediment analysis. II. Certified reference materials; Trends Anal. Chem. 1998, 17, 632–642.
[49] Quevauviller, Ph., Operationally defined extraction procedures for soil and sediment analysis. Part 3. New CRMs for trace-element extractable contents. Trends Anal. Chem. 2002, 21, 774–785.

[50] Meers, E.; Du Laing, G.; Unamuno, V.G.; Lesage, E.; Tack, F.M.G.; Verloo, M.G., Water extractability of heavy metals from soils: some pitfalls; Water Air Soil Pollut. 2006, 176, 21–35.
[51] Sinaj, S.; Machler, F.; Frossard, E., Assessment of isotopically exchangeable zinc in polluted and unpolluted soils; Soil Sci. Soc. Am. J. 1999, 63, 1618–1625.
[52] McLaren, R.G. and Crawford, D.V., Fractionation of copper in soils; J. Soil Sci. 1973, 24, 172–181.
[53] Gibbs, R.J., Transport phases of transition-metals in Amazon and Yukon rivers; Geol. Soc. Am. Bull. 1977, 88, 829–843.
[54] Engler, R.M.; Brannon, J.M.; Rose, F., A practical selective extraction procedure for sediment characterization; in: Yen, T.F. (ed.), Chemistry of Marine Sediments; Ann Arbor Science, Ann Arbor, USA, 1977, pp. 163–171.
[55] Tessier, A.; Campbell, P.G.C.; Bisson, M., Sequential extraction procedure for the speciation of particulate trace metals; Anal. Chem. 1979, 51, 844–851.
[56] Meguellati, N.; Robbe, D.; Marchandise, P.; Astruc, M., A new chemical extraction procedure in the fractionation of heavy metals in sediments-interpretation; in: Müller, G. (ed.), Proc. Int. Conf. Heavy Metals in the Environment, vol. 2; CEP Consultants, Edinburgh, UK, 1983, pp. 1090–1903.
[57] Shuman, L.M., Separating soil iron-oxide and manganese-oxide fractions for micro-element analysis; Soil Sci. Soc. Am. J. 1982, 46, 1099–1102.
[58] Salomons, W.; Förstner, U., Metals in the Hydrocycle; Springer-Verlag; Berlin, 1984.
[59] Miller, W.P.; Martens, D.C.; Zelazny, L.W., Effect of sequence in extraction of trace metals from soils; Soil Sci. Soc. Am. J. 1986, 50, 598–601.
[60] Elliott, H.A.; Dempsey, B.A.; Maille, M.J., Content and fractionation of heavy-metals in water-treatment sludges; J. Environ. Qual. 1990, 19, 330–334.
[61] Krishnamurti, G.S.R.; Huang, P.M.; Vanrees, K.C.J.; Kozak, L.M.; Rostad, H.P.W., Speciation of particulate-bound cadmium of soils and its bioavailability; Analyst 1995, 120, 659–665.
[62] Sahuquillo, A.; López-Sánchez, J.F.; Rubio, R.; Rauret, G.; Thomas, R.P.; Davidson, C.M. et al., Use of certified reference material for extractable trace metals to assess sources of uncertainty in the BCR three-stage sequential extraction procedure; Anal. Chim. Acta 1999, 382, 317–327.
[63] Gibson, M.J.; Farmer, J.G., Multistep sequential chemical extraction of heavy metals from urban soils; Environ. Pollut. B 1986, 11, 117–135.
[64] Ahnstrom, Z.S.; Parker, D.R., Development and assessment of a sequential extraction procedure for the fractionation of soil cadmium; Soil Sci. Soc. Am. J. 1999, 63, 1650–1658.
[65] Arunachalam, J.; Emons, H.; Krasnodebska, B.; Mohl, C., Sequential extraction studies on homogenized forest soil samples; Sci. Total Environ. 1996, 181, 147–159.
[66] Hall, G.E.M.; Gauthier, G.; Pelchat, J.C.; Pelchat, P.; Vaive, J.E., Application of a sequential extraction scheme to ten geological certified reference materials for the determination of 20 elements; J. Anal. Atom. Spectrom. 1996, 11, 787–796.
[67] Du Laing, G.; Vanthuyne, D.R.J.; Vandecasteele, B.; Tack, F.M.G.; Verloo, M.G., Influence of hydrological regime on pore water metal concentrations in a contaminated sediment-derived soil; Environ. Pollut. 2007, 147, 615–625.
[68] Zachara, J.M.; Smith, S.C.; Resch, C.T.; Cowan, C.E., Cadmium sorption to soil separates containing layer silicates and iron and aluminum oxides; Soil Sci. Soc. Am. J. 1992, 56, 1074–1084.
[69] Tack, F.M.G.; Van Ranst, E.; Lievens, C.; Vandenberghe, R.E., Soil solution Cd, Cu and Zn concentrations as affected by short-time drying or wetting: the role of hydrous oxides of Fe and Mn; Geoderma 2006, 137, 83–89.
[70] Tipping, E.; Hetherington, N.B.; Hilton, J.; Thompson, D.W.; Bowles, E.; Hamilton-Taylor, J., Artifacts in the use of selective chemical-extraction to determine distributions of metals between oxides of manganese and iron; Anal. Chem. 1985, 57, 1944–1946.
[71] Krasnodebska-Ostrega, B.; Emons, H.; Golimowski, J., Selective leaching of elements associated with Mn-Fe oxides in forest soil, and comparison of two sequential extraction methods; Fresen. J. Anal. Chem. 2001, 371, 385–390.
[72] Chao, T.T.; Zhou, L., Extraction techniques for selective dissolution of amorphous iron oxides from soils and sediments; Soil Sci. Soc. Am. J. 1983, 47, 225–232.

[73] Hall, G.E.M.; Vaive, J.E.; Beer, R.; Hoashi, M., Selective leaches revisited, with emphasis on the amorphous Fe oxyhydroxide phase extraction; J. Geochem. Explor. 1996, 56, 59–78.
[74] Cabral, A.R.; Lefebvre, G., Use of sequential extraction in the study of heavy metal retention by silty soils; Water Air Soil Pollut. 1998, 102, 329–344.
[75] Slavek, J.; Wold, J.; Pickering, W.F., Selective extraction of metal-ions associated with humic acids; Talanta 1982, 29, 743–749.
[76] Du Laing, G.; Van Ryckegem, G.; Tack, F.M.G.; Verloo, M.G., Metal accumulation in intertidal litter through decomposing leaf blades, sheaths and stems of *Phragmites australis*; Chemosphere 2006, 63, 1815–1823.
[77] Xiang, H.F.; Tang, H.A.; Ying, Q.H., Transformation and distribution of forms of zinc in acid, neutral and calcareous soils of China; Geoderma 1995, 66, 121–135.
[78] Norrstrom, A.C.; Jacks, G., Concentration and fractionation of heavy metals in roadside soils receiving de-icing salts; Sci. Total Environ. 1998, 218, 161–174.
[79] Kersten, M.; Förstner, U., Chemical fractionation of heavy-metals in anoxic estuarine and coastal sediments; Water Sci. Technol. 1986, 18, 121–130.
[80] Keon, N.E.; Swartz, C.H.; Brabander, D.J.; Harvey, C.; Hemond, H.F., Validation of an arsenic sequential extraction method for evaluating mobility in sediments; Environ. Sci. Technol. 2001, 35, 2778–2784.
[81] Wenzel, W.W.; Kirchbaumer, N.; Prohaska, T.; Stingeder, G.; Lombi, E.; Adriano, D.C., Arsenic fractionation in soils using an improved sequential extraction procedure; Anal. Chim. Acta 2001, 436, 309–323.
[82] Ure, A.; Davidson, C., Chemical speciation in soils and related materials by selective chemical extractions; in: Ure, A. and Davidson, C. (eds), Speciation in the Environment. 2nd edn; John Wiley & Sons, Ltd, Chichester, 2002, pp. 265–300.
[83] Bacon, J.R.; Davidson, C.M., Is there a future for sequential chemical extraction?; Analyst 2008, 133, 25–46.
[84] Ure, A.M.; Davidson, C.M.; Thomas, R.P., Single and sequential extractions schemes for trace metal speciation in soil and sediment; in: Quevauviller, P.; Maier, E.A.; Griepink, B. (eds), Quality Assurance for Environmental Analysis; Elsevier, Amsterdam, 1995, pp. 505–523.
[85] Huang, J.H.; Ilgen, G., Factors affecting arsenic speciation in environmental samples: sample drying and storage; Int. J. Environ. Anal. Chem. 2006, 86, 347–358.
[86] Bordas, F.; Bourg, A.C.M., A critical evaluation of sample pretreatment for storage of contaminated sediments to be investigated for the potential mobility of their heavy metal load; Water Air Soil Pollut. 1998, 103, 137–149.
[87] Davidson, C.M.; Ferreira, P.C.S.; Ure, A.M., Some sources of variability in application of the three-stage sequential extraction procedure recommended by BCR to industrially-contaminated soil; Fresen. J. Anal. Chem. 1999, 363, 446–451.
[88] Clevenger, T.E., Use of sequential extraction to evaluate the heavy-metals in mining wastes; Water Air Soil Pollut. 1990, 50, 241–254.
[89] Hirner, A., Trace element speciation in soils and sediments using sequential chemical extraction methods; Int. J. Environ. Anal. Chem. 1992, 46, 77–85.
[90] Whalley, C.; Grant, A., Assessment of the phase selectivity of the European Community Bureau of Reference (BCR) sequential extraction procedure for metals in sediment; Anal. Chim. Acta 1994, 291, 287–295.
[91] Nirel, P.; Morel, F., Technical note – Pitfalls of sequential extractions; Water Res. 1990, 24, 1055–1056.
[92] Luo, Y.; Christie, P., Choice of extraction technique for soil reducible trace metals determines the subsequent oxidisable metal fraction in sequential extraction schemes; Int. J. Environ. Anal. Chem. 1998, 72, 59–75.
[93] Gomez-Ariza, J.; Giraldez, I.; Sanchez-Rodas, D.; Morales, E., Metal readsorption and redistribution during the analytical fractionation of trace elements in oxic estuarine sediments; Anal. Chim. Acta 1999, 399, 295–307.
[94] Perez-Cid, B.; Lavilla, I.; Bendicho, C., Speeding up of a three-stage sequential extraction method for metal speciation using focused ultrasound; Anal. Chim. Acta 1998, 360, 35–41.

[95] Campos, E.; Barahona, E.; Lachica, M.; Mingorance, M.D., A study of the analytical parameters important for the sequential extraction procedure using microwave heating for Pb, Zn and Cu in calcareous soils; Anal. Chim. Acta 1998, 369, 235–243.

[96] Shiowatana, J.; McLaren, R.G.; Chanmekha, N.; Samphao, A., Fractionation of arsenic in soil by a continuous-flow sequential extraction method; J. Environ. Qual. 2001, 30, 1940–1949.

[97] Chomchoei, R.; Shiowatana, J.; Pongsakul, P., Continuous-flow system for reduction of metal readsorption during sequential extraction of soil; Anal. Chim. Acta 2002, 472, 147–159.

[98] Fedotov, P.S.; Fitz, W.J.; Wennrich, R.; Morgenstern, P.; Wenzel, W.W., Fractionation of arsenic in soil and sludge samples: continuous-flow extraction using rotating coiled columns versus batch sequential extraction; Anal. Chim. Acta 2005, 538, 93–98.

[99] Miro, M.; Hansen, E.H.; Chomchoei, R.; Frenzel, W., Dynamic flow-through approaches for metal fractionation in environmentally relevant solid samples; Trends Anal. Chem. 2005, 24, 759–771.

[100] Nakazato, T.; Akasaka, M.; Tao, H., A rapid fractionation method for heavy metals in soil by continuous-flow sequential extraction assisted by focused microwaves; Anal. Bioanal. Chem. 2006, 386, 1515–1523.

[101] Tongtavee, N.; Shiowatana, J.; McLaren, R.G.; Buanuam, J., Evaluation of distribution and chemical associations between cobalt and manganese in soils by continuous-flow sequential extraction; Comm. Soil Sci. Plant. Anal. 2005, 36, 2839–2855.

[102] Buanuam, J.; Shiowatana, J.; Pongsakul, P., Fractionation and elemental association of Zn, Cd and Pb in soils contaminated by Zn minings using a continuous-flow sequential extraction; J. Environ. Monit. 2005, 7, 778–784.

[103] Ure, A.M., Trace-element speciation in soils, soil extracts and solutions; Microchim. Acta 1991, 2, 49–57.

[104] Buffle, J.; Leppard, G.G., Characterization of aquatic colloids and macromolecules. 2. Key role of physical structures on analytical results; Environ. Sci. Technol. 1995, 29, 2176–2184.

[105] Bouyssiere, B.; Szpunar, J.; Potin-Gautier, M.; Lobinski, R., Sample preparation techniques for elemental speciation studies; in: Cornelis, R., Crews, H., Caruso, J.; Heumann, K. (eds), Handbook of Elemental Speciation: Techniques and Methodology; John Wiley & Sons, Ltd, Chichchester, 2003, pp. 95–118.

[106] Olivas, R.M.; Cámara, C., Sample preparation. Sample treatment for speciation analysis in biological samples; in: Cornelis, R., Crews, H., Caruso, J., Heumann, K. (eds), Handbook of Elemental Speciation: Techniques and Methodology; John Wiley & Sons, Ltd, Chichester, 2003, pp. 119–146.

[107] Bayona, J.M., Supercritical fluid extraction in speciation studies; Trends Anal. Chem. 2000, 19, 107–112.

[108] LopezAvila, V.; Liu, Y.; Beckert, W.F., Interlaboratory evaluation of an off-line supercritical fluid extraction and gas chromatography with atomic emission detection method for the determination of organotin compounds in soil and sediments; J. Chromatogr. A 1997, 785, 279–288.

[109] Wai, C.M.; Wang, S.F., Supercritical fluid extraction: metals as complexes; J. Chromatogr. A 1997, 785, 369–383.

[110] Foy, G.P.; Pacey, G.E., Specific extraction of chromium(VI) using supercritical fluid extraction; Talanta 2000, 51, 339–347.

[111] Foy, G.P.; Pacey, G.E., Supercritical fluid extraction of mercury species; Talanta 2003, 61, 849–853.

[112] Encinar, J.R.; Rodriguez-Gonzalez, P.; Fernandez, J.R.; Alonso, J.I.G.; Diez, S.; Bayona, J.M. et al., Evaluation of accelerated solvent extraction for butyltin speciation in PACS-2 CRM using double-spike isotope dilution-GC/ICPMS; Anal. Chem. 2002, 74, 5237–5242.

[113] Welter, E., Direct speciation of solids: X-ray absorption fine structure spectroscopy for species analysis in solid samples; in: Cornelis, R., Crews, H., Caruso, J.; Heumann, K.; (eds), Handbook of Elemental Speciation: Techniques and Methodology; John Wiley & Sons, Ltd, Chichester, 2003, pp. 526–546.

[114] Parsons, J.G.; Aldrich, M.V.; Gardea-Torresdey, J.L., Environmental and biological applications of extended X-ray absorption fine structure (EXAFS) and X-ray absorption near edge structure (XANES) spectroscopies; Appl. Spectrosc. Rev. 2002, 37, 187–222.
[115] Meitzner, G., Experimental aspects of X-ray absorption spectroscopy; Catal. Today 1998, 39, 281–291.
[116] Rehr, J.J.; Albers, R.C., Theoretical approaches to X-ray absorption fine structure; Rev. Mod. Phys. 2000, 72, 621–654.
[117] Furnare, L.J.; Vailionis, A.; Strawn, D.G., Polarized XANES and EXAFS spectroscopic investigation into copper(II) complexes on vermiculite; Geochim. Cosmochim. Acta. 2005, 69, 5219–5231.
[118] Chang, S.H.; Wei, Y.L,; Wang, H.P., Zinc species distribution in EDTA-extract residues of zinc-contaminated soil; J. Electron Spectrosc. 2007, 156, 220–223.
[119] Manning, B., Arsenic speciation in As(III)- and As(V)-treated soil using XANES spectroscopy; Microchim. Acta 2005, 151, 181–188.
[120] Ghabbour, E.A.; Scheinost, A.C.; Davies, G., XAFS studies of cobalt(II) binding by solid peat and soil-derived humic acids and plant-derived humic acid-like substances; Chemosphere 2007, 67, 285–291.
[121] LoPresti, V.; Conradson, S.D.; Clark, D.L., XANES identification of plutonium speciation in RFETS; J. Alloy. Compd. 2007, 444, 540–543.
[122] Voegelin, A.; Weber, F.A.; Kretzschmar, R., Distribution and speciation of arsenic around roots in a contaminated riparian floodplain soil: Micro-XRF element mapping and EXAFS spectroscopy; Geochim. Cosmochim. Acta. 2007, 71, 5804–5820.

5
Fractionation and Speciation of Trace Elements in Soil Solution

Gijs Du Laing

Ghent University, Laboratory of Analytical Chemistry and Applied Ecochemistry, Faculty of Bioscience Engineering, Ghent, Belgium

5.1 Introduction

Solution speciation is a critical parameter controlling the bioavailability, solution–solid phase distribution and transport of trace elements in soils. Most trace elements are mainly taken up by soil biota as soluble, free ions via the soil solution [1–6]. A wide range of natural metal-complexing ligands also exist in soil solution, which include inorganic anions, inorganic colloids, organic humic substances, amino acids (notably phytosiderophores and bacterial siderophores) and low-molecular-mass organic acids. The latter two groups are of particular significance in the soil surrounding plant roots (the rhizosphere). As a result, a number of analytical methodologies, encompassing computational, spectroscopic, physicochemical and separation techniques, have been applied to the measurement of speciation of trace elements in soil solution. However, perhaps with the exception of the determination of the free metal cation, the majority of these techniques rarely provide species-specific information [7]. Moreover, some preconditions complicate the use of some speciation techniques in soil solutions, such as the high amounts of suspended solids, organic matter and/or salt contents and the variable redox potentials. This chapter describes the most important separation and detection techniques which have recently been used for trace element speciation and fractionation in soil solution. Some sample preparation steps, for example filtration, extraction, preconcentration, clean-up and/or derivatization, and detection techniques are discussed individually, whereas other techniques involve a combination of such sample preparation steps and detection techniques. The techniques discussed can also be used for fractionation and speciation of trace elements in soil extracts, which were discussed in Chapter 4.

Most attention is given to techniques which are used for the determination of free ion activities in soil solution, as only the free ions are considered to interact with biological ligands [8]. These are often seen as a surrogate for the determination of metal-, species- and soil type-dependent bioavailable and bioaccessible metal pools. Soil solution speciation techniques are mainly described from a technical point of view in this chapter; their suitability for assessing bioavailability is discussed in Chapter 11. In addition, these analytical methods can be used for assessing the origin of trace elements in soils, as well as their leaching potential from the soils and the extent of soil contamination [9].

5.2 Soil Solution Sampling, Storage and Filtration

For a wide range of trace element speciation procedures, the soil solution is first physically separated from the soil solid fraction. Bufflap and Allen [10] distinguished four means of extracting soil solution. Two of the methods, squeezing and centrifugation, are *ex situ*, requiring the removal of soil. The other two, vacuum filtration and dialysis, have the advantage that they can be performed *in situ* and allow repeated sampling. Gravity-driven zero-tension lysimeters have proved to be effective tools for *in situ* sampling of the freely draining soil solution. High temporal and spatial resolution is difficult to achieve especially within the transient rhizosphere environment [11]. To overcome these problems, tension lysimeters have been miniaturized in recent years. In this context, micro suction cup techniques have gained widespread acceptance for gathering interstitial liquids and root exudates. Consisting of porous cups or plates with a thickness of a few millimetres, they can be easily deployed within specific locations, and allow periodic collection of the soil pore water via mostly hand-powered or automatic vacuum pumping systems [11–13].

If samples are not filtered during or after extraction, the residual particles can cause interferences during analysis and can adsorb metals from the solution, altering the pore water concentration [14]. If filters are used, these filters can, however, also be a source of error as they can adsorb, exclude or release trace elements [13]. Moreover, they can affect the original species distribution [15]. For example, ceramic suction cups for vacuum filtration are known to suffer from adsorption effects [16,17]. Such effects may be particularly important when filtering soil solutions with low overall ionic strength, compared with stronger salt solutions for which the filters are usually designed. Under these circumstances, testing the filter material prior to use is recommended, for example by filtering and analysing standard solutions of trace element species with a similar ionic background and pH as the studied samples. Sample acidification reduces adsorption of trace elements by the filters and sample containers while also inhibiting microbial activity. Unfortunately, acidification changes speciation conditions (acid–base equilibria and coupled redox and complex formation equilibria) and results in the solubilization of metals previously precipitated or adsorbed to colloidal soil particles [15]. Soil solution acidification should therefore only be employed for total analysis.

In addition to individual practical advantages and disadvantages and the role of filtration, several other general sources of error can alter the pore water chemical concentrations and affect the species integrity of the trace elements, for example redox and pH changes, microbial activity, metal contamination, and temperature artefacts. Especially the oxidation of metal sulphides and subsequent co-precipitation of trace elements with Fe and Mn oxides have been found to induce pH and metal speciation changes in pore waters during handling and storage of initially reduced soil and sediment samples. To avoid redox changes during sampling, *in situ* vacuum filtration and dialysis are preferred over squeezing and centrifugation to remove the soil solution. Removal of interstitial water changes the local soil moisture conditions and potentially perturbs the chemical equilibrium. So a re-equilibration of dissolved species at the soil-solution interface is needed prior to the next sampling. It should also be stressed that the collected fractions are not free from microbial activity, so that metal speciation changes can be expected during transport and storage. To reduce the above mentioned shortcomings, flow-through microdialysis sampling systems have been developed in recent years, which are based on the diffusive transport of dissolved target analytes from the environment probed through a perm-selective dialysis membrane into a continuously flowing perfusion liquid [18]. Dedicated microdialysers are currently being used for *in situ* investigation of the influence of rhizospheric processes on trace element dynamics in soils [11].

Temperature control is also critical for trace element speciation analysis. Temperature changes during sampling, sample transport, storage and analysis can induce shifts in chemical equilibria and the activity and structure of microbial communities, which both can affect the speciation. Often a compromise is sought between refrigerating the samples to limit microbial growth during storage and keeping them at their initial temperature to prevent shifts in chemical equilibria. Overall the validation of sampling and storage techniques with respect to species-retaining operations must be performed with reference to the target species, as the use of general approaches may not preserve the species [15]. Therefore, the most promising approach for speciation analysis of dissolved species consists of the application of direct *in situ* measurements.

Size-selective fractionation enables information on the distribution of trace elements between different sized particles. Fractionation using centrifugation and (micro)filtration with particle size thresholds of, for example, <0.20 and <0.45 μm, combined with subsequent analysis of trace elements in the collected filtrates (or the residue retained on the filter) can easily be conducted. These size fractions coincide with the distinction between dissolved and particulate loads in many studies and represent the two most common operational fractions isolated by traditional membrane filtration [19]. However, it is well known that in natural waters, many trace elements are present in colloidal (that is between 1 nm and 1 μm) rather than truly dissolved forms. Methods to establish a further fractionation below 0.2 μm have therefore been developed in recent years and are discussed in the next section.

5.3 Particle Size Fractionation

Methods to establish a fractionation for particles below 0.2 μm include ultrafiltration, field-flow fractionation and size-exclusion chromatography, which can all be coupled to inductively coupled plasma mass spectrometry (ICP-MS) as a detection technique.

84 Basic Principles, Processes, Sampling and Analytical Aspects

Ultrafiltration (UF) uses pressure to force sample liquid through a semipermeable membrane. Suspended solids and solutes of high molecular size are retained, while water and low molecular size solutes pass through the membrane. Different ultrafiltration membranes are available with size cutoff ranging from 0.001 to 0.1 µm, so they can remove proteins. Ultrafiltration is not fundamentally different from microfiltration, nanofiltration or reverse osmosis, except in terms of the size of the molecules it retains. A typical pore size range for microfiltration is 0.1–10 µm, whereas nanofiltration and reverse osmosis remove even smaller molecules than does ultrafiltration. The latter are mainly applied in drinking water purification process steps, such as water softening, decolourizing and pollutant removal. Ultrafiltration has, however, been used in many studies for the fractionation of trace elements associated with dissolved organic material, including fulvic and humic acids, and the study of colloidal-material-associated trace elements, both in soil extracts/solutions and in groundwater [20–24].

Field-flow fractionation (FFF) is a chromatographic-like family of methods particularly suitable for the purification and characterization of macromolecules, colloids and particles in a size range from 10^{-3} to 10^2 µm. The primary separation is effected by the action of an external field (Figure 5.1). The force induced by that field moves molecules and particles toward the wall of a channel and, since they cannot penetrate the wall, an exponential concentration distribution is built up. The thickness of the gradient layer depends solely on the diffusion coefficient of the molecule or particle. Slowly diffusing large molecules and particles lie closer to the wall than smaller molecules and particles with higher diffusion coefficients. The separation is amplified by the parabolic flow profile of the liquid pumped through the channel. The highest velocity is observed in the middle of the channel, while closer to the wall the flow is slowed down by the frictional drag. Large molecules near to the wall are in slower streamlines than the smaller ones and will therefore be more retained than small molecules [25]. If sample particles have a radius of the same magnitude as their layer thickness (>1 µm), the particles will be in contact with the wall, which results in larger particles eluting at the same time as smaller particles in this 'steric mode', compared with the normal mode. To investigate the lower size ranges the samples therefore need to be fractionated to <1 µm before FFF analysis. Sedimentation FFF (SdFFF) uses either the

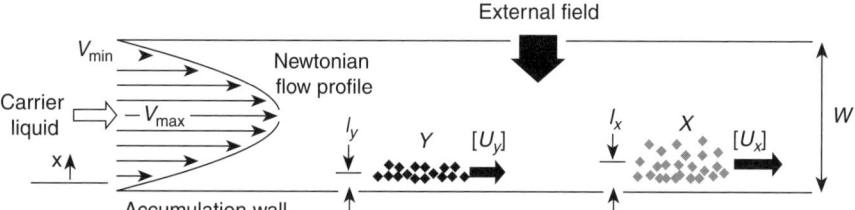

Figure 5.1 Principle of field-flow fractionation (FFF). The separation effect is reached by a combined action of the nearly parabolic Newtonian flow velocity profile of a carrier liquid in a capillary tube (50–500 µm diameter) and a transverse physical field applied perpendicularly to the flow of this carrier liquid (Reprinted from T. Kowalkowski, B. Buszewski, C. Cantado and F. Dondi, Field-flow fractionation: Theory, techniques, applications and the challenges, Critical Reviews in Analytical Chemistry, 2006, 36, 129–135. Reprinted by permission of Taylor & Francis Ltd [26].)

Earth's gravitational field or centrifugal force to fractionate a variety of particulate analytes in the micrometre size range. This technique has been shown to be successful for the separation and further characterization of cells, microorganisms and the large particles [26]. In flow field-flow fractionation (FlFFF), the wall elements of the flow FFF channel are made of semipermeable ceramic frit, which allows an additional flow to penetrate the channel perpendicularly to the main flow. FlFFF systems have several advantages over SdFFF in being easier to maintain, with the ability to determine molecular weight distributions and particle size distributions. FlFFF has greater resolution at the lower size limit (1 nm or 1000 daltons), whereas the lower size boundary in SdFFF is approximately 50 nm. While FlFFF requires assumptions to be made about particle shape, SdFFF also relies on deductions with respect to particle density. Field-flow fractionation can be used to obtain information on particle size or molar mass distributions in complex environmental matrices over the entire colloidal size range (0.001–1 μm). The output from a FFF instrument is a fractogram of UV absorbance (254 nm) against elution time and this is converted into a plot of relative mass against particle size [27]. Trace elements associated with different particle sizes can be assessed by analysing their contents in collected fractions. ICP-MS can be directly coupled to FFF techniques for the direct quantitative measurement of trace elements in the specific, submicrometre size-fraction particulates [28]. As a result, field-flow fractionation has already been used for the study of colloidal material associated trace elements in soil extracts, soil solutions and groundwater and for the fractionation of trace elements associated with dissolved organic material [27,29–33].

Size-exclusion chromatography (SEC) can also be used to separate different sizes of particles and uses a chromatographic column tightly packed with porous polymer beads designed to have pores of different sizes. Smaller molecules penetrate more deeply into the pores of the stationary phase and thus their pathway is longer, which results in a longer retention time (greater elution volume). Larger molecules by-pass the stationary phase pores and reach the end of the chromatographic column in a shorter time (with a smaller elution volume). It is evident that the separation is based rather on differences in molecular size (hydrodynamic size, effective diameter) than on differences in molecular mass. Although the term 'molecular mass (weight) distribution' is widely used in the literature, the term 'molecular size distribution' would be more appropriate for the values determined by gel permeation chromatography/SEC [34]. It should however be mentioned that all SEC columns exhibit also ionic properties as a result of the residual organic functional groups (primarily COOH and OH) remaining after manufacture of the stationary phase. Although this interaction may occasionally be minimized with the use of high-ionic-strength mobile phases, the process of separation in SEC is actually a combination of size, ion exclusion and ion exchange effects [7]. When an aqueous solution is used to transport the sample through the column, the technique is known as gel filtration chromatography. Gel permeation chromatography (GPC) uses an organic solvent mobile phase. Compared to FFF, the working mass range of these chromatographic techniques is much narrower [26]. However, SEC can also be directly coupled to ICP-MS (SEC-ICP-MS), which allows a direct fractionation of trace elements into different associated particle size fractions. As a result, this technique has also been used for the analysis of trace elements associated with particles and molecules of different sizes, mainly metalloproteins, but also dissolved organic material, both in soil extracts/solutions and groundwater [35–38]. Although SEC is most commonly used for the separation of relatively large compounds such as polymers and biopolymers, it can be used for the separation of compounds of intermediate molecular weight as well. For example, Shibata *et al.* [39] determined diphenylarsinic acid in urine, groundwater, and soil extract samples by SEC-ICP-MS. The analysis was performed with a sub-parts-per-billion detection limit.

For successful determination of a specific metal–organic complex using chromatographic techniques it is, however, required that this metal–organic complex is sufficiently kinetically inert to remain intact, as the perturbation of equilibrium conditions is unavoidable during the chromatographic separation [40]. Collins [7] reviewed the separation of low-molecular-mass organic acid–metal complexes by different types of chromatographic techniques and concluded that only those metal–organic complexes whose dissociation rate is insignificant during the separation process are amenable for analysis by chromatographic techniques (that is only those complexes which are not at thermodynamic equilibrium during separation). On the basis of thermodynamic equilibrium and complex dissociation kinetics, it would appear that among the low-molecular-mass organic acid–metal complexes especially the Fe, Ni and Co complexes of some di- and tricarboxylic acids may have the necessary characteristics for analysis by chromatographic techniques [7].

5.4 Liquid–Liquid Extraction

Liquid–liquid extraction can be directly applied to nonfiltered samples with complex matrices and allows the direct transfer of analytes into a nonpolar organic solvent. Nonpolar species, for example tetraalkyllead, can be quantitatively extracted into a small volume of hexane. The same applies to some 'ionic species' with a marked covalent character (for example, triphenyltin, tributyltin, triethyllead, methylmercury), but the use of water-immiscible solvents with polar character (toluene, ethyl acetate) is advised. The properties of organometallic species with a smaller number of organic substituents (for example monobutyltin) do not allow their quantitative extraction by any organic solvent. In order to transfer such compounds into an organic phase, the formation of extractable chelate complexes or nonpolar covalent compounds in the aqueous solution prior to extraction is necessary. Diethyldithiocarbamate is the most commonly used reagent for chelation, which is followed by derivatization (see Section 5.6). Inorganic interferents can be efficiently masked with EDTA [41]. Supercritical fluid extraction, SFE (see Chapter 4) of ionic compounds from aqueous samples cannot be accomplished directly with CO_2 because of the charge neutralization requirement and the weak solute–supercritical fluid interaction. Thus, the direct extraction of organometallic species from aqueous samples with supercritical fluids has not been attempted, but the combination of solid-phase extraction (see Section 5.5) followed by SFE of the alkylated organotin derivatives has already been successful [42].

5.5 Ion-Exchange Resins and Solid-Phase Extraction

Ion exchange is frequently used to differentiate between metal species, based on the electrostatic charge or on the lability of complexes, as it is rapid, sensitive, and not affected by interferences of high concentrations of electrochemically active compounds such as humic substances [43]. If a known amount of well-characterized resin is supplied to a soil solution sample, its ligands will compete with those naturally present in the sample for sorption of the free ions. After equilibration, the resin is separated from the soil solution and the adsorbed trace element species can be eluted and quantified. The type and quantity of resin should be selected carefully, so that the equilibrium in the natural water is not excessively perturbed, but sufficient binding occurs to allow measurement in the resin phase. Moreover, adequate time must be allowed for the equilibrium to be established [44]. Resins have been successfully applied for the analysis of trace cations in the soil solution or leachates from soil, compost and wastes [45–48].

Solid phase extraction (SPE) followed by elution with a small amount of organic solvent and often derivatization, is becoming increasingly popular for sample preparation in organic analysis, and has also found application in speciation analysis for organometallics in recent years. Its advantages over liquid–liquid extraction include a higher enrichment factor, lower solvent consumption and risk of contamination, and the ease of application to field sampling and automation. However, using SPE, organometallics can only be extracted from filtered samples, which constitutes a considerable drawback [41]. In solid-phase microextraction (SPME), organometallic compounds are derivatized *in situ* in the aqueous phase and simultaneously extracted onto a coated silica fibre. The adsorptive-type polydimethylsiloxane (PDMS) coatings are the most popular. The sorbed organometals are subsequently released from the fibre in a GC injection liner by thermal desorption. Derivatization, extraction, preconcentration and injection into the GC take only 10 minutes with a minimum of handling steps, and the technique is also solvent-free. Owing to the very low detection limits (0.13–3.7 ng l^{-1} as metal), only small sample amounts are needed for analysis [49]. Most studies which used SPME in recent years involved the sampling of organometallic species of lead, arsenic, mercury, tin, or selenium, followed by a chromatographic separation (see Section 5.7) and atomic detection technique of the different species [50]. Direct coupling of solid-phase microextraction (SPME) with inductively coupled plasma mass spectrometry (ICP-MS) by using thermal desorption after extraction has also been described, for example for the detection of organoselenium species [51] and methylmercury [52].

5.6 Derivatization Techniques to Create Volatile Species

Derivatization is the process of controlled conversion of species originally present in a sample into forms with improved chromatographic yield or separation coefficient. The most popular type of derivatization is the conversion of ionic or highly polar species into nonpolar species which can be readily separated by gas chromatography (see Section 5.7). Hydride generation (HG) is a derivatization procedure which is applicable to compounds of several elements such as As, Sb, Bi, Ge, Pb, Se, Te and Sn. The inorganic forms of these elements react with sodium borohydride with the formation of simple hydrides, but not all species of a given element react equally fast and in the same solution conditions. Selenium, As and Sb readily form hydrides only in their lower oxidation states. This behaviour can in some cases be used for the separation of different species. Ethylation by using sodium tetraethylborate is another derivatization procedure which can be applied for Sn, Se, Hg, Pb. Liquid–liquid extraction remains necessary, however, to extract the derived compounds from their matrix and, therefore, organic solvents are required. Moreover, sometimes there is no discrimination among the different species of an element; for example, in the case of lead, Pb^{2+}, Et_3Pb^+, or Et_2Pb^{2+} all react with the reagent $NaEt_4B$. Derivatization using Grignard reagents (alkyl- or arylmagnesium chlorides) has a more universal character. It has been applied for Sn, Hg, and Pb speciation, although sample preparation is laborious and time-consuming as these reagents are atmospherically unstable and can be hydrolysed in the presence of water [53]. Preconcentration, for example by cryotrapping, is required after extraction and/or derivatization in some cases.

5.7 Chromatographic Separation of Trace Element Species

Chromatographic techniques involve passing a mixture dissolved in a 'mobile phase' through a stationary phase in a column, which separates the analyte to be measured from other molecules in the mixture and allows it to be isolated. The mixtures are separated as a function of the affinity of the analyte for the current mobile phase composition relative to the stationary phase. This partitioning process is similar to that which occurs during a liquid–liquid extraction but is continuous, not stepwise. The mobile phase can be either a gas (gas chromatography, GC) or a liquid (liquid chromatography, LC).

Liquid chromatography has emerged as one of the most popular separation techniques for elemental speciation analysis. LC is an extremely versatile technique since both the stationary phase and the mobile phase may be altered to achieve the desired separation, and an enormous variety of stationary phases are commercially available. Moreover, separations can be further enhanced by the addition of additives to the mobile phase. Usually, minimum sample preparation is required. When the mobile phase is pumped at high pressure through the stationary phase, liquid chromatography is called HPLC (high-performance liquid chromatography). HPLC columns differ from earlier LC columns because the stationary-phase particles have smaller diameters, typically 3–5 μm. To build an LC separation method, the nature of the chemical species to be separated must first be considered. The interactions between an analyte and the stationary and mobile phases are based upon dipole forces, electrostatic interactions, and dispersion forces. Generally, compounds that are quite different in polarity or chemical structure will be much easier to separate than compounds that are more similar. A wide variety of variables may be adjusted to optimize a liquid chromatography separation, such as the choice of the stationary phase, the mobile phase composition, the flow rate of the mobile phase and the temperature [54].

Liquid chromatographic separations can be further subdivided by the general type of stationary phase that is used in the separation. Four chromatographic techniques have principally been used for separating metals in aqueous solutions that have different oxidation states (for example As, Cr, Fe and Se), metal–inorganic anion complexes (for example Al–F complexes), metals covalently bound to organic molecules (for example As, Hg, Pb and Se) and those metal–organic molecule complexes formed by chelation reactions (for example metal–EDTA complexes). These techniques are reversed-phase liquid chromatography (RP-HPLC), reversed-phase ion-pair chromatography (RP-IPC), size-exclusion chromatography

(SEC) and ion-exchange chromatography (IEC) [7]. In SEC, discussed in Section 5.3, a fractionation is established based on the size of the particles the trace elements are associated with. Separation using reversed-phase chromatography is achieved through the partitioning of a trace element-containing compound or complex between a nonpolar stationary phase and a polar liquid mobile phase. Continuous partitioning between the mobile and stationary phases, which depends upon the polarity of the analyte, results in differential migration through the analytical column. As a result, trace element-containing compounds of higher polarity elute earlier than those with lower polarity. In reversed-phase ion-pair chromatography, RP-HPLC stationary phases are used, but an ion pair reagent is added to the mobile phase. An ion pair reagent is a salt with a cation or anion having a polar head group and a nonpolar tail, for example sodium alkyl sulfonates or tri- and tetraalkylammonium salts. RP-IPC is able to simultaneously separate anions, cations and noncharged species [7,54]. In ion-exchange chromatography, the stationary phase contains ions on its surface, which interact with ions of the opposite charge in the mobile phase. As many trace element containing compounds are charged, IEC has formed the basis of many HPLC separation procedures for trace element species. The separation of compounds can be optimized by variation of the pH of mobile phase, the identity and concentration of the buffer added to the mobile phase, and the temperature [55]. As was the case for size-exclusion chromatography, metal complexes should be sufficiently kinetically inert to remain intact during chromatographic separation despite the perturbation of equilibrium conditions that necessarily occurs during the chromatographic separation (see Section 5.3). Other liquid chromatography types exist, but have been used less commonly for trace element speciation in the soil solution.

Compared to liquid chromatography, gas chromatography has been less commonly used for trace element speciation in soils because compounds must be sufficiently volatile and thermally stable to form gaseous molecules. Within the field of speciation only very few compounds fulfil these requirements directly, for example dimethylmercury, tetramethyltin, trimethylantimony, trimethylbismuth, dimethylselenium and tetraalkylated lead compounds. The analyst can, however, resort to chemical reactions to transform nonvolatile compounds into volatile thermally stable compounds (see Section 5.6).

After chromatographic separation, the total amount of trace elements in each collected fraction is subsequently detected using inductively coupled plasma mass spectrometry (ICP-MS), atomic absorption (AAS) or atomic emission (AES) spectrometric techniques, which have been discussed in Chapter 4. Direct online coupling of liquid chromatography separation to these detection techniques offers enhanced selectivity and sensitivity. Results of a chromatographic separation of different arsenic species in extracts of different soils using HPLC-ICP-MS are given in Figure 5.2. Interfacing of the LC system to ICP-based detectors is easily achieved using inert tubing to connect the end of the LC column with the sample nebulizer. Care should be taken to minimize the length and internal diameter of the tubing to reduce band broadening. Moreover, when introducing high concentrations of organic solvents into the plasma, some precautions have to be taken to avoid the plasma escaping and blocking of the injector due to carbon deposition. These measures can be the addition of oxygen to the plasma to stimulate the oxidation of the solvent and a reduction of the temperature inside the spray chamber to reduce the solvent loading into the plasma. Coupling to hydride generation systems (see Section 5.6) or GF-AAS (see Chapter 4) is also possible, but is less easy. When coupling GC directly to ICP-based detectors, the nebulizer and spray chamber have to be removed and a heated transfer line is necessary to

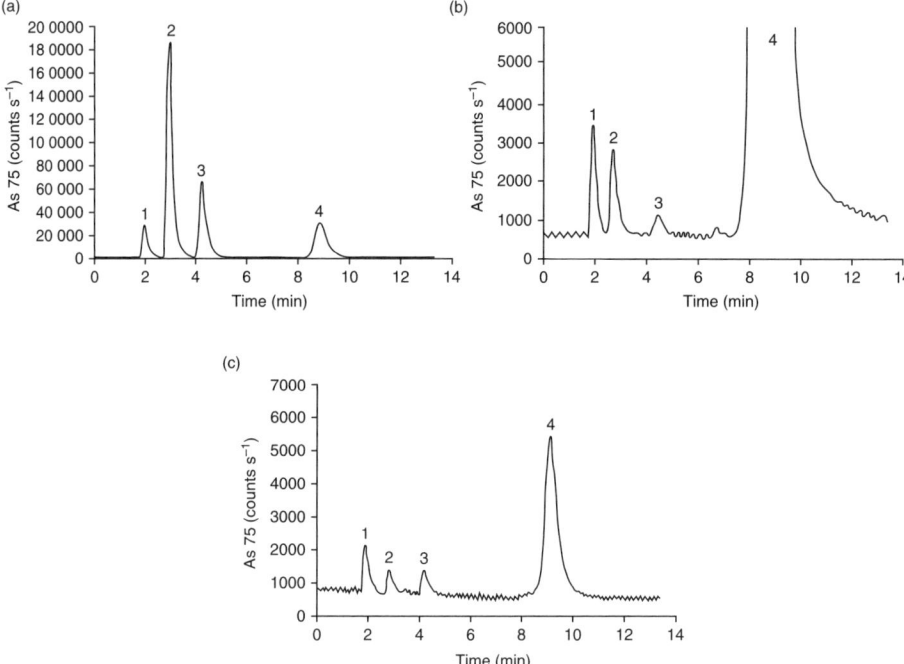

Figure 5.2 *HPLC-ICP-MS arsenic speciation profiles of soil water extracts: (a) arsenic-spiked soil mixture; (b) SRM 2711 moderate levels Montana soil; (c) SRM 2709 San Joaquin soil. 1, arsenite (As(III)); 2, dimethylarsonic acid (DMA); 3, monomethylarsonic acid (MMA); 4, arsenate (As(V)). Column: PRP X-100 anion-exchange column. Mobile phase: 10 mM ammonium dihydrogenphosphate (pH 5.8). Flow rate 1.0 ml min^{-1}. Injection volume 100 µl (Reproduced from C. Kahakachchi, P.C. Uden, J.F. Tyson, Extraction of arsenic species from spiked soils and standard reference materials, Analyst, 2004, 129, 714–718. Reproduced by permission of The Royal Society of Chemistry [56].)*

avoid condensation in the connecting tubing. The development of interfaces between chromatographic techniques and ICP-MS has been extensively reviewed by Vanhaecke and Köllensperger [57].

For less commonly studied organometallic compounds, no pure standard solutions are commercially available nor can they be synthesized and purified in a reliable manner. In that case, it is impossible to use the chromatographic retention time to identify the species, and chromatographic techniques should be coupled to other detection techniques which allow structural identification of the complete organometallic compound, for example electrospray ionization mass spectrometry (ESI-MS).

5.8 Capillary Electrophoresis

Techniques based on the use of the migration of electrically charged particles or ions in solution due to an applied electric field between an anode and a cathode (electrophoretic methods) are well established and a wide range of electrophoretic methods are currently available. The real explosion in electrophoresis started, however, in the 1990s with the introduction of capillary electrophoresis (CE) which involves the application of an electric potential of around 30 kV across a capillary between two buffer solutions. Compounds in the capillary migrate at different rates through the capillary according to the size and charge of the compounds, the length of the capillary, and the magnitude of the electric field. CE is able to separate cationic, anionic and neutral species; it shows potential to handle labile complexes and colloidal systems; it is rapid; and the columns are relatively simple and cheap. The method represents an interesting alternative to HPLC-ICP-MS when 'gentle' separation schemes are required to preserve the true chemical information in a real-life sample [57]. Although CE offers some advantages, its application for trace element speciation has been relatively limited because of interfacing difficulties with atomic spectrometry detection systems and its relatively low concentration sensitivity [55]. However, since CE-ICP-MS can combine high separation efficiency with the sensitivity of ICP-MS, it can be considered as the most important recent development in the field of hyphenated techniques [58]. Capillary electrophoresis has been used for arsenic speciation in soil extracts [59].

5.9 Diffusive Gradients in Thin Films

The technique of diffusive gradients in thin films (DGT) was originally developed for measuring metal speciation in natural waters [60], but it has now been consolidated as an attractive tool for predicting bioavailable element fractions in the soil environment [61,62]. The DGT sampler consists of a simple, implantable device that accumulates the mobilized elements on a binding reagent after well-defined diffusional transport through a hydrogel layer [63]. A plastic device, which contains a layer of gel, generally a chelex cation exchange resin selective for trace metals, is overlain by a layer of pure gel, where molecular diffusion can occur (diffusive gel, usually polyacrylamide), and a protective 0.45 μm filter membrane (Figure 5.3). Trace element ions diffuse through the filter and diffusive gel layer and are removed at the resin layer [44]. Shortly after submersion in a solution, a steady-state concentration gradient is established from the solution to the resin layer. Because the diffusive gel layer is relatively thick (typically 0.8 mm), the diffusive boundary layer in stirred solutions and many natural waters is usually negligible. The device is normally deployed for a time (t) in excess of one hour. The DGT sampler is retrieved and the resin layer is removed. The trace elements are eluted with acid and measured by any convenient technique (AAS, ICP-OES or ICP-MS, see Chapter 4), enabling calculation of the accumulated mass of trace element. Knowing the geometry of the device (diffusion layer thickness, Δg, and exposure area, A) and the diffusion coefficient (D) of the ion through the diffusive layer, the concentration in solution, C, can be simply calculated: $C = M\Delta g/DAt$ [44]. This methodology is founded on kinetic rather than equilibrium principles, thereby allowing the measurement of trace element fluxes and monitoring of plant–soil interfacial concentrations under dynamic conditions.

The DGT sampler can be used in direct contact with the soil, which constitutes a major advantage. The DGT measurement provides the mean concentration at the surface of the

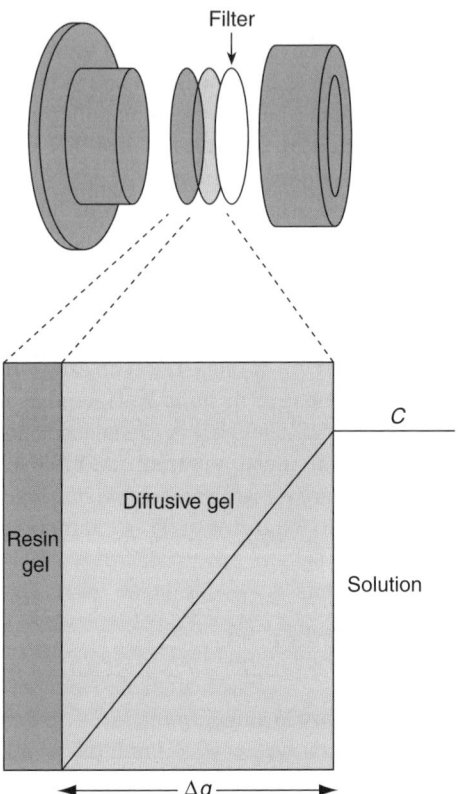

Figure 5.3 Schematic representation of a DGT device and cross-section through the device showing the concentration gradient of a trace element in a DGT assembly in contact with a solution. C is the concentration in the solution and Δg is the thickness of the diffusive gel (Reproduced from H. Zhang and S.D. Young, Characterizing the availability of metals in contaminated soils. II. The soil solution, Soil Use Manage, 2005, 21, 459–467. Reproduced with permission from Blackwell Publishing Ltd [44].)

DGT sampler during the deployment time, which could be related to the soil solution concentration [64]. As convection in the soil solution is very limited, the surface concentration also depends on the rate of diffusive trace element supply and/or the release of trace elements from the solid phase to the soil solution. If the trace element diffuses rapidly in the soil solution and the equilibrium between the solid phase and the soil solution is restored quickly, the concentration changes in the soil solution due to the uptake of trace elements by the sampler are continuously compensated from the bulk soil. DGT is then considered to measure the bulk concentration of the soil solution. If not, the DGT measurements provide information on the rate of trace element supply, which can be used to identify kinetically active trace element pools [65], to estimate the size of the pools and kinetics of trace element supply [66,67]. In this context, dynamic models of the soil-DGT system have been developed [68,69]. The above illustrates that plants and DGT samplers both continuously remove trace elements from the soil, accumulating elements which originate from both the soil solution and part of the solid phase. As a result, the effective solution concentration experienced by the DGT sampler, defined as the concentration needed in the soil solution for the DGT sampler to accumulate the same amount of trace element as if the supply originated solely from the soil solution without any compensation from the solid phase, has been found to correlate well with plant uptake for a wide range of trace elements and soil types [61,63,70–72].

Hooda et al. [62] studied the feasibility of DGT *in situ* application in soils and found that metal fluxes to the DGT sink are highly reproducible over a range of soil moisture contents. Where the moisture content is in excess of saturation, the measured DGT fluxes reflect metal concentrations in solution. At moisture contents less than field capacity, however, the measured DGT fluxes are less than would be expected from soil solution concentrations. As DGT responds systematically at moisture contents as low as 27 %, *in situ* deployment in relatively wet soils should not be a problem. Moreover, Hooda et al. [62] concluded that the role of gel layer thickness is complex and requires further studies, as the gel membrane thickness and/or its pore size may change below the field capacity moisture contents. However, for a given geometry and deployment time the results are consistent. Therefore, measurements made by different laboratories in different soils should provide fluxes that are fully comparable. This contrasts with experiments that simply use resin in bags in which the geometry is poorly controlled so that results can be compared only within experiments [62].

Although DGT samplers can be implemented in the soil body or at fixed distances to the root plane, DGT-mediated analyses invariably call for in-laboratory extraction or digestion to recover the elements entrapped in the active gel [11], which constitutes a disadvantage compared with techniques in which microelectrodes can be used *in situ* (see Section 5.12). The large volumes of solution needed for standard laboratory experiments (about 50 ml for 4 h [44]) constitute another disadvantage. The latter contributes to the fact that its use for measurements of trace elements in soil solutions has so far been limited [44] and has promoted the development of smaller devices.

Due to the structure of the DGT probe, this technique will measure only dissolved species with molecular sizes significantly smaller than the pore size of the hydrogel that allows them to diffuse freely through it, and which are sufficiently labile to bind on the resin's functional groups. Normally, the Chelex-100 resin is used, because the functional group, iminodiacetic acid, competes effectively with natural ligands for divalent and trivalent metal ions. However, alternative resins have been described, such as the AG50W-X cation-exchange resin for radioactive Cs and Sr [73] and the Spheron-Thiol resin for mercury [74,75]. Thus DGT measures all species that can dissociate during their transport through the diffusion layer, which may include metal fulvic and humic complexes. A single device is unable to quantify the exact contribution of organic complexes. However, use of two or more DGT devices with different gel layers of different pore sizes can enable fully quantitative discrimination between small (inorganic) species and larger (organically complexed) species [44].

5.10 Ion-Selective Electrodes

Ion-selective electrodes (ISEs) respond to free ion activities. The ISE is immersed in the solution containing the ions to be measured, together with a reference electrode, usually a calomel or silver/silver chloride electrode. Both electrodes could be built into the same body and are connected to a very high-impedance voltmeter. A potential difference is developed across an ion-selective membrane of the ISE when the target ions diffuse through from the high-concentration side to the lower-concentration side, which is related to the activity of the free ion in solution. The use of ion-selective electrodes for trace element speciation purposes has remained somewhat limited due to their low sensitivity. Moreover, trace element analyses using ion-selective electrodes often suffer from interferences [76–78]. Ion-selective Cu electrodes are generally more sensitive and selective than other ion-selective metal electrodes (for example Cd, Pb and Zn). Rachou *et al.* [79] recently developed and validated a method for the determination of free Cu^{2+} at environmentally relevant activities, ranging between 10^{-14} and 10^{-4} M, in low-salinity and low-ionic-strength matrices, whereas Lu *et al.* [80] and Yang *et al.* [81] reported detection limits for Pb(II) ion-selective membrane electrodes of around 10^{-6} M. In recent years research efforts have focused on improving the detection limits and selectivity of ISEs. As a result, a series of new interesting perspectives on ISEs application in environmental analysis has been identified [82]. For research concerning trace elements in soils, ion-selective electrodes are currently mainly used to measure free metal ion (mainly Cu^{2+}) activities for the assessment of the complexation of trace elements by organic matter [83–86], their sorption by clays [87], their mobility, availability and toxicity in soils [88–90] and their chemical behaviour in rhizosphere microenvironments [91]. They are also frequently used in procedures for validation of speciation models [92]. As the Cu ISE has a good selectivity, direct determination of Cu^{2+} in soil extracts is possible [3,61,93,94]. At very low activities (down to 10^{-15} M), the equilibration times can be very long (hours) and the relationship between log[Cu^{2+}] and the potential can be nonlinear due to the dissolution of the ion-sensitive membrane. Moreover, the metal ion buffers should have the same ionic strength as the solution being measured. This is relatively simple to arrange when measuring in soil extracts, whereas the ionic strength of pore waters can be quite variable [44]. In this context, calibration should also be done carefully. The use of metal ion buffers in combination with a speciation programme to calculate free metal ion activities is recommended. Due to their lower selectivity, the other ISEs (for example Cd, Pb and Zn ISEs) are primarily used for potentiometric titrations of soil extracts [95–98]. In numerous studies, ion-selective electrodes have been used in combination with other techniques such as DGT (for example [99]) or anodic stripping voltammetry (for example [100]) to assess the trace element speciation.

5.11 Donnan Membrane Technique

One of the techniques that can also measure the free metal ion concentration is the Donnan membrane technique (DMT). This technique was developed to reduce the interference effects from which some other speciation procedures are known to suffer [101]. A DMT cell consists of an acceptor compartment, which is separated from the sample solution by a negatively charged cation-exchange membrane. The membrane allows exchange only of 'free' cations between the sample, which is referred to as the donor solution, and the acceptor solution, while anions and metal complexes with larger molecular weight are

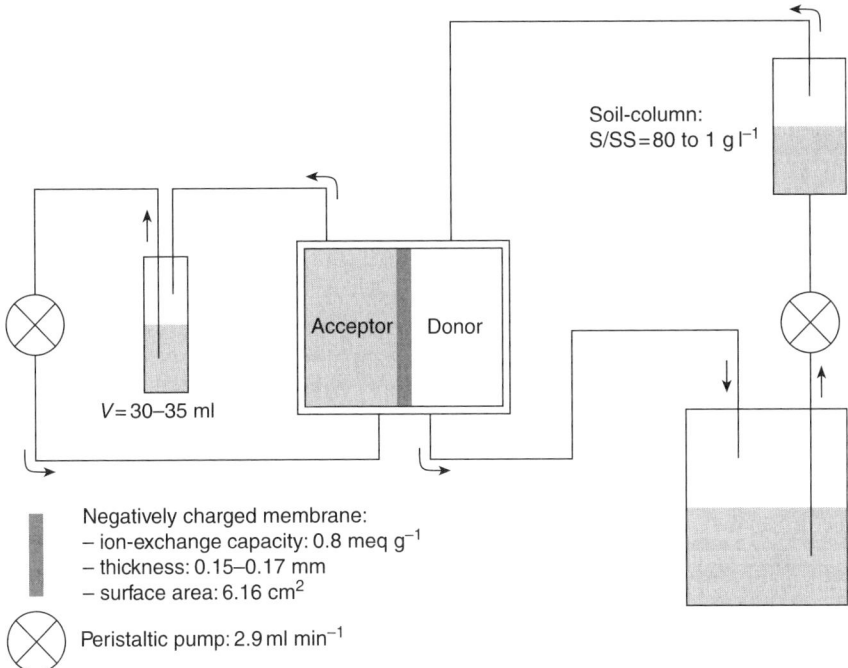

Figure 5.4 Coupling of a soil column to a Donnan cell (Reprinted from Geoderma, 113, B. Cancès et al., Metal ions speciation in a soil and its solution: experimental data and model results, 341–355. Copyright 2003, with permission from Elsevier [105].)

excluded on the basis of charge or size. After some time, a 'Donnan equilibrium' is reached between free cation concentrations in the acceptor solution and the free cation concentrations in the sample solution [102,103]. The driving force is the attempt by DMT to achieve an equal charge between the ions in the sample solution and the ions in the acceptor solution which diffuse in the opposite direction.

The Donnan membrane technique was originally developed mainly for use in surface waters, but has been further developed to make it applicable for analysing soil samples by direct linkage of the soil column with the Donnan cell (Figure 5.4) [104, 105]. The technique has the advantage that it can measure many different free metal ion concentrations simultaneously in a multicomponent sample while also minimizing the disturbance of substrate [104,106]. However, it has also limitations in terms of a long equilibration time. Helmke and co-workers [107] originally designed the system with a 30–50 ml volume donor solution and a very small volume (for example 200 μl) acceptor solution. This allowed for an equilibration within a few hours, although metal analysis and pH measurements were difficult due to the very small acceptor solution volume. Nowadays, the development of microelectrodes and small volume sample introduction systems for ICP-MS helps to overcome these problems [44]. The need for expensive analytical equipment capable of conducting sensitive DMT measurements can, however, be seen as a disadvantage. A larger acceptor solution volume can be achieved by the use of flowing donor and acceptor solutions, as in the recently developed Wageningen Donnan membrane technique (WDMT). However, this implies longer equilibration times of about 2 days [108]. The developments of similar dynamic modifications of the standard DMT have recently been reported, such as the Flux Donnan membrane (FDM) technique [109]. The requirement of relatively large soil solution volumes as donor solution is also seen as a disadvantage, but as the soil solutions can be easily exposed *in situ* to the DMT device, speciation artefacts which may occur during sampling and sample handling can be avoided [110]. As the Donnan membrane technique is also still in its development stage, so far it has rarely been used for routine monitoring purposes in soil research.

5.12 Voltammetric Techniques

Voltammetric techniques are based on the recording of a current, i, which flows between a working electrode and an auxiliary electrode, due to the reduction or oxidation of the test element, as function of the potential imposed on the working electrode and expressed with respect to that of a reference electrode. A large number of voltammetric techniques with variable imposed modulations of potential have been developed, in particular to increase sensitivity [111]. The most widely used voltammetric technique for trace metal analysis is anodic stripping voltammetry (ASV), which basically consists of two steps: a deposition step and a stripping step. Metal ions from the solution are reduced at a negative potential and deposited onto an electrode in the deposition step. In this way, the species of interest is preconcentrated, which is the main feature for the success of voltammetric stripping techniques in trace element analyses. Then the potential is gradually made more positive. At characteristic potentials, the metals are reoxidized to free metal ions and stripped back into solution, generating a current which is proportional to the amount of metals on the electrode. The latter is proportional to the concentration of metal ions in solution for a given deposition time. When the potential is varied linearly with time during the stripping step, a capacitative current is present, which limits the sensitivity. More complicated waveforms of potential versus time, coupled with measuring the current after perturbations, have been developed to minimize the capacitative current and maximize the current due to the electrode reaction. Examples of these include square wave (SWASV), differential pulse (DPASV) and staircase techniques [44].

The ASV device consists of a cell in which the sample is placed, a stirrer and three electrodes: the working electrode, at which the reaction of interest occurs, the reference electrode, which provides a known, steady potential and an auxiliary or counter electrode, which minimizes errors from cell resistance in controlling the potential of the working electrode. Parameters which should be set are the deposition potential and time, and the stripping speed and waveform. As adsorption of natural organic matter on the electrode can change the sensitivity, calibration can be difficult. Standard solutions with a composition which simulates the composition of the sample (for example pH, ionic strength) are recommended. Different working electrodes can be used, such as a thin film mercury electrode (TFME), a hanging mercury drop electrode (HMDE), and a composite carbon electrode. The HMDE has an advantage over the other two electrodes since a new mercury drop is generated for each run. This is important as the accuracy and reproducibility of ASV depends strongly on the quality of the surface of the electrode. In turn, composite carbon electrodes provide an extended anodic potential range, which implies that they can be used for a wider range of metals. In addition, carbon electrodes allow the system to be used in the field and they are safer to use than mercury-containing electrodes. Anodic stripping voltammetry measures all metal species that can dissociate during the time it takes for them to diffuse from the bulk solution to the electrode surface, typically 0.1 s. This usually includes all inorganic complexes and most metals complexed by fulvic and humic acids. The generated current is proportional to the concentration of each species and its diffusion coefficient. All inorganic species have the same diffusion coefficient, but fulvic and humic complexes have diffusion coefficients which are typically three to five times smaller. As a result, they contribute to the total measured current with less sensitivity and their contribution cannot be fully quantified [44]. Specific reduction potentials for each element should be applied when discerning free metal forms from labile complexes [112]. Too low reduction potentials result in increasingly less labile complexes being reduced; this causes an overestimation of the free ionic form [43].

An alternative to ASV for analysing positively charged metal ions is (cathodic) adsorptive stripping voltammetry (AdSV) [113–115]. In AdSV, a ligand is added to form a complex with the metal ion, which is preconcentrated by adsorption at the electrode surface, whereas the reduction peak of this adsorbed complex is measured in the second step. This technique has broadened the application range of voltammetry to a large number of elements; however, more work is still needed to evaluate its reliability for routine and multielement analysis [111]. Other promising stripping techniques for speciation analyses are anodic stripping chronopotentiometry (SCP) and SCP at scanning deposition potential, SSCP [116,117] and the 'absence of gradients and Nernstian equilibrium stripping' (AGNES) technique. Anodic stripping chronopotentiometry is somewhat less sensitive than SWASV, but also less sensitive to redox reversibility, and much less prone to artefacts due to adsorption, compared with ASV techniques [111]. The AGNES technique was developed for direct free metal ion measurements [118].

The strength of stripping voltammetric techniques are the low detection limits (10^{-10}–10^{-12} M), the multielement and speciation capabilities, and the suitability for on-site and *in situ* applications [111,119]. They have mainly been used to measure trace element concentrations in surface waters, but also have numerous applications in soil science. In some studies, the test sample is deliberately perturbed by addition of a synthetic ligand forming well-known soluble voltammetrically labile or inert complexes with the test metal or complexes adsorbed on the electrodes, as in the AdSV technique, after which the free ion concentrations are computed from the known stability constants of the test metals with the synthetic ligands [111]. In another approach, trace metals are directly analysed in the unperturbed soil solution or a soil extract [120–122], or in an acid soil digest when the determination of (pseudo)total trace element contents in the soil solid fraction is targeted [114]. A simplified soil extraction procedure in combination with DPASV measurement which can yield semiquantitative data for rapid field-based screening of contaminated sites was recently developed [123]. As a result, portable ASV instruments capable of detecting and identifying trace metals *in situ* in soil have also been developed [124]. Soil extractions using various concentrations of KNO_3 are often conducted when ASV is used to assess metal speciation in the soil solution [44,125].

Although the extraction procedure is operationally defined and not completely selective, it has the advantage of supplying a good electrolyte which is needed for the voltammetric measurement. The free ion activity is calculated assuming that only inorganic species are measured, which is often the case when measuring trace elements which are not substantially complexed by humic substances, for example Cd and Zn. For Cu substantial complexation by humic substances often results in a significant overestimation of the free ion activity. The importance of this interference depends not only on the abundance of these complexes and their contribution to the measured current, but also on their lability. Taking this uncertainty into account, ASV measurements are often operationally-defined as 'ASV-labile metal concentrations' [91].

5.13 Microelectrodes and Microsensors

Recent work has focused on eliminating interferences in voltammetric techniques. As a result, for example, rotating disc electrodes were developed which can minimize the diffusion layer thickness and thus the time available for dissociation of labile complexes [126]. In this context, the development of microelectrodes has been an important step. Although definitions of microelectrodes are ambiguous [127], these are mainly considered as the electrodes with dimensions of tens of micrometres or less [128]. They include single and composite types and can be produced in diverse geometries, with different electrode materials, and using various fabrication techniques [129]. While planar diffusion occurs on the surface of macroelectrodes, radial diffusion occurs around microelectrodes, as the electrode radius is much smaller than the diffusion-layer thickness. Microelectrodes have a few unique characteristics that make them preferable to macroelectrodes for voltammetric environmental monitoring [111,129]:

1. They have a low ohmic drop due to their small size, which enables direct measurements in low ionic strength freshwater without adding electrolyte.
2. Because of the increased mass-transport due to radial diffusion, a non-zero, steady-state reduction current is quickly established at constant potential, even in quiescent solution. Thus, stirring the solution is unnecessary during the preconcentration step of stripping techniques, which greatly improves the reliability of analysis and is anyway a prerequisite for performing stripping voltammetry *in situ* in soil and sediment pore waters.
3. Finally, due to their increased mass-transport compared to planar electrodes, combined with their lower capacitance, a significantly larger signal-to-noise (S/N) ratio is obtained which allows the determination of subnanomolar concentrations with short preconcentration times.

In addition to the advantages of eliminating interferences, microelectrodes are also very suitable for the development of portable analytical instruments and *in situ* analyses. As already mentioned, these have several advantages compared with the conventional analytical methods used in central laboratories, including elimination of artefacts due to sample handling (for example transportation, storage and pretreatment); rapid and real-time monitoring of pollutants, allowing quick remedial action to minimize their effects; the capacity to measure metal speciation; and the possibility to obtain detailed spatial and temporal data [129–131].

Especially for *in situ* measurements, it might be necessary to protect the microelectrodes from fouling by suspended matter and organic compounds. This can be done using a gel which can be also used as a dialysis membrane. The latter allows diffusion of metal ions and small metal complexes, while hindering the access of colloids and macromolecules to the surface of the electrode [132]. The measured current intensities are a function of the diffusion coefficients of the analyte inside the gel and are little affected by the composition of the test medium, as opposed to measurements made with bare electrodes; this is considered the most important feature of such gel-integrated microsensors. The simplest reported micro-analytical system with voltammetric detection for water analysis is the gel integrated microelectrode (GIME). This consists of Hg-plated Ir microelectrodes, covered by an agarose gel a few hundred micrometres thick. The GIME sensor is left to equilibrate with the test solution, and the test ions are analysed inside the gel by a direct or an anodic stripping voltammetric technique. The complexing gel integrated microelectrode (CGIME) is a modified GIME, which specifically determines the free metal-ion concentration at sub-nanomolar concentrations as it contains a thin layer of complexing resin (such as Microchelex) below the antifouling agarose gel layer. In a first step the trace metals of interest are accumulated on the chelating resin in proportion to their free ion concentration in solution, then they are released in acidic solution and detected simultaneously using square-wave anodic stripping voltammetry [133]. In the even more complicated PLM-GIME-μTAS device, the flux through a hydrophobic permeation liquid membrane and binding to a carrier in a stripping solution can be used to determine the speciation of a trace element in the test solution [134]. As trace element diffusion through the gels of gel-integrated microsensors can be seen as physicochemical analogues of various migration steps of trace elements into microorganisms, these microsensors are often also referred to as 'bioanalogical sensors'. This principle is illustrated in Figure 5.5. Capabilities and limitations of all these voltammetric techniques have been extensively discussed by Buffle and Tercier-Waeber [111].

The further development of microelectrodes in terms of keeping intact their sensitivity and speciation capabilities over long periods of time (weeks) without maintenance is currently a major challenge, which requires competence in the physical chemistry of membrane and gel layers. Until recently, the application of microelectrodes to environmental analysis has largely been limited by the fact that they were not reliable enough as they were manually fabricated using classical mechanical processes. These problems have been overcome by using microtechnology-based techniques, which allow automated mass production of low-cost, reliable, microelectrode arrays, with a wide flexibility in the choice of electrode material and geometry [111].

Several studies report the development of Hg-plated Ir-based microelectrode array sensors for the determination of trace elements using square wave anodic stripping voltammetry, SWASV [129,135–137]. Specifically in soil science, microelectrode arrays have already been used in combination with ASV for on-site analysis of arsenic (As^{3+}) in groundwater. A nafion-coated, mercury-free, Au-based microelectrode array including 564 microelectrodes with a diameter of 12 μm and centre-to-centre spacing of 58 μm was used [138]. The As^{3+} was analysed after deposition of an As^0 monolayer by SWASV. This resulted in a detection limit of 0.05 μg l^{-1} and a precision of 2.5 % relative standard deviation. Anodic stripping voltammetry with Hg-coated microelectrodes has recently also been used to measure concentration profiles of trace metal ions at the soil/water interface of a natural sediment and a bauxitic soil sample [139] and for the analysis of Cd, Cu and Pb in soil extracts [140]. Gold microelectrode arrays were developed for the analysis of Cu in soil extracts [141].

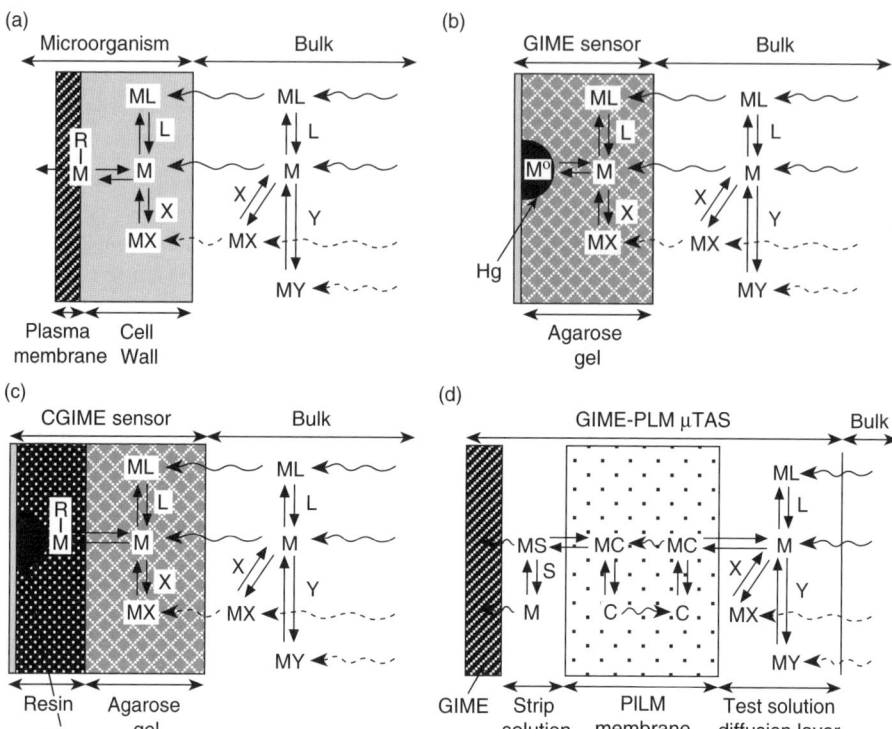

Figure 5.5 Schematic representation of (a) a microorganism/solution interface; (b) a GIME/solution interface; (c) a CGIME/solution interface; and (d) a PLM-GIME-μTAS/solution interface. (Note: thicknesses of the various layers are not to scale). The test solution includes free-metal ion (M), small ligands (L), small complexes (ML), colloidal complexants (X,Y) and colloidal complexes (MX, MY). Only one species of each type is shown for the sake of simplicity. X and MX are small enough to penetrate the gel, unlike Y and MY (Reproduced from Trends Anal. Chem. 24, J. Buffle and M.L. Tercier-Waeber, Voltammetric environmental trace metal analysis and speciation: from laboratory to in situ measurements, *172–191*. Copyright 2005, with permission from Elsevier [111].)

5.14 Models for Predicting Metal Speciation in Soil Solution

Various computer models can be used to calculate the speciation of trace elements in soil solutions and soil extracts (for example, GEOCHEM, SOILCHEM, NICA-DONNAN, MINTEQ2, WHAM, PHREEQC, and others). The most common input variables for such models are pH, redox potential, dissolved organic matter (DOM), the concentrations of major inorganic anions and cations, and trace elements.

The core of such chemical equilibrium models resides in the use of the ion-association approach, also called the Debye–Hückel model, to calculate the activity coefficient of charged aqueous species in dilute waters. Some selected aqueous chemical equilibrium models also propose the ion-interaction approach, which relies on empirical coefficients to describe ion complexing at high ionic strength (>0.3 M), when the ion-association description loses accuracy. In most models, acid–base and oxidoreduction reactions, trace element complexation and mineral dissolution/precipitation are described as aqueous phase equilibria, although some also consider these reactions by using various kinetic expressions. The input value of the solution pH in conjunction with the thermodynamic equilibrium constants registered in the database for acid–base reactions (pK_a) is used to calculate the speciation of acids and bases. Similarly, the redox potential (E_h) is used to distribute the members of redox couples according to the thermodynamic constants of half reactions. Complexation and chelation, as ion pairing, are also treated as aqueous phase equilibrium between a ligand and a free ion, or simply between two ions present in the solution. After the speciation of the initial solution, an ion activity product (IAP) is calculated for each mineral registered in the database (with possible exclusions by the user) and compared with its equilibrium constant (K) through a saturation index. If mass transfer from the mineral to the solution (dissolution) or vice versa (precipitation) is required to reach equilibrium, the aqueous speciation is recalculated in response to corresponding variations in total concentrations [142].

Most chemical equilibrium models also take into account some of the interactions occurring between the aqueous-phase ions and surface sites. Sorption of aqueous ions onto specific surface sites can be described using nonelectrostatic or electrostatic models. Nonelectrostatic sorption is described using either a simple equilibrium distribution coefficient (K_d) of the component between the aqueous phase and the solid phase or the Langmuir or Freundlich models, whereas electrostatic sorption models describe the charge-based attraction of anions and cations by positively and negatively charged surface sites, respectively. The widely used double-layer model considers a first layer of counterions covering the surface and a second diffuse layer of ions balancing the charge difference created by the first layer. Ion-exchange reactions between aqueous phase cations and exchanger organic or mineral phases can also be modelled using various selectivity coefficient descriptions for the major cations [142].

Which model is applied has very little influence on the end result when mainly inorganic salts are involved. Unfortunately, there is some difficulty in applying these speciation programmes to real soil solutions as assumptions must be made regarding the complexing characteristics of humic substances and inorganic colloidal material. The actual stability constants used and model assumptions can indeed have a large influence on the results. A majority of the data available for the constants are derived from titration data where large concentrations are added to purified organic acids of mostly aquatic origin. The purification protocol, which is desired to produce consistent experimental results, transforms the properties of the fulvic and humic acids. Furthermore, removing cations not included in the study can render the results unrealistic, misrepresenting dissolved organic matter (DOM) occurrence under field conditions. However, various chemical models have been used recently to describe adsorption in which molecular features, specific surface complexation, chemical reactions and charge balances are considered. The prediction of adsorption on the soil using such models has been successful for B, Mo, and As [143–145]. Dudal and Gerard [142] reviewed six chemical equilibrium models, namely NICA-Donnan, EQ3/6, GEOCHEM, MINTEQA2, PHREEQC and WHAM, in light of the account they make of natural organic matter (NOM) with the objective of helping potential users in choosing a modelling approach. Various approaches have been used to incorporate the role of natural organic matter, mainly the addition of specific molecules within the existing model databases (EQ3/6, GEOCHEM, and PHREEQC) or by using either a discrete (WHAM) or a continuous

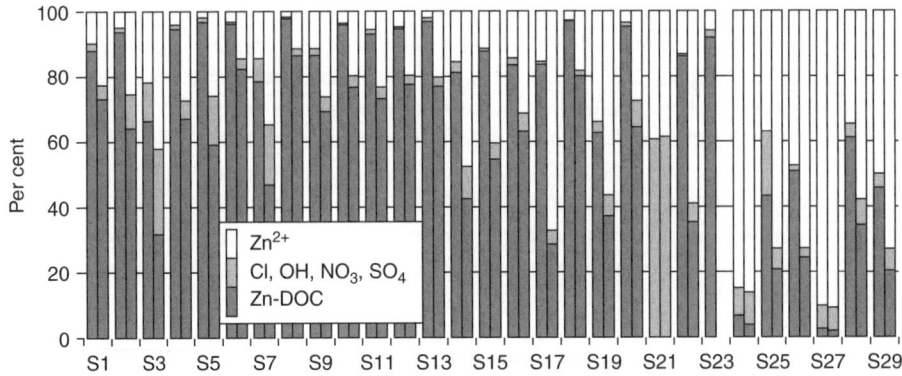

Figure 5.6 Free Zn^{2+} (white), inorganic Zn-complexes (light grey) and Zn complexed by dissolved organic matter (DOC, dark grey) in the soil solution of 29 soils, according to Visual Minteq 2.23 and WHAM VI. For each soil, speciation according to Visual Minteq 2.23 is presented on the left, speciation according to WHAM VI is presented on the right (Reproduced from Geoderma, 136, E. Meers et al., Zn in the soil solution of unpolluted and polluted soils as affected by soil characteristics, 107–119. Copyright 2006, with permission from Elsevier [147].)

(NICA-Donnan and MINTEQA2) distribution of the deprotonated carboxylic and phenolic groups. WHAM and NICA-Donnan, which are considered as the two most advanced models, were found to reasonably describe most of the experimental results. Dudal and Gerard [142], however, concluded that more nonhumic compounds should also be introduced into the model databases, notably the ones that readily interact with the soil microorganisms, such as low-molecular-weight organic acids. Even more difficult to integrate in the currently used chemical equilibrium models is that, by definition, they often fail to account for the slow kinetics of some of the reactions. This is an area of research that needs to be addressed, especially in soil environments where complexation of a particular trace element may be slowed and hindered by the need to displace cations, such as calcium or magnesium, from the ligands before complexation can take place [146]. Speciation differences which might occur when using different models are illustrated in Figure 5.6.

So, although significant progress has been made in modelling trace element complexation by DOM and colloidal material, there is still some accuracy uncertainty in application to real soil solutions and soil extracts [7,101,148]. Regardless of these current limitations, computer models are often used in combination with data obtained by instrumental speciation techniques when assessing trace element speciation in the soil solution or in soil extracts (for example [147]), or to cross-validate the instrumental speciation techniques (for example [149]) or the models themselves (for example [150]) in various development stages.

5.15 Conclusions

Numerous techniques which can be used for trace element speciation analysis in soil solutions have been developed in recent years. Most attention has been given to the development of computer models and analytical techniques which aim at the assessment of free ion activities in the soil solution, as only the free ions are considered to interact directly with biological ligands. Other techniques were originally developed for the purpose of assessing trace element speciation in biological materials, but can also be used in soil science. The low analyte concentrations involved, the need for preservation of the species throughout the procedure, and the strict requirements in terms of sample volume, polarity and matrix acceptable by the instrumental set-up make the sample preparation protocol for speciation analysis critically important. Before carrying out any speciation procedure, the stability of the chemical species during sampling and sample preparation needs to be critically addressed. A distinct trend is automation; this may result in the development of more efficient, faster, and possibly single-step speciation techniques.

References

[1] Allen, H.E., The significance of trace metal speciation for water, sediment and soil quality standards; Sci. Total Environ. 1993, 134(S1), 23–45.
[2] Sauvé, S.; Dumestre, A.; McBride, M.; Hendershot, W., Derivation of soil quality criteria using predicted chemical speciation of Pb^{2+} and Cu^{2+}, Environ. Toxicol. Chem. 1998, 17, 1481–1489.
[3] McGrath, S.P.; Knight, B.; Killham, K.; Preston, S.; Paton, G.I., Assessment of the toxicity of metals in soils amended with sewage sludge using a chemical speciation technique and a lux-based biosensor; Environ. Toxicol. Chem. 1999, 18, 659–663.
[4] Vulkan, R.; Zhao, F.-J.; Barbosa-Jefferson, V. et al., Copper speciation and impacts on bacterial biosensors in the pore water of copper-contaminated soils; Environ. Sci. Technol. 2000, 34, 5115–5121.
[5] Di Toro, D.M.; Allen, H.E.; Bergman, H.L.; Meyer, J. S.; Paquin, P.R.; Santore, R.C., Biotic ligand model of the acute toxicity of metals. 1. Technical basis; Environ. Toxicol. Chem. 2001, 20, 2383–2396.
[6] Pampura, T.; Groenenberg, J.E.; Lofts, S.; Priputina, I., Validation of transfer functions predicting Cd and Pb free metal ion activity in soil solution as a function of soil characteristics and reactive metal content; Water Air Soil Pollut. 2007, 184, 217–234.
[7] Collins, R.N., Separation of low-molecular mass organic acid-metal complexes by high-performance liquid chromatography; J. Chromatogr. A 2004, 1059, 1–12.
[8] Playle, R.C., Modelling metal interactions at fish gills; Sci. Total Environ. 1998, 219, 147–163.
[9] Peijnenburg, W.J.G.M.; Zablotskaja, M.; Vijver, M.G., Monitoring metals in terrestrial environments within a bioavailability framework and a focus on soil extraction; Ecotoxicol. Environ. Saf. 2007, 67, 163–179.
[10] Bufflap, S.E.; Allen, H.E., Comparison of pore water sampling techniques for trace metals; Water Res. 1995, 29, 2051–2054.
[11] Miró, M.; Hansen, E.H., Recent advances and perspectives in analytical methodologies for monitoring the bioavailability of trace metals in environmental solid substrates; Microchim. Acta 2006, 154, 3–13.
[12] Göttlein, A.; Hell, U.; Blasek, R., A system for microscale tensiometry and lysimetry; Geoderma 1996, 69, 147–156.
[13] Meers, E.; Du Laing, G.; Unamuno, V.G.; Lesage, E.; Tack, F.M.G.; Verloo, M.G.; Water extractability of trace metals from soils: some pitfalls; Water Air Soil Pollut. 2006, 176, 21–35.
[14] Bufflap, S.E.; Allen, H.E., Sediment pore water collection methods for trace metal analysis: a review; Water Res. 1995, 29, 165–177.
[15] Emons, H., Sampling: collection, processing and storage of environmental samples; in: R. Cornelis, H.; Crews, J.; Caruso, K.; Heumann (eds), Handbook of Elemental Speciation: Techniques and Methodology; John Wiley & Sons, Ltd, Chichester, 2003, pp. 7–22.

[16] Wenzel, W.W.; Wieshammer, G., Suction cup materials and their potential to bias trace metal analyses of soil solutions: a review; Int. J. Environ. Anal. Chem. 1995, 59, 277–290.
[17] Grossman, J.; Udluft, P., The extraction of soil water by the suction-cup method: a review; J. Soil Sci. 1991, 42, 83–93.
[18] Miró, M.; Frenzel, W., Implantable flow-through capillary-type microdialyzers for continuous in situ monitoring of environmentally relevant parameters; Anal. Chem. 2004, 76, 5974–5981.
[19] Buffle, J.; Leppard, G.G., Characterization of aquatic colloids and macromolecules .1. Structure and behavior of colloidal material; Environ. Sci. Technol. 1995, 29, 2169–2175.
[20] Burba, P.; Aster, B.; Nifanteva, T.; Shkinev, V.; Spivakov, B.Y., Membrane filtration studies of aquatic humic substances and their metal species: a concise overview Part 1. Analytical fractionation by means of sequential-stage ultrafiltration; Talanta 1998, 45, 977–988.
[21] Paulenova, A.; Rajec, P.; Kandrac, J. et al., The study of americium, yttrium and lead complexation by humic acids of different origin; J. Radioan. Nucl. Ch. 2000, 246, 617–622.
[22] Van Hees, P.A.W.; Lundstrom, U.S.; Giesler, R., Low molecular weight organic acids and their Al-complexes in soil solution - composition, distribution and seasonal variation in three podzolized soils; Geoderma 2000, 94, 173–200.
[23] Wang, W.S.; Wen, B.; Zhang, S.Z.; Shan, X.Q., Distribution of heavy metals in water and soil solutions based on colloid-size fractionation; Int. J. Environ. Anal. Chem. 2003, 83, 357–365.
[24] Pokrovsky, O.S.; Dupre, B.; Schott, J., Fe-Al-organic colloids control of trace elements in peat soil solutions: Results of ultrafiltration and dialysis; Aquat. Geochem. 2005, 11, 241–278.
[25] Hlavay, J.; Polyák, K., Sample preparation – fractionation (sediments, soils, aerosols, and fly ashes); in: R. Cornelis, H.; Crews, J.; Caruso, K.; Heumann (eds), Handbook of Elemental Speciation: Techniques and Methodology; John Wiley & Sons, Ltd, Chichester, 2003, pp. 119–146.
[26] Kowalkowski, T.; Buszewski, B.; Cantado, C.; Dondi, F., Field-flow fractionation: theory, techniques, applications and the challenges; Crit. Rev. Anal. Chem. 2006, 36, 129–135.
[27] Gimbert, L.J.; Haygarth, P.M.; Beckett, R.; Worsfold, P.J., The influence of sample preparation on observed particle size distributions for contrasting soil suspensions using flow field-flow fractionation; Environ. Chem. 2006, 3, 184–191.
[28] Taylor, H.E.; Garbarino, J.R.; Murphy, D.M.; Beckett, R., Inductively coupled plasma mass-spectrometry as an element-specific detector for field-flow fractionation particle separation; Anal. Chem. 1992, 64, 2036–2041.
[29] Von der Kammer, F.; Forstner, U., Natural colloid characterization using flow-field-flow-fractionation followed by multi-detector analysis; Water Sci. Technol. 1998, 37, 173–180.
[30] Ranville, J.F.; Chittleborough, D.J.; Shanks, F. et al., Development of sedimentation field-flow fractionation-inductively coupled plasma mass-spectrometry for the characterization of environmental colloids; Anal. Chim. Acta 1999, 381, 315–329.
[31] Amarasiriwardena, D.; Siripinyanond, A.; Barnes, R.M., Trace elemental distribution in soil and compost-derived humic acid molecular fractions and colloidal organic matter in municipal wastewater by flow field-flow fractionation-inductively coupled plasma mass spectrometry (flow FFF-ICP-MS); J. Anal. Atom. Spectrom. 2001, 16, 1344–1344.
[32] Ranville, J.F.; Chittleborough, D.J.; Beckett, R., Particle-size and element distributions of soil colloids: Implications for colloid transport; Soil Sci. Soc. Am. J. 2005, 69, 1173–1184.
[33] Siripinyanond, A.; Barnes, R.M.; Amarasiriwardena, D., Flow field-flow fractionation-inductively coupled plasma mass spectrometry for sediment bound trace metal characterization; J. Anal. Atom. Spectrom. 2002, 17, 1055–1064.
[34] Janos, P., Separation methods in the chemistry of humic substances; J. Chromatogr. A 2003, 983, 1–18.
[35] De Leon, C.A.P.; DeNicola, K.; Bayon, M.M.; Caruso, J.A., Sequential extractions of selenium soils from Stewart Lake: total selenium and speciation measurements with ICP-MS detection; J. Environ. Monit. 2003, 5, 435–440.
[36] Mueller-Spitz, S.R.; Vonderheide, A.P.; Shann, J.R.; Caruso, J.A.; Kinkle, B.K., Use of SEC-ICP-MS with a collision cell for determining the interaction of chromium with DNA extracted from metal-contaminated soils; Anal. Bioanal. Chem. 2006, 386, 142–151.

[37] Newton, K.; Amarasiriwardena, D.; Xing, B.S., Distribution of soil arsenic species, lead and arsenic bound to humic acid molar mass fractions in a contaminated apple orchard; Environ. Pollut. 2006, 143, 197–205.

[38] Steely, S.; Amarasiriwardena, D.; Xing, B.S., An investigation of inorganic antimony species and antimony associated with soil humic acid molar mass fractions in contaminated soils; Environ. Pollut. 2007, 148, 590–598.

[39] Shibata, Y.; Tsuzuku, K.; Komori, S.; Umedzu, C.; Imai, H.; Morita, M., Analysis of dephenylarsinic acid in human and environmental samples by HPLC-ICP-MS; Appl. Organomet. Chem. 2005, 19, 276–281.

[40] Piatina, T.B.; Hering, J.G. Direct quantification of metal-organic interactions by size-exclusion chromatography (SEC) and inductively coupled plasma mass spectrometry (ICP-MS); J. Environ. Qual. 2000, 29, 1839–1845.

[41] Bouyssiere, B.; Szpunar, J.; Potin-Gautier, M.; Lobinski, R., Sample preparation techniques for elemental speciation studies; in: R. Cornelis, H.; Crews, J.; Caruso, K.; Heumann (eds), Handbook of Elemental Speciation: Techniques and Methodology; John Wiley & Sons, Ltd, Chichester, 2003, pp. 95–118.

[42] Bayona, J.M., Supercritical fluid extraction in speciation studies; Trends Anal. Chem. 2000, 19, 107–112.

[43] Tack, F.M.G.; Verloo, M.G., Chemical speciation and fractionation in soil and sediment heavy-metal analysis – a review; Int. J. Environ. Anal. Chem. 1995, 59, 225–238.

[44] Zhang, H.; Young, S.D., Characterizing the availability of metals in contaminated soils. II. The soil solution; Soil Use Manag. 2005, 21, 459–467.

[45] Holm, P.E.; Christensen, T.H.; Tjell, J.C.; McGrath, S.P., Speciation of cadmium and zinc with application to soil solutions; J. Environ. Qual. 1995, 24, 183–190.

[46] Holm, P.E.; Andersen, S.; Christensen, T.H., Speciation of dissolved cadmium – interpretation of dialysis, ion-exchange and computer (GEOCHEM) methods; Water Res. 1995, 29, 803–809.

[47] Lorenz, S.E.; Hamon, R.E.; Holm, P.E. et al., Cadmium and zinc in plants and soil solutions from contaminated soils; Plant Soil 1997, 189, 21–31.

[48] Christensen, J.B.; Christensen, T.H., The effect of pH on complexation of Cd, Ni and Zn by dissolved organic carbon from leachate-polluted groundwater; Water Res. 2000, 34, 3743–3754.

[49] De Smaele, T.; Moens, L.; Sandra, P.; Dams, R., Determination of organometallic compounds in surface water and sediment samples with SPME-CGC-ICPMS; Microchim. Acta 1999, 130, 241–251.

[50] Kaur, V.; Malik, A.K.; Verma, N., Applications of solid phase microextraction for the determination of metallic and organometallic species; J. Sep. Sci. 2006, 29, 333–345.

[51] Dietz, C.; Perez-Corona, T.; Madrid-Albarran, Y.; Camara, C., SPME for on-line volatile organo-selenium speciation; J. Anal. Atom. Spectrom. 2003, 18, 467–473.

[52] Mester, Z.N.; Lam, J.; Sturgeon, R.; Pawliszyn, J., Determination of methylmercury by solid-phase microextraction inductively coupled plasma mass spectrometry: a new sample introduction method for volatile metal species; J. Anal. Atom. Spectrom. 2000, 15, 837–842.

[53] Olivas, R.M.; Cámara, C., Sample preparation. Sample treatment for speciation analysis in biological samples; in: R. Cornelis, H.; Crews, J.; Caruso, K.; Heumann (eds), Handbook of Elemental Speciation: Techniques and Methodology; John Wiley & Sons, Ltd, Chichester, 2003, pp. 73–94.

[54] Ackley, K.L.; Caruso, J.A., Separation techniques. 4.1 Liquid chromatography; in: R. Cornelis, H.; Crews, J.; Caruso, K.; Heumann (eds), Handbook of Elemental Speciation: Techniques and Methodology; John Wiley & Sons, Ltd, Chichester, 2003, pp. 147–162.

[55] Butcher, D.J., Environmental applications of arsenic speciation using atomic spectrometry detection; Appl. Spectrosc. Rev. 2007, 42, 1–22.

[56] Kahakachchi, C.; Uden, P.C.; Tyson, J.F., Extraction of arsenic species from spiked soils and standard reference materials; Analyst 2004, 129, 714–718.

[57] Vanhaecke, F.; Köllensperger, G., Detection by ICP–mass spectrometry; in: R. Cornelis, H.; Crews, J.; Caruso, K.; Heumann (eds), Handbook of Elemental Speciation: Techniques and Methodology; John Wiley & Sons, Ltd, Chichester, 2003, pp. 281–312.

[58] Sutton, K.; Sutton, R.M.C.; Caruso, J.A., Inductively coupled plasma mass spectrometric detection for chromatography and capillary electrophoresis; J. Chromatogr. A 1997, 789, 85–126.
[59] Kutschera, K.; Schmidt, A.C.; Kohler, S.; Otto, M., CZE for the speciation of arsenic in aqueous soil extracts; Electrophoresis 2007, 28, 3466–3476.
[60] Zhang, H.; Davison, W., Performance characteristics of the technique of diffusion gradients in thin-films (DGT) for the measurement of trace metals in aqueous solution; Anal. Chem. 1995, 67, 3391–3400.
[61] Zhang, H.; Zhao, F.J.; Sun, B.; Davison, W.; McGrath, S.P., A new method to measure effective soil solution concentration predicts copper availability to plants; Environ. Sci. Technol. 2001, 35, 2602–2607.
[62] Hooda, P.S.; Zhang, H.; Davison, W.; Edwards, A.C., Measuring bioavailable trace metals by diffusive gradients in thin films (DGT): soil moisture effects on its performance in soils; Eur. J. Soil Sci. 1999, 50, 285–294.
[63] Fitz, W.J.; Wenzel, W.W.; Zhang, H. *et al.*, Rhizosphere characteristics of the arsenic hyperaccumulator Pteris vittata L. and monitoring of phytoremoval efficiency; Environ. Sci. Technol. 2003, 37, 5008–5014.
[64] Degryse, F.; Smolders, E.; Oliver, I.; Zhang, H., Relating soil solution Zn concentration to DGT measurements in contaminated soils; Environ. Sci. Technol. 2003, 37, 3958–3965.
[65] Zhang, H.; Davison, W.; Knight, B.; McGrath, S., In situ measurements of solution concentrations and fluxes of trace metals in soils using DGT; Environ. Sci. Technol. 1998, 32, 704–710.
[66] Ernstberger, H.; Davison, W.; Zhang, H.; Tye, A.; Young, S., Measurement and dynamic modelling of trace metal mobilization in soils using DGT and DIFS; Environ. Sci. Technol. 2002, 36, 349–354.
[67] Ernstberger, H.; Zhang, H.; Davison, W.; Tye, A.; Young, S., Trace metal partitioning and sorption kinetics in relation to soil properties; Environ. Sci. Technol. 2005, 39, 1591–1597.
[68] Harper, M.; Davison, W.; Zhang, H.; Tych, W., Solid phase to solution kinetics in sediments and soils interpreted from DGT measured fluxes; Geochim. Cosmochim. Acta. 1998, 62, 2757–2770.
[69] Sochaczewski, L.; Tych, W.; Davison, W.; Zhang, H., 2D DGT-induced fluxes in sediments and soils (2D DIFS); Environ. Modell. Softw. 2007, 22, 14–23.
[70] Song, J.; Zhao, F.J.; Luo, Y.M., McGrath, S.P.; Zhang, H., Copper uptake by *Elsholtzia splendens* and *Silene vulgaris* and assessment of copper phytoavailability in contaminated soils; Environ. Pollut. 2004, 128, 307–315.
[71] Nolan, A.; Zhang, H.; McLaughlin, M.J., Prediction of Zn, Cd, Pb and Cu availability to wheat in contaminated soils using chemical speciation, DGT, extraction and isotopic dilution techniques; J. Environ. Qual. 2004, 34, 469–507.
[72] Zhang, H.; Lombi, E.; Smolders, E.; McGrath, S., Kinetics of Zn release in soils and prediction of Zn concentration in plants using diffusive gradient in thin films; Environ. Sci. Technol. 2004, 38, 3608–3613.
[73] Chang, L.Y.; Davison, W.; Zhang, H.; Kelly, M., Performance characteristics for the measurement of Cs and Sr by diffusive gradients in thin films (DGT); Anal. Chim. Acta 1998, 368, 243–253.
[74] Divis, P.; Leermakers, M.; Docekalova, H.; Gao, Y., Measurement of mercury concentrations in sediment pore waters using centrifugation and diffusive gradients in thin film sampling techniques; Anal. Bioanal. Chem. 2005, 382, 1715–1719.
[75] Gao, Y.; Leermakers, M.; Gabelle, C. *et al.*, High-resolution profiles of trace metals in the pore waters of riverine sediment assessed by DET and DGT; Sci. Total Environ. 2006, 362, 266–277.
[76] Lund, W., Speciation analysis – why and how; Fresen. J. Anal. Chem. 1990, 337, 557–564.
[77] Langford, C.H.; Gutzman, D.W., Kinetic-studies of metal-ion speciation; Anal. Chim. Acta 1992, 256, 183–201.
[78] Domergue, F.L.; Védy, J.C., Mobility of heavy-metals in soil profiles; Int. J. Environ. Anal. Chem. 1992, 46, 13–23.
[79] Rachou, J.; Gagnon, C.; Sauve, S., Use of an ion-selective electrode for free copper measurements in low salinity and low ionic strength matrices; Environ. Chem. 2007, 4, 90–97.

[80] Lu, X.; Chen, Z.; Hall, S.B.; Yang, X., Evaluation and characteristics of a Pb(II) ion-selective electrode based on aquatic humic substances; Anal. Chim. Acta 2000, 418, 205–212.
[81] Yang, X.H.; Kumar, N.; Hibbert, D.B.; Alexander, P.W., Lead(II)-selective membrane electrodes based on 4,7,13,16-tetrathenoyl-1,10-dioxa-4,7,13,16-tetraazacyclooctadecane; Electroanalysis 1998, 10, 827–831.
[82] De Marco, R.; Clarke, G.; Pejcic, B., Ion-selective electrode potentiometry in environmental analysis; Electroanalysis 2007, 19, 1987–2001.
[83] Logan, E.M.; Pulford, I.D.; Cook, G.T.; MacKenzie, A.B., Complexation of Cu^{2+} and Pb^{2+} by peat and humic acid; Eur. J. Soil Sci. 1997, 48, 685–696.
[84] Wu, J.; West, L.J.; Stewart, D.I., Effect of humic substances on Cu(II) solubility in kaolin-sand soil; J. Hazard. Mater. 2002, 94, 223–238.
[85] Gondar, D.; Fiol, S.; Lopez, R.; Ramos, M.A.; Antelo, J.M.; Arce, F., Determination of intrinsic complexation parameters for Cu^{2+} and a soil fulvic acid by ion selective electrode; Chem. Spec. Bioavailab. 2000, 12, 89–96.
[86] Gondar, D.; Iglesias, A.; Lopez, R.; Fiol, S.; Antelo, J.M.; Arce, F., Copper binding by peat fulvic and humic acids extracted from two horizons of an ombrotrophic peat bog; Chemosphere 2006, 63, 82–88.
[87] Ashworth, D.J.; Alloway, B.J., Complexation of copper by sewage sludge-derived dissolved organic matter: effects on soil sorption behaviour and plant uptake; Water Air Soil Pollut. 2007, 182, 187–196.
[88] Evangelou, V.P.; Marsi, M., Stability of Ca^{2+}, Cd^{2+}, and Cu^{2+}-illite complexes; J. Environ. Sci. Health. A 2002, 37, 811–828.
[89] Sauvé, S.; McBride, M.B.; Norvell, W.A.; Hendershot, W.H., Copper solubility and speciation of in situ contaminated soils: effects of copper level, pH and organic matter; Water Air Soil Pollut. 1997, 100, 133–149.
[90] McBride, M.B., Cupric ion activity in peat soil as a toxicity indicator for maize; J. Environ. Qual. 2001, 30, 78–84.
[91] Courchesne, F.; Kruyts, N.; Legrand, P., Labile zinc concentration and free copper ion activity in the rhizosphere of forest soils; Environ. Toxicol. Chem. 2006, 25, 635–642.
[92] Cloutier-Hurteau, B.; Sauvé, S.; Courchesne, F., Comparing WHAM 6 and MINEQL+4.5 for the chemical speciation of Cu^{2+} in the rhizosphere of forest soils; Environ. Sci. Technol. 2007, 41, 8104–8110.
[93] Sauvé, S.; McBride, M.B.; Hendershot, W.H., Ion-selective electrode measurments of copper(II) activity in contaminated soils; Arch. Environ. Contam. Tox. 1995, 29, 373–379.
[94] Temminghoff, E.J.M.; Van der Zee, S.E.A.T.M.; de Haan, F.A.M., Copper mobility in a copper-contaminated sandy soil as affected by pH and solid and dissolved organic matter; Environ. Sci. Technol. 1997, 31, 1109–1115.
[95] Morley, G.F.; Gadd, G.M., Sorption of toxic metals by fungi and clay minerals; Mycol. Res. 1995, 99, 1429–1439.
[96] Christl, I.; Milne, C.J.; Kinniburgh, D.G.; Kretzschmar, R., Relating ion binding by fulvic and humic acids to chemical composition and molecular size. 2. Metal binding; Environ. Sci. Technol. 2001, 35, 2512–2517.
[97] Estives da Silva, J.C.G.; Oliveira, C.J.S., Metal ion complexation properties of fulvic acids extracted from composted sewage sludge as compared to a soil fulvic acid; Water Res. 2002, 36, 3404–3409.
[98] Kaschl, A.; Romheld, V.; Chen, Y., Cadmium binding by fractions of dissolved organic matter and humic substances from municipal solid waste compost; J. Environ. Qual. 2002, 31, 1885–1892.
[99] Kraal, P.; Jansen, B.; Nierop, K.G.J.; Verstraten, J., Copper complexation by tannic acid in aqueous solution; Chemosphere 2006, 65, 2193–2198.
[100] Cao, J.; Lam, K.C.; Dawson, R.W.; Liu, W.X.; Tao, S., The effect of pH, ion strength and reactant content on the complexation of Cu^{2+} by various natural organic ligands from water and soil in Hong Kong; Chemosphere 2004, 54, 507–514.
[101] Nolan, A.L.; McLaughlin, M.J.; Mason, S.D., Chemical speciation of Zn, Cd, Cu, and Pb in pore waters of agricultural and contaminated soils using Donnan dialysis; Environ. Sci. Technol. 2003, 37, 90–98.

[102] van der Stelt, B.; Temminghoff, E.J.M.; van Riemsdijk, W.H., Measurement of ion speciation in animal slurries using the Donnan Membrane Technique; Anal. Chim. Acta 2005, 552, 135–140.

[103] Van Laer, L.; Smolders, E.; Degryse, F.; Janssen, C.; De Schamphelaere, K., Speciation of nickel in surface waters measured with the Donnan membrane technique; Anal. Chim. Acta 2006, 578, 195–202.

[104] Weng, L.; Temminghoff, E.J.M.; Van Riemsdijk, W.H., Determination of the free ion concentration of trace metals in soil solution using a soil column Donnan membrane technique; Eur. J. Soil Sci. 2001, 52, 629–637.

[105] Cancès, B.; Ponthieu, M.; Castrec-Rouelle, M.; Aubry, E.; Benedetti, M.F., Metal ions speciation in a soil and its solution: experimental data and model results; Geoderma 2003, 113, 341–355.

[106] Kalis, E.J.; Weng, L.; Temminghoff, E.J.; van Riemsdijk, W.H., Measuring free metal ion concentrations in multicomponent solutions using the Donnan membrane technique; Anal Chem. 2007, 79, 1555–1563.

[107] Fitch, A.; Helmke, P.A., Donnan equilibrium graphite-furnace atomic-absorption estimates of soil extract complexation capacities; Anal. Chem. 1989, 61, 1295–1298.

[108] Temminghoff, E.J.M.; Plette, A.C.C.; Van Eck, R.; Van Riemsdijk, W.H., Determination of the chemical speciation of trace metals in aqueous systems by the Wageningen Donnan Membrane Technique; Anal. Chim. Acta 2000, 417, 149–157.

[109] Marang, L.; Reiller, P.; Pepe, M.; Benedetti, M.F., Donnan membrane approach: from equilibrium to dynamic speciation; Environ. Sci. Technol. 2006, 40, 5496–5501.

[110] Sigg, L.; Black, F.; Buffle, J. et al., Comparison of analytical techniques for dynamic trace metal speciation in natural freshwaters; Environ. Sci. Technol. 2006, 40, 1934–1941.

[111] Buffle, J.; Tercier-Waeber, M.L., Voltammetric environmental trace metal analysis and speciation: from laboratory to in situ measurements; Trends Anal. Chem. 2005, 24, 172–191.

[112] Larsen, J.; Svensmark, B., Labile species of Pb, Zn and Cd determined by anodic-stripping staircase voltammetry and their toxicity to Tetrahymena; Talanta 1991, 38, 981–988.

[113] Abu Zuhri, A.Z.; Voelter, W., Applications of adsorptive stripping voltammetry for the trace analysis of metals, pharmaceuticals and biomolecules; Fresen. J. Anal. Chem. 1998, 360, 1–9.

[114] Gunkel, P.; Fabre, B.; Prado, G.; Baliteau, J.Y., Ion chromatographic and voltammetric determination of heavy metals in soils. Comparison with atomic emission spectroscopy; Analysis 1999, 27, 823–828.

[115] Opydo, J., Cathodic adsorptive stripping voltammetry for estimation of the forest area pollution with nickel and cobalt; Microchim. Acta 2001, 137, 157–162.

[116] Town, R.M.; van Leeuwen, H.P., Significance of wave form parameters in stripping chronopotentiometric metal speciation analysis; J. Electroanal. Chem. 2002, 535, 11–25.

[117] Town, R.M.; van Leeuwen, H.P., Depletive stripping chronopotentiometry: A major step forward in electrochemical stripping techniques for metal ion speciation analysis; Electroanalysis 2004, 16, 458–471.

[118] Galceran, J.; Companys, E.; Puy, J.; Cecilia, J.; Garces, J.L., AGNES: a new electroanalytical technique for measuring free metal ion concentration; J. Electroanal. Chem. 2004, 566, 95–109.

[119] Wang, J., Electrochemical Preconcentration; in: Kissinger, P.T.; Heineman, W.R. (eds), Laboratory Techniques in Electroanalytical Chemistry; Marcel Dekker, New York, 1996, pp. 719–738.

[120] Andrews, P.; Town, R.M.; Hedley, M.J.; Loganathan, P., Measurment of plant-available cadium in New Zealand soils; Aust. J. Soil Res. 1996, 34, 441–452.

[121] Columbo, C. and van den Berg, C.M.G., Determination of trace metals (Cu, Pb, Zn and Ni) in soil extracts by flow analysis with voltammetric detection; Int. J. Environ. Anal. Chem. 1998, 71, 1–17.

[122] Nedeltcheva, T.; Atanassova, M.; Dimitrov, J.; Stanislavova, L., Determination of mobile form contents of Zn, Cd, Pb and Cu in soil extracts by combined stripping voltammetry; Anal. Chim. Acta 2005, 528, 143–148.

[123] Cooper, J.; Bolbot, J.A.; Saini, S.; Setford, S.J., Electrochemical method for the rapid on site screening of cadmium and lead in soil and water samples; Water Air Soil Pollut. 2007, 179, 183–195.

[124] Christidis, K.; Robertson, P.; Gow, K.; Pollard, P., Voltammetric in situ measurements of heavy metals in soil using a portable electrochemical instrument; Measurement 2007, 40, 960–967.
[125] Martinez, C.E.; Jacobson, A.R.; McBride, M., Aging and temperature effects on DOC and elemental release from a metal contaminated soil; Environ. Pollut. 2003, 122, 135–143.
[126] Donat, J.R.; Lao, K.A.; Bruland, K.W., Speciation of dissolved copper and nickel in San Franscisco Bay: a multi-method approach; Anal. Chim. Acta 1994, 284, 547–571.
[127] Aoki, K., Theory of ultramicroelectrodes; Electroanalysis 1993, 5, 627–639.
[128] Stulik, K.; Amatore, C.; Holub, K.; Marecek, V.; Kutner, W., Microelectrodes. Definitions, characterization, and applications; Pure Appl. Chem. 2000, 72, 1483–1492.
[129] Xie, X.; Stueben, D.; Berner, Z., The application of microelectrodes for the measurement of trace metals in water; Anal. Lett. 2005, 38, 2281–2300.
[130] Buffle, J.; Tercier, M.L.; Parthasarathy, N.; Wilkinson, K.J., Analytical techniques for the in situ measurement and speciation of trace compounds in natural waters; Chimia 1997, 51, 690–693.
[131] Buffle, J.; Wilkinson, K.J.; Tercier, M.L.; Parthasarathy, N., In situ monitoring and speciation of trace metals in natural waters; Ann. Chim.-Rome 1997, 87, 67–82.
[132] Tercier, M.L.; Buffle, J., Antifouling membrane-covered voltammetric microsensor for in situ measurements in natural waters; Anal. Chem. 1996, 68, 3670–3678..
[133] Noel, S.; Tercier-Waeber, M.L.; Lin, L.; Buffle, J.; Guenat, O.; Koudelka-Hep, M., Integrated microanalytical system for simultaneous voltammetric measurements of free metal ion concentrations in natural waters; Electroanalysis 2006, 18, 2061–2069.
[134] Salaun, P.; Buffle, J., Integrated microanalytical system coupling permeation liquid membrane and voltammetry for trace metal speciation. Theory and applications; Anal. Chem. 2004, 76, 31–39.
[135] Belmont, C.; Tercier, M.L.; Buffle, J.; Fiaccabrino, G.C.; Koudelka-Hep, M., Mercury-plated irridum-based microelectrode arrays for trace metals detection by voltammetry: optimum conditions and reliability; Anal. Chim. Acta 1996, 329, 203–214.
[136] Feeney, R.; Herdan, J.; Nolan, M.A.; Tan, S.H.; Tarasov, V.V.; Kounaves, S.P., Analytical characterization of microlithographically fabricated iridium-based ultramicroelectrode arrays; Electroanalysis 1998, 10, 89–93.
[137] Silva, R.P.M.; Khakani, M.A.E.L.; Chaker, M. et al., Development of Hg-electroplated-iridium based microelectrode arrays for heavy metal traces analysis; Anal. Chim. Acta 1999, 385, 249–255.
[138] Feeney, R.; Kounaves, S.P. Voltammetric measurement of arsenic in natural waters; Talanta 2002, 58, 23–31.
[139] Daniele, S.; Ciani, I.; Baldo, M.A.; Bragato, C., Application of sphere cap mercury microelectrodes and scanning electrochemical microscopy (SECM) for heavy metal monitoring at solid/solution interfaces; Electroanalysis 2007, 19, 2067–2076.
[140] Silva, P.R.M.; El Khakani, M.A.; Chaker, M.; Dufresne, A.; Courchesne, F., Simultaneous determination of Cd, Pb, and Cu metal trace concentrations in water certified samples and soil extracts by means of Hg-electroplated-Ir microelectrode array based sensors; Sensor Actuat. B-Chem. 2001, 76, 250–257.
[141] Berduque, A.; Lanyon, Y.H.; Beni, V. et al., Voltammetric characterisation of silicon-based microelectrode arrays and their application to mercury-free stripping voltammetry of copper ions; Talanta 2007, 71, 1022–1030.
[142] Dudal, Y.; Gerard, F., Accounting for natural organic matter in aqueous chemical equilibrium models: a review of the theories and applications; Earth-Sci. Rev. 2004, 66, 199–216.
[143] Goldberg, S.; Lesch, S.M.; Suarez, D.L., Predicting molybdenum adsorption by soils using soil chemical parameters in the constant capacitance model; Soil Sci. Soc. Am. J. 2002, 66, 1836–1842.
[144] Goldberg, S., Modeling boron adsorption isotherms and envelopes using the constant capacitance model; Vadose Zone J. 2004, 3, 676–680.
[145] Goldberg, S.; Lesch, S.M.; Suarez, D.L.; Basta, N.T., Predicting arsenate adsorption by soils using soil chemical parameters in the constant capacitance model; Soil Sci. Soc. Am. J. 2005, 69, 1389–1398.

[146] Carrillo-González, R.; Jirka Simunek, J.; Sauvé, S.; Adriano, D., Mechanisms and pathways of trace element mobility in soils; Adv. Agron. 2006, 91, 111–178.
[147] Meers, E.; Unamuno, V.R.; Du Laing, G. *et al.*, Zn in the soil solution of unpolluted and polluted soils as affected by soil characteristics; Geoderma 2006, 136, 107–119.
[148] Weng, L.P.; Temminghoff, E.J.M.; Lofts, S.; Tipping, E.; Van Riemsdijk, W.H., Complexation with dissolved organic matter and solubility control of heavy metals in a sandy soil; Environ. Sci. Technol. 2002, 36, 4804–4810.
[149] Ge, Y.; Sauve, S.; Hendershot, W.H., Equilibrium speciation of cadmium, copper, and lead in soil solutions; Comm. Soil Sci. Plant Anal. 2005, 36, 1537–1556.
[150] Ge, Y.; MacDonald, D.; Sauve, S.; Hendershot, W., Modeling of Cd and Pb speciation in soil solutions by WinHumicV and NICA-Donnan model; Environ. Modell. Softw. 2005, 20, 353–359.

Section 2

Long-Term Issues, Impacts and Predictive Modelling

6

Trace Elements in Biosolids-Amended Soils

Weiping Chen[1], Andrew C. Chang[2], Laosheng Wu[2], Albert L. Page[2] and Bonjun Koo[3]

[1] State Key Lab of Urban and Regional Ecology, Research Center for Eco-Environmental Sciences, Chinese Academy of Sciences, Beijing, China
[2] Department of Environmental Science, University of California, Riverside, California, USA
[3] Department of Natural and Mathematical Sciences, California Baptist University, California, USA

6.1 Introduction

Sewage sludge is an inevitable by-product of wastewater treatment processes. Being the concentrate of impurities removed from the spent water, it contains compounds of agricultural value that mainly include nitrogen, phosphorus, and organic matter, and potential harmful substances that include trace elements (for example, Cd and Pb), organic chemicals (for example, pesticides, polychlorinated biphenyls (PCBs)) and pathogens. In the urban settings and from the environmental sanitation perspective, sewage sludge is a 'waste'. Its final disposal, however, has been problematic as it lacks appropriate outlets and there is not an entirely satisfactory resolution. Land application is by far the most common approach employed by treatment works around the world.

'Biosolids', in the United States, refers to municipal sewage sludge that has been treated to meet product quality standards for environmentally safe land applications. The term biosolids is employed to emphasize the beneficial nature of the products. The United States Environmental Protection Agency (USEPA) defines biosolids as 'the nutrient-rich organic materials resulting from the treatment of domestic sewage in a treatment facility' that may be recycled and applied as fertilizers to improve and maintain productivity of soils and stimulate plant growth [1].

Given the time-tested experiences of using human excrement, sewage, and animal manure on croplands worldwide, applying biosolids to agricultural lands was a logical extension. According to a US National Research Council report [2], approximately 60 % of the biosolids generated in the nation was land applied in 2002. With proper treatment, biosolids had been applied on croplands, forests, and reclamation sites, parks, golf courses, lawns, and home gardens with negligible adverse impacts to the exposed populations [1,3,4]. The USEPA expected that the use of biosolids would increase to 70 % in 2010 [5]. In a survey of its production, use, and disposal, the United Kingdom reported that nearly 45 % of the nation's sewage sludge was spread on approximately 0.3 % of the agricultural land [6]. The situations in other parts of the world are expected to be similar [7,8].

Biosolids when land applied provides soils with organic matter and offers the opportunity of recycling plant nutrients that improve their fertility and physicochemical properties [6–13]. On the average, biosolids contains 3.2 % N, 2.3 % P, and 0.3 % K [14]. Biosolids however may contain disease-causing organisms and/or chemical pollutants that are potentially hazardous in land applications. In the soil, pathogens rapidly die off and do not present long-lasting detrimental environmental effects. Chemical pollutants such as trace elements, however, may persist in receiving soils and be absorbed in sufficient quantities by growing plants to adversely affect the health of consumers and/or growing plants. Long-term, repeated land applications of biosolids may result in significant build-up of trace elements in cropland soils [15]. The enrichments in the soils in turn may lead to loss in soil fertility and to accelerate transfer of trace elements through the food chain.

A conservative approach to dealing with the uncertainties of the potential harm is to minimize accumulation of potentially toxic trace elements in the receiving soils by matching the inputs from biosolids application to the outputs through crop removal, soil erosion, and leaching. When the input equals the output, trace element concentrations of the receiving soils remain at the background levels [16]. Sweden, Norway, the Netherlands, and Denmark all have adopted standards based on this principle. The EU sewage sludge disposal initiative as outlined in the *Working Document on Sludge* (3rd draft, Brussels, 2000, http://europa.eu.int/comm/environment/sudge/report10.htm) embraced a similarly restrictive approach that limited the trace element loading comparably to the no observed adverse effect level applicable to the most sensitive soil biota. In the spirit of resource conservation and recycling, the land application regulations of the US maximized the pollutant attenuation capacities of the receiving soils and permitted the pollutants (trace elements) to accumulate in the receiving soils to the maximum levels that considered safe to the exposed populations, that is, the growing plants, foraging animals, and human consumers [1]. Because of the differences in the underlying regulatory philosophy and goal, the resulting numerical limits for the trace elements were significantly different (see Chapter 13). There have been calls for caution since the promulgation of the regulations [17–20].

Are USEPA regulations protective over the long term with respect to accumulation of potentially toxic trace elements from cropland application of biosolids? Conversely, is the approach adopted in the EU sewage sludge disposal initiatives too restrictive to limit the beneficial use of biosolids as fertilizer supplements on croplands? It is imperative that the environmental behavior of trace elements in biosolids-amended soils be objectively evaluated.

This chapter focuses on the concepts and approaches of assessing long-term behavior of trace elements in biosolids-amended soils. A root exudates-based model to assess long-term availability of biosolids-borne trace elements in receiving soils is elaborated.

6.2 Biosolids-Borne Trace Elements in Soils

Biosolids contains a wide range of trace elements. Biosolids from industrial catchments generally have higher trace elements contents than those from mainly suburban domestic areas. In the US part 503 Rule, 10 elements (that is, As, Cd, Cr, Cu, Hg, Mo, Ni, Pb, Se, and Zn) are regulated. The elements of primary concern are Cd, Zn, Cu, Pb and Ni, which, when applied to soils in excessive amounts, may depress plant yields or degrade the quality of food or fiber produced [21].

6.2.1 Land Application and Trace Element Loading

Trace elements such as Cd, Cr, Cu, Pb, Ni, and Zn may accumulate in the receiving soils following the land application of biosolids. The extent and the rate of accumulation are dependent on their concentrations in the biosolids, frequency of applications, and the mass loading. Kabata-Pendias and Pendias [22] reported that the Cd concentration of biosolids that were cropland applied varied from less than 2 mg kg^{-1} to as high as 1500 mg kg^{-1}. The quality of biosolids has improved as the community wide industrial wastewater pretreatment processes were implemented [18,23–24]. Chang et al. [8] showed that for biosolids in the United States the concentration of trace elements (for example, Cd, Cr, Cu, Pb, Ni, and Zn) from point source discharges steadily decreased from 1976 through 1996 while the concentration of trace elements from non-point source discharges (for example, As, Hg, Mo, and Se) remained unchanged over the same time period (Table 6.1).

It is generally agreed that biosolids should be applied on land in accordance with the nitrogen (N) and/or phosphorus (P) requirements of the growing plants. The trace element inputs through land application could be substantial, even when the biosolids are applied in

Table 6.1 Trends in trace element concentrations (mg kg^{-1}) of biosolids produced by wastewater treatment plants in the United States

Year[a]	As	Cd	Cr	Cu	Pb	Hg	Mo	Ni	Se	Zn
1976	—	110	2620	1210	1360	—	—	320	—	2790
1979	7	69	429	602	369	3	18	135	7	1594
1987	12	26	430	711	308	3	109	167	6	1540
1988	10	7	119	741	134	5	9	43	5	1202
1996	12	6	103	506	111	2	15	57	6	830

[a]Note: the data were based on average from 150, 40, 364, 200, and >200 treatment plants in United States.
Reproduced with permission from A.C. Chang, G. Pan, A.L. Page and T. Asano, Developing human health-related chemical guidelines for reclaimed water and sewage sludge applications in agriculture, World Health Organization, Geneva, Switzerland, 2002 [8].

accordance with agronomic rates. Assuming the biosolids contain 2 % of N, an annual application of 10 Mg ha^{-1} provides 200 kg N ha^{-1}, which meets the requirement of many crops. Based on most current data in Table 6.1, the annual loading of Cd in this case will be 60 g ha^{-1}. For a 20-year continuous application, the cumulative loading of Cd is 1200 g ha^{-1}. It would increase the Cd concentration of the plow layer (top 20 cm) by an estimated 0.4 mg kg^{-1}. The estimated accumulation of other trace elements may be obtained in the same manner. However, total metal concentration does not furnish sufficient information regarding the potential availability of elements for plant uptake.

6.2.2 Trace Element Availability in Biosolids-Amended Soils – A Time Bomb?

The environmental fate and phytoavailability of trace elements in the soils are dependent upon the dynamic equilibrium between those present in solid phase and solution phase through sorption and desorption, precipitation and dissolution, oxidation and reduction, plant absorption, and organic matter mineralization [25–30]. The dynamic and interactive processes of trace element transformations are best demonstrated by a mass balance model [31,32] and the local specific conditions such as total element concentration, soil pH, soil organic matter, soil texture, and concentrations of the competing elements could influence the outcomes. Chang et al. [33] concluded that soil texture did not appear to influence sorption of trace elements as two soils, the Domino loam and Greenfield sandy loam, under comparable land application conditions exhibited the identical trace element accumulation characteristics. Sauerbeck and Hein [34] reported that trace element concentration in soils and pH were more important than organic matter and texture in regulating the availability of Cd, Cu, Ni, and Zn to several plant species. Loganathan et al. [35] concluded that if factors such as topsoil pH and organic matter content were equal, then the amount of Cd taken up by plants was proportional to the amount accumulated in the topsoil. Hooda et al. [36] noted that total element concentration, pH, and clay fraction of the soil affected trace element concentrations in plants. Plants that were grown at neutral soil pH and fine-textured soil tended to accumulate fewer trace elements.

There appeared to be element-specific chemical barriers that limited the plant transfer of selected potentially toxic elements from soil to plants. Chromium and lead exhibit a low potential for plant absorption compared with Cd, Cu, Ni, and Zn under the same conditions. These two elements do not usually translocate in significant amounts to the above-ground tissues of leaves, fruit or seed even if they are accumulated at the root–soil interface [20]. The potentials of trace elements for plant absorption in ascending order are Pb, Cr, and Hg < Cu < Ni, Zn, and Cd < Mo and Tl [20].

In land applications, the nature of biosolids might also affect the availability of biosolids-borne trace elements in receiving soils. Chen et al. [37] illustrated that the biosolids dependent on the treatment exhibited distinctively different particle morphology and size distributions. When biosolids were size classified and applied at the same mass loading, the plant absorptions of Cd, Cu, Ni, and Zn were inversely proportion to the particle-size of the applied material while the plant absorption of Cr and Pb was not affected by the particle size. For the same trace element in different biosolids, plants absorbed greater amounts of Cd, Cu, Ni, and Zn from the biosolids that exhibited porous than those with blocky solid morphology. The root–biosolids interfacial contact appeared to influence the trace element uptake. Again, the particle morphology did not significantly affect the plant uptake of Cr and Pb [37]. The inherent chemical nature of the metal solid phases appeared to govern the dissolution of Cr and Pb.

The chemical forms of trace elements in biosolids and thus their availability for plant absorption might change prior to the soil incorporation due to the dewatering and composting processes [38–40]. It had been postulated that during the course of wastewater treatment and sewage sludge processing, the biosolids-borne trace elements formed complexes with the organic solids, limiting their availability for plant absorption upon land application [17]. When the organic matter decomposed over time, the organically bound trace elements would become readily available and cause an environmental calamity – thus the advent of the 'time bomb' hypothesis [17,41,42]. Inorganic minerals such as Fe and Al oxides, silicates, phosphates, and carbonates may constitute 30–60 % of the finished biosolids. Trace elements in composted biosolids were found to be associated primarily with fine and dense inorganic fractions [43,44]. Li *et al.* [45] and Hettiarachchi *et al.* [46] demonstrated that both organic carbon and hydrous oxides of Fe, Mn, and Al could significantly contribute toward the increased trace element sorption capacity in biosolids-amended soils. Upon land application, the biosolids-borne trace element-organic matter complexes may be further stabilized when they react with the inorganic minerals in the receiving soils [47]. Brown *et al.* [48] found that the loss of organic matter in biosolids-amended soils did not increase the uptake of Cd and suggested that organic carbon loss was not the primary factor limiting Cd phytoavailability. They concluded that inorganic mineral components played a more significant and persistent role in limiting phytoavailability than organic matter. Additional reports showed that the phytoavailability of trace elements in biosolids-amended soils did not increase for long after the land application ceased [44,49]. In fact, the highest uptake by plants most likely took place during the period immediately following the termination [49–52].

Wiseman and Zilbilske [53] reported that the biosolids-borne organic matter rapidly mineralized within days following its soil incorporation. The organic matter decomposition greatly decreased once the land application was discontinued [49,54] and in some cases plant adsorption of trace elements decreased as the organic matter decomposed over time to more stable chemical forms [49,55]. McGrath [56] observed that the DTPA (diethylenetriamine pentaacetic acid)-extractable trace elements of the soils decreased with time following cessation of the biosolids application. Parkapin *et al.* [57] reported that biosolids-borne trace elements existed in readily mobile forms, namely, the soluble and exchangeable forms, when they were first incorporated into the soil and reverted to less-mobile pools after the biosolids–soil mixtures were incubated for 12 weeks in the laboratory setting. McGrath *et al.* [54] showed that 24 years after termination of biosolids application, the soil organic matter decreased by up to 40 % but plant available Cd in the soil remained unchanged over this period of time. Following the rapid decomposition of the accumulated organic matter in the initial years, it might take the remaining recalcitrant fraction hundreds of years to slowly degrade and the organic matter content of the soil to revert to the baseline level. Hyun *et al.* [44] found no evidence that soluble Cd concentration or phytoavailability of Cd in soils previously amended with biosolids increased because the organic carbon content of the soils decreased over time. Frost and Ketchum [58] reported that tissue concentrations of trace elements of wheat grown on soils amended with heavy loads of trace element laden biosolids that was stockpiled for over 20 years were comparable to those of wheat grown on soils receiving simply the commercial fertilizers.

In summary, the trace element phytoavailability in biosolids-amended soils is difficult to generalize since it is strongly dependent on the nature of the element, soil properties, properties of the biosolids, and types of crop. Current researches tend to point out that phytoavailability of trace elements in biosolids-amended soils is low due to the specific trace element adsorption capacity added with biosolids and this capacity will persist following termination of biosolids application. However, due to a lack of accurate and consistent methods of assessing the phytoavailability of trace elements (see Chapter 11), the often observed low phytoavailability does not guarantee the safety of the food chain.

6.2.3 Plant Response to Trace Elements in Biosolids-Amended Soils – Is There a Plateau?

The mobility and plant absorption of biosolids-borne trace elements appeared limited in soils. Years after the introduction, the added trace elements remained in the near surface soil horizons [59]. The plant absorption of biosolids-borne trace elements would be influenced by the amounts accumulated in the receiving media. Street *et al.* [60] showed that the Cd concentrations of corn seedlings gradually rose and approached a 'plateau' as the concentrations of Cd in biosolids-amended soil rose. Dowdy *et al.* [61] reported that the Cu and Zn concentrations of soybean leaf tissue continued to increase with amounts of biosolids applied up to a point and then gradually levelled off, after which the uptake no longer further responded to continued applications. Logan *et al.* [62] observed that the uptake of Cd, Cu, and Zn by corn grown on biosolids-amended soils would gradually level off with respect to the inputs, yet uptake of Cd, Cu, and Zn in lettuce growing under the same ranges of biosolids treatments rose linearly in proportion to those present in the soils. Chang et al [63] demonstrated that among the 12 species of plants grown on biosolids-amended soils with Cd contents ranging from the background level of 0.25 mg kg^{-1} to 20 mg kg^{-1}, the majority of the species tested exhibited the Cd 'plateau' in plant tissue that asymptotically approached the upper limits as the soil Cd concentrations increased. Hamon *et al.* [64] found Cd and Zn in the plants (*Raphanus sativus* L.) grown in a soil historically amended at different rates of biosolids displayed a 'plateau' response. There are other examples in which 'plateaus' in trace element uptake have been described [65–67].

The 'plateau' in this case referred to the phenomenon that plants' absorption of trace elements from biosolids-amended soils might be restricted by factors inherent to biosolids, soils, or plant species. Ryan and Chaney [68] postulated that the biosolids were intrinsically rich in organic matter and amorphous oxides and hydroxides of iron and manganese that possess high capacities to react with the trace elements during the course of wastewater treatment, rendering the biosolids-borne trace elements less likely to dissolve into soil solution phase. They reasoned that phytoavailability of elements in the biosolids-amended soils would rise with the inputs until the threshold governed by biosolids and/or soil was reached, hence the trace element plant uptake 'plateau' (see Figure 6.1).

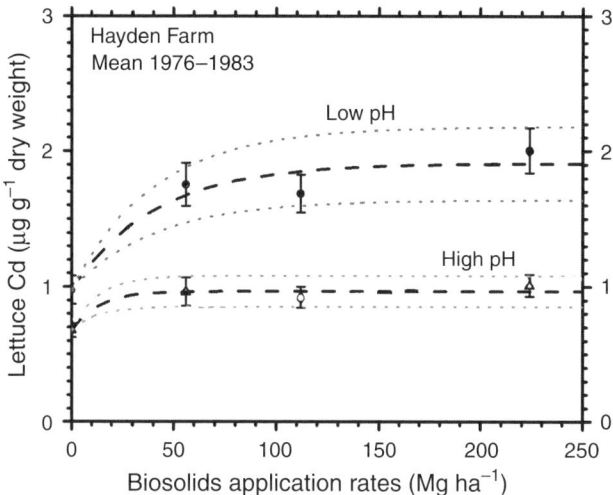

Figure 6.1 Example showing the trace element plant uptake plateau in response to different biosolids application rates, at low and high soil pH (based on data provided by Dr. Rufus L. Chaney at USDA). The data were fitted to plateau regression, the lines were for ±95 % confidence interval and mean, respectively

Hypothetically, the trace element concentrations of plants grown in biosolids-amended soils might exhibit different outcomes if the trace element loading rate was not restrictive [64]. There could be a barrier through which the trace element uptake by plants is limited by the chemical characteristics of soils and/or biosolids. In this regard, uptake of trace elements by a plant species would be held at a threshold even when total trace element contents of the biosolids-amended soils reached extraordinary high levels. The physiological mechanisms of plants could regulate trace element uptake as well thus blocking the uptake or translocation of the trace element to the shoot tissue [69]. Plant mechanisms that can account for the observed decrease in uptake at high trace element loadings include exclusion of trace elements, limited transport from root to shoot, saturation of the carrier system responsible for element transport into root globes and root avoidance [17,70]. Either process could result in the plateaus as described by Chaney and Ryan (Figure 6.1).

The plant tissue concentration plateau for plant grown in biosolids-amended soils might be temporary as the phenomenon could be caused by short-term increases in soil pH, growth dilution, and ion competition effects due to excess application of macronutrients, or root toxicity effects that limit the crop's ability to take up trace elements [20]. Alternatively, the trace element uptake by roots extended beyond the trace element-laden surface soil layer would be significantly reduced, thus offsetting for the enhanced uptake by roots remaining in the zone of biosolids incorporation. The trace element concentration in the plant tissue of short-season and shallow-root species would be expected to rise continuously as their roots are not likely to extend beyond the influence of biosolids. McGrath *et al.* [54] showed that uptake of Cd and Zn by beets, carrots, and barley increased in proportion to the total soil Cd and Zn up to 20 and 600 mg kg^{-1}, respectively, 20-plus years after the last biosolids application. Based on 10 years of data on plant uptake of Cd, Chang *et al.* [50] reported that neither a 'time bomb' nor a 'plateau' was imminent in previously biosolids-amended soils even when the biosolids mass loading reached 2880 Mg ha^{-1} over a 16-year period.

Regardless of how the plants responded to the trace elements in the biosolids-receiving soils, there is no definitive answer about the long-term phytoavailability of biosolids-borne trace elements.

6.3 Assessing Availability of Trace Elements in Biosolids-Amended Soils

The phytoavailability of trace elements in soils may be assessed, by determining the amounts of their available forms in the soil, the source assessment, and/or determining the amounts absorbed by the growing plants, the end measurement.

6.3.1 Source Assessment

In soils, trace elements are present in both the solid phase and the aqueous (that is, soil solution) phase. In solution, trace elements can exist either as free ions or as various complexes with organic and inorganic ligands. In the solid phase, trace element ions can either be retained on organic and inorganic soil components by various sorption processes (for example, ion exchange or surface complexation), be coprecipitated with other minerals in the soil (for example, carbonates, oxides, phosphates, sulfides), or exist as ions in crystal lattices of primary and secondary minerals. Ions in solution generally are more available for plant uptake and transport; however, trace elements associated with a number of physicochemical forms may become available if environmental conditions change. The degree of trace element association with these soil components is strongly dependent upon soil pH, the redox conditions prevailing, extent of organic matter degradation, and grain size distribution (see Chapter 2).

Trace elements present in biosolids-amended soils may be partitioned through sequential extractions employing incrementally stronger chemical reagents into various fractions (see Chapter 4). Conceptually, as the strength of the sequential extractant increases, the trace element reactivity and mobility in soils decreases and therefore potential availability to plants also decreases [71,72].

Conventionally, chemical reagents that extract the soil-borne trace elements have been employed to assess the potentially available trace element pool of the soil (see Chapter 11). However, the phytoavailability is related not only to the chemical forms and solubility of the trace element but also to a series of reactions, specifically, the rate at which other trace element pools in the soil system can replenish the available pool of the trace element. Ideally, the soil extractant used to indicate the availability of the trace element would account for contributions of all applicable physical, chemical, and biological processes in the soil related to the uptake of trace elements. Actually, no static extraction process can simulate these dynamic processes. Thus, the use of trace element extractants to evaluate the soil's phytoavailability potential is derived from observations on the amounts extracted and the corresponding plant responses (the amount of uptake).

In source assessment, the plant available elements (pool) of the soils would be obtained from one-time extraction by a reagent consisting of weak acids, neutral electrolytes, and chelating agents [73–80]. Marchi *et al.* [81] compared the amount of elements extracted from biosolids-amended soils by four organic acid mixtures that mimic the compositions of organic acids in the rhizosphere and three reagents commonly used for trace element extraction, namely DTPA-TEA [73], Mehlich-I [77], and ammonium acetate [82]. They showed that the one-time extraction did not remove all of the extractable elements in the soils. The total extractable trace elements (pool) might be obtained by fitting the cumulative element removal through successive extractions to an exponential decay model [83]. The rhizosphere-based organic acid mixtures extracted more trace elements than the commonly used chemical extracting agents. In reality, the static extractions could not simulate trace element release through the dynamic processes taking place over a growing season or, if not exhausted in one growing season, over multiple seasons.

6.3.2 End Measurement

If the plant growth is not affected by the presence of trace elements, the amounts accumulated in the end products, that is, the plant tissues, would be an indication of the available trace element in the growing medium, that is, soils. The availability is represented simply by the amounts of trace elements the plant absorbed or the concentrations in the harvested plant tissues. Greater amounts in the plant tissues denote greater availability of the soil trace elements. The trace elements in plants vary according to the species and cultivar, the part of the plant sampled as well as its stage of development, the position of the tissue on fruiting versus nonfruiting branches, and the substrate. For example, Sauerbeck and Hein [84] found that Ni uptake in 13 crops depended on plant species and organs. Nickel concentrations were higher in grain and storage organs than in vegetative plant parts. Barley accumulated low amounts of Ni, lettuce was a medium accumulator, and radish absorbed high amounts of Ni. The seasonal and cultivar effects on grain Cd content in wheat grown on a sandy loam soil were reported by Chaudri *et al.* [85]. While these authors [84,85] indicated the extents of trace element uptake by the growing plants, none was able to probe the pool of potentially available trace elements in the soils. The trace element concentration of plant tissue would be better suited to diagnosing deficiencies and phytotoxicities of trace elements than to assessing the potential plant available trace elements of the soils.

Under the same circumstances, the trace elements in the plant tissue (end products) are expected be related to those present in the soils (source). The statistically significant linear regressions between trace elements extracted and trace element concentrations of the plant tissue were frequently used to predict what would be expected in the end products based on what was present in the source and vice versa [49,54,71,86–88]. However, the empirical relationships were largely case specific and not transferable to other situations. As a dependent variable, the trace element concentration of plant tissue is not a mathematically valid parameter for estimating the trace element content of the soil through the same regression. While the outcomes of the trace element extractions might be adequate to diagnose whether the soil is deficient in essential micronutrients or has trace element levels potentially harmful to plant growth, they failed to produce the interrelationship that predicts the amounts of trace elements available for plant absorption and how rapidly the growing plants absorb them.

6.4 Long-Term Availability Pool Assessment through a Root Exudates-Based Model

6.4.1 Rationale for Root Exudate-Based Trace Element Phytoavailability

The phytoavailability of an element in soil customarily is defined by the amount of element absorbed by growing plants or by concentration in the harvested plant tissue. Plants grown on biosolids-amended soils, typically absorb less than 1 % of the trace elements present in the soil. It would take a long time to deplete the biosolids-borne trace elements that are available for plant absorption in biosolids-amended soils. Therefore, phytoavailability of biosolids-borne trace elements must account for not only the plant uptake in one growing season, but also the total amounts available over time, that is, their total plant available pool. In this regard, neither the soil-based source assessment nor the plant-based end product measurements is able to delineate the true nature of the potential plant available trace element pool of a soil. To assess the availability of trace elements to plants in the biosolids-receiving soils, the rate at which the biosolids-borne trace element is absorbed by plant and total amounts of plant available trace elements in biosolids-amended soils must be addressed. The former describes the plant uptake for a given time period such as one growing season. The latter defines the total available trace element pool of the receiving soil. Together, they track the time dependent changes of plant available trace element pool and uptake of plants over time.

In land application of biosolids, the majority of the introduced trace elements are associated with the solid phase [68,89]. These biosolids-borne trace elements tend to be discrete particulates intermingled with masses of soil aggregates. The trace elements in solution phase are likely present in the vicinity of trace element-containing particulates and their transport through soils to the plant root would be short-range mass transfers. Therefore, plants will not absorb these trace elements unless the root is near the proximity of the trace element containing-particles. Those trace elements in particulate forms may be mobilized by the root exudates and their derivatives, such as acetic, citric, oxalic, fumaric, tartaric, and succinic acids [90–93]. Researchers had found that these low-molecular-weight organic acids which were often found in root exudates could increase availability of trace elements. Mench and Martin [94] found that soil Cd, Cu, Fe, Mn, Ni, and Zn were solubilized by root exudates of *Nicotiana tabacum* L., *N. rustica* L., and *Zea mays* L. Veeken and Hamelers [95] showed that the pH of the rhizosphere soils was relatively consistent between pH 3 and 5 regardless of the pH in the bulk soils, which may vary from less than 4 to greater than 8. Under the circumstances, the organic acids found in the rhizosphere are capable of forming complexes with trace element ions in solution [96–98], thus enhancing their mobilization and uptake by plants [99,100].

Koo *et al.* [101] studied the production and composition of organic acids in the rhizosphere as influenced by different biosolids amendments, plant species and varieties, and length of growing period. The results showed the following.

- When plants were not present, only acetic, lactic, butyric, and propionic acids were recovered from the cultivating media. When the plants were present, lactic, acetic propionic, butyric, glutaric, maleic, valeric, succinic, tartaric, oxalic, and citric acids were found in the rhizosphere of the cultivating media. Among them, lactic, acetic, butyric, and oxalic acids were always present and were invariably the most abundant,

accounting for 60–80 % of the total organic acids recovered. Other organic acids were also found some of the time but they were in considerably lower quantities.
- The volumes of organic acids recovered were enhanced by the biosolids treatment in both planted and unplanted cultivating media.
- The organic acids in the rhizosphere, while subject to biodegradation were constantly replenished through the microbial metabolism. During the growing season, the compositions dynamically changed depending on the availability of the substrate (that is, C source for microbial activity) and the environmental conditions (for example, soil type and pH). The total amounts of organic acids recovered did not change significantly due to the stage of plant growth, plant species and varieties, or origin of biosolids.

The dissolution by the root exudates may be a significant pathway through which plants take up trace elements from biosolids-amended soils. Therefore, in the root exudates-based long-term availability model proposed by Chang et al [102], it is hypothesized that the plant available trace elements in biosolids-amended soils may be determined as amounts of trace elements dissolved by organic acids in the rhizosphere. Typically, the rhizosphere consists of the 1- to 5-mm layer of soils surrounding the root cylinder. As such, in a given growing season all of the trace element-containing particles in the soil may not be under the influence of organic acids in the rhizosphere. Over time, however, all of the trace element-containing particles in the growth medium will be exposed to the influence of rhizosphere. The trace element uptake by plants, which determines the plant tissue concentration and phytotoxicity, is determined by the kinetics of trace element released into solution by organic acids in the rhizosphere. In this manner, the availability of biosolids-borne trace elements to plants may be defined in terms of a capacity factor (that is, trace elements solubilized by root exudates) which defines 'how much' of the trace elements are available and an intensity factor that defines the rate at which trace elements are solubilized.

Assuming the plant uptake of trace elements from biosolids-amended soils follows exponential decay kinetics, the long-term availability of biosolids-borne trace elements in the soil can be expressed as:

$$M_t = C \times [1 - e^{-kt}] \quad (6.1)$$

where M_t (mg kg^{-1}) is the cumulative trace element removal from the soils by growing and harvesting plants for t years, C is the capacity factor representing the plant available trace element pool (mg kg^{-1}) which is determined as the total rhizosphere organic acids-soluble trace elements, k is a function of the trace element dissolution rate (y^{-1}) of the soil.

The total amount of plant available trace elements in biosolids-amended soils, C in Equation (6.1), can be estimated by extracting the soils using a mixture of organic acids that mimic the composition and strength of organic acids found in the rhizosphere of plants grown on biosolids-amended medium and has the chemical matrix and ionic strength of the soil solution. Chang et al. [102] formulated a synthetic root exudates solution based on the organic acid composition outlined in Table 6.2. In laboratory extraction, it is necessary that all of the synthetic root exudates soluble trace elements in the biosolids-amended soils must be accounted for, thus the soils should be extracted successively by the synthetic root exudates. Figure 6.2 illustrates the cumulative trace element extracted from a biosolids-amended soil with the cumulative volume of synthetic root exudates used. The cumulative

Table 6.2 Synthetic organic acid mixtures based on composition of organic acids found in the rhizosphere of corn

Organic acid	Molecular weight	Formula	Mole fraction (COO$^-$)	Number of COOH functional groups
Acetic	60.05	$C_2H_4O_2$	0.287	1
Butyric	88.11	$C_4H_8O_2$	0.209	1
Glutaric	132.12	$C_5H_8O_4$	0.004	2
Lactic	90.08	$C_3H_6O_3$	0.366	1
Maleic	116.07	$C_4H_4O_4$	0.042	2
Oxalic	90.04	$C_2H_2O_4$	0.043	2
Propionic	74.08	$C_3H_6O_2$	0.010	1
Pyruvic	88.06	$C_3H_4O_3$	0.0004	1
Succinic	118.09	$C_4H_6O_4$	0.006	2
Tartaric	150.09	$C_4H_6O_6$	0.032	2
Valeric	102.13	$C_5H_{10}O_2$	0.001	1

Data based on A.C. Chang, D.E. Crowley and A.L. Page, Assessing bioavailability of metals in biosolids-treated soils, WERF Report 97-REM-5, IWA, 2004 [102].

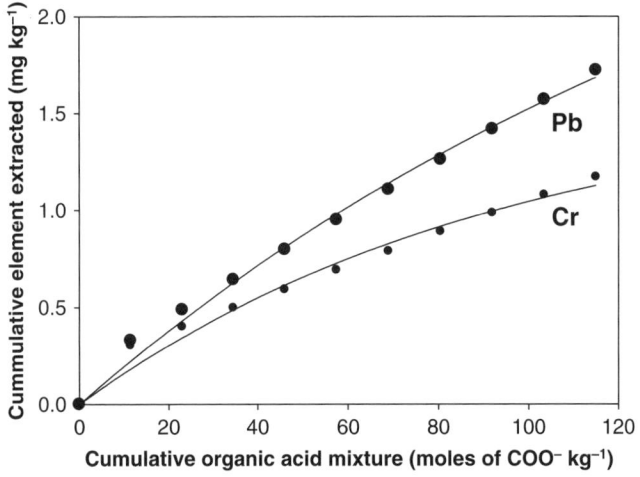

Figure 6.2 Cumulative Cr and Pb extracted from the long-term biosolids-amended soil by the successive extraction, using synthetic organic acid mixture (Data from on A.C. Chang, D.E. Crowley and A.L. Page, Assessing bioavailability of metals in biosolids-treated soils, WERF Report 97-REM-5, IWA, 2004 [102].)

trace element release has not yet approached the plateau after 10 time successive extractions. This indicates the soil can continuously supply the labile trace elements for a long-term period under field conditions. The synthetic root exudates extractable trace elements may be estimated by fitting the data to the exponential decay model:

$$M_q = a \times [1 - e^{-bq}] \qquad (6.2)$$

where M_q represents the cumulative trace elements removed (mg kg^{-1}) following successive extractions using q volumes of synthetic organic acid (moles of COO$^-$ kg^{-1} soil); parameter a represents the ultimate trace element removal (mg kg^{-1}); parameter b is the reaction rate constant (kg soil per mole of COO$^-$).

Conceptually, the constant a in Equation (6.2) is comparable to the trace element capacity factor of soil, C, in Equation (6.1). While both represent the trace element release rates, b in Equation (6.2) is based on laboratory synthetic root exudates extraction and k in Equation (6.1) is based on field plant removal rate. These two equations can be reconciled by assuming a new constant A which is the effective annual root exudates production factor of the rhizosphere with the unit of moles of root exudates per kg of soil per year (mole COO$^-$ kg^{-1} y^{-1}). For plants in the field, it produces an equivalent of q volume of organic acids during a period of t years:

$$q = A \times t \tag{6.3}$$

Substituting q into Equation (6.2):

$$M_q = a \times [1 - e^{-bAt}] \tag{6.4}$$

Let $M_q = M_t$; the field-based trace element uptake coefficient, k, is mathematically related to the successive extraction trace element release rate constant, b, through the effective field organic acid production, A.

$$k = b \times A \tag{6.5}$$

6.4.2 Case Studies

The protocol was employed to extract trace elements from soils at two long-term biosolids land application sites, namely, Rosemont Experiment Station at University of Minnesota and the Fulton County land reclamation farm in Metropolitan Water Reclamation District of Greater Chicago (MWRDGC). The long-term phytoavailability of trace elements in biosolids-amended soil was assessed based on the data collected from these two sites using the above-described root exudate-based phytoavailability model (Section 6.4.1).

The three fields of MWRDGC received comparable total biosolids mass loadings of 522–710 Mg ha^{-1} dry weight from 1974 to 1984. The cumulative loading of Cd, Cu, Ni, and Zn ranged from 130 to 177, 839 to 1142, 179 to 267, and 1583 to 2154 kg ha^{-1}, respectively. Following the termination of biosolids applications in 1984, the cultivations continued. Based on the successive root exudes extraction results, the percentages of the total trace elements in the receiving soils that would be available to plants were comparable in the three fields of MWRDGC at the time of termination (Table 6.3). The availability, however, varied greatly between the trace elements. For example, approximately 0.2 % of the Cr in receiving soils would be available (assuming the availability in soil does not change with time), while the total available Zn pool accounted for 50 %+ of the total Zn in the receiving soils. Using the percentage of the total trace elements of the receiving soils as a benchmark, the phytoavailability of trace elements in biosolids-amended soil follows a descending order of Zn > Ni > Cd > Cu > Pb > Cr.

Table 6.3 Estimated plant available trace element pool, C (mg kg^{-1}), and its percentage of the total trace element of biosolids-amended soils at the MWRDGC sites

	Cd		Cr		Cu		Ni		Pb		Zn	
	C	%	C	%	C	%	C	%	C	%	C	%
Field#1	42.7	39	1.62	0.22	35.5	11	36.1	42	3.44	1.7	372	50
Field#2	30.9	38	1.16	0.20	29.9	13	33.5	46	2.39	1.4	299	49
Field#3	21.8	31	0.70	0.12	22.9	13	28.5	43	1.28	1.2	275	53

Adapted with permission from A. C. Chang, D. E. Crowley and A.L. Page, Assessing bioavailability of metals in biosolids-treated soils, WERF Report 97-REM-5, IWA, 2004 [102].

At the Rosemount Experiment Station, University of Minnesota, biosolids were applied annually on experimental plots at rates of 0, 20, 40, and 60 Mg ha^{-1} y^{-1} from 1975 to 1977, where the plots were maintained in a three-year rotation of corn, alfalfa, and oat. The mass loadings of Cd, Cr, Cu, Ni, Pb, and Zn ranged from 9 to 25, 48 to 141, 41 to 127, 16 to 43, 54 to 97, and 106 to 343 kg ha^{-1}, respectively. Following the termination of biosolids application, the trace element capacity factors (C, mg kg^{-1}) of the receiving soils determined by the successive root exudes extraction were from 21 to 24, 1.4 to 2.0, 7.6 to 13, 25 to 39, 2.1 to 3.3, and 43 to 47 % of the total Cd, Cr, Cu, Ni, Pb, and Zn in the receiving soils, respectively (Table 6.4) and for each trace element increased in proportion to the mass loading rates.

Following the termination of biosolids application, corn was grown and harvested annually in these two field sites. The field trace element uptake was tracked for 13 years at the Fulton County land reclamation site of MWRDGC and for 7 years at Rosemont Experiment Station of University of Minnesota. The removal constant k of a trace element was estimated according to Equation (6.1). Since it is a first-order kinetic equation, the duration for removing half of the plant available trace element pool can be estimated as:

$$t_{1/2} = 0.693/k \tag{6.6}$$

The results are summarized in Table 6.5 and Table 6.6. Under these specific field conditions, it will require 141–178, 158–187, 546–597, and 147–158, years of cultivation to remove 50 % of the plant available pool of Cd, Cu, Ni, and Zn by

Table 6.4 Estimated plant available trace element pool, C (mg kg^{-1}), and its percentage of the total trace element of biosolids-amended soils at the Minnesota sites

	Cd		Cr		Cu		Ni		Pb		Zn	
	C	%	C	%	C	%	C	%	C	%	C	%
Control	0.01	18	0.38	1.4	0.36	7.4	3.59	24	0.41	2.5	11.5	43
Low	1.07	22	0.68	1.6	3.73	7.6	4.26	25	0.68	2.1	30.7	43
Medium	2.76	21	1.13	1.4	9.56	9.6	14.3	30	0.99	2.1	68.1	45
High	3.70	24	1.72	2.0	14.9	13	23.4	39	1.59	3.3	97.3	47

Adapted with permission from A. C. Chang, D. E. Crowley and A.L. Page, Assessing bioavailability of metals in biosolids-treated soils, WERF Report 97-REM-5, IWA, 2004 102].

Table 6.5 Estimated removal rate constants (k, y^{-1}) and corresponding time ($t_{1/2}$, y) to remove half of the plant available trace element pools for biosolids-amended soils at the MWRDGC sites

	Cd		Cu		Ni		Zn	
	k	$t_{1/2}$	k	$t_{1/2}$	k	$t_{1/2}$	k	$t_{1/2}$
Field#1	0.0039	178	0.0044	158	0.0013	546	0.0046	151
Field#2	0.0041	169	0.0037	187	0.0012	597	0.0047	147
Field#3	0.0049	141	0.0037	187	0.0012	559	0.0044	158

Adapted with permission from A. C. Chang, D. E. Crowley and A.L. Page, Assessing bioavailability of metals in biosolids-treated soils, WERF Report 97-REM-5, IWA, 2004 [102].

Table 6.6 Estimated removal rate constants (k, y^{-1}) and corresponding time ($t_{1/2}$, y) to remove half of the plant available trace element pools for biosolids-amended soils at the Minnesota sites

	Cd		Cr		Cu		Ni		Pb		Zn	
	k	$t_{1/2}$	k	$t_{1/2}$	k	$t_{1/2}$	k	$t_{1/2}$	k	$t_{1/2}$	k	$t_{1/2}$
Low	0.0053	130	0.0040	171	0.0077	90	0.0012	586	0.0062	113	0.0085	82
Medium	0.0056	124	0.0046	150	0.0061	114	0.0010	703	0.0062	113	0.0081	86
High	0.0049	142	0.0032	214	0.0045	155	0.0010	680	0.0039	177	0.0077	89

Adapted with permission from A. C. Chang, D. E. Crowley and A.L. Page, Assessing bioavailability of metals in biosolids-treated soils, WERF Report 97-REM-5, IWA, 2004 [102].

plant harvests respectively at the Fulton County land reclamation sites of MWRDGC (assuming the cultivation remain unchanged). The estimated time to remove half of the plant available trace element pools for the biosolids-amended soils at Rosemont Experiment Station (University of Minnesota) would be 124–142, 150–214, 90–155, 586–703, 173–177, and 82–89 years for Cd, Cr, Cu, Ni, Pb, and Zn, respectively. Definitely, these numbers are site-specific. Plants vary greatly in their ability to absorb trace elements and trace element availability in soils is affected by many soil-specific factors as reviewed in the previous sections. Nevertheless, these results illustrate the long-term persistence of these biosolids-borne trace elements in soils.

Under realistic conditions, there is often not enough data to obtain the trace element removal rate k through Equation (6.1). In this case, the model parameter k can be estimated through parameter b obtained through the synthetic root exudates-based successive trace element extraction protocol and the effective annual organic acids production of the rhizosphere, A (Equation 6.5). The statistical inferences indicate that the effective organic acid production factor for a plant species (*Zea mays* in this case), A, is not significantly different ($p < 0.05$) between trace element loading, origin of biosolids, the nature of the element, and locations. The data from the two field studies were therefore pooled to determine the probabilistic characteristics of the population. The A values followed a normal distribution (Figure 6.3). It appeared that for a plant species, A could be treated as a universal parameter

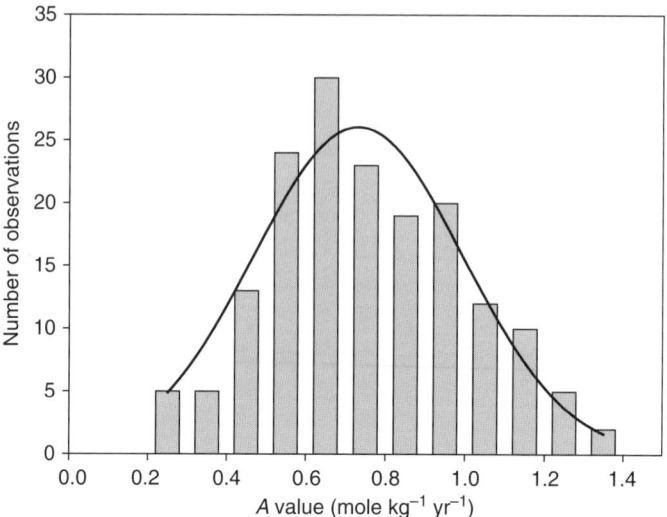

Figure 6.3 Distribution of the effective annual organic acid production (A) of the rhizosphere of corn

with distinctive probabilistic attributes. The cumulative probability distribution can be predicted based on the mean, x_0 and standard deviation, σ, of the samples as:

$$F(x) = 0.5 + 0.5 erf\left(\frac{(x-x_0)}{\sigma\sqrt{2}}\right) \qquad (6.7)$$

where $F(x)$ denotes the probability that $A \leq x$. Based on the probability distribution function, the uncertainty associated with estimating trace element availability due to the A values may be expressed in terms of risk. For example: where $A = 1.2$ mol COO^- kg^{-1} y^{-1}, there is 96.6 % probability according to the probability density function (for corn in this case) that k will be equal to or less than 0.012 y^{-1}, assuming a laboratory determination of the successive extraction trace element release rate constant, $b = 0.01$ kg soil per mole of COO^-. Consequentially, there is 96.6 % probability that it would take 58 years or more to remove 50 % of the plant available trace element in the biosolids-amended soil. Similar evaluation can be made if the A value probability density functions for other plant species are known.

6.5 Conclusions

The soil is the entry point for potentially harmful substances to enter the human food chain. When municipal biosolids are applied on croplands, potentially harmful trace elements are introduced into the root zone along with beneficial plant nutrients. How much of the biosolids-borne trace element are available for plant absorption? And how long will the trace element remain available in the soils? The conventional approaches of assessing the availability of trace elements in biosolids-amended soils relied on empirical relationships between trace element concentrations in the soil vs. trace element concentrations in plants that were developed through data obtained usually from short-term small-scale observations. They have relatively weak fundamental basis in describing the dynamic changes of plant available trace element in the biosolids-amended soils in the long term.

Organic acids present in the rhizosphere of growing plants are widely recognized to be responsible for dissolving the solid-phase trace elements in the soil and making them available for plant absorption. Based on successive extraction results with a synthetic organic acid mixture, the root exudates-based model protocol proposed by Chang *et al.* [102] provides a way to evaluate the long-term phytoavailability of trace element in biosolids-amended soils. The model was proved successful in two specific field cases. The feasibility of the protocol needs to be further validated, however, especially for different type of crops (the two field cases were both for corn).

References

[1] US Environmental Protection Agency (USEPA); Standards for the use and disposal of sewage sludge (40 Code of federal Regulation Part 503); Office of Wastewater Management, Washington, DC, 1993.
[2] National Research Council (NRC); Biosolids Applied to Land: Advancing Standards and Practices; National Academy Press, Washington, DC, 2002.
[3] US EPA; Land application of Biosolids; EPS/832/F-00-064, Office of Water, Washington, DC, 2000.
[4] Water Environmental Research Foundation; Proceedings from the Biosolids Research Summit; WERF/03-HHE-1, Consensus Building Institute, Cambridge, 2004.
[5] US EPA; Biosolids Generation, Use, and Disposal in the United States; EPA530-R-99-009, Municipal and Industrial Solid Waste Division, Office of Solid Waste, Washington, DC, 1999.
[6] UK Department of Environment; UK Sludge Survey; London, UK, 1993.
[7] Bastian, R.K., Biosolids management in the United States, a state-of-the-nation overview; Water Environ. Technol. 1997, 9, 45–50.
[8] Chang, A.C.; Pan, G.; Page, A.L; Asano, T., Developing human health-related chemical guidelines for reclaimed water and sewage sludge applications in agriculture; World Health Organization, Geneva, Switzerland, 2002.
[9] Tsadilas, C.D.; Matsi, T.; Barbayiannis, N.; Dimoyiannis, D., Influence of sewage sludge application on soil properties and on the distribution and availability of heavy metal fractions; Comm. Soil Sci. Plant Anal. 1995, 26, 2603–2619.
[10] Walter, I.; Cuevas, G.; Chemical fractionation of heavy metals in a soil amended with repeated sewage sludge application; Sci. Total Environ. 1999, 226, 113–119.
[11] Wei, Y.J.; Liu, Y.S., Effects of sewage sludge compost application on crops and cropland in a 3-year field study; Chemosphere 2005, 59, 1257–1265.
[12] Su, D.C.; Wong, J.W., Chemical speciation and phytoavailability of Zn, Cu, Ni and Cd in soil amended with fly ash-stabilized sewage sludge; Environ. Int. 2003, 29, 895–900.
[13] Illera, V.; Walter, I.; Souza, P.; Cala, V., Short-term effects of biosolids and municipal solid waste applications on heavy metals distribution in a degraded soil under a semi-arid environment; Sci. Total Environ. 2000, 255, 29–44.
[14] Metcalf and Eddy, Inc.; Wastewater Engineering, 3rd Edition, New York, McGraw Hill, 1991.
[15] Chang A.C.; Page, A.L., Trace elements slowly accumulating, depleting in soils; Calif. Agric. 2000, 54, 49–55.
[16] Witter, E., Toward zero accumulation of heavy metals in soils; Fert. Res. 1996, 43, 225–233.

[17] McBride, M.B., Toxic metal accumulation from agricultural use of sewage sludge: are USEPA regulations protective? J. Environ. Qual. 1995, 24, 5–18.
[18] Harrison, E.Z.; McBride, M.B.; Bouldin, D.R., The case for caution: Recommendations for land application of sewage sludges and an appraisal of the USEPA Part 503 sludge rules; Working Paper, Cornell Waste Management Institute; Rice Hall, Ithaca, NY, 1997 (revised 1999).
[19] Schmidt, J.P., Understanding phytotoxicity thresholds for trace elements in land applied sewage sludge; J. Environ. Qual. 1997, 23, 50–57.
[20] McBride, M.B., Toxic metals in sewage sludge-amended soils: has promotion of beneficial use discounted the risks? Adv. Environ. Res. 2003, 8, 5–19.
[21] Alloway, B.J., Heavy Metals in Soils, 2nd edn; Blackie Academic & Professional, London, New York, 1995.
[22] Kabata-Pendias A.; Pendias, H., Trace Elements in Soils and Plants; CRC Press, Boca Raton, FL, 2001.
[23] Walker, J., Biosolids management, use and disposal; in: Meyers, R.A. (ed.), Encyclopedia of Environmental Analysis and Remediation, John Wiley & Sons, Inc., New York, 1998.
[24] Stehouwer, R.; Wolf, A., Pennsylvania sewage sludge quality survey: 1978–1998; Pennsylvania State University, 1998.
[25] Welch, J.E.; Lund, L.J., Soil properties, irrigation water quality, and soil moisture level influences on the movement of nickel in sewage sludge-treated soils; J. Environ. Qual. 1987, 16, 403–410.
[26] Baldwin, K.R.; Shelton, J.E., Availability of heavy metals in compost-amended soil; Bioresource Technol. 1999, 69, 1–14.
[27] Zorpas, A.A.; Constantinides, T.; Vlyssides, A.G.; Haralambous, I.; Loizidou, M., Heavy metal uptake by natural zeolite and metals partitioning in sewage sludge compost; Bioresource Technol. 2000, 72, 113–119.
[28] Benítez, E.; Romero, E.; Gómez, M.; Gallardo-Lara, F.; Nogales, R., Biosolids and biosolids-ash as sources of heavy metals in a plant-soil system; Water Air Soil Pollut. 2001, 132, 75–87.
[29] Wong, J.W.; Lai, K.M.; Su, D.S.; Fang, M., Availability of heavy metals for *Brassica Chinesis* grown in an acidic loamy soil amended with a domestic and an industrial sewage sludge; Water Air Soil Pollut. 2001, 128, 339–353.
[30] Silveira, M.L.; Chang, A.C.; Alleoni, L.R.F.; O'Connor, G.A.; Berton, R., Metal-associated forms and speciation in biosolids amended oxisols; Comm. Soil Sci Plant Anal. 2007, 38, 851–869.
[31] Chen, W.P.; Chang, A.C.; Wu, L.S., Assessing long-term environmental risks of trace elements in phosphate fertilizers; Ecotoxi. Environ. Saf. 2007, 67, 48–58.
[32] Chen, W.P.; Wu, L.S.; Chang, A.C.; Hou, Z.A., Assessing the effect of long-term crop cultivation on distribution of Cd in the root zone. Ecol. Model. 2009, 220, 1836–1843.
[33] Chang, A.C.; Warneke, J.E.; Page, A.L.; Lund, L.J., Accumulation of heavy metals in sewage sludge treated soils; J. Environ. Qual. 1984, 13, 87–91.
[34] Sauerbeck, D.R.; Hein, A., The nickel uptake from different soils and its prediction by chemical extractants; Water Air Soil Pollut. 1991, 57–58, 861–871.
[35] Loganathan, P.; Mackay, A.D.; Lee J.; Hedley M.J., Cadmium distribution in hill pastures as influenced by 20 years of phosphate fertilizer application and sheep grazing; Aust. J. Soil Res. 1995, 33, 859–871.
[36] Hooda, P.S.; McNaulty, D.; Alloway, B.J.; Aitken, M.N., Plant availability of heavy metals in soils previously amended with heavy applications of sewage sludge; J. Sci. Food Agric. 1997, 73, 446–454.
[37] Chen, W.P.; Chang, A.C.; Wu, L.S.; Zhang, Y.S., Metal uptake by corn grown on media treated with particle-size fractionated biosolids; Sci. Total Environ. 2008, 392, 166–173.
[38] Leita, L.; De Nobili, M., Water-soluble fractions of heavy metals during composting of municipal solid waste; J. Environ. Qual. 1991, 20, 73–78.
[39] Amir, S.; Hafidi, M.; Merlina, G.; Revel, J.C., Sequential extraction of heavy metals during composting of sewage sludge; Chemosphere 2005, 59, 801–810.

[40] Liu, Y.S.; Ma, L.L.; Li, Y.Q.; Zheng, L.T., Evolution of heavy metal speciation during the aerobic composting process of sewage sludge; Chemosphere 2007, 67, 1025–1032.
[41] McBride, M.B., Richards, B.K.; Steenhuis, T.; Russo, J.J.; Sauve, S., Mobility and solubility of toxic metals and nutrients in soil fifteen years after sludge application; Soil Sci. 1997, 162, 487–500.
[42] McBride, M.B.; Richards, B.K.; Steenhuis, T.; Spiers, G., Long-term leaching of trace elements in a heavily sludge-amended silty clay loam soil; Soil Sci. 1999, 164, 613–623.
[43] Essington, M.E.; Mattigod, S.V.; Element partitioning in size-fractioned and density-fractioned sewage sludge and sludge-treated soil; Soil Sci. Soc. Am. J. 1990, 54, 385–390.
[44] Hyun, H.; Chang, A.C.; Parker, D.R.; Page, A.L., Cadmium solubility and phytoavailability in sludge treated soil: effects of organic carbon; J. Environ. Qual. 1998, 27, 329–334.
[45] Li, Z.; Ryan, J.A.; Chen, J.L.; Al-Abed, S.R., Adsorption of cadmium on biosolids-amended soils; J. Environ. Qual. 2001, 30, 903–911.
[46] Hettiarachchi, G.M.; Ryan, J.A.; Chaney, R.L.; La Fleur, C.M., Sorption and desorption of cadmium by different fractions of biosolids-amended soils; J. Environ. Qual. 2003, 32, 1684–1693.
[47] Corey, R.B.; King, L.D.; Lue-Hing, C.; Fanning, D.S.; Street, J.J.; Walker, J.M., Effects of sludge properties on accumulation of trace elements by crops; in: Page, A.L. et al. (eds), Land Application of Sludge, Food Chain Implications; Lewis Pub., Chelsea, MI, 1987.
[48] Brown, S.; Chaney, R.L.; Angle, J.S.; Ryan, J.A., The phytoavailability of cadmium to lettuce in long-term biosolids-amended soils; J. Environ. Qual. 1998, 27, 1071–1078.
[49] Bidwell, A.M.; Dowdy, R.H., Cadmium and zinc availability to corn following termination of sewage sludge applications; J. Environ. Qual. 1987, 16, 438–442.
[50] Chang, A.C.; Hae-nam, H.; Page, A.L., Cadmium uptake by swiss chard grown on composted sewage sludge treated field plots: Plateau or time bomb? J. Environ. Qual. 1997, 26, 11–19.
[51] Chang, A.C.; Hinsely, J.D.; Bates, T.E.; Doner, H.E.; Dowdy, R.H.; Ryan, J.A., Effects of long-term sludge application on accumulation of trace elements by crops; in: Page, A.L. et al. (eds), Land Application of Sludge: Food Chain Implications. Lewis Pub. Chelsea, MI, 1987.
[52] Sommers, L.E.; Page, A.L.; Logan, T.J.; Ryan, J.A., Optimum use of sewage sludge on agricultural land; Western Regional Res. Pub. W-124, Colorado Agric. Exp. Stn, Fort Collins, 1991.
[53] Wiseman, J.T.; Zilbilske, L.M., Effect of sludge application sequence on carbon and nitrogen mineralization in soil; J. Environ. Qual. 1988, 17, 334–339.
[54] McGrath, S.P.; Zhao, F.J.; Dunham, S.J.; Crosland, A.R.; Coleman, K., Long term changes in extractability and bioavailability of zinc and cadmium after sludge application; J. Environ. Qual. 2000, 29, 875–883.
[55] Walter, I.; Martınez, F.; Alonso, L.; De Gracia, J.; Guevas, G., Extractable soil heavy metals following the cessation of biosolids application to agricultural soil; Environ. Pollut. 2002, 117, 315–321.
[56] McGrath, S.P., Metal concentration in sludges and soil from a long-term field trial; J. Agric. Sci. 1984, 103, 25–35.
[57] Parkapin, P.; Sreesai, S.; Delaune, R.D., Bioavailability of heavy metals in sewage sludge-amended Thai soils; Water Air Soil Pollut. 1999, 122, 163–182.
[58] Frost, H.L.; Ketchum, L.H., Trace metal concentration in durum wheat from application of sewage sludge and commercial fertilizer; Adv. Environ. Res. 2000, 4, 347–355.
[59] Dowdy, R.H.; Latterell, J.J.; Hinesly, T.D.; Grossman, R.B.; Sullivan, D.L., Trace metal movement in an Aeric Ochraqualf following 14 years of annual sludge applications; J. Environ. Qual. 1991, 20, 119–123.
[60] Street, J.J.; Sabey, B.R.; Lindsay, W.L., Influence of pH, phosphorus, cadmium, sewage sludge, and incubation time on the solubility and plant uptake of cadmium; J. Environ. Qual. 1978, 7, 286–290.
[61] Dowdy, R.H.; Larson, W.E.; Titrud, J.M.; Latterell J.J., Growth and metal uptake of snap beans grown on sewage sludge amended soil: a four-year field study; J. Environ. Qual. 1978, 7, 252–257.

[62] Logan, T.J.; Lindsay, B.J.; Goins, L.E.; Ryan, J.A., Field assessment of sludge metal bioavailability to crops: sludge rate response; J. Environ. Qual. 1997, 26, 534–550.

[63] Chang, A.C.; Page, A.L.; Warneke, J.E., An experimental evaluation of the Cd uptake by 12 plant species grown in a sludge-treated soil: Is there a plateau?; in: Adriano, D.C.; Chen, Z.S.; Yang, S.S.; Islandar, K. (eds.); Biogeochemistry of Trace Metals, Science Reviews, Northwood, UK, 1997.

[64] Hamon, R.E.; Holm, P.E.; Lorenz, S.E.; McGrath, S.P.; Christensen, T.H., Metal uptake by plants from sludge-amended soils: caution is required in the plateau interpretation; Plant Soil 1999, 216, 53–64.

[65] Barbarick, K.A.; Ippolito, J.A.; Westfall, D.G., Biosolids effect on phosphorus, copper, zinc, nickel, and molybdenum concentrations in dry land wheat; J. Environ. Qual. 1995, 24, 608–611.

[66] Chaney, R.L.; Ryan, J.A., Heavy metals and toxic organic pollutants in MSW-composts: research results on phytoavailability, bioavailability, fate, etc.; in: Hoitink, H.A.J.; Keener, H.M. (eds.); Science and Engineering of Composting: Design, Environmental, Microbiological and Utilization Aspects; Renaissance Pub., Worthington, OH, 1993.

[67] Chaney, R.L.; Ryan, J.A., Regulating residue management practices; Water Environ. Technol. 1992, 4, 36–41.

[68] Ryan, J.A.; Chaney, R.L., Development of limits for land application of municipal sewage sludge: Risk assessment; In: Transactions, 15th World Congress of Soil Science, Acapulco, Mexico, 1994.

[69] Baker, A.J.M., Accumulators and excluders-strategies in the response of plants to heavy metals; J. Plant Nutr. 1981, 3, 643–654.

[70] Hamon, R.E., Identification of factors governing cadmium and zinc bioavailability in polluted soils; PhD Thesis, University of Nottingham, UK, 1995.

[71] Ahnstrom, Z.S.; Parker, D.R., Development and assessment of a sequential extraction procedure for the fractional of soil cadmium; Soil Sci. Soc. Am. J. 1999, 63, 1650–1658.

[72] Chen, W.P.; Chang, A.C.; Wu, L.S.; Page, A.L., Modeling dynamic sorption of cadmium in cropland soils; Vadose Zone J. 2006, 5, 1216–1221.

[73] Lindsay, W.L.; Norvell, W.A., Development of a DTPA micronutrient soil test; Agron. Abstr. 1969, pp 84.

[74] Soltanpour, P.N.; Schwab, A.P., A new soil test for simultaneous extraction of macro- and micro-nutrients in alkaline soils; Commun. Soil Sci. Plant Anal. 1977, 8, 195–207.

[75] Norvell, W.A., Comparison of chelating agents as extractants for metals in diverse soil materials; Soil Sci. Soc. Am. J. 1984, 48, 1285–1292.

[76] Novozamski, I.; Lexmond, T. M.; Houba, V.J.G.; A single extraction procedure of soil for evaluation of uptake of some heavy metals by plants; Int. J. Environ. Anal. Chem. 1993, 51, 47–58.

[77] Allen, E.R.; Johnson, G.V.; Unruh, L.G., Current approaches to soil testing methods: problems, and solutions; in: Havlin, J.L.; Jacobsen, J.S. (eds.), Soil testing: Prospects for improving nutrient recommendations; Spec. Publ. No. 40., Soil Science Society of America, Madison, WI, 1994.

[78] Hooda, P.S.; Alloway, B.J., The plant availability and DTPA extractability of trace metals in sludge-amended soils; Sci. Total Environ. 1994, 149, 39–51.

[79] Esnaola, M.V.; Bermond, A.; Millan E., Optimization of DTPA and calcium chloride extractants for assessing extractable metal fraction in polluted soils; Commun. Soil Sci. Plant Anal. 2000, 31, 13–29.

[80] Alvarez, J.M., Influence of soil type on the mobility and bioavailability of chelated zinc; J. Agric. Food Chem. 2007, 55, 3568–3576.

[81] Marchi, G.; Guilherme, L.R.G.; Chang, A.C.; do Nascimento, C.W.A., Heavy metals extractability in a soil amended with sewage sludge; Sci. Agric. 2009, 66, 643–649.

[82] Soltanpour, P.N., Use of ammonium bicarbonate DTPA soil test to evaluate elemental availability and toxicity; J. Plant Nutr. 1985, 8, 323–338.

[83] Pires, A.M.M.; Marich, G.; Mattiazzo, M.E.; Guilherme, L.R.G.; Organic acids in the rhizosphere and phytoavailability of sewage sludge-borne trace elements; Pesq. Agropec. Bras. 2007, 42, 917–924.

[84] Sauerbeck, D.R.; Hein, A., The nickel uptake from different soils and its prediction by chemical extractants; Water Air Soil Pollut. 1991, 57–58, 861–871.

[85] Chaudri, A.M.C.; Allain, M.G.; Badawy, S.H.; Adams, M.L.; McGrath, S.P., Chambers, B.J.; Cadmium content of wheat grain from a long-term field experiment with sewage sludge; J. Environ. Qual. 2001, 30, 1575–1580.

[86] King, L.D., Effect of selected soil properties on cadmium content of tobacco; J. Environ. Qual. 1988, 17, 251–255.

[87] Berti, W.R.; Jacobs, L.W., Chemistry and phytotoxicity of soil trace elements from repeated sludge application; J. Environ. Qual. 1996, 25, 1025–1032.

[88] Zhao, F.J.; Dunham, S.J.; McGrath, S.P., Lessons to be learned about soil-plant metal transfers from the 50-year sewage sludge experiment at Woburn, U.K.; in: Extended Abstracts of the Fourth International Conference on the Biogeochemistry of Trace Elements, University of California, Berkeley, USA. 1997.

[89] Candalaria, L.M.; Chang, A.C., Cadmium activities, solution speciation and solid phase distribution in cadmium nitrate and sewage sludge treated soil systems sewage sludge treated soil systems; Soil Sci. 1997, 162, 722–732.

[90] Chang, A.C.; Page, A.L.; Koo, B.J., Biogeochemistry of phosphorus, iron, and trace elements on soils as influenced by soil-plant-microbial interactions; in: Violante, A. *et al.* (eds); Soil Mineral-Organic Matter-Microorganism Interactions and Ecosystem Health; Developments in Soil Science, Vol. 28B, Elsevier, New York, 2002.

[91] Krishnamurti, G.S.R.; Cieslinski, G.; Huang, P.M.; Van Rees, K.C.J., Kinetics of cadmium release from soils as influenced by organic acids: Implication in cadmium availability; J. Environ. Qual. 1997, 26, 271–277.

[92] Gao, Y.; He, J.; Ling, W.; Hu, H.; Liu, F., Effects of organic acids on copper and cadmium desorption from contaminated soils; Environ. Int. 2003, 29, 613–618.

[93] Chen, Y.X.; Lin, Q.; Luo, Y.M. *et al.*, The role of citric acid on the phytoremediation of heavy metal contaminated soil; Chemosphere 2003, 50, 807–811.

[94] Mench, M.; Martin, E., Mobilization of cadmium and other metals from two soils by root exudates of *Zea mays* L., *Nicotiana tabacum* L. and *Nicotiana rustica* L.; Plant Soil 1991, 132, 187–196.

[95] Veeken, A.H.M.; Hamelers, H.V.M., Removal of heavy metals from sewage sludge by extraction with organic acids; Water Sci. Technol. 1999, 40, 129–136.

[96] Merckx, R.; Ginkel, J.H.; Sinnaeve, J.; Cremers, A., Plant-induced changes in the rhizosphere of maize and wheat. II. Complexation of cobalt, zinc and manganese in the rhizosphere of maize and wheat; Plant Soil 1986, 96, 95–107.

[97] Inskeep, W.P.; Comfort, S.D., Thermodynamic predictions for the effects of root exudates on metal speciation in the rhizosphere; J. Plant Nutr. 1986, 9, 567–586.

[98] Mench, M.; Morel, J.L.; Guckert, A.; Guillet, B., Metal binding with root exudates of low molecular weight; Soil Sci. 1988, 39, 521–527.

[99] Treeby, M.; Marschner, H.; Romheld, V., Mobilization of iron and other micronutrient cations from a calcareous soil by plant-borne, microbial, and synthetic metal chelators; Plant Soil 1989, 114, 217–226.

[100] Awad, F.; Romheld, V., Mobilization of heavy metals from a contaminated calcareous soil by plant borne and synthetic chelators and its uptake by wheat plants; J. Plant Nutr. 2000, 23, 1847–1855.

[101] Koo, B.J.; Chang, A.C.; Crowley, D.E.; Page, A.L., Characterization of organic acids recovered from rhizosphere of corn grown on biosolids-treated medium; Comm. Soil Sci. Plant Anal. 2006, 37, 871–887.

[102] Chang, A.C.; Crowley, D.E.; Page, A.L., Assessing bioavailability of metals in biosolids-treated soils; WERF Report 97-REM-5, IWA, 2004.

7
Fertilizer-Borne Trace Element Contaminants in Soils

Samuel P. Stacey[1], *Mike J. McLaughlin*[1,2] *and Ganga M. Hettiarachchi*[3]

[1]*Soil and Land Systems, School of Earth and Environmental Sciences, The University of Adelaide, South Australia, Australia*
[2]*CSIRO Land and Water, Glen Osmond, South Australia, Australia; Australia*
[3]*Department of Agronomy, Kansas State University, Manhattan, Kansas, USA*

7.1 Introduction

Fertilizers play an essential role in maintaining and/or increasing world food production. In many parts of the world spectacular crop yield increases have been recorded following fertilizer application. Additionally, inorganic fertilizers have become an integral part of sustainable farming systems, by allowing farmers to supply specific nutrients to crops that are required for optimum and long-term crop or pasture production.

Some inorganic fertilizers are known to contain contaminant metals, metalloids and radionuclides. Concentrations of contaminants in fertilizers are either derived from the source materials from which fertilizers are manufactured or are introduced into the fertilizer during the manufacturing process. The latter is very uncommon and the dominant source of contaminants in manufactured fertilizers is the raw material used for manufacturing. Pure nitrogenous and potassic fertilizers generally contain few trace elemental contaminants (for a review see [1]) so these will not be considered further in this chapter. The dominant sources of fertilizer contaminants are the raw materials used to manufacture phosphatic and micronutrient fertilizers [2].

A considerable research effort has attempted to assess the environmental and toxicity risks that may be associated with long-term fertilizer use. This chapter reviews the levels of metal, metalloid and radionuclide contaminants in fertilizers with a particular emphasis on inorganic phosphate and trace element sources. The chapter then assesses the likelihood of long-term soil accumulation of the most important contaminants in fertilizers and the risk of food-chain transfer to humans and animals.

7.2 Phosphatic Fertilizers

Phosphate rock is a general and imprecise term that is usually used to describe naturally occurring mineral assemblages that contain a relatively high concentration of P-bearing minerals. There are two types of phosphate deposits: sedimentary and igneous. Sedimentary deposits make up about 80% of the world's production of phosphate rock [3] and often contain high concentrations of contaminant metals. Phosphate rock is used to produce phosphate fertilizers, such as ammonium and calcium phosphates. The most common contaminant metals, metalloids and radionuclides found in phosphate rock are arsenic (As), cadmium (Cd), chromium (Cr), mercury (Hg), lead (Pb), selenium (Se), uranium (U) and vanadium (V). Fluoride (F) may also be present. Therefore, phosphatic fertilizers are considered as one of the most important diffuse sources of trace element contamination in agricultural soils [1].

Phosphorus fertilizers made from sedimentary phosphate deposits tend to have higher levels of Cd compared to fertilizers made from igneous phosphate deposits [4]. Van Kauwenbergh [5] reviewed contaminant concentrations in sedimentary phosphate rock from 25 countries (41 regions) and found that contaminant concentrations varied widely between deposits, even within countries (Table 7.1). For example, Cd concentrations varied between 3 and 150 mg kg^{-1} in samples from Florida and Idaho, respectively [5].

This diverse range of contaminant concentrations in phosphate rocks means that manufactured fertilizers can have a wide range of contaminant concentrations depending on the source of the material used for fertilizer manufacture [2,6–8]. Where products in a country

Table 7.1 Summary of contaminant concentrations in sedimentary phosphate rock

Contaminant	Contaminant concentration (mg kg^{-1})				
	Lowest		World average	Highest	
Arsenic	3	Algeria, Nauru	13.2	79	India
Cadmium	0.5	Australia	20.8	150	Idaho, USA
Chromium	16	Tanzania	129.2	1000	Idaho, USA
Lead	<1	North Carolina, USA	8.4	55	Florida, USA
Mercury	0.01	Egypt	0.4	9.9	China
Uranium	10	China	96.1	390	Tanzania

Adapted from S.J. van Kauwenbergh, Cadmium and Other Minor Elements in World Resources of Phosphate Rock, The International Fertiliser Society – Proceedings 400, 1997 [5].

or region are manufactured from the same source of raw materials, variability in contaminant concentrations is usually low [9].

The manufacturing process may reduce the concentration of contaminant elements in fertilizer products. The manufacture of phosphoric acid produces a phosphogypsum by-product [10], which may remove up to 30–40% of Cd from the phosphate rock [11,12]. However, single superphosphate is manufactured by reacting sulfuric acid with phosphate rock. Therefore, all of the metal contaminants present in phosphate rock are transferred to the final product [11]. Further contaminant removal has been investigated but not adopted for fertilizer production; at present, there is no cost-effective or environmentally sound method of further reducing metal contaminants during the manufacturing process [13,14]. Experimental metal removal processes have generally been based on solvent extractions [15]. Two alternative strategies have been proposed to reduce contaminant levels in phosphate fertilizers. They are to preferentially mine low-contaminant phosphate rock or to blend high- and low-quality rock to reduce contaminant levels to more acceptable levels. The European Fertilizer Manufacturers Association (EFMA), which is dependent on phosphate rock imports into Europe, has argued that while purchasing low-Cd phosphate rock is desirable, the world supply is insufficient to meet food production needs [14]. Moreover, the association acknowledged that phosphate rock exports represent an important source of revenue for developing countries and that controlling trade may have an adverse economic impact in these regions [14]. However, some countries have adopted this approach and, in Australia, concentrations of Cd in manufactured phosphatic fertilizers have dropped markedly over the last two decades due to different sources of phosphate rock being used in manufacturing or the import of finished fertilizers made from low-Cd rock sources [16] (Figure 7.1). However,

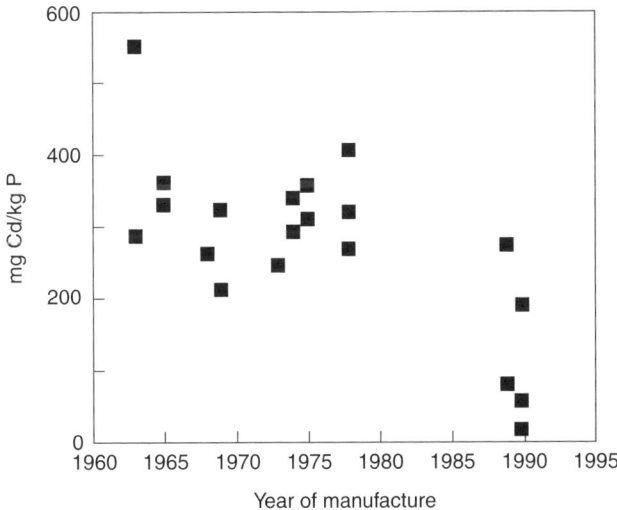

Figure 7.1 Change in concentrations of Cd in manufactured phosphatic fertilizers in Australia between 1960 and 1990 [16] (Reproduced with permission from B.A. Zarcinas and R.O. Nable, Boron and Other Impurities in South Australian Fertilizers and Soil Amendments, CSIRO, Divisional Report No. 118, 1992.)

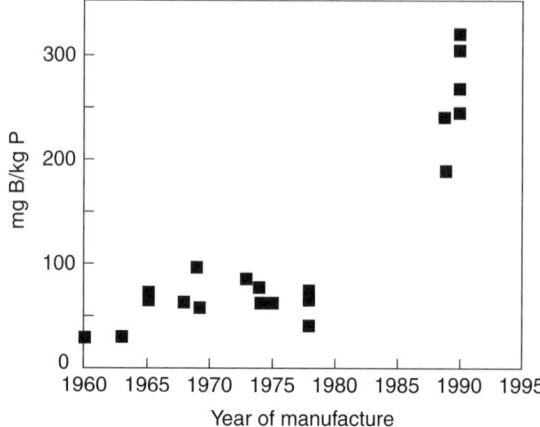

Figure 7.2 Change in concentrations of B (boron) in manufactured phosphatic fertilizers in Australia between 1960 and 1990 [16] (Reproduced with permission from B.A. Zarcinas and R.O. Nable, Boron and Other Impurities in South Australian Fertilizers and Soil Amendments, CSIRO, Divisional Report No. 118, 1992.)

Zarcinas and Nable [16] found that concentrations of other impurities in fertilizers may rise due to changes in source of rocks to control Cd. For example, boron (B) concentrations in manufactured phosphatic fertilizers were found to rise over the same period (Figure 7.2).

The EFMA used an analysis of the world demand for phosphate rock to justify its position. World demand for phosphate rock is between 145×10^6 and 151×10^6 Mg y^{-1} (Table 7.2). At current rates of consumption, world phosphate reserves (both high and low contaminant levels) would support only 88–120 years of fertilizer production [17,18] (Table 7.2). The phosphate rock reserve base could potentially support over 300 years of fertilizer manufacture. However, extraction is not economically or technologically viable at present. For example, large reserve bases of phosphate rock are located on the continental shelves and in the Atlantic Ocean [18]. As the world's reserves of high-grade phosphate rocks are depleted, contaminant levels in phosphate fertilizers will probably increase, particularly if phosphate rock with low contaminant levels is preferentially mined.

Table 7.2 Summary of world phosphate rock consumption, economic reserves and total reserve base

Data source	Consumption (10^6 Mg)	Reserves (10^6 Mg)	Reserve base (10^6 Mg)
Roberts and Stewart [17]	151 (1997–2001)	13 224	51 794
Jasinski [18]	145 (2006)	18 000	50 000

Table 7.3 Limits on contaminants in Zn fertilizers in USA

Element	Maximum allowed concentration per 1% Zn (mg kg^{-1})
As	0.3
Cd	1.4
Cr	0.6
Hg	2.8
Pb	0.3

Zinc Fertilizers Made From Recycled Hazardous Secondary Materials, United States Environmental Protection Agency, US Federal Register 2002, 67, 40 CFR, Parts 261, 266, 268 and 271 [FRL-7248-3] RIN 2050-AE69 [22].

7.3 Micronutrient Fertilizers

Some potentially toxic trace elements may be present in micronutrient fertilizers depending on the source of the raw material used to manufacture the fertilizer [19,20]. Concentrations vary widely depending on the product and the source of raw material used. For example, Cd and Pb concentrations have been reported at up to several hundred to thousands of milligrams per kilogram levels in some micronutrient fertilizers [19,20].

There have been several cases where industrial wastes have been improperly used as raw materials for the manufacture of micronutrient fertilizer, and use of these products is therefore a potential exposure pathway for soils to accumulate potentially toxic elements. In the early 1990s, a shipment of contaminated fertilizer was exported from the United States to Australia and Bangladesh. The fertilizer was manufactured from baghouse dust, a waste material derived from air purification systems at a copper recycling operation. High concentrations of Cd and Pb were found in the product, and the material was banned from sale in Australia. This prompted Australia to review its fertilizer quality regulations and the monitoring of trace elements in fertilizers at importation [21]. In the late 1990s a series of articles in the *Seattle Times* in the united States highlighted the issue of industrial wastes being used as fertilizers, and this also prompted a review of fertilizer regulations in the United States by the US Environmental Protection Agency (US EPA) and The Fertilizer Institute (for a review see [20]). As a result, the US EPA set maximum limits (Table 7.3) for As, Cd, Cr, Hg and Pb in zinc fertilizers made from hazardous industrial materials [22].

7.4 Long-Term Accumulation of Fertilizer-Borne Trace Element Contaminants

Schroeder and Balassa [23] were the first to alert the scientific community to the possibility of a potentially toxic metal (cadmium) accumulating in soils and crops due to applications of phosphatic fertilizers. This stimulated a large number of studies in the 1970s and 1980s on metal accumulation in soil from fertilizers and the possible transfer to food crops, with most effort focussing on Cd [24–37]. For example, a study at Rothamsted, England, showed that 60 years of phosphate rock and superphosphate application increased soil Cd levels in pasture soils, on average, by 4 g ha^{-1} y^{-1} above that contributed from atmospheric deposition [37]. However, under arable crops, inorganic P fertilizer did not significantly increase soil Cd. The authors hypothesized that Cd leaching was higher under arable crops due to reduced soil organic matter and, hence, reduced Cd sequestration. The study also showed that farmyard manure was a more important source of soil Cd than inorganic P fertilizers; manure application increased soil Cd levels by 12.8 g ha^{-1} y^{-1} above atmospheric deposition [37]. It should be noted that Cd concentrations in the superphosphate were relatively low, between 3 and 8 mg kg^{-1} (mean 5.2). The relative importance of different sources of Cd varies widely between countries, with atmospheric deposition often being as significant as fertilizer inputs in some highly populated and industrialized countries, for example the United Kingdom [38], while fertilizer Cd dominates in other countries with lower levels of population and industrialization, for example Australia or New Zealand [1,39].

While Cd has been the main focus of studies of fertilizer contaminants, the potential accumulation of other elements in soils through use of fertilizers has also been evaluated. Sauerbeck [40] performed a simple hazard ranking of potential contaminants in phosphatic fertilizers by examining the range of contaminant concentrations in phosphate rocks and comparing this to average elemental concentrations in the earth's crust. This analysis identified As, Cd, Cr, F, strontium (Sr), thorium (Th), U and Zn as deserving of greater investigation [40]. Data for accumulation of F in fertilized soils have been reported [30,41,42], as have data for U [32,43,44].

Nziguheba and Smolders [45] measured contaminant levels in phosphate fertilizers sold in Europe and calculated annual metal inputs in European agricultural soils. Inputs of As, Cd, Cr and Pb were, on average, 2.3, 1.6, 20.7 and 1.0 g ha^{-1} y^{-1}, respectively. The average Cd concentration in the European fertilizers surveyed by Nziguheba and Smolders [45] was 7.4 mg kg^{-1}. However, Nziguheba and Smolders [45] also showed that, in Europe, the average annual soil application of Cd, Ni, Pb and Zn from P fertilizers was below that contributed via atmospheric deposition. In a long-term fertilizer trial in Nigeria, 50 years of inorganic N, P and K application contributed less Pb to soil than cow manure; however, all Pb concentrations were within permissible levels [46]. One trace contaminant in fertilizer that has more recently caused concern is perchlorate anion (ClO_4^-), which can interfere with iodine metabolism in humans. Perchlorate may be found in trace concentrations in nitrate-based fertilizers and may be a significant contaminant in some fertilizer sources [47,48].

Obviously, measurable changes in soil trace element concentrations are a function of inputs, mixing depth and removal in produce or leaching. Thus for detection of accumulation of potentially toxic trace elements in fertilized soils, cumulative fertilizer addition rates must be large, coupled with high concentrations of trace elements in the fertilizer. In addition, measurement of elements at trace concentrations in soils is not straightforward and, coupled with variability of elemental concentrations in field soils, requires significant fertilizer inputs to soil before changes are measurable and significantly different from field variability. Long-term trials, such as those at Rothamsted, are invaluable in this regard. Studies that conclude that fertilizer-derived contaminants do not accumulate in soil need to provide good information on sampling and analytical quality control, as well as reasons why an element added to soil can no longer be found there (leaching losses, soil or product removal, etc.).

Removal of trace element contaminants from fertilized soils may be large for elements that do not bind strongly to the soil solid phase (that is have a low partition coefficient), and hence are prone to leaching or to remain in soil solution and therefore are more readily available for plant uptake, for example B or $HClO_4^-$ [49,50]. Elements that bind strongly to soils will remain near the zone of fertilization (often the surface soil) and removal by leaching or plant uptake will often be low, for example for Hg or Pb [51,52]. It is therefore evident that the physicochemical properties of the trace element, coupled with climatic and agronomic factors relating to leaching, crop uptake and product removal, will determine the propensity for trace element contaminants to accumulate in fertilized soils. Thus, prediction of accumulation of contaminants in fertilized soils is more complex than suggested by Sauerbeck [40], and requires the use of modelling approaches to determine potential accumulation related to fertilizer quality and fertilization practices [53].

7.5 Trace Element Contaminant Transfer to Crops and Grazing Animals

Trace element contaminants from fertilizers may enter the food chain via two main pathways; direct uptake by crops and translocation to harvested grains, shoots or tubers, or via pastures and inadvertent soil ingestion by grazing animals. McLaughlin *et al.* [54] summarized the main risk pathways for food-chain transfer and ecotoxicology for selected trace elements (Table 7.4). Of these elements, As, Cd and Pb were identified as posing the highest risk of food-chain transfer.

Table 7.4 Primary and secondary risk pathways for selected trace-elemental contaminants in fertilizers

Element	Primary risk pathway	Secondary risk pathway	Most important predictor required
As	Soil ingestion by animals/humans	Food-chain transfer	Oral bioavailability assessment, speciation
Cd	Food-chain transfer	Phyto- and ecotoxicity	Soil–plant uptake
Cr	Phyto- and ecotoxicity	Leaching	Toxic threshold definition, speciation
Cu	Phyto- and ecotoxicity	Soil ingestion by animals/humans	Toxic threshold definition
F	Soil ingestion by animals	Phyto- and ecotoxicity	Oral bioavailability assessment, speciation
Ni	Phyto- and ecotoxicity	Soil ingestion by animals/humans	Toxic threshold definition
Pb	Soil ingestion by animals/humans	Phyto- and ecotoxicity	Oral bioavailability assessment
Zn	Phyto- and ecotoxicity	Food-chain transfer	Soil–plant uptake

Adapted with permission from M.J. McLaughlin, B.A. Zarcinas, D.P Stevens et al., Soil testing for heavy metals, Communications in Soil Science and Plant Analysis, 31, 1661–1700. Copyright 2000, Taylor & Francis Ltd [54].

7.5.1 Arsenic

The main predicators of As solubility are soil mineralogy, organic matter, soil pH and As oxidation state. Arsenic is primarily found as the anions, arsenate (As(V)) and arsenite (As(III)) under oxidized and reduced conditions, respectively [55,56]. Inorganic As can be relatively mobile in soils in both oxidation states [55,56], particularly in alkaline soils. Arsenic may form insoluble compounds with Fe and Al oxides [56,57], be adsorbed to organic matter or hydroxyl groups on clay minerals [58], and be desorbed from these sites by phosphate fertilizers [57]. However, Huang and Matzner [55] argued that sorption by organic matter was of minor importance in most soils. Arsenic can also be readily absorbed by a large range of crop and pasture species [59]. The fate of As in soil and its absorption by crop and pastures has important implications for food-chain transfer of As.

Californian models have suggested that 100 years of intensive cultivation would not increase the concentration of As in sandy-loam soils [60]. In their study, the concentration of arsenic in fertilizer was 13 mg kg^{-1}, the median level of As measured in P fertilizers used in California. The concentration of As in fertilizer was also similar to the world average measured in sedimentary phosphate rock (Table 7.1). Less than 5% of the total As inputs into the system came from P fertilizers. Atmospheric deposition and irrigation water contributed 79% and 16% of annual soil As inputs, respectively [60]. Due to As mobility in the sandy loam soil, modelling suggested that 92% of the total As inputs were likely to be lost by leaching below the root zone. Thus, in this study, the contribution of root-zone As from P fertilizers was considered to be negligible. A recent survey suggested that As inputs to European soils from inorganic P fertilizers were generally similar to inputs from atmospheric deposition. Arsenic inputs averaged 2.3 and 2 g ha^{-1} y^{-1} from P fertilizers and atmospheric deposition, respectively [45].

In a field experiment, Szabo and Fodor [61] showed that As did not accumulate in wheat, corn or sunflower seeds at As application rates as high as 270 kg ha^{-1} applied to the soil as $NaAsO_2$. Up to 5.2, 1.8 and 12.8 mg As kg^{-1} was detected in wheat, maize and sunflower shoots, respectively. However, As was not readily translocated into the harvested grains. Gulz et al. [62] found that the concentration of As in sunflower and canola seeds rose above the Swiss tolerance value for food, only when the soil was spiked with 110 mg As kg^{-1} soil as $Na_2HAsO_4 \cdot 7H_2O$. In both these studies, the soil concentrations of As reflected highly contaminated soil, more likely from industrial contamination than from agronomic use of fertilizers. Therefore, these studies suggest that the average contribution of As from inorganic P fertilizers [45] would have a negligible effect on grain and seed As concentrations. Furthermore, surveys of As in European foods have shown that seafood contributes significantly more As to human diets than agricultural produce [63,64]. A recent survey of Spanish foods indicated that 60% of the populations daily As consumption was derived from seafood [64]. Approximately 13% of the daily As consumption came from meat and animal organs, 11% from vegetables, 9.5% from dairy products and the remaining 6.5% came from cereals, legumes, fruits and drinking water. In both surveys, the contribution of As from agricultural produce was well below the provisional tolerable weekly intake (0.015 mg kg^{-1} body weight) recommended by the Food and Agriculture Organization and the World Health Organization [63–65]. The low concentrations of As present in inorganic P fertilizers, therefore, appears not to pose a significant threat to human health.

7.5.2 Cadmium

Cadmium is almost exclusively present in fertilized soils as the divalent cation Cd^{2+} and associated organic (Cd-dissolved organic carbon) and inorganic complexes ($CdCl_n^{2-n}$, $CdSO_4^0$, $CdHCO_3^+$). Being cationic, Cd solubility is greatest in acidic soils, those of low CEC and soils having high concentrations of complexing anionic ligands, either organic or inorganic [66–69]. Cadmium is one of the trace elements that accumulates in plants to concentrations that could potentially cause animal toxicity but have no effect on plant growth, unlike Cu and Zn for which phytotoxicity occurs at concentrations in plants much lower than those that affect humans or grazing animals [70]. Of all the trace element contaminants in fertilizers, Cd is perhaps the most widely studied.

Cadmium present in granular fertilizers rapidly dissolves [71,72] and the solid-phase Cd in the fertilizer, likely present as the phosphate, moves into the soil as free Cd^{2+}. The propensity for Cd to complex with Cl [66] means that Cd is relatively mobile in saline soils and therefore more available for uptake by plants [73,74]. In contrast to other trace metals such as Cu and Zn, organic ligands do not bind strongly with free Cd^{2+} in soil solutions. Many studies have now found that either the free Cd^{2+} ion or Cd complexed by inorganic ligands are the dominant Cd species present in soil solution of most soils [75–81].

Cadmium uptake by plants is influenced by a wide range of factors – plant cultivar, soil pH, salinity, mineralogy and cation exchange capacity (CEC), organic matter content, and concentrations of other nutrients, especially N, P and Zn [82–85]. These are discussed below.

Generally, Cd concentrations are greater in roots, tubers or leaves of plants, so that leafy and root vegetables generally have higher Cd concentrations than fruits or grains [86]. Due to the many factors affecting Cd uptake by plants, it is difficult to predict soil-plant transfer of Cd by examining concentrations (total or extractable) of Cd in soil alone [54,85], and where good relationships have been found this is commonly because a small range of soils has been examined, usually under laboratory settings.

Effects of pH on plant Cd accumulation are mostly through the strong pH-dependency of Cd sorption to soil. At low pH, Cd partitions to the solution phase (low K_d) and despite increased competition of H^+ ions for Cd uptake by plant roots [87], plant Cd uptake increases due to the 2- to 4-fold increase in solution Cd concentrations for each unit decline in soil pH [88]. It should be noted however, that plant Cd uptake can sometimes be increased by liming of soils [89–91], and hypotheses have been developed to explain this phenomenon – increased Ca competition with Cd for soil sorption surfaces, or induced micronutrient deficiencies increasing Cd uptake.

The enhancement of plant Cd uptake by salinity was first noted by Bingham *et al.* [92] but was attributed to Na^+ effects on Cd^{2+} displacement from soil. Later work verified the effect of salinity was important under field conditions [73,93,94], and more mechanistic work has found that effects of Cl on Cd uptake are mainly due to the increased diffusion of Cd to plant roots induced by chloro-complexation of free Cd^{2+} ions [95–98]. This Cl effect may also be evident when KCl fertilizers are used [99].

The effects of mineralogy, CEC and organic mater content of soil on fertilizer-Cd accumulation by plants are all explained through Cd partitioning. Soils high in CEC, organic matter or having minerals with high surface (negative) charge exhibit strong Cd sorption [68,100,101], so soil solution Cd concentrations are low and therefore plant uptake is also low (per unit Cd input).

Fertilizer-derived macronutrients can also affect plant uptake of Cd. Addition of N generally tends to increase crop Cd uptake [102–105], with the effect of N attributed to variable pH changes due to N fertilizer in the rhizosphere (although the effects are often not consistent with this hypothesis), and counter ions added with the N source (for example calcium in nitrate-based N fertilizers) affecting fertilizer Cd partitioning in soil. Addition of P, excluding the effect of additional added Cd with the P, may increase plant Cd accumulation [106], decrease it [107], or have no effect [108]. There is some evidence in the literature that fertilizer type plays an important role in determining Cd availability, where DAP (diammonium phosphate) reduces Cd solubility due to pH effects around the fertilizer granule [107], but field evidence for the importance of this to crops under agronomic conditions is lacking [108]. The effect of Zn on plant availability of fertilizer Cd is generally to decrease plant uptake and accumulation in edible plant parts [83,103,109]. This is both through competition for uptake sites in the root, and in reducing Cd loading from xylem to phloem [110]. In some situations addition of Zn fertilizer could increase Cd uptake by plants – where Zn–Cd uptake competition at the root is low, and where Zn^{2+} displaces Cd^{2+} from soil sorption sites into soil solution [111,112].

Thus, it has been clearly demonstrated that fertilizer Cd accumulates through the food chain and the propensity for Cd to move through the food chain is governed by a complex array of plant, soil and environmental conditions. Many countries already impose controls on Cd concentrations in fertilizers (Table 7.5) and some (for example Australia) have instituted more holistic Cd management strategies in agriculture (see www.cadmium-management.org.au/). There is much debate in the toxicology literature regarding safe limits for Cd in foods [113–115], but the World Health Organization has retained the human provisional tolerable weekly intake (PTWI) figure for Cd unchanged since the 1970s. The PTWI is used by various health jurisdictions to set food Cd standards, and farmers must meet these standards in the produce they sell. Thus, it is evident that the interest in minimizing Cd in fertilizers, and minimizing Cd transfer through the food chain, will continue into the future.

Table 7.5 Proposed or implemented Cd limits for P fertilizers[11]

Country	Cd limit (mg kg^{-1} P)	Effective year
Australia	300	2000
Austria	120	In effect
Belgium	200	Voluntary
Denmark	110	1995
Finland	50	In effect
Germany	200	Voluntary
Japan	343	In effect
Norway	100	In effect
Sweden	100	In effect
Switzerland	50	In effect
New Zealand	280	2000

7.5.3 Fluorine

While not a trace element in rocks, soils and fertilizers, F is a trace element in plants and in animals. Fluorine constitutes 0.078% of the earth's crust with the most important F-containing mineral being fluorite (CaF_2). Fluorine concentrations in soil depend on atmospheric input; variation in soil parent material, the rate of F translocation through the soil profile and the rate of biological cycling of F. The total concentration of F in normal agricultural soils averages 150–360 mg kg^{-1}, and concentrations in fertilizers vary, ranging from negligible in nitrogenous, potassic and trace element fertilizers to levels of a few per cent in phosphatic fertilizers [1,116].

Fluorine is usually present in soils as the fluoride ion (F^-), usually retained on soil mineral surfaces by ligand exchange reactions, or incorporated into soil minerals, for example fluorite. Fluorine in soil solution is either present as the free F^- ion or as complexes with Al or H [117,118]. The free F^- ion is highly reactive with soil components and is retained strongly by soils [119], thus fertilizer-derived F accumulates at the soil surface [42,118]. The strong retention of F by soil also results in low soil–plant transfer coefficients [120], unless F is complexed by H [121] or Al [122]. Thus in acidic soils, herbage F concentrations may increase due to plant uptake of complexed F species. The key risk pathway for fertilizer-derived F in agricultural systems is animal ingestion of soil in neutral and alkaline soils, and ingestion of both soil and herbage in acidic soils. Several studies have also suggested that adverse effects of fertilizer F on crop growth could also occur in the future [121–123].

At low concentrations, F is essential for human and animal nutrition. However, at higher concentrations, F may accumulate in bone tissue, reduce bone and tooth strength and immobilize joints [124]. Clark *et al.* [125], O'Hara *et al.* [126] and East [127] have all reported acute F poisoning in livestock by direct ingestion of top-dressed phosphate fertilizers. Thus, accumulation of fertilizer-F in soils could, in the long-term, lead to problems with grazing livestock if F concentrations build up to concentrations similar, or near, those in fertilizers. Again, accurate dynamic models are needed to assess the risk to grazing animals from fertilizer F under various agronomic scenarios.

7.5.4 Lead

The average concentration of Pb in the soils around the world is about 29.2 mg kg^{-1} with the range of <1 to 888 mg kg^{-1} [128]. Lead is a chalcophile element that mainly occurs as sulfides. Lead also occurs in a large number of common rock forming minerals (as a lithophile element) substituted for potassium, strontium, barium, and calcium [129]. Lead ($6s^2\ 6p^2$) has three naturally occurring oxidation states: Pb(0), Pb(II) and Pb(IV). However, only Pb^{+2} is important over the wide range of environmental conditions that exist in soils [130].

Although past usage of pesticide sprays and other agrochemicals were a significant source of Pb in agricultural land, inputs of Pb in phosphatic fertilizers to agricultural soils do not appear to be significant according to the current estimates. It has been estimated that inputs of Pb from fertilizers are about 2 g ha^{-1} y^{-1} or less [1]. A summary of Pb concentrations in some phosphate rocks and fertilizers is shown in Table 7.6. Lead

Table 7.6 Lead and/or phosphorus concentrations of some phosphate rocks, selected fertilizers and soil amendments

Material source	Pb (mg kg^{-1})	P (%)	Reference
Phosphate rock			
Morocco (Khouribga)	2.0	14.4	[131]
Tunisia (Gafsa)	2.0	13.4	[131]
USSR (Kola)	4.0	17.2	[131]
Senegal	4.0	15.8	[131]
Morocco (Boucraa)	7.5	15.7	[1]
Togo	14.5	16.0	[1]
Florida 74/75%	18.0	15.0	[1]
Florida 72%	18.0	14.4	[1]
Peru (Sechura)	17.6	13.1	[1]
South Africa (Phalaborwa)	28.0	17.2	[1]
South Africa	35.0	—	[132]
Christmas Island	25	15.3	[1]
Australia (Duchess)	29.3	13.9	[1]
Nauru	3.3	15.6	[1]
P fertilizers			
Single superphosphate	2–71	9.7^a	[133]
Single superphosphate	4–40	9.1^b	[1]
Diammonium phosphate	5–20	20.0^b	[1]
NPK mixes	13–225	5.0–9.0	[134]
PK mixes	1.6–3.7	6.0–10.0	[134]
P fertilizers	1.9–14.0	5.0–11.0	[134]
Others			
Copper sulfate	63–117	—	[1]
Zinc oxide	3800–23 750	—	[1]
Zinc sulfate (California, USA)	0–1430.0	—	[135]
Manganese sulfate	4.0–8.0	—	[1]

^{a}Mean P concentration reported.
^{b}Registered P concentration.

concentrations in soils using phosphatic fertilizers (assuming a mean lead value of 15 mg kg^{-1} fertilizer) will not exceed the upper limit of natural soil background values (about 100 mg kg^{-1} soil) for five to ten thousand years, and this will be longer if fertilizers of lower Pb concentration are used.

Lead is toxic to humans, especially to young children, and to animals. Plant uptake of Pb is very limited. Therefore, soil containing Pb would need to be ingested in order for substantial exposure to occur. Thus the primary route of Pb exposure to humans or animals from soil is by direct ingestion of soil particles or fertilizer rather than via food-chain transfer. Lead that is absorbed by plants is largely retained within the roots, with only minimal translocation to above-ground parts [136]. We conclude that fertilizer-derived Pb poses negligible risk of accumulating to toxic concentrations in agricultural food crops.

7.5.5 Uranium

About 80–90% of U is present in the soil primarily in the +6 oxidation state as the uranyl (UO_2^{2+}) cation [137]. It exists in solution predominantly as the stable linear ion UO_2^{2+} and as soluble complexes. For example, U may form a range of soluble complexes with carbonate, such as $(UO_2)_2CO_3(OH)^{-3}$, $UO_2CO^0{}_3$, $UO_2(CO_3)_2{}^-$, $UO_2(CO_3)_4{}^{-3}$, and possibly $(UO_2)_3(CO_3)_6{}^{-6}$ [138]. Complexes with sulfate, such as $UO_2(SO_4)_2{}^{2-}$, may dominant at low pH and in soils with high concentrations of sulfate [139]. In addition, within the pH range of 4.0 and 7.5 and in the absence of dissolved inorganic ligands such as carbonate, fluoride, sulfate, and phosphate, the hydrolysis species UO_2OH^+, $UO_2(OH)^0{}_2$, and $(UO_2)_2(OH)_2{}^{+2}$ dominate U(VI) speciation [140].

The concentration of U in fertilizers is dependent on levels present in phosphate rock, which varies according to the deposit mined. Kratz et al. [141] showed that mean U concentrations in P fertilizers of different origins varied enormously, between 2–325, 50–362, 2.9–188, 43–205, 82–99, and 0.5–66 mg kg^{-1} for superphosphates, triple superphosphates, NP fertilizers, soft/ground rock phosphates, PK and NPK fertilizers, respectively.

There is disagreement in the literature about the mobility of U in the soil environment. On one hand, Spalding and Sackett [142] attributed excess U measured in rivers flowing into the Gulf of Mexico to U in phosphate fertilizers applied to agricultural soils. In contrast, Rothbaum et al. [32] and Taylor [44], using P fertilizer applied long-term field plots, showed that almost all added U remained in the surface soils and did not leach to waters or was not taken up by plants in significant amounts. Rothbaum et al. [32] used soils from three sites at Rothamsted. The sites had superphosphate applied between 1889 to 1976 at the rate of 33 kg P ha^{-1} y^{-1} and 15 g U ha^{-1} y^{-1}. А fourth site, located at Papatoetoe, New Zealand, had superphosphate applied between 1889 and 1975/76 at the rate of 37 kg P ha^{-1} y^{-1} and 16 g U ha^{-1} y^{-1}. Two neutral to slightly alkaline soils from Barfield, Rothamsted showed averaged annual increase of U in soils from 0.003 and 0.004 mg kg^{-1}, while two acid soils from Park Grass, Rothamstead and Papatoetoe, New Zealand showed increases of 0.008 and 0.014 mg kg^{-1}y^{-1}, respectively. These long-term trials suggest that U levels in phosphate fertilizers are low enough not to cause significant risk of food-chain transfer.

7.6 Conclusions

Phosphatic and micronutrient fertilizers are the main sources of trace-elemental contaminants in fertilized soils, with phosphatic fertilizers being the dominant source. The contaminants are derived from the phosphate rock used to manufacture the fertilizer. A hazard ranking of contaminant concentrations in phosphate rocks relative to concentrations in the earth's crust identified As, Cd, Cr, F, Sr, Th, U and Zn as likely to accumulate in P-fertilized soils [40] and research data to date has confirmed this for Cd, F and U. Much is already known regarding the chemistry and soil-plant transfer of Cd, with much less information being available for fertilizer-derived F and U. Research to date also suggests that fertilizer-derived As and Pb pose negligible risk for food-chain transfer. Cadmium remains the fertilizer contaminant of most concern from a human health perspective and

F from an animal health perspective. Accurate dynamic models are needed to evaluate under which soil and agronomic conditions, and over what time, these elements will reach critical concentrations in foods or soils.

References

[1] McLaughlin, M.J.; Tiller, K.G.; Naidu, R. *et al.*, Review: the behaviour and environmental impact of contaminants in fertilizers; Aust. J. Soil Res. 1996, 34, 1–54.

[2] Raven, K.P.; Reynolds, J.W.; Loeppert, R.H., Trace element analyses of fertilizers and Soil amendments by axial-view inductively-coupled plasma atomic emission spectrophotometry; Commun. Soil Sci. Plant Anal. 1997, 28, 237–257.

[3] Stewart, W.M.; Hammond, L.L.; Van Kauwenbergh, S.J., Phosphorus as a natural resource; in Sims, J.T.; Sharpley, A.N. (eds); Phosphorus: agriculture and the environment; Agron. Monogr. 2005, 46; ASA, CSSA, and SSSA, Madison, WI, pp. 3–22.

[4] Louekari, K.; Mäkelä-Kurtto, R.; Pasanen, J. *et al.*, Cadmium in fertilizers – risks to human health and the environment; Ministry of Agriculture and Forestry in Finland, Publications 4/2000, PrintLink, Helsinki.

[5] van Kauwenbergh, S.J., Cadmium and other minor elements in world resources of phosphate rock; The International Fertiliser Society – Proceeding 400; The International Fertiliser Society, London, 1997.

[6] Charter, R.A.; Tabatabai, M.A.; Schafer J.W., Arsenic, molybdenum, selenium, and tungsten contents of fertilizers and phosphate rocks; Commun. Soil Sci. Plant Anal. 1995, 26, 3051–3062.

[7] Raven, K.P.; Loeppert, R.H., Microwave digestion of fertilizers and soil amendments; Commun. Soil Sci. Plant Anal. 1996, 27, 2947–2971.

[8] Raven, K.P.; Loeppert, R.H., Trace element composition of fertilizers and soil amendments; J. Environ. Qual. 1997, 26, 551–557.

[9] Charter, R.A.; Tabatabai, M.A.; Schafer, J.W., Metal contents of fertilizers marketed in Iowa; Commun. Soil Sci. Plant Anal. 1993, 24, 961–972.

[10] Windridge, K., Mineral Fertilizer Production and the Environment Part 1. The Fertilizer Industry's Manufacturing Processes and Environmental Issues; United Nations Environment Programme and United Nations Industrial Development Organization, Technical Report No. 26 Part 1 1998, United Nations Publication.

[11] Mortvedt, J.J., Heavy metal contaminants in inorganic and organic fertilizers; Fert. Res. 1996, 43, 55–61.

[12] Wakefield, Z.T., Distribution of Cadmium and Selected Heavy Metals in Phosphate Fertilizer Processing; Bull. Y-159, National Fertilizer Development Center, Tennessee Valley Authority, Muscle Shoals, AL, USA, 1980.

[13] US EPA, Profile of the Agricultural Chemical, Pesticide, and Fertilizer Industry, Sector Notebook Project EPA/310-R-00–003 2000; United States Environmental Protection Agency, Washington DC.

[14] EFMA (The Fertilizer Industry of the European Union), The Issues of Today, the Outlook for Tomorrow; European Fertilizer Manufacturers Association, Brussels, 1997.

[15] Scorovarov, J.I.; Ruzin, L.I.; Lomonosov, A.V. *et al.*, Solvent extraction for cleaning phosphoric acid in fertilizer production; J. Radioanal. Nucl. Chem. 1998, 229, 111–116.

[16] Zarcinas, B.A.; Nable, R.O., Boron and Other Impurities in South Australian Fertilizers and Soil Amendments; South Australia: CSIRO; Divisional Report No. 118, 1992.

[17] Roberts, T.L.; Stewart, W.M., Inorganic phosphorus and potassium production and reserves; Better Crops 2002, 86, 6–7.

[18] Jasinski, S.M., Phosphate rock; in Mineral Commodity Summaries, US Geological Survey, United States Government Printing Office, Washington DC, 2007, pp. 120–121.

[19] Gabe, U.; Rodella, A.A., Trace Elements in Brazilian agricultural limestones and mineral fertilizers; Commun. Soil Sci. Plant Anal. 1999, 30, 605–620.
[20] Franklin, R.E.; Duis, L.; Brown, R. et al., Trace element content of selected fertilizers and micronutrient source materials; Commun. Soil Sci. Plant Anal. 2005, 36, 1591–1609.
[21] Rippon, G.D.; Angel, R.; Patel, P. et al., Rechargeable potatoes – cadmium contaminated trace nutrient fertilisers; Presentation to the Fenner Conference on the Environment, Canberra, 4–5 July, 2002, http://www.csu.edu.au/special/fenner/abstracts4.html#Rippon.
[22] US EPA, Zinc Fertilizers Made From Recycled Hazardous Secondary Materials; United States Environmental Protection Agency, US Federal Register 2002, 67, 40 CFR, Parts 261, 266, 268 and 271 [FRL-7248-3] RIN 2050-AE69.
[23] Schroeder, H.A.; Balassa, J.J., Cadmium: uptake by vegetables from superphosphate in soil; Science 1963, 140, 819–820.
[24] Williams, C.H.; David, D.J., The effect of superphosphate on the cadmium content of soils and plants; Aust. J. Soil Res. 1973, 11, 43–56.
[25] Vahter, M.; Stenström, T., Cadmium and lead in Swedish commercial fertilizers; Ambio 1974, 3, 91–92.
[26] Williams, C.H., Heavy metals and other elements in fertilizers – environmental considerations; in Leece, D.R. (ed.), Fertilizers and the Environment: Proceedings of a Symposium Dealing with Ecological Aspects of Fertilizer Technology and Use, Wesley College, Univeristy of Sydney, 13–15 May 1974; Australian Institute of Agricultural Science, Sydney NSW, 1974.
[27] Williams, C.H.; David, D.J., The accumulation in soil of cadmium residues from phosphate fertilizers and their effect on the cadmium content of plants; Soil Sci. 1974, 121, 86–93.
[28] Jaakkola, A., Effect of fertilizers, lime and cadmium added to soil on the cadmium content of spring wheat; J. Sci. Agric. Soc. Finl. 1977, 49, 406–414.
[29] Jaakkola, A.; Korkman, J.; Juvankoski, T., The effect of cadmium contained in fertilizers on the cadmium content of vegetables; J. Sci. Agric. Soc. Finl. 1979, 21, 158–162.
[30] Omueti, J.A.I.; Jones, R.L., Fluorine content of soil from morrow plots over a period of 67 years.; Soil Sci. Soc. Am. J. 1977, 41, 1023–1024.
[31] Reuss, J.O.; Dooley, H.L.; Griffis, W., Uptake of cadmium from phosphate fertilizers by peas, radishes, and lettuce; J. Environ. Qual. 1978, 7, 128–133.
[32] Rothbaum, H.P.; McGaveston, D.A.; Wall, T. et al., Uranium accumulation in soils from long continued applications of superphosphate; J. Soil Sci. 1979, 30, 147–153.
[33] Mulla, D.J.; Page, A.L.; Ganje, T.J., Cadmium accumulations and bioavailability in soils from long-term phosphorus fertilization; J. Environ. Qual. 1980, 9, 408–412.
[34] Andersson, A.; Hahlin, M., Cadmium effects from phosphorus fertilization; Swed. J. Agric. Res. 1981, 11, 3–10.
[35] Mortvedt, J.J.; Mays, D.A.; Osborn, G., Uptake by wheat of cadmium and other heavy metal contaminants in phosphate fertilizers; J. Environ. Qual. 1981, 10, 193–197.
[36] Rothbaum, H.P.; Goguel, R.L.; Johnston, A.E. et al., Cadmium accumulation in soils from long-continued applications of superphosphate; J. Soil Sci. 1986, 37, 99–107.
[37] Jones, K.C.; Symon, C.J.; Johnston, A.E., Long-term changes in soil and cereal grain cadmium: studies at Rothamsted Experimental Station; Trace Substances in Environmental Health – XXI. Proceedings of the University of Missouri's 21st Annual Conference on Trace Substances in Environmental Health, University of Missouri Press, Columbia, 1987, pp. 450–460.
[38] Nicholson, F.A.; Smith, S.R.; Alloway, B.J. et al., Quantifying heavy metal inputs to agricultural soils in England and Wales; Water Environ. J. 2006, 20, 87–95.
[39] Bramley, R.G.V., Cadmium in New Zealand agriculture; N.Z. J. Agric. Res. 1990, 33, 505–519.
[40] Sauerbeck, D., Conditions controlling the bioavailability of trace elements and heavy metals derived from phosphate fertilizers in soils; Proceedings of the International IMPHOS Conference on Phosphorus, Life and Environment, Casablanca, Morocco; Institute Mondial du Phosphate, 1993, pp. 419–448.
[41] McLaughlin, M.J.; Simpson, P.G.; Fleming, N. et al., Effect of fertiliser type on cadmium and fluorine concentrations in clover herbage; Aust. J. Exp. Agric. 1997, 37, 1019–1026.

[42] Loganathan, P.; Hedley, M.J.; Wallace, G.C. et al., Fluoride accumulation in pasture forages and soils following long-term applications of phosphorus fertilisers; Environ. Pollut. 2001, 115, 275–282.

[43] Takeda, A.; Tsukada, H.; Takaku, Y. et al., Accumulation of uranium derived from long-term fertilizer applications in a cultivated Andisol; Sci. Total Environ. 2006, 367, 924–931.

[44] Taylor, M.D., Accumulation of uranium in soils from impurities in phosphate fertilisers; Landbauforsch. Volk. 2007, 57, 133–139.

[45] Nziguheba, G.; Smolders, E., Inputs of trace elements in agricultural soils via phosphate fertilizers in European countries; Sci. Total. Environ. 2008, 390, 53–57.

[46] Ogunwole, J.O.; Ogunleye, P.O., Surface soil aggregation, trace, and heavy metal enrichment under long-term application of farm yard manure and mineral fertilizers; Commun. Soil Sci. Plant Anal. 2004, 35, 1505–1516.

[47] Hunter, W.J., Perchlorate is not a common contaminant of fertilizers; J. Agron. Crop Sci. 2001, 187, 203–206.

[48] Urbansky, E.T.; Brown, S.K.; Magnuson, M.L. et al., Perchlorate levels in samples of sodium nitrate fertilizer derived from Chilean Caliche; Environ. Pollut. 2001, 112, 299–302.

[49] Harada, T.; Tamai, M., Some factors affecting behaviour of boron in soil. i. some soil properites affecting boron adsorption of soil; Soil Sci. Plant Nut. 1968, 14, 215–224.

[50] Urbansky, E.T.; Brown, S.K., Perchlorate retention and mobility in soils; J. Environ. Monit. 2003, 5, 455–462.

[51] Gerritse, R.G.; van Driel, W., The relationship between adsorption of trace metals, organic matter, and pH in temperate soils; J. Environ. Qual. 1984, 13, 197–204.

[52] Lumsdon, D.G.; Evans, L.J.; Bolton, K.A., The influence of pH and chloride on the retention of cadmium, lead, mercury, and zinc by soils; J. Soil Contam. 1995, 4, 137–150.

[53] Lofts, S.; Chapman, P.M.; Dwyer, R. et al., Critical loads of metals and other trace elements to terrestrial environments; Enviro. Sci. Technol. 2007, 41, 6326–6331.

[54] McLaughlin, M.J.; Zarcinas, B.A.; Stevens, D.P. et al., Soil testing for heavy metals; Commun. Soil Sci. Plant Anal. 2000, 31, 1661–1700.

[55] Huang, J.H.; Matzner, E., Mobile arsenic species in unpolluted and polluted soils; Sci. Total Environ. 2007, 377, 308–318.

[56] Frentiu, T.; Vlad, S.N.; Ponta, M. et al., Profile distribution of As(III) and As(V) species in soil and groundwater in Bozanta Area; Chemical Papers 2007, 61, 186–193.

[57] Codling, E.E.; Dao, T.H., Short-term effect of lime, phosphorus, and iron amendments on water-extractable lead and arsenic in orchard soils; Commun. Soil Sci. Plant Anal. 2007, 38, 903–919.

[58] Mitchell, P.; Barr, D., The nature and significance of public exposure to arsenic: A review of its relevance to South West England; Environ. Geochem. Health 1995, 17, 57–82.

[59] Raab, A.; Williams, P.H.; Meharg, A. et al., Uptake and translocation of inorganic and methylated arsenic species by plants; Environ. Chem. 2007, 4, 197–203.

[60] Chen, W.; Chang, A.C.; Wu, L., Assessing long-term environmental risks of trace elements in phosphate fertilizers; Ecotoxicol. Environ. Saf. 2007, 67, 48–58.

[61] Szabo, L.; Fodor, L., Uptake of microelements by crops grown on heavy metal-amended soil; Commun. Soil Sci. Plant Anal. 2006, 37, 2679–2689.

[62] Gulz, P.A.; Gupta, S.K.; Schulin, R., Arsenic accumulation of common plants from contaminated soils; Plant Soil 2005, 272, 337–347.

[63] van Dokkum, W.; de Vos, R.H.; Muys, T.H. et al., Minerals and trace elements in total diets in The Netherlands; Br. J. Nutr. 1989, 61, 7–15.

[64] Delgado-Andrade, C.; Navarro, M.; Lopez, H. et al., Determination of total arsenic levels by hydride generation atomic absorption spectrometry in foods from South-East Spain: estimation of daily dietary intake; Food Addit. Contam. 2003, 20, 923–932.

[65] FAO/WHO, Thirty-third Report of the Joint FAO/WHO Expert Committee on Food Additives; World Health Organization Technical Report Series 1989 No. 776, Food and Agriculture Organization of the United Nations/World Health Organization, Geneva, 1989.

[66] Hahne, H.C.H.; Kroontje, W., Significance of pH and chloride concentration on behaviour of heavy metal pollutants: mercury (II), cadmium (II), zinc (II), and lead (II); J. Environ. Qual. 1973, 2, 444–450.

[67] Garcia-Miragaya, J.; Page, A.L., Influence of ionic strength and inorganic complex formation on the sorption of trace amounts of Cd by montmorillonite; Soil Sci. Soc. Am. J. 1976, 40, 658–663.
[68] Garcia-Miragaya, J.; Page, A.L., Sorption of trace quantities of cadmium by soils with different chemical and mineralogical composition; Water Air Soil Pollut. 1978, 9, 289–299.
[69] Chubin, R.G.; Street, J.J., Adsorption of cadmium on soil constituents in the presence of complexing ligands; J. Environ. Qual. 1981, 10, 225–228.
[70] Chaney, R.L., Crop and food chain effects of toxic elements in sludges and effluents; Proceedings of the Joint Conference on 'Recycling Municipal Sludges and Effluents on Land' 1973. National Association of State Universities and Land-Grant Colleges, Washington DC, 1973, pp. 129–141.
[71] Williams, C.H.; David, D.J., The Accumulation in soil of cadmium residues from phosphate fertilizers and their effect on the cadmium content of plants; Soil Sci. 1976, 121, 86–93.
[72] Williams, C.H.; David, D.J., Some effects of the distribution of cadmium and phosphate in the root zone on the cadmium content of plants; Aust. J. Soil Res. 1977, 15, 59–68.
[73] McLaughlin, M.J.; Tiller, K.G.; Beech, T.A. et al., Soil salinity causes elevated cadmium concentrations in field-grown potato tubers; J. Environ. Qual. 1994, 23, 1013–1018.
[74] Smolders, E.; Lambrechts, R.M.; McLaughlin, M.J. et al., Effect of soil solution chloride on cadmium availability to Swiss chard; J. Environ. Qual. 1998, 27, 426–431.
[75] Mahler, R.J.; Bingham, R.T.; Sposito, G. et al., Cadmium-enriched sewage sludge application to acid and calcareous soils: relation between treatment, cadmium in saturation extracts, and cadmium uptake; J. Environ. Qual. 1980, 9, 359–364.
[76] Emmerich, W.E.; Jund, L.J.; Page, A.L. et al., Predicted solution phase forms of heavy metals in sewage sludge-treated soils; J. Environ. Qual. 1982, 11, 182–186.
[77] Fujii, R.; Hendrickson, L.L.; Corey, R.B., Ionic activities of trace metals in sludge-amended soils; Sci. Total Environ. 1983, 28, 179–190.
[78] Tills, A.R.; Alloway, B.J., The use of liquid chromatography in the study of cadmium speciation in soil solutions from polluted soils; J. Soil Sci. 1983, 34, 769–781.
[79] Hirsch, D.; Banin, A., Cadmium speciation in soil solutions; J. Environ. Qual. 1990, 19, 366–372.
[80] Holm, P.E.; Futtrup, J.; Christensen, T.H. et al., Determination of cadmium species in soil solution samples; International Conference on Heavy Metals in the Environment: Proceedings 1993, 1, 290–293.
[81] Nolan, A.L.; McLaughlin, M.J.; Mason, S.D., Chemical speciation of Zn, Cd, Cu, and Pb in pore waters of agricultural and contaminated soils using Donnan dialysis; Environ. Sci. Technol. 2003, 37, 90–98.
[82] Chaney, R.L.; Hornick S.B., Accumulation and effects of cadmium on crops; cadmium 77: Proc. 1st Int. Cd. Conf. 1978, San Francisco, Metal Bulletin Ltd, London, pp 125–140.
[83] Oliver, D.P.; Hannam, R.; Tiller, K.G. et al., The effects of zinc fertilization on cadmium concentration in wheat grain; J. Environ. Qual. 1994, 23, 705–711.
[84] Grant, C.A.; Buckley, W.T.; Bailey L.D. et al., Cadmium accumulation in crops; Can. J. Plant Sci. 1998, 78, 1–17.
[85] McLaughlin, M.J.; Maier, N.A.; Correll, R.L. et al., Prediction of cadmium concentrations in potato tubers (*Solanum Tuberosum* L.) by pre-plant soil and irrigation water analyses; Aust. J. Soil Res. 1999, 37, 191–207.
[86] Wolnik, K.A.; Fricke, F.L.; Capar, S.G. et al., Elements in major raw agricultural crops in the United States. I. Cadmium and lead in lettuce, peanuts, potatoes, soybeans, sweetcorn and wheat; J. Agri. Food Chem. 1983, 31, 1240–1244.
[87] Jarvis, S.C.; Jones, L.H.P.; Hopper, M.J., Cadmium uptake from solution by plants and its transport from roots to shoots; Plant Soil 1976, 44, 179–191.
[88] Christensen, T.H., Cadmium soil sorption at low concentrations: I. Effect of Time, cadmium load, pH, and calcium; Water Air Soil Pollut. 1984, 21, 105–114.
[89] Andersson, A.; Siman, G., Levels of Cd and some other trace elements in soils and crops as influenced by lime and fertilizer level; Acta Agric. Scand. 1991, 41, 3–11.
[90] Oborn, I.; Jansson, G., Effects of liming on cadmium contents of spring wheat and potatoes; Proc. World Congress Soil Sci. 1998, Montpellier, France, SR2025.

[91] Maier, N.A.; McLaughlin, M.J.; Heap, M. *et al.*, Effect of current-season application of calcitic lime and phosphorus fertilization on soil pH, potato growth, yield, dry matter content, and cadmium concentration; Commun. Soil Sci. Plant Anal. 2002, 33, 2145–2165.

[92] Bingham, F.T.; Strong, J.E.; Sposito, G., Influence of chloride salinity on cadmium uptake by Swiss chard; Soil Sci. 1982, 135, 160–165.

[93] Li, U.-M.; Chaney, R.L.; Schneiter, A.A., Effect of soil chloride level on cadmium concentration in sunflower kernels; Plant Soil 1994, 167, 275–280.

[94] Norvell, W.A.; Wu, J.; Hopkins, D.G. *et al.*, Association of cadmium in durum wheat grain with soil chloride and chelate-extractable soil cadmium; Soil Sci. Soc. Am. J. 2000, 64, 2162–2168.

[95] Smolders, E.; McLaughlin, M.J., Chloride increases cadmium uptake in Swiss chard in a resin-buffered nutrient solution; Soil Sci. Soc. Am. J. 1996, 60, 1443–1447.

[96] Smolders, E.; McLaughlin, M.J., Effect of Cl and Cd uptake by Swiss chard in nutrient solution; Plant Soil 1996, 179, 57–64.

[97] Degryse, F.; Smolders, E.; Parker, D.R., An agar gel technique demonstrates diffusion limitations to cadmium uptake by higher plants; Environ. Chem. 2006, 3, 419–423.

[98] Degryse, F.; Smolders, E.; Merckx, R., Labile Cd complexes increase Cd availability to plants; Environ. Sci. Technol. 2006, 40, 830–836.

[99] Salardini, A.A.; Sparrow, L.A.; Holloway, R.J., Effects of potassium and zinc fertilizers, gypsum and leaching on cadmium in the seeds of poppies (*Papaver somniferum* L.); in: Barrow N.J. (ed.), Plant Nutrition – From Genetic Engineering to Field Practice; Proceedings of the Twelfth International Plant Nutrition Colloquium; Kluwer Academic Publishers, Dordrecht, 1993, pp. 795–798.

[100] Haghiri, F., Plant Uptake of cadmium as influenced by cation exchange capacity, organic matter, zinc, and soil temperature; J. Environ. Qual. 1974, 3, 180–183.

[101] MacLean, A.J., Cadmium in different plant species and its availability in soils as influenced by organic matter and additions of lime, P, Cd and Zn; Can. J. Soil Sci. 1976, 56, 129–138.

[102] Willaert, G.; Verloo, M., Effects of various nitrogen fertilizers on the chemical and biological activity of major and trace elements in a cadmium contaminated soil; Pedologie 1992, 43, 83–91.

[103] Grant, C.A.; Bailey, L.D., Nitrogen, phosphorus and zinc management effects on grain yield and cadmium concentration in two cultivars of durum wheat; Can. J. Plant Sci. 1998, 78, 63–70.

[104] Maier, N.A.; McLaughlin, M.J.; Heap, M. *et al.*, Effect of nitrogen source and calcitic lime on soil pH and potato yield, leaf chemical composition, and tuber cadmium concentrations; J. Plant Nutr. 2002, 25, 523–544.

[105] Wangstrand, H.; Eriksson, J.; Oborn I., Cadmium concentration in winter wheat as affected by nitrogen fertilization; Eur. J. Agron. 2007, 26, 209–214.

[106] Sparrow, L.A.; Salardini, A.A.; Bishop, A.C., Field studies of cadmium in potatoes (*Solanum tuberosum* L.) II. Response of cvv. Russet Burbank and Kennebec to two double superphosphates of different cadmium concentration; Aust. J. Agric. Res. 1993, 44, 855–861.

[107] Levi-Minzi, R; Petruzzelli, G., The influence of phosphate fertilizers on Cd solubility in soil; Water Air Soil Pollut. 1984, 23, 423–429.

[108] McLaughlin, M.J.; Maier, N.A.; Freeman, K. *et al.*, Effect of potassic and phosphatic fertilizer type, phosphatic fertilizer cadmium content, and additions of zinc on cadmium uptake by commercial potato crops; Fert. Res. 1995, 40, 63–70.

[109] Abdel-Sabour, M.F.; Mortvedt, J.J.; Kelsoe, J.J., Cadmium-zinc interactions in plants and extractable cadmium and zinc fractions in soil; Soil Sci. 1988, 145, 424–431.

[110] Hart, J.J.; Welch, R.M.; Norvell, W.A. *et al.*, Zinc effects on cadmium accumulation and partitioning in near-isogenic lines of durum wheat that differ in grain cadmium concentration; New Phytol. 2005, 167, 391–401.

[111] Christensen, T.H., Cadmium soil sorption at low concentrations: VI. A model for zinc competition; Water Air Soil Pollut. 1987, 34, 305–314.

[112] Moraghan, J.T., Accumulation of cadmium and selected elements in flax seed grown on a calcareous Soil; Plant Soil 1993, 150, 61–68.

[113] Satarug, S.; Baker, J.R.; Urbenjapol, S. *et al.*, A global perspective on cadmium pollution and toxicity in non-occupationally exposed population; Toxicol. Lett. 2003, 137, 65–83.

[114] Chaney, R.L.; Reeves, P.G.; Ryan, J.A. *et al.*, An improved understanding of soil Cd risk to humans and low cost methods to phytoextract cd from contaminated soils to prevent soil Cd risks; Biometals 2004, 17, 549–553.
[115] Satarug, S.; Moore, M.R., Adverse health effects of chronic exposure to low-level cadmium in foodstuffs and cigarette smoke; Environ. Health Perspect. 2004, 112, 1099–1103.
[116] Omueti, J.A.I.; Jones, R.L., Regional distribution of fluorine in Illinois soils; Soil Sci. Soc. Am. J. 1997, 41, 771–774.
[117] Munns, D.N.; Helyar, K.R.; Conyers, M., Determination of aluminum activity from measurements of fluoride in acid soil solutions; J. Soil Sci. 1992, 43, 441–446.
[118] McLaughlin, M.J.; Stevens, D.P.; Keerthisinghe, D.G. *et al.*, Contamination of soil with fluoride by long-term application of superphosphates to pastures and risk to grazing animals; Aust. J. Soil Res. 2001, 39, 627–640.
[119] Barrow, N.J.; Ellis, A.S., Testing a mechanistic model. III. The effects of pH on fluoride retention by soil; Soil Sci. 1986, 37, 287–293.
[120] Stevens, D.P.; McLaughlin, M.J.; Randall, P.J. *et al.*, Effect of fluoride supply on fluoride concentrations in five pasture species: levels required to reach phytotoxic or potentially zootoxic concentrations in plant tissue; Plant Soil 2000, 227, 223–233.
[121] Stevens, D.P.; McLaughlin, M.J.; Alston, A.M., Phytotoxicity of hydrogen fluoride and fluoroborate and their uptake from solution culture by Lycopersicon esulentum and Avena sativa; Plant Soil 2000, 200, 175–184.
[122] Stevens, D.P.; McLaughlin, M.J.; Alston, A.M., Phytotoxicity of aluminium- fluoride complexes and their uptake from solution culture by Avena sativa and Lycopersicon esculentum; Plant Soil 1997, 192, 81–93.
[123] Manoharan, V.; Loganathan, P.; Tillman, R.W. *et al.*, Interactive effects of soil acidity and fluoride on soil solution aluminium chemistry and barley (*Hordeum vulgare* L.) root growth; Environ. Pollut. 2007, 145, 778–786.
[124] Clark, R.G.; Stewart, D.J., Fluorine (F); in Grace, N. D. (ed.), Mineral Requirements of Grazing Ruminants; THL Publishers; Singapore, 1983, pp. 129–134.
[125] Clark, R.G.; Hunter, A.C.; Stewart, D.J., Deaths in cattle suggestive of sub-acute fluorine poisoning following ingestion of superphosphate; NZ. Vet. J. 1976, 24, 193–197.
[126] O'Hara, P.J.; Fraser, A.J.; James M.P., Super-phosphate poisoning of sheep – the role of fluoride; N.Z. Vet. J. 1982, 30, 199–201.
[127] East, N.E., Accidental superphosphate fertilizer poisoning in Pregnant ewes; J. Am. Vet. Med. Assoc. 1993, 203, 1176–1177.
[128] Ure, A.M.; Berrow, M.L., The elemental constituents of soils; in Bowen, H. J. M. (ed.), Environmental Chemistry 1982, Vol. 2; Royal Society of Chemistry, London, 1982.
[129] Garrels, R.M.; Christ, C.L., Solutions, Minerals, and Equilibria; Harper and Row, New York, 1965.
[130] Lindsay, W.L., Chemical Equilibria in Soils; John Wiley & Sons, Inc., New York, 1979.
[131] Singh, B. R., Unwanted components of commercial fertilizers and their agricultural effects; Proceedings of the Fertilizer Society, London, December 1991, pp. 2–28.
[132] Kongshaug, G.; Bøckman, O.C.; Kaarstad, O. *et al.*, Paper presented at the International Symposium for Chemical Climatology and Geomedical Problems, Oslo, Norway, 21–22 May 1992.
[133] Williams, C.H., Trace metals and superphosphate. Toxicity problems; J. Aust. Inst. Agric. Sci. 1977, 43, 99–109.
[134] Stenstrom, T.; Vahter, M., Cadmium and lead in Swedish commercial fertilizers; Ambio 1974, 3, 91–92.
[135] Pierzynski, G.M.; Sims J.T.; Vance, F., Soils and Environmental Quality, 2nd edn.; CRC Press, Boca Raton, 2000.
[136] Koeppe, D.E., Lead, understanding the minimal toxicity of lead in plants; in: Lepp, N.W. (ed.), Effect of Heavy Metal Pollution on Plants, Vol. 1; Applied Science, London, 1981, pp. 55–76.
[137] Ebbs, S.D.; Brady, D.J.; Kochian, L.V., Role of uranium speciation in the uptake and translocation of uranium by plants; J. Exp. Bot. 1998, 49, 1183–1190.
[138] Duff, M.C.; Amrhein, C., Uranium(VI) adsorption on goethite and soil in carbonate solutions; Soil Sci. Soc. Am. J. 1996, 60, 1393–1400.

[139] Hennig, C.; Schmeide, K.; Brendler, V. et al., EXAFS investigation of U(VI), U(IV), and Th(IV) sulfato complexes in aqueous solution; Inorg. Chem. 2007, 46, 5882–5892.
[140] Meinrath, G.; Kato, Y.; Kimura; T. et al., Solid-aqueous phase equilibria of uranium(VI) under ambient conditions; Radiochim. Acta 1996, 75, 159–167.
[141] Kratz, S.; Knappe, F.; Schick, J. et al., U balances in agroecosystems, International Symposium, Protecting Water Bodies from Negative Impacts of Agriculture – Loads and Fate of Fertilizer Derived Uranium. 4–5 July 2007, Braunschweig, Germany.
[142] Spalding, R.F.; Sackett, W.M., Uranium in runoff from the Gulf of Mexico distributive province: Anomalous concentrations; Science 1972, 175, 69–631.

8
Trace Metal Exposure and Effects on Soil-Dwelling Species and Their Communities

David J. Spurgeon

Centre for Ecology and Hydrology, Monks Wood, Abbots Ripton, Huntingdon, Cambridgeshire, UK

8.1 Introduction

The metal content of soil is derived either naturally from the geogenic weathering of base rock during soil formation or anthropogenically through point source or diffuse pollution and/or solid waste disposal (see Chapter 1). The presence of these trace metals in soil is essential for life as they are needed as micronutrients by plants, microbes and soil-dwelling invertebrates. When the trace metal concentration of soil is below the essential requirements of a given species, symptoms attributable to micronutrient deficiency can result. When, however, concentrations exceed an upper tolerance limit, symptoms associated with toxicity can result [1]. This 'window of essentiality' varies between different species as a result of differences in metal toxicokinetics and toxicodynamics (metal ion handling, target site sensitivity) and comprises a component part of the ecological niche controlling species biogeography and habitat preferences [2].

This chapter considers the impacts of exposure to trace metal concentrations that fall outside the niche preferences. The focus is on exposures above the upper threshold (that is on toxicity). Toxic effects of trace metal are known to be important for terrestrial ecosystems because it has been demonstrated that exposure to high concentrations of trace metals in soils can have adverse impacts on soil communities [3–5]. These effects arise for two main reasons. First, it is well known that a number of soil taxa are sensitive to the toxic effects of trace metals, with high concentrations known to alter the abundance and diversity of earthworms [6,7], microarthropods [8], nematodes [9], microbes [10,11] and plants [12] and also soil functions, such as the breakdown of organic matter and turnover of essential nutrients [13,14]. Secondly, unlike organic contaminants, metals are not subject to degradation over time. This means that the loss of metals from soils is effected only through the relatively slow process of erosion, leaching and in some areas cropping [15,16]. The natural presence and additional augmentation of the metal content of soils, thus, offers one of the most serious and long-term threats to soil communities.

8.2 Hazards and Consequences of Trace Metal Exposure

8.2.1 Effects on Individuals, Risk Assessment and the Prediction of Population Effects

Among the first approaches to be used to assess the effects of metals on soil communities was the use of single-species toxicity tests conducted to generate concentration–response profiles for acute and chronic life-cycle end points. This focus on direct ecotoxicological effects has resulted in the generation of a considerable database of metal toxicity studies for soil species (see www.epa.gov/ecotox; also [17,18]). Among studies, the majority focus on assessing the toxicity of the most common pollutant metals and metalloids such as Cd, Pb, Cu and Zn and to a lesser extent Ni, Hg, Cr, Co, Se and As. For remaining metals, only a limited number of toxicity studies exist, although for certain metals this situation may change should further data be requested or made available as a result of studies conducted in support of European Council Regulation 793/93/EEC on Existing Chemicals. Over the coming years, further systematic studies may address these gaps in knowledge, and thereby improve the basis for metals risk assessment at sites polluted with single and multiple metals.

Consideration of the toxicity database for soil species indicates that studies on only a few taxa dominate. These are principally those groups for which toxicity tests initially developed and standardized for pesticide registration have been adopted for metals testing. These include earthworms, enchytraeids, springtails, terrestrial (crop) plants and soil microbial function such as carbon and nitrogen mineralization and enzyme activities [19]. The availability of the standardized procedures for these groups means that it is relatively simple to undertake comparative studies of between-taxa sensitivity for individual metals [20–24]. Further, by slight modification, each test procedures can be fine tuned to allow testing with nonstandard species from the same taxa, thereby allowing within-taxon sensitivity variation to be estimated [25]. For the remaining soil groups, a focused effort was made (in the EU funded SECOFASE project) to develop standardized test procedures to support the expansion of the available metal toxicity database. This work resulted in the publication of test protocols for nematodes, mites, staphylinid beetles, centipedes, millipedes and woodlice that have been used to assess metal toxicity for representative species within these taxa [26].

While single-species toxicity studies are a valuable resource for effect prediction in their own right, it was the development of method for the 'meta-analysis' of the metal toxicity database that proved key to exploiting this available data resource for metals risk assessment. This species sensitivity distribution (SSDs) approach was initially developed from principles outlined by Kooijman [27] and later refined by Van Straalen and Denneman [28] and Aldenberg and Slob [29]. The SSD approach requires that toxicity data for a given metal for multiple species is available. This interspecies variability in sensitivity among tested species is used as a surrogate for the interspecies variation in ecological communities. By plotting the distribution of tested sensitivities, as defined by established toxicity measures such as the no observed effect concentration (NOEC) or particular effect concentrations (EC_x), the mean and the standard deviation of these toxicity metrics can be used to estimate a concentration within the left-hand tail of the SSD that results in toxicity for only a small proportion (usually 5%) of exposed species. This concentration is designated as HCp, 'hazardous concentration for p per cent of species' [30].

Although controversial from the outset [31,32], the SSD approach has found wide application in metals risk assessment [17,33]. Take-up of the SSD approach for risk assessment was facilitated by a validation exercise that suggested that an HC_5 derived from the soil species toxicity database for Zn, provided the requisite level of protection for soil microbial functions (no effect on 95% of all functions) as measured by community physiological profiling using BIOLOG^{TM} plates [34,35]. From this point, the SSD approach has became formalized as a key part of risk assessment guidance for existing substances, including for the assessment of trace metals in a series of risk assessments that have recently been conducted within the European Union under European Council Regulation 793/93/EEC.

A recent example of the use of SSDs in metals risk assessment under EU existing substance guidelines has been the assessment for Zn [33]. In this appraisal, a large volume of toxicity data was available from the literature and this was further augmented with data generated in a series of targeted studies. One of the experiments conducted compared the toxicity of Zn in freshly spiked soil with that in a naturally contaminated soils collected along a transect from galvanized pylons [36]. Results indicated that toxicity was greater in the freshly spiked soil by at least a factor of 3. This supported previous work that also identified a similar difference between freshly spiked and long-term metal smelter contaminated soil [37]. The discrepancies between toxicity in spiked and long-term contaminated soils was attributed to differences in metal speciation linked to the slow kinetics of high-stability binding between metal ions and soil constituents such as organic matter and Al and Fe oxides. These results led directly to the introduction of a 3-fold laboratory-field correction factor on HC_5 values for Zn in soil that were derived from the available laboratory toxicity tests, since the studies were almost exclusively in freshly spiked soil.

While effect measures based on individual traits (for example survival, reproduction, maturation) provide information on the relative sensitivities to metals and between species and also invaluable input into SSDs, it is widely recognized that measurements of the toxic effect of a metal on a single life-history trait alone is insufficient to predict the fitness of populations. This is because, as has been demonstrated, the most sensitive trait at the life history level may not have the largest impact on population growth rate [38]. To translate toxicant effects on individual to consequences for populations, demographic models can be used [39]. These models allow calculation of the Malthusian parameter, the intrinsic rate of population increase or r (an estimate of the ability of a population to increase in an unlimited environment) from Euler's equation (for further information see [39]).

To use the demographic approach to assess the toxic effects of metals on population growth rate requires that survival and fecundity schedules are quantified throughout the life cycle. This focus on life-cycle testing means that to date most studies conducted to assess the fitness effect of metals have used short-lived soil species. These include particularly nematodes [40,41] and springtails [42,43]. While less common due to the technical complexity of acquiring the necessary data, demographic models can also be developed for longer-lived species such as snails and earthworms [44,45]. Such models can be parameterized using data derived from partial life-cycle tests. A key aspect of most demographic studies, is the use of sensitivity analysis to identify the extent of fitness effect arising from a particular level of change in a given life-cycle trait. This information can be used to focus future testing on assessing toxic effects on the most important traits.

A further approach to analysing the potential population effects of exposure, also based on measurement of changes in life-cycle traits, are physiologically based model such as the dynamic energy budget (DEB) model [46]. This model has been adapted in a number of ways for metals risk assessment. For example, a physiological based models was used to assess the population-level consequences of changes in growth and reproduction of earthworms in metal-polluted soils [47]. These polluted field soils resulted in a 23% lower growth rate of exposed earthworms and a change in population demography toward younger individuals. Field data on population composition of earthworms were used to support the laboratory results; however, despite this and other advances, the challenge for demographic studies remains how to link laboratory data on the toxicity effect of metals on multiple traits with the effect of extrinsic factors that also affect population size and performance in the field.

8.2.2 Populations and Communities

While measurement of toxic effects of metals on life-history traits of individual species provides the foundation for much of modern metals ecotoxicology and risk assessment, it is worth remembering that these studies are not an end in their own right. Instead, they are designed to provide data that can be used to predict the effects of metals on populations and communities at metal-polluted field sites. Alternative approaches to predicting field effects from laboratory data, which can be undertaken when moving from generic to site specific studies, are direct population and community measurements. In one of his classic contributions to the field, Hopkin [48] outlined four approaches to assessing the field effects of pollutants in studies of the impact of point source or diffuse pollution. These were:

- assessing impacts on community structure and function;
- measuring pollutant concentrations in the tissues of sentinel species;
- quantifying pollutant effects on individual performance; and
- detecting the presence of genetically altered races that are resistant to pollutants.

All of these strategies can play a role in any study of the effects of metals on soil ecosystems. For example, measuring tissue residue concentrations can provide information on metal transfer to predators [49,50]; measuring effects on individuals provides information on specific trait responses; while adapted populations can provide understanding of the mechanisms that allow species to maintain viable populations in long-term polluted mine and smelter contaminated soils [51–53]. Of the outlined approaches, however, perhaps the one with the most obvious relevance to the detection of metal effects is the assessment of metal-induced change in soil community structure. This is both because the absence of species from areas where they would normally be expected to occur provides a strong indication of the deleterious effects of metals and also because loss of species can be expected to compromise the multi-functionality of the soil communities present at polluted sites [54,55].

There are numerous studies that have assessed the effect of metals on the dominance structure and diversity of different soil communities. These include studies undertaken to quantify community change in major invertebrate groups along metal contamination gradients [7,56], as well as studies with specific soil taxa like earthworms [57–59], springtails [8,60], mites [61], ground beetles [62] and spiders [63]. Changes observed in these studies included reduced overall abundance (including the presence of deserts for some groups), lower diversity as assessed using traditional metrics, altered community composition and changes in species dominance structure [64]. Structural characterization of the microbial community by physiological and genetic methods has also revealed that metal can reduce the community diversity of the microflora [65–67].

One of the most extensive studies conducted to date of the extent and effects of metals on soil ecosystems is that conducted for surface-active and soil/litter dwelling invertebrates at site along a transect from the Avonmouth smelter in South West England. Over the 100-year operational period of this plant, emission of Cd, Cu, Pb, Zn and to a lesser extent of other metals has resulted in the contamination of a wide area of seminatural or pasture land around the factory. This has resulted in the presence of a contamination gradient that extends for over 10 km over which soil concentrations range from grossly polluted ($>20\ 000$ mg Zn kg^{-1}, $>10\ 000$ mg Pb kg^{-1}, >1000 mg Cu kg^{-1}, >200 mg Cd kg^{-1}) to the regional background. Studies undertaken at the Avonmouth site have included an assessment of the distribution of major invertebrate groups [68,69]; a carabid beetle communities diversity study that identified effect on community structure associated with altered seasonal patterns of reproduction and resulting year class recruitment [62,70]; and an earthworm communities study that identified the sensitivity of the weakly calciferous *Allolobophora* and *Aporrectodea* species compared with the calciferous *Lumbricus* worms [6,71].

The most detailed study conducted to date of the community effects of the metals present at Avonmouth was conducted for macro-arthropod communities by Sandifer [72]. Results indicated that the most severe effects impacts of metals occurred for decomposers taxa such as earthworms, isopods, molluscs, myriapods, springtails and mites (Table 8.1). These groups were all absent or reduced in abundance and/or diversity at the two most metal-polluted sites (and in some case at more distant locations). The sensitivity of these groups accords with results from community studies conducted in areas such as vineyards, where the addition of Cu as a pesticide has lead to an elevation of soil Cu levels. Here reductions in the diversity and abundance of

Table 8.1 Invertebrate groups at sites located along a pollution transect from the Avonmouth Cd/Pb/Zn smelter: ▓▓ = group present at 'normal' abundance; ■■ = group hyper-abundant; ▒▒ = group present at low numbers or reduced diversity; ░░ = group absent from site. Data summarized from Spurgeon and Hopkin, 1999 and Sandifer 1996.

Class	Order	Family	Most polluted	Polluted	Moderately polluted	Slightly polluted
Arachnida						
	Araneida					
		Lycosidae	■	▓	▓	▓
		Linyphiidae	░	▒	▓	▓
	Opiliones		░	▒	▓	▓
	Acari		░	▒	▓	▓
Insecta						
	Collembola		░	▒	▓	▓
	Orthoptera		▓	▓	▓	▓
	Hemiptera		▓	▓	▓	▓
	Coleoptera					
		Staphylinidae	■	▓	▓	▓
		Carabidae	░	▓	▓	▓
	Hymenoptera					
		Formicidea	▓	■	▓	▓
Crustacea						
	Isopoda		░	░	▒	▓
Myriapoda						
	Chilopoda		▓	▓	▓	▓
	Diplopoda		░	░	▒	▓
Mollusca						
	Slugs		░	▓	▓	▓
	Snails		░	▒	▓	▓
Oligochaeta			░	▒	▓	▓

similar decomposer taxa has been recorded, with earthworms in particular proving to be a sensitive group [73,74].

Although the changes occurring when soil communities are exposed to metals have been characterized at a number of sites, uncertainties in the taxonomy of some soil groups and the general lack of taxonomic expertise for some fauna has meant that the widespread use of community cenosis approaches has not taken off in soils as it did in the 1980s for riparian monitoring [75]. As an alternative to characterizing community structural change, it has been proposed that the effect of metal on the functional activity of the soil community can be used to provide a surrogate indication of community response. With respect to the range of functions undertaken by soil communities, perhaps the single most important is litter decomposition. This activity is course vital for carbon and nitrogen cycling, the knowledge of which is emerging as one of the key needs for understanding the process and effects associated with long-term environmental change.

Studies that describe the effects of metals on litter decomposition (or surrogate measures) are widely reported in the literature [14,76]. In many cases it has been shown that exposure to high concentrations of metals (for example in smelter contaminated areas or vineyards) results in a reduction in litter turnover. These results explain previous observations of a build-up of undecomposed litter in heavily metal-polluted soils and suggests that toxic metal exposure can have a profound effect on soil community-mediated carbon turnover [77,78]. Beyond studies at point sources and vineyards, some evidence of wider effect of metals on litter degradation has been suggested by bait lamina assessments made in a metal-contaminated flood plain [79]. In this work, key decomposers (millipedes, woodlice and earthworms) inhabiting these contaminated areas were also found to have higher tissue metal burdens. Since it has been suggested that increased concentrations of metals in tissues may have a physiological effect on feeding rate [40,80], the observed reduction in bait lamina feeding may be linked to an individual-based effect associated with high tissue metal levels rather than any community-level change. This concept of a physiological feedback to ecosystem functioning is a relatively new development.

Assays that can be used to further investigate the effects of metals on soil functions range from measurement of community rates of litter removal to more focused assays of the activity of specific soil enzymes and utilization of carbon substrates. At the coarsest level, studies with litter bags and similar approaches have been used to identify a general effect of metals on litter removal that can explain the accumulation of litter that is frequently seen at grossly polluted sites [81]. Focusing more specifically on the microbial community, measurement of functions including elements of the nitrogen cycle and enzyme activity can be used to provide useful information on metal effects on soil systems. For nitrification, which is the second step in the mineralization of nitrogen from organic matter, so far studies of the effects on metals of this process have not produced a consistent picture. Dusek [82] found that Cd disrupted nitrification in spiked soils; however, high concentrations (100 and 500 mg Cd kg^{-1}) were required, suggesting that the function had low sensitivity. Murray *et al.* [83], in contrast, found no effect of Cd, Cu, Ni and Pb, while Kandeler *et al.* [84] found that nitrification was influenced more by the supply of organic substrate in sludge than by the heavy metal content. Like nitrification, nitrogen fixation can also respond to metals, with population sizes of nitrogen fixers, nodulation, nitrogen fixation and growth of legumes all impacted by metal exposure [85,86].

A range of soil enzymes, such as dehydrogenase, glucose oxidase, catalase, peroxidase, phosphatases, cellulase, proteinase and urease can also be impacted by metal exposure [87]. Chander and Brookes [88] for example, found that both dehydrogenase and phosphatase activities were reduced by Cu, with the former enzyme being more sensitive. In industrially contaminated soils, containing significant amounts of hydrocarbons and metals, Brohon *et al.* [89] found that urease and dehydrogenase were inhibited in the most contaminated soil. Trasar-Cepeda *et al.* [90], however, found that activities of individual enzymes were apparently unaffected in three soils contaminated by tanning effluent, hydrocarbons or landfill effluent.

Focusing more on carbon utilization, community-level physiological profiling, CLPP (for example using BIOLOG plates) has shown the effects of metals on individual carbon substrate use that signify changes in underlying soil microbial community structure. One particular application of the BIOLOG system has been the analysis of pollution-induced community tolerance (PICT) in microbial communities. PICT is the process by which a stressor (for example metals) alters microbial assemblage structure through adaptation and extinction, thereby, leading to a more resilient community. Rutgers *et al.* [35] used the PICT approach to study the effects of Zn in a field plot. For most substrates, the metabolic activities showed an increased community tolerance with increasing Zn concentration, indicating PICT. The data were used to demonstrate a community change at soil Zn concentrations close to the soil protection guideline value and this was taken as supporting evidence for the validity of these regulatory limits.

8.3 Routes of Exposure, Uptake and Detoxification

8.3.1 Uptake Routes and Speciation Models

In concert with efforts to describe the toxic effects of metals on individuals and scale these up to predict effects on populations and communities, an effort has also been made to understand the routes of metal exposures, the toxicokinetic of uptake and the principal molecular cellular and molecular pathways involved in detoxification. For plants and microbes, a well-supported assumption is that the main route of metal exposure is by passive or active uptake from soil pore water. For invertebrates, however, a debate continues about the balance of uptake between the dermal and feeding routes. In a study of organic molecule uptake in earthworms, Jager *et al.* [91] concluded that it was only for very hydrophobic molecules that the gut was an important route of accumulation. Although similar systematic work is lacking for metals, it is widely anticipated that the passive diffusion route from the soil solution will be dominant in those soil species (microorganisms, protozoa, annelids, nematodes, microarthropods) that live in close association with the soil pore water [92,93]. For surface-active species, and in particular for predators like beetles, centipedes, harvestmen and spiders, exposure from food is likely to be an important route of (secondary) exposure [68,94].

Identification of the soil solution as the likely major route of metal exposure has allowed soil chemistry-based modelling approaches originally developed in aquatic ecotoxicology to be used for exposure prediction in soil-dwelling species. Such models were initially based on the free ion activity model and then later modified to utilize the Biotic Ligand Model (BLM) framework [95]. There are two main postulates of BLM theory:

1. For organisms primarily exposed from their surrounding medium (soil pore water), the free metal ion is the chemical species which is taken up at the site of toxic action.
2. The rates of free ion uptake are determined by two factors: the concentration of the free metal ion in soil solution (as measured or determined using chemical speciation models) and the composition of the medium including specifically the concentrations of major cations (predominantly H^+, Ca^{2+}, Mg^{2+}) that compete with the free ion for binding at the uptake site (the biotic ligand).

In aquatic ecotoxicology the adoption of the BLM has greatly improved the theoretical basis for predicting the effect of metals on individual species [96–99]. Transfer of the approach to predicting sediment toxicity [100] opened up the way to the development of BLMs for terrestrial species. To date BLMs have been developed for species within a number of taxa including microbes [101,102], plants [102,103], macroinvertebrates [104–106] and microarthropods [92], and for metals including Cu, Zn, Ni, Co and Cd.

While the development of BLM theory has undoubtedly allowed a major advance in our understanding of metal exposure in soil species, there has been some debate about how the approach can be used for risk assessment. As outlined previously, the state of the art for metal risk assessment had moved away from a single-species approach to a multispecies assessment based on the use of SSDs. This presented the problem of how to integrate BLM approaches, which are parameterized for an individual species and metals into a generic approach that exploits information on differential species sensitivity.

One simplified approach to linking metal speciation and BLM theory with the SSD approach was developed by Lofts *et al.* [87,107]. This work proposed an alternative empirical approach to the BLM, where the major cations are considered to protect the organism against the toxic effects of the free metal ion but no specific protection mechanism is used to derive expressions quantifying this protective effect. For soils, only the protective effect of the proton (H^+) could be studied, since concentration of other competing cations were not reported in the literature studies from which data for the assessment of sensitivity was derived. Based on a comprehensive search of literature data, the approach used for assessing each metal followed a three-phase methodology. Initially, functions were derived to predict free metal ion concentrations from the physiochemical properties of a given soil. Next, toxicity test data (NOEC, EC_x from laboratory studies) were collated and the derived functions were used to convert these values to equivalent free metal ion concentrations based on reported soil properties (pH, organic matter content). A weighted linear regression of the toxicity value expressed as log free metal ion concentration was then plotted against soil pH and used to derive a relationship for the protective (competition) effect of H^+. Finally, toxicity values for individual studies were plotted on this relationship and their residual from the regression was taken as a measurement of species sensitivity for input into an SSD (Figure 8.1). While limited in the extent that soil properties are taken into account, the approach has the advantage that information of species sensitivity can be captured and used generically to predict sensitivity in different soil types for combinations of metals and soil species for which no BLM exists [108].

8.3.2 Toxicokinetics and Compartment Models

For predicting internal concentration of metals in the tissue of exposed species, the balance between routes of exposure and the influence of soil properties on uptake at any potential biotic ligand represents only a part of the equation. Mechanisms that control rates of uptake (across any biotic ligand) and elimination are also of fundamental importance. The kinetics of metal transport can be addressed by combining time series measurement of tissue metal concentration with compartment models that describe uptake and elimination from internal pools. The most commonly used model of this type to describing metal uptake by soil invertebrates has been the one-compartment model [109–111]. This is a simple pharmacokinetic model that assumes uptake and elimination from a single undifferentiated whole body compartment [109]. Recent work, however, has suggested that for some metals, a

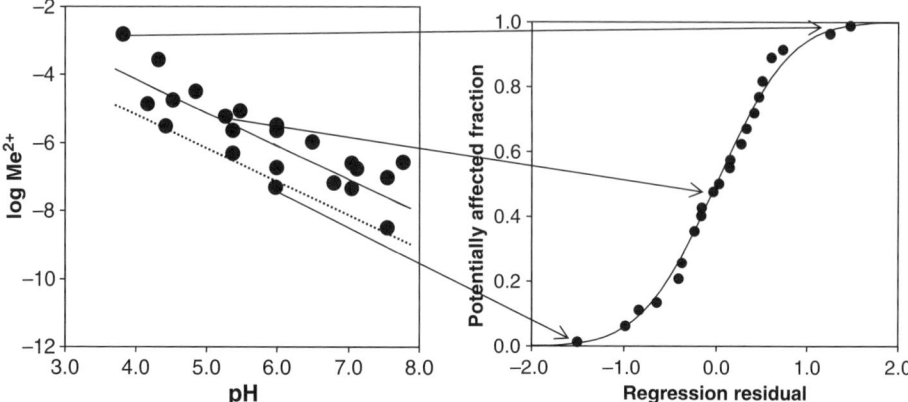

Figure 8.1 Schematic representation of the procedure used by Lofts et al. [106] to derive environmental quality standards for soil organisms based on toxicity data collected in soils of diverse physicochemical properties

model that describes both a loosely bound and rapidly eliminated compartment and a storage bound and slowly eliminated compartment (that is a two-compartment model) may provide a better description of observed accumulation patterns [112]. The two-compartment concept can be used to explain the results of two previous studies of metal kinetics in soil species as outlined below.

In the first study, the accumulation kinetics of four metals (Cd, Cu, Pb, and Zn) were assessed in the earthworm *Eisenia fetida* exposed to a smelter polluted soil [113]. Different kinetics were found, with fast uptake and elimination rates found for the essential metals Cu and Zn and slow rates for the nonessential metals Cd and Pb (Figure 8.2). These differences in uptake could be explained by a number of factors associated with the essentiality and chemistry of the four metals; however, in the context of two-compartment model; the distribution of these four metals into two types of granules found in cells of the chloragogenous tissue can provide a biological explanation. One of the granule types contained Ca, Pb, and a limited amount of Zn bound to one or more oxygen-containing, probably phosphate-rich, ligands; the second type was sulfur-rich and contained Cd that was most probably bound to compartmentalized metallothionein [114,115]. For Pb and Cd, the presence of these granules represents a storage pool from which release kinetics are slow. In contrast to the two nonessential metals, for both Cu and Zn no major storage pool is known. Instead it has been suggested that these metals when accumulated become associated with cytosolic metal-binding protein or small-metabolite chelators [116,117] from which they can be readily released and so rapidly eliminated.

In the second study, Janssen *et al.* [109] used a one-compartment model to describe the uptake and excretion of Cd from four soil invertebrate species, the mite *Platynothrus peltifer*, the springtail *Orchesella cincta*, the carabid *Neobisium muscorum*, and the harvestman *Notiophilus biguttatus*. Results indicated that excretion kinetics were rapid in the insects *Orchesella cincta* and *Neobisium muscorum* and low in the arachnids *Platynothrus peltifer* and *Notiophilus biguttatus*. The conclusion from this work in the

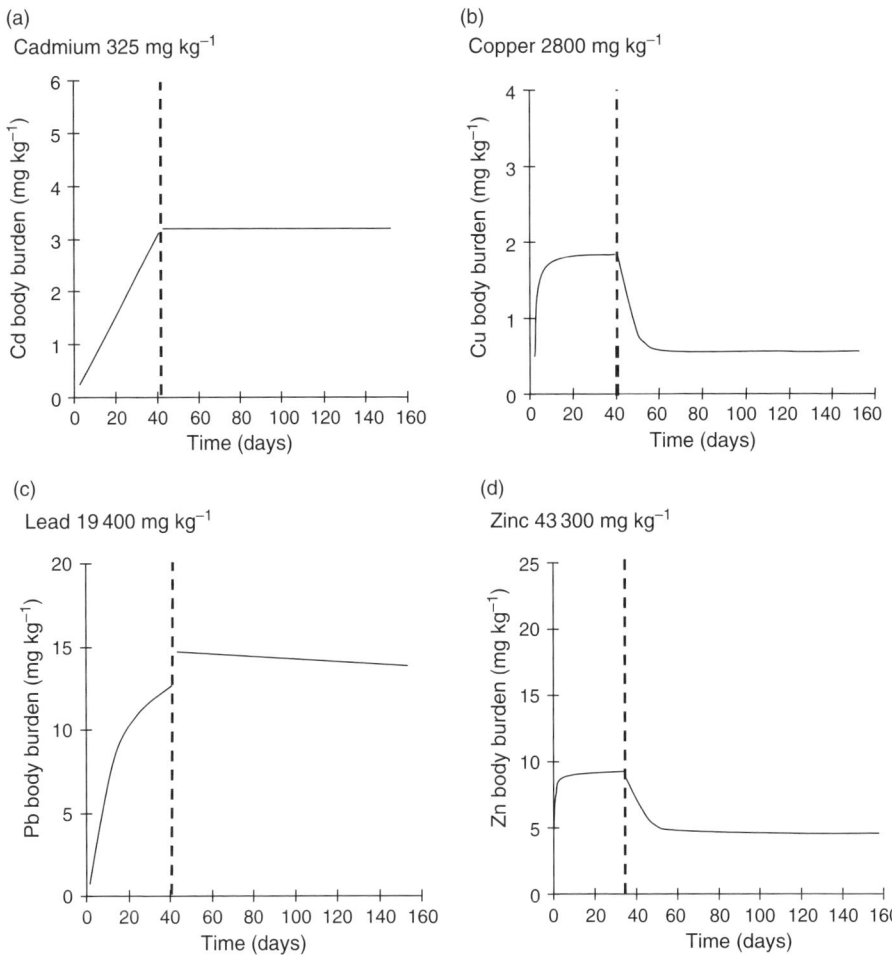

Figure 8.2 *Accumulation and elimination of (a) cadmium, (b) copper, (c) lead, (d) zinc in Eisenia fetida exposed to field soils collected from a highly metal-polluted site located close to the Avonmouth cadmium/lead/zinc smelter*

light of the study of Vijver *et al.* [112] and associated work is that the species that show low rates of elimination are those that rely on storage detoxification of Cd into pools from which kinetics are slow, while species with rapid elimination utilize small-molecule chelators, from which the metal can be rapidly excreted.

A further insight to be gained from toxicokinetics studies with metals is an insight into the effects of exposure duration on species sensitivity. Within general ecotoxicology it is often assumed that the probability of a toxic effect occurring following exposure is related to the internal concentration of the chemical [118]. This idea has led to the development of the 'critical body residue' approach. This postulates that a particular adverse response will occur when the body concentration reaches the critical level for that effect [119,120]. The link between body residue concentrations and the likelihood of toxicity has implications for the effect of toxicity exposure duration on studies conducted to compare toxicity between different species or metals for which toxicokinetics differ. In cases where accumulation to equilibrium is rapid (such as Cu and Zn in earthworms and Cd in some insects); the probability and extent of toxic effect initially increases rapidly, but thereafter remains constant. This means that toxicity parameters (NOEC, EC_x) calculated from toxicity tests will be time dependent only for this initial short period. In cases where accumulation to equilibrium is slow (for example Cd and Pb in earthworms and Cd in some arachnids); toxicity statistics will tend to decrease with extended exposure. Since standardized test are time limited and usually conducted without prior knowledge of toxicokinetics, this means that for some cases where the uptake is slow, routine (short-term) laboratory tests may underestimate long-term field toxicity.

8.3.3 Molecular Mechanisms of Detoxification and Effect

Much of the classic toxicology on metals was, of course, conducted in non-soil-dwelling species. This means that well-known conclusions about the mechanisms through which metals exert their toxicity, such as the fact that Cd is a hepatotoxin and Hg a neurotoxin, are not necessarily relevant for all the diversity of soil life. While it seems likely that molecular mechanisms linked to nonspecific modes of action, such as binding to thiol groups in proteins and cofactor substitution, are important as causes of metal toxicity, the focus of ecotoxicology on measuring effects on individuals means that the underlying mechanism causing toxicity frequently remain to be established in many soil species.

Recognizing that changes in the abundance of gene transcripts and proteins are the first indications of toxicity, measurements of expression changes have become core 'tools' for assessing the biochemical response of soil species to metal exposure [4,121,122]. Within these studies, the focus is most usually on individual gene products, with three generalized groups of markers usually the focus: there are proteins linked to exposure, physiological compensation, and/or toxic effect.

Metal exposure-associated molecular responses known in soil species include a number of proteins involved in metal handling and detoxification. Metallothionein-like proteins, for example, have been found to be upregulated in metal-exposed earthworms [123,124], enchytraeids [125], molluscs [126,127], and springtails [128]. High metallothionein expression phenotypes have been found in springtail populations collected from metal-contaminated sites, indicating a likely role of these proteins in inherited metal tolerance [129]. A further gene product that is directly responsive to metal exposure is phytochelatin synthase. This thiol-rich peptide has been shown to play a key role in the development of Cd adaptation in *Caenorhabditis elegans* [130]. The globin-related protein ferritin, plays a specific key role in iron trafficking and has been found to show altered expression following metal exposure of both plants and animals [131,132].

Adaptive responses of soil species to metals include differential regulation of proteins involved in common compensatory stress-responsive pathways. Expression of heat shock proteins, for example, has been found to alter expression following metal exposures in earthworms [133,134], oribatid mites [135], woodlice [136,137], and nematodes [136]. Other gene products associated with compensatory responses to metals include lysosomally associated proteins that are responsive to toxicity mediated by radical oxygen species [138,139] and genes encoding components of the antioxidant system [140]. Finally, a widely observed compensatory response that also has a direct link to the observed toxic effects of metals is expression change of mitochondrially associated transcripts [123]. Changes in the expression of these markers indicate a change in energy metabolism in response to metal exposure, such as has also been identified from the analysis of small-metabolite markers in earthworms inhabiting long-term metal-polluted mine sites [141].

Building on the use of the targeted pathway approach for mechanistic assessment has been the use of toxicogenomics (transcriptomics, proteomics and metabolomics). The relatively early availability of an extensive database of genome and/or expressed sequence tag (EST) information initially for the soil-dwelling nematode *C. elegans* (www.wormbase.org) and latterly of ESTs in two earthworms (*Lumbricus rubellus*, *Eisenia andrei*) [142] (www.earthworms.org) and two springtails (*Folsomia candida*, *Orchesella cincta*) (www.collembase.org), as well as publication of genome sequences for a range of plants and soil-dwelling microbes has opened up the possibilities of toxicogenomic studies in a variety of soil taxa.

An example of the type of toxicogenomic study that is possible and the insight into the mechanisms of metal toxicity that these can provide is given by Bundy *et al.* [143]. This work monitored both the molecular (cDNA transcript microarrays and nuclear magnetic resonance, NMR-based metabolic profiling – metabolomics) and ecological/functional responses (reproduction rate and weight change) of the earthworm *L. rubellus* exposed to Cu. Measured transcript and metabolite profiles indicated that Cu disrupted a number of aspects of energy metabolism including oxidative phosphorylation and carbohydrate metabolism. By treating both enzymes and metabolites as functional cohorts, clear inferences were gained about changes in energetic metabolism (carbohydrate use and oxidative phosphorylation), which would not have been possible by taking a targeted (gene-by-gene) approach. The cohort of metabolic changes identified could further be linked to life-history response to Cu toxicity, including reductions in feeding, growth rate, reproduction and survival. Thus, by adopting a mechanistic approach the fundamental link between the mechanisms of metal exposure and the responses of individuals that are the currency of mainstream ecotoxicology and risk assessment could be gained. Studies of the type outlined above have the potential to provide a greater level of mechanistic understanding of the toxic effects and adaptive change that can occur when soil species are exposed to metals over long exposure durations in the field.

8.4 Conclusions

In industrialized regions trace metal are near ubiquitous contaminants of soils. This has led to a long-standing research interest in assessing the effects of these metals (if any) on soil communities. Starting with the utilization of standardized toxicity tests (originally developed for pesticides) to establish the effects of single metals on acute and chronic life-cycle response, the field has broadened to focus on both higher-tier effects and mechanistic causes. The most prominent developments include the use of demographic approaches to integrate life-history effects for predicting effects on populations; community structure and functional analysis at metal-polluted sites; development of models to account for soil physiochemical property effects on availability and toxicity; studies of the major routes of exposure; and the use of species sensitivity distribution to support more scientifically robust risk assessment. The research undertaken in these areas has meant that some of the mechanisms and processes linking metal exposure on individual species and communities are now much better understood. With the recent addition of a systems toxicology aspect, it is now possible to link the biochemical responses of soil species to metal exposure to their higher-tier consequences in more exquisite detail than previously achieved.

References

[1] Hopkin, S.P., Ecophysiology of Metals in Terrestrial Invertebrates; Elsevier Applied Science, London, 1989.
[2] Van Straalen, N.M.; Roelofs, D., An Introduction to Ecological Genomics; Oxford University Press, Oxford, 2007.
[3] Walker, C.H.; Hopkin, S.P.; Sibly, R.M. et al., Principles of Ecotoxicology, 3rd edn; Taylor & Francis, London, 2006.
[4] Weeks, J.M.; Spurgeon, D.J.; Svendsen, C. et al., Critical analysis of soil invertebrate biomarkers: a field case study in Avonmouth, UK; Ecotoxicology 2004, 13, 817–822.
[5] Dallinger, R.; Rainbow, P.S. (eds), Ecotoxicology of Metals in Invertebrates; Lewis, Boca Raton, FL, 1993.
[6] Spurgeon, D.J.; Hopkin, S.P., Seasonal variation in the abundance, biomass and biodiversity of earthworms in soils contaminated with metal emissions from a primary smelting works; J. Appl. Ecol. 1999, 36, 173–183.
[7] Hunter, B.A.; Johnson, M.S.; Thompson, D.J., Ecotoxicology of copper and cadmium in a contaminated grassland ecosystem. 2. Invertebrates; J. Appl. Ecol. 1987, 24, 587–599.
[8] Pedersen, M.B.; Axelsen, J.A.; Strandberg, B. et al., The impact of a copper gradient on a microarthropod field community; Ecotoxicology 1999, 8, 467–483.
[9] Korthals, G.W.; Alexiev, A.D.; Lexmond, T.M. et al., Long-term effects of copper and pH on the nematode community in an agroecosystem; Environ. Toxicol. Chem. 1996, 15, 979–985.
[10] Smit, E.; Leeflang, P.; Wernars, K., Detection of shifts in microbial community structure and diversity in soil caused by copper contamination using amplified ribosomal DNA restriction analysis; FEMS Microb. Ecol. 1997, 23, 249–261.
[11] Oorts, K.; Bronckaers, H.; Smolders, E., Discrepancy of the microbial response to elevated copper between freshly spiked and long-term contaminated soils; Environ. Toxicol. Chem. 2006, 25, 845–853.
[12] Strandberg, B.; Axelsen, J.A.; Pedersen, M.B. et al., Effect of a copper gradient on plant community structure; Environ. Toxicol. Chem. 2006, 25, 743–753.
[13] Salminen, J.; van Gestel, C.A.M.; Oksanen, J., Pollution-induced community tolerance and functional redundancy in a decomposer food web in metal-stressed soil; Environ. Toxicol. Chem. 2001, 20, 2287–2295.
[14] Filzek, P.D.B; Spurgeon, D.J.; Broll, G. et al., Metal effects on soil invertebrate feeding: measurements using the bait lamina method; Ecotoxicology 2004, 13, 807–816.
[15] Tipping, E.; Lawlor, A.J.; Lofts, S. et al., Simulating the long-term chemistry of an upland UK catchment: heavy metals; Environ. Pollut. 2006, 141, 139–150.
[16] Hall, J.R.; Ashmore, M.; Fawehinmi, J. et al., Developing a critical load approach for national risk assessments of atmospheric metal deposition; Environ. Toxicol. Chem. 2006, 25, 883–890.
[17] Posthuma, L.; Traas, T.P.; Suter, G.W., The Use of Species Sensitivity Distributions in Ecotoxicology; Lewis, Boca Raton, FL, 2001.

[18] Edwards, C.A.; Bohlen, P.J., The effects of toxic chemical on earthworms; Rev. Environ. Contam. Toxicol. 1992, 125, 23–99.
[19] Spurgeon, D.J.; Svendsen, C.; Hankard, P.K. et al., Review of sublethal ecotoxicological tests for measuring harm in terrestrial ecosystems; Technical Report P5-063/TR1, Bristol, Environment Agency, 2002.
[20] Lock, K.; Janssen, C.R., Ecotoxicity of mercury to *Eisenia fetida*, *Enchytraeus albidus* and *Folsomia candida*; Biol. Fertil. Soils. 2001, 34, 219–221.
[21] Lock, K.; Janssen, CR., Ecotoxicity of nickel to *Eisenia fetida*, *Enchytraeus albidus* and *Folsomia candida*; Chemosphere 2002, 46, 197–200.
[22] Lock, K.; Janssen, C.R., Toxicity of arsenate to the compost worm *Eisenia fetida*, the potworm *Enchytraeus albidus* and the springtail *Folsomia candida*; Bull. Environ. Contam. Toxicol. 2002, 68, 760–765.
[23] Lock, K.; Janssen, C.R., Ecotoxicity of chromium (III) to *Eisenia fetida*, *Enchytraeus albidus*, and *Folsomia candida*; Ecotox. Environ. Saf. 2002, 51, 203–205.
[24] Lock, K.; Janssen, C.R., Comparative toxicity of a zinc salt, zinc powder and zinc oxide to *Eisenia fetida*, *Enchytraeus albidus* and *Folsomia candida*; Chemosphere 2003, 53, 851–856.
[25] Spurgeon, D.J.; Svendsen, C.; Rimmer, V.R. et al., Relative sensitivity of life-cycle and biomarker responses in four earthworm species exposed to zinc; Environ. Toxicol. Chem. 2000, 19, 1800–1808.
[26] Løkke, H.; Van Gestel, C.A.M., Handbook of Soil Invertebrate Toxicity Tests; John Wiley & Sons. Ltd, Chichester, 1998.
[27] Kooijman, S.A.L.M., A safety factor for LC_{50} values allowing for differences in the sensitivity among species; Ecotox. Environ. Saf. 1987, 18, 241–252.
[28] Van Straalen, N.M.; Denneman, C.A.J., Ecotoxicological evaluation of soil quality criteria; Ecotox. Environ. Saf. 1989, 18, 241–251.
[29] Aldenberg, T.; Slob, W., Confidence limits for hazardous concentrations based on logistically distributed NOEC toxicity data; Ecotox. Environ. Saf. 1993, 25, 48–63.
[30] Van Straalen, N.M., Threshold models for species sensitivity distributions applied to aquatic risk assessment for zinc; Environ.Toxicol. Pharmacol. 2002, 11, 167–172.
[31] Hopkin, S.P. Ecological implications of '95% protection levels' for metals in soils; Oikos 1993, 66, 137–141.
[32] Forbes, V.E.; Forbes, T.L., A critique of the use of distribution-based extrapolation models in ecotoxicology; Funct. Ecol. 1993, 7, 249–254.
[33] Bodar, C.M.; Pronk, M.E.J.; Sijm, D.T.H.M., The European Union risk assessment on zinc and zinc compounds: the process and the facts; Int. Environ. Assess. Manag. 2005, 1, 301–319.
[34] Posthuma, L.; Van Gestel, C.A.M.; Smit, C.E. et al., Validation of toxicity data and risk limits for soils: final report. Bilthoven, The Netherlands: Rijksinstituut voor Volksgezondheid en Milieu; 1998. Report No.: 607505004.
[35] Rutgers, M.; van't Verlaat, IM; Wind, B. et al., Rapid method for assessing pollution-induced community tolerance in contaminated soil; Environ. Toxicol. Chem. 1998, 17, 2210–2213.
[36] Smolders, E.; McGrath, S.P.; Lombi, E. et al., Comparison of toxicity of zinc for soil microbial processes between laboratory-contaminated and polluted field soils; Environ. Toxicol. Chem. 2003, 22, 2592–2598.
[37] Spurgeon, D.J.; Hopkin, S.P., Extrapolation of the laboratory based OECD earthworm toxicity test to metal contaminated field sites; Ecotoxicology 1995, 4, 190–205.
[38] Kammenga, J.E.; Riksen, J.A.G., Comparing differences in species sensitivity to toxicants: phenotypic plasticity versus concentration-response relationships; Environ. Toxicol. Chem. 1996, 15, 1649–1653.
[39] Kammenga, J.; Laskowski, R., Demography in Ecotoxicology; John Wiley & Sons, Ltd, Chichester, 2000.
[40] Alvarez, O.A.; Jager, T.; Redondo, E.M. et al., Physiological modes of action of toxic chemicals in the nematode *Acrobeloides nanus*; Environ. Toxicol. Chem. 2006, 25, 3230–3237.
[41] Kammenga, J.E.; Busschers, M.; Van Straalen, N.M. et al., Stress induced fitness reduction is not determined by the most sensitive life-cycle trait; Funct. Ecol. 1996, 10, 106–111.

[42] Son, J.; Ryoo, M.I.; Jung, J. et al., Effects of cadmium, mercury and lead on the survival and instantaneous rate of increase of *Paronychiurus kimi* (Lee) (Collembola); Appl. Soil Ecol. 2007, 35, 404–411.

[43] Crommentuijn, G.H.; Doodeman, C.J.A.M.; Doornekamp, A. et al., Life-table study with the springtail *Folsomia candida* (Wilem) exposed to cadmium, chlorpyrifos and triphenyl tin hydroxide. Ecological Principles for Risk Assessment of Contaminants in Soil. Chapman and Hall, London, 1997, pp. 275–291.

[44] Laskowski, R.; Hopkin, S.P., Effects of Zn, Cu, Pb and Cd on fitness in snails (*Helix aspersa*); Ecotox. Environ. Saf. 1996, 34, 59–69.

[45] Spurgeon, D.J.; Svendsen, C.; Weeks, J.M. et al., Quantifying copper and cadmium impacts on intrinsic rate of population increase in the terrestrial oligochaete *Lumbricus rubellus*; Environ. Toxicol. Chem. 2003, 22, 1465–1472.

[46] Kooijman, S.A.L.M., Dynamic Energy Budget Models in Biological Systems: Theory and Application in Ecotoxicology; Cambridge University Press, Cambridge, 1993.

[47] Klok, C.; Van der Hout, A.; Bodt, J., Population growth and development of the earthworm *Lumbricus rubellus* in a polluted field soil: possible consequences for the godwit (*Limosa limosa*); Environ. Toxicol. Chem. 2006, 25, 213–219.

[48] Hopkin, S.P., In situ biological monitoring of pollution in terrestrial and aquatic ecosystems; in: Calow P. (ed.), Handbook of Ecotoxicology. Blackwell Scientific Publications, Oxford, 1993, pp. 397–427.

[49] Veltman, K.; Huijbregts, M.A.J.; Hamers, T. et al., Cadmium accumulation in herbivorous and carnivorous small mammals: meta-analysis of field data and validation of the bioaccumulation model optimal modeling for ecotoxicological applications; Environ. Toxicol. Chem. 2007, 26, 1488–1496.

[50] Rogival, D.; Scheirs, J.; Blust, R., Transfer and accumulation of metals in a soil-diet-wood mouse food chain along a metal pollution gradient; Environ. Pollut. 2007, 145, 516–528.

[51] Spurgeon, D.J.; Hopkin, S.P., Tolerance to zinc in populations of the earthworm *Lumbricus rubellus* from uncontaminated and metal-contaminated ecosystems; Arch. Environ. Contam. Toxicol. 1999, 37, 332–337.

[52] Spurgeon, D.J.; Hopkin, S.P., Life-history patterns in reference and metal-exposed earthworm populations; Ecotoxicology 1999, 8, 133–141.

[53] Posthuma, L., Genetic differentiation between populations of *Orchesella cincta* (Collembola) from heavy metal contaminated sites; J. Appl. Ecol. 1990, 27, 609–622.

[54] Moore, J.C.; De Ruiter, P.C.; Hunt, H.W., Soil invertebrate/micro-invertebrate interactions: disproportionate effects of species on food web structure and function; Veterinary Parisitol. 1993, 48, 247–260.

[55] De Ruiter, P.C.; Wolters, V.; Moore, J.C. et al., Food web ecology: playing Jenga and beyond; Science. 2005, 309, 68.

[56] Hågvar, S.; Abrahamsen, G., Microarthropoda and Encytraeidae (Oligochaeta) in a naturally lead contaminated soil: a gradient study; Environ. Entomol. 1990, 19, 1263–1277.

[57] Bengtsson, G.; Nordström, S.; Rundgren, S., Population density and tissue metal concentration of Lumbricids in forest soils near a brass mill; Environ. Pollut. 1983, 30, 87–108.

[58] Bisessar, S., Effect of heavy metals on the micro-organisms in soils near a secondary lead smelter; Water Air Soil Pollut. 1982, 17, 305–308.

[59] Pizl, V.; Josens, G., Earthworm communities along a gradient of urbanization; Environ. Pollut. 1995, 90, 7–14.

[60] Bengtsson, G.; Rundgren, S., The Gusum case: a brass mill and the distribution of soil Collembola; Can. J. Zool. 1988, 66, 1518–1526.

[61] Stamou, G.P.; Argyropoulou, M.D., A preliminary-study on the effect of Cu, Pb and Zn contamination of soils on community structure and certain life-history traits of Oribatids from urban areas; Exp. Appl. Acarol. 1995, 19, 381–390.

[62] Read, H.J.; Martin, M.H.; Rayner, J.M.V., Invertebrates in woodlands polluted by heavy metals – An evaluation using canonical correspondence analysis; Water Air Soil Pollut. 1998, 106, 17–42.

[63] Tanasevitch, A.V., The effects of metallurgic smelter pollution on spider communities (Arachnida, Araneae): preliminary observations; in: Butovsky RO (ed.), Pollution-Induced

Changes in Soil Invertebrate Food-Webs. Netherlands Organization for Scientific Research, Amsterdam, 1999.
[64] Van Straalen, N.M., Community structure of soil arthropods, bioindicators of soil quality; in: Pankhurst, C.E., Doube, B.M., Gupta, V.V.S.R. (eds), Bioindicators of Soil Health; CAB International, Wallingford, 1997, pp. 235–264.
[65] Rutgers, M.; Breure, A.M., Risk assessment, microbial communities, and pollution-induced community tolerance; Hum. Ecol. Risk Assess. 1999, 5, 661–670.
[66] Kozdroj, J.; van Elsas, JD., Structural diversity of microbial communities in arable soils of a heavily industrialised area determined by PCR-DGGE fingerprinting and FAME profiling; Appl. Soil Ecol. 2001, 17, 31–42.
[67] Muller, A.K.; Westergaard, K.; Christensen, S. et al., The effect of long-term mercury pollution on the soil microbial community; FEMS Microb. Ecol. 2001, 36, 11–9.
[68] Hopkin, S.P.; Martin, M.H., Assimilation of zinc, cadmium, lead, copper and iron by the spider *Dysdera crocata*, a predator of woodlice; Bull. Environ. Contam. Toxicol. 1985, 34, 183–187.
[69] Martin, M.H.; Duncan, E.M.; Coughtrey, P.J., The distribution of heavy metals in a contaminated woodland ecosystem; Environ. Pollut. (Series B) 1982, 3, 147–157.
[70] Read, H.J.; Wheater, C.P.; Martin, M.H., Aspects of the ecology of Carabidae; Environ. Pollut. 1987, 48, 61–76.
[71] Spurgeon, D.J.; Hopkin, S.P., The effects of metal contamination on earthworm populations around a smelting works – quantifying species effects; Appl. Soil Ecol. 1996, 4, 147–160.
[72] Sandifer, R.D., The effects of cadmium, copper, lead and zinc contamination on arthropod communities in the vicinity of a primary smelting works; PhD thesis; University of Reading; UK, 1996.
[73] Eijsackers, H.; Beneke, P.; Maboeta, M. et al., The implications of copper fungicide usage in vineyards for earthworm activity and resulting sustainable soil quality; Ecotox. Environ. Saf. 2005, 62, 99–111.
[74] Paoletti, M.G.; Sommaggio, D.; Favretto, M.R. et al., Earthworms as useful bioindicators of agroecosystem sustainability in orchards and vineyards with different inputs; Appl. Soil Ecol. 1998, 10, 137–150.
[75] Wright, J.F.; Sutcliffe, D.W.; Furse, M.T. (eds), Assessing the Biological Quality of Fresh Waters RIVPACS and other techniques. The Freshwater Biological Association, Ambleside, UK, 2000.
[76] Rantalainen, M.L.; Torkkeli, M.; Strommer, R. et al., Lead contamination of an old shooting range affecting the local ecosystem – a case study with a holistic approach; Sci. Total Environ. 2006, 369, 99–108.
[77] Streit, B., Effects of high copper concentrations on soil invertebrates (earthworms and oribatid mites): experimental results and a model; Oecologia (Berlin). 1984, 64, 381–388.
[78] Filzek, P.D.B.; Spurgeon, D.J.; Broll, G. et al., Pedological characterisation of sites along a transect from a primary cadmium/lead/zinc smelting works; Ecotoxicology 2004, 13, 725–737.
[79] Hobbelen, P.H.F.; Koolhaas, J.E.; vanGestel CAM. Effects of heavy metals on the litter consumption by the earthworm *Lumbricus rubellus* in field soils; Pedobiologia 2006, 50, 51–60.
[80] Hobbelen, P.H.F.; Van Gestel, C.A.M., Using dynamic energy budget modeling to predict the influence of temperature and food density on the effect of Cu on earthworm mediated litter consumption; Ecol. Model. 2007, 202, 373–384.
[81] Geissen, V.; Brümmer, G.W., Decomposition rates and feeding activities of soil fauna in deciduous forest soils in relation to soil chemical parameters following liming and fertilization; Biol. Fertil. Soils. 1999, 29, 335–342.
[82] Dusek, L., Activity of nitrifying populations in grassland soil polluted by polychlorinated-biphenyls (PCBs); Plant Soil 1995, 176, 273–282.
[83] Murray, P.; Ge, Y.; Hendershot, W.H., Evaluating three trace metal contaminated sites: a field and laboratory investigation; Environ. Pollut. 2000, 107, 127–135.
[84] Kandeler, E.; Luftenegger, G.; Schwarz, S., Soil microbial processes and *Testacea* (Protozoa) as indicators of heavy-metal pollution; Zeitschrift Fur Pflanzenernahrung Und Bodenkunde. 1992, 155, 319–322.

[85] Simon, T., The effect of nickel and arsenic on the occurrence and symbiotic abilities of native rhizobia; Rostlinna Vyroba 2000, 46, 63–68.

[86] McGrath, S.P., Effects of heavy metals from sewage sludge on soil microbes in agricultural ecosystems, in: Ross SM (ed.), Toxic Metals in Soil Plant systems. John Wiley & Sons, Inc., New York, 1994.

[87] Lofts, S.; Spurgeon, D.; Svendsen, C., Fractions affected and probabilistic risk assessment of Cu, Zn, Cd, and Pb in soils using the free ion approach; Environ. Sci. Technol. 2005, 39, 8533–8540.

[88] Chander, K.; Brooks, P.C., Is the dehydrogenase assay invalid as a method to estimate microbial activity in copper-contaminated soils? Soil Biol. Biochem. 1991, 23, 909–915.

[89] Brohon, B.; Delolme, C.; Gourdon, R., Complementarity of bioassays and microbial activity measurements for the evaluation of hydrocarbon-contaminated soils quality; Soil Biol. Biochem. 2001, 33, 883–891.

[90] Trasar-Cepeda, C.; Leiros, M.C.; Seoane, S. et al., Limitations of soil enzymes as indicators of soil pollution; Soil Biol. Biochem. 2000, 32, 1867–1875.

[91] Jager, T.; Fleuren, R.; Hogendoorn, E.A. et al., Elucidating the routes of exposure for organic chemicals in the earthworm, *Eisenia andrei* (Oligochaeta); Environ. Sci. Technol. 2003, 37, 3399–3404.

[92] Van Gestel, C.A.M.; Koolhaas, J.E., Water-extractability, free ion activity, and pH explain cadmium sorption and toxicity to *Folsomia candida* (Collembola) in seven soil–pH combinations; Environ. Toxicol. Chem. 2004, 23, 1822–1833.

[93] Vijver, M.G.; Vink, J.P.M.; Miermans, C.J.H. et al., Oral sealing using glue: a new method to distinguish between intestinal and dermal uptake of metals in earthworms; Soil Biol. Biochem. 2003, 35, 125–132.

[94] Scheifler, R.; Gomot-de Vaufleury, A.; Toussaint, M.L. et al., Transfer and effects of cadmium in an experimental food chain involving the snail *Helix aspersa* and the predatory carabid beetle *Chrysocarabus splendens*; Chemosphere 2002, 48, 571–579.

[95] Di Toro, D.M.; Allen, H.E.; Bergman, H.L. et al., Biotic ligand model of the acute toxicity of metals. 1. Technical basis; Environ. Toxicol. Chem. 2001, 20, 2383–2396.

[96] De Schamphelaere, K.A.C.; Janssen, C.R., A biotic ligand model predicting acute copper toxicity for *Daphnia magna*: the effects of calcium, magnesium, sodium, potassium, and pH; Environ. Sci. Technol. 2002, 36, 48–54.

[97] DeSchamphelaere, K.A.C.; Janssen, C.R., Bioavailability models for predicting copper toxicity to freshwater green microalgae as a function of water chemistry; Environ. Sci. Technol. 2006, 40, 4514–4522.

[98] Heijerick, D.G.; De Schamphelaere, K.A.C.; Janssen, C.R., Predicting acute zinc toxicity for *Daphnia magna* as a function of key water chemistry characteristics: development and validation of a biotic ligand model. Environ; Toxicol. Chem. 2002, 21, 1309–1315.

[99] Santore, R.C.; Di Toro, D.M.; Paquin, P.R. et al., Biotic ligand model of the acute toxicity of metals. 2. Application to acute copper toxicity in freshwater fish and *Daphnia*; Environ. Toxicol. Chem. 2001, 20, 2397–2402.

[100] Di Toro, D.M.; McGrath, J.A.; Hansen, D.J. et al., Predicting sediment metal toxicity using a sediment biotic ligand model: methodology and initial application; Environ. Toxicol. Chem. 2005, 24, 2410–2427.

[101] Mertens, J.; Degryse, F.; Springael, D. et al., Zinc toxicity to nitrification in soil and soilless culture can be predicted with the same biotic ligand model; Environ. Sci. Technol. 2007, 41, 2992–2997.

[102] Thakali, S.; Allen, H.E.; DiToro, D.M. et al., A Terrestrial biotic ligand model. 1. Development and application to Cu and Ni toxicities to barley root elongation in soils; Environ. Sci. Technol. 2006, 40, 7085–7093.

[103] Antunes, P.M.C.; Berkelaar, E.J.; Boyle, D. et al., The biotic ligand model for plants and metals: technical challenges for field application; Environ. Toxicol. Chem. 2006, 25, 875–882.

[104] Koster, M.; deGroot, A.; Vijver, M. et al., Copper in the terrestrial environment: verification of a laboratory-derived terrestrial biotic ligand model to predict earthworm mortality with toxicity observed in field soils; Soil Biol. Biochem. 2006, 38, 1788–1796.

[105] Peijnenburg, W.; Zablotskaja, M.; Vijver, M.G., Monitoring metals in terrestrial environments within a bioavailability framework and a focus on soil extraction; Ecotox. Environ. Saf. 2007, 67, 163–179.

[106] Thakali, S.; Allen, H.E.; DiToro, D.M. *et al.*, Terrestrial biotic ligand model. 2. Application to Ni and Cu toxicities to plants, invertebrates, and microbes in soil; Environ. Sci. Technol. 2006, 40, 7094–7100.

[107] Lofts, S.; Spurgeon, D.J.; Svendsen, C. *et al.*, Deriving soil critical limits for Cu, Zn, Cd and Pb: a method based on free ion concentrations; Environ. Sci. Technol. 2004, 38, 3623–3631.

[108] Spurgeon, D.J.; Lofts, S.; Hankard, P.K. *et al.*, Effect of pH on metal speciation and resulting uptake by and toxicity to earthworms; Environ. Toxicol. Chem. 2006, 25.

[109] Janssen, M.P.M.; Bruins, A.D.V.T.H.; Van Straalen, N.M., Comparison of cadmium kinetics in four soil arthropod species; Arch. Environ. Contam. Toxicol. 1991, 20, 305–312.

[110] Sterenborg, I.; Vork, N.A.; Verkade, S.K. *et al.*, Dietary zinc reduces uptake but not metallothionein binding and elimination of cadmium in the springtail, *Orchesella cincta*; Environ. Toxicol. Chem. 2003, 22, 1167–1171.

[111] Spurgeon, D.J.; Hopkin, S.P., Risk assessment of the threat of secondary poisoning by metals of predators of earthworms in the vicinity of a primary smelting works; Sci. Total Environ. 1996, 187, 167–183.

[112] Vijver, M.G.; Vink, J.P.M.; Jager, T. *et al.*, Biphasic elimination and uptake kinetics of Zn and Cd in the earthworm *Lumbricus rubellus* exposed to contaminated floodplain soil. Soil Biol. Biochem. 2005, 37, 1843–1851.

[113] Spurgeon, D.J.; Hopkin, S.P., Comparisons of metal accumulation and excretion kinetics in earthworms (*Eisenia fetida*) exposed to contaminated field and laboratory soils; Appl. Soil Ecol. 1999, 11, 227–243.

[114] Vijver, M.G.; Van Gestel, C.A.M.; Van Straalen, N.M., *et al.*, Biological significance of metals partitioned to subcellular fractions within earthworms (*Aporrectodea caliginosa*); Environ. Toxicol. Chem. 2006, 25, 807–814.

[115] Morgan, A.J.; Kille, P.; Sturzenbaum, S.R., Microevolution and ecotoxicology of metals in invertebrates; Environ. Sci. Technol. 2007, 41, 1085–1096.

[116] Gibb, J.O.T.; Svendsen, C.; Weeks, J.M. *et al.*, H-1 NMR spectroscopic investigations of tissue metabolite biomarker response to Cu(II) exposure in terrestrial invertebrates: identification of free histidine as a novel biomarker of exposure to copper in earthworms; Biomarkers 1997, 2, 295–302.

[117] Stürzenbaum, S.R.; Kille P.; Morgan, A.J., The identification, cloning and characterization of earthworm metallothionein; FEBS Lett. 1998, 431, 437–442.

[118] Kooijman, S.; Bedaux, J.J.M., Analysis of toxicity tests on *Daphnia* survival and reproduction; Water Res. 1996, 30, 1711–1723.

[119] Lanno, R.; Wells, J.; Conder, J. *et al.*, The bioavailability of chemicals in soil for earthworms; Ecotox. Environ. Saf. 2004, 57, 39–47.

[120] Meador, J., Rationale and procedures for using the tissue-residue approach for toxicity assessment and determination of tissue, water, and sediment quality guidelines for aquatic organisms; Human and Ecological Risk Assessment 2006, 12, 1018–1073.

[121] Triebskorn, R.; Adam, S.; Casper, H. *et al.*, Biomarkers as diagnostic tools for evaluating effects of unknown past water quality conditions on stream organisms; Ecotoxicology. 2002, 11, 451–465.

[122] Forbes, V.E.; Palmqvist, A.; Bach, L., The use and misuse of biomarkers in ecotoxicology; Environ. Toxicol. Chem. 2006, 25, 272–280.

[123] Galay-Burgos, M.; Spurgeon, D.J.; Weeks, J.M. *et al.*, Developing a new method for soil pollution monitoring using molecular genetic biomarkers; Biomarkers 2003, 8, 229–239.

[124] Sturzenbaum, S.R.; Georgiev, O.; Morgan, A.J. *et al.*, Cadmium detoxification in earthworms: from genes to cells; Environ. Sci. Technol. 2004, 38, 6283–6289.

[125] Willuhn, J.; Schmittwrede, H.P.; Greven, H. *et al.*, cDNA cloning of a cadmium inducible messenger RNA encoding a novel cysteine rich, nonmetallothionein 25 KDa protein in an Enchytraeid earthworm; J. Biol. Chem. 1994, 269, 24688–24691.

[126] Dallinger, R.; Chabicovsky, M.; Berger, B., Isoform-specific quantification of metallothionein in the terrestrial gastropod *Helix pomatia* I. Molecular, biochemical, and methodical background; Environ. Toxicol. Chem. 2004, 23, 890–901.

[127] Dallinger, R.; Chabicovsky, M.; Lagg, B. et al., Isoform-specific quantification of metallothionein in the terrestrial gastropod *Helix pomatia*. II. A differential biomarker approach under laboratory and field conditions; Environ. Toxicol. Chem. 2004, 23, 902–910.

[128] Hensbergen, P.J.; Donker, M.H.; Van Velzen, M.J.M. et al., Primary structure of a cadmium-induced metallothionein from the insect *Orchesella cincta* (Collembola); Eur. J. Biochem. 1999, 259, 197–203.

[129] Sterenborg, I.; Roelofs, D., Field-selected cadmium tolerance in the springtail *Orchesella cincta* is correlated with increased metallothionein mRNA expression; Insect Biochem. 2003, 33, 741–747.

[130] Vatamaniuk, O.K.; Bucher, E.A.; Sundaram, M.V. et al., CeHMT-1, a putative phytochelatin transporter, is required for cadmium tolerance in *Caenorhabditis elegans*; J. Biol. Chem. 2005, 280, 23684–23690.

[131] Kumar, T.R.; Prasad, M.N.V., Ferritin induction by iron mediated oxidative stress and ABA in *Vigna mungo* (L.) hepper seedlings: role of antioxidants and free radical scavengers; J. Plant Physiol. 1999, 155, 652–655.

[132] Lee, M.S.; Cho, S.J.; Tak, E.S. et al., Transcriptome analysis in the midgut of the earthworm (*Eisenia andrei*) using expressed sequence tags; Biochem. Biophys. Res. Commun. 2005, 328, 1196–1204.

[133] Marino, F.; Winters, C.; Morgan, A.J., Heat shock protein (hsp60, hsp70, hsp90) expression in earthworms exposed to metal stressors in the field and laboratory; Pedobiologia 1999, 43, 615–624.

[134] Nadeau, D.; Corneau, S.; Plante, I. et al., Evaluation for Hsp70 as a biomarker of effect of pollutants on the earthworm *Lumbricus terrestris*; Cell Stress Chaperones 2001, 6, 153–163.

[135] Kohler, H.R.; Alberti, G.; Seniczak, S. et al., Lead-induced hsp70 and hsp60 pattern transformation and leg malformation during postembryonic development in the oribatid mite, *Archegozetes longisetosus* Aoki; Comp. Biochem. Physiol. C Toxicol. Pharmacol. 2005, 141, 398–405.

[136] Arts, M-J.; Schill, R.O.; Knigge, T. et al., Stress proteins (hsp70, hsp60) induced in isopods and nematodes by field exposure to metals in a gradient near Avonmouth, UK; Ecotoxicology 2004, 13.

[137] Knigge, T.; Kohler, H.R., Lead impact on nutrition, energy reserves, respiration and stress protein (hsp 70) level in *Porcellio scaber* (Isopoda) populations differently preconditioned in their habitats; Environ. Pollut. 2000, 108, 209–217.

[138] Liao, V.H.C.; Dong, J.; Freedman, J.H., Molecular characterization of a novel, cadmium-inducible gene from the nematode *Caenorhabditis elegans* – A New gene that contributes to the resistance to cadmium toxicity; J. Biol. Chem. 2002, 277, 42049–42059.

[139] Kille, P.; Sturzenbaum, S.R.; Galay, M. et al., Molecular diagnosis of pollution impact in earthworms - Towards integrated biomonitoring; Pedobiologia 1999, 43, 602–607.

[140] Brulle, F.; Mitta, G.; Cocquerelle, C. et al., Cloning and real-time PCR testing of 14 potential biomarkers in *Eisenia fetida* following cadmium exposure; Environ. Sci. Technol. 2006, 40, 2844–2850.

[141] Bundy, J.G.; Keun, H.C.; Sidhu, J.K. et al., Metabolic profile biomarkers of metal contamination in a sentinel terrestrial species are applicable across multiple sites; Environ. Sci. Technol. 2007, 41, 4458–4464.

[142] Stürzenbaum, S.R.; Blaxter, M.; Parkinson, J. et al., The earthworm EST sequencing project; Pedobiologia 2004, 47, 447–451.

[143] Bundy, J.G.; Sidhu, J.K.; Rana, F. et al., Systems toxicology' approach identifies coordinated metabolic responses to copper in a terrestrial non-model invertebrate, the earthworm *Lumbricus rubellus*. BMC Biol. 2008 Jun 3, 6, 25.

9
Trace Element-Deficient Soils

Rainer Schulin[1], Annette Johnson[2] and Emmanuel Frossard[3]

[1]*Institute of Terrestrial Ecosystems, ETH Zurich, Switzerland*
[2]*Eawag, Dübendorf, Switzerland*
[3]*Institute of Plant Sciences, ETH Zurich, Switzerland*

9.1 Introduction

A soil is called deficient in a certain trace element if it cannot supply sufficient amounts of that element to plants or other organisms that depend on such delivery for normal growth and health. Trace element-deficient soils are a widespread problem in agriculture. In many parts of the world, deficiencies in essential trace elements are limiting the growth and healthy development of crop plants and animals. Although it may seem straightforward at first glance to identify and delineate trace element-deficient soils, relating concentrations of trace elements in soil to their deficiencies in plants or animals is not at all a simple task. The problem is rarely the lack of a trace element *per se*. Even under conditions of severe deficiency the root zone usually contains many times the amount of a trace element that is needed by a plant. In general, deficiencies of trace elements in soils are due only to limitations in bioavailability. This chapter, therefore, begins by developing the concept of trace element-deficient soils in relation to the essentiality and bioavailability of trace elements for plant and animal nutrition. The concept provides the framework required for the discussion of methodological questions concerning the detection and identification of micronutrient deficiencies and the delineation and mapping of soils deficient in essential trace elements. A brief general account of the main soil factors associated with trace element deficiencies completes this discussion and provides the transition to the main part of this chapter, which is a review of the relevant physiological and chemical properties of the essential micronutrient metals and metalloids and of the soil conditions under which deficiencies in these elements are likely to develop. As they have very specific and widely different characteristics, each element is presented individually in this part. The chapter concludes with an overview of agricultural approaches to cope with trace element-deficient soils.

9.2 The Concept of Trace Element-Deficient Soils

9.2.1 The Role of Trace Elements as Essential Micronutrients

Providing organisms with nutrients is one of the most important functions of soils. For normal growth and health all higher plants require the nine macronutrient elements carbon (C), hydrogen (H), oxygen (O), nitrogen (N), potassium (K), calcium (Ca), magnesium (Mg), phosphorus (P) and sulfur (S) and the seven micronutrient elements (≤ 100 mg kg^{-1} dry mass) chlorine (Cl), manganese (Mn), iron (Fe), zinc (Zn), boron (B), copper (Cu) and molybdenum (Mo). At least some plants also need nickel (Ni) as an essential element. While cobalt (Co) is not an essential plant nutrient [1], it is required in trace amounts by microorganisms for atmospheric nitrogen fixation, which makes it essential for effective N_2-fixation by microorganisms in symbiotic association with leguminous and other plants. Furthermore, some authors reported that sodium (Na) is required by C4 plants [2,3]. While this essentiality of Na is cited by Asher [4] and indirectly referred to by Welch [5], no role of Na in plant nutrition is mentioned in the more recent review of Fageria *et al.* [6].

In addition to that for higher plants, the list of micronutrient elements that are essential to animals also includes Co, chromium (Cr), fluorine (F), iodine (I) and selenium (Se) [7–9], while it does not include Cl, B, and Ni. For animals Cl is a macronutrient together with Na and the nine macronutrient elements of the higher plants. Boron and Ni are suspected to be essential in animals and human nutrition, but their essentiality has not been conclusively proven [9–11].

Deficiencies of essential micronutrients in crop plants are a widespread problem in agriculture [6,12]. Only Cl deficiency is not known to be a problem in crop production. In recent years micronutrient deficiency problems have become more acute for the following reasons:

1. Increase in micronutrient demands due to intensified agricultural crop production.
2. Crop production being extended to marginal soils with low fertility in many countries with high population growth and urbanization pressure on cultivated fertile land.
3. Increased use of high-quality macronutrient fertilizers that are poor in trace elements.
4. Decreased recycling of nutrients in animal manures, composts and crop residues.
5. Depletion of low-nutrient soils by over-use.
6. Decrease in micronutrient bioavailability by inappropriate soil management, for example over-liming [6].

Micronutrient deficiencies are also becoming increasingly important in livestock farming as marginal land becomes more and more important for grazing and forage production due to increasing demand for food and increasing competition of agricultural food production with other land uses [13].

9.2.2 Soil Trace Element Concentrations and Micronutrient Deficiencies

Soil is the primary source of micronutrient elements under natural conditions, but as mentioned in the introduction, the relationship between the deficiency of an element in a plant and its concentration in soil in general is not straightforward. The reason is that the uptake of an element by a plant does not simply depend on its total concentration in soil, but depends also on the availability of that element for uptake as well as on other soil and environmental factors such as climatic conditions [14]. Furthermore, plant micronutrient requirements vary between plant species, between different genotypes within species, and with age and physiological status [5,15–17].

Low nutrient concentrations in plants can translate into mineral deficiencies of humans and animals. Deficiencies in Co, Cu, I, Fe, Mn, Se and Zn resulting from low accumulation in feed plants are common occurrences in grazing livestock. Thornton [13] compiled several examples of close relationships between trace element concentrations in soils or sediments and mineral deficiencies in livestock, for example between low Co concentrations in stream sediments and a high degree of pine (wasting) in sheep and cattle grazing on soils that had developed on granite bedrocks in SW England, and between Cu deficiency in cattle and high concentrations of Mo in pasture soils that had developed on marine black shales. Such correlations have led to the proposal of using geochemical mapping to delineate areas associated with a high risk of insufficient micronutrient supply to grazing agricultural livestock and wildlife [13,18,19].

The establishment of such relationships is generally not easy or straightforward, even in regions where grazing livestock rely on indigenous forage for their mineral nutrition. Fordyce *et al.* [20], for example, analysed trace element concentrations in stream sediments, soils, forage plants and cattle serum from an area with a wide range in sediment Zn in northeast Zimbabwe. Comparing districts with different average geochemical Cu and Zn status, they found significant correlations between the respective metal (that is Cu and Zn) concentrations of the sediment, soil and forage samples, but not between the respective metal concentrations of these samples and those of the serum samples. In fact, the Zn concentrations in the serum samples even tended to decrease with increasing Zn in the forage samples. No mineral nutrient supplementation was practised in the area, so grazing including some direct soil ingestion was the only source of mineral nutrient supply for the cattle. The lack of correlations was attributed to several factors. Apart from variability in the age, gender and health of the animals and uncertain contribution of individual plant species and plant parts with different mineral contents to the feed, the main reason was assumed to be that total Zn and Cu do not adequately represent the bioavailability of nutrients, not only with respect to their uptake by forage plants from soil, but also for the absorption of nutrient elements by the animals from feed-stuffs. In addition, Fordyce *et al.* [20] considered antagonistic relationships in the absorption of nutrient elements as contributing factors to the observed lack of a clear relationship between trace element concentrations in the serum samples and that in the sediments, soils and forage plants. This hypothesis was supported by the finding that serum Cu was negatively correlated with the concentrations of Zn, Fe and Mn in forage samples and the finding of similar negative correlations also between serum Zn and forage Cu, Fe and Mn concentrations. The latter three elements have all been reported to interact antagonistically with the absorption and metabolism of Zn in animals, while Cu absorption has been found reduced by high dietary Zn and Fe contents [7,21].

When relationships between trace elements in soils or plants and their uptake by animals and humans are investigated, it should not be assumed that the concentration of an element in serum samples is the best indicator of its supply to animals or humans. López Alonso et al. [22] found that the concentrations of Cu and Zn in the liver of calves in the province of Galicia, Spain, progressively increased with their subsoil concentrations in the grazing areas of the animals. This, however, was not reflected in blood Cu and Zn concentrations when subsoil Cu and Zn concentrations were <100 mg kg^{-1} and <200 mg kg^{-1}, respectively. Blood Cu tended to increase with subsoil Cu only at high subsoil Cu concentrations (>100 mg kg^{-1}), whereas Zn even tended to decrease with increasing soil Zn at high subsoil Zn concentrations (>200 mg kg^{-1}).

9.2.3 Trace Element Deficiency as a Disturbance-Related Concept

Nutrient deficiency can be considered as an opposite condition to toxicity. Deficiency results from the lack or shortage of an essential nutrient, toxicity from excessive exposure of an organism to a harmful substance. Deficiency and toxicity are analogous phenomena in so far as both refer to the effect of a substance on a receptor or target organism and both depend on the bioavailability of that substance rather than on its total concentration in the medium from which it is taken up. Moreover, in both cases the physiological state of the recipient is assessed, as they refer to effects that are considered not to be 'normal' or 'healthy' for the organisms. Thus, the notion of nutrient deficiency represents an anthropocentric concept. As such it is useful to characterize soils with respect to human land use for plant and animal production, but there is no definition of trace element deficiency in a soil that has a useful meaning for the characterization of a soil *per se*, without referring to a function for which the element under consideration is used and without defining what is considered 'sufficient'. In plant and animal production the definition of a reference state is, at least in principle, straightforward: the state in which a further increase in supply of the trace element under consideration does not produce any further increase in the respective production. Under natural ecosystem conditions, populations adapt to environmental conditions through evolutionary processes and tend to evolve to a state in which their use of the available nutrients is optimal under the existing constraints. Changes in nutrient availability will in general change the balance of competitive advantages and disadvantages between different organisms. As a result, some populations or species will increase (that is 'win'), while others will decrease (that is 'lose') relative to each other. With such a shift in the community composition/response the ecosystem is no longer the same and it is not possible to define a soil condition of 'nutrient sufficiency' in which the ecosystem has the same population composition as under the condition of a postulated 'deficiency'. Without a meaningful reference state, however, any definition of deficiency is arbitrary and of no scientific or practical value.

This does not mean that micronutrient deficiencies cannot occur in wildlife populations. Firstly, there may always be individuals suffering from such conditions. Secondly, deficiencies may arise as a result of disturbances. Migratory herbivores in Africa present an example illustrating this point. Maskall and Thornton [18] cite various studies in which the spatial distribution and seasonal movements of grazing animals were found to be related to the mineral content of forages, indicating that the animals had to cope with strongly limited supplies of some mineral nutrients. In such a case, restricting the migrations by enclosure in small national parks may cause mineral deficiency problems in the affected migratory grazer populations as they no longer have access to sufficiently nutritious feed throughout the seasons.

As the notion of deficiency relates to a nutrient requirement that is not met by the available supply, a soil cannot be deficient in an element as such, but only in relation to the function of soil to provide nutrient elements to organisms in quantities sufficient for their normal growth and functioning. As nutrient requirements differ for different organisms, depending in particular on species, genotype, developmental stage and physiological conditions, a classification of soils by their status with respect to element deficiency or sufficiency that is independent of a specific recipient organism would require the definition of standard conditions. Standards are also needed with respect to the environmental factors governing the bioavailability of micronutrient elements. Unfortunately, such standards have not been established. However, there is a large volume of literature on micronutrient uptake by plants and animals, covering a wide range of species, elements and experimental conditions, so that ranges of critical and sufficient trace element concentrations in soils and plants can be given for many important agricultural crop plants (for example [14,17,23–29]) as well as ranges for the amounts of trace elements required in feedstuffs for livestock [7,30].

9.3 Methods to Identify and Map Soil Trace Element Deficiencies

9.3.1 Detection and Diagnosis of Trace Element Deficiency

The derivation of criteria that can be used to assess trace-element deficiencies in soils in the sense outlined above requires the establishment of relationships between the degree of deficiency of a particular trace element in a selected target organism and soil properties that can be used as reliable indicators of the soil's capacity to supply that element to the organism. In the case of crop plants, experiments are usually performed in which the plant is grown in test soils with different concentrations of the individual element, and the response of crop yield or growth is then used to delimit ranges of deficiency, sufficiency and toxicity by critical limits [16,31].

A growth response upon addition of a nutrient is ultimate proof that insufficient availability of the nutrient was indeed a limiting factor under the given conditions. As such trials are very time-consuming and laborious, deficiencies are usually diagnosed by comparing nutrient concentrations measured in plant or soil samples with experimentally determined critical values. Furthermore, bioassays, enzyme and other biomarker tests or visual symptoms are used for the diagnosis of micronutrient deficiencies [31].

The analysis of the chemical composition of plant tissue (see for example [32] for an overview on methods for the analysis of metals and metalloids in plant samples) is the most direct way to assess nutrient deficiency, but usually provides less information on the cause of a deficiency than the analysis of the soil. As the content of a micronutrient can vary considerably with growth and between different organs, the selection of the plant parts to be used for the analysis needs to be given careful consideration. For trace elements with low mobility such as Cu and Zn, young leaves and shoot tips are to be preferred [31]. Critical levels also depend on interactions with other elements. Interpretations of analytical results, therefore, should be based not only on the absolute concentration of the element under consideration, but also on its concentration relative to other elements, in particular those known for antagonistic effects. Analogous considerations apply to the analysis of tissue samples for the diagnosis of micronutrient deficiencies in animals.

Chemical analyses of soil samples (see Chapter 4 for an overview on methods for the analysis of trace elements in soil samples) are more difficult to interpret with respect to the potential risk of micronutrient deficiencies in receptor organisms. However, if appropriate methods are chosen, they can provide valuable information on the availability of trace elements for uptake by an organism. In general, such information is based on the analysis of a soil extract supposed to approximately represent the fraction of the trace element that can be accessed and absorbed by the organism, for example a crop plant. For example, diethylenetriamine pentaacetate (DTPA) is a common extractant used for micronutrients such as Cu, Fe, Mn and Zn, which are often deficient in calcareous soils [33]. While chemical analyses of soil samples can be very useful as indicators of potential deficiency problems, the information they can give on the underlying processes and mechanisms is very limited. Macroscopic parameters representing average properties of entire test samples cannot adequately account for the complex interactions that occur between soil and plant or other organisms involved in micronutrient uptake processes on the microscale. In particular, such analyses cannot capture the strong local influence of water and solute uptake, root exudates and the decomposition of root debris on the composition of the soil solution adjacent to absorbing roots, that is in the rhizosphere. As these interactions vary from plant to plant, between as well as within species, due to differences in genotypes, developmental stage and physiological conditions, the bioavailability of a trace element invariably also depends on the receptor organism and not only on the soil conditions. Therefore, no chemical soil test can be a universal predictor of the bioavailability of a trace element in soil (see Chapter 11), let alone deficiency. The latter would mean that the test would also have to account for a plant's requirement for the element. Nonetheless, soil tests that indicate the amounts and the association of trace elements with certain soil components can be very useful for the identification of soils where deficiency disorders may be expected for certain crops or grazing livestock [34].

Visual symptoms of the phenotype of a plant can give valuable clues about a nutrient deficiency condition [35]. Such diagnoses are inexpensive, but generally limited to situations where a deficiency is already severe. Moreover, while some mineral deficiencies produce quite characteristic symptoms, others are less specific. The diagnosis of visual symptoms becomes even more complicated in the case of multiple deficiencies and with the interference of responses to other stresses such as pests, diseases, frost and drought. Nonetheless, visual symptoms are extremely helpful in detecting plant nutrition problems and in giving a quick first hint of the type of problem and as an aid in the selection of more expensive and time-consuming tests.

Monitoring a metabolic function for which a particular trace element is required is a much more sensitive, but also a methodologically more difficult way to assess a deficiency. Usually, a specific enzyme activity is chosen for this purpose. To determine whether the activity of an enzyme is really limited by a suspected micronutrient deficiency, Steward and Durzan [36] proposed to assess the response of the enzyme activity to a re-supply of the micronutrient under consideration. Enzyme tests, for example, are based on nitrate reductase activity to test for Mo deficiency [37] and on ascorbic acid oxidase to test for Cu deficiency [38]. There are also metabolic tests that are not directly assaying the activity of a specific enzyme. For example the phloroglucin test used by Rahimi and Bussler [39] to diagnose Cu deficiency in plants is based on the Cu-dependence of lignin production. The main advantages of enzyme bioassays and metabolic product indicators are that they are capable of detecting even slight deficiencies and that the diagnosis is possible at an early stage in the development of a deficiency.

9.3.2 Mapping of Trace Element Deficiencies

The fact that micronutrient deficiencies in crops and livestock are widespread and often appeared to be related to specific soil conditions has motivated scientists to analyse geographical distribution patterns in order to relate the occurrence of deficiencies to soil properties, climate factors and epidemiological data, to delineate problem areas, to identify locations for more detailed surveys, and to provide a basis for the planning of fertilizer and animal feed applications and related marketing strategies. The investigation of the geographic patterns of malnutrition disorders can also lead to important progress in our understanding of micronutrients. The essential role of iodine was discovered by studying the geographical distribution of goitre [12]. Similarly, the essential role of cobalt was detected through investigations into a particular condition of livestock health problems in certain regions of Australia [16].

Mapping of trace element problems started in the 1940s [40]. The first maps were prepared primarily on the basis of literature surveys, personal communications, some plant analyses and hypotheses [41,42]. Subsequently, more detailed maps of Co, Mo and Se deficiencies and toxicities were produced in the United States for areas where related nutritional problems had been observed, based on the analysis of plant and soil samples. Sampling sites were selected that were considered representative for specific geological strata and soil units. The plant species chosen were agriculturally important crop and forage plants as well as accumulator plants considered to indicate the phytoavailability of the trace element under study to crop plants [43]. In the United Kingdom, geochemical mapping related to trace element nutrition focused on Co and Mo-induced Cu deficiency. Problem areas were identified first by sampling of stream sediments and soils; then detailed maps were produced on the basis of systematic soil sampling using geostatistical interpolation methods [13]. As mentioned before, stream sediment samples were also found to reliably indicate regional Cu and Zn levels in soil and forage samples in central Africa, although not in samples of cattle serum [20]. Extensive geochemical mapping was also performed by national geological services, in particular in the United States. These maps are based on extensive surveys of rocks, soils, plants, sediments and waters performed on systematic grids. The primary objective of these surveys was usually mineral prospecting, but they are also valuable sources of information on the geographical distribution of micronutrients and their availability in soils. For more detailed accounts of micronutrient deficiency and geochemical trace element mapping the reader is referred to Welch *et al.* [12] and White and Zasoski [40].

Most maps generated from the analysis of soil samples are based on total element concentrations and thus relate to the scarcity of the mapped elements in terms of absolute amounts, while the intended use of geochemical deficiency maps for agriculture is to indicate the *inability* of soils to supply an element at rates sufficient for crop growth or for the nutrition of grazing livestock. Some soils indeed lack this ability due to absolute paucity in a trace element. This condition may arise if the parent material from which the soil had formed was poor in the respective element, as for example for lithogenic elements in weathered sandy soils and organic soils derived from peat bogs. It may also result from depletion due to excessive leaching or erosion. Most often, however, as has been mentioned before, deficiency is not related to absolute scarcity of a nutrient, but largely due to its low bioavailability.

9.4 Soil Factors Associated with Trace Element Deficiencies

9.4.1 General Relationships between Soil Factors and Micronutrient Deficiencies

Key factors governing the bioavailability of trace elements are pH, redox conditions, ionic strength, sorption sites, and the presence of other solutes, in particular complex-forming ligands and competing ions (see Chapter 2). Furthermore, environmental conditions such as temperature and water regime can have an important influence. The relationship of these factors to the occurrence of trace element deficiencies is not limited to their influence on bioavailability. Soil factors that are closely linked to certain conditions and processes governing the formation of a soil, as for example the increase in soil acidity that results from mineral weathering and leaching in the pedogenesis of podzolic soils, also are indicators of the trace element status associated with specific soil conditions.

Parent material is a key factor for the presence of most trace elements. It is particularly important for those elements that enter soil primarily through weathering and are of low mobility, for example Co, Cu, Fe, Mn and Zn. The water regime has a dominant influence on weathering, as solutes are removed by leaching and atmospheric acidity is imported by infiltration, while alkalinity and salinity are accumulated as residues of evapotranspiration. Furthermore, it has a dominant influence on redox conditions, as waterlogging leads to anaerobiosis. The soil solution is a medium of intense chemical reactivity, and water availability is a precondition for all life in soil.

Soil pH is a master variable of the soil solution and its interactions with other phases. With increasing pH, metal hydrolysis, dissolved carbonate concentrations and sorption of cations to deprotonating surface groups increases as competition with protons decreases (see Chapter 2). Eventually the increasing concentrations of hydroxide and carbonate anions associated with increasing soil pH may lead to the precipitation of metal hydroxides, oxides and carbonates. At the same time, increasing competition by hydroxy anions reduces binding of anionic species of trace elements (see Chapter 2). Consequently, metal cations such as Cu^{2+}, Co^{2+} and Zn^{2+} solubilize with decreasing pH, while the solubility of (hydr)oxyanions such as molybdate and borate increases with pH.

Redox potential is the second most important chemical master variable. It directly influences the speciation of those trace elements that occur in different oxidation states, such as Fe, Co, I, Mn and Mo, but indirectly also affects the solubility/stability of other minerals/components. Most dramatic is the impact of changing redox potentials on the solubility of Mn and Fe. Reductive dissolution of Mn and Fe (hydr)oxides causes not only a corresponding increase in Fe and Mn cations in solution, but also the disappearance of an important sorbent for many micronutrient elements. On the other hand, the precipitation of Fe and Mn hydroxides can result not only in deficiencies of these elements, but also of other micronutrient elements that strongly sorb to the precipitates.

Apart from sesquioxides (Al, Fe and Mn hydroxides as oxides), organic matter and clay minerals are the main sorbents in soils, and thus play an important role in micronutrient retention–release in soil (see Chapter 2). For some micronutrient elements, including Fe, Ni, Zn and in particular Cu, dissolved organic carbon (DOC) plays a major role in their solubility and mobility, with potential implications for their distribution, fate and plant availability in soil.

Other important factors are the presence of other inorganic cations and anions, competing with trace elements for sorption sites. In calcareous soils, the competition between calcium and micronutrient cations can be an important factor, whereas among anions phosphate and carbonate often exert a major influence on the solubility of many trace elements.

As the various micronutrients have widely differing properties, a collective review would be rather cumbersome. Instead, therefore, a brief account of their properties and conditions involved in the occurrence of deficiencies is given in the following sections for each element individually. Table 9.1 gives an overview of the micronutrient deficiencies that are typical for certain soil groups. Table 9.2 presents a compilation of the concentration ranges of trace elements essential to higher plants that are typically found in rocks, soils and plants, as well as typical ranges of their critical concentrations in soils and plants below which there are deficiency risks for crop plants.

9.4.2 Boron

Although the metalloid boron (B) is required by higher plants in concentrations that are high in comparison with other micronutrient elements, its physiological role is still not very well understood. Most B in plants is associated with polysaccharides and bound to a substantial fraction in the cell walls, due to the extraordinary capacity of boric acid to form stable complexes with *cis*-diols [16]. Playing an essential role in cell wall synthesis, B is in particular required for root elongation and pollen tube growth. It also plays a key role in plasma membrane functioning.

Table 9.1 Association of trace element (TE) deficiencies with major soil groups

Soil group	Limiting property	Deficiency	Additional TE problems
Acrisols	Leached, strong acidity	Most micronutrients	Fe and Mn toxicity
Andosols	Amorphous hydrated oxides	B, Mo	Al toxicity
Arenosols	Sandy texture	Cu, Fe, Mn, Zn	
Chernozems	High pH, calcium carbonate	Fe, Mn, Zn	
Ferralsols	Strongly weathered, Al and Fe oxide accumulation	Mo	Al, Fe and Mn toxicity
Gleysols	Waterlogging	Mn	Fe and Mo toxicity
Histosols	Accumulated organic matter, water-logging	Cu	
Kastanozems	High pH, calcium carbonate	Cu, Mn, Zn	
Planosols	Strongly leached, water-logging	Most micronutrients	
Podsols	Acid, leached, strong organocomplexes	Co, Cu and most others	
Rendzic Leptosols	Shallow, calcareous	Fe, Mn, Zn	
Solonetz	High pH, Na accumulation	Cu, Fe, Mn, Zn	
Yermosols	Arid, high pH, Ca carbonate	Co, Fe, Zn	

After Srivastava and Gupta [31], and Fageria et al. [6].

Boron is unique among the essential trace elements in being normally present as an uncharged molecule in soil solution, given that undissociated boric acid is the predominant dissolved boron species at soil pH values below 8. Borate anions become relevant only under extremely alkaline conditions. Complexes of boric acid with other cations are generally weak. Retention in soil is due to surface complexes formed with organic matter, clay minerals and metal hydroxides, in particular Al hydroxides. The stability of these complexes increases up to a pH between 9 and 10, and decreases above that range, as borate anions become dominant. Due to its low retention as a nonionic species, B is very mobile, especially in coarse-textured acid soils, and is easily leached in humid climates, limiting its availability for plant uptake. Binding of B by soil organic matter is stronger than by minerals due to the capability of boric acid to form complexes with polyhydroxylic compounds [16]. The slow release of B from organic matter in the course of microbial decomposition appears to be an important source for plant nutrition [14].

Boron uptake by plants is considered to be primarily a passive, that is non-metabolically driven process [5]. Due to the high mobility of B, its transport from soil to roots occurs mainly by mass flow, as long as conditions permit sufficient water flow to the roots. Also, translocation of B within plants primarily follows the transpirational water stream. Mobility in the phloem is medium to low [16,31]. Boron accumulated in older leaves is

Table 9.2 Trace element concentrations (mg kg^{-1}) in rocks and soils and critical levels in soils and plants associated with deficiency conditions

Element	Typical concentrations				Critical levels			Sufficient level in plants
	Rocks		Soils		Soils		Plants	
	Acid	Basic	Sedimentary	Total	Soluble	Extractant[a]		
Boron	3–10	1–5	500	2–150	15–25	0.1–2	Hot water	<10	10–100
Chlorine	100	200	160	20–900	70–200			<2000	2000–20 000
Copper	10–20	100–200	30–40	2–250		0.1–2.5	DTPA[b]	2–5	5–30
Iron				200–500 000		2.5–5	DTPA	<50	50–500
Manganese	200–12 000	1000–2000	200–1200	7–10 000	1000–3000	1–5	DTPA	10–25	20–300
Molybdenum	1.9	1.4	2	0.1–40	2–13	0.1–0.3	Ammonium oxalate	0.03–0.15	0.1–5
Nickel	5–10	1500	20–100	3–1000				<0.1	0.01–10
Zinc	50–60	70–130	20–120	1–900		0.1–2	DTPA	10–20	20–150

After Brown [71]; Alloway [69]; Jones [15]; Srivastava and Gupta [31]; and Fageria et al. [6].
[a] Used for the determination of the critical soluble concentration in soils.
[b] DTPA = diethylenetriamine pentaacetate.

poorly available for the growth of new leaves. Plant species vary considerably in their requirement for B and also in their capacity for uptake of this microelement [16]. This explains, at least partially, why the geographical pattern of B deficiency is more related to the cultivation of crops with high B demand than to soils and their parent materials [12]. Boron demand was found to be enhanced by high K supply in corn [44], while Graham *et al.* [45] reported that B uptake was increased under Zn deficiency in barley.

Boron deficiency is often associated with low-boron parent materials, such as many granitic and other igneous rocks (Table 9.2), but may also occur on limed peat soils [14]. It is typical for Andosols, which are rich in B-sorbing amorphous Al hydroxides (Table 9.1), but low B bioavailability may also result from its sorption by organic matter and coprecipitation with calcium carbonate in humous calcareous soils [31]. Soils with low pH, sandy texture and low organic matter content have a low retention capacity for B and are thus prone to suffer B deficiency as a result of leaching [46,47]. Given the high mobility of B, the water regime is a decisive factor. Under arid conditions B is often found to accumulate with salinity to toxic concentrations, despite the decrease in relative bioavailability with alkaline conditions. As B is not redox sensitive in soils and sorption by soil Fe and Mn hydroxides usually does not seem to play a major role, its availability is not significantly affected by soil aeration. Boron supply can become critical under dry conditions, due to reduced mass flow, but possibly also due to polymerization of boric acid [16].

9.4.3 Cobalt

Cobalt is an essential component of cobalamin (the coenzyme form of vitamin B_{12}) and of some specific enzymes found in N-fixing bacteria [16,48]. For ruminants, Co intake with fodder is indirectly essential, because Co is required by their rumen bacteria. The vitamin B_{12} synthesized by these bacteria is absorbed in the lower intestine to meet the nutritional demand of the animals for this nutrient. Nonruminant animals, including humans, depend on direct supply of preformed vitamin B_{12} and therefore do not need to take in Co. Apart from the dependence of legume and some nonlegume plants on symbiotic N-fixation by root nodule bacteria, Co is not known to be essential for plants [49], although some Co-deficiency symptoms have been reported, for example some chlorosis of young leaves and slightly reduced growth in Co-deficient wheat seedlings [50].

Cobalt is a siderophile element, that is it has only a weak affinity for oxygen and sulfur. Analogous to Fe, it is found in the environment in oxidation states Co(+2) and Co(+3). Geogenic Co is primarily associated with ferromagnesian minerals and can replace Fe in iron oxides and hydroxides. In soils, it is strongly sorbed by Fe and Mn oxides. Sorption by mineral surfaces and organic matter decreases with pH in a similar way to that of Zn and Ni [51]. Due to the high solubility under acid conditions, Co is easily leached from acid soils, similarly to other metal cations. Sandy texture and low organic matter content are conditions that favour leaching. Thus, Co deficiency is frequent on sandy podzols (Table 9.1) and also occurs on very acidic leached ferralitic and peaty soils. As the amounts of Co required by N-fixing bacteria are generally low in comparison with the availability of Co in most soils, Co deficiency is a rather rare condition even in Co-sensitive leguminous plants. Grasses generally accumulate less Co than legumes. Thus, pasture on soils with low Co may not provide sufficient amounts of Co to grazing ruminants depending on this source. As a consequence Co deficiency is a potential problem primarily for grazing livestock in areas with soils of low plant-available Co concentrations. In fact, the role of Co as an essential micronutrient was first detected in field investigations into malnutrition of grazing livestock in Australia in the 1930s [16]. Similar observations were also made in the United States, and maps were produced to delineate Co-deficient soils [52,53]. Some plants that accumulate high concentrations of Co are good indicators of Co availability in soil and can be used to identify areas in which Co deficiency is a potential problem for grazing livestock [54].

9.4.4 Copper

Copper is an essential structural component of many enzymes and other proteins involved in electron transfer reactions, for example in the mitochondrial enzyme cytochrome oxidase and in the plastocyanin protein of the chloroplasts in plants [16]. It is also required for N-fixation. As in the case of Zn, total soil Cu concentrations are largely determined by the parent materials. Soils derived from acid rocks and sediments are particularly low in Cu (Table 9.2). In soil, Cu is strongly associated with soil organic matter and hydroxides. Except under very acid conditions, the fraction of free copper (Cu^{2+}) is very low due to the formation of complexes with DOC. While free Cu concentrations decrease with increasing pH, the total concentration of Cu in solution is often higher at neutral than at acid pH due to increased DOC concentrations. Cu uptake by plants is under tight metabolic control. Cu transport to the roots is limited by the formation and mobility of the organic Cu-complexes present in the solution. Substantial mobilization of copper has been observed in the rhizosphere of Fe-deficient grasses and attributed to the exudation of phytosiderophores [16].

Copper deficiency in plants is not a problem on most soils. It is primarily observed on podzolic sandy soils due to inherently low Cu contents and in organic soils due to low bioavailability (Table 9.1). Due to the frequent occurrence of Cu-deficiency on peat soils reclaimed for agriculture, the term "reclamation disease" was coined for this condition [12]. Low bioavailability can also be a problem in calcareous and clayey soils. Copper deficiency may be aggravated by antagonistic interactions with dissolved Fe and Mo in waterlogged soils. Also N and P fertilization can accentuate Cu deficiency, an effect that partially at least can be attributed to increased Cu demand resulting from enhanced plant growth [31]. Antagonism between Cu and Mo is found not only in plants [55], but also in animals [7]. Molybdenum-induced Cu deficiency (molybdenosis) is a widespread problem in ruminant farm animals feeding on forages grown on poorly drained soils that formed from parent materials rich in Mo, particularly under alkaline conditions. Patterns of Mo-induced Cu deficiency are closely related to the geographical distribution of soils [13,53]. Cu deficiency in ruminants is induced if the Cu:Mo ratio of the fodder is <2. Another dietary factor interfering with Cu metabolism is sulfur (S), which aggravates Mo-induced Cu deficiency in ruminants, while alleviating Mo toxicity in nonruminants [30,7].

9.4.5 Iron

Iron is in general not a trace element in soils (Table 9.2). In fact, it is one of the most abundant elements in the lithosphere (around 5%), primarily in the form of ferromagnesium silicates, and also a major element in most soils, averaging 3–4%. It is included here, because of its importance as a micronutrient and because Fe deficiency is considered the most widespread micro-element malnutrition problem in many crops, animals and human populations throughout the world [6–8,12,56,57]. The physiological functions of Fe relate to the ease of this transition element to change valency between +2 (ferrous Fe) and +3 (ferric Fe) and its capability to form complexes with many organic and inorganic ligands. As a component of haemoglobin, myoglobin and many enzymes, Fe is involved in electron transfer reactions of photosynthesis and the respiratory energy metabolism as well as in defence and detoxification mechanisms against free oxygen radicals [16,31].

Given its abundance in terrestrial environments, scarcity is not the reason for the widespread problem of Fe deficiency, but the generally very low bioavailability of Fe in soil. When Fe is released from primary minerals by weathering, it is readily precipitated in form of ferric oxides and hydroxides under aerobic conditions. Due to the extremely low solubility of these secondary minerals the total concentrations of aquo- and hydroxy-cations of Fe are far below the levels required for a sufficient supply to crop plants, in particular at neutral to alkaline pH. Apart from pH, Fe solubility primarily depends on the oxidation state and the presence of soluble organic ligands. Under waterlogged conditions, Fe is reduced to the ferrous form, which is three orders of magnitude more soluble than ferric iron. Plants have developed various strategies to mobilize Fe in the rhizosphere by the exudation of acidity and chelating compounds, some of which, the so-called phytosiderophores, are extremely specific for this element [16]. Given the strong dependence of Fe availability on pH and oxidation state, Fe deficiency is commonly observed on well-aerated calcareous or saline soils such as Chernozems, Rendzic Leptosols, Yermosols and Solonetz (Table 9.1). Thus it is a common occurrence in arid and semi-arid regions. It is also common in leached acid sandy soils. High organic matter content may aggravate Fe deficiency by further decreasing its availability [6]. Plants vary widely in their efficiency for Fe uptake and their sensitivity to Fe deficiency, depending also on interference by other elements. In particular the heavy metals Zn, Mn, Cu, Co, Cr, Ni are known for antagonistic effects, but also high concentrations of Ca may reduce Fe uptake and transport [31]. This often obscures geographical patterns of iron deficiency, in combination with the variability of soils in water regimes and redox conditions [12].

9.4.6 Manganese

Manganese is much less abundant than Fe in the earth crust, but still markedly above trace concentrations in many soils (Table 9.2). Its biogeochemistry bears much analogy to Fe, and it is included here for similar reasons as Fe. Manganese plays an essential role in photosynthesis and in catalysing the conversion of free oxygen radicals to hydrogen peroxide as part of the enzyme superoxide dismutase, due to its capability to easily change oxidation states. Manganese also is a cofactor activating many other enzymes, but is often replaceable in this role by other cations, in particular Mg [16].

Manganese in soils derives primarily from ferromagnesium minerals. It accumulates in the form of oxides and hydroxides as secondary soil minerals, as coprecipitates with Fe oxides and also in complexes with organic matter. Uptake by plants is under metabolic control, although passive uptake may occur at high concentrations leading to toxicity. Manganese deficiency is found worldwide in soils of many different types. Deficiency results from factors that reduce the concentration of free divalent Mn in the soil solution, as this is the form in which Mn is taken up by plants. For a given total concentration, the free Mn(II) cation concentration primarily depends on the redox potential, pH and organic ligands. Solubility as a free ion decreases with increasing pH by 2 orders of magnitude per unit of pH [31]. Therefore, Mn is often deficient due to low bioavailability in calcareous and alkaline soils (Table 9.1). Manganese deficiency due to leaching is common in podzols and other highly weathered acidic soils, in particular acidic sandy soils low in native Mn [12]. Manganese complexes with organic matter are less stable than, for example, Zn complexes, but at high pH humic acids can fix Mn so that Mn deficiency is also found in peaty soils with calcareous subsoil. As waterlogging leads to the dissolution of Mn oxides by reduction of Mn(IV) to Mn(II), it also promotes loss of Mn by leaching, in particular in Gleysols, Planosols and other soils with fluctuating redox conditions (Table 9.1). On the other hand, dissolution of Mn oxides and subsequent accumulation of Mn^{2+} ions in solution often results in Mn toxicity, which may induce Ca, Mg and Fe deficiency [31]. Plant uptake of Mn may in turn be reduced by competition with Fe and Zn [14]. However, Zn has also been found to increase Mn uptake [58].

9.4.7 Molybdenum

As a cofactor of some important enzymes in plants and N-fixing microorganisms, Mo is directly involved in the catalysis of redox reactions in the metabolism of N [16]. Two key enzymes for which Mo is an essential component are nitrate reductase and nitrogenase.

Molybdenum differs from the other essential trace metals in occurring as an oxyanion or hydroxyanion, that is as negatively charged species, in soil solution except for very acidic conditions (pH <5), where it occurs as molybdic acid. As a consequence, sorption decreases with increasing pH in contrast to trace elements that occur as cations. Maximum sorption occurs around pH 4, coinciding with the pK_a value of molybdic acid (H_2MoO_4). Low Mo solubility at low pH is attributed to precipitation of ferric molybdates ($Fe_2(MoO_4)_3$) and formation of ferric oxide molybdate surface complexes [12]. Organic matter also sorbs molybdate, but is much less important for Mo availability than pH and iron oxides. Under anaerobic conditions increased molybdate (MoO_4^{2-}) concentrations in the soil solution may result from the reduction of ferric iron (Fe(III)) to ferrous iron (Fe(II)) and the concomitant dissolution of Fe oxides. Moreover, ferrous molybdates or molybdites are much more soluble than ferric molybdates. Organic matter promotes the mobilization of Mo in waterlogged soils [6].

Molybdate and sulfate anions strongly compete for uptake by roots [16]. As molybdate is the form in which Mo is mainly taken up by roots, this may explain why high soil sulfate concentrations can aggravate Mo deficiency [59]. There is also a strong antagonism between Mo and Cu that may exacerbate Mo deficiency. In plants, Cu interferes with the role of Mo in the enzymatic reduction of nitrate [31]. On the other hand, phosphorus generally increases Mo availability, which appears to be at least partially due to competition between phosphate and molybdate for sorption sites [14].

Shale, clay schists and granite are the primary geogenic sources of Mo in soil [60]. Molybdenum deficiency is mostly found in well-aerated acid soils. Although it is most pronounced on soils derived from low-Mo parent materials, pedogenic factors, in particular soil drainage and pH, are generally more relevant for the development of Mo deficiency than Mo contents of soil parent material [12]. Under acid conditions this is due to the effects of drainage on soil aeration and thus iron chemistry, while well-drained alkaline soils tend to impoverish in Mo in high-rainfall areas due to molybdate leaching.

9.4.8 Selenium

While all trace elements become toxic above certain element- and organism-specific threshold concentrations, selenium (Se) is particular in that the margin between its deficiency and toxicity is very narrow [61]. A metabolic role of Se was first identified in 1973 when it was found to be a constituent of the antioxidant enzyme glutathione peroxidase. Today a variety of other selenoproteins are known to play a critical role in many processes including thyroid hormone metabolism, redox balance and the protection against cardiovascular disorders, chemically induced carcinogenesis and mercury poisoning [61,62]. In livestock Se deficiency is known to produce muscular degeneration (white muscle disease), a myopathy found especially in young animals, primarily lambs and calves [7]. Severe deficiency in humans was also found to be the prime cause of the Keshan disease, an endemic cardiac myopathy in the Keshan region in China [8].

Selenium is chemically very similar to sulfur. It is highly redox-sensitive and occurs in the environment in various oxidation states ranging from -2 to $+6$. Uptake of Se appears to vary more widely among plants than that of most other trace elements [23,63]. Selenate (SeO_4^{2-}) is actively and more easily accumulated by plants than selenite (SeO_3^{2-}). Selenate strongly competes with sulfate for uptake. Selenium speciation in soil solution is governed primarily by pH and redox potential. Selenide (Se^{2-}) dominates at low, selenate at high and selenite at intermediate pe+pH values [64]. Selenium solubility increases with pH and redox potential [65]. Under acidic conditions, it is limited by poorly soluble ferric oxide–selenite complexes, at low redox potentials by metal selenides. Organic matter and iron oxides have a higher sorption capacity for selenium than clay minerals and mostly control Se solubility in alkaline and well-aerated conditions. Binding to soluble organics and competition with other anions for sorption sites increase Se in the soil solution.

Selenium is not an abundant micronutrient element in the lithosphere and in most soils (<10 mg kg^{-1}), but its total contents can exceed 1 g kg^{-1} in so-called seleniferous soils. Selenium concentrations of soils are closely related to parent rock and water regime. Low concentrations are common in highly weathered acidic soils, derived from igneous rocks such as granites. In high-rainfall areas, leaching from aerobic soils is considerable, leading to Se concentrations in forage plants that are insufficient as sole Se source for livestock. Selenium deficiency is expected in livestock when the Se concentration of forage plants is <0.04 mg kg^{-1} dry matter [31].

9.4.9 Zinc

Zinc deficiency is a major problem for crop production as well as in animal and human nutrition in many countries all over the globe. After Fe deficiency, deficiency in Zn is now recognized as the most serious mineral deficiency disorder in human populations worldwide [66]. In crop production the occurrence of Zn deficiency was found to be more frequent than that of any other micronutrient deficiency in field experiments performed on 190 sites in 15 countries, with some degree of Zn deficiency being observed in about 50% of the investigated sites [27]. In a previous study, the same author found particularly low Zn concentrations in wheat and maize grown on soils from 10 countries out of 30 countries sampled [24].

Like Cu and Mo, Zn is essential to plants and animals as well. Its metabolic functions are manifold. It is a key component of a large number of enzymes such as carbonic anhydrase, alcohol dehydrogenase and other dehydrogenases, superoxide dismutase and RNA polymerase, and it activates or modulates many other enzymes [16]. Through the formation of polypeptide loops, so-called Zn fingers, which bind to specific DNA sequences, it also plays a crucial role in DNA replication and transcription. Zinc complexes with polypeptides are also important for the stability of membranes, apparently protecting them against oxidative damage [16].

Under deficiency conditions diffusion becomes the limiting process of Zn supply to plant roots. Zinc uptake by plants is primarily an active process, involving biochemical cross-membrane pumping of free Zn cations. The availability of Zn in soil solution mainly depends on pH, the presence of competing cations, sorption sites provided by metal hydroxides, clay minerals, Ca carbonate and organic matter. It may also be limited by precipitation with phosphate. Zinc does not interact very strongly with organic substances in comparison to other trace metals. Complexes with DOC therefore do not usually play a major role for the speciation of Zn in soil solution. Zinc is not reduced under anaerobic conditions in soil, but Zn concentrations may be increased, for example through the reductive dissolution of sorbing Fe and Mn oxides under waterlogging, as well as decreased, for example through precipitation as franklinite ($ZnFe_2O_4$) or sulfide, depending on the soil chemical conditions. Synthesis of high-affinity transporters and exudation of acidity by roots into the rhizosphere are common strategies of plants to enhance Zn uptake. Also mycorrhizal fungi can be very important for supplying Zn to plants. Phosphorus can induce Zn deficiency not only through the precipitation of Zn phosphate or binding of Zn to phosphate sorbed to Fe oxides [67], but also by Zn dilution through enhanced growth or by inactivation of Zn in plants through binding to phytate [14]. Other antagonistic interactions with Zn have been reported for Cd, Cu and Fe. However they are usually less important for the availability of Zn in field soils than those with P.

Total Zn contents in soils generally depend more on parent material than on other pedogenic factors [60]. Soils developed from siliceous acid rocks such as sandstones and gneiss are mostly poor, whereas those from basic igneous rocks and limestones are comparatively rich in Zn [31]. Much more than other trace elements, Zn is retained in noncropped soils through recycling by vegetation, as it is accumulated in the biomass and subsequently released into the soil from decomposing litter [68]. Zinc deficiency is most common on leached acid sandy soils and on soils with low Zn availability due to high pH (calcareous and alkali soils) or high organic matter content (for example peat soils). The risk of

Zn-deficiency is also high on soils developed on phosphatic rock as parent material [12] and waterlogged soils such as paddy rice fields [69]. The geographic distribution of Zn deficiency is often very patchy due to high local variability in soil conditions affecting available Zn and large variations in Zn demand by crops [12].

9.4.10 Other Micronutrients

Deficiencies in Cr and Ni are not known to be major problems. Nickel is a component of various bacterial enzymes, but in higher plants only of urease. It is, thus, required by plants that depend on urea as nitrogen source. Nickel was found to be essential for some higher plants such as barley – independently of the nitrogen source, however, suggesting that there may be also other functions of this transition metal [70]. Nickel is likely an essential micronutrient also for animals [11,71]. The difficulty in establishing the essentiality of Ni derives from the fact that it is required only in ultra-trace amounts (Table 9.2). As Ni is comparatively abundant in most soils, its deficiency is a very rare condition, even where the bioavailability is low due to high carbonate, sesquioxide or organic matter content. In fact, field occurrence of Ni deficiency has been demonstrated only recently in pecan trees in the south-eastern United States [72].

While Cr is essential for the glucose metabolism in mammals and also plays an important role in the lipid and protein metabolism of animals [31], nutritional deficiency of Cr is not known to be a problem [7]. The total content of this element varies widely in soils, from trace concentrations to as high as 0.4% Cr, primarily dependent on the parent material.

9.5 Treatment of Soils Deficient in Trace Elements

Low contents or limited bioavailability of essential trace elements are constraints that affect agricultural crop production on many soils throughout the world. Many strategies have been developed to cope with these limitations [6]. One way is to select crop plant species and cultivars that are adapted to the site-specific soil conditions. In particular, this includes breeding and growing crop plants that are efficient in the use of the limiting micronutrients. Equally important is to optimize physical, chemical and biological soil conditions for the development of root systems and the supply of water and nutrients. Soil compaction due to the use of excessively heavy machinery, for example, has become a major physical constraint to agricultural soil productivity in recent years, impeding root growth by increased mechanical resistance and reduced soil aeration. Another factor strongly influenced by agricultural management practices is soil organic matter. Providing a sufficient input of easily decomposable organic substrates through inputs of crop residues, manures, composts and other biomaterials is important for the formation of soil aggregate structure, water retention, binding of toxic compounds and nutrient supply. An important soil biological factor in plant nutrition is root colonization by mycorrhizal fungi. Mycorrhizae are primarily known for their beneficial effects on the P nutrition of plants, but they can also increase the acquisition of Cu and Zn by roots [73]. Inoculation of soils with mycorrhizal fungi can be particularly helpful where vegetation is to be newly established on nutrient-poor soils under difficult soil chemical conditions, for example on acid, alkaline or saline soils. Liming of soils is another technique to improve plant growth conditions. It is used to ameliorate soils with excessive acidity. Raising the pH can in particular improve the availability of Mo and reduce Mn toxicity in acid soils. On the other hand, care has to be taken in the application of lime in order to avoid an excessive increase in pH as this may cause deficiencies in Zn, Cu, Mn or Fe.

By far the most direct and effective measure to correct micronutrient deficiencies in agricultural soils is fertilization, using both organic and inorganic sources. Soil, foliar and seed application of micronutrient fertilizers are practised. When applied to soil, inorganic or synthetic organic micronutrient fertilizers are often blended with granular N, P, and K fertilizers or mixed with fluid macronutrient fertilizers. Application rates need to be carefully determined, as overdoses may result in toxic concentration levels. Remediation of soils with toxic levels of trace elements is generally difficult and costly, if feasible at all. Toxicity risks are particularly high in foliar applications and seed priming. Soil application has the advantage of a comparatively large duration of the fertilization effect for most micronutrients, whereas foliar applications have little residual effects and may have to be repeated several times during a growing season because limited amounts of nutrients can be supplied at a time, without causing toxic effects. The advantage of foliar application is that the response of the crops is almost immediate and that lower rates need to be applied. The lack of residual effects, however, also means that the response of crops to foliar fertilization is in general closely related to the time of application. Foliar application of micronutrients is usually the method of choice in emergency situations, whereas soil applications of micronutrients are advantageous in achieving a long-term improvement of soil fertility, depending on the rate at which residual fertilizer is converted into plant-unavailable forms. According to Martens and Westermann [74] residual Cu, for example, was found to increase wheat yields for up to 12 years after soil application of $CuSO_4$. Similarly, residual Zn from broadcast application of $ZnSO_4$, the standard form of inorganic Zn fertilizer, was reported to be effective for correcting Zn deficiency over 5–7 years. On the other hand, soil applications of inorganic Fe sources are generally ineffective for correcting Fe deficiency, as these sources are too insoluble or become rapidly immobilized in well-aerated agricultural soils. Synthetic Fe chelates are usually more effective than inorganic Fe sources due to their stability and mobility, but their application is usually limited to crops of high value due to the high costs of these synthetic fertilizers [74]. Foliar sprays of synthetic organic Fe fertilizers, such as FeEDDHA (Fe-ethylenediamine-di-*o*-hydroxyphenylacetate), are usually more economic. Also soil application of Mn fertilizers is generally not recommended for similar reasons as in the case of Fe, while soil application is the preferred application method for B due to the risk that direct application of B fertilizers on leaves and seeds may cause toxicity [6].

Enhancing the acquisition of micronutrients by crops using genetic or agronomic techniques can also be very efficient in fighting mineral deficiencies in humans and animals by providing micronutrient-enriched food or feed, respectively. This strategy, which is called biofortification, is considered a cost-effective and sustainable approach to alleviating micronutrient deficiencies in human populations that depend on plant-derived food for an adequate supply of these micronutrients. Increasing Se concentrations in food crops by application of Se-supplemented NPK fertilizers effectively improved the nutritional status of the Finnish population with respect to this element [62,75]. In Turkey a similar agronomic biofortification strategy was successfully implemented in the mid-1990s to increase the yield and Zn concentration of wheat grains grown on the Zn-deficient alkaline soils of Central Anatolia [76]. In comparison with other strategies such as dietary diversification, supplementation of diets with pharmaceutical micronutrient preparations or food fortification during commercial food crop processing, biofortification does not require populations to change their dietary habits, is attractive for the farmers because of increased yields and reduced seedling losses, and is also practicable in countries that are not in a position to implement costly supplementation programmes.

Acknowledgement

We thank Dr Richard Hurrell for reviewing the chapter with respect to the nutritional role of trace elements.

References

[1] Arnon, D.I.; Stout, P.R., The essentiality of certain elements in minute quantity for plants with special reference to copper; Plant Physiol. 1939, 14, 371–385.
[2] Brownell, P.F., Sodium as an essential micronutrient element for plants and its possible role in metabolism; Adv. Bot. Res. 1979, 7, 117–224.
[3] Nable, R.O.; Brownell, P.F., Effect of sodium and light upon the concentrations of alanine in leaves of C4 plants; Aust. J. Plant Physiol. 1984, 11, 314–324.
[4] Asher, C.J., Beneficial elements, functional nutrients and possible new essential elements; in: J.J. Mortvedt, F.R. Cox, L.M. Shuman, R.M. Welch (eds), Micronutrients in Agriculture, SSSA Book Series 4, 2nd edn; Soil Science Society of America, Madison WI, 1991, pp. 703–723.
[5] Welch, R.M., Micronutrient nutrition of plants; Crit. Rev. Plant Sci. 1995, 14, 49–82.
[6] Fageria, N.K.; Baligar, V.C.; Clark, R.B., Micronutrients in crop production; Adv. Agron. 2002, 77, 185–268.
[7] Miller, E.R.; Lei, X.; Ullrey, D.E., Trace elements in animal nutrition; in: J.J. Mortvedt, F.R. Cox, L.M. Shuman, R.M. Welch (eds), Micronutrients in Agriculture, SSSA Book Series 4, 2nd edn; Soil Science Society of America, Madison WI, 1991, pp. 593–662.
[8] Van Campen, D.R., Trace elements in human nutrition; in: J.J. Mortvedt, F.R. Cox, L.M. Shuman, R.M. Welch (eds), Micronutrients in Agriculture, SSSA Book Series 4, 2nd edn, Soil Science Society of America, Madison WI, 1991, pp. 663–701.
[9] Selinus, O.; Alloway, B.J.; Centeno, J.A. *et al.* (eds), Essentials of Medical Geology. Impacts of the Natural Environment on Public Health; Elsevier Academic Press, Amsterdam, 2005.
[10] Nielsen, F.H., Ultratrace elements in nutrition; Annu. Rev. Nutr. 1984, 4, 21–41.
[11] Maroney, M.J., Structure/function relationships in nickel metallo-biochemistry; Curr. Opin. Chem. Biol. 1999, 3, 188–199.
[12] Welch, R.M., Allaway, W.H.; House, W.A.; Kubota, J., Geographic distribution of trace element problems; in: J.J. Mortvedt, F.R. Cox, L.M. Shuman, R.M. Welch (eds), Micronutrients in Agriculture; SSSA Book Series 4, 2nd edn; Soil Science Society of America, Madison WI, 1991, pp. 31–57.
[13] Thornton, I., Geochemistry and the mineral nutrition of agricultural livestock and wildlife; Appl. Geochem. 2002, 17, 1017–1028.
[14] Moraghan, J.T.; Mascagni Jr., H.J., Environmental and soil factors affecting micronutrient deficiencies and toxicities; in: J.J. Mortvedt, F.R. Cox, L.M. Shuman, R.M. Welch (eds),

Micronutrients in Agriculture; SSSA Book Series 4, 2nd edn; Soil Science Society of America, Madison WI, 1991, pp. 371–425.
[15] Jones, J.B., Plant tissue analysis in micronutrients; in: J.J. Mortvedt, F.R. Cox, L.M. Shuman, R.M. Welch (eds), Micronutrients in Agriculture, SSSA Book Series 4, 2nd edn; Soil Science Society of America, Madison WI, 1991, pp. 477–521.
[16] Marschner, H., Mineral Nutrition of Higher Plants, 2nd edn; Academic Press, London, 1995.
[17] Barker, A.V.J. Pilbeam, D.J., Handbook of Plant Nutrition; CRC Press, Taylor & Francis, Boca Raton FL, 2007.
[18] Maskall, J.; Thornton, I., The distribution of trace and major elements in Kenyan soil profiles and implications for wildlife nutrition; in: J.D. Appleton, R. Fuge, G.J.H. Hall (eds), Environmental Geochemistry and Health; Geological Society Special Publication No. 113, 1996, pp. 47–62.
[19] Mills, C.F., Geochemical aspects of the aetiology of trace element related diseases; in: J.D. Appleton, R. Fuge, G.J.H. Hall (eds), Environmental Geochemistry and Health; Geological Society Special Publication No. 113, 1996, pp. 1–5.
[20] Fordyce, F.M.; Masara, D.; Appleton, J.D., Stream sediment, soil and forage chemistry as indicators of cattle mineral status in northeast Zimbabwe; in: J.D. Appleton, R. Fuge, G.J.H. Hall (eds), Environmental Geochemistry and Health, Geological Society Special Publication No. 113, 1996, pp. 23–37.
[21] Mertz, W., Trace Elements in Human and Animal Nutrition, 5th edn; Academic Press, London, 1987.
[22] López Alonso, M.; Benedito, J.L.; Miranda, M.; Castillo, C.; Hernandez, J.; Shore, R.F., Cattle as biomonitors of soil arsenic, copper, and zinc concentrations in Galicia (NW Spain); Arch. Environ. Contam. Toxicol. 2002, 43, 103–108.
[23] Chapman, H.D. (ed.), Diagnostic Criteria for Plants and Soils; Department of Soils and Plant Nutrition, University of California, Riverside CA, 1966.
[24] Sillanpää, M., Micronutrients and the nutrient status of soils: a global study; FAO Soils Bulletin 48; FAO, 1982.
[25] Katyal, J.C.; Randhawa, N.S., Micronutrients; FAO Fertilizer and Plant Nutrition Bulletin 7; FAO, 1983.
[26] Mengel, K.; Kirkby, E.A., Principles of Plant Nutrition; International Potash Research Institute, Worblaufen-Bern, Switzerland, 1987.
[27] Sillanpää, M., Micronutrients assessment at the country level: an international study; FAO Soils Bulletin 63; FAO, 1990.
[28] Benton, Jr. J., Plant tissue analysis in micronutrients; in: J.J. Mortvedt, F.R. Cox, L.M. Shuman, R.M. Welch (eds), Micronutrients in Agriculture, SSSA Book Series 4, 2nd edn; Soil Science Society of America, Madison WI, 1990, pp. 477–521.
[29] Kabata-Pendias, A.; Mukherjee, A.B., Trace Elements from Soil to Humans; Springer, Berlin, 2007.
[30] Underwood, E.J., The Mineral Nutrition of Livestock, 2nd edn; Commonwealth Agricultural Bureaux, London, 1981.
[31] Srivastava, P.C.; Gupta, U.C., Trace Elements in Crop Production; Science Publishers, Lebanon NH, 1996.
[32] Prasad, M.N.V. (ed.), Heavy Metal Stress in Plants – From Biomolecules to Ecosystems, 2nd edn; Springer, Berlin, 2004.
[33] Lindsay, W.L.; Norvell, W.A., Development of a DTPA soil test for zinc, iron, manganese, and copper; Soil Sci. Soc. Am. J. 1978, 42, 421–428.
[34] de Abreu, C.A.; van Raij, B.; de Abreu, M.F.; Gonzalez, A.P., Routine soil testing to monitor heavy metals and boron; Scientia Agricola 2005, 62, 564–571.
[35] Bergmann, W. (ed.), Nutritional Disorders of Plants: Development, Visual and Analytical Diagnosis; Fischer, Jena, 1992.
[36] Steward, F.C.; Durzan, D.J., Metabolism of nitrogenous compounds; in: F.C. Steward (ed.), Plant Physiology – A Treatise, Volume IVa; Academic Press, New York, 1965.
[37] Shaked, D.; Bar-Akiva, A., Nitrate reductase activity as an indicator of molybdenum level and requirement of citrus plants; Phytochemistry 1967, 6, 347–350.

[38] Bar-Akiva, A.; Lavon, R.; Sagiv, J., Ascorbic acid oxidase activity as a measure of the copper nutrition requirements of citrus plants; Agrochimica 1967, 14, 47–54.
[39] Rahimi, A.; Bussler, W., Kupfermangel in höheren Pflanzen und sein histochemischer Nachweis; Landwirtschaftliche Forschung 1974, 30, 101–111.
[40] White, J.C.; Zasoski, R.J., Mapping soil micronutrients. Field Crops Res. 1999, 60, 11–26.
[41] Beeson, K.C., The occurrence of mineral nutritional diseases of plants and animals in the United States; Soil Sci. 1945, 60, 9–13.
[42] Berger, K.C., Micronutrient deficiencies in the United States; Agric. Food Chem. 1962, 10, 178–181.
[43] Kubota, J., Soils and plants in the geochemical environment; in: I. Thorrnton (ed.), Applied Experimental Geochemistry; Academic Press, New York, 1983, pp. 103–122.
[44] Woodruff, J.R.; Moore, F.W.; Musen, H.L., Potassium, boron, nitrogen and lime effects on corn yield and earleaf nutrient concentrations; Agron. J. 1987, 79, 520–524.
[45] Graham, R.D.; Welch, R.M.; Grunes, D.L.; Cary, E.E.; Norvell, W.A., Effect of zinc deficiency on the accumulation of boron and other mineral nutrients in barley; Soil Sci. Soc. Am. J. 1987, 51, 652–657.
[46] Bradford, G.R., Boron; in: H. D. Chapman (ed.), Diagnostic Criteria for Plants and Soils, University of California, Berkeley, 1966, pp. 33–61.
[47] Gupta, U.C., Factors affecting boron uptake by plants; in: U.C. Gupta (ed.), Boron and Its Role in Crop Production; CRC Press, Boca Raton, FL, 1993, pp. 87–123.
[48] Dilworth, M.J.; Robson, A.D.; Chatel, D.L., Cobalt and nitrogen fixation in *Lupinus angustifolius* L. II. Nodule formation and function; New Phytol. 1979, 83, 63–79.
[49] Talukder, G.; A. Sharma, A., Cobalt; in: A.V. Barke ; D.J. Pilbeam (eds), Handbook of Plant Nutrition; Taylor & Francis, CRC Press, Boca Raton FL, 2007, pp. 499–514.
[50] Wilson, S.B.; Nicholas, D.J.D., A cobalt requirement for non-nodulated legumes and for wheat; Phytochemistry 1967, 6, 1057–1060.
[51] Sposito, G., The Surface Chemistry of Soils; Oxford University Press, New York, 1984.
[52] Underwood, E.J., Trace Elements in Human and Animal Nutrition, 4th edn; Academic Press, New York, 1977.
[53] Kubota, J.; Welch, R.M.; van Campen, D.R., Soil-related nutritional problem areas for grazing animals; Adv. Soil Sci. 1987, 6, 189–215.
[54] Vanselow, A.P. 1966. Cobalt; in: H.D. Chapman (ed.), Diagnostic Criteria for Plants and Soils; University of California, Berkeley, 1966, pp. 142–156.
[55] Mackay, D.C.; Chipman, E.W.; Gupta, U.C., Copper and molybdenum nutrition of crops grown on acid sphagnum peat soil; Soil Sci. Soc. Am. Proc. 1966, 30, 755–759.
[56] Frossard, E.; Bucher, M.; Mächler, F.; Mozafar, A.; Hurrell, R., Potential for increasing the content and bioavailability of Fe, Zn and Ca in plants for human nutrition; J. Sci. Food Agric. 2000, 80, 861–879.
[57] Vose, P.B., Iron nutrition in plants: a world overview; J. Plant Nutr. 1982, 5, 233–249.
[58] White, M.C.; Decker, A.M.; Chaney, R.L., Differential cultivar tolerance in soybean to phytotoxic levels of soil Zn. I. Range of cultivar response; Agron. J. 1979, 71, 121–126.
[59] Gupta, U.C.; MacLeod, J.A., The effects of sulfur and molybdenum on the molybdenum, copper, and sulfur concentrations of forage crops; Soil Sci. 1975, 119, 441–447.
[60] Aubert, H.; Pinta, M., Trace Elements in Soils; Elsevier, Amsterdam, Netherlands, 1977.
[61] Reilly, C., Selenium in Food and Health, 2nd edn; Springer, New York, 2006.
[62] Lyons, G.; Stangoulis, J.; Graham, R., High-selenium wheat: biofortification for better health; Nutr. Res. Rev. 2003, 16, 45–60.
[63] Kopsell, D.A.; Kopsell, D.E., Selenium; in: A. V. Barker, D.J. Pilbeam (eds), Handbook of Plant Nutrition; Taylor & Francis, CRC Press, Boca Raton, FL, 2007, pp. 515–549.
[64] Elrashidi, M.A.; Adriano, D.C.; Workman, M.; Lindsay, W.L., Chemical equilibria of selenium in soils: a theoretical development; Soil Sci. 1987, 144, 141–152.
[65] Adriano, D.C., Trace Elements in the Terrestrial Environment; Springer, New York, 1986.
[66] Welch, R.M.; Graham, R.D., Breeding for micronutrients in staple food crops from a human nutrition perspective; J. Exp. Bot. 2004, 55, 353–364.

[67] Xie, R. J.; Mackenzie, A.F., Effects of sorbed orthophosphate on zinc status in three soils of eastern Canada; J. Soil Sci. 1989, 40, 49–58.
[68] Nowack, B.; Obrecht, J.-M.; Schluep, M.; Schulin, R.; Hansmann, W.; Köppel, V., Elevated lead and zinc contents in remote alpine soils of the Swiss National Park; J. Environ. Qual. 2001, 30, 919–926.
[69] Alloway, B.J., Zinc in Soils and Crop Nutrition, Online book published by the International Zinc Association, Brussels, Belgium, http://www.zinc-crops.org/, 2006.
[70] Brown, P.H.; Welch, R.M.; Cary, E.E., Nickel: a micronutrient essential for higher plants; Plant Physiol. 1987, 85, 801–803.
[71] Brown, P.H., Nickel; in: A.V. Barker, D.J. Pilbeam (eds), Handbook of Plant Nutrition; Taylor & Francis, CRC Press, Boca Raton, FL, 2007, pp. 395–409.
[72] Wood, B.W.; Reilly, C.C.; Nyczepir, A.P., Mouse-ear of pecan: a nickel deficiency; HortScience 2004, 39, 1238–1242.
[73] Marschner, H.; Römheld, V., Root-induced changes in the availability of micronutrients in the rhizosphere; in: Y. Waisel, A. Eshel, U. Kafkafi (eds), Plant Roots – The Hidden Half; Marcel Dekker, New York, 1996, pp. 557–579.
[74] Martens, D.C.; Westermann, D.T., Fertilizer applications for correcting micronutrient deficiencies; in: J.J. Mortvedt, F.R. Cox, L.M. Shuman, R.M. Welch (eds), Micronutrients in Agriculture, SSSA Book Series 4, 2nd edn; Soil Science Society of America, Madison WI, 1991, pp. 549–592.
[75] Aspila, P., History of selenium supplemented fertilization in Finland; Agrifood Research Reports 2005, 69, 8–13.
[76] Cakmak, I., Enrichment of cereal grains with zinc: Agronomic or genetic biofortification?; Plant Soil 2008, 302, 1–17.

10
Application of Chemical Speciation Modelling to Studies on Toxic Element Behaviour in Soils

Les J. Evans[1], *Sarah J. Barabash*[1], *David G. Lumsdon*[2] *and Xueyuan Gu*[3]

[1]*School of Environmental Sciences, University of Guelph, Ontario, Canada*
[2]*Macaulay Institute, Aberdeen Scotland*
[3]*School of the Environment, Nanjing University, China*

10.1 Introduction

Soils are very important in the attenuation of toxic elements in the environment because they contain mineral and humic constituents that are involved in elemental retention. These surface-active sites include both constant negative charge sites associated with phyllosilicate clays as well as variable charge sites associated with oxide minerals, the edges of clays and soil organic material [1]. The mobility of many elements in soils is also influenced by a number of other important chemical factors that include the pH of the aqueous phase and the type and content of soluble complexing ligands, such as chloride, sulfate and carbonate ions, as well as organic acids and low-molecular- weight humic substances [2]. The magnitude of these effects will depend on both the binding constants of the aqueous element–ligand species and the intrinsic binding constants for the various complexes of the element with the reactive particle surfaces [3,4]. To fully model soil systems, a combination of both aqueous speciation and surface complexation models need to be applied.

To be able to predict the behaviour of trace elements in contaminated soils, it is necessary to understand both the soluble forms of the elements in solution and the nature and amount of solid phases with which the element is, or could be, associated. This information will aid in predicting the bioavailability of toxic elements; help in describing their fate in both terrestrial and aquatic ecosystems; and aid in the development of scientifically sound remediation options and guidelines for the safe disposal of wastes onto land.

The concentration of toxic elements in contaminated soils is generally measured by digesting the soil in aqua regia, a mixture of HCl and HNO_3. There are, however, many studies which show poor correlations between the toxic element content measured by aqua regia digests and the content and response of the elements in various bioreceptors, such as plants and meso- and microfauna [5]. This is most likely because not all of the forms of the metal dissolved by the digestion procedure are readily bioavailable. To overcome this problem, attempts have been made to estimate the potentially bioavailable fraction of toxic elements in soils using computer models [6] or develop analytical techniques to estimate trace element speciation [7]. Over the past few years, however, aqueous speciation and surface complexation models have become important tools in understanding how toxic elements behave in aqueous environments in general and specifically for soils, see for example, Gustaffson [8], Lumsdon [9], Bonten et al. [10].

Other examples of equilibrium model applications include studies of soil solution-phase speciation [11–13], the influence of solution speciation on ion exchange/adsorption [14,15], the effects of speciation on precipitation and dissolution phenomena [16,17], the influence of kinetics on chemical speciation [18,19], and the influence of reactive transport on chemical speciation [20,21].

Many different models have been developed over the years for solving problems involving chemical speciation. Commonly used models in the literature with regards to metal complexation include MINTEQA2, PHREEQC, WHAM, CHEAQS, ECOSAT, and ORCHESTRA (Table 10.1). The majority of these models were developed from earlier versions of MINEQL [33] and MICROQL [34]. The models differ in the level of sophistication with which they handle adsorption of elements to mineral and humic surfaces. To date, none of the commercially available models can adequately handle all the various adsorbing surfaces found in soils, although some models now include multiple adsorbing phases [4,35–37] and the original WHAM model for organic surfaces has been extended to include oxide surfaces [38].

Table 10.1 Some commonly used chemical speciation models

Program name	Author	Program name	Author
ECOSAT	Keizer and van Reimsdijk [22]	CHEAQS	Verweij [27]
MINETQA2	Allison et al. [23]	WHAM	Tipping [28,29]
PHREEQC	Parkhurst and Appelo [24]	EQ3/6	Wolery [30]
MINEQL+	Schecher [25]	CHESS	Santore and Driscoll [31]
Geochemists Workbench	Benthke [26]	ORCHESTRA	Meeussen [32]

Current models can manage many geochemical processes, including precipitation/dissolution reactions, oxidation/reduction reactions, surface complexation reactions, cation exchange, coprecipitation and kinetics [39]. Calculations are performed to take into account varied environmental conditions, such as pH, ionic strength, temperature and redox conditions. The program MINTEQA2 contains enthalpy values within its database which allows the user to vary the temperature from 25 °C to the temperature of the solution of interest [23].

10.2 The Structure of Chemical Speciation Models

Because of the large number of simultaneous equations that need to be solved in both aqueous and surface complexation models, chemical equilibrium problems are reduced to an algebraic form in order to calculate the variables required to solve the problem. To do this, a number of constraining equations are needed to define the system. To solve any problem the number of equations needed must equal the number of unknowns. Many of the current aqueous speciation and surface complexation models use the algorithms originally described in the program MICROQL [40]. The algorithms contained within the program MICROQL include Newton–Raphson and Jacobian iterations and codes for Gaussian elimination and back substitution (Table 10.2).

This traditional approach for solving chemical systems may be slow for large chemical systems. Significant improvements in computing speed have been achieved in the program ORCHESTRA [32] whose solver functions are implemented in JAVA, a computer language that contains standard language constructs for parallel processing allowing the efficient combination of both chemical speciation and transport of solutes.

The major inputs into the programs are the total concentrations of each component, T_i, as measured by various analytical techniques. For each species in the chemical problem, the activity, a_i, and concentration, c_i, of each species are related through the activity coefficient, γ, $a_i = \gamma\, c_i$. Most commercially available computer software program contains a database that contains many formation constants, log β, that are usually reported at 25 °C, 1 atmosphere pressure and zero ionic strength. The success of geochemical models is very dependent on the quality and internal consistency of the database. The most recent and internally consistent data base for aqueous species and for precipitates is that compiled by the National Institute of Standards and Technology (NIST) [41]. Ionic strength can also be varied within most programs with activity coefficients calculated using the Davies equation [42]. Revised formation constants at the new ionic strength are recalculated from their values at zero ionic strength. In the PHREEQC program the Pitzer equation [43] is included to calculate activity coefficients for solutions with very high ionic strengths.

Formation constants of both soluble and surface species are most commonly determined from potentiometric titration data and batch adsorption experiments using least-squares fitting programs such as FITEQL [44]. FITEQL is an iterative, nonlinear least-squares optimization program that can be used to adjust parameters in a chemical equilibrium model to fit experimental data. FITEQL version 3.1 [45] not only allows calculations for aqueous speciation but also includes four surface complexation models: the constant capacitance model (CCM), the diffuse layer model (DLM), the Stern model (SM) and

Table 10.2 Algorithms used in the computer program MICROQL

Mass action

$$c_i = \beta_{p,q,r} \prod_j X_j^a$$

where c_i is the concentration of species I, $\beta_{p,q,r}$ is the formation constant for species I, X_j is the concentration of component j and a is the stoichiometric coefficient of component j in species I.

The mass action equation written in logarithmic form

$$\log c_i = \log \beta_{p,q,r} + \sum_j a \log X_j$$

which in matrix format is $C = AX + K$, where C is a column vector of log c_i, A is the matrix of stoichiometric coefficients a, X is the column vector of log X_j and K is the column vector of log $\beta_{p,q,r}$.

Mass balance

$$[X]_T = \sum_i a c_i$$

where $[X]_T$ is the total concentration of component X in the system, c_i is the concentration of species I and a is the stoichiometric coefficient of component j in species I.

Newton–Raphson iteration

From an initial guess for the concentration of each component, the concentration of each species can be calculated. The sum of the calculated concentration of each species, multiplied by the stoichiometric amount of each component in the species is compared with the known total concentration of each component to give the following equation:

$$Y_j = \sum_i a c_i \cdot [X]_T$$

where Y_j is the difference in the known total concentration and $[X]_T$, from the calculated total concentration $\sum_j a c_j$ of each species.

An iterative technique is used to find improved values of X such that the error term Y becomes smaller. The equilibrium problem is solved when $Y = 0$. Using the Newton–Raphson iterative technique, improved values for X may be found from the matrix equation $Z\Delta X = Y$ where Z is a square matrix, the Jacobian of Y with respect to X, calculated at iteration n, $\Delta X = X - X_{n-1}$ is a column vector, the change in X from iteration $n+1$ to n, Y is a column vector, the remainder, or error, in the mass balance equation at iteration n.

Jacobian iteration

The Jacobian element for components j and k is given by:

$$z_{jk} = \frac{\partial Y_j}{\partial X_k} = \sum_i a_{ij} a_{ik} c_i / X_k$$

---

the triple layer model (TLM). FITEQL is also able to determine the formation constants of complexes, the total concentrations of components, the solubility products of solids as well as the partial pressures of gases. The program includes ionic strength calculations using the Davies equation to calculate activity coefficients. Input of data into the program is generally based on an equilibrium model and a titration dataset. FITEQL has been used extensively throughout the literature to elucidate surface complexation constants for the adsorption of a variety of different metals onto various mineral surfaces [for example 6,15,46].

The algorithm used in FITEQL is based on the Gauss method, which is commonly used in optimization programs for chemical equilibrium models [47]. Basic steps in the use of FITEQL for the optimization of surface complexation constants (K) are: (1) input the chemical equilibrium model; (2) input the total component concentrations and known K values, including guesses for unknown K values; (3) input experimental equilibrium data; (4) compute the equilibrium concentrations; (5) compute the residuals for all components where both free and total concentrations are known; (6) test for convergence (minimization of the least squares of the residuals); (7) compute improved estimates for the unknown K values and continue until convergence is achieved [48,49]. An indicator of the goodness of fit is the overall variance, V, in Y.

$$V_y = \frac{\sum(y/s_y)^2}{(n_p n_c - n_u)}$$

where y is the difference function, s_y is the error estimate for each experimental data point, n_p is the number of experimental points, n_c is the number of components for which both free and total concentrations are known and n_u is the number of adjustable parameters [43].

10.3 The Species/Component Matrix

A chemical tableau is a convenient and systematic way of chemically organizing a chemical problem [48]. Construction of the tableau is based on the linear algebraic nature of the chemical problem. A tableau consists of a matrix of both species, i, and components, j. Species are composed of all the soluble, adsorbed and solid chemical entities in the system. A component is a set of chemical entities that permits a complete description of the stoichiometry of the system [50]. Each component must be independent and not a product of a reaction involving other components. It is usually more convenient to use the overall equilibrium constant, β, when solving problems numerically because every species is then defined with a minimum set of unknowns rather than as a function of several other species, as would be the case if the stepwise constants, K, are used.

The speciation of Hg(II) in a soil solution in equilibrium with atmospheric $CO_2(g)$ will be considered as an example of the construction of a species/component tableau for a chemical problem. At the concentration of Hg usually found in contaminated soils ($<10^{-6}$ M), there are eight possible Hg species present in solution, the 'free' metal plus three hydroxo complexes and four carbonate species: Hg^{2+}, $Hg(OH)^+$, $Hg(OH)_2^0$, $Hg(OH)_3^-$, $HgHCO_3^+$, $HgCO_3^0$, $HgCO_3OH^-$, $Hg(CO_3)_2^{2-}$. In addition to these Hg-containing species, the carbonate species, $H_2CO_3^0$, HCO_3^-, CO_3^{2-} and the hydroxyl ion, OH^- and the proton, H^+, also need to be included to complete the mass balance equations for inorganic C and H, respectively. The chemical problem can be described using three components: Hg^{2+}, $CO_2(g)$ and H^+. Although the OH^- ion could also be chosen as a component, the H^+ ion is more convenient as pH is a variable in many chemical problems. Similarly, $CO_2(g)$ is chosen as a component instead of the carbonate ion, CO_3^{2-}, since the system is open to the atmosphere. Chemical equations describing the formation of the mononuclear

Table 10.3 Chemical reactions involved in the speciation of Hg(II) in water

$Hg^{2+} + H_2O \rightleftharpoons HgOH^+ + H^+$	$Hg^{2+} + CO_2(g) + H_2O \rightleftharpoons HgHCO_3^+ + H^+$
$Hg^{2+} + 2H_2O \rightleftharpoons Hg(OH)_2^0 + 2H^+$	$Hg^{2+} + CO_2(g) + H_2O \rightleftharpoons HgCO_3{}^0 + 2H^+$
$Hg^{2+} + 3H_2O \rightleftharpoons Hg(OH)_3^- + 3H^+$	$Hg^{2+} + CO_2(g) + 2H_2O \rightleftharpoons HgCO_3OH^- + 3H^+$
	$Hg^{2+} + 2CO_2(g) + 2H_2O \rightleftharpoons Hg(CO_3)_2^{2-} + 4H^+$

Table 10.4 Species/component tableau for Hg(II) in water

Components		Hg^{2+}	$CO_2(g)$	H^+
Species 1	Hg^{2+}	1	0	0
2	$HgOH^+$	1	0	−1
3	$Hg(OH)_2{}^0$	1	0	−2
4	$Hg(OH)_3^-$	1	0	−3
5	$HgHCO_3^+$	1	1	−1
6	$HgCO_3{}^0$	1	1	−2
7	$HgCO_3OH^-$	1	1	−3
8	$Hg(CO_3)_2{}^{2-}$	1	2	−4
9	$H_2CO_3{}^0$	0	1	0
10	HCO_3^-	0	1	−1

Hg hydroxo-complexes and carbonate complexes used to compile the species/component tableau are shown in Table 10.3.

The species/component tableau constructed for this problem contains the names of each of the species, the stoichiometric coefficients, a, of each component in the formation of each species and the formation constant, $\beta_{p,q,r}$, for each species (Table 10.4).

10.4 Aqueous Speciation Modelling

The soil solution is the universal medium for soil chemical reactions and because of its movement within the soil profile is directly involved in many translocation processes. Its chemical composition is determined by both equilibration reactions with the mineral and humic phases in the horizon within which it resides; by its chemical evolution through overlying horizons; and by the rates of reaction of these processes.

Toxic elements in solution may exist either as 'free' ionic species, such as Ni^{2+} and Cd^{2+}, or as complexes with various inorganic and organic ligands [2]. Common inorganic complexant ligands in soil solutions include the hydroxyl, OH^-, carbonate, CO_3^{2-}, chloride, Cl^- and sulfate, SO_4^{2-}, ions. Examples of metal complexes with inorganic ligands include $ZnOH^+$, $Hg(OH)_2{}^0$, $CuCO_3{}^0$ and $CdCl^+$. Inner-sphere complexes form when a covalent bond is formed between an electron-donating ligand and a metallic ion.

Outer-sphere complexes form when oppositely charged cations and anions approach close enough to form a loose association in which the coordinated water of one or both of the ions is retained. Some elements are fully hydrolysed in water and exist as anionic species. These elements include Mo as MoO_4^{2-}, Cr as CrO_4^{2-} and As as $HAsO_4^{2-}$. These anionic species may form soluble complexes with the major cationic components of soil solutions, such as Ca^{2+} and Mg^{2+} ions.

Organic complexes may be formed between metallic cations and dissolved organic carbon (DOC). These complexes include those with relatively simple aliphatic or aromatic acids, such as citric, oxalic or gallic acids, or with more structurally complex humic substances. As it is very difficult to distinguish between these various complexes, formation constants are usually reported as complexes with DOC.

As there are few formation constants for metals with DOC in the literature, these formation constants can be estimated by assuming that the DOC can be represented as a simple diprotic acid, H_2L^0 and the metal complex as MeL^{z-2} [51,52]. Using a procedure described by Gunneriusson and Sjöberg [53] the value for the formation constant of any metal is calculated by assuming that the metal binding constant is similar to the complexation constant of the metal with oxalic acid. Although this procedure may be empirical, it does provide a means of estimating the formation constants with DOC for metals for which no reported formation constants exist. An alternative approach is to assume that the DOC behaves as a fulvic acid fraction and use the WHAM [28,29] or the NICA-Donnan [54] models. Both these metal-organic binding models contain an extensive data base for metal-fulvic acid interactions. The concentration of DOC in a soil solution is however not constant over a wide range of pH, generally increasing with an increase in soil pH. Changes in DOC concentrations as a function of pH can be calculated using the WHAM model, which can partition the organic C into both soluble and colloidal phases [9, 55].

10.4.1 Calculating the Concentration of Soluble Species of Toxic Elements

The mass balance equation for the soluble species of a cationic element in a soil solution is the sum of the concentration of the 'free' element plus all its complexant species. An example of the mass balance equation for the aqueous species of Hg(II) in a soil solution is shown below and the species/component tableau for Hg species is shown in Table 10.5.

$$[Hg]_T = [Hg^{2+}] + [HgOH^+] + [Hg(OH)_2^0] + [Hg(OH)_3^-] + [HgHCO_3^+] + [HgCO_3^0]$$
$$+ [HgCO_3OH^-] + [Hg(CO_3)_2^{2-}] + [HgCl^+] + [HgClOH^0] + [HgCl_2^0] + [HgCl_3^-]$$
$$+ [HgCl_4^{2-}] + [HgSO_4^0] + [Hg(SO_4)_2^{2-}] + [HgL^0]$$

The complete species/component tableau will also contain protons and hydroxyls, in addition to carbonate, sulfate and DOC species.

A speciation diagram showing the relative proportions of 'free' and complexed Hg as a function of pH in a representative soil solution is shown in Figure 10.1.

The mass balance equation for anionic species includes the undissociated acid plus its conjugate bases and any soluble complexes formed with major cations, such as Ca^{2+} ions, found in the soil solution. For example, the mass balance for As(V) includes arsenic acid,

Table 10.5 Species/components matrix for Hg species in a soil solution

Components		Hg^{2+}	Cl^-	SO_4^{2-}	H_2L	$CO_2(g)$	H^+
Species 1	Hg^{2+}	1	0	0	0	0	0
2	$HgOH^+$	1	0	0	0	0	−1
3	$Hg(OH)_2^0$	1	0	0	0	0	−2
4	$Hg(OH)_3^-$	1	0	0	0	0	−3
5	$HgHCO_3^+$	1	0	0	0	1	−1
6	$HgCO_3^0$	1	0	0	0	1	−2
7	$HgCO_3OH^-$	1	0	0	0	1	−3
8	$Hg(CO_3)_2^{2-}$	1	0	0	0	2	−4
9	$HgCl^+$	1	1	0	0	0	0
10	$HgClOH^0$	1	1	0	0	0	−1
11	$HgCl_2^0$	1	2	0	0	0	0
12	$HgCl_3^-$	1	3	0	0	0	0
13	$HgCl_4^{2-}$	1	4	0	0	0	0
14	$HgSO_4^0$	1	0	1	0	0	0
15	$Hg(SO_4)_2^{2-}$	1	0	2	0	0	0
16	HgL^0	1	0	0	1	0	−2

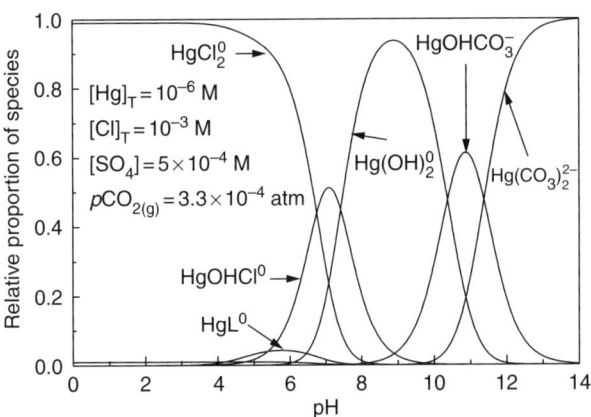

Figure 10.1 Speciation of mercury(II) in a representative soil solution

$H_3AsO_4^0$ and its conjugate bases, $H_2AsO_4^-$, $HAsO_4^{2-}$ and AsO_4^{3-}, and three possible soluble complexes with Ca^{2+} ions, $CaH_2AsO_4^+$, $CaHAsO_4^0$ and $CaAsO_4^-$. The mass balance equation for aqueous species of arsenic(V) is shown below and the species/component tableau is shown in Table 10.6.

$$[As(V)]_T = [H_3AsO_4^0] + [H_2AsO_4^-] + [HAsO_4^{2-}] + [AsO_4^{3-}] + [CaH_2AsO_4^+] \\ + [CaHAsO_4^0] + [CaAsO_4^-]$$

A speciation diagram for As(V) as a function of pH for a representative soil solution is shown in Figure 10.2.

Table 10.6 Species/components matrix for As(V) and Ca(II) species in a soil solution

Components		AsO_4^{3-}	Ca^{2+}	$CO_2(g)$	H^+
Species 1	$H_3AsO_4^0$	1	0	0	3
2	$H_2AsO_4^-$	1	0	0	2
3	$HAsO_4^{2-}$	1	0	0	1
4	AsO_4^{3-}	1	0	0	0
5	$CaH_2AsO_4^+$	1	1	0	2
6	$CaHAsO_4^0$	1	1	0	1
7	$CaAsO_4^-$	1	1	0	0
8	$CaOH^+$	0	1	0	−1
9	$CaHCO_3^+$	0	1	1	−1
10	$CaCO_3^0$	0	1	1	−2

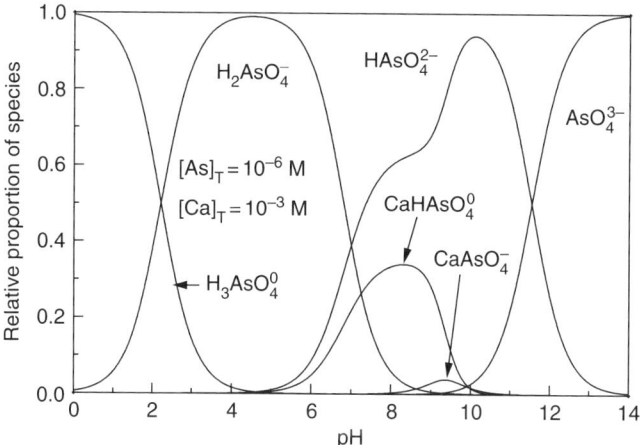

Figure 10.2 Speciation of arsenic(V) in a representative soil solution

10.5 Modelling of Surface Complexation to Mineral Surfaces

10.5.1 Proton and Toxic Element Binding to Oxide Minerals

A variable pH-dependent charge, the net proton charge, σ_H, can arise from proton association and dissociation reactions with both the surfaces and edges of inorganic colloids. They differ, however, from the permanent structural charges in the phyllosilicate clay minerals in that the magnitude and sign of the charge depend on pH. These charges are associated with the edges of clay minerals, such as kaolinite and vermiculite; with the edges and surfaces of oxides, hydroxides and oxyhydroxides of Al, Fe and Mn, such as gibbsite, $Al_2(OH)_6$, hematite, Fe_2O_3, and pyrolusite, MnO_2; and with the surfaces of amorphous and poorly crystalline aluminosilicates, such as allophanes, $2SiO_2 \cdot Al_2O_3 \cdot 3H_2O$ and imogolite, $SiAl_2O_3(OH)_4$.

At the surfaces and edges of oxide minerals there is no cation to completely neutralize the charge on edge oxygens or hydroxyls and so a negatively charged site will result, $\equiv$S–O$^-$, where $\equiv$S represents a surface atom, such as Fe, Al, Mn or Si. The edges of 1:1 and 2:1 clay minerals also have octahedral hydroxyls not shared by adjacent Al^{3+} or Mg^{2+} or tetrahedral oxygens not shared with adjacent Si^{4+}. As these negative charges on both oxide minerals and the edges of phyllosilicate clays have a great affinity for protons and, depending on the pH of the system, they may adsorb protons to become neutral sites, $\equiv$S–OH^0, or positively charged sites, $\equiv$S–OH$_2^+$. The intrinsic reactions leading to the dissociation of protons from variable charge surfaces can be regarded as analogous to the dissociation reactions that occur for weak diprotic acids in solution. This approach has been called the 2-pK model [56]. Intrinsic surface proton binding constants are generally determined by potentiometric titrations but can also be calculated from crystallographic considerations [57].

The intrinsic surface proton binding reactions and their associated equilibrium constants for these reactions are:

$$\equiv S\text{–}OH_2^+ \rightleftharpoons \equiv S\text{–}OH^0 + H^+; \text{ at equilibrium,}$$

$$^cK_{a1}^{int} = \frac{[\equiv S\text{–}OH^0]\{H^+\}}{[S\text{–}OH_2^+]} \exp\left(\frac{-\psi_s F}{RT}\right)$$

$$\equiv S\text{–}OH^0 \rightleftharpoons \equiv S\text{–}O^- + H^+; \text{ at equilibrium,}$$

$$^cK_{a2}^{int} = \frac{[\equiv S\text{–}OH^-]\{H^+\}}{[S\text{–}OH^0]} \exp\left(\frac{-\psi_s F}{RT}\right)$$

where $^cK_{a1}^{int}$ and $^cK_{a2}^{int}$ are the first and second conditional intrinsic acidity constants for the surface acidity reactions and ψ_s is the electrical potential at the charged surface.

The total mass balance for these surface sites, $[\equiv S\text{–}OH]_T$, is given by:

$$[\equiv S\text{–}OH]_T = [\equiv S\text{–}OH_2^+] + [\equiv S\text{–}OH^0] + [\equiv S\text{–}O^-]$$

A number of surface complexation models have been developed to calculate the electrical charge at the surfaces of variable charged minerals. These include the diffuse layer model, the constant capacitance model, the triple layer model and the four layer model. These models differ in the number of adsorption planes considered and the equations used to describe the relationships between the surface charge density and the surface electrical potential. Two commonly used models, the constant capacitance model (CCM) and the diffuse layer model (DLM) differ in that the DLM contains two adsorption planes, whereas the CCM has one adsorption plane. They also differ in the formal relationships between the surface electrical charge, ψ_s, and the surface charge density, σ_s.

In the DLM, these parameters are related through the Gouy–Chapman equation, which for a 1:1 binary background electrolyte is given by:

$$\sigma_s = (8RT\varepsilon_0 \varepsilon c_0)^{0.5} \sinh\left(\frac{z\psi_s F}{2RT}\right)$$

where ϵ is the dielectric constant of water, 78.5, ε_0 is the permittivity in a vacuum, 8.854×10^{-12} C^2 J^{-1} m^{-1}, c_0 is the concentration of the background electrolyte, mol m^{-3} and z is the valency of the ion. This equation simplifies to:

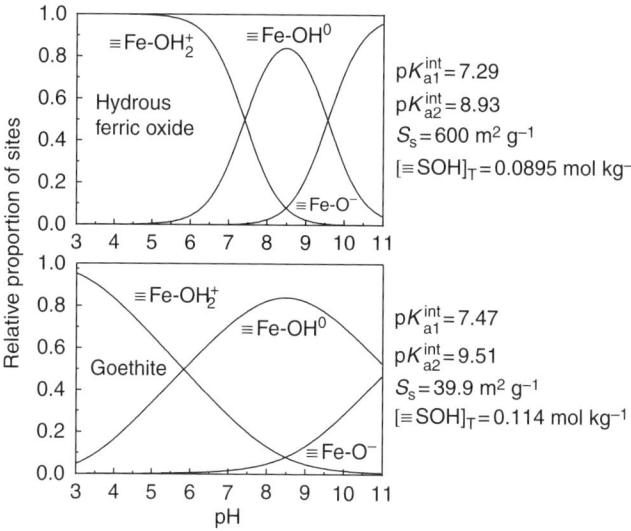

Figure 10.3 Relative proportion of surface sites on hydrous ferric oxide and goethite. (Suspension density $S_s = 0.089$ g l^{-1}.)

$$\sigma_s = 0.1174 \, I^{0.5} \sinh(19.46 z \psi_s)$$

where I is the ionic strength in mol l^{-1}.

In the CCM, these parameters are related through the capacitance, κ (C^2 J^{-1} m^{-2}), $\sigma_s = \kappa \psi_s$.

The relative proportions of the three surface sites, the positively charged site, $\equiv$S–OH$_2^+$, the neutral site, $\equiv$S–OH^0, and the negatively charged site, $\equiv$S–O$^-$, for hydrous ferric oxide, Fe$_2$O$_3$·H$_2$O, and goethite, α-FeOOH, as a function of pH in the absence of adsorbed cations and anions and using the CCM are shown in Figure 10.3 and the species/component tableau used to create this figure is shown in Table 10.7. The figure for hydrous ferric oxide was constructed using the DLM and the data of Dzombak and Morel [48] and for goethite using the data of Lövgren et al. [58].

In the charge-distribution multisite ion complexation model (CD-MUSIC), the charges on the various types of reactive groups on the surfaces of minerals are based on

Table 10.7 Species/components tableau for protons on variable charge surfaces

		Components		
		$\equiv$S–OH^0	$e^{-\psi F/RT}$	H$^+$
Species	$\equiv$S–OH$_2^+$	1	1	1
	$\equiv$S–OH^0	1	0	0
	$\equiv$S–O$^-$	1	−1	−1
	OH$^-$	0	0	−1
	H$^+$	0	0	1

the application of the Pauling bond valence concept [56,59]. The basic charging of the surfaces of the mineral goethite, for example, is due to the protonation of singly, $\equiv \text{FeOH}^{-1/2}$, and triply, $\equiv \text{Fe}_3\text{O}^{-1/2}$, coordinated surface groups [60]. The pK values for both surface protonation reactions are commonly set equal to the experimentally determined PZC of goethite [61].

$$\equiv \text{FeOH}^{-1/2} + \text{H}^+ \rightleftharpoons \equiv \text{FeOH}_2^{+1/2}$$
$$\equiv \text{Fe}_3\text{O}^{-1/2} + \text{H}^+ \rightleftharpoons \equiv \text{Fe}_3\text{OH}^{+1/2}$$

Conditional intrinsic acidity constants for minerals are usually determined from data obtained from potentiometric titrations in a similar manner to that for those used to determine dissociation constants for weak acids or metal complexes in solution. Suspensions are titrated with acids and bases in the presence of an indifferent background electrolyte, such as LiClO_4 or NaNO_3. Nitrate or perchlorate salts of sodium or lithium are preferred as they do not form or form very weak soluble complexes with other cations and anions in solution.

The majority of metallic cations, such as the heavy metals, are retained by the surface sites on variably charged minerals, and they are specifically adsorbed through covalent bonds and form inner-sphere complexes with the mineral surface. Specific adsorption of cations can occur on mineral surfaces for those metals that are involved in hydrolysis reactions with water. These metals include Zn, Cu, Co, Ni, Mn and Pb. The reactions of divalent metals at these variable charged surfaces to form monodentate surface complexes can be described by:

$$\equiv \text{S}-\text{OH}^0 + \text{Me}^{2+} \rightleftharpoons \equiv \text{S}-\text{OMe}^+ + \text{H}^+; \text{ at equilibrium,}$$

$$^c K_{\text{Me}}^{\text{int}} = \frac{[\equiv \text{S}-\text{OMe}^0]\{\text{H}^+\}}{[\equiv \text{S}-\text{O}^-]\{\text{Me}^{2+}\}} \exp\left(\frac{\psi_{is} F}{RT}\right)$$

where $^c K_{\text{Me}}^{\text{int}}$ is the intrinsic formation constant for the surface complex and ψ_{is} is the electrical potential at the plane of adsorption of the inner-sphere complexes.

The total content of a divalent metal in solution and adsorbed onto a variably charged mineral in a soil solution with negligible amounts of Cl^-, SO_4^{2-} and DOC can be described by:

$$[\text{Me}]_T = [\text{Me}^{2+}] + [\text{MeOH}^+] + [\text{Me(OH)}_2^0] + [\text{Me(OH)}_3^-] + [\text{MeHCO}_3^+] + [\text{MeCO}_3^0] + [\equiv \text{S}-\text{OMe}^+]$$

and the relative proportion, α, of adsorbed species by $[\equiv \text{S}-\text{OMe}^+]/[\text{Me}]_T$.

The extent of the adsorption reaction is dependent on pH, and increases to a maximum as the pH is raised. The maximum amount of adsorption occurs at a pH about 3 units below that of the first hydrolysis constant of the metal. The shape of the adsorption curve is described as an 'adsorption edge'. Some typical results describing the adsorption of metals onto hydrous ferric oxide (HFO) using the data from Dzombak and Morel [48] and the DLM are shown in Figure 10.4. Since the initial studies of Dzombak and Morel [48], more comprehensive structural models for HFO have been developed for metal adsorption [62].

The tendency of a metal to complex with oxide surfaces is related to its ability to react with other oxygen atoms, such as those in water molecules. It is to be expected therefore

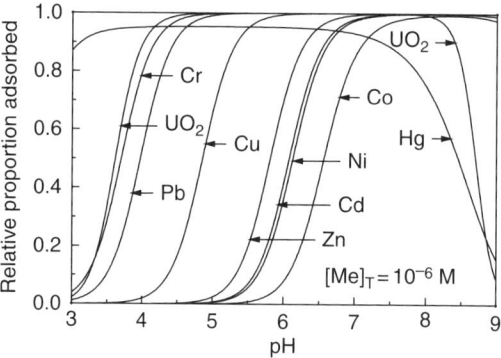

Figure 10.4 Adsorption of metals onto hydrous ferric oxide. (Suspension density = 0.089 g l^{-1}.)

that there should be a linear free energy relationship between the negative logarithm of the first hydrolysis constants of metals, p$\beta_{-1,1,1}$, and the negative logarithm of their intrinsic complexation constants, p$^c K_{Me}^{int}$ [48,63].

The amount of metal complexed by a variable charged surface can be reduced or enhanced if there is present in the soil solution a complexant ligand, such as Cl$^-$ ions. Reduction in adsorption can occur if the metal forms strong soluble chloride complexes, as is the case for Hg and Cd, which form complexes, such as HgCl$_2^0$ and CdCl$_2^0$. Enhanced adsorption in the presence of Cl$^-$ ions can, however, occur if ternary complexes, such as ≡S–OMeCl^0, form with the variable charge surface that are stronger than the soluble complexes. The effect of Cl$^-$ ions on the retention of Hg onto goethite is shown in Figure 10.5 using data from Gunneriusson and Sjoberg [64]. Similar studies for the effect of Cl$^-$ ions on metal retention to goethite have been reported for Cd(II) [65] and Pb(II) [66].

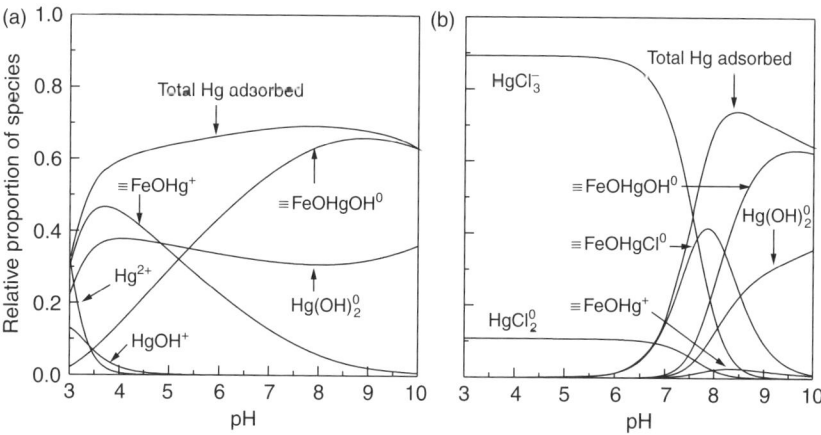

Figure 10.5 Modelled Hg(II) speciation on goethite ([Hg]$_T$ = 0.0001 M and a suspension density of 0.089 g l^{-1}). (a) No Cl$^-$ ions present; (b) [Cl]$_T$ = 0.01 M

Anionic species also can form inner-sphere complexes with the charged surfaces of variably charged mineral colloids. These complexes generally involve the formation of covalent bonds and the adsorbed species are not readily displaced. As with metallic cations, most anions are specifically adsorbed. Examples of such anions include CrO_4^{2-}, SeO_4^{2-} and AsO_4^{2-}. This type of adsorption process is often termed ligand exchange or chemisorption. As most anions are the conjugate bases of weak acids, their content in soil solutions is dependent upon pH and will increase as the pH is raised. Similarly, the content of positively charged sites, $\equiv S-OH_2^+$, decreases as the pH is raised. The extent of adsorption increases to a maximum and then declines – the shape of the adsorption curve being described as an 'adsorption envelope' – and the maximum amount of retention of the conjugate bases occurs at a pH at or near the pK_a of the weak acid. The extent of the deviation from the pK_a value(s) is dependent on the point of zero net proton charge (PZNPC) of the mineral.

The mechanism of adsorption involves the replacement of surface $-O^-$, $-OH^0$ or $-OH_2^+$ groups by the adsorbing ligand. The process of ligand exchange for a monovalent anion, L^-, can be described by the equation:

$$\equiv S-OH^0 + L^- + H^+ \rightleftharpoons \equiv S-L^0 + H_2O; \text{ at equilibrium,}$$

$$^c K_L^{int} = \frac{[\equiv S-L^0]}{[\equiv S-OH^0]\{H^+\}\{L^-\}}$$

and for diprotic acids to form monodentate surface complexes by:

$$\equiv S-OH^0 + HL^- + H^+ \rightleftharpoons \equiv S-HL^0 + H_2O; \text{ at equilibrium,}$$

$$^c K_{HL}^{int} = \frac{[\equiv S-HL^0]}{[\equiv S-OH^0]\{H^+\}\{HL^-\}}$$

$$\equiv S-OH^0 + L^{2-} + H^+ \rightleftharpoons \equiv S-L^- + H_2O; \text{ at equilibrium,}$$

$$^c K_L^{int} = \frac{[\equiv S-L^-]}{[\equiv S-OH^0]\{H^+\}\{L^{2-}\}} \exp\left(\frac{-\psi_{is}F}{RT}\right)$$

where $^c K_{HL}^{int}$ and $^c K_L^{int}$ are the intrinsic formation constant for the two surface complexes and ψ_{is} is the electrical potential at the plane of adsorption of the inner-sphere complexes.

The mass balance equation for a diprotic acid, for example, in the presence of a variably charged surface and in the absence of complexant cations would be given by:

$$[L]_T = [H_2L^0] + [HL^-] + [L^{2-}] + [\equiv S-HL^0] + [\equiv S-L^-]$$

and the relative proportion, α, of adsorbed species by $[\equiv S-HL^0] + [\equiv S-L^-]/[L]_T$.

As an example of the adsorption of anions onto variable charge surfaces the adsorption envelopes for some anions adsorbed on hydrous ferric oxide calculated using the DLM and the data given in Dzombak and Morel [48] are shown in Figure 10.6.

In the CD-MUSIC model for goethite [56,59], the surface sites containing protons are contained in an adsorption plane at the mineral surface and surface inner-sphere complexes and outer-sphere complexes are contained in two adsorption planes adjacent to this surface plane. Specifically adsorbed inner-sphere complexes of cations, Me^{z+}, or anions, L^{l-}, are

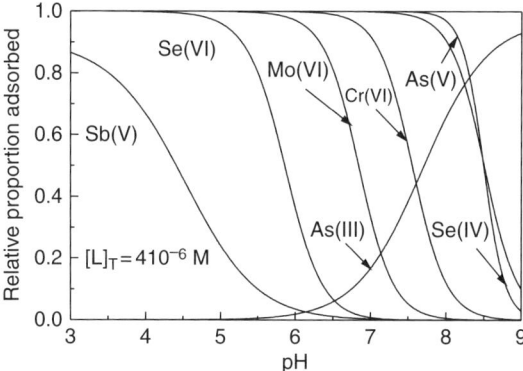

Figure 10.6 Adsorption of anions adsorbed onto hydrous ferric oxide

contained in an adsorption plane adjacent to the surface as either as monodentate complexes:

$$\equiv FeOH^{-1/2} + Me^{z+} \equiv \rightleftharpoons FeO-Me^{z-1/2}$$
$$\equiv FeOH_2^{+1/2} + L^{l-} \rightleftharpoons \equiv Fe-L^{l-1/2} + H_2O$$

and/or bidentate complexes:

$$2 \equiv SOH^{-1/2} + Me^{z+} \rightleftharpoons 2 \equiv SO-Me^{z-1}$$
$$2 \equiv SOH_2^{+1/2} + L^{l-} \rightleftharpoons 2 \equiv S-L^{l-1} + 2H_2O$$

The charge for these specifically adsorbed complexes is distributed over both the surface and the adsorption plane for inner-sphere surface complexes as described by the charge distribution (CD) model [56].

The retention of ions in the background electrolyte is considered as nonspecifically adsorbed outer sphere complexes in a plane coincident with the start of the diffuse layer. For a background electrolyte containing, for example, $NaNO_3$:

$$\equiv FeOH^{-1/2} + Na^+ \rightleftharpoons \equiv FeOH^{-1/2} \cdot Na$$
$$\equiv Fe_3O^{-1/2} + Na^+ \rightleftharpoons \equiv Fe_3O \cdot Na^{+1/2}$$
$$\equiv FeOH^{+1/2} + NO_3^- \rightleftharpoons \equiv FeOH^{+1/2} \cdot NO_3^-$$
$$\equiv Fe_3O^{-1/2} + H^+ + NO_3^- \rightleftharpoons \equiv Fe_3OH \cdot NO_3^{-1/2}$$

These adsorption planes are similar to those described in the TLM.

Recent application of the CD-MUSIC model to studies on toxic elements in the presence of goethite include retention of Se(VI) [67] and F(–1) [68] and surface speciation and competitive adsorption of As(V) and As(III) [69,70].

10.5.2 Proton and Toxic Element Binding to Clay Minerals

Unlike the oxide minerals, clay minerals have both permanent and variably charged surfaces which are involved in trace metal and proton binding. Permanent charges, $\equiv X^-$, occur on the mineral faces in the interlayer space, whereas variable charge sites,

$\equiv$S–OH$_2^+$, $\equiv$S–OH^0 and $\equiv$S–O$^-$, occur on the clay mineral edges, similar to those found on the oxide minerals [71–73]. The edge adsorption sites on clay minerals generally include both the silanol, $\equiv$SiOH, and aluminol, $\equiv$AlOH, hydroxyl groups, but may also include ferrinol, $\equiv$FeOH groups [74]. The protonation and deprotonation reactions on the edges of clay minerals are treated in the same way as those on variably charged oxide minerals and the intrinsic equilibrium reaction and constants for these reactions are given by:

$$\equiv\text{S–OH}_2^+ \rightleftharpoons \equiv\text{S–OH}^0 + \text{H}^+; \text{at equilibrium}, \quad ^cK_{a1}^{int} = \frac{[\equiv\text{S–OH}^0]\{\text{H}^+\}}{[\text{S–OH}_2^+]} \exp\left(\frac{-\psi_s F}{RT}\right)$$

$$\equiv\text{S–OH}^0 \rightleftharpoons \equiv\text{S–O}^- + \text{H}^+; \text{at equilibrium}, \quad ^cK_{a2}^{int} = \frac{[\equiv\text{S–O}^-]\{\text{H}^+\}}{[\text{S–OH}^0]} \exp\left(\frac{-\psi_s F}{RT}\right)$$

where $^cK_{a1}^{int}$ and $^cK_{a2}^{int}$ are the first and second conditional intrinsic acidity constants for the surface acidity reactions and ψ_s is the electrical potential at the charged surface.

Permanent negative charges, $\equiv$X$^-$, on clay surfaces arise from isomorphous substitution or nonideal octahedral occupancy. The relative importance of edge sites versus surface sites varies greatly amongst the clay minerals. Edge sites are most important for kaolinite, of intermediate importance for illites and least important for smectites and vermiculites. Representative distributions of surface sites for kaolinite [75] and montmorillonite [76] are shown in Figure 10.7.

Metal adsorption to clay minerals takes place on both the permanent charge sites and on the variably charged sites [77]. When metal ions are adsorbed to the permanent charges sites, $\equiv$X$^-$, outer-sphere complexes are formed by displacing protons and cations from the background electrolyte through cation exchange reactions. To determine the surface acidity properties associated with the permanent charge sites of clay minerals, potentiometric titrations are usually used in the presence of an indifferent background electrolyte, such as NaNO$_3$ or LiClO$_4$. The surface species present at any given pH value in the absence of a contaminant metal are $\equiv$X$^-\equiv$H$^+$ and that involving the cation in the indifferent background electrolyte, such Na$^+$, to give $\equiv$X$^-\cdot$Na$^+$.

$$\equiv\text{X}^-\cdot\text{H}^+ + \text{Na}^+ \rightleftharpoons \equiv\text{X}^-\cdot\text{Na}^+ + \text{H}^+; \text{at equilibrium} \quad ^cK_{Na/H}^{int} = \frac{[\equiv\text{X}^-\cdot\text{Na}^+]\{\text{H}^+\}}{[\equiv\text{X}^-\cdot\text{H}^+]\{\text{Na}^+\}}$$

On the variably charge edge sites, inner-sphere surface complexes, $\equiv$SOMe^{z-1}, are formed between the metal ion and the surface sites. The specific adsorption of metals, Me^{z+}, at the edges of clay minerals can be described by:

$$\equiv\text{SOH}^0 + \text{Me}^{z+} \rightleftharpoons \equiv\text{SOMe}^{z-1} + \text{H}^+; \text{at equilibrium},$$

$$^cK_{Me}^{int} = \frac{[\equiv\text{SOMe}^{z-1}]\{\text{H}^+\}}{[\equiv\text{SOH}^0]\{\text{Me}^{z+}\}} \exp\left(\frac{\psi_s F}{RT}\right)$$

where $^cK_{Me}^{int}$ is the intrinsic formation constant for the surface complexes.

The surface species present on the permanent charged sites at any given pH value are those involving the proton, $\equiv$X$^-\cdot$H$^+$, the major soluble cation, usually calcium,

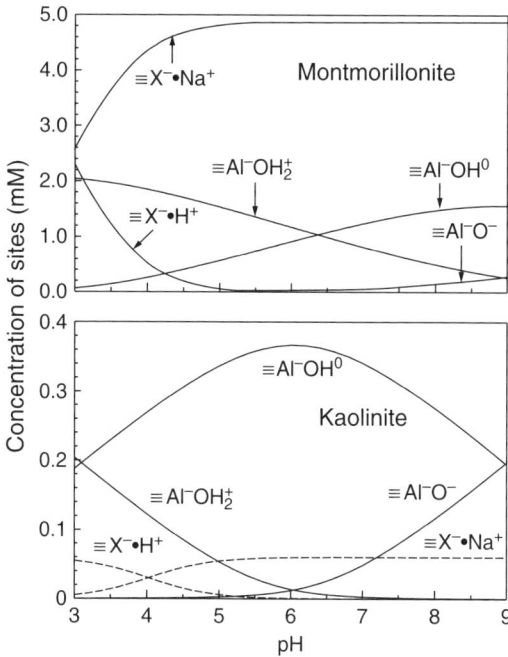

Figure 10.7 Concentration of surface species on montmorillonite and kaolinite. (Suspension density $= 7.8$ g l^{-1}.)

$\equiv X_2^- \cdot Ca^{2+}$, and the contaminant metal of interest, $\equiv X_2^- \cdot Me^{z+}$. To determine the binding constant, log $\equiv X_2^- \cdot Me^{z+}$, of the contaminant metal of interest to a particular clay mineral or extracted clay fraction, batch adsorption reactions are usually employed in an indifferent background electrolyte.

$$z \equiv X^- \cdot Na^+ + Me^{z+} \rightleftharpoons \equiv X_z^- \cdot Me^{z+} + z\, Na^+; \text{ at equilibrium,}$$

$$^c K^{int}_{Me/Na} = \frac{[\equiv X_z^- \cdot Me^{z+}]\{Na^+\}^z}{[\equiv X^- \cdot H^+]^z \{Me^{z+}\}}$$

The species/component tableau used to determine the binding constant from the data from the batch adsorption experiments is shown in Table 10.8.

Some results for the adsorption of Ni and Zn onto kaolinite [75] at three ionic strengths are shown in Figure 10.8. From Figure 10.8 it can be seen that as the ionic strength increases, the relative proportion of cations adsorbed onto the exchange sites, $\equiv X_2^- \cdot Me^{z+}$, decreases because of increased competition from the background electrolyte, whereas the amount of the metal adsorbed onto the variably charged edge sites, $\equiv AlOMe^{z-1}$, increases. The effect of ionic strength on the adsorption of metallic cations is greatest for smectites and vermiculites, intermediate for illites and least for kaolinite because of the different relative proportions of permanent and edge adsorption sites in these clay minerals. This is illustrated in Figure 10.9, where the adsorption of Cd and Pb are compared at three ionic strengths for kaolinite, illite and montmorillonite [44,75,76].

Table 10.8 Species/component matrix for binding constants of divalent metal for clay minerals

Species		≡AlOH⁰	$e^{(-\psi F/RT)}$	≡X⁻·H⁺	Me^{2+}	Na^+	$CO_2(g)$	H^+
1	Me^{2+}	0	0	0	1	0	0	0
2	$MeOH^+$	0	0	0	1	0	0	−1
3	$Me(OH)_2^0$	0	0	0	1	0	0	−2
4	$Me(OH)_3^-$	0	0	0	1	0	0	−3
5	$MeHCO_3^+$	0	0	0	1	0	1	−1
6	$MeCO_3^0$	0	0	0	1	0	1	−2
7	$Me(CO_3)_2^{2-}$	0	0	0	1	0	2	−4
8	$\equiv SOH_2^+$	1	1	0	0	0	0	1
9	$\equiv SOH^0$	1	0	0	0	0	0	0
10	$\equiv SO^-$	1	−1	0	0	0	0	−1
11	$\equiv X^-\cdot H^+$	0	0	1	0	0	0	0
12	$\equiv X^-\cdot Na^+$	0	0	1	0	1	0	−1
13	$\equiv SOMe^+$	1	1	0	0	0	0	−1
14	$2\equiv X^-\cdot Me^{2+}$	0	0	2	1	0	0	−2
15	Na^+	0	0	0	0	1	0	0
16	OH^-	0	0	0	0	0	0	−1
17	H^+	0	0	0	0	0	0	1

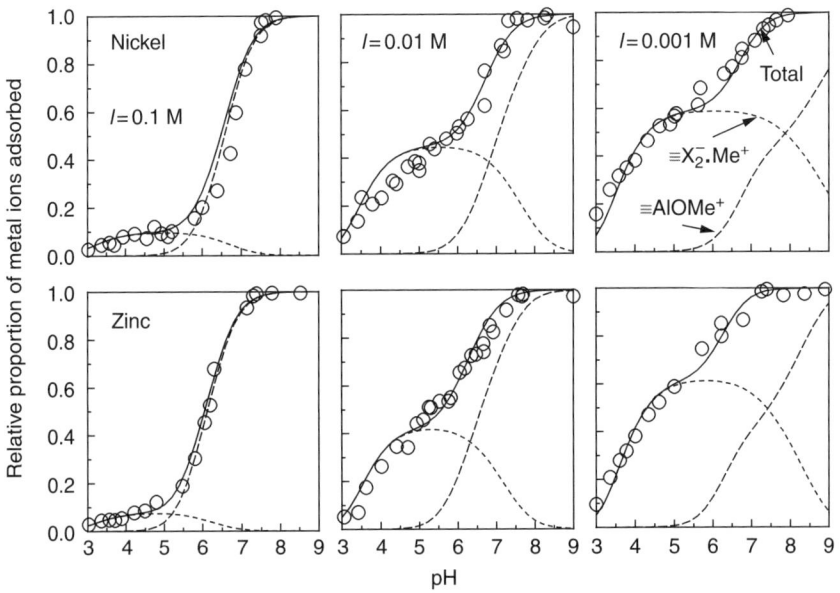

Figure 10.8 Binding of nickel and zinc to kaolinite at three ionic strengths

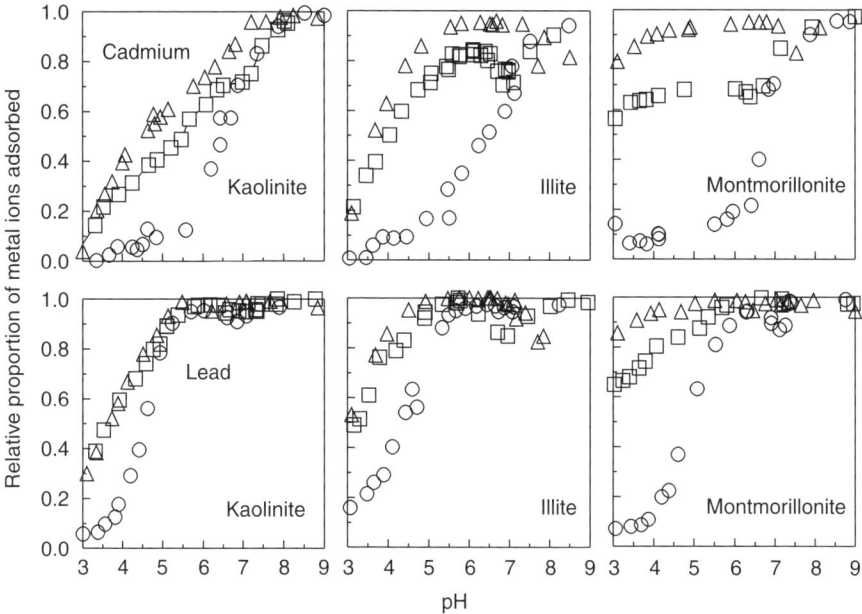

Figure 10.9 Adsorption of Cd and Pb onto kaolinite, illite and montmorillonite in 0.1 M (○), 0.01 M (□) and 0.001 M (△) NaNO$_3$

10.6 Modelling of Surface Complexation to Soil Organic Matter

Soil organic matter is composed of a highly complex mixture of humic and fulvic acid fractions as well as nonhumic constituents, such as carbohydrates, proteins and long-chain fatty acids. The humic and fulvic acid fractions contain a wide range of functional groups whose metal-complexing abilities can be expected to vary considerably. The nature of the complexant ligand, the type of complex formed and the electronic configuration of the metal ion affect the magnitude of the stability constant.

Multiligand representations of humic material have been used to empirically fit titration data for organic colloids using both discrete and continuous multiligand models [78,79]. Examples of continuous distribution models include the Windermere humic aqueous model, WHAM [28,29] and the nonideal competitive adsorption (NICA) model [80–82] and the NICA-Donnan model [83,84]. Both the WHAM and NICA-Donnan models have successfully modelled humic chemistry and the relative merits of the two are of continuing interest [85]. The complexity of humic materials suggests that a large number of different sites may be involved in metal binding and that these sites are best characterized by a continuous distribution of complexation constants. However, it has been shown that titration data can often be adequately described using only two or three discrete complexant ligands. Such models have been used by Bolton *et al.* [86] and Evans *et al.* [87] to describe Cd adsorption to a soil humic acid fractions using a site-specific adsorption model.

The NICA model is a continuous adsorption model in which the overall adsorption of a species is described by its adsorption per group of identical sites integrated over the affinity distribution. The original NICA model has been extended to include nonspecific binding in the Donnan phase [83,84] and has been developed to model both cation and proton binding by organic matter and now has an extensive database for metal binding by a generic fulvic and humic acid [54,88]. The adsorption process is also ion specific in that the affinity distributions of different ions for the same surface may differ. The NICA-Donnan model accounts for both this intrinsic heterogeneity and ion-specific nonideality [80]. To model ion binding by soil humic material a bimodal distribution of binding sites is assumed [83]. The two binding sites of this bimodal distribution represent weak proton affinity sites associated with carboxyl groups and those sites with stronger proton affinity associated with phenolic sites. Due to the presence of hydrogen ions in aqueous environments, the binding of metals is always competitive. Both metallic and H^+ ions are assumed to form monodentate binding sites with the humic material, which is assumed to be composed of two monoprotic acids, HA to represent binding to a carboxylic acid site and HB to represent binding to a phenolic site. The total adsorption of component i, in the NICA-Donnan model is expressed as:

$$Q_{i,t} = Q_{max1} \frac{(\tilde{K}_{i,1} c_{D,1})^{ni,1}}{\sum_j (\tilde{K}_{j,1} c_{D,j})^{nj,1}} \cdot \frac{\left\{ \sum_j (\tilde{K}_{j,1} c_{D,j})^{nj,1} \right\}^{p1}}{1 + \left\{ \sum_j (\tilde{K}_{j,1} c_{D,j})^{nj,1} \right\}^{p1}}$$

$$+ Q_{max2} \frac{(\tilde{K}_{i,2} c_{D,i})^{ni2}}{\sum_j (\tilde{K}_{j,2} c_{D,j})^{nj,2}} \cdot \frac{\left\{ \sum_j (\tilde{K}_{j,2} c_{D,j})^{nj,2} \right\}^{p2}}{1 + \left\{ \sum_j (\tilde{K}_{j,2} c_{D,j})^{nj,2} \right\}^{p2}}$$

where $Q_{i,t}$ is the total amount of component i specifically bound to the humic material, Q_{max} is the total site density, $\tilde{K}_i$ is the median affinity constant for component i, $c_{D,j}$ is the concentration or activity of i in the Donnan volume, n_j accounts for component-specific nonideality, p accounts for the intrinsic surface heterogeneity, and the subscripts 1 and 2 refer to carboxylic and phenolic groups, respectively.

The mass balance for divalent metals specifically bound to humic colloids in the NICA-Donnan model is given by:

$$[Me]_T = [Me^{2+}] + [MeOH^+] + [Me(OH)_2^0] + [Me(OH)_3^-] + [MeHCO_3^+] + [MeCO_3^0]$$
$$+ [Me(CO_3)_2^{2-}] + [MeA^+] + [MeB^+] + [Me_{Don}]$$

where $[MeA^+]$ and $[MeB^+]$ are the concentrations of metal bound to the carboxylic and phenolic groups, respectively, and $[Me_{Don}]$ is the concentration of the metal in the Donnan layer.

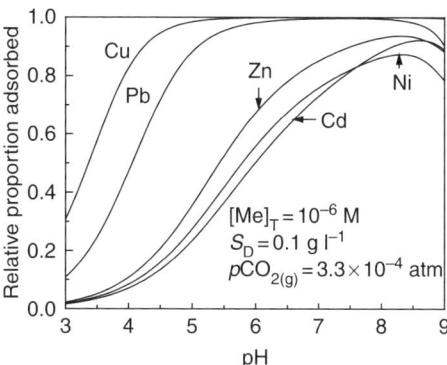

Figure 10.10 *Metal retention by humic colloids as predicted by the NICA-Donnan model*

The extent of metal binding as a function of pH using the NICA-Donnan model and the generic parameters provided by Milne et al. [54,88] using the computer program ORCHESTRA [33] are shown in Figure 10.10.

Both the WHAM and the NICA-Donnan model contain generic fitting parameters suitable for many studies in soil contamination studies, but site-specific models usually give better fits to experimental data than generic models [87]. However, the development of site-specific organic binding models involves much more work in that both the proton and metal binding constants have to be developed experimentally for a particular organic soil or a separated humic or fulvic acid fraction. These binding constants are usually determined by potentiometric titrations and batch adsorption experiments. In a study by Bolton et al. [86], the binding of Cd to a soil humic acid fraction was best described by assuming that the humic fraction behaved as two discrete diprotic acids H_2A^0 and H_2B^0, with H_2A^0 forming mono-, bi- and tridentate complexes with Cd^{2+} ions and the second diprotic acid, H_2B^0, forming only a monodentate complex. In a follow-up study, Cd binding by an organic soil was better described by assuming both mono- and bidentate complexes to H_2A^0, and a monodentate complex to the second diprotic acid, H_2B^0, assuming the same proton binding constants as in the previous study [87] (Figure 10.11).

10.7 Discussion

The prediction of the chemical speciation of toxic elements using aqueous speciation and surface complexation calculations and computer programs has been discussed previously [89]. Most computer programs used in studies in soil chemistry contain extensive databases which include thermodynamic data for soluble complexes, mineral solids, gas solubilities and redox couples [23,27]. Inputs for the chemical problem being investigated require the total analytically measured concentrations of the selected components as well as temperature, ionic strength, pH and E_h. Soluble complexes are selected from the database, with those that are kinetically slow or irrelevant to the problem can be removed from the calculation. Databases also include surface complexation constants for both

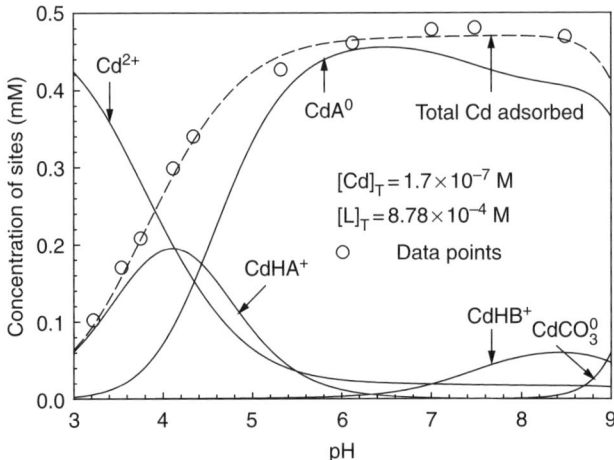

Figure 10.11 Modelled Cd adsorption to an organic soil using a two diprotic acid model

mineral and organic surfaces with many different toxic elements. Recent additions to geochemical equilibrium programs include both kinetic reactions and transport. Both PHREEQC and ECOSAT include a program which allows the user to define rate expressions for kinetic reactions within its input file. The program PHREEQC also contains the algorithms for one-dimensional solute transport [24], the program ORCHESTRA for one- and two-dimensional solute transport [32], and the program ECOSAT the algorithms for one-, semi-two- and semi-three-dimensional solute transport [22]

Whereas most models use similar databases for aqueous speciation, complexation constants, particularly for organic materials vary widely depending on the model chosen. For example in the program CHEAQS [27], organic complexation was initially represented by DOC and used the method of Canbaniss and Shuman [90] to extrapolate the metal binding constants. The most recent version of CHEAQS now incorporates WHAM V in its data base. Conversely, the program MINTEQA2 gives the user three options to model DOC; Gaussian [91], NICA-Donnan [79] and the Stockholm humic acid model [8,23].

Recently there has been much activity to validate the predicted amounts of both 'free' and 'labile' forms of toxic elements generated by aqueous speciation models. Techniques used include diffusive gradients in thin films [92–94] and Donnan membrane techniques [95,96]. Results from these studies generally show good agreement between the measured and the model predicted forms of the toxic elements for both organic and inorganic forms of the toxic element.

There are very few models that can adequately handle all the various adsorbing surfaces found in soils and sediments. Some additive adsorption models include organic and oxide surfaces [8,38], while others include soil organic matter, oxide minerals and clays [35–37]. These models are generic and not specifically written for a particular contaminated site. While much is known about the surface chemistry of oxide minerals, such as hydrous ferric oxide and goethite, the corresponding chemistry of humic materials and clay fractions is much less known.

To develop models that are site specific, both humic and clay fractions need to separated from soils and their proton and metal binding constants determined by techniques, such as potentiometric titrations and batch adsorption experiments. In addition, the chemical composition of the soil solution needs to be known with respect to its content of major cations and anions and the concentrations of the contaminant metals of interest. One such model has been developed for nickel in soils of the Sudbury area, Ontario, Canada [97]. This model was written in Visual Basic^{TM} and includes aqueous speciation and surface complexation parameters and solves the various chemical problems using the algorithms given in MICROQL [34]. The model provided a significant relationship between the soluble Ni^{2+} concentrations in the soil and the Ni content in various vegetable crops (Figure 10.12).

Although much progress has been made in the development of adsorption models for minerals and for humic fractions, much more work is needed in the development of these models for application to soil systems. Specifically, internally consistent data bases are needed for the adsorption of toxic elements with commonly occurring soil components, such as hydrous ferric oxide, goethite, gibbsite, birnessite, etc. The data base developed by Dzombak and Morel [48] for hydrous ferric oxide has been incorporated into a number of computer models but was developed over twenty years ago for a relatively small number of

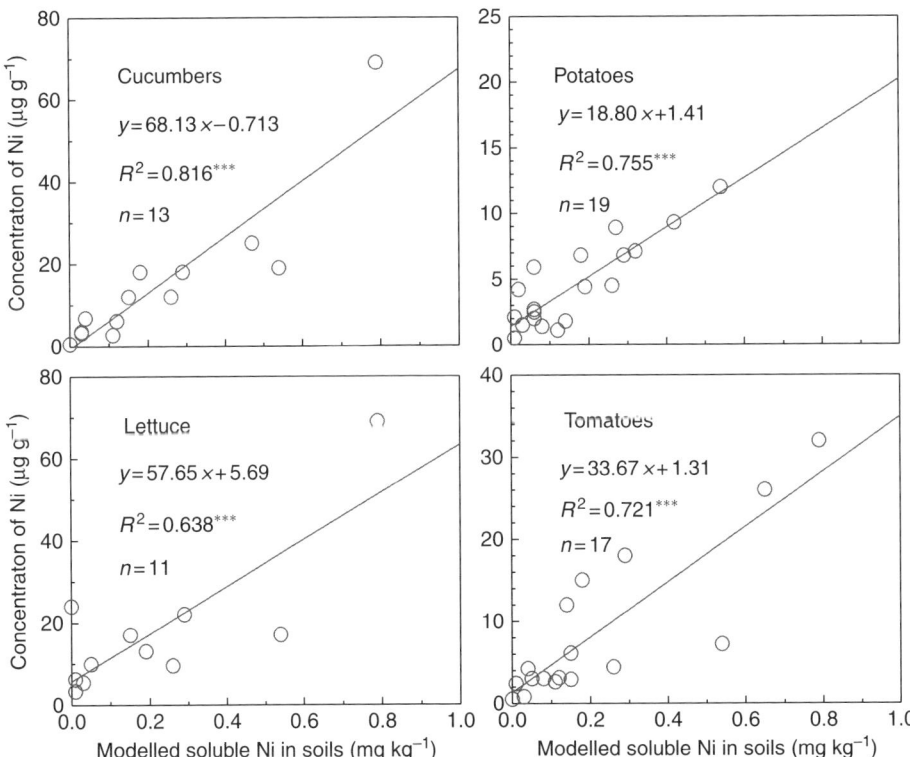

Figure 10.12 Regressions between modelled soluble soil Ni concentrations and Ni contents in four vegetable crops for soils with pH values below 6.5 (*** significant at $p < 0.01$)

toxic elements and only uses the diffuse layer model to describe the surface adsorption processes. In addition, emerging toxic elements of concern, such as antimony, silver, thallium, vanadium, have not been extensively studied.

The soil adsorption models that have currently been developed to include clay, soil organic matter and oxide minerals assume that the adsorption capacity of the soil is the sum of the adsorption capacities of the individual components. These models are referred to as 'additive models'. To date they do not account for interactions between the various components. While these interactions may be insignificant for cationic ions, for anionic ions these interactions may be significant. Although there are a number of reports of the binding constants of toxic elements with pure clay minerals, such as kaolinite, illite and montmorillonite, soil clay fractions generally contain a suite of clay minerals that are specific to particular soil zones. Site-specific adsorption models will need to be able to predict the adsorption of toxic elements to soil clay fractions either through direct measurement or from predictions based on the clay mineralogy of the soil.

In conclusion, 'what we must remember is that modelling is a tool – a useful tool to be sure, but not something that can ever replace the experience gained from directly working on a hydrogeochemical problem for several years' [98]. Finally 'we must admit that a model may confirm our biases and support incorrect intuitions. Therefore, models are most useful when they are used to challenge existing formulations, rather than to validate or verify them' [99].

References

[1] McBride, M.B.; 1989. Reactions controlling heavy metal solubility in soils; Adv. Soil Sci. 1989, 10, 1–56.
[2] Evans, L.J.; Chemistry of metal retention by soils; Environ. Sci. Technol. 1989, 23, 1046–1056.
[3] Benjamin, M.M.; Leckie, L.O.; Effects of complexation by Cl, SO_4, and S_2O_3 on adsorption behaviour of Cd on oxide surfaces; Environ. Sci. Technol. 1982, 16, 162–170.
[4] Dijkstra, J.J.; Meeussen, J.C.L.; Comans, R.N.J., Leaching of heavy metals from contaminated soils: an experimental and modelling study; Environ. Sci. Technol. 2004, 38, 4390–4395.
[5] Peijnenburg, W.J.G.M.; Zablotskaja, M.; Vijver, M., Monotoring metals in terrestrial environments within a bioavailability frame work and focus on soil extraction; Ecotox. Environ. Safety, 2007, 67, 163–179.
[6] Goldberg, S.; Suarez, D.L., Prediction of anion adsorption and transport in soil systems using the constant capacitance model.; in: J. Lützenkirchen (ed.), Surface Complexation Modelling. pp. 491–517. Academic Press, London, 2006.
[7] Sigg, L.; Black, F.; Buffle.; Cao, J. *et al.*, Comparison of analytical techniques for dynamic trace metal speciation in natural waters; Environ. Sci. Technol. 2006, 40, 1934–1941.
[8] Gustaffson, J.P., Modelling competitive anion adsorption on oxide minerals and an allophane-containing soil; Eur. J. Soil. Sci. 2001, 52, 639–653.
[9] Lumdson, D.G.; Partitioning of organic carbon, aluminum and cadmium between solid and solution in soils: application of a mineral–humic additivity model; Eur. J. Soil. Sci. 2004, 55, 271–285.
[10] Bonten, L.T.C.; Groenenberg, J.E.; Weng, L.; van Riemsdijk, W.H., Use of speciation and complexation models to estimate heavy metal sorption in soils. Geoderma, 2008, 146, 303–310.
[11] Turner, A.; Martino, M., Modelling the equilibrium speciation of nickel in the Tweed Estuary, UK: Voltammetric determinations and simulations using WHAM; Mar. Chem. 2006, 102, 198–207

[12] Ge, Y.; Sauvé, S.; Hendershot, W.H., Equilibrium speciation of cadmium, copper, and lead in soil solutions; Comm. Soil Sci. Plant Anal. 2005, 36, 1537–1556.
[13] Fest, P.M.J.E., Temminghoff, E.J.M; Comansa, R.N.J.; van Riemsdijk, W.H., Partitioning of organic matter and heavy metals in a sandy soil: effects of extracting solution, solid to liquid ratio and pH; Geoderma. 2008, 146, 66–74.
[14] Arai, Y.; McBeath, M.; Bargar, J.R.; Joye, J.; Davis, J.A., Uranyl adsorption and surface speciation at the imogolite–water interface: self-consistent spectroscopic and surface complexation models; Geochim. Cosmochim. Acta. 2006, 70, 2492–2509.
[15] Gustafsson, J.P., Modelling molybdate and tungstate adsorption to ferrihydrite; Chem. Geol. 2003, 200, 105–115.
[16] Wingenfelder, U.; Hansen, C.; Furrer, G.; Schulin, R., Removal of heavy metals from mine waters by natural zeolites; Environ. Sci. Technol. 2005, 39, 4606–4613.
[17] Lin, Y.P.; Singer, P.C., Inhibition of calcite precipitation by orthophosphate: speciation and thermodynamic considerations; Geochim. Cosmochim. Acta. 2006, 70, 2530–2539.
[18] Halim, C.E.; Short, S.A.; Scott, J.A.; Amal, R.; Low, G, Modelling the leaching of Pb, Cd, As and Cr from cementitious waste using PHREEQC; J. Haz. Mat. 2005, 125, 45–61.
[19] Stollenwerk, K.G.; Coleman, J.A., Natural remediation potential of arsenic-contaminated ground water; in: A.H. Welch and K.G. Stollenwerk (eds), Arsenic in Ground Water: Geochemistry and Occurrence, Kluwer Academic Publishers, Boston, MA, 2003, 351–379.
[20] van Beinum, W.; Meeussen, J.C.L.; Edwards, A.C.; van Riemsdijk, W.H., Modeling transport of protons and calcium ions in an alginate gel bead system: the effects of physical non-equilibrium and nonlinear competitive sorption; Environ. Sci. Tech. 2000, 34, 4902–4907.
[21] Brown, J.G.; Basset, R.L.; Pierre, D.G., Reactive transport of metal contaminants in alluvium-model comparison and column simulation; App. Geochem. 2000, 15, 35–49.
[22] Keizer, M.G.; van Riemsdijk, W.H., ECOSAT User Manual; Wageningen Agricultural University, Wageningen, 1998.
[23] Allison, J.D.; Brown, D.S.; Novo-Gradac, K.J., MINTEQA2/PRODEFA2, a Geochemical Assessment Model for Environmental Systems, Ver 3 User's Manual; US Environmental Protection Agency, Athens, GA, 1991.
[24] Parkhurst, D.L.; Appelo. C.A.J. User's guide to PHREEQC (version 2) – A computer program for speciation, batch-reaction, one-dimensional transport, and inverse goechemcial calculations; US Geological Survey, Water Resources Investigation Report 99-4259, 1999, 312 pp.
[25] Schecher W.D.; McAvoy, D.C., MINEQL+ chemical equilibrium modeling system, Version 4.5 for Windows. Environmental Research Software. Hallowell, ME, 2001.
[26] Bethke, C.M., The Geochemists Workbench; University of Illinois, 2002.
[27] Verweij, W.; Equilibria and constants in CHEAQS: selection criteria, sources and assumptions. Version 5. 2003. (http:home.tisali.nl/cheaqs/)
[28] Tipping, E.; WHAM – a chemical equilibrium model and computer code for waters, sediments and soils incorporating a discrete/electrostatic model of ion-binding by humic substances; Comp. Geosci. 1994, 20, 973–1023.
[29] Tipping, E., Humic ion-binding model VI: An improved description of the interactions of protons and metal ions with humic substances; Aqua. Geochem. 1998, 4, 3–48.
[30] Wolery, T. J., EQ3/6: A software package for geochemical modeling of aqueous systems: package overview and installation guide (version 7.0); Report UCRL-MA-110662 PT I, Lawrence Livermore National Laboratory, Livermore, CA, 1992.
[31] Santore, R.C.; Driscoll, C.T., 1995. The CHESS model for calculating chemical equilibria in soils and solutions; Chemical Equilibrium and Reaction Models; SSSA Special Publication 42, Soil Society of America, American Society of Agronomy, 1995.
[32] Meeussen, J.C.L., ORCHESTRA, a new object-oriented framework for implementing chemical equilibrium models; Environ. Sci. Technol. 2003, 37, 1175–1182.
[33] Westall J.C.; Zachary, J.L.; Morel, F.M.M., MINEQL, a computer program for the calculation of chemical equilibrium; MIT, 1976.
[34] Westall, J.C., MICROQL: I. A chemical equilibrium program in BASIC; EAWAG CH-8600. Swiss Federal Institute of Technology, Duebendorf, 1979.

[35] Weng, L.P.; Temminghoff, E.J.M.; van Riemsdijk, W.H., Contribution of individual sorbents to the control of heavy metal activity in sandy soil; Environ. Sci. Technol. 2001, 35, 4436–4443.
[36] Weng, L.P.; Temminghoff, E.J.M.; Lofts, S.; Tipping, E.; van Riemsdijk, W.H., Complexation with dissolved organic matter and solubility control of heavy metals in a sandy soil; Environ. Sci. Technol. 2002, 36, 4804–4810.
[37] Weng, L.P.; Wolthoorn, A.; Lexmond, T.M.; Temminghoff, E.J.M.; van Riemsdijk, W.H., Understanding the effects of soil characteristics on phytotoxicity and bioavailability of nickel using speciation models; Environ. Sci. Technol. 2004, 38, 156–162.
[38] Lofts, S., Tipping, E., An assemblage model for cation-binding by natural particles; Geochim. Cosmochim. Acta, 1998, 62, 2609–2625.
[39] van der Lee, J.; de Windt, L., Present state and future directions of modelling geochemistry in hydro-geochemical systems; J. Contam. Hydrol. 2001, 47, 265–282.
[40] Westall, J.C.; A chemical equilibrium program in Basic; Report 86–02; Department of Chemistry, Oregon State University, Corvallis, OR, 1986.
[41] Martell, A.E.; Smith, R.M.; NIST Standard Reference Database 46, version 8.0; Gaithersburg, MD, 2004.
[42] Davies, C.W., Ion Association; Butterworth, Washington, DC, 1962.
[43] Pitzer, K.S., Thermodynamics of electrolytes. V. Effects of high-order electrostatic terms; J. Soln. Chem. 1975, 4, 249–265.
[44] Westall, J.C., FITEQL: A computer program for determination of equilibrium constants from experimental data; Report 82–01. Department of Chemistry, Oregon State University, Corvallis, OR, 1982.
[45] Herbelin, A.L.; Westall, J.C., FITEQL: A computer program for determination of chemical equilibrium constants from experimental data; Report 96–01, Version 3.2; Department of Chemistry, Oregon State University, Corvallis, OR, 1996.
[46] Gu, X.; Evans, L.J., Modelling the adsorption of Cd(II), Cu(II), Ni(II), Pb(II) and Zn(II) onto Fithian Illite; J. Coll. Interf. Sci. 2007, 307, 317–325.
[47] Gaizer, F., Computer evaluation of complex equilibria; Coord. Chem. Rev. 1979, 27, 195–222.
[48] Dzombak, D.A.; Morel, F.M.M.; Surface Complexation Modeling: Hydrous Ferric Oxide. John Wiley & Sons Canada, Toronto, 1990.
[49] Goldberg, S.; The use of surface complexation models in soil chemical systems. Adv. Agron. 1992, 47, 233–329.
[50] Morel, F.M.M.; Hering, J.G.; Principles and Applications of Aquatic Chemistry. John Wiley & Sons Canada, Toronto, 1993.
[51] Lövgren, L., Hedlund, T.; Öhman, L., Sjöberg, S., Equilibrium approaches to natural water systems. 6: Acid–base properties of a concentrated bog-water and its complexation reactions with aluminium(III); Wat Res. 1987, 21, 1401–1407.
[52] Lövgren, L.; Sjöberg, S.; Equilibrium approaches to natural water systems. 7. Complexation reactions of copper (II), cadmium (II) and mercury (II) with dissolved organic carbon in a bog-water; Wat. Res. 1989, 23, 327–332.
[53] Gunneriusson, L.; Sjöberg, S.; Equilibrium speciation models for Hg, Cd and Pb in the Gulf of Bothnia and its catchment area; Nordic Hydr. 1991, 22, 67–80.
[54] Milne, C.J.; Kinniburgh, D.G.; van Riemsdijk, W. H.; Tipping. E., Generic NICA-Donnan model parameters for metal-ion binding by humic substances; Environ. Sci. Technol. 2003, 37, 958–971.
[55] Lumsdon, D.G.; Stutter, M.I.; Cooper, R.J.; J. R.Manson, Model assessment of biogeochemical controls on dissolved organic carbon partitioning in an acid organic soil; Environ. Sci. Technol. 2005, 39, 8057–8063.
[56] van Riemsdijk, W.H.; Heimstra, T., The CD-MUSIC model as a framework for interpreting ion adsorption on metal (hydr) oxide surfaces; in J. Lützenkirchen (ed.), Surface Complexation Modelling; Academic Press, London, 2006, pp. 251–268.
[57] Sverjensky, D.A.; Sahai, S., Theoretical predictions of single-site surface-protonation equilibrium constants for oxides and silicates in water. Geochim. Cosmochim. Acta. 1996, 60, 3773–3979.
[58] Lövgren, L.; Sjöberg, S.; Schindler, P.W., Acid/base reactions and Al(III) complexation at the surface of goethite. Geochim. Cosmochim. Acta. 1990, 54, 1301–1306.

[59] Hiemstra, T.; van Riemsdijk, W.H. A surface structural approach to ion adsorption: the charge distribution (CD) model; J. Coll. Interf. Sci. 1996, 179, 488–508.
[60] Weng, L.; van Reimsdijk, W.H.; Heimstra, T., Cu^{2+} and Ca^{2+} adsorption to goethite in the presence of fulvic acids; Geochimica Cosmochimica Acta, 2008, 72, 5857–5870.
[61] Tadanier, C.J.; Eick. M.J., (2002). Formulating the charge-distribution multisite surface complexation model using FITEQL; Soil Sci. Soc. Amer. J. 2002, 66, 1505–1517.
[62] Spadini, L.; Schindler, P.W.; Charlet, L.; Manceau, A.; Ragnarsdottir, K.V., Hydrous ferric oxide: evaluation of Cd-HFO surface complexation models combining Cd_K EXAFS data, potentiometric titration results and surface site structures identified from mineralogical knowledge; J. Coll. Interf. Sci. 2003, 266, 1–18.
[63] Mathur, S.S.; Dzombak, D.A., Surface complexation modelling: Goethite; in: J. Lützenkirchen (ed.), Surface Complexation Modelling; Academic Press, London, 2006, pp. 443–468.
[64] Gunneriusson, L.; Sjöberg, S., Surface complexation in the H^+-goethite(α-FeOOH)–Hg(II)–chloride system; J. Coll. Interf. Sci. 1993, 156, 121–128.
[65] Gunneriusson, L. Composition and stability of Cd(II) - chloro and -hydroxo complexes at the goethite(α-FeOOH)/water interface; J. Coll. Interf. Sci. 1994, 163, 484–492.
[66] Gunneriusson, L.; Lövgren, L.; Sjöberg, S., Complexation of Pb(II) at the goethite (α-FeOOH) water interface – the influence of chloride; Geochim. Cosmochim. Acta 1994, 58, 4973–4983.
[67] Rene, P. J.; Rietra, J.; Hiemstra, T.; van Riemsdijk, W.H., Sulfate adsorption on goethite; J. Coll. Interf. Sci. 1999, 218, 511–521.
[68] Hiemstra, T.; van Riemsdijk, W.H., Fluoride adsorption on goethite in relation to different types of surface sites; J. Coll. Interf. Sci. 2000, 225, 94–104.
[69] Stachowicz, M.; Hiemstra, T.; van Riemsdijk W.H., Surface speciation of As(III) and As(V) in relation to charge distribution; J. Coll. Interf. Sci, 2006, 302, 62–75.
[70] Stachowicz, M.; Hiemstra, T.; van Riemsdijk W.H., Multi-competitive interaction of As(III) and As(V) oxyanions with Ca^{2+}, Mg^{2+}, $PO4^{3-}$ and $CO3^{2-}$ ions on goethite; J. Coll. Interf. Sci, 2008, 320, 400–414.
[71] Angove, M.J.; Johnson, B.B.; Wells, J.D., Adsorption of cadmium (II) on kaolinite; Colloid Surf. 1997, A, 126, 137–147.
[72] Ikhsan, J.; Johnson, B.B.; Wells, J.D., A comparative study of the adsorption of transition metals on kaolinite; J. Coll. Interf. Sci. 1999, 217, 403–410.
[73] Lackovic, K.; Angove, M.J.; Wells, J.D.; Johnson, B.B., Modeling the adsorption of Cd(II) onto Muloorina illite and related clay minerals; J. Coll. Interf. Sci. 2003, 257, 31–40.
[74] Lackovic, K.; Wells, J.D.; Johnson, B.B.; Angove, M.J., Modeling the adsorption of Cd(II) onto kaolinite and Muloorina illite in the presence of citric acid; J. Coll. Interf. Sci. 2004, 270, 86–93.
[75] Gu, X.; Evans, L.J., Surface complexation modelling of Cd(II), Cu(II), Ni(II), Pb(II) and Zn(II) adsorption onto kaolinite; Geochim. Cosmochim. Acta 2008, 72, 267–276.
[76] Gu, X.; Evans, L.J., Barabash, S. J., Surface complexation modelling of Cd(II), Cu(II), Ni(II), Pb(II) and Zn(II) adsorption onto montmorillonite; Geochim. Cosmochim. Acta (in press).
[77] Bradbury, M.H.; Baeyens, B., A quasimechanistic non-electrostatic approach to metal sorption on clay minerals; in: J. Lützenkirchen (ed.), Surface Complexation Modelling; Academic Press, London, 2006, pp. 518–538.
[78] Perdue, E.M.; Lytle, C.R., Distribution model for binding protons and metal ions by humic substances; Environ. Sci. Technol. 1983, 17, 654–60.
[79] Dzombak, D.A.; Fish, W.; Morel, F.M.M., Metal-humate interactions. I. Discrete ligand and continuous distribution models. Environ. Sci. Technol. 1986, 20, 669–75.
[80] Koopal, L.K.; van Riemsdijk, W.H.; de Wit, J.C.M.; Benedetti, M.C., Analytical isotherm equations for multi-component adsorption to heterogeneous surfaces. J. Coll. Interf. Sci. 1994, 166, 51–60.
[81] Benedetti, M.F.; Milne, C.J.; Kinniburgh, D.G.; van Riemsdijk, W.H.; Koopal, L.K., Metal ion binding to humic substances: application of the non-ideal competitive adsorption model; Environ. Sci. Technol. 1995, 29, 446–457.
[82] Benedetti, M. F.; van Riemsdijk, W. H.; Koopal, L. K. Humic substances considered as a heterogeneous Donnan gel phase; Environ. Sci. Technol. 1996, 30, 1805–1813.

[83] Kinniburgh, D. G.; Milne, C. J.; Benedetti, M. F.; Pinheiro, J. P.; Filius, J. D.; Koopal, L. K.; van Riemsdijk, W. H. Metal ion binding by humic acid: application of the NICA-Donnan model; Environ. Sci. Technol. 1996, 30, 1687–1698.

[84] Kinniburgh, D.G.; van Riemsdijk, W.H.; Koopal, L.K.; Borkovec, M.; Benedetti, M.F.; Avena, M.J., Ion binding to natural organic matter: competition, heterogeneity, stoichiometry and thermodynamic consistency; Coll. Surf. 1999, 151, 147–166.

[85] Christensen, J.B.; Tipping, E.; Kinniburgh, D.G.; Gron, C.; Christensen, T.H., Proton binding by groundwater fulvic acids of different age, origins, and structure modeled with the model V and NICA-Donnan model. Environ. Sci. Technol. 1998, 32, 3346–3355.

[86] Bolton, K.A.; Sjöberg, S.; Evans, L.J.; Proton binding and cadmium complexation constants for a soil humic acid using a quasi-particle model, Soil Sci. Soc. Amer. J. 1996, 60, 1064–1072.

[87] Evans, L.J., Sengdy, B.; Lumsdon, D.G.; Stanbury. D.; Cadmium adsorption by an organic soil: a comparison of some humic - metal complexation models; Chem. Spec. Bioavail. 2003, 15, 92–100.

[88] Milne, C.J.; Kinniburgh, D.G.; Tipping. E., Generic NICA-Donnan model parameters for proton binding by humic substances; Environ. Sci. Technol. 2001, 35, 2049–2059.

[89] Lumsdon, D.G.; Evans, L.J., Predicting chemical speciation and computer simulation; in: A.M. Ure, C.M. Davidson (eds), Chemical Speciation in the Environment; Blackie Academic, London, 1995, pp. 86–134.

[90] Canbaniss, S.E.; Shuman, M.S., Copper binding by dissolved organic matter. Geochim. Cosmochim. Acta, 1988, 52, 185–200.

[91] Dobbs, J.C.; Stuseyo, W.; Carreria, L.A.; Azarraga, L.V., Competitive binding of protons and metal ions in humic substances by lanthanide ion probe spectroscopy; Anal. Chem. 1989, 61, 1519–1524.

[92] Zhang, H.; Davidson, W., In situ measurement of labile inorganic and organically bound metals in synthetic solutions and natural waters using diffusive gradients in thin films; Anal. Chem. 2000, 72, 4447–4457.

[93] Unsworth, E.R.; Zhang, H.; Davidson. W., Use of diffusive gradients in thin films to measure cadmium speciation on solutions with synthetic and natural ligands: comparison with model predictions; Environ. Sci. Technol. 2005, 39, 624–630.

[94] Warnken, K.W.; Davidson, W.; Zhang, H., Interpretation of in situ measurements of inorganic and organically complexed trace metals in freshwater by DGT; Environ. Sci. Technol. 2008, 42, 6903–6909.

[95] Kalis, E.J.J.; Weng, L.P.; Dousma, F.; Temminghoff, E.J.M.; van Riemsdijk, W.H., Measuring free metal ion concentrations in situ in natural waters using the Donnan membrane technique; Environ. Sci. Technol. 2006, 40, 955–961.

[96] Ge, Y.; Wang, W.; Zhang, C.; Zhou, Q., Determination of speciation and bioavailability of Cd in soil solution using a modified soil column Donnan membrane technique; Chem. Spec. Bioavail. 2009, 21, 7–13.

[97] Evans, L.J.; Bolton, K.A.; Barabash, S.J; Gu, X., Development and application of an adsorption model for nickel in Sudbury area soils. Final report, research project conducted as part of the Sudbury Area Risk Assessment (SARA), Ontario Ministry of the Environment, Toronto. 58 pp. 2006.

[98] Nordstrom, D.K., Modeling low-temperature geochemical processes; in: J. I. Drever (ed), Surface and Groundwater, Weathering and Soils. Vol. 5, Treatise on Geochemistry, H.D., Holland, K.K. Turekin (eds); Elsevier–Pergamon, Oxford, 2004, pp. 1–64.

[99] Oreskes, N.; Shrader-Frechette, K.; Belitz, K., Verification, validation, and conformation of numerical models in the earth sciences; Science, 1994, 263, 641–645.

Section 3

Bioavailability, Risk Assessment and Remediation

11
Assessing Bioavailability of Soil Trace Elements

Peter S. Hooda

School of Geography, Geology and the Environment, Kingston University London, Kingston upon Thames, Surrey, UK

11.1 Introduction

Elevated accumulation of trace elements (TEs) in soils as a consequence of anthropogenic activities represents a range of risks to human and ecosystem health, including bio- and water resources. Risks to human health may occur due to excess exposure to trace elements either via the food chain or through direct exposure by dermal contact or ingestion of soil-borne TEs, for example soil–human or soil–water–human pathways [1]. Environmental risks may arise due to the toxicity of TE contaminants to plants and soil-dwelling organisms (for example [2,3]); also see Chapter 8.

Trace elements in soil occur in various chemical species or forms distributed across solution and solid phases: Broadly, they can be grouped into the following compartments [4–6]; also see Chapter 3:

- free ions and organic and inorganic complexes (present in soil solution);
- adsorbed ions and compounds (clay and organic colloids);
- those occluded in secondary minerals and precipitated oxides of Fe and Mn, present as carbonates and phosphates or complexed with organic matter;
- those present in lattice of primary minerals.

The equilibrium between compartments 1 and 2 above is typically fast and the two compartments together represent the labile pool of TEs in soil [4,6,7]. Plants absorb nutrients, including TEs from soil essentially via the soil solution, which represents the most mobile and plant-available fraction. The solid phase-associated labile pool resupplies TEs to the solution in response to their removal, for example by plant uptake or leaching, and this can effectively buffer their solution concentrations [7]. Trace elements associated with compartments 3 and 4 have limited environmental significance. However, while the pool of elements associated with compartment 3 above is only slowly labile and may not contribute much to the soil solution, it can be mobilized in response to a major change in soil pH or redox potential [8].

It is now widely accepted that the risk of spread of TE contaminants into the wider environment increases with their soil concentrations (for example [9–14]). Solubility, mobility and 'bioavailability' depend on the forms or chemical species, which, in turn, are controlled by a number of soil characteristics such as pH, the amounts of clay, organic matter and secondary minerals (Fe and Mn oxides), and redox potential (see Chapter 3). The importance of these factors in controlling the solubility, the partition between the solid and solution phases or the distribution of TEs among the various soil compartments has been extensively studied and reviewed (for example [1,15–19]). Chapter 3 provides a further review of the key soil processes that control the partitioning of TEs between the labile and nonlabile phases, and ultimately their behaviour in soil.

The determination of 'total' concentrations of TEs in soils, for example after aqua regia digestion, does not necessarily provide an indication of their potential bioavailability or toxicity. Total TE analysis includes labile as well as nonlabile fractions; the latter, being strongly bound within the soil solid phases, are not available for plant uptake or for transport from the soil [1,20,21]. An evaluation of the wider environmental risks that TEs pose, therefore, requires reliable measurements of their labile species or forms in both solution and solid phases. The realization of this has led to the development of a plethora of selective soil tests; however, there is no general consensus on how the chemical species, fractions or 'bioavailable' concentrations can best be measured. This chapter presents an overview of TE speciation, bioavailability and risk relationships in soils. The range of tests and procedures that are currently used together with their reliability and general usefulness, that is, for assessing the risks TE-contaminated soils pose to the wider environment, will be reviewed. For sampling, procedural and analytical aspects of such tests, see Chapters 3–5.

11.2 Speciation, Bioavailability and Bioaccumulation: Definitions and Concepts

A number of definitions of speciation, bioavailability, bioaccessibility and bioaccumulation exist in the literature. This has created a considerable degree of ambiguity within scientific and regulatory communities, as often these terms have been used interchangeably. Other similarly expressing the concept of bioavailability include 'availability', 'plant availability' or 'extractability'. There is an obvious connection between these terms; for example, free ionic species of TEs influence their bioavailability and perhaps their uptake by plants or other soil biota. However, these are different concepts (Figure 11.1). It is generally accepted that the bioavailability of contaminants should be considered in environmental risk assessment and setting up risk-based contaminated land clean-up targets. This, however, has not been possible, because there is no general consensus on how speciation or bioavailability should inform risk assessment [23–25].

The term bioavailability was first used in pharmacology to express the amount of drug absorbed as a fraction of intravenous or oral dosage. In a similar way the amount of soil nutrients present in a form (as estimated by a given soil test) that is potentially 'available' for plant uptake is often termed plant-available. The term bioavailability in soil/environmental science was later coined to encompass availability of soil-borne nutrients and chemicals to soil fauna and flora and other organisms, including humans, which directly or indirectly can be exposed to such chemicals. This clearly puts bioavailability in a much wider context as it includes chemical availability to a variety of biota [26]. Authors have used different definitions to describe contaminants/nutrients bioavailability at least partly due to receptor species being different:

1. Adriano [27] distinguishes between 'availability' and 'bioavailability'; the former is defined as the rate and extent at which a chemical is released from a medium (soil) of concern, while the bioavailability of the chemical is to living receptors, through direct contact or uptake, for example plant roots. This encompasses both the supply (from soil solid and solution phases) and absorption aspects (see Figure 11.1) and appears to include the exposure and uptake of soil-borne chemicals by plants as well as other soil organisms.
2. Naidu *et al.* [28] defined bioavailability with respect to the uptake of metals by plants, as the fraction of extractable metals which correlates with their plant concentrations.
3. Alexander [29] described bioavailability to microorganisms as the accessibility of a chemical for assimilation and possible toxicity.
4. Ruby *et al.* [30,31] described bioavailability to higher organisms as the fraction of an administered dose that reaches the bloodstream from the gastrointestinal (GI) tract. This is the same definition as originally used in pharmacology. These authors also define 'bioaccessibility' as the fraction of a contaminant that becomes soluble in the GI tract and hence is available for uptake into the circulatory system. However, it should be noted that the amount of chemical (in this case soil-borne trace element) that becomes soluble in the GI tract does not mean that all will be absorbed in the body. Bioaccessibility has been applied in the context of inadvertent soil ingestion, and in essence represents potential bioavailability (absorption) of soil-borne contaminants via the soil ingestion route.
5. According to Sposito [32] 'a chemical element is bioavailable to plants if it is present as, or can be readily transformed to, the free ion species, if it can move to plant roots on a time scale that is relevant to plant growth and development and if once absorbed by the root, it affects the life cycle of the plant'. This seems to suggest that elements are only taken up by plants if they are in the form of their free ionic aqueous species (Figure 11.1). While free ionic species are considered to be most easily absorbed, there is evidence the plants can take up complexed species [33,34]. Secondly, here an effect on the life cycle of the plant is implicit, which may not be the case when the amount absorbed meets the plant's requirement (as in the case of essential TEs) or is well below its tolerance limit as for nonessential TEs.

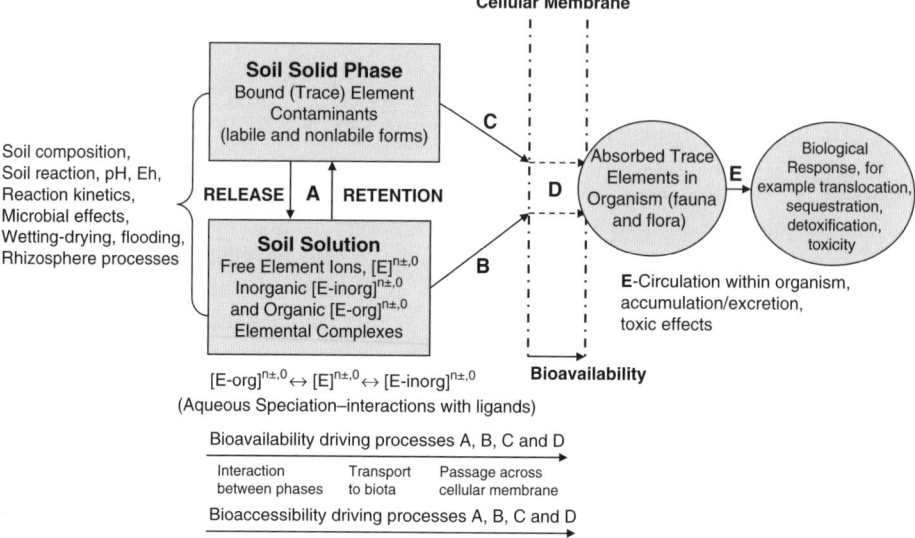

Figure 11.1 Bioavailability processes of soil trace element partitioning between soil solid and solution phases, including speciation in soil solution and how they can be influenced by soil properties and environmental conditions (A), transport from soil solution to plant roots (or soil fauna) cell membrane (B), direct contact of soil-bound TEs with cell membrane (C), and absorption following entry through cell membrane (D). In the case of soil ingestion, process D represents bioaccessibility – this may not lead to absorption. Organisms' response after absorption (E) is as such not part of bioavailability processes (Adapted from Semple, K.T.; Doick, J.K.; Jones, K.C.; Bureauel, P.; Craven, A.; Harms, H., Defining bioavailability bioaccessibility of contaminated soil and sediments is complicated; Environ. Sci. Technol. 2004, 38, 228A–231A [22].)

While there are some similarities between the various definitions, there is still no universal definition. For environmental risk assessment of TEs in soils, bioavailability assessment should include their soluble (pore water) as well as solid-phase-associated labile fractions. This may not best correlate to TE uptake by a specific plant species but is likely to be applicable to a wider variety of receptors, including different plant species. It makes sense when the solid-phase-associated labile pool of elements is directly in equilibrium with those in the solution phase (Figure 11.1), and the transformation between those species present in the latter phase is common knowledge [4,19,35], including that mediated by chemical and biological processes in the rhizosphere [36,37]. The measurement of either total or free ionic species present in soil solution will exclude the potential contribution from aqueous complexed species and the labile solid phase, and will underestimate the risk that trace element-contaminated soils pose to the wider environment.

Given the complexity of the processes (dissolution–desorption, transport and uptake) involved together with the lack of general consensus regarding the definition of bioavailability, the Committee on Bioavailability of Contaminants in Soils and Sediments termed bioavailability as a series of processes [38], as shown in Figure 11.1. Some authors consider bioavailability as a dynamic process (for example [39]). This approach includes at least two distinct phases [40]: (1) physicochemically driven desorption process (A, Figure 11.1), and (2) physiologically driven uptake process (D, Figure 11.1). However, diffusion, advection or mass flow driven transport of a contaminant is essential before uptake can occur (B, Figure 11.1), and in some cases, for example dermal contact or soil ingestion by earthworms, contaminants can be supplied by direct contact/ingestion to the site of uptake (C, Figure 11.1).

Bioavailability for all practical purposes therefore is the fraction of the total concentration of contaminants present in soil (solid and solution phases) which is potentially available for plant uptake or absorption by soil-dwelling organisms. Bioaccessibility, on the other hand, is the fraction of a contaminant that becomes soluble in the GI tract and hence is available for potential absorption in the circulatory system. Bioaccessibility explicitly implies intake/ingestion of solid-bound contaminants, and is an integral part of bioavailability – that is, in such situations a contaminant must be bioaccessible before uptake/absorption can occur.

Bioaccumulation and uptake are also often used interchangeably, though the latter term generally refers to uptake via plant roots. Nolan *et al.* [24] define bioaccumulation as the total accumulation of any chemical in the tissue of any biota via any route of exposure. The use of biological tissues for assessing bioaccumulation reduces uncertainties associated with bioavailability as a measure of risk assessment, but certain organs/tissues can concentrate TEs.

Speciation of trace elements in soil is important for their uptake by biota and hence can help asses their bioavailability. While the concentration/speciation of elements in soil solution has direct relevance to uptake by soil fauna and flora, the solid-phase-associated elements, particularly those in labile forms are also important (Figure 11.1) for bioavailability/bioaccessibility assessment purposes. In soil science, the term 'speciation' until recently has been used vaguely to describe fractions of elements associated with various soil components, particularly in so called solid-phase speciation procedures (SPs). Parker *et al.* [41] described speciation more specifically as the distribution of elements among their various chemical and physical forms, including their free ions, complexes, ion pairs, and chelates in solution, and their amorphous and crystalline solid phases, and stated that all of these forms may influence the reactivity, mobility and bioavailability of elements in soil.

The International Union of Pure and Applied Chemistry (IUPAC) recommends 'chemical species' as the form of an element defined as to isotopic composition, electronic or oxidation state and/or complex of molecular structure. Accordingly, Templeton *et al.* [42] define speciation as the distribution of an element among the IUPAC-defined chemical species, including inorganic compounds and complexes, organometallic compounds, and organic and macromolecular complexes. The speciation obtained using traditional techniques such as partial extractions by chemical reagents (for example exchangeable or plant-available), sequential extractions or soil solution (total) concentration does not conform to the IUPAC-defined chemical speciation criteria [43]. The determination of well-defined chemical species [42] of TEs in soil solution is possible using a range of analytical techniques (see Chapter 5). Partial and sequential chemical extraction procedures cannot determine discrete chemical species of TEs [43]; such 'species' are operationally defined and do not hold any significant chemical meaning [24], and therefore should be termed by the procedure used, for example 0.01 M $CaCl_2$ extractable fraction.

11.3 Bioavailability Assessment Approaches

A number of approaches have been used for the purpose of assessing bioavailability of soil TEs. These approaches include partial extractions using chemical reagents, solid-phase extractions into various fractions, total soil solution concentrations or speciation, isotopic dilution for lability assessment, diffusive membranes, biosensors and biological indicators. This section provides a brief overview of the main approaches and their appropriateness in risk measurements, particularly in terms of how effective they are in assessing the bioavailability of TEs. There are a number of seminal reviews available for the procedural as well wider aspects of these tests (for example [1,7,24,44,45]); also see Chapters 4 and 5.

11.3.1 Single Chemical Extraction Procedures

A variety of single extraction tests have been used as surrogate measures for some form of bioavailability of soil TEs. The underpinning premise is that such extractions are capable of mobilizing and isolating certain 'species' or forms of TEs that represent their mobile pool in soil. TEs extracted by such chemical leaching have often been functionally defined as exchangeable, mobile, plant-available or bioavailable. In reality they are operationally defined and extract a fraction of TEs that is potentially plant-available, mobile or bioavailable. An ideal test should extract only the labile fraction (that is water-soluble and those associated with the solid phases) and should not alter soil-specific element partitioning controls between the phases. Single extraction procedures have been widely used and are still commonly applied because they are relatively simple from laboratory-operational aspects, are cost-effective and generally provide reproducible results. The commonly used partial extraction tests can be grouped as follows:

- Strong chelating solutions, for example ethylenediamine tetraacetic acid (EDTA), diethylenetriamine pentaacetic acid (DTPA).
- Unbuffered neutral salt solutions, for example $CaCl_2$, $MgCl_2$, NH_4NO_3, $NaNO_3$, KNO_3.
- Organic (weak) acids alone or combined acid–salt solutions, for example acetic acid (CH_3COOH), oxalic acid ($H_2C_2O_4$), citric acid ($C_6H_8O_7$), ammonium acetate (CH_3COONH_4), sodium acetate (CH_3COONa).
- Diluted mineral acid solutions, for example HCl, HNO_3, $HCl + HNO_3$ or $HCl + H_2SO_4$ (Mehlich-1).

The general approach has been to establish linear correlations between TEs extracted by a particular chemical solution (for example EDTA, DTPA or $CaCl_2$) or extracted by a group of such extractants with the elements accumulation in the whole or parts of plants or other biota. The extent of correlation between the plant element content and that extracted by a given chemical extraction test is used to assess the suitability of the test for bioavailability assessment purposes. These extractants have also been used as secondary tests, that is without establishing correlations with plant or other biota metal contents, and the concentrations extracted have often been termed as being bioavailable, plant-available, extractable or mobile fractions. Here, the use of these tests is limited to a general assessment of a potentially mobile or soluble TE fraction in the soil.

Initially the DTPA and EDTA tests, originally designed to assess micronutrient deficiency/availability in soils, were preferred for estimating availability of metals in contaminated soils for plant uptake. In part this was due to their ability in extracting large concentrations suitable for the early analytical techniques. Lindsay and Norvell [46] developed the DTPA test, which is an aqueous mixture of 0.005 M DTPA, 0.1 M TEA (triethanolamine) and 0.01 M $CaCl_2$, buffered to pH 7.3. It was developed for calcareous soils, specifically for Cu, Fe, Mn and Zn. The appropriateness of extractants such as DTPA in contaminated soils was questioned by O'Connor [47]. Nevertheless, the test continues to be widely used on contaminated soils and regardless of soil type, for example acidic, calcareous, noncalcareous (for example [11,48–52]). Likewise, the use of EDTA was initially proposed for assessing soil micronutrient supply [53]. Subsequently, Clayton and Tiller [54] adapted the EDTA test for TEs in contaminated soils. Despite this early adaptation work, several variants of EDTA (for example Na or NH_4 EDTA, 0.005–1.0 M) have been used extensively for assessing plant availability of soil TEs [11,51,55–58]. Both DTPA and EDTA as strong chelating agents are known to extract labile and nonlabile element fractions [1,44,59,60]; and while they may correlate with plant metal uptake (for example [11]), are likely to overestimate bioavailability in contaminated soils. This has led to the suggestion that less aggressive chelating agents, such as nitrilotriacetic acid (NTA) and pyrophosphate may provide a better estimate of plant-available soil metals. The use of these 'milder' chelating agents for establishing their relationship with plant metal uptake, however, has been to a limited extent [61–63] compared with the DTPA and EDTA tests. These tests extract smaller amounts of TEs than do DTPA and EDTA, and while being better suited, they have not received wide enough use to test their suitability for predicting mobility and bioavailability of TEs in contaminated soils.

Despite reports of good correlations between elements extracted by chelating agents (for example DPTA and EDTA) and their uptake by plants, there is a general acceptance that milder chemical extractants, particularly unbuffered neutral salt solutions can provide a more realistic estimation of potential bioavailability of TEs in contaminated soils [39,64–68]. Examples of such aqueous chemical solutions used for bioavailability assessment purposes include $CaCl_2$, NH_4NO_3, $Ca(NO_3)_2$, $NaNO_3$, KNO_3, $MgCl_2$ [11,51,66,68–72]. Lebourg et al. [73], while studying the speciation of Cd, Cu, Pb and Zn in different neutral salt extracts observed similarities between the composition of these extracts and the soil solution. These unbuffered neutral salt solutions are therefore not expected to alter soil conditions/processes and thus can provide a reasonable estimate of the labile TE pool. Table 11.1 provides some examples where TEs extracted by a range of chemical extractants have been compared with their plant uptake. Generally TEs extracted by unbuffered neutral salt solutions correlated better with their plant contents compared to those extracted with EDTA, DTPA or acetic acid, although the relationships varied across the metals and studies. This may be due to the expected differences in TE lability; for example, Pb is expected to be less labile than Cd or Zn. A study involving soils from 14 vegetable gardens with variable source of contamination, while not showing an advantage of 0.01 M $CaCl_2$ over 0.005 M DTPA, had poor correlations (Table 11.1). The authors observed that using multiple regression ($CaCl_2$-extractable element content, and soil pH and cation-exchange capacity, CEC), the correlations were improved significantly. Similarly, a study of 15 field sites by Mench et al. [74] produced mixed results, where 0.1 M $Ca(NO_3)_2$-extractable

Table 11.1 *Some examples of commonly used chemical extractants for predicting plant metal uptake. The data show linear correlation coefficients (r) between metals in plant tissues with those extracted using chemical extraction tests*

Field grown wheat, a large area contaminated by Cu smelter/refinery ($n = 100$ from 216 km^2), metal content in wheat shoots of unknown growth stage [71]

	KNO$_3$, 0.05 M	NaNO$_3$, 0.1 M	MgCl$_2$, 1 M	EDTA, 0.05 M	CH$_3$COOH, 2% w/v
As	0.711	0.822	0.686	0.425	0.495
Cd	0.695	0.915	0.711	0.486	0.499
Cr	0.702	0.352	0.611	0.473	0.421
Pb	0.422	0.795	0.775	0.325	0.584

Field soils ($n = 13$), pot-grown lettuce [66]

	KNO$_3$, 0.1 M	CaCl$_2$, 0.05 M	NaNO$_3$, 0.1 M	NH$_4$NO$_3$, 0.01 M
Cu	0.64	0.65	0.68	0.63
Zn	0.89	0.84	0.87	0.87
Cd	0.49	0.61	0.85	0.51
Pb	0.23	0.22	0.18	0.38

Plants (Chinese cabbage and spinach) and soils collected from vegetable gardens, with various contamination history from 14 cities, China [52]

	Chinese cabbage		Spinach	
	0.01 M CaCl$_2$	0.005 M DTPA	0.01 M CaCl$_2$	0.005 M DTPA
Cr	0.385	−0.576	0.410	−0.741
Zn	0.202	−0.394	0.303	0.012
Cu	0.235	−0.286	0.823	−0.044
Cd	0.223	0.047	−0.222	0.124

Cd correlated well (r = 0.919) with wheat grain-Cd whereas Cd in the shoots was best (r = 0.883) predicted by 0.005 M EDTA-extracted Cd.

Differences in the ability of chemical extractants to correlate with plant uptake are not uncommon, at least partly due to the differences in the chemical behaviour of individual TEs, and a single test is unlikely to be suitable for all TEs. Also, plant tissue composition varies with age, although sometimes this is not considered; for example, the study by Chojnacka *et al.* [71] (see Table 11.1) does not mention the growth stage at which wheat shoots were sampled. The good correlations obtained by using 1 M MgCl$_2$ for all four elements in wheat shoots may not hold when compared with their wheat grain content. Also, some elements are known to be poorly translocated to shoots; for example, Pb tends to accumulate in roots, particularly in soils with ample supply of phosphate, and excess soil available Zn can inhibit plant Cd uptake (see Chapter 17).

Single chemical extractants have been widely used, but often this has been limited only to TE extractions for making bioavailability inferences. There is a general lack of systematic testing/comparison for establishing the robustness of such tests in predicting element concentrations in the plant tissues of interest.

Hooda et al. [11] compared four extractants (0.005 M DTPA, 0.05 M EDTA-$(Na)_2$, 1 M NH_4NO_3 and 0.05 M $CaCl_2$) in 13 soils where wheat, carrots and spinach were grown to maturity. The plants were grown in field-like conditions in large tubs kept in a rural field (each containing about 50 kg soil). Cadmium, Cu, Ni, Pb and Zn contents in the edible portions of the plants were compared with those extracted by the test chemical solutions before planting (Table 11.2). The soils varied considerably in terms of their pH, CEC, organic matter and clay contents [11]. While metals extracted by all tests correlated well with the plant metal concentrations, best predictor test varied across the metals and plant species studied. Given the strong extractability of the DTPA and EDTA tests, it would appear sensible to use unbuffered neutral salt solutions such as NH_4NO_4 and $CaCl_2$ (Table 11.2).

The use of unbuffered neutral salt solutions as extractants for assessing the bioavailability of TEs is gaining wider acceptance. However, it is important that these tests are standardized for their concentration, extraction protocols and precision. Fortunately, some efforts have already been made to resolve the standardization issues; for example, the Standards Measurements and Testing Programme of the EU (formerly BCR) carried out

Table 11.2 Linear regression equations of the form y = a + bx where y represents element concentration in the edible plant tissues (mg kg^{-1}) and x is the soil element concentration (mg kg^{-1}) extracted by one of the four extractants used (0.005 M DTPA, 0.05 M EDTA-$(Na)_2$, 1 M NH_4NO_3 and 0.05 M $CaCl_2$). Similar results were obtained for Pb, where the EDTA test best predicted its concentration in the crop tissues (data, not shown). The study was carried out using 13 soils with past history of sewage sludge application and carrot, spinach and wheat were grown in 'field-like conditions' in large tubs, kept in the open

Y	a	SE	b	SE	X	R^2	P
Cadmium							
Carrot	0.653	0.297	0.469	0.672	EDTA	0.781	<0.001
Spinach	2.41	1.91	3.92	0.672	DTPA	0.734	<0.001
Wheat	0.069	0.125	0.224	0.029	EDTA	0.855	<0.001
Copper							
Carrot	5.23	0.79	0.052	0.008	DTPA	0.769	<0.001
Spinach	12.67	1.81	1.93	0.543	NH_4NO_3	0.493	0.004
Wheat	2.40	0.187	1.21	0.218	$CaCl_2$	0.752	<0.001
Nickel							
Carrot	1.50	0.770	3.48	0.307	$CaCl_2$	0.914	<0.001
Spinach	4.52	1.37	0.301	0.047	EDTA	0.763	<0.001
Wheat	0.453	0.137	1.52	0.080	NH_4NO_3	0.973	<0.001
Zinc							
Carrot	26.62	5.04	0.130	0.026	EDTA	0.663	<0.001
Spinach	299.10	56.50	16.86	5.59	$CaCl_2$	0.403	0.012
Wheat	37.78	5.42	2.60	0.494	$CaCl_2$	0.728	<0.001

SE = standard error.
Plant availability of heavy metals in soils previously amended with heavy applications of sewage sludge, PS. Hooda, D. McNulty, B.J. Alloway and M.N. Aitken, Journal of the Science of Food and Agriculture, 73, 446–454. 1997, © Society of Chemical Industry, first published by John Wiley & Sons Ltd [11].

several interlaboratory exercises and has provided indicative values for $CaCl_2$, $NaNO_3$ and NH_4NO_3 extractable metals in two sludge amended soils (BCR CRM 483 and BCR CRM 484) as reference materials [75]. The $CaCl_2$, $NaNO_3$, NH_4NO_3 and KNO_3 (see Tables 11.1 and 11.2) seem similar, but they have not been robustly compared. Nevertheless, 0.1 M $NaNO_3$ [76] and 1.0 M NH_4NO_3 [77] have been adopted as standard national protocols in Switzerland and Germany, respectively (cf. [68]), and 0.01 M $CaCl_2$ [67] has been recommended in the Netherlands for similar metal testing protocols (cf. [68]). $CaCl_2$ was also suggested by Houba *et al.* [78] as the most suitable for a universal test for assessing risks from TEs in soils. Peijnenburg *et al.* [44] also support this test based upon findings and rationale provided by Novozamsky *et al.* [65] and Houba *et al.* [78].

Recently there has been some interest in developing 'rhizosphere-based' methods for assessing TE bioavailability. These methods generally require element extractions using a mixture of low-molecular-weight organic acids. For example, Feng *et al.* [51,58] used a mixture of organic acids (acetic, lactic, citric, malic and formic acids), and compared their 'rhizosphere-based test' results with other traditional extractants (for example DTPA, EDTA, $NaNO_3$ and $CaCl_2$) for assessing the accumulation of Cd, Cr, Cu and Zn in pot-grown barley and wheat plants (roots and shoots). The mixture of organic acids used by these authors was found a somewhat better predictor of plant metal uptake compared with the other tests, but results were not consistent. The 'rhizosphere-based' methods may simulate soil–root interactions that influence nutrient/metal acquisition and uptake processes [79]; however, more rigorous testing is required to evaluate their effectiveness in predicting TE bioavailability in soils. Chang *et al.* [80] have used a similar approach together with modelling to predict changes over time in the total bioavailability pool of metals in sewage sludge amended soils (see Chapter 6).

11.3.2 Sequential Extraction Procedures

The characterization of TE species present in the solid phases of soil or sediments has been the focus of much research during the past 40 years. Earlier work on speciation included either assessing the association of TEs with various mineral and organic phases present in the solid matrix or quantification of associations between TEs and organic and inorganic phases, resulting in the development of numerous 'speciation' techniques. As noted previously, the sequential extraction procedures (SEPs) that are typically used to isolate species of TEs associated with the various soil solid phases do not meet the IUPAC criteria of chemical species or chemical speciation [42]; henceforth such species and speciation determined by using SEPs will be referred to as TE forms/fractions and fractionation, respectively.

As discussed previously, within the soil solid phase TEs exist in different forms; some solid forms are relatively stable (for example bound to carbonates or Fe and Mn oxides) while other forms (for example water-soluble and easily exchangeable) can significantly increase the risk of leaching and plant uptake. As a result, information about the forms of TEs has been considered potentially valuable for predicting their availability to plants, their movement in the soil profile and transformation between different forms in soils (for example [35,81,82]).

Among the numerous SEPs, most widely used are those proposed by Tessier *et al.* [83] and Stover *et al.* [84] or certain variants. Some of the SEPs involve elaborate extraction protocols to isolate up to 10 identifiable element fractions [85, cf. 7]. The sequential extraction procedure proposed by Tessier *et al.* [83] fractionates TEs into the following forms: (a) exchangeable, (b) bound to carbonates, (c) bound to Fe and Mn oxides, (d) bound to organic matter, and (e) residual, that is bound to primary mineral structure. All such schemes, however, are fraught with a number of problems, for example: lack of specificity of the extractants as a result overlapping of the fractions (for example [86]); incomplete dissolution or solubilization of multiple phases in one extract (for example [87]); extractants changing pH during the extraction, which may result in re-adsorption [88]; or redistribution as well as re-adsorption [89,90]. The extent of such problems or at least the possibility of their occurrence generally increases with the number of stages in the sequential extraction procedure. Needless to say, these fractionation procedures are hugely labour-intensive. Another added problem is that these so-called functionally defined schemes were never standardized, resulting in a plethora of procedures and making it impossible to compare the findings or the validity of the procedures used [86,91].

There are a number of excellent reviews (for example [1,7,85,92,93]) on the procedural aspects as well as the difficulties in fractionating TEs in clearly defined fractions (also see Chapter 4). The realization of these problems led to a major standardization programme by the Standards, Measurements and Testing Programme of European Community, formerly known as BCR [75], eventually developing an operationally defined sequential extraction procedure as well as producing certified reference materials for its quality assurance purposes [94,95].

The BCR fractionation procedure is a three-step procedure that fractionates TEs into three operationally defined fractions (CH_3COOH, $NH_2OH \cdot HCl$ and H_2O_2/CH_3COONH_4 extractable), and the residual fraction can either be determined directly or by the difference between the sum of the TEs extracted in the three steps and their total amounts determined separately. Figure 11.2 shows an example application of the BCR scheme [35]. The main aim in this case was to assess how sewage sludge borne-Cd is distributed among the forms and to examine subsequent transformations. It shows that the sludge application contributed Cd to all three forms, and as expected H_2O_2/CH_3COONH_4-soluble Cd decreased as the sludge mineralized. Cadmium released from the sludge (H_2O_2/CH_3COONH_4-soluble; SS-C, Figure 11.2) was largely in the most mobile pool, that is CH_3COOH-extractable. Clearly the application of the BCR fractionation procedure here provides additional information on the dynamics of Cd forms during sewage sludge mineralization. Such information can be useful in understanding both short- and long-term behaviour of TEs in soils. Further examples of SEPs use include studies on potential changes in TE fractions due to ageing effects [96,97]; however, these are limited to a few weeks following metal inputs and thus such short-term studies may not provide much useful information.

It should be noted that such fractionation schemes seldom serve any purpose for assessing TE bioavailability, largely because plant uptake generally correlates only with that extracted in the first step of any sequential extraction procedure, which often includes their most labile fraction. Despite the ineffectiveness of the strongly bound fractions in assessing the bioavailability of soil TEs, a number of authors have used SEPs for examining either the relationship of the various metal fractions with plant uptake (for example [98–102]) or with microbial activities (for example [103]).

Notwithstanding the numerous criticisms, fractionation by SEPs can be useful for understanding the chemical reaction of TEs or their transformation in soils [35,104,105], provided

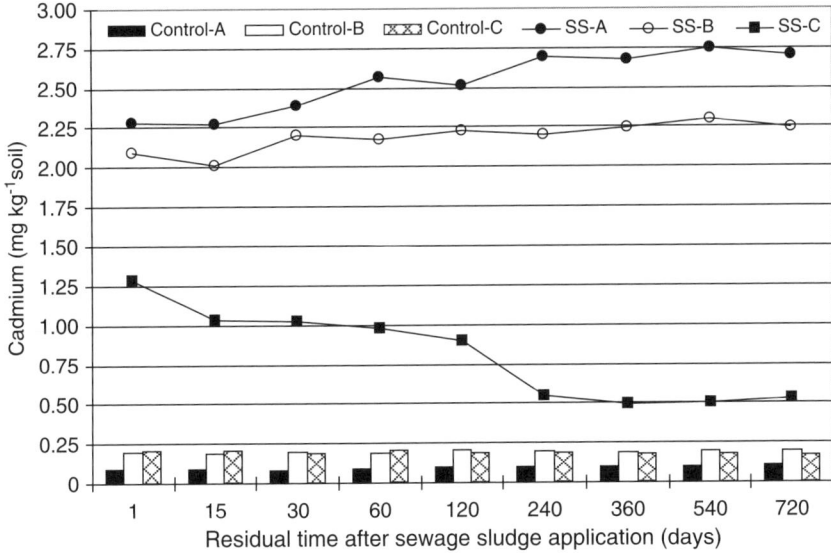

Figure 11.2 An example of the BCR trace element fractionation scheme. A, B and C represent 0.11 M CH_3COOH, 0.10 M $NH_2OH \cdot HCl$ (pH 2) and 30 % H_2O_2/1.0 M CH_3COONH_4 (pH 5) extractable Cd, respectively. Both control and sewage sludge (SS) amended (150 Mg SS ha^{-1}) soils were incubated at 25 ± 2 °C for 2 years under greenhouse conditions (Data from P.S. Hooda and B.J. Alloway, Changes in operational fractions of trace metals in two soils during a two years reaction time following treatment with sewage sludge, *Intern. J. Anal. Environ. Chem.* 1994, 57, 289–311 [35].)

the validity of such procedures can be demonstrated as in the case of the BCR procedure [91,94,95]. Information obtained from such fully validated procedures can be useful in terms of making long-term risk assessment, as well as in providing supplementary information for remediation decisions. For example, in the United Kingdom, remediation is not obligatory if the concentrations of TEs measured in samples from the site is greater than the regulatory or guide value for a specific land use, provided no significant risk is determined. The use of SEPs for the purpose of assessing the risk of uptake of TEs by plants is unnecessary [106], which can be better assessed using other less time-consuming means, for example using $CaCl_2$ or other similar tests (Table 11.1).

11.3.3 Soil Solution Concentration and Speciation

Plants and other soil biota take up nutrients and contaminants, including TEs mainly from the soil solution. This indeed is the primary exposure route of plants, microorganisms and soft-bodied organisms to TEs, while soil-ingesting invertebrates are exposed to TEs that are associated with the soil solid phase. Nevertheless, it has been suggested that the dissolved TEs are most readily available to soil biota, including plants (for example [107,108]). The determination of total dissolved elements in soils as a result gained recognition as a means of assessing the bioavailability and toxicity of TEs in soils (for example [109–111]). Within soil solution TEs occur in a variety of chemical forms, for example free hydrated ions, complexed with a wide range of naturally occurring organic (humic substances, phyto- and bacterial siderophores, and low-molecular-mass organic acids) and inorganic ligands and bound to colloidal material.

Later work recognized the need for solution phase speciation when the role of chemical species, particularly free ionic forms, in delineating TE toxicity was demonstrated [112,113]. Soil solution speciation may also be important from a contaminant bioavailability standpoint, as metal cationic species are considered to be more readily absorbed compared to their complexed counterparts because of the negative charge potential in plant roots, favouring preferential transport of the former species. The importance of free ionic species in determining the bioavailability of TEs in soil is examined later in this section. Speciation of TEs in soil solution is also considered useful because some forms of the element can be essential for life while other forms can be highly toxic. For example, while Cr(VI) is highly toxic, Cr(III) is an essential/beneficial element (see Chapter 19); likewise As(III) is much more toxic than As(V) or its organic species (see Chapter 16).

A number of analytical methodologies and computational tools have been applied to the measurement of TE speciation in soil solution, mainly for the purpose of determining concentrations/activities of free TE ions (for example Cd^{2+}, Zn^{2+} and Ni^{2+}). Free element activity can be determined using analytical or computational approaches. The analytical techniques and models have evolved and improved considerably in recent years. A detailed review of their application, capabilities and limitations can be found in Chapter 5. Briefly, techniques such as ion-selective electrodes, ion-exchange resins, Donnan membrane systems, and anodic stripping voltametry have been used for the determination of free metal ions in soil solution [45]. However, chemical separation and quantification of different free metal species is complex despite the improvements (see Chapter 5). Methods such as the Donnan membrane system, ion-exchange resins and voltametry suffer from either chemical interferences, poor limits of detection or disturbance of solution equilibrium [1,24 and references therein]. The successful use of ion-selective electrodes (ISEs) for the determination of free metal activity is mainly limited to Cu (for example [110,114,115]); attempts have been made to use ISEs for other TEs, for example Cd [116], Pb [115] and Zn [117], but these ISEs are not sensitive or reliable enough. Voltametry-based techniques are impossible to calibrate and are not species specific because they measure the electrochemically labile species, which can include both inorganic and easily dissociable organic-metal complexes [118,119]. As a result, voltametry (for example anodic stripping voltametry) provides a measure of all labile species in the soil solution. Speciation in terms free ion activity by ion-exchange resins is operationally defined, as organic and inorganic metal complexes can also contribute to the measurement [120]. The Donnan dialysis system uses ion exchange in the form of a membrane, with separation being controlled by the membrane pore size, and is also known to minimize the problem of equilibrium disturbance [121,122].

More recent analytical techniques such as chromatography coupled with ICP-MS, LC-MS or capillary zone electrophoresis-ICP-MS are indeed highly advanced techniques but are not free from problems. The coupled chromatography is generally limited to thermodynamically stable and kinetically inert species because more kinetically labile species are likely to change as they pass through the chromatographic column [24,123]. This will inevitably disturb the solution equilibrium during analysis. In essence, soil solution speciation, particularly the determination of free TE activity, is mostly operationally defined and is clearly not a trivial analytical task [24].

An alternative approach is to compute the free metal activity or metal speciation by using measured pH, concentrations of major cations, anions, dissolved organic carbon and TEs in soil solution. Over the years many models have been developed and used for the purpose of estimating TE speciation, for example GEOCHEM, MINTEQA2, WHAM, ECOSAT, ORCHESTRA and many more (see Chapter 10). The use of these models has made significant advances in terms of describing the mechanism of TE adsorption on soil primary and secondary minerals as well as organic matter and hence their distribution between the solid and soil phases. Most of the currently available models, however, cannot effectively account for TE retention simultaneously onto all binding surfaces found in soils, that is primary and secondary minerals, and dissolved and solid organic matter [124–129]. This is despite major improvements in some of the models which take into account competitive binding to both soil components and dissolved organic matter, for example WHAM VI [130] and NICA-Donnan [131]. Nevertheless, a number of studies have demonstrated the use of computational models in predicting soil solution metal speciation (for example [129,132,133]).

A major limitation of all such computational techniques in terms of predicting TE speciation in soil solution is their inability to distinguish between TE complexes formed with simple organic ligands (for example citric acid or oxalic acid) or more complex humic substances (see Chapter 10). As a result, significant assumptions are made with respect to the extent of TE interactions as well their formation constants [24]. Most of these models rely on default databases, derived from rather simple systems, with highly variable speciation output depending upon the assumptions made. Thus, depending on what stability constants one chooses, the speciation output can be entirely different. The geochemical modelling of soil solution is clearly fraught with difficulties given the uncertainties in the metal-organic complex stability constants. Theses problems were further demonstrated by a study of 29 soils where for the same data input (including metal-organic complex formation constants) there were major differences in the distribution of organic and inorganic Zn speciation as predicted by two independent models, MINTEQ 2.23 and WHAM VI [134].

11.3.3.1 Free Ion Activity and Bioavailability

Until recently it was generally accepted that plant uptake of nutrients or TEs is related to their free ion activities in the solution (for example [109,110,112,113,135,136]), although this was largely based upon uptake from metal ion buffer solutions. This resulted in the development of the free ion activity model, FIAM [137], and its subsequent adaptation, the biotic ligand model (BLM); the latter was developed for describing uptake by fish gills and is based on the assumption that the free ions interact with ligands at the membrane surface [138]. The BLM has also been used in ecotoxicity studies of soil-dwelling organisms (see Chapter 8). Both the FIAM and the BLM consider uptake by biota essentially in the form of free ions. There is some evidence for a link between uptake of free metal ionic species and toxicity as well as soluble organic metal–ligand complexes not being readily taken up and being less toxic compared to free metal ion species (for example [111,139,140]). The concept of FIAM and its capability to handle metal–organic interactions, for example uptake and toxicity, is supported by plenty of evidence that arose mainly from aquatic environments involving algae or plankton (see Morel [110]). Nolan *et al.* [24] provide a critical analysis of the FIAM and why it may not be able to predict TE uptake by terrestrial plants. The FIAM or BLM may apply if the free ion activity at the plant root or biota membrane is effectively buffered from other labile species, which may be possible in situations where uptake is relatively slow and rhizosphere processes do not alter TE speciation. However, there is considerable evidence which challenges the free ion activity concept. For example, Krishnamurti *et al.* [141] demonstrated organic complex-bound Cd species contributing significantly to the bioavailability of Cd. Likewise, McLaughlin *et al.* [142] and Weggler *et al.* [143] found enhanced Cd uptake with increasing amount of chloride salinity in the soils, possibly due to uptake of cadmium–chloride complexes. Also, enhanced uptake of Cd by alfalfa [34] and by Indian mustard [33] following EDTA addition demonstrates, although indirectly, that Cd uptake occurs via its nonionic species as well in theses cases, possibly as EDTA–Cd complexes.

Findings from three major studies which compared plant metal uptake with the various soil tests, including the total soil solution concentration and the free metal ion concentration (Table 11.3) demonstrate unequivocally that the concentration/activity of the free metals is no better than their total concentration in the soil solution or that extracted by using traditional soil tests, and in some cases the total soil metal content was better in predicting plant uptake (Table 11.3).

The relationship between the solution and solid phases is intrinsically important, as in response to local removal by plants the solid-phase associated labile pool contributes TEs to the solution pool. It is important to recognize that transformation between species is a dynamic process; that is, both organic (particularly those associated with simple organic acids) and inorganic complexes can exchange rapidly with the free ion species (Figure 11.1). Also, inorganic and weak organic TE complexes can easily dissociate, for example at the root membrane surface. The free ion activity concept is further complicated by the excretion products of plant roots, and organic compounds and their metabolites from rhizosphere microorganisms which can include a variety of soluble and insoluble organic substances [147,148]. Such excretion products (for example acetic, oxalic, fumaric, citric, tartaric acids, uronic acids, and polysaccharides) can form complexes and chelates with TE ions (for example [148–152]), which inevitably will modify their solubility in soils and their speciation in soil solution. Furthermore the production of protons, exudates and metabolites can modify the pH by as much as 2 units [153], with profound implications for element solubility and speciation. More importantly, these changes in the rhizosphere can be soil and plant species specific and can immobilize free metal ions and mobilize less-labile species.

Speciation in soil solution is therefore a highly complex and dynamic process and it may not add much to the measurement of total soluble (soil solution) metal concentration. However, speciation is important from toxicity viewpoints for elements like arsenic, mercury, chromium and tin.

Table 11.3 Examples of soil solution (total dissolved), free metal ion concentration/activity and other soil tests used for predicting plant metal uptake. The data show linear correlation coefficients (r) between metal concentrations in the plant tissues and that measured in the soils

Urban soils and grass and herbaceous vegetation, n = 23 [144]

	Free metal ion	Total soil solution	Total soil metal
Cd	−0.203	−0.142	0.281
Cu	−0.120	−0.057	−0.076
Ni	0.465*	0.379	0.503*
Pb	−0.125	−0.289	0.223
Zn	0.381	0.465*	0.405

Pot-grown *Silene vulgaris* (SV) and *Elsholtzia splendens* (EP) in 29 soils with a range of Cu levels [145]

	Total Cu	EDTA	NH_4NO_3	Soil solution total	Soil solution free	$C_E(Cu)^a$
SV-shoot Cu	0.77***	0.78***	0.86***	0.85***	0.72***	0.82***
SV-root Cu	0.75***	0.76***	0.92***	0.92***	0.77***	0.92***
EP-shoot Cu	0.60***	0.62***	0.79***	0.76***	0.72***	0.78***
EP-root Cu	0.74***	0.75***	0.93***	0.90***	0.80***	0.93***

Pot-grown radish in 10 arable or grassland soils with long history of contamination collected from six European countries [146]. Data show linear regression coefficient values (adjusted R^2)^b

	Soil Cd_T	Solution Cd (NR)		Solution Cd (R)		Soil Zn_T	Solution Zn (NR)		Solution Zn (R)	
		Total	Free	Total	Free		Total	Free	Total	Free
Leaf	0.79***	0.47*	0.24	0.98***	0.90***	0.61**	0.00	0.00	0.25	0.25
Tuber	0.80***	0.51*	0.28	0.98***	0.92***	0.15	0.07	0.07	0.53**	0.55**

*, ** and *** represent the level of significance at $P < 0.05$, < 0.01 and < 0.001, where reported.
^{a}DGT (diffusive gradient through thin films) effective concentration
^{b}R and NR represent rhizosphere and nonrhizosphere soil. Subscript T refers to total metal concentration.

11.3.4 Other Approaches

There are difficulties in the application of the FIAM to real soil situations because of a number of reasons noted in Section 11.3.3. Most of the studies supporting relationships of plant metal uptake to free metal ion activity are based on controlled aqueous conditions. This section provides brief overviews of X-ray based spectroscopic and isotopic dilution techniques that can help examine TE reactions, their fixation and transformation with ageing, and hence are capable of providing information which can be useful in evaluating risks from TEs in soils, including the effectiveness of remediation treatments such as formation of stable metal minerals. Secondly, a relatively new technique called DGT (diffusive gradients in thin films) is discussed, particularly in terms of its effectiveness in predicting plant TE uptake.

11.3.4.1   X-ray-Based Spectroscopic Techniques

A number of solid-phase nondestructive techniques such as particle induced X-ray emission (PIXE), X-ray fluorescence (XRF), energy dispersive X-ray fluorescence (EDXRF), X-ray absorption spectroscopy (XAS), and extended X-ray absorption fine structure spectroscopy (see Chapter 5) can allow quantitative analysis of TEs at the molecular level in terms of their distribution, reaction with soil primary and secondary minerals, speciation, and formation and identification of metal minerals [154–158]. These X-ray-based spectroscopic techniques can provide much accurate insight into the physical and chemical state of TEs, as well as mechanisms associated with their retention in soils, compared with traditional methods (for example adsorption isotherms, SEPs). While this can provide information that can be useful for assessing risks from TEs in soils, these techniques have not been used *per se* for the purpose of bioavailability assessment and thus are outside the scope of this chapter.

11.3.4.2   Isotopic Dilution Techniques

Historically, isotopic dilution (ID) techniques were used initially to study the reactions, fixation and residual availability of fertilizer nutrients, particularly phosphorus and potassium, using radioisotopes of P and K. Subsequently these techniques have been used for TEs, for example Ni, Zn, Cd, Cd and As (for example [159–164]). Isotopic dilution techniques using radioisotopes of TEs allow examination of their partitioning into isotopically and nonisotopically exchangeable pools, which are described as labile and nonlabile, respectively. TEs in the nonlabile pool are considered as 'fixed', while those in the labile pool are potentially available for plant uptake or soil reactions/processes such as precipitation, coprecipitation with secondary soil minerals or attenuation/fixation. As any other form or fraction of TEs in soils, exchange between labile and nonlabile pools occurs. In other words, nonlabile pool can be mobilized, for example following a change in soil pH and/or E_h and reaction with plant root exudates organic compounds and their metabolites [161–163,165]. In essence, the partitioning of TEs using isotopic dilution methods is similar to SEPs but there are two major differences: (a) instead of fractionating TEs in several pools as in SEPs (for example exchangeable, bound to carbonates, organic matter and metal oxides), this fractionates them into two distinct pools, labile and nonlabile; and (b) being less intrusive and more or less direct measurements the ID techniques do not suffer from procedural artefacts as is common in SEPs. Also, these techniques have similarities with chemical extraction-based soil tests used for assessing bioavailability of TEs in soils, as they are intended to provide an estimate of their labile pool, although some strong chemical reagents (for example EDTA) are known to attack nonlabile TE fractions as well. So an isotopic dilution approach in essence is a combined solution and solid-phase fraction technique where the labile pool is comprised of isotopically exchangeable TEs in the solution and those associated with the solid phase, which like chemical extraction-based tests reflects their potential bioavailability.

The technique typically requires mixing an isotope of an element of interest in a soil suspension; after a period of equilibration the elements distribute between its labile and nonlabile pools, which can be determined by measuring the specific activity and concentration of the element for estimating the labile element pool such that:

$$A^*_{\text{Solution}}/[M]_{\text{Solution}} = A^*_{\text{Total}}/[M]_{\text{Labile}} \quad (11.1)$$

$$[M]_{\text{Labile}} = [M]_{\text{Solution}} \times A^*_{\text{Total}}/A^*_{\text{Solution}} \quad (11.2)$$

where A^*_{Total} and A^*_{Solution} are the total activities of the radioisotope mixed in the soil and that measured in the soil solution phase, respectively, and $[M]_{\text{Labile}}$ and $[M]_{\text{Solution}}$ are the concentrations of the labile element in the soil and solution phases, respectively. Equation (11.2) allows the calculation of unknown labile pool of the element, that is $[M]_{\text{Labile}}$, usually known as the E-value. The bioavailability of an element can be assessed by growing plants in the labelled soil and measuring its activity in the plant material; this plant-accumulated labile pool of an element is traditionally known as its L-value. In principle the E- and L-values represent the same pool and should be equal but this is rarely the case as E-values change with time and exchange between soil labile and nonlabile pools is not uncommon due to environmental conditions (E_h, pH and root exudates) as well natural attenuation, as noted above. While the isotopic dilution techniques have been used for assessing TE fixation and ageing effects on their lability (for example [166–168]), the use of such techniques for the purpose of assessing the bioavailability of TEs in soils has largely been limited to a few studies (for example [161,163,169]). This is at least partly due to the potential health/environmental hazards associated with the use of radioisotopes. Also, there is no concrete or consistent evidence of isotopic dilution techniques being superior to other tests, such as DGT, chemical extractions or speciation-based bioavailability soil tests (for example [169–171]). With the availability and analysis of stable isotopes, for example using advanced versions of ICP-MS, it is possible to apply this technique in the field, but it remains to be seen if this offers significant advantages over more widely used techniques such as chemical extraction-based tests, specifically in terms of its ability to predict the bioavailability of TEs in soil.

11.3.4.3 Diffusive Gradient in Thin Films

It has been suggested that measurement of concentrations of TEs present in soil solution can be a more rational and sensitive approach for evaluating their bioavailability [172] compared with the chemical extraction-based procedures. This, however, is not without problems as it provides only a snapshot for the sampling time and cannot account for the dynamic nature of the partition between the solid and solution phases, and has not proved better than conventional soil tests (see Section 11.3.3). Furthermore, as plants continually take up TEs, local depletion in the solution may induce replenishment from the soil, effectively giving the plant access to the labile components associated with the solid phase. Therefore, the labile TE species available to plants or other soil biota can best be measured using a test that is capable of measuring contributions from both soil solution and solid phases and as well can account for any change in TE lability due to prevailing field conditions.

The bioavailability of TEs is related to their flux into the plant roots or cell membrane of soil-dwelling (non-soil-ingesting) organisms, which is dependent on both their concentration in soil solution and their transport rate through the soil. Where there is active removal of TEs, the local soil solution equilibrium concentration is further dependent on the resupply from the solid phase. Quantitative interpretation of this flux of TEs from the solid phase is central to considerations of soil testing for bioavailability purposes, yet it is not assessed by conventional soil testing procedures. Several studies (for example [173,174]) have investigated the bioavailability of nutrients and TEs in soils using ion-exchange resin directly or resin-impregnated membranes and resin imbedded in capsules or dialysis bags. All these procedures respond to a flux from the soil to the resin-sink, but the geometry of these sinks with respect to the soil and the diffusion layer between soil and resin-sink is so poorly defined that this flux from soil solution to the sink cannot be quantitatively interpreted.

Davison and Zhang [175] developed an *in situ* technique capable of quantitatively measuring labile metal species in waters. Further work on this technique, known as DGT, has shown its application for the measurement of labile metal species in soils [176,177]. The application of DGT for the purpose of assessing metal bioavailability has been developed on the premise that in conventional methods of testing soil solution (a) metal speciation may change during sampling and extraction, and (b) the kinetics of metal re-supply from solid phase to solution are not considered. The theoretical background to DGT as well the protocols of its application in soil and the interpretation of the measurements can be found in Zhang *et al.* [176,178,179], Song *et al.* [145], and Hooda and Zhang [180]. A brief description of the DGT technique is also given in Chapter 5.

For shorter deployment periods of 2–3 days, the DGT accumulated TE mass increases linearly with time, and soil moisture in excess of the maximum water holding capacity can prolong this linear response by facilitating in-soil element transport by diffusion [177], effectively operating as a sink, and is therefore somewhat analogous to root uptake. The length of time it can be deployed depends on the capacity of the resin-gel [177,180].

Depletion of metals in soil solution, as a consequence of plant uptake or leaching, results in the re-supply from the solid phase. DGT can provide information about this solid phase to solution re-supply of metals. DGT devices with different diffusive gel thickness can be deployed to create various diffusion gradients and hence different rates of demand for metal ions [180]; also see Chapter 5. The results suggest that the flux from soil to solution is able to maintain a steady-state concentration in the soil solution, providing the DGT demand is kept relatively low by having thicker gel membranes (see Chapter 5). That is, the concentration of metal in the soil solution is well buffered by rapid re-supply from the solid phase [176] as long as the DGT TE uptake is not fast (by having thinner diffusive gel membrane) and the local supply from the solid phase has not been exhausted. Hooda *et al.* [177] found a similar relationship between DGT-measured metal fluxes and the thickness of the diffusive membrane. They further observed that DGT-measured metal fluxes are influenced by soil moisture content. The response was maximal for moisture contents between field capacity, FC and maximum water holding capacity, MWHC (42 % and 53 %, respectively). For moisture contents greater than the MWHC, the DGT-measured fluxes decreased, consistent with a dilution-associated decrease in metal concentrations in soil solution [177,181]. The results also showed an unexpectedly lower response at soil moisture contents lower than FC, which suggested that diffusional transport may be restricted in gel and/or soil at low moisture contents.

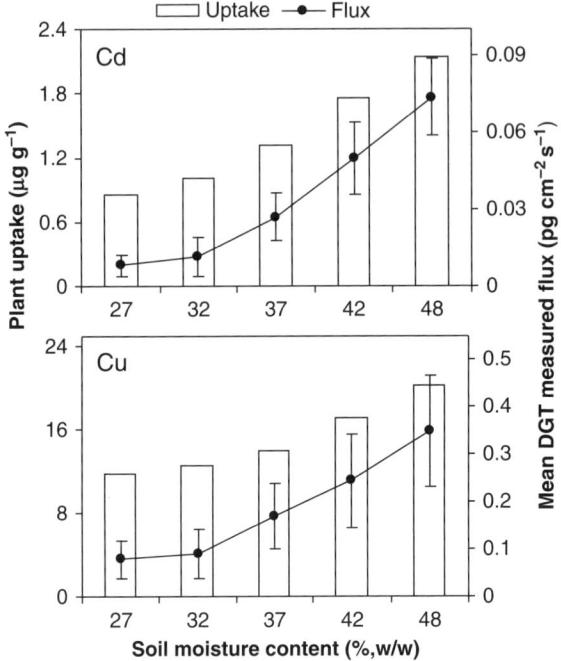

Figure 11.3 Dependence of metal uptake and DGT measured fluxes on soil moisture content (Reproduced with permission from W. Davison, P.S. Hooda, H. Zhang, A.C. Edwards, DGT measured fluxes as surrogates for uptake of metals by plants, Adv. Environ. Res. 2000, 3, 550–555 [181].)

This effect of soil moisture on metal fluxes measured by DGT should also be reflected by plant metal uptake, if the controlling process in soil for the uptake is diffusion. This was demonstrated by a small pot experiment in which soil moisture content was maintained at 27 %, 32 %, 37 %, 42 % and 48 %; the MWHC being 53 %. Cress (*Lepidium sativum* L.) was grown in pots. During the 18-day experiment, DGT deployments were made consecutively, each lasting for 48 hours. Mean DGT metal fluxes obtained during this 18-day period were compared with the concentration of metals in the whole cress plants harvested at the end of the experiment [181]. The integrated DGT flux increased with moisture content, as did the amount of Cd and Cu in the plants (Figure 11.3). The general increase in metal uptake with soil moisture is consistent with other work on nutrient uptake by plants [182]. The results, though based on limited data, demonstrated that the response of fluxes measured by DGT to changes in moisture content is similar to the response of plant uptake. The results from this one simple experiment should not be overinterpreted, but it is instructive to consider the mechanistic similarities of uptake by DGT and plants.

DGT Measurements and Plant Uptake
Zhang *et al.* [178] developed a new concept of expressing DGT measurements as effective soil solution concentration, C_E which incorporates re-supply from soil solid phase. In a greenhouse study of 29 soils, *Lepidium heterophyllum* was grown in pots. Concentrations

of Cu in the plants were plotted against the total concentration of Cu in soil solution, free Cu^{2+} activity, EDTA-extractable Cu and DGT-measured C_E. The best predictor of the Cu concentration in plant shoots was C_E ($r^2 = 0.98$) and it was demonstrated that C_E, rather than free Cu^{2+} activity in soil solution determines uptake of Cu by *L. heterophyllum* under glasshouse/pot conditions [178].

A similar study assessed the bioavailability to wheat (*Triticum aestivum* L.) of Zn, Cd, Pb and Cu in 13 metal-contaminated soils. The TE concentrations in wheat shoots were compared with their 'total' (aqua regia-extractable) soil content, total soil solution concentrations, free activities in soil solution, 0.01 M $CaCl_2$-extractable concentrations, E values measured by isotopic dilution, and DGT-measured C_E. While associations tended to be element specific, DGT-measured C_E generally best predicted accumulation of TEs in wheat shoots, particularly for Cd and Zn [169]. These findings support earlier findings where the C_E best predicted Zn uptake by *Lepidium sativum* [179], and Cu uptake by two plant species, *Elsholtzia splendens* and *Silene vulgaris* [145].

The above findings [169,178,179] clearly demonstrate that kinetically labile solid-phase metal contributes to plant uptake, which, unlike soil solution, free metal activity or isotopic dilution, is included in the DGT measurement. The DGT technique has proved a consistently better predictor of plant available metal since it provides a measure of the labile metals in soil solution as well as their kinetically labile pool associated with the soil solid phase.

Degryse *et al.* [183] provided a seminal review of the application of DGT for TE studies. According to these authors, strong correlations between DGT fluxes and plant uptake as seen in the above-mentioned studies are mostly likely in situations where the diffusive transport of TEs from soil to the plant roots is rate-limiting, and the fluxes may still correlate when plant uptake is not limited by TE diffusive transport. Other competitive ions, however, may affect the plant uptake in the latter situations, which would have no bearing on the DGT-measured fluxes (see Degryse *et al.* [183]). Furthermore, the application of the DGT technique has largely been limited to soils which were previously homogenized either in the laboratory or in pot experiments. Clearly, field application of DGT is important in order to further establish and validate the technique.

11.3.5 Bioaccessibility

Soils contaminated with TEs can also pose risk to human health via direct exposure pathways such as soil ingestion, and dermal contact and inhalation of air-borne particulate material, particularly in urban environments. Among these direct exposure pathways, incidental ingestion of soil is considered a major concern compared with dermal and respiratory pathways [184]. Ingestion of soil may be either deliberate or inadvertent. In the latter case, children, especially toddlers, mouth dirty toys which may be soiled with TE-contaminated soil. Children may also develop soil-pica behaviour – an eating disorder of intentionally and repeatedly ingesting soil, although this is not limited only to children, with potential health implications [185,186].

The risk from soil ingestion to human health is, however, dependent on the amount of the TEs that become available from the ingested soil for potential absorption (bioaccessibility) or their amount that actually enters the bloodstream, that is bioavailability [184].

Furthermore, the mineral nutrition status of soil-ingesting children may also affect the bioavailability of soil-borne TEs. For example, marginal nutritional (Fe, Zn and Ca) status rats when fed with Cd-spiked sunflower or rice-based diets were found to accumulate increased amounts of Cd compared with those with normal nutritional status [187,188].

The risk from ingestion of contaminated soil has traditionally been assessed using *in vitro* and *in vivo* approaches. *In vivo* testing involving the use of animals (for example pigs, rabbits, rats, monkeys) is time consuming and costly and also raises ethical issues, making it less acceptable as a method for assessing potential bioavailability (bioaccessibility) or bioaccumulation of TE contaminants [189]. On the other hand, *in vitro* approaches do not represent the entire physiological process controlling the absorption of nutrients/contaminants in the gastrointestinal (GI) tract or their dissolution from the soil, but are considered useful in assessing broad potential impacts of soil ingestion on human health (for example [30,186,190,191]).

A variety of *in vitro* gastrointestinal methods have been developed and proposed. Rodriguez *et al.* [190] used a two-phase test which involved pH 1.8 at the first phase and pH 5.5 for the second phase, to correspond to the stomach and the small intestine, respectively. These pH values do not reflect a close approximation of the human gastrointestinal system, especially at the intestinal phase. The prevailing pH in the small intestine of the human gastrointestinal tract is about 7 [30], particularly towards the top end.

Ruby *et al.* [30] proposed a physiologically based extraction test (PBET) which is the most used method since it represents a closer resemblance to the GI tract than the previously used tests. The PBET involves two extraction steps that represent the stomach and the small intestine phases with the use of enzymes to stimulate both phases as in the human body. The PBET proposed by Ruby *et al.* [30] and later used by others (for example [192]), has no background nutrients to represent already available nutrients. However, there exists a possible interaction between the ingested soil and the food-borne nutrients that may be already present in the GI tract. Also, Maddaloni *et al.* [193] showed that the inclusion of food lowered the potential bioavailability of soil-borne Pb compared with when the test included no food. The presence or absence of background nutrients in an *in vitro* gastrointestinal test can influence the release–retention processes of soil-borne mineral nutrients and metals [194]. The presence of nutrients is likely to shift the quasi equilibrium towards decreased desorption of contaminants from the soil solid phase; that is, a reduction in potential bioavailability is expected. This means that a PBET with no background nutrients to represent already available nutrients in the GI tract is likely to overestimate the potential bioavailability (bioaccessibility) of ingested soil-borne TEs.

The potential bioavailability, that is, bioaccessibility of metal contaminants is also likely to be influenced by the soil:solution ratio in the test used. The dissolution of the contaminants increases with increase in the soil:solution ratio to a certain extent [189]. Hence the potential bioavailability of ingested soil-borne-TEs is likely to be overestimated if a larger soil:solution ratio than the normal is used in a test [194]. A large soil:solution ratio, for example a soil to test solution ratio of 1:100 (for example [30,191,192]) is larger than the typical human gastrointestinal system solid:fluid ratio of about 1:20, that is by about 5 times [194]. Even though soil:solution ratios ranging from 1:2.5 to 1:5000 have been used in physiologically based extractions [189], it is necessary to use a ratio which closely resembles that of the human gastrointestinal system.

The lack of background nutrient or food representation, combined with the use of much larger soil:solution ratios in currently used physiologically based extraction tests is likely to provide an overestimation of risk from soil ingestion. Furthermore, ingested soil may also release TEs in the small intestine fluid following their absorption, since absorption from the fluid may create a disequilibrium which could result in additional dissolution/desorption of the soil-bound TEs [30]. None of the currently used PBETs can adequately represent this process fully. Assessing the risk of gastrointestinal absorption of TEs from soil ingestion remains a significant challenge, regardless of the approach used (*in vitro* or *in vivo*) (see Chapter 12).

11.3.6 Bioassays, Biosensors and Bioavailability

Environmental risk from exposure to TEs in soils may be expressed as toxicity to plants and soil-dwelling organisms (for example microorganisms, earthworms, nematodes and other invertebrates). High concentrations of TEs in soils can adversely affect soil organisms, and the resulting ecotoxicity can manifest in changes in their community structure and diversity and as well can impact their soil functional activities, for example nutrient mineralization, nitrogen fixation and litter decomposition (for example [3,195,196]). Ecotoxicity tests to assess the effects of chemicals on soil organism are fairly standard techniques, for example for worms [197] or BIOLOG plates for microbial and cell biology analysis.

Considerable progress has been made in understanding the effects of TEs on soil organisms as well as interpretation and application of the various approaches. Chapter 8 reviews TE exposure and effects on soil organisms and their communities. The toxicity or adverse effects of TEs, however, will ultimately be determined by their bioavailability. This section briefly considers bioavailability/bioaccumulation aspects of ecotoxicity. The bioavailability of TEs to soil organisms, like plants, depends upon a number of factors as discussed in previous sections (see Figure 11.1), that is the influence of soil properties as well as the exposure time, and also the overall impacts are often organism specific and metal specific (for example [39,198]).

Bioavailability of TEs to soil organisms can be assessed using indirect and direct approaches. Indirect assessment of potential bioavailability can be made by external chemical measurements, that is by measuring the labile TEs concentrations in soil, and by measuring the functional activities, for example biomass specific-respiration (for example [199,200]) or soil enzyme activities [103,201]. Direct measurements of TEs in soil organisms (for example body concentrations) or their biological effects (for example growth, reproduction, life cycle) can provide a more accurate assessment of bioavailability in terms of the bioaccumulation of an element and its impact compared to indirect approaches. Also, exposure time is important in determining uptake of TEs and their effect on soil organisms. Ecotoxicological studies often involve short exposure time, for example 2–3 weeks for earthworms. This may present difficulties in extrapolation of the findings of such laboratory-based tests to field situations, as it has been found that soil worms may continue to accumulate TEs for their lifespan [202]. Often such studies have been carried out using soils freshly spiked with soluble metal salts. This does not represent real contamination situations, and like plants, such approaches overestimate TE bioavailability to soil organisms (for example [203,204]). Short-term bioassays and those involving freshly spiked soils therefore may not help in assessing the risk that TEs pose to soil organisms under real field situations.

Other work shows that empirical models that incorporate the main soil-related metal partitioning controls [205] can be used to predict TE bioavailability to soil organisms, for example uptake by the compost worm *Eisenia andrei* [206]. As with plants, 'total' concentrations of metals in soil are not likely to provide a reliable measure of their bioavailability to soil organisms or ecotoxicity. Hobbelen *et al.* [198] demonstrated this in a field study of metal (Cd, Cu, Pb and Zn)-polluted floodplain soils where, despite the high level of contamination, concentrations of metals in biota (millipedes, isopods, earthworms and plants) were generally similar to those in uncontaminated reference sites. In this study the concentrations of TEs in the soils clearly overestimated the risk of metal exposure to the soil biota, while the 0.01 M $CaCl_2$-extractable and pore water concentrations were similar to those in uncontaminated areas.

The use of whole-cell microbial biosensors is another approach to assessing bioavailability of TEs. This approach entails the use of genetically engineered microorganisms, designed to produce specific output/signal upon exposure to target chemicals. Biosensors are considered useful alternative to chemical extraction-based assays, at least for assessing bioavailability or toxicity of chemicals to microorganisms, as they offer a number of advantages, for example short exposure time, small sample requirement and ease of use. Such biosensors can potentially also be used for rapid screening purposes prior to more detailed analysis of the samples identified in the bioassay.

A variety of whole-cell bacterial biosensors have been developed for TE and organic contaminants. Currently used biosensors are generally of two broad types: (a) those which provide a general measure of toxicity, and (b) those which are designed to detect and respond to specific chemicals. The first group of biosensors are general effect/toxicity bioassays where the response is based upon overall bioavailable amount of chemicals present in the soil, water or sediment. Microtox^{TM} is one of the most popular biosensors of this type, which utilizes naturally bioluminescent marine bacterium *Vibrio fischeri*. Exposure of the bacterium to chemicals present in the sample results in the reduction of luminescence. More recent variants of this type of biosensors are based on genetically engineered bacteria in which the luminescence increases in the presence of toxic chemicals (for example [207]). The sensor is comprised of two genetic elements – a regulatory element and a reporter element; the former is induced by the target group of toxic chemicals, while the latter element is a gene or a group of genes with readily measurable activity. The two elements are coupled together through a genetic fusion into a desired microbial host cell. In the presence of the target group of chemicals, the amount of reporter protein in the (biosensor) cell increases. Bioluminescence (*lux* genes), green fluorescent protein (*gfp*)-related genes or colour production are some of the most commonly used reporter elements in such general toxicity biosensors [208–210].

Biosensor development for the detection and effects of specific TEs or chemicals has made significant progress. Metal-specific biosensors are based on a similar genetic engineering technique as used for general toxicity biosensors, with a metal-specific regulatory element as the sensing element. Several metal-specific biosensors for the detection of bioavailable TEs have been developed (for example [211,212]). Upon sensing the target element these biosensors produce luminance in a dose–response manner.

Despite certain inherent limitations (see below), bacterial sensors have shown their potential for the purpose of assessing TE bioavailability/toxicity, although most studies have been carried out under laboratory conditions. Ritchie *et al.* [213], using *Escherichia coli* as a biosensor found that EC_{50} (half maximal effective concentration) values were similar to free metal activities (Pb and Zn), supporting the FIAM, albeit in artificial soil solutions. Similar response of *Escherichia coli* and *Pseudomonas fluorescens* biosensors to free Cu ion activity was observed by Vlukan *et al.* [214]. Everhart *et al.* [72] demonstrated successful use of a Ni-specific bacterial biosensor in predicting Ni plant uptake. Ivask *et al.* [215], on the other hand, found that the biosensor's estimated bioavailability of Cd and Pb was several orders of magnitude greater than that based on their pore-water concentrations from 50 agricultural soils sampled near Zn and Pb smelters in northern France.

The use of bacterial biosensors for the purpose of assessing metal bioavailability/toxicity has several inherent and procedural limitations, including adsorption on cell membranes [213], soil matrix interferences [215], luminescence without the presence of the analyte or no luminescence in the presence of the analyte [216], procedural conditions (for example temperature, pH, incubation time and medium) all can affect biosensors' performance [216]. Also, the relatively short lifetime of bacterial biosensors together with limited genetic stability as well as (in some cases) a lack of sensitivity of such biosensors causes significant variability in the response which is difficult to resolve [217,218]. Despite their being-based on genetically engineered microorganisms and the response being specific to their membranes, it is perhaps reasonable to assume that soil microorganisms may respond to metal contaminants in a similar manner. The main difficulty lies in transposing information gathered from biosensors and applying it directly to higher soil organisms and plants [216].

Despite the above-mentioned challenges, whole-cell bacterial biosensors have the potential to become a useful tool for at least initial site screening and post-remediation monitoring. This, however will require further work, including testing in the field.

11.4 Discussion and Conclusions

The existing legislation on TE contaminants in soils is primarily based on their total concentrations. Such maximum permissible (total) concentrations (MPC) in soil are indeed based on good science; for example, the EU Sewage Sludge Directive permits 100 times more Pb than Cd in agricultural soils – a reflection of the limited solubility and hence bioavailability of Pb although both elements are equally toxic. Biological response (bioavailability and bioaccumulation or toxicity) of a TE, however, can be entirely different for the same level of MPC, depending upon soil type and prevailing physical, chemical and biological processes. Furthermore, the recognition that a large fraction of the total soil contaminant content may be present in forms which may not be available to plants, microorganisms or leaching processes has led to much scientific interest in speciation and bioavailability studies. Consequently significant progress has been made in terms of understanding the mechanism of TE bioavailability, the overriding influence of soil properties and environmental conditions, and the significance of the level of contamination. Also, studies on TE bioavailability have created a valuable knowledge base in terms of their uptake, translocation and bioaccumulation behaviours across a wide range of plant species as well as the significance of individual elements and plants species and soil-dwelling organisms.

In addition to measures of the total amount of TE contaminants, secondary tests (bioavailability, bioaccessibility and ecotoxicity) capable of predicting reliable biological response to soil TEs are clearly required. Indeed a plethora of approaches for assessing TE bioavailability to plants (phytoavailability), bioaccessibility in human/animal gut, and toxicity to soil organisms have been proposed and used, but there is no universal test for either of the biological/environmental endpoints. While a single test may provide some qualitative risk assessment, it is highly unlikely that such a test can accurately predict bioavailability of TEs to soil micro- and macroorganisms and plants, at least partly due the route of exposure and uptake mechanism being different.

Bioassays and biosensors for ecotoxicity, functional or life-cycle effects on soil organisms have made considerable progress [208–213]. While bioassays are fairly standard procedures, biosensors are not without problems – so further work is required for their reliability, specificity, and standardization and validation [216–218]. Also, it is perhaps not unreasonable that (genetically engineered) whole-cell biosensors can be surrogates for other, similar indigenous microorganisms. It is difficult to see how such short-lived biosensors' response can be applicable to higher soil organisms and plants. Biosensors can potentially be used for general bioavailability monitoring purposes, for example to assess the effectiveness of a remediation treatment. The work so far, however, has largely been limited to laboratory-based studies. There exist bioassay-derived ecotoxicity/ bioavailability databases for a large number of soil organisms, largely based on the use of soluble metal salts and often with no soil matrix effects. Clearly such measurements overestimate many-fold the risks from soil TEs when compared with field situations [203,204,215]. There is an obvious need to revisit such ecotoxicity database for necessary corrections.

Bioavailability (*in vivo*) or bioaccessibility (*in vitro*) assessment of TE exposure via soil ingestion (animal/human) is problematic. Determination of bioavailability using animal models is time-consuming and difficult to conduct. Currently used bioaccessibility procedures provide some conservative estimate of potential bioavailability, as they lack proper representation of the gastrointestinal tract (see Section 11.3.5), and have not been validated. A further problem arises from the difficulties in estimating the amount of soil ingestion (see Chapter 12), without which it is not possible to assess the risk that exposure to TEs via this route poses to human/animal health.

Assessing bioavailability or toxicity of soil TEs to crop plants faces many challenges. Many chemical extraction based tests have been used. While all of them provide some operationally defined estimate of potentially plant-available soil TEs, some of them extract element fractions which may not have much relevance to plant uptake. Furthermore, they have not been properly standardized/validated, resulting in several variants of any given test in terms of chemical treatment and leaching conditions.

Among the non-extraction-based or new techniques (isotopic dilution, Donnan dialysis system, DGT and solution speciation or rhizosphere-based methods), DGT is the only procedure which is capable of measuring soil solution labile TEs as well their kinetically labile solid-phase-associated pool (for example [169,178,179]). It is important to stress that any test that does not account for the total labile pool (solution + solid phase) is not likely to be a reliable predictor of TE plant uptake. DGT so far has produced the most consistent predictions of plant uptake when compared with other traditional or new techniques (for example [145,169,178,179]). However, DGT is yet to be tested in real situations.

Soil solution speciation, particularly measurement of free ion concentration or activity, is pointless because of its dynamic nature together with the influence of rhizosphere processes and changes in redox potential mediated by environmental conditions. It should also be noted that free metal ion activity/concentration is largely operationally defined. Furthermore, nonionic (complex) species also contribute to uptake perhaps either by dissociation at the root–soil interface or directly as such [33,141–143].

The use of unbuffered neutral salt solutions, particularly $CaCl_2$, $NaNO_3$ and NH_4NO_3, at present offers an easy-to-use and pragmatic approach. These tests generally correlate better with the uptake of TEs by plants than other chemical extraction tests. $NaNO_3$ (0.1 M), NH_4NO_3 (1.0 M) and $CaCl_2$ (0.01 M) have already been adopted in Germany, Switzerland and the Netherlands, respectively [67,76,77]. The $CaCl_2$ test generally has a wider support among researchers (for example [44,65,78]) due to its similar concentration to soil solution and thus lower likelihood of modifying soil-specific existing chemical, physical and biological controls on soil TEs. Also, divalent Ca is better suited to release soil solid-phase associated labile TEs. The tests have been standardized as discussed previously [75]. They have, however, not been robustly tested, so there is need to validate these tests in real field conditions across many different soils and crop plant species. It is possible that such a validation exercise may provide a reliable test capable of predicting soil TE bioavailability to crop plants. Unlike the current situation where suitability or otherwise of a test is decided mainly on the basis of correlation between uptake of TEs and that extracted or measured by a test, researchers in the field need to develop empirical or semi-empirical models, capable of predicting the probability of exceeding TE- and plant/biota-specific threshold or safe concentrations. Since soil characteristics (for example pH, CEC, and clay, organic matter and Fe–Mn oxides) play a key role in determining bioavailability of soil TEs, incorporation of these soil characteristics in such a model, for example based on 0.01 M $CaCl_2$-extractable TEs could further improve bioaccumulation or toxicity predictions [52,219]. Without such a robust model, it is difficult to see how a bioavailability risk-based regulatory framework for soil TEs could be developed.

References

[1] McLaughlin, M.J.; Zarcinas, B.A.; Stevens, D.P.; Cook, N., Soil testing for metals; Commun. Soil Sci. Plant Anal. 2000, 31, 1661–1700.
[2] Hunter, B.A.; Johnson, M.S.; Thompson, D.J., Ecotoxicology of copper and cadmium in a contaminated grassland ecosystem. 2. Invertebrates; J. Appl. Ecol. 1987, 24, 587–599.
[3] Spurgeon, D.J.; Hopkin, S.P., Seasonal variation in the abundance, biomass and biodiversity of earthworms in soils contaminated with metal emissions from a primary smelting works; J. Appl. Ecol. 1999, 36, 173–183.
[4] West, T.S., Soil as a source of trace elements; Phil. Trans., R. Soc. Lond. 1981, B294, 19–30.
[5] Tessier, A.; Campbell, P.G.C., Partitioning of trace metals in sediments; in: J.R. Kramer, H.E. Allen (eds), Metal Speciation: Theory, Analysis and Application; Lewis Publisher, Chelsea, MI, 1988, pp. 183–199.

[6] Thornton, I., Metals in the Global Environment: Facts and Misconceptions; Ottawa, International Council on Metals and the Environment, 1995.
[7] Young, S.D.; Zhang, H.; Tye, A.M.; Maxted, A.; Thums, C.; Thornton, I., Characterizing the availability of metals in contaminated soils. 1. The solid phase: sequential extractions and isotopic dilution; Soil Use Manage. 2006, 21: 450–458.
[8] Lindsay, W.L., Chemical Equilibria in Soils; John Wiley & Sons, Inc., New York, 1979.
[9] Alloway, B.J.; Jackson, A.P., The behaviour of heavy-metals in sewage sludge-amended soils; Sci. Total Environ. 1991, 100, 151–176.
[10] Alloway B.J., Heavy Metals in Soils, 2nd edn; Blackie Academic and Professional, London, 1995.
[11] Hooda, P.S.; McNulty, D.; Alloway, B.J.; Aitken, M.N., Plant availability of heavy metals in soils previously amended with heavy applications of sewage sludge; J. Sci. Food Agric. 1997, 73, 446–454.
[12] Xue, H.; Nhat, P.H.; Gachter, R.; Hooda, P.S, The transport of Cu and Zn from agricultural soils to surface water in a small catchment; Adv. Environ. Res. 2003, 8, 69–76.
[13] Alvarez-Ayuso, E.; Garcia-Sanchez, A.; Querol, X.; Moyano, A., Trace element mobility in soils seven years after the Aznalcollar mine spill; Chemosphere 2008, 73, 1240–1246.
[14] Zhao, L.Y.L.; Schulin, R.; Nowack, B., Copper and Zn mobilization in soil columns percolated by different irrigation solutions; Environ. Pollut. 2009, 157: 823–833.
[15] McLaren, R.G.; Williams, J.G.; Swift, R.S., Some observations on the desorption and distribution behaviour of copper with soil components; J. Soil Sci. 1983, 32: 325–331.
[16] McBride, M.B., Reactions controlling heavy metal solubility in soil; Adv. Soil Sci. 1989, 10, 1–56.
[17] Schindler, P.W.; Sposito, G., Surface complexation at (hydro)oxide surfaces; in: G.H.E.A. Bolt (ed.), Interactions at the Soil Colloid–Soil Solution Interface; Kluwer Academic Publishers, Dordrecht, 1991, pp 115–145.
[18] Naidu, R.; Harter, R.D., Effect of different organic ligands on cadmium sorption and extractability from soils; Soil Sci. Soc. Am. J. 1998, 62, 644–650.
[19] Rieuwerts, J.J.; Thornton, I.; Farago, M.E.; Ashmore, M.R., Factors influencing metal bioavailability in soils: preliminary investigations for development of a critical loads approach for metals; Chem. Spec. Bioavail. 1998, 10, 61–75.
[20] Harrison, R.M., Chemical association of Pb, Cd, Cu and Zn in street dusts and roadside soils; Environ. Sci. Technol. 1981, 15, 1378–1383.
[21] Xian X., Effects of pH on chemical forms and plant availability of cadmium, zinc and lead in polluted soils; Water Air Soil Polut. 1989, 45: 265–273.
[22] Semple, K.T.; Doick, J.K.; Jones, K.C.; Bureauel, P.; Craven, A.; Harms, H., Defining bioavailability bioaccessibility of contaminated soil and sediments is complicated; Environ. Sci. Technol. 2004, 38, 228A–231A.
[23] NEPC, National Environmental Protection Measures for the Assessment of Site Contamination; National Environmental Protection Council (NEPC) Service Corporation, Adelaide, 1999.
[24] Nolan, A.L.; Lombi, E.; McLaughlin, M.J., Metal bioaccumulation and toxicity in soils – why bother with speciation; Aust. J. Chem. 2003, 56, 77–91.
[25] Naidu, R.; Semple, K.T; Megharaj, M. et al., Bioavailability: definition, assessment, and implications for risk assessment; in: R. Naidu et al. (eds), Chemical Bioavailability in Terrestrial Environments, Development in Soil Science, Vol. 32; Elsevier, London, 2008, Chapter 3.
[26] Naidu, R.; Bolan, N.S.; Megharaj, M. et al. (eds), Chemical Bioavailability in Terrestrial Environments, Development in Soil Science, Vol. 32; Elsevier, London, 2008.
[27] Adriano, D.C., Trace Elements in Terrestrial Environments: Biogeochemistry, Bioavailability and Risk of Metals; Springer-Verlag, New York, 2001.
[28] Naidu, R.; Rogers, S.; Gupta, V.V.S.R.; Kookana, R.S.; Bolan, N.S.; Adriano, D., The bioavailability of metals in soil-plant environment and its potential role in risk assessment: an overview; in: R. Naidu, S. Rogers, V.V.S.R. Gupta, R.S. Kookana, N.S. Bolan, D. Adriano (eds), Bioavailability and Its Potential Role in Risk Assessment; Science Publishers, New York, 2003, pp. 21–59.

[29] Alexander, M., Ageing, bioavailability, and overestimation of risk from environmental pollutants; Environ. Sci. Technol. 20, 2000, 4259–4265.
[30] Ruby, M.V.; Schoof, W.B.; Eberle, R. *et al.*, Estimation of lead and arsenic bioavailability using a physiologically based test; Environ. Sci. Technol. 1996, 30, 422–430.
[31] Ruby, M.V.; Schoof, W.B.; Brattin, W. *et al.*, Advances in evaluating the oral bioavailability of inorganics in soil for use in human health risk assessment; Environ. Sci. Technol. 1999, 33, 3697–3705.
[32] Sposito, G. The Chemistry of Soils; Oxford University Press, New York, 1989.
[33] Jiang, X.J.; Luo, Y.M.; Zhao, Q.G.; Baker, A.J.M.; Christie, P.; Wong, M.H., Soil Cd availability to Indian mustard and environmental risk following EDTA addition to Cd-contaminated soil; Chemosphere 2003, 50, 813–818.
[34] López, M.L.; Peralta-Videa, J.R.; Benitez, T.; Gardea-Torresdey, J.L., Enhancement of lead uptake by alfalfa (*Medicsgo sativa*) using EDTA and a plant growth promoter; Chemosphere 2005, 61, 595–598.
[35] Hooda, P.S.; Alloway, B.J., Changes in operational fractions of trace metals in two soils during a two years reaction time following treatment with sewage sludge; Int. J. Anal. Environ. Chem. 1994, 57, 289–311.
[36] Lin, Q.; Chen, Y.X.; He, ; He, Y.F.; Tian, G.M., Root induced changes of lead availability in the rhizosphere of *Oryza sativa* L.; Agric. Ecosyst. Environ. 2004, 104, 605–613.
[37] Ruiz, E.; Rodríguez, L.; Alonso-Azcárate, J., Effects of earthworms on metal uptake of heavy metals from polluted mine soils by different crops; Chemosphere 2009, 75, 1035–1041.
[38] NRC (National Research Council), Bioavailability of Contaminants in Soils and Sediments: Processes, Tools and Applications; National Academic Press, Washington DC, 2002.
[39] Peijnenburg, W.J.G.M.; Baerselman, R.; de Groot, A.C.; Jager, T.; Posthuma, L.; Van Veen, R.P.M., Relating environmental availability to bioavailability: soil type-dependent metal accumulation in the *Oligochaete Eisenia Andrei*; Ecotoxicol. Environ. Saf. 1999, 44, 294–310.
[40] McCarty, L.S.; Mackay, D., Enhancing ecotoxicological modeling and assessment, body residues and modes of action; Environ. Sci. Technol. 1993, 27, 1719–1728.
[41] Parker, D.R.; Chaney, R.L.; Norvell, W.A., Chemical equilibrium models: applications to plant nutrition research; in: R.H. Leoppert, A.P. Schwab, S, Goldberg (eds), Chemical Equilibrium and Reaction Models; American Society of Agronomy, Madison, WI, 1995, pp. 163–200.
[42] Templeton, D.M.; Ariese, F.; Cornerlis, R. *et al.*, Guidelines for terms related to chemical speciation and fractionation of elements. Definitions, structural aspects and methodological approaches (IUPAC Recommendations); Pure Appl. Chem. 2000, 72, 1453–1470.
[43] Ure, A.M.; Davidson, C.M., Chemical Speciation in the Environment, *2nd edn*; Blackwell Science Ltd, Oxford, 2002.
[44] Peijnenburg, W.G.M.; Zablotskaja, M.; Vijver, M.G, Monitoring metals in terrestrial environments within a bioavailability framework and a focus on soil extraction; Ecotoxocol. Environ. Saf. 2007, 67, 163–179.
[45] Zhang, H.; Young, S.D., Characterizing the availability of metals in contaminated soils. II. The soil solution; Soil Use Manage. 2006, 21, 459–467.
[46] Lindsay, W.L.; Norvell, W.A., Development of a DTPA soil test for zinc, iron, manganese and copper; Soil Sci. Soc. Am. J. 1978, 42, 421–428.
[47] O'Connor, G.A., Use and misuse of the DTPA soil test; J. Environ. Qual. 1988, 17, 715–718.
[48] Korcak, R.F.; Fanning, D.S., Extractability of cadmium, copper, nickel and zinc by double acid versus DTPA and plant content at excessive soil levels; J. Environ. Qual. 506–512.
[49] Merry, R.H.; Tiller, K.G.; Alston, A.M., The effects of soil contamination with copper, lead and arsenic on the growth and composition of plants; Plant Soil 1986, 95, 255–269.
[50] Hooda, P.S.; Alloway, B.J., The plant availability and DTPA extractability of trace metals in sludge-amended soils; Sci. Total Environ. 1994, 149, 39–51.
[51] Feng, M.-H.; Shan, X.-O.; Zhang, S.; Wen, B., A comparison of the rhizosphere-based method with DTPA, EDTA, $CaCl_2$ and $NaNO_3$ extraction methods to for prediction of bioavailability of metals in soil to wheat; Chemosphere 2005, 59, 939–949.
[52] Wang, X-P.; Shan, X-Q.; Zhang, S-Z.; Wen, B., A model foe evaluation of phytoavailability of trace elements to vegetables under field conditions; Chemosphere 2004, 55, 811–822.

[53] Viro, P.J., Use of ethylenediaminetetraacetic acid in soil analysis. 1. Experimental; Soil Sci. 1955, 79, 459–465.
[54] Clayton, P.M.; Tiller, K.G., A chemical method for the determination of heavy metal content in soils in environmental studies; CSIRO Australian Division of Soils Technical Paper 41, 1979, pp. 1–17.
[55] Anonymous; The Analysis of Agricultural Materials; Ministry of Agriculture, Food and Fisheries Reference Book 427; HMSO, London, 1986.
[56] Hammer, D.; Keller, C., Changes in the rhizosphere of metal accumulating plants evidenced by chemical extractants; J. Environ. Qual. 2002, 31, 1561–1569.
[57] Chaignon, V.; Sanchez-Neira, I.; Herrmann, P.; Jaillard, B.; Hinsinger, P., Copper bioavailability and extractability as related to chemical properties of contaminated soils from a vine growing area; Environ. Pollut. 2003, 123, 229–238.
[58] Feng, M.-H.; Shan, X.-O.; Zhang, S.-Z.; Wen, B., Comparison of a rhizosphere-based method with other one-step extraction methods to for assessing the bioavailability of metals in soil to barley; Environ. Pollut. 2005, 137, 231–240.
[59] Payá-Pérez, A.; Sala, J.; Mousty, F., Comparison of ICP-AES and ICP-MS for the analysis of trace elements in soil extracts; Int. J. Environ. Anal. Chem. 1993, 51, 223–230.
[60] Bermond, A.; Yousfi, I.; Ghestem, J.P., Kinetic approach to the chemical speciation of trace metals in soils; Analyst 1998, 123, 785–789.
[61] Haq, A.U.; Bates, T.E.; Soon, Y.K., Comparison of extractants for plant-available zinc, cadmium, nickel, and copper in contaminated soils; Soil Sci. Soc. Am. J. 1980, 44, 772–777.
[62] Krishnamurti, G.S.R.; Huang, P.M.; van Rees, K.C.J.; Kozak, L.M.; Rostad, H.P.W., A new soil tests method for determination of plant-available cadmium in soils; Commun. Soil Sci. Plant Anal. 1995, 26, 2857–2867.
[63] Garrett, R.G.; MacLaurin, A.I; Gawalko, E.J.; Tkachuk, R.; Hall, G.E.M., A predction model for estimating the cadmium content of durum wheat from soil chemistry; J. Geochem. Explor. 1998, 64, 101–110.
[64] Houba, V.J.G.; Novozamsky, I.; Lexmond, T.M.; vander Lee, J.J., Applicability of 0.01 m $CaCl_2$ as a single extraction solution for the assessment of the nutrient status of soils and other diagnostic purposes; Commun. Soil Sci. Plant Anal. 1990, 21, 2281–2290.
[65] Novozamsky, I.; Lexmond, T.M.; Houba, V.J.G, A single extraction procedure of soil for evaluation of uptake of some heavy-metals by plants; Int. J. Environ. Anal. Chem. 1993, 51, 47–58.
[66] Aten, C.F.; Gupta, S.K., On heavy metals in soil; rationalization of extractions by dilute salt solutions, comparison of the extracted concentrations with uptake by ryegrass and lettuce, and the possible influence of pyrophosphate on plant uptake; Sci. Total Environ. 1996, 178: 45–53.
[67] Houba, V.J.G; Lexmond, T.M.; Novozamsky, I.; vander Lee, J.J., State of the art and future developments in soil analysis for bioavailability assessment; Sci. Total Environ. 1996, 178, 21–28.
[68] Pueyo, M.; López-Sánchaez, J.F.; Rauret, G., Assessment of $CaCl_2$, $NaNO_3$ and NH_4NO_3 extraction procedures for the study of Cd, Cu, Pb and Zn extractability in contaminated soils; Anal. Chim. Acta 2004, 504, 217–226.
[69] Ullrich, S.M.; Ramsey, M.H.; Helois-Rybicka, E., Total and exchangeable concentrations of heavy metals in soils near Bytom, an area in Upper Silesia, Poland; Appl. Geochem. 1999, 14, 187–196.
[70] Bhogal, A.; Nicholson, F.A.; Chambers, B.J.; Shepherd, M.A., Effects of past sewage sludge additions on heavy metal aavailability in light textured soils: implications for crop yields and metal uptake; Environ. Pollut. 2003, 121, 413–423.
[71] Chojnacka, K.; Chojnacki, A.; Górecki, H., Bioavailability of heavy metals from polluted soils to plants; Sci. Total Environ. 2005, 337: 175–182.
[72] Everhart, J.L.; McNear Jr., D.; Peltier, E.; van der Leile, D.; Chaney, R.L.; Sparks, D.L, Asessing nickel bioavailability in smelter-contaminated soils; Sci. Total Environ. 2006, 367, 732–744.

[73] Lebourg, A.; Sterckeman, T.; Ciesielski, H.; Proix, N., Trace metal speciation in three unbuffered salt solutions used to assess their bioavailability in soil; J. Environ. Qual. 1998, 27, 584–590.
[74] Mench, M.; Baize, D.; Mocquot, B., Cadmium availability to wheat in five soil series from the Yonne District, Burgundy, France; Environ. Pollut. 1997, 95, 93–103.
[75] Queavauviller, P.H.; Rauret, G.; Ure, A.; Bacon, J.; Muntau, H., EUR Report 17127 EN, European Commission, Brussels 1997.
[76] Bo, V.S., Verordnung über Schadstoffgehalt im Boden, Swiss Ordinance on Pollution in Soils Nr.814.12; Publ. Eidg. Drucksachen und Materialzentrale (EDMZ), 3000, Bern, 1986, p. 1.
[77] DIN, Deutsches Institut für Normung, Bodenuntersuchungs; Extraktion von Spurenelemente mit Ammonium-nitralösung. Vornorm DINV 19730; DIN Boden-Chemische Bodenuntersuchungs – verfahren, Berlin, 1995.
[78] Houba, V.J.G.; Temmingoff, E.J.M.; Gaikhorst, G.A.; Van Vark, W., Soil analysis procedures using 0.01M calcium chloride as extraction reagent; Commun. Soil Sci. Plant Anal. 2000, 31, 1299–1396.
[79] Sauvé, S.; Cook, N.; Hendershot, W.H.; McBride, M.B., Linking plant tissue concentrations and soil copper pools in urban contaminated soils; Environ. Pollut. 1996, 94, 153–157.
[80] Chang, A.C.; Crowley, D.E.; Page, A.L., Assessing bioavailability of metals in biosolids-treated soils; WERF Report 97-REM-5, IWA, 2004.
[81] McBride, M.B.; Tyler, L.D.; Hovde, D.A., Cadmium adsorption by soils and uptake by plants as affected by soil chemical-properties; Soil Sci. Soc. Am. J. 1981, 45, 739–744.
[82] Soon, S.K.; Bates, T.E., Chemical pools of cadmium, nickel and zinc in polluted soils and some preliminary indications of their availability to plants; J. Soil Sci. 1982, 33, 477–488.
[83] Tessier, A.; Campbell, P.G.C.; Bisson, M., Sequential extraction procedure for the speciation of trace metals; Anal. Chem. 1979, 51, 844–851.
[84] Stover, R.C.; Sommers, L.E.; Silviera, D.J., Evaluation of metals in wastewater-sludge; J. Water Pollut. Control Fedn. 1976, 48, 2165–2175.
[85] Hlavay, J.; Prochasks, Y.; Weisz, M.; Wenzel, W.M.; Stringerder, G.J., Determination of trace elements bound to soils and sediment fractions, IUPAC Technical Report; Pure Appl. Chem. 2004, 76, 415–442.
[86] Sheppard, M.I.; Stephenson, M., Critical evaluation of selective extraction methods for soils and sediments; in: R. Prost (ed.), Contaminated Soils, 3rd International Conference on the Biogeochemistry of Trace Elements; Institut National de al Recherche Agronomique, 1997, pp. 69–97.
[87] Miller, W.P.; Martens, D.C.; Zelazny, L.W., Effect of sequence in extraction of trace-metals from soils; Soil Sci. Soc. Am. J. 1986, 50, 598–601.
[88] Bermond, A., Limits of sequential extraction procedures re-examined with emphasis on H^+ ion activity; Anal. Chim. Acta, 2001, 445, 79–88.
[89] Kheboian, C.; Bauer, F., Accuracy of selective extraction procedures for metal special in model aquatic sediments; Anal. Chem. 1987, 59, 1417–1423.
[90] Bermond, A.; Yousfi, I., Reliability of comparisons based on sequential extraction procedures applied to soil samples: the thermodynamic point of view; Environ. Technol. 1997, 18, 219–224.
[91] Queavuviller, P.H., Rauert, G.; Griepink, B., Single and sequential extraction in sediments and soils; Int. J. Anal. Chem. 1993, 51, 231–235.
[92] Ure, A.M., Single extraction schemes for soil analysis and related applications; Sci. Total Environ. 1996, 178, 3–10.
[93] Clark, M.W.; Davies-McConchie, F.; McConchie, D.; Birch, G.F., Selective chemical extraction and grain size normalisation for environmental assessment of anoxic sediments: validation of an integrated procedure; Sci. Total Environ. 2000, 258, 149–170.
[94] Queavuviller, P.H., Operationally defined extraction procedures for soil and sediment analysis. 1. Standardisation; Trends Anal. Chem. 1998, 17, 289–298.
[95] Queavuviller, P.H., Operationally defined extraction procedures for soil and sediment analysis. 2. Certified reference materials; Trends Anal. Chem. 1998, 17, 632–642.

[96] Lu, A.; Zhang, S.; Shan, X.-Q., Time effect on the fractionation of heavy metals in soils; Geoderma 2005, 125, 225–234.
[97] Jalali, M.; Khanlari, Z.V.; Effect of aging processes on the fractionation of heavy metals in some calcareous soils of Iran; Geoderma 2008, 143, 26–40.
[98] Qian, J.; Wang, Z.-J.; Shan, X.-Q.; Tu, Q.; Wen, B.; Chen, B., Evaluation of plant availability of soil trace metals chemical reaction and multiple regression analysis; Environ. Pollut. 1996, 91, 309–315.
[99] Maiz, I.; Arambarri, I.; Garcia, R.; Millán, E., Evaluation of heavy metal availability in polluted soils by two sequential extraction procedures using factor analysis; Environ. Pollut. 2000, 110, 3–9.
[100] Li, J.X.; Yang, X.E.; He, Z.L.; Jilani, G.; Sun, C.Y.; Chen, S.M.; Fractionation of lead in paddy soils and its bioavailability to rice plants; Geoderma 2007, 141, 174–180.
[101] Obrador, A.; Alvarez, J.M.; Lopez-Valdivia, L.M.; Gonazalez, D.; Novillo, J.; Rico, M.I., Relationship of sol properties with Mn and Zn distribution in acidic soils and their uptake by a barley crop; Geoderma 2007, 137, 423–443.
[102] Guerra, P.; Ahumada, I.; Carrasco, A., Effects of biosolid incorporation to mollisol soils on Cr, Cu, Ni, Pb and Zn fractionation and relationship with their bioavailability; Chemosphere 2007, 68, 2021–2027.
[103] Bhattacharyya, P.; Tripathy, S.; Chakrabarti, K.; Chakraborty, A.; Banik, P., Fractionation and bioavailability of metals and their impacts on microbial properties in sewage irrigated soil; Chemosphere 2008, 72, 543–550.
[104] Filgueiras, A.V.; Lavilla, I.; Bendicho, C., Chemical sequential extraction for metal partitioning in environmental solid samples; J. Environ. Monit. 2002, 823–857.
[105] Bacon, J.R.; Hewitt, I.J.; Cooper, P., Reproducibility of the BCR sequential extraction procedure in a long-term study of the association of heavy metals with soil components in an upland catchment in Scotland; Sci. Total Environ. 2005, 337, 191–205.
[106] Singh, S.P.; Tack, F.M.; Verloo, M.G., Heavy metal fractionation and extractability in dredged sediment derived surface soils; Water Air Soil Pollut. 1998, 102, 313–328.
[107] Barber, S.A., Soil Nutrient Bioavailability: A Mechanistic Approach; John Wiley & Sons, Inc., New York, 1984.
[108] Plette, A.C.C.; Nederlof, M.M.; Temminghoff, E.J.M.; van Riemsdijk, W.H; Bioavailability of heavy metals in terrestrial and aquatic systems: a quantitative approach; Environ. Toxicol. Chem. 1999, 18, 1882–1890.
[109] Pavan, M.A.; Bingham, F.T.; Pratt, P.F., Toxicity of aluminium to coffee in ultisols and oxisols ammended with $CaCO_3$, $MgCO_3$ and $CaSO_4.2H_2O$; Soil Sci. Soc. Am. J. 1982, 46, 1201–1207.
[110] Morel, F.M.M, Principles of Aquatic Chemistry; John Wiley & Sons, Inc., New York, 1983.
[111] Parker D.R.; Pedler, J.F., Reevaluating the free-ion activity model of trace metal availability to higher plants; Plant Soil 1997, 196, 223–228.
[112] Allen, H.E.; Hall, R.H.; Brisbin, T.D., Metal speciation. Effects on aquatic toxicity; Environ. Sci. Technol. 1980, 14, 441–443.
[113] Minnich, M.M.; McBride, M.B.; Chaney, R.L., Copper activity in soil solution: II. Relation to copper accumulation in young snapbeans; Soil Sci. Soc. Am. J. 1987, 51, 573–578.
[114] Sauvé, S.; McBride, M.B.; Hendershot, W.H., Ion-selective electrode measurements of Cu(II) actiovity in contaminated soils; Arch. Environ. Contam. Toxicol. 1995, 29, 373–379.
[115] Christ, I.; Miline, C.J.; Kinnibiergh, D.K.; Knetzschmar, R., Relating ion binding by fulvic and humic acids to chemical composition and molecular size. 2. Metal binding; Environ. Sci. Technol. 2001, 35, 2512–2517.
[116] Kaschl, A.; Romheld, V.; Chen, Y., Cadmium binding by fractions of dissolved organic matter and humic substances from municipal solid waste compost; J. Environ. Qual. 2002, 31, 1885–1892.
[117] Morley, G.F.; Gudd, G.M., Sorption of toxic metals by fungi and clay minerals; Mycol. Res. 1995, 99, 1429–1438.
[118] Florence, T.M., Electrochemical approaches to trace element speciation in waters: a review; Analyst 1986, 111, 489–505.

[119] Sauvé, S.; Norvell, W.A.; McBride, M.B.; Hendershot, W.H., Speciation and complexation of cadmium in extracted soil solutions; Environ. Sci. Technol. 2000, 34, 291–296.
[120] Persaud, G.; Cantwell, F.F, Determination of free magnesium-ion concentration in aqueous-solution using 8-hydroxyquinoline immobilized on a nonpolar adsorbent; Anal. Chem. 1992, 64, 89–94.
[121] Helmke, P.A.; Salam, A.K.; Li, Y., Measurement and behavior of indigenous levels of the free hydrated cations of Cu, Zn and Cd in the soil-water system; in: R. Prost (ed.), Contaminated Soils; INRA Editions number 85, Versailles, 1997.
[122] Temminghoff, E.J.M.; Plette, A.C.C.; Van Eck, R.; Van Riemsdijk, W.H., Determination of the chemical speciation of trace metals in aqueous systems by the Wageningen Donnan Membrane Technique; Anal. Chim. Acta 2000, 417, 149–157.
[123] Hill, S.J.; Pitts, L.J.; Fisher, A.S., High-performance liquid chromatography-isotope dilution inductively coupled plasma mass spectrometry for speciation studies: an overview; TrAC-Trends Anal. Chem. 2000, 19, 120–126.
[124] Weng, L.P.; Temminghoff, E.J.M.; Lofts, S.; Tipping, E.; van Riemsdijk, W.H., Complexation with dissolved organic matter and solubility control of heavy metals in a sandy soil; Environ. Sci. Technol., 2002, 36, 4804–4810.
[125] Weng, L.P.; Temminghoff, E.J.M.; Lofts, S.; Tipping, E.; van Riemsdijk, W.H., Complexation with dissolved organic matter and solubility control of heavy metals in a sandy soil; Environ. Sci. Technol., 2002, 36, 4804–4810.
[126] Bolton, K.A.; Sjöberg, S.; Evans, L.J.; Proton binding and cadmium complexation constants for a soil humic acid using a quasi-particle model; Soil Sci. Soc. Am. J., 1996, 60, 1064–1072.
[127] Tipping, E., Berggren, D., Mulder, J., Woof, C., Modelling the solid-solution distributions of protons, aluminium, base cations and humic substances in acid soils; Eur. J. Soil Sci. 1995, 46, 77–94.
[128] Benedetti, M.F., van Riemsdijk, W.H., Koopal, L.K., Kinniburgh, D.G., Gooddy, D.C., Milne, C.J., Metal ion binding by natural organic matter: from the model to the field; Geochim. Cosmochim. Acta 1996, 60, 2503–2513.
[129] Lumsdon, D.G.; Evans, L.J., Predicting chemical speciation and computer simulation; in: A.M. Ure, C.M. Davidson (eds), Chemical Speciation in the Environment; Blackie Academic, London, 1995, pp. 86–134.
[130] Tipping, E., 1998. Humic Ion-binding Model VI: an improved description of the interactions of protons and metal ions with humic substances; Aqu. Geochem. 4, 3–48.
[131] Kinniburgh, D.G., Van Riemsdijk, W.H., Koopal, L.K., Borkovec, M., Benedetti, M.F., Avena, M.J., Ion binding to natural organic matter: competition, heterogeneity, stoichiometry and thermodynamic consistency; Colloids Surf., A 199, 151, 147–166.
[132] Weng, L.P,; Temminghoff, E.J.M.; van Riemsdijk, W.H., Contribution of individual sorbents to the control of heavy metal activity in sandy soil; Environ. Sci. Technol., 2001, 35, 4436–4443.
[133] Weng, L.P.; Wolthoorn, A.; Lexmond, T.M.; Temminghoff, E.J.M.; van Riemsdijk, W.H., Understanding the effects of soil characteristics on phytotoxicity and bioavailability of nickel using speciation models; Environ. Sci. Technol., 2004, 38, 156–162.
[134] Meers, E.; Unamuno, V.R.; Du Laing, G. *et al.*, Zn in the soil solution of unpolluted and polluted soils as affected by soil characteristics; Geoderma 2006, 136, 107–119.
[135] Anderson, M.A.; Marel, F.M.M.; Guillard, R.R.L., Growth limitation of coastal diatoms by low zinc ion activity; Nature 1978, 276, 70–77.
[136] Sparks, D.L., Ion activities: an historical and theoretical overview; Soil Sci. Soc. Am. J. 1983, 48, 514–518.
[137] Campbell, P.G.C., Interactions between trace metals and aquatic organisms: a critique of the free ion activity model; in: A. Teessier, D.R. Turner (eds), Metal Speciation and Bioavailability in Aquatic Systems; John Wiley & Sons, Ltd, Chichester, 1995, pp 45–102.
[138] Playle, R.C.; Modelling metal interactions at fish gills; Sci. Total Environ. 1998, 219, 147–163.

[139] Bell, P.F.; Chaney, R.L.; Angle, J.S., Free metal activity and total metal concentrations as indices of micronutrient availability to barley [*Hordeum vulgare* (L.) 'Klages']; Plant Soil 1991, 130, 51–62.

[140] Cheng, T.; Allen, H.E., Prediction of uptake of copper from solution by lettuce (*Lactuca sativa* Romance; Environ. Toxicol. Chem. 2001, 20, 2544–2551.

[141] Krishnamurti, G.S.R.; Cieslinski, G.; Haung, P.M.; Rees, K.C.J., Kinetics of cadmium release from soils as influenced by organic acids: implication of cadmium availability; J. Environ. Qual. 1997, 26, 271–277.

[142] McLaughlin, M.J; Tiller, K.G.; Beech, T.A.; Smart, M.K., increased soil-salinity causes elevated cadmium concentrations in field-grown potato-tubers; J. Environ. Qual. 1994, 23, 1013–1018.

[143] Weggler, K.; McLaughlin, M.J.; Graham, R.D., Effect of chloride in soil solution on the plant availability of biosolid-borne cadmium; J. Environ. Qual., 2004, 33, 496–504.

[144] Ge, Y.; Murray, P.; Hendershot, W.H., Trace metal speciation and bioavailability in urban soils; Environ. Pollut. 2000, 107, 137–144.

[145] Song, J.; Zhao, F.-J.; Luo, Y.-M.; McGrath, S.P.; Zhang, H., Copper uptake by *Elsholtzia splendens* and *Silene vulgaris* and assessment of copper phytoavailability in contaminated soils; Environ. Pollut. 2004, 128, 307–315.

[146] Lorenz, S.E.; Hammon, R.E.; Holm, P.E.; Domingues, H.C.; Sequeira, E.M., Cadmium and zinc in plants and soil solution from contaminated soils; Plant Soil 1997, 189, 21–31.

[147] Leyval, C.; Berthelin, J., Rhizodeposition and net release of soluble organic compounds by pine and beech seedlings inoculated with rhizobacteria and ectomycorrhizal fungi; Biol. Fertil. Soil 1993, 15, 259–267.

[148] Morel, J.L.; Mench, M.; Guckert, A., Measurement of Pb^{2+}, Cu^{2+}, and Cd^{2+} binding with mucilage exudates from maize (*Zea mays* L.) roots; Biol. Fertil. Soil 1986, 2, 29–34.

[149] Mench, M.; Morel, J.L.; Guchert, A., Metal binding properties of high molecular weight soluble exudates from maize (*Zea mays* L.) roots; Biol. Fertil. Soil 1987, 3, 165–169.

[150] Mench, M.; Morel, J.L.; Guckert, A.; Guillet, B., Metal binding with root exudates of low molecular weight; J. Soil Sci. 1988, 39, 521–527.

[151] McLaughlin, M.J.; Smolders, E.; Merckx, R., Soil root interface: physicochemical processes; in: Chemistry and Ecosystem Health; Special Publication No. 52, Soil Science Society of America, Madison, USA, 1998, pp. 233–277.

[152] Lombi, E.; Wenzel, W.W.; Gobran, G.R.; Adriano, D.C., Dependency of phytoavailability of metals on indigenous and induced rhizosphere processes: a review; in: G.R. Gobran, W.W. Wenzel, E. Lombi (eds.), Trace Elements in the Rhizosphere; CRC Press, Boca Raton, FL, 2001, pp. 4–23.

[153] Marschner, H.; Romheld, V., In vivo measurement of root-induced pH changes at the soil–root interface: effect of plant species and nitrogen source; Z. Pflanzenphysiol. 1983, 111, 241–251.

[154] Sun, .XH.; Doner, H.E., An investigation of arsenate and arsenite bonding structures on goethite by FTIR; Soil Sci. 1996, 161, 865–872.

[155] Manceau, A.; Lanson. B.; Schlegel, M.L. *et al.*, Quantitative Zn speciation in smelter-contaminated soils by EXAFS spectroscopy; Am. J. Sci. 2000, 300, 289–343.

[156] Sparks, D.L., Elucidating the fundamental chemistry of soils: past and recent achievements and future frontiers; Geoderma 2001, 100, 303–319.

[157] Arai, Y.; Sparks, D.L., Residence time effects on arsenate surface speciation at the aluminium oxide-water interface; Soil Sci. 2002, 167, 303–314.

[158] Roberts, D.R.; Scheinost, A.C.; Sparks, D.L., Zinc speciation in a smelter-contaminated soil profile using bulk and microspectroscopic techniques; Environ. Sci. Technol. 2002, 36, 1742–1750.

[159] Tiller, K.G.; Honeysett, J.L.; de Vries, M.P.C., Soil zinc and uptake by plants. 1. Isotopic exchange equilibria and the application of tracer techniques; Aus. J. Soil Res. 1972, 10, 151–164.

[160] Fuji, R.; Corey, R.B., Estimation of isotopically exchangeable cadmium and zinc in soils; Soil Sci. Soc. Am. J. 1986, 50, 306–308.

[161] Hamon, R.E.; Wundke, J.; McLaughlin, M.J.; Naidu, R., Availability of zinc and cadmium to different plant species; Aus. J. Soil Res. 1997, 35, 1267–1277.
[162] Smolders, E.; Brans, K.; Foldi, A.; Merkx, R., Cadmium fixation in soils measured by isotopic dilution; Soil Sci. Soc. Am. J. 1999, 63, 78–85.
[163] Hutchinson, J.J.; Young, S.D.; McGrath, S.P.; West, H.W.; Black, C.R.; Baker, A.J., Determining uptake of 'non-labile' soil cadmium by *Thlaspi caerulescens* using dilution techniques; New Phytol. 2000, 146, 453–460.
[164] Young, S.D.; Tye A., Carstensen, A.; Resende, L.; Crout, N., Methods for determining labile cadmium and zinc in soil; Eur. J. Soil Sci. 2000, 51, 129–136.
[165] Stanhope, K.G.; Young, S.D.; Hutchinson, J.J.; Kamath, R., Use of isotopic dilution techniques to assess the mobilization of non-labile Cd by chelating agents in phytoremediation; Environ. Sci. Technol. 2000, 4123–4127.
[166] Hamon, R.E.; McLaughlin, M.J.; Naidu, R.; Correll, R., Long-term changes in cadmium bioavailability in soil; Environ. Sci. Technol. 1998; 32, 3699–3703.
[167] Tye, A.M.; Young, S.D.; Crout, N.M.J. *et al.*, Predicting arsenic solubility in contaminated soils using isotopic dilution techniques; Environ. Sci. Technol. 2002, 36, 982–988.
[168] Tye, A.M.; Young, S.D.; Crout, N.M.J. *et al.*, Predicting the activity of Cd^{2+} and Zn^{2+} in soil pore water from the radio-labile metal fraction; Geochim. Cosmochim. Acta 2003, 67, 375–385.
[169] Nolan, A.L.; Zhang, H.; McLaughlin, M.J., 2005. Prediction of zinc, cadmium, lead, and copper availability to wheat in contaminated soils using chemical speciation, diffusive gradients in thin films, extraction, and isotopic dilution techniques; J. Environ. Qual. 2005, 34, 496–507.
[170] Gérard, E.; Echevarria, G.; Sterckeman, T.; Morel, J.L., Cadmium availability to three plant species varying in cadmium accumulation pattern; J. Environ. Qual. 2000, 29, 1117–1123.
[171] Degryse, F.; Buekers, J.; Smolders, E. Radio-labile cadmium and zinc in soils as affected by pH and source of contamination; Eur. J. Soil Sci. 2004, 55, 113–121.
[172] McGrath, S.P.; Knight, B.; Kilham, K.; Preston, S.; Paton, G.I., Assessment of the bioavailability of heavy metals in soils amended with sewage sludge using a chemical speciation technique and a lux-based biosensor; J. Environ. Toxicol. Chem. 1999, 18, 659–663.
[173] Cooperband, L.R.; Logan, T.J., Measuring in situ changes in labile soil phosphorus with anion-exchange membranes; Soil Sci. Soc. Am. J. 1994, 58, 105–114.
[174] Yang, J.E.; Skogley, E.O.; Schaff, B.E., Nutrient flux to mixed-bed ion-exchange resin: Temperature effects; Soil Sci. Soc. Am. J. 1991, 55, 762–767.
[175] Davison, W.; Zhang, H., In-situ speciation measurements of trace components in natural waters using thin-film gels; Nature 1994, 367: 546–548.
[176] Zhang, H.; Davison, W.; Knight, B.; McGrath, S., 1998. In situ measurements of solution concentrations and fluxes of trace metals in soils using DGT; Environ. Sci. Technol. 1998, 32, 704–710.
[177] Hooda, P.S.; Zhang, H.; Davison, W.; Edwards, A.C., Measuring bioavailable trace metals by diffusive gradients in thin films (DGT): soil moisture effects on its performance in soils; Eur. J. Soil Sci. 1999, 50, 285–294.
[178] Zhang, H.; Zhao, F.J.; Sun, B.; Davison, W.; McGrath, S.P., 2001. A new method to measure effective soil solution concentration predicts copper availability to plants; Environ. Sci. Technol. 2001, 35, 2602–2607.
[179] Zhang, H.; Lombi, E.; Smolders, E.; McGrath, S., Zinc availability in field contaminated and spiked soils; Environ. Sci. Technol. 2004, 38, 3608–3613.
[180] Hooda, P.S.; Zhang, H., DGT measurements to predict metal bioavailability in soils; in: R. Naidu *et al.* (eds), Chemical Bioavailability in Terrestrial Environments, Development in Soil Science, Vol. 32; Elsevier, London, 2008, pp. 169–185.
[181] Davison, W.; Hooda, P.S.; Zhang, H.; Edwards, A.C., DGT measured fluxes as surrogates for uptake of metals by plants; Adv. Environ. Res. 2000, 3, 550–555.
[182] Singh, R.; Chandel, J.D.; Bhandari, A.R., 1998. Effect of soil moisture regime on plant, growth, fruiting, quality and nutrient uptake by mango (*Magnifera indica*); J. Indian Agric. Sci. 1998, 68, 135–138.

[183] Degryse, F.; Smolders, E.; Zhang, H.; Davison, W., Predicting availability of mineral elements to plants with the DGT technique: a review of experimental data and interpretation by modelling; Environ. Chem. 2009, 6, 198–218.

[184] Paustenbach, D., The practice of exposure assessment: a state-of-the-art review; J. Toxicol. Environ. Health Part B, 2000, 3, 179–291.

[185] Reid, R.M., Cultural and medical perspective on geophagia. Med. Anthropol. 1992, 13, 337–351.

[186] Hooda, P.S.; Henry, C.J.K.; Seyoum, T.A.; Armstrong, D.M.; Fowler, M.B., The potential impact of geophagia on bioavailability of iron, zinc and calcium in human nutrition; Environ. Geochem. Health 2002, 24, 305–319.

[187] Reeves, P.G.; Chaney, R.L., Mineral status of female rats affects the absorption and organ distribution of dietary cadmium derived from edible sunflower kernels (*Helianthus annuus* L.); Environ. Res. 2001, 85, 215–225.

[188] Reeves, P.G.; Chaney, R.L., Marginal nutritional status of zinc, iron, and calcium increases cadmium retention in the duodenum and other organs of rats fed rice-based diets; Environ. Res. 2004, 96, 311–322.

[189] Wragg, J.; Cave, M.R., In-vitro methods for the measurement of the oral bioaccessibility of selected metals and metalloids in soil; Critical Review R&D Technical Report P5-062/TR/01, 2002.

[190] Rodriguez, R.R.; Basta, N.T.; Casteel, S.W.; Pace, L.W., An in vitro gastrointestinal method to estimate bioavailable arsenic in contaminated soils and solid media; Environ. Sci. Technol. 1999, 33, 642–649.

[191] Poggio, L.; Vrscaj, B.; Schulin, R.; Hepperle, E.; Marsan, F.A., Metals pollution and human bioaccessibility of topsoil in Grugliasco (Italy); Environ. Pollut. 2009, 157, 680–689.

[192] Smith B.; Rawlins, B.G.; Cordeiro, M.J.A.R. *et al.*, The bioaccessibility of essential and potentially toxic trace elements in tropical soils from Mukono district, Uganda; J. Geol. Soc. 2000, 57, 885–91.

[193] Maddaloni, M.; Lolacono, N.; Manton, W.; Blum, C.; Drexler, J.; Graziano, J., Bioavailability of soil-borne lead in adults by stable isotopes dilution; Environ. Health Perspect. 1998, 106, 1589–1594.

[194] Hooda, P.S.; Henry, C.J.K.; Seyoum, T.A.; Armstrong, D.M.; Fowler, M.B., The potential impact of soil ingestion on human mineral nutrition; Sci. Total Environ. 2004, 333, 75–87.

[195] Smit, E.; Leeflang, P.; Wernars, K., Detection of shifts in microbial community structure and diversity in soil caused by copper contamination using amplified ribosomal DNA restriction analysis; FEMS Microb. Ecol. 1997, 23, 249–61.

[196] Filzek, P.D.B; Spurgeon, D.J.; Broll, G. *et al.*, Metal effects on soil invertebrate feeding: Measurements using the bait lamina method; Ecotoxicol. 2004, 13, 807–816.

[197] OECD, Guidelines for the testing of chemicals, No. 207, Earthworm Acute Toxicity Test. Adopted 4/4/1984; OECD, Paris, France, 1984.

[198] Hobbelen, P.H.F.; Koolhaas, J.E.; van Gestel, C.A.M., Risk assessment of metal pollution for detritivores in floodplain soils in the Biesbosch, the Netherlands, taking bioavailability into account; Environ. Pollut. 2004, 129, 409–419.

[199] Chander, K.; Brookes, P.C., Effects of heavy metals from past applications of sewage sludge on microbial biomass and organic matter accumulation in a sandy loam and silty loam UK soil; Soil Biol. Biochem. 1991, 23, 927–932.

[200] Nannipieri, P.; Gregos, S.; Ceccanti, B., Ecological significance of the biological activity in soil; in: J.L. Smith, E.A. Paul (eds.), Soil Biochemistry; Marcel Dekker, New York, 1990, vol. 6, 293–355.

[201] Dick, R.P., Soil enzyme activities as integrative indicators of soil health; in: C.E. Pankhurst, B.M. Doube, V.V.S.R. Gupta (eds.), Biological Indicators of Soil Health; CAB International, New York, 1997, pp. 121–156.

[202] Sheppard, S.; Evenden, W.G.; Cornwell, T. C., Depuration and uptake kinetics of I, Cs, Mn, Zn, and Cd by the earthworm (*Lumbricus terrestris*) in radiotracer-spiked litter; Environ. Toxicol. Chem. 1997, 16, 2106–2112.

[203] Spurgeon, D.J.; Hopkin, S.P., Extrapolation of the laboratory based OECD earthworm toxicity test to metal contaminated field sites; Ecotoxicology 1995, 4, 190–205.
[204] Smolders, E.; McGrath, S.P.; Lombi, E. et al, Comparison of toxicity of zinc for soil microbial processes between laboratory-contaminated and polluted field soils; Environ. Toxicol. Chem. 2003, 22, 2592–2598.
[205] Janssen, R.P.T., Peijnenburg, W.J.G.M., Posthuma, L.; Van der Hoop M.A.G.T., Equilibrium partitioning of heavy metals in Dutch field soils. I. Relationships between metal partition coefficients and soil characteristics; Environ. Toxicol. Chem. 1997, 16, 2470–2478.
[206] Janssen, R.P.T.; Posthuma, L.; Baerselman, R.; Den Hollander, H.A.; Van Veen, R.P.M.; Peijnenburg, W.J.G.M., Equilibrium partitioning of heavy metals in Dutch field soils. II. Prediction of metal accumulation in earthworms; Environ. Toxicol. Chem. 1997, 16, 2479–2488.
[207] Belkin, S.; Smulski, D.R.; Dadon, S.; Vollmer, A.C.; Van Dyk, T.K.; LaRossa, R.A., A panel of stress-responsive luminous bacteria for toxicity detection; Wat. Res. 1997, 31, 3009–3016.
[208] Chalfie, M.; Tu, Y.; Euskirchen, G.; Ward, W.W.; Prasher, D.C., Green fluorescent protein as a marker for gene expression; Science 1994, 263, 802–805.
[209] Kohler, S.; Belkin, S.; Schmid, R.D., Reporter gene bioassays in environmental analysis (review); Fresen. J. Anal. Chem. 2000, 366, 769–779.
[210] Daunert, S.; Barrett, G.; Feliciano, J. S.; Shetty, R. S.; Shrestha, S.; Smith-Spencer W., Genetically engineered whole-cell sensing systems: coupling biological recognition with reporter genes; Chem. Rev. 2000, 100, 2705–2738.
[211] Riether, K.B.; Dollard, M.A.; Billard, P., Assessment of heavy metal bioavailability using *Escherichia coli* zntAp::lux and copAp::lux-based biosensors; Appl. Microbiol. Biotechnol. 2001, 57, 712–716.
[212] Shetty, R.S.; Deo, S.K.; Shah, P.; Sun, Y.; Rosen, B.P.; Daunert, S., Luminescence-based whole-cell-sensing systems for cadmium and lead using genetically engineered bacteria; Anal. Bioanal. Chemi. 2003, 376, 11–17.
[213] Ritchie, J.M., Cressser, M.; Cotter-Howells, J., Toxicological response of a bioluminescent microbial assay to Zn, Pb and Cu in artificial soil solution: relationship with total metal concentrations and free ion activities; Environ. Pollut. 2001, 114, 129–136.
[214] Vulkan, R.; Zhao, F.-J.; Barbara-Jefferson, V.; Preston, S. *et al.*, Copper speciation and impacts on bacterial biosensors in the pore water of copper-contaminated soils; Environ. Sci. Technol. 2000, 34, 5114–5121.
[215] Ivask, A.; François, M.; Kahru, A.; Dubourguier, H.-C.; Virta, M.; Douay, F., Recombinant luminescent bacterial sensors for the measurement of bioavailability of cadmium and lead in soils polluted by metal smelters; Chemosphere 2004, 55, 147–156.
[216] Strosnider, H., Whole-cell bacterial biosensors and the detection of bioavailable arsenic – a report prepared for US EPA, Washington, DC, 2003.
[217] NRC (National Research Council), Bioavailability of Contaminants in Soils and Sediments: Processes, Tools, and Applications; National Academic Press, Washington DC, 2003.
[218] Tauriainen, S.; Virta, M.; Chang, W.; Karp, M., Measurement of firefly luciferase reporter gene activity from cells and lysates using *Escherichia coli* arsenite and mercury sensors; Anal. Biochem. 1999, 272, 191–198.
[219] Rieuwerts, J.S.; Ashmore, M.R.; Farago, M.E.; Thornton, I., The influence of soil characteristics on the extractability of Cd, Pb and Zn in upland and moorland soils; Sci. Total Environ. 2006, 366, 864–875.

12
Bioavailability: Exposure, Dose and Risk Assessment

Rupert L. Hough
The Macaulay Land Use Research Institute, Aberdeen, Scotland, UK

12.1 Introduction

'Risk' is the probability that a specific outcome will occur. The term 'risk' is usually associated with adverse outcomes. A risk assessment, when properly conducted, provides an indication of the magnitude of the threat and of the certainty of this estimate [1]. For trace elements in soils, there may be risks associated with both deficiency, for example goitre associated with soils that have inherently low concentrations of iodine [2] and toxicity, for example decreased bone density and increased risk of fractures associated with elevated concentrations of cadmium in soils [3].

Risk may be estimated from either the perspective of the outcome or the perspective of the causative agent [4]. For trace elements in soils, we may be interested in, for example, the risk of kidney cancer associated with elevated concentrations of arsenic in the soil of a residential area. To study this problem from the perspective of the outcome will require identification of individuals with kidney cancer and those without kidney cancer or any other health outcomes associated with exposure to arsenic. For both groups of people, exposure to arsenic through soil ingestion/inhalation and via consumption of food grown in garden soil would be ascertained. Finally, risk would be determined by comparing the incidence of cancer in the two groups of individuals. Conversely, to look at the problem from the perspective of the

268 Bioavailability, Risk Assessment and Remediation

causative agent would require an initial geochemical survey to ascertain concentrations of arsenic in garden soils. Assumptions regarding peoples' activity budgets would be used to estimate levels of exposure. A further complication associated with assessing risks from trace elements in soils, is that the 'total' concentration of a trace element measured in a soil matrix may provide little indication of the 'active' or bioavailable concentration [5]. Geochemical modelling may be applied to make estimates of uptake by food crops and/or bioavailability of arsenic ingested with the soil matrix. Final inferred doses would be compared with 'safe' doses from published toxicological studies.

Although addressing the problem from the perspective of the outcome is widely considered the 'gold standard', the choice of approach really depends on the availability of information and which questions we are trying to answer. Studying risks from the perspective of the causative agent is more likely to improve our understanding of environmental mechanisms, while studying risks from the perspective of the outcome usually increases our understanding of the mechanisms of exposure.

12.1.1 The 'Classical' Risk Assessment Model

The approach to assessing risks to human health or the environment can vary widely depending on a variety of factors including both scientific and political reasoning. Risk assessment has its roots in a number of disciplines including, among others, the biomedical sciences, environmental sciences, statistics and engineering. The classical model involves four stages, namely 'hazard identification', 'dose–response assessment', 'exposure assessment', and 'risk characterization' (Figure 12.1). This model was developed primarily for the application of biomedical and public health research, as such it is most suited to extrapolating the findings of observational studies conducted from the perspective of the outcome (that is where existing health issues are being studied in existing populations) to a wider context. To this extent, it is primarily favoured by the research, rather than the regulatory, community. The hazard-based approach outlined in 1983 by the US National

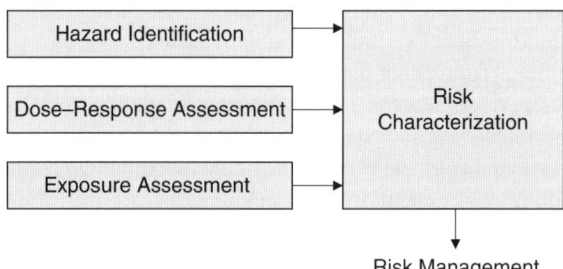

Figure 12.1 *The four stages of the 'classical model of risk assessment'. The first three stages ('hazard identification', 'dose–response assessment', and 'exposure assessment') feed into the final derivation of risk ('risk characterization'). The risk characterization may then be used to inform 'risk management', that is management of risk factors in order to reduce impacts of causative agents*

Research Council (NRC) in their report known as the 'Red Book' [6] is also based to a large extent on the classical model.

In the initial stages of the risk assessment process, hazard identification aims to identify whether there is cause for concern, that is 'what is the hazard?' This is usually a desk-based exercise with information gathered from toxicological and epidemiological studies, as well as more specific information from case and accident reports. Apart from determining the toxicological potential of the trace element of interest, the hazard identification also determines (conceptually) whether pathways exist by which receptors may become exposed to the agent of interest; and whether the receptor is sensitive to a given exposure. The hazard identification is essentially a screening phase, identifying areas requiring prioritization and/or further investigation.

When a trace element is deemed to be a hazard, then the next stage is to determine what dose of a substance will cause what response in a receptor (human, ecological). This process is termed 'dose–response assessment'. Data may be derived experimentally, observationally, or using modelling approaches. In most cases, it will be difficult to find data that matches the situation of concern – namely a specific trace element (or species of that element) within a specific soil matrix under specific conditions, with a specific route of exposure, with a specific mode of action on (a) specific receptor(s). Usually, the most suitable (available) data are selected; these are typically experimental/toxicological data where animals have been exposed to the trace element of interest under laboratory conditions.

Once the effect of dose magnitude has been determined, it is important to find the actual dose(s) that the receptors are being exposed to. This is a key stage in the risk assessment process, especially where trace elements are concerned, and is termed the 'exposure assessment'. This stage is, however, often poorly characterized using proxy measures of exposure such as distance from a source of pollution, or time spent working with soil. For trace elements in soils, bioavailability is a key consideration when estimating exposure. Bioavailability may be thought of as the proportion of an ingested/inhaled/contacted causative agent that actually crosses into the tissues and cells of the body. Bioavailability of a trace element varies with the matrix of delivery (soil, soil type, food, water, etc.), and the chemical and physical form of the element (solid, liquid, gas, inorganic, organic, adsorbed, free ion, etc.). The nature of the matrix of delivery can also reduce absorption of already bioavailable elements, which may lead to deficiencies [7].

Finally, 'risk characterization' pools the information from the previous three stages in order to make some estimation of 'risk'. In many cases this is a quantitative estimate of the probability (risk) of the receptor experiencing a deleterious effect as a result of being exposed to the trace element(s) present in the soil at a particular site/circumstance, or over a given period. A true 'risk-based approach' attempts to quantify the probability of being exposed to specific trace element(s) and experiencing an adverse effect(s) compared to the probability of receiving the same exposure(s) and *not* experiencing an adverse effect(s).

This chapter reviews the four stages of the classical risk assessment process, in the context of assessing risks to human health and ecosystems from trace elements in the soil environment. There are many comprehensive tomes describing the risk assessment process (for example [8]) and as such this chapter will refer to these rather than enter into excessive detail regarding the process itself. Instead, in line with the overall remit of this book, this chapter considers areas of contention within the risk assessment approach, and where improvements may be made in the future.

12.2 Hazard Identification

12.2.1 Approaches, Uncertainties, Issues for Discussion

Hazard identification is an evaluation of the potential effects and concerns related to the intrinsic properties of a substance [9–12]. Thus the goal of hazard identification is to identify all situations or substances that can, under any (reasonable) circumstances, pose a risk to human health or the environment [13]. In most circumstances, the hazard identification takes the form of a desk-based study where the analyst collates as much information as available regarding the trace element of interest. The overall aim is to answer the following questions: (i) what are the main sources of the trace element (geogenic, anthropogenic – past or present industrial activity, agricultural practice, waste disposal or recycling, etc.)? (ii) what are the pathway(s) by which the element(s) may 'flow' from the source to the receptor(s) (aeolian transport, via surface run-off, leaching into ground or surface water, uptake into food crops, etc.)? (iii) how do the receptors become exposed (inhalation, dermal contact, ingestion, etc.)?, and (iv) if there is the potential for exposure, should we be worried – that is, is the trace element considered toxic at the expected doses? Often it is useful to use a filtering process that narrows a long list of hazards down to a short list of those hazards that are of most pressing concern. The Department of Environment [14] risk assessment provides a sound rationale for the identification and analysis of hazardous agents. The approach (Figure 12.2 [15]) identifies hazardous agents and then applies a successive series of screens, or filters with the aim of focusing on hazards of greatest concern:

(**Filter 1**)   Does the hazardous agent have potentially serious health effects, or environmental impact?
(**Filter 2**)   For those with a potentially serious health effect, is the hazardous agent likely to evade destruction if contamination is not contained?
(**Filter 3**)   For those agents with serious health effects that may evade destruction, might the hazardous agent present itself at a point of exposure in a quantity sufficient to be of concern?

Sources of information may include toxicological studies where experimental animals or cell cultures have been purposefully exposed to the trace element of interest; epidemiological studies where humans (or sometimes animals or microorganisms) that have previously been (or are still being) exposed to the element of interest are studied; one-off case reports (medical and veterinary) that postulate a link between the trace element and an adverse outcome; and information regarding potential pathways through which receptors may become exposed, for example data that demonstrate the circumstances under which a specific trace element may become mobile within the soil matrix and hence increase its potential impact on controlled waters.

Although a large amount of information is collated, it is not the aim of the hazard identification to enter into quantification of these relationships. Rather, the analyst simply

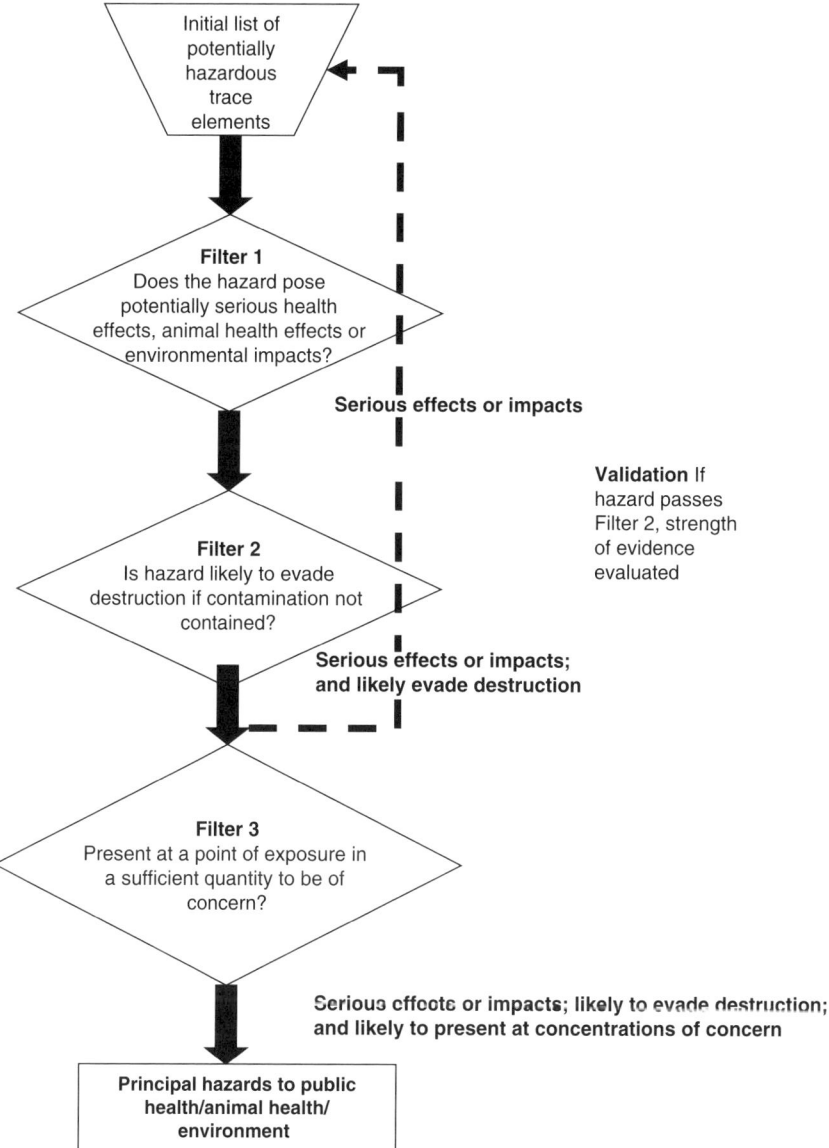

Figure 12.2 Flow chart for identifying principal public health hazards associated with trace elements from contaminated land (Adapted with permission from S.J.T. Pollard, G.A.W. Hickman, P. Irving et al., Exposure assessment of carcase disposal options in the event of a notifiable exotic animal disease—application to avian influenza virus, Environ. Sci. Technol. 42, 3145–3154. Copyright 2008 American Chemical Society [15].)

'flags up' potential scenarios that may be of concern. In most cases, the output from the hazard identification step is some form of conceptual model, an example of which is shown in Figure 12.3 [16]. The general structure of the conceptual model is based around a series of source–pathway–receptor linkages.

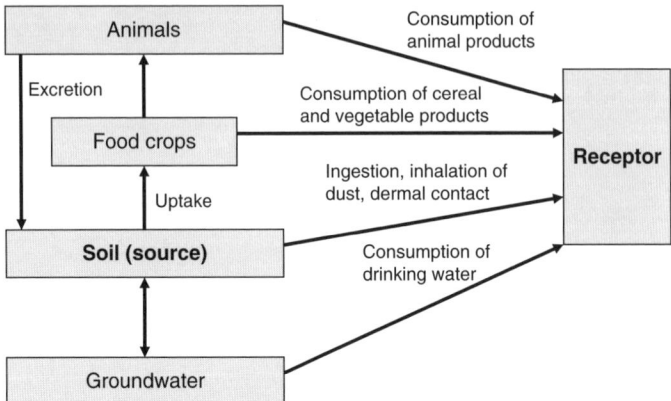

Figure 12.3 *Example of a conceptual model: schematic representation of the exposure pathways deriving dose from a smallholding situated on contaminated soil (Adapted from C. Stephens et al., Assessing the Health Risk and Impact Associated with Living on the Thames View Estate: a Rapid Literature Review of methods, Barking & Dagenham Primary Care Trust,* London, 2004 [16].*)*

12.3 Exposure Assessment

'Exposure' is the amount or dose of a specific substance that may enter or act upon the target organism [17]. It is important to note that in most cases, *exposure* is different from the concentrations of, and the spatial distribution of the substance in the environment. Exposure is the combination of the amount of a trace element entering the body of the target organism, and the bioavailability of that trace element once ingested. This is very important, and often overlooked by studies assessing risk from the perspective of the causative agent. In the Cd and bone density example, concentrations of Cd may be highest in soils where the exposure is actually relatively low. If the soil has been sealed, exposure through ingestion and/or inhalation will be much reduced. If the people living on the most contaminated soils spent most of their time indoors and had little interest in gardening, exposure would be minimized. If the form of Cd in the soil, or the soil chemistry, meant that the majority of Cd ingested was not absorbed over the length of the digestive tract, exposure would be much less than expected.

12.3.1 Approaches to Estimating Exposure from Trace Elements in Soils

The exposure pathways identified in Figure 12.3 can, allowing for assumptions, all be quantified. In all cases, the key information required is the concentration of the contaminant of interest in each exposure medium (soil, dust, home-grown food) and the amount of the exposure medium ingested/inhaled. However, once inside the body, not all of a specific contaminant is absorbed.

12.3.1.1 Home-Grown Produce

For many trace elements (with the possible exceptions of As and Pb) soil is the principle source of contamination of food plants [18]. Contamination of garden soils is widespread in urban areas mainly due to past industrial activity, from the burning of fossil fuels and from automotive fuels [19–23]. A comprehensive survey of urban gardens and allotments [24] measured elevated levels of Pb and Cd in over 4100 gardens in 48 towns and city boroughs across the United Kingdom. In the United Kingdom, fresh vegetables provide around 17% of the average weekly intake of Pb and Cd [21].

Assessing exposure from home-grown food may be laborious because different crops take up contaminants to different extent [25,26]. Also, the uptake of contaminants by crops is not necessarily related to the concentration of the contaminants in the soil, but is also dependent on the chemical characteristics of that soil [27–32]. For example, many authors have reported weak associations between the 'total' concentration of metal in the soil, and the concentration measured in plants. More significant relationships have been reported between various proxy measures of bioavailability, such as soil solution chemistry, and plant uptake (Figure 12.4).

There is a vast body of literature measuring the uptake of specific soil contaminants by garden food plants [21,25,29,30,33–38]. To estimate exposure from the perspective of the causative agent, it is possible to use information gathered during a soil survey of the site (pH, *total* concentration of contaminants, soil organic matter and clay contents, etc.) to predict the uptake of metal contaminants by different food plants [6,27,31,39]. These predictions can then be used in combination with statistics on the amount of home-grown produce generally consumed by different demographics in society. Alternatively, when

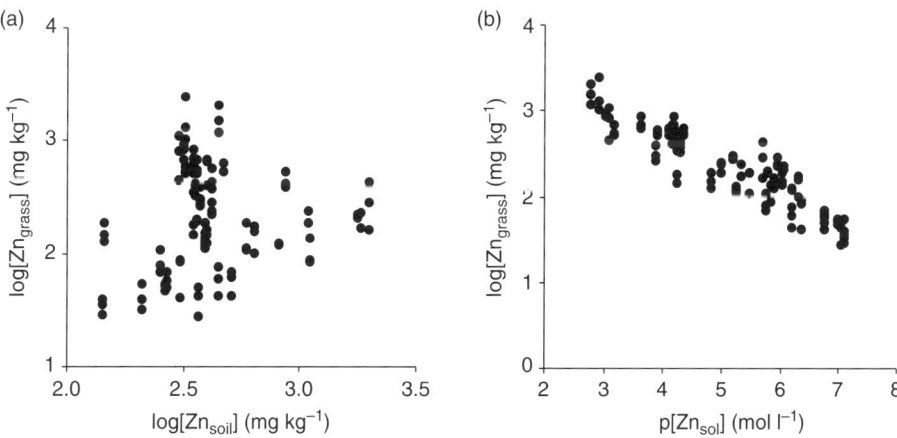

Figure 12.4 The concentration of Zn in ryegrass, $[Zn_{grass}]$, as a function of (a) total (aqua regia) Zn concentration in the soil, $[Zn_{soil}]$, and (b) the concentration of Zn in the soil solution $[Zn_{sol}]$ expressed similarly to pH as a negative $\log_{10}(p)$ with low values indicating the greatest concentrations (Reproduced with kind permission from Springer Science + Business Media: Plant and Soil, Evaluating a 'Free Ion Activity Model' applied to metal uptake by Lolium perenne L. grown in contaminated soils, 270, 2005, R.L. Hough et al., 1–12 [4].)

estimating risk from the perspective of the outcome, exposure is likely to be determined using direct measurements. Samples of vegetables and fruits from participants' gardens may be gathered and concentrations of contaminants measured in these samples. The amount of home-grown produce consumed by each individual in the study would then be determined using a food frequency questionnaire.

12.3.1.2 Ingestion of Soil and Dust

A number of scientists consider ingestion of soil and dust to be the single most important exposure pathway for inorganic pollutants (Table 12.1). This is considered especially true for younger children who ingest larger quantities than other sections of the population, especially in terms of ingested quantity relative to body mass [40–43].

Early studies of soil ingestion were based on observations of mouthing behaviour coupled with some measurement of dust or soil on the hands of the participant. Early studies were primarily concerned with evaluating soil ingestion by children in order to estimate the amount of Pb ingested. For example, Duggan and Williams [44] investigated the amount of street dust a child might consume by observing the number of times per day a 'typical' child sucked his or her finger or thumb. Davies [45] used video recording to more accurately describe hand-to-mouth behaviour of two-year-old children. Observation methods, however, can not estimate soil and dust ingestion reliably without some knowledge of how much soil or dust is removed from the hands during the mouthing action. The amount of soil retained on the hands and available for ingestion from hand-to-mouth activities probably depends on soil type [46], although this is rarely taken into account. One method that has been employed is the use of hand wipes [47–49], which is a cheap and easy method to execute. Concentrations of contaminants in the soil adhering to the skin are enriched relative to the concentrations of contaminants present in the bulk soil. This may be due to fine particles, on which contaminants tend to be concentrated, preferentially adhering to the skin [50].

Overall, estimation of soil ingestion using hand-to-mouth observations is fraught with uncertainties. Not only is it an indirect way of estimating soil ingestion, the method focuses on only one single route by which soil may enter the body. For example, such a study cannot take into account the ingestion of soil adhering to food items such as vegetables.

12.3.1.3 Tracer studies

The obvious uncertainties associated with hand-to-mouth soil ingestion estimates stimulated a different approach based on mass balance of tracer elements. Mass balance studies are labour-intensive; require specialist equipment and also a high degree of cooperation from participants. The basic concept is that some elements ubiquitously present in soils are very poorly absorbed by the digestive tract, these include: aluminium (Al), cerium (Ce), lanthanum (La), neodymium (Nd), silicon (Si), titanium (Ti), yttrium (Y), and zirconium (Zr). In simple terms, if the soil (input) contains $10 \mu g\ Y\ g^{-1}$, and $20 \mu g\ Y$ is recovered in a subject's faeces (output), then it could be concluded that the subject consumed 2 g soil. However, all sources of the tracer element must be taken into account (food, medicines, vitamin pills, toothpaste, etc.), the activity budget of the person, and the concentration of the tracer in outside soil and indoor dust. It may also be important to consider the time lag between ingestion and excretion. The mass balance equation for a tracer element can therefore be written as:

$$I_a + I_{fo} + I_s + I_w = O_f + O_u \qquad (12.1)$$

where I represent the intakes of a specific tracer element (from air (I_a), food (I_{fo}), soil (I_s), and water (I_w)), and O the outputs of that same tracer (in faeces (O_f) and urine (O_u)). The most suitable tracer will be the one that displays negligible inputs from air, water, and possibly food. The recovery of some tracers in urine is also negligible. In some cases it is possible to reduce Equation (12.1) to Equation (12.2):

$$I_{fo} + I_s = O \qquad (12.2)$$

Tracers that display the behaviour described by Equation (12.2) should provide more accurate estimates of soil ingestion as there are fewer possible sources of error.

In order to satisfy the terms of Equation (12.2), it is usual to use more than one tracer and employ the 'best tracer method' [51] so as to reduce input–output error and improve tracer recovery. The 'best tracer' is therefore the tracer that displays the best percentage recovery and whose inputs and outputs are fewest as in Equation (12.2).

Risk assessments tend to use mean soil ingestion estimates from recommended tracer studies [42,52,53]. Large variation in soil ingestion, with standard deviations several times greater than the mean ingestion estimate, however, is a common problem. A number of studies [42,52,53] have concentrated on children aged 1–4 years old. These age groups are of particular interest as they ingest the largest quantities of soil, both purposely and inadvertently, compared with other age groups. Pica behaviour has been defined as the repeated eating of nonnutritive substances, including clay, paint, plaster, hair cloth, matches, paper and various other items [40,54]. Deliberate ingestion of soil is termed 'geophagy' [46,55] and is characterized by ingestion of >1 g soil per day [52]. There is considerable variation in geophagy between children. One child was observed to ingest 20–25 g soil per day on two out of eight days [56,57], while a second child was observed to ingest 1–3 g soil per day on four days out of seven [42]. In total, 62% of children aged 1–4 years will ingest >1 g soil per day [51].

12.3.1.4 Inhalation of Dust

There are two main aspects to estimating exposure due to dust inhalation. Firstly there is a need to estimate the amount of dust that a person is exposed to, that is the concentration of dust particles in the air that can be inhaled and the chemical composition of those particles. The second stage is to estimate the amount of dust that is actually respired by individuals; this usually involves gathering information about peoples' activity budgets and hence inhalation rates.

The simplest devices available for estimating the amount of suspended particulate matter are collectively referred to as passive samplers. Typically these consist of deposition plates that can be used to quantify the contribution of indoor and outdoor sources of particulate matter onto indoor surfaces [58]. However, this type of sampling regime does not consider the movement of contaminants, or the fraction of contaminated dust which is inhalable.

Recently researchers have begun to consider human exposure in residential neighbourhoods and in doing so designed field studies that focused on quantifying the movement of resuspended, contaminated soil into indoor locations [59,61]. These studies led to parallel investigations looking at the concentrations of contaminants in house dust once it has been transported indoors [24].

Techniques for the sampling of indoor dust were developed prior to 1970, the majority of these being wipe-sampling. A major review of dust sampling techniques, focusing on wipe-sampling, was published in 1992 by McArthur [62]. Although surface sampling is relatively low cost, McArthur [62] stated that surface sampling by wipe-sampling had limited reliability for use in exposure assessment. During the late 1980s and early 1990s further work has improved methods of surface sampling. These include the use of size-selective vacuum samplers as a means of sampling the fraction of dust that may be inhalable [63]. Once dust samples have been collected from both indoor and outdoor locations, the concentrations of the contaminants of interest may then be determined. These measurements can then be combined with information about the activity budgets of the individuals. Usually, when risk is being estimated from the perspective of the causative agent, the activity budgets (and corresponding inhalation rates) used are theoretical [4,40]. However, for a population-specific health study directed from the perspective of the outcome, it may be beneficial to construct activity budgets for subsets of the population. It should be noted that the estimation of activity budgets are inherent with uncertainty and there is also significant between-subject variation. It is therefore desirable to validate or look at the reproducibility of such studies. For example, an activity budget estimated with diaries for a large group of people could be validated using a smaller subset whose inhalation rates are measured, or whose activity budgets are determined using a more in-depth diary tool. Alternatively, activity budget diaries could be filled in on more than one occasion to look at the reproducibility of the diary tool.

A further refinement of dust sampling strategies is the application of personal samplers. These are devices that are worn by the participant, typically on their shirt just below their head. The personal sampler provides a more refined estimate of a person's actual dust inhalation. It is important to use a size-selective sampler that measures the inhalable dust fraction. A wide range of European personal samplers are available for this purpose, among which the British IOM Sampler, the Dutch PAS6 Sampler, and the German GSP Sampler are considered the most reliable, having undergone extensive validation under both wind tunnel and field conditions [64]. The use of such samplers requires a high level of cooperation with the community, but the resulting dust measurements are a closer representation of the actual dust which is inhaled by the individuals.

12.3.2 Environmental Measurements and Influence of Bioavailability

Absorption of contaminants across the gastrointestinal tract can be highly variable in human populations because it is influenced by a variety of factors that include the chemical form of the contaminant of interest, the environmental matrix in which the ingested contaminant is contained, the gastrointestinal contents, diet, nutritional status, age, and, in some cases, genotype [16]. The bioavailability of contaminants ingested in water tends to be close to 100%, while absorption of contaminants from soil and dust tends to be much lower. The role of the gut is to extract nutrients from ingested material through rigorous mixing, extreme shifts in pH and enzymatic attack. Even so, soil will remain a competitive sink for strongly sorbed contaminants. Bioavailability varies enormously depending upon the environmental matrix and the contaminant of interest. For example, bioavailability of Cd and Pb in ingested soil has been found to be as little as 2% [67] and as much as 76% [68].

Estimating gastrointestinal absorption of contaminants remains a significant challenge. There are three main approaches to estimating the bioavailability of different contaminants in food, soil and water. The first is an *in vitro* approach using synthetic gastric fluids that act as a model for the human gastrointestinal tract. The second approach is to use laboratory animals as a model for humans, and the third is to use a human 'whole-body burden' biomarkers of exposure.

There are a large number of studies looking at estimating bioavailability using *in vitro* systems, for example [67–70]. These studies try to reconstruct the gastrointestinal system and typically follow a two-step model comprising an acidic gastric phase followed by a neutral or slightly alkaline intestinal phase. The *in vitro* method is probably the quickest, and logistically least difficult, way of estimating bioavailability. However, it has to be assumed that the *in vitro* system is a close representation of the *in vivo* system. Also, the assumption has to be made that the bioavailability of each metal contaminant from each specific environmental matrix is constant across the whole population. Making this assumption is probably more valid than not including a measure of bioavailability.

Estimating bioavailability in human and animal populations has been reviewed by Diamond *et al.* [71]. Balance studies measure daily intake and excretion of a specific contaminant over periods of days or weeks. The difference between intake and faecal excretion can then provide an estimate of absorption. This is very similar to mass balance studies used to estimate soil ingestion, except in this case the tracers of interest are absorbed by the gastrointestinal tract. Balance studies are widely used to estimate absorption in humans because they are noninvasive.

Gastrointestinal absorption may also be estimated using the relationship between the intake of a metal and the concentration of that metal in the tissues of the body, or body burden [72]. Such methods may be employed using experimental animals [73–75] or with human subjects [76–82]. In human studies, subjects are dosed with a radiolabelled isotope of the contaminant in question. The activity of the isotope is measured in faeces and absorption is estimated as the difference between the activity in the original dose and the activity measured in the faeces. A number of researchers have found that the isotope recovery in the faeces is dependent on the food currently in the gastrointestinal tract [76–78,80–81]. Because of this it is common practice for subjects to fast for a period of time before they ingest the dosed material. For example, James et al [82] measured the fraction retained in the whole body of adult subjects for seven days after they ingested a dose of radioactive Pb after a 19-hour fast. These methods are more suited to studying the bioavailability of contaminants from food products rather than from soils or dusts as it is usually difficult to recruit people willing to ingest a dosed quantity of soil.

Table 12.1 Selected ingestion studies outlining methodologies to assess the health impacts of soil contamination

Author	Investigation of:	Country, source, time period	Study design/ focus of paper	Measurement of interest	Remarks
Calabrese et al. [42]	Soil ingestion: children	USA, Superfund site	Tracer	8 tracer elements, 5 selected under 'best tracer methodology'	Estimates of 20–500 mg soil d^{-1}
Calabrese et al. [52]	Soil ingestion: children	USA, Superfund	Modelling	13 contaminant concentrations, quantity of soil ingested during pica episode	Acute human lethal dose exceeded for four of the chemicals
Charlesworth et al. [65]	Dust exposure	UK, Birmingham and Coventry	Descriptive	Heavy metal distribution	Large concentrations associated with main roads and junctions, especially those where traffic stops, for example traffic lights
Casteel et al. [73]	Bioavailability	USA	Laboratory	Mass balance of soil Pb fed to swine	Relative bioavailability estimated at 58–74%
Davies et al. [45]	Soil ingestion	USA, urban housing, 1987	Tracer	Aluminium, silicon, titanium	Inconsistencies between tracers reported
Diamond et al. [71]	Bioavailability	USA	Review	Determining bioavailability of metals for risk assessment	Comprehensive overview
Ferguson and Marsh [46]	Human health risks from soil ingestion	UK	Review	Mass of soil ingested per day	Highlights poor quantification in other studies and need to differentiate between deliberate and inadvertent soil ingestion
Fouchecourt et al. [74]	Bioavailability	France	Laboratory	Mass balance of soil PAH in rats exposed by natural routes	Interesting use of natural exposure routes
Kenny et al. [64]	Dust inhalation	Europe	Review	Personal sampler performance	Provides overview of the market samplers and suitability to different situations

Table 12.1 (continued)

Author	Investigation of:	Country, source, time period	Study design/ focus of paper	Measurement of interest	Remarks
Gasser et al. [68]	Bioavailability	USA	Laboratory	Synthetic gastric conditions with Pb waste from mining	>50% of Pb bioavailable
Georgopoulos et al. [61]	Copper exposure	USA	Review	Soil ingestion, inhalation, dietary	Overview of environmental dynamics of copper and exposure to copper
Hammel et al. [67]	Bioavailability	USA	Laboratory	Synthetic gastric conditions and metals	Bioavailability of metals in soil is affected by solid: solution ratio in 'stomach'
Lanphear et al. [83]	Indoor dust inhalation	USA	Meta-analysis	Pb in children's blood	Age, race, social status influenced dust inhalation
Linden et al. [75]	Bioavailability	Sweden	Laboratory	Bioavailability of soil and vegetable Cd in pigs using biomarkers	Highlights huge variation in bioavailability estimates
McArthur [62]	Dust ingestion/ inhalation	USA	Review	Dermal measurement and wipe sampling	Methodological review
Meyer et al. [84]	Contaminant levels in housedust	Former E. Germany 1992–1993	Passive dust sampling	Pb, Cd and As	Mean dust loading 8.9 mg m^{-2} with 1.14, 0.02 and 0.02 μg m^{-2} for Pb, Cd and As respectively
Pakkanen et al. [66]	Dust exposure	Finland, 1997	Passive sampling	Particle size and chemical composition	Chemical concentrations higher at roof level than at street level
Roberts and Dickey [85]	Dust exposure: children	USA	Review	Metals, allergens, pesticides, PAHs, PCBs[a]	Most significant exposures occur in infants and toddlers
Ruby et al. [69]	Bioavailability	USA	Laboratory	Synthetic gastric conditions and Pb	Lower bioavailability of Pb from mining sites compared with urban areas

(continued overleaf)

Table 12.1 (continued)

Author	Investigation of:	Country, source, time period	Study design/ focus of paper	Measurement of interest	Remarks
Sheppard and Evenden [50]	Ingestion/ inhalation of Pb	Canada	Laboratory	Contaminant enrichment due to weathering processes	Adhesion of soil particles to leaf surfaces provides as great concentration of contaminants as uptake from soil
Sheppard [55]	Soil ingestion	Canada	Review	Mass of soil ingested per day and potential exposures	Up to 50 g soil d^{-1} ingested purposefully
Stanek et al. [41]	Soil ingestion: children	USA, superfund site	Tracer	8 tracer elements, 4 selected under 'best tracer methodology'	No significant differences between male and female, but age-related increase noted
Thatcher and Layton [59]	Dispersion of dust within the home	USA, California, Summer 1993	Passive dust sampling	Volume of dust and particle size distribution	Re-suspension due to activity of inhabitants only occurred with super-micrometre particles

^{a}PAH, polycyclic aromatic hydrocarbon; PCB, polychlorinated biphenyl.
Adapted with permission from C. Stephens *et al.*, Assessing the Health Risk and Impact Associated with Living on the Thames View Estate: A Rapid Literature Review of methods, London School of Hygiene & Tropical Medicine, 2004[16].

12.4 Dose–Response

'Dose (concentration)–response (effect)' is the estimation of the relationship between dose, or level of exposure to a substance, and the incidence and severity of an effect. In their most basic form, dose–response relationships can be determined by exposing groups of experimental animals to specific agents of interest. In a toxicological study, the general objective of dose selection is to select, at one extreme, a dose sufficiently high to produce serious adverse effects without causing the early deaths of the animals, and, at the other, one that should produce minimal or, ideally, no observable toxicity – termed the No Observed adverse Effects Level (NOEL) [17]. The other three groups of animals are exposed to doses between the two extremes in order to construct a dose–response function (Figure 12.5).

12.4.1 High- to Low-Dose Extrapolation

In order to estimate risks to human health, it is often necessary to predict risk from exposures or doses below the range of observed data. For example, in Figure 12.5, the

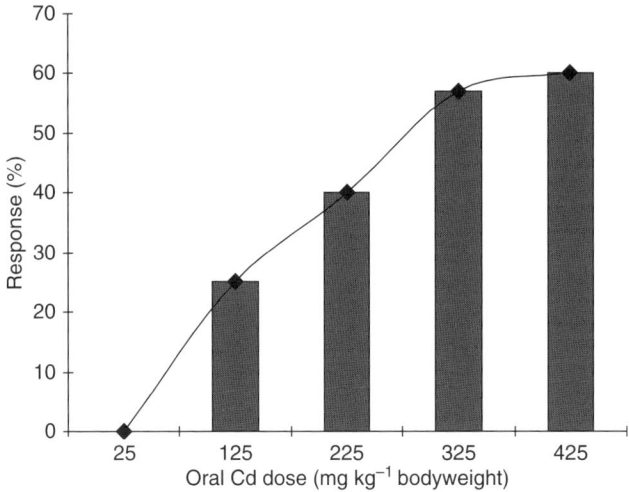

Figure 12.5 *Dose–response curve derived from a hypothetical animal study. Large doses have been used in order to 'see' a response in small populations of experimental animals, in effect increase the statistical power of the experiment. A series of extrapolations are then required to derive a 'safe' daily exposure/dose for humans*

lowest dose used was 25 mg Cd kg^{-1} bodyweight. Environmental exposures are likely to be one to three orders of magnitude lower. This presents a number of specific challenges, for example a number of US EPA programmes require regulation at the 10^{-6} to 10^{-4} exposure level – risks (and by implication, exposures) well below those observable in experimental or even epidemiological settings. Studies dealing with high-dose exposures are most plentiful and include toxicological studies using experimental animals and epidemiological approaches in occupational settings. Studies looking at exposures that are of more relevance to environmental regulation are confined to a small number of environmental epidemiology and exposure biomarker studies that cover a limited number of trace elements from an even more limited range of exposure scenarios. Certainly, the number of environmental epidemiology studies examining health outcomes associated with exposure to soil is surprisingly small [4]. Out of these, only a few studies consider specific trace elements [4].

Methodologies for high- to low-dose extrapolation can be broadly categorized as being 'model-independent' or 'model-dependent'. Model-independent approaches are most common in the regulatory setting. They are defined by an explicit separation of the dose–response continuum into an 'observed range' and an 'extrapolated range'. Since they are model-independent, no attempt is made to quantify any detailed biological associations below the observed range of data. Instead a point of departure (POD) is specified that defines where the observed data end. Traditionally, the POD has been identified as the No Observed adverse Effects Level (NOEL) or, when that is unavailable, a Lowest Observed adverse Effects Level (LOEL). Extrapolation to lower doses is then performed using either a linear or a nonlinear model. The linear model is simply a straight line from the POD down to the origin. The slope of the line can be used to determine a 'slope factor' that is used to represent a plausible upper bound on the potency [86]. This approach is usually employed when assessing carcinogens. Nonlinear extrapolation relies on a number of uncertainty factors (UFs) that are applied in order to reach a reference dose (RfD) that represents a daily dose that would result in no appreciable deleterious health effects over an individual's lifetime. Reference doses are commonly used when assessing noncancer risks from potentially toxic elements in, for example contaminated soils [31].

Uncertainty factors used in the estimation of a 'safe' or threshold dose are necessary reductions to account for the lack of data and inherent uncertainty in these extrapolations. The majority of UFs commonly employed have little biological basis, and are more of a precautionary approach to ensuring thresholds are 'safe' for the majority of the human population. For example, the majority of agencies use a default 100-fold factor to address the extrapolation of a NOEL found in a chronic animal study to the subthreshold dose for humans. This 100-fold factor reflects a 10-fold factor for experimental animal-to-human extrapolation and a 10-fold factor for extrapolation of an average human NOEL to a sensitive human NOEL. For the majority of trace elements, data on their toxicity in humans are very limited. Therefore, data from toxicological experiments using laboratory animals are used as the basis of the assessment, and a 10-fold UF is applied to the data. The basic assumptions made in order to employ this UF are that the results seen in experimental animals are relevant to humans, that toxicokinetic and toxicodynamic differences exist between species, and that humans are more sensitive than animals when exposed to a given dose or concentration. Again, there is a paucity of information regarding sensitive individuals in the human population and the toxicity of specific trace elements to different population subgroups. A further 10-fold UF is used to address this uncertainty. The use of this UF assumes that variability in response from one human to the next occurs and, if dealing with epidemiological data, that this variability could not be detected in the study, for example due to low statistical power. Use of this UF also assumes that subgroups exist in the human population that are more sensitive to the toxicity of the trace element than the average population.

To address the limitations of UFs, probabilistic approaches have been proposed. Each UF is characterized by a probability distribution, with the aim to characterize the dose–response as a dose versus risk function. A dose associated with a *de minimus* risk can then be derived from this function. The underlying uncertainty in this estimate can be characterized using the uncertainty components of the input distributions. This approach is illustrated in Figure 12.6 (after Evans *et al.* [87]). For each individual animal in a toxicological experiment, a threshold dose is assumed (Figure 12.6a). Thresholds are assumed to vary for the different animals in the critical experiment. Assuming that animal sensitivity follows a lognormal distribution, a probit model may be fitted to the animal threshold data (Figure 12.6b) providing a straight-line dose–response curve. To construct the human dose–response relationship, the animal dose is scaled to a human equivalent dose by taking a single value (that is the median value) from the interspecies extrapolation distribution (Figure 12.6c). A distribution assumed to reflect the variability among humans is used to define the spread of the human dose–response curve, which in this example is described by a lognormal distribution. A dose associated with a *de minimus* risk level (for example 1 per 100 000) can be established from the resulting human dose–response relationship. Quantitative measures of uncertainty in the *de minimus* dose estimate can be derived from the probabilistic description of uncertainty. This description is a

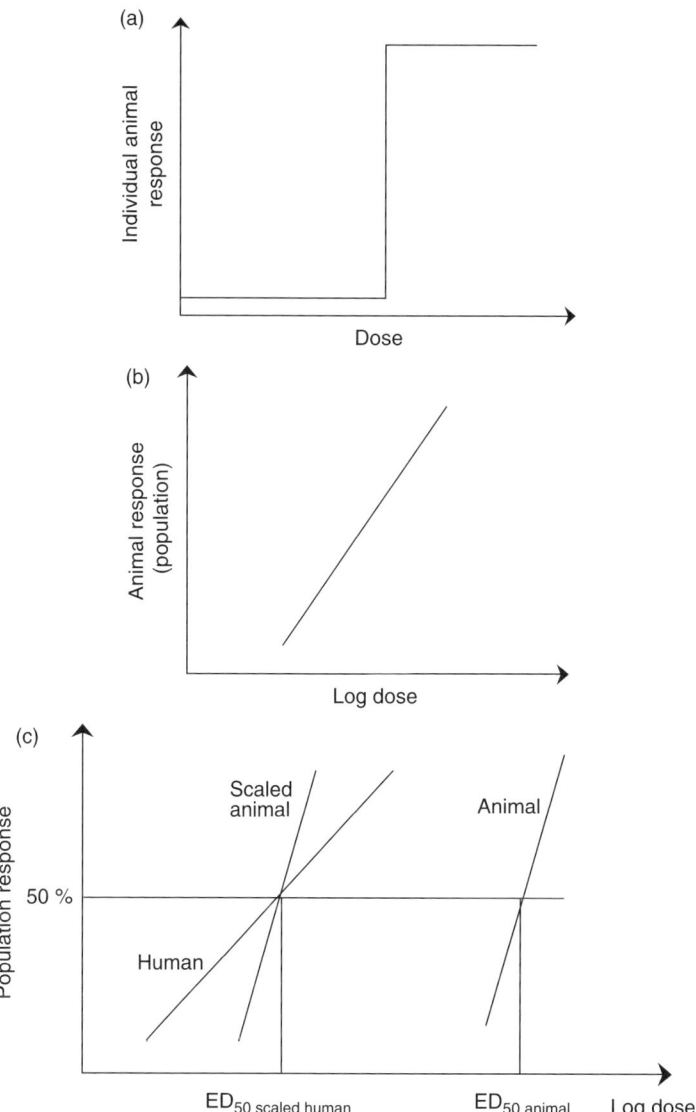

Figure 12.6 (a) A threshold is assumed to exist for each individual animal in a toxicological experiment, the threshold varies between individuals. The distribution of these responses may be assumed to follow a log-normal distribution (common for biological responses). (b) Assuming the log-normal assumption is correct, a probit model can be fitted to the individual animal responses described in (a), thus providing a straight-line dose–response relationship. (c) A human dose–response relationship is then constructed. The animal dose is scaled to a human equivalent dose by taking a single value (that is the median value, in this case the effective dose that results in the outcome of interest being realized on 50% of the study population, ED_{50}) from the interspecies extrapolation distribution. (Modified from Evans, J. S.; Rhomberg, L.R.; Williams, P.L. et al., Reproductive and developmental risks from ethylene oxide: a probabilistic characterization of possible regulatory thresholds, *Risk Anal.* 2001, 21, 697–717 [87].)

convolution of distributions representing the uncertainties in interspecies extrapolation, human variability and other factors such as quality of data. This distribution is assumed to portray the confidence in the dose estimate associated with the *de minimus* risk.

Model-dependent approaches employ statistical or biological-based models to perform the extrapolation down to low doses. In many cases, a number of models will provide satisfactory fit to the experimental data. However, implicit in this approach is the presumption that all underlying assumptions in the model are valid below the observed range of data. This approach, assuming the model is 'true', enables statistical analysis of estimates of 'safe' doses as well as estimating bounds of uncertainty – essential for evaluating the model itself. Estimates from extrapolation models may also be compared to estimates made using model-independent approaches. This may be achieved by treating model-dependent estimates as mean values of a distribution of possible values, and model-independent estimates as bounding values of the same distribution.

One of the main issues with model-dependent approaches is the accurate characterization of model uncertainty. For example, it is well known that different empirical cancer dose–response models (for example Weibull, One-hit) may fit experimental data equally well while differing by orders of magnitude when extrapolated to low doses [88]. More recently, the increased use of physiologically based pharmacokinetic (PBPK) models which provide little or no characterization of uncertainty or variability, presents similar difficulties. Different PBPK models for the same toxicant can provide 'safe' estimates that differ by as much as 10-fold [89].

12.5 Risk Characterization

Risk characterization is the final stage in the risk assessment process. At this point, the information gathered in the three previous stages is combined to make an estimate of risk. For potentially toxic elements in soils, it is most likely that some form of quantitative estimate of risk is required. Any such estimate will be sensitive to the inherent uncertainties in the processes described in the previous sections of this chapter. It is important, if risks are to be communicated, that some indication of this uncertainty is also presented. In this section, those uncertainties within the control of the analyst are discussed, and when using information from other studies, how uncertainties can be recognized.

Once a causal association between a risk factor for a specific potentially toxic element in soil and an adverse outcome is considered likely, interest often focuses on the quantitative relationship with risk. Inaccuracy of exposure assessment will impact both on the final outcome of epidemiological studies, and also how this relates to risk [90]. For final risk characterization, it is important that the analyst has an understanding of the different types of potential measurement error in order to evaluate the weight of evidence provided by a single study.

Improving the accuracy of exposure measurement will in general improve statistical power and the accuracy of the estimate of the effect of the exposure. However, improving measurement accuracy usually requires resources that could be used to enhance other aspects of the study or simply to increase study size, a measure that will increase statistical power. Due to this trade-off, it is usual to accept a certain level of error in exposure measurements, including the use of proxy measures of exposure. It is important to understand the different types of error associated with different measurements of exposure and how these may impact on final estimates of risk.

As with most forms of measurements, errors in exposure estimates may be differential or nondifferential, random or systematic, and of varying magnitude. Additionally, as measurements of exposure can either be direct or 'proxy', errors can also be described as being 'classical' or 'Berkson'. Classical error occurs when the average of many replicate measurements of the same exposure would approximately equal the true exposure. Berkson-distributed error occurs when the same approximate exposure ('proxy') is used for many subjects (for example 'near', 'medium', 'far' from source of contamination, with each category being assigned a value or concentration of the trace element of interest); the true exposures vary randomly around this proxy, with mean equal to it. Prediction models also tend to result in predominantly Berkson-distributed error. In general terms, random measurement error leads to bias in effect measures (estimates of relative risk, regression coefficients, differences in means, etc.). This bias is usually towards the null and therefore tends to underestimate effects associated with exposure. Classical errors bias linear regression coefficients towards zero. For logistic and log-linear (Poisson) regression coefficients the same qualitative result is true. Berkson errors, however, lead to no bias in linear regression coefficients, and little or no bias in logistic or log-linear regression coefficients. Therefore, the distinction between classical and Berkson error is very important when assessing studies for inclusion within the assessment.

All types of nondifferential random measurement error will reduce the probability that a study will find a statistically significant association if one is present (that is reduce study power). This is true for both Berkson and classically distributed errors. Differential error can bias estimations of effect either to under- or overestimate the effect. This depends on whether adverse outcomes are associated with over- or underestimation of exposure. Significance tests are invalid in the presence of differential error.

When selecting data for use in the risk characterization process, it is often not possible from the information available to estimate the magnitude of error associated with measures of exposure. However, the basic designs of most studies are widely reported. The analyst needs to judge whether detection of a causal relationship is reliant primarily on statistical power (for example for rare outcomes), or on the quality of the effect measure. Where bias in effect measure is the major consideration, rather than study power, a study design that utilizes the fact that Berkson error causes little bias is usually more appropriate. Berkson-type error will bias effect measures little if at all, but this design will reduce overall study power.

Once data for exposure estimation have been selected, a central issue is 'what can we expect from estimates of exposure?' Quantitative exposure assessment has been a routine tool for the assessment of land contamination for over 20 years. International jurisdictions have introduced exposure tools based, to a greater or lesser extent, on the USEPA exposure factors handbook and risk assessment guidance for Superfund released in the late 1980s. Exposure assessments estimate (i) an intake – a dose (mg kg^{-1} body weight d^{-1}) – that can be presented, following dose–response assessment, as a risk estimate (excess lifetime or annual risk) or unitless hazard quotient; or (ii) by reverse estimation, a soil guideline – a concentration (mg contaminant kg^{-1} soil) informed by regulatory interpretation of acceptable risk through interpretation of the baseline toxicology. In both cases, a quantitative surrogate for the risk is presented. These estimates and guideline values are not only subject to the uncertainties of environmental fate assumptions and receptor characteristics captured within, but also to the policy decisions of governments and agencies seeking to uphold a consistency of approach to risk management. Unsurprisingly, these factors have marked influences on the characterization of risk (Figure 12.7).

(a)
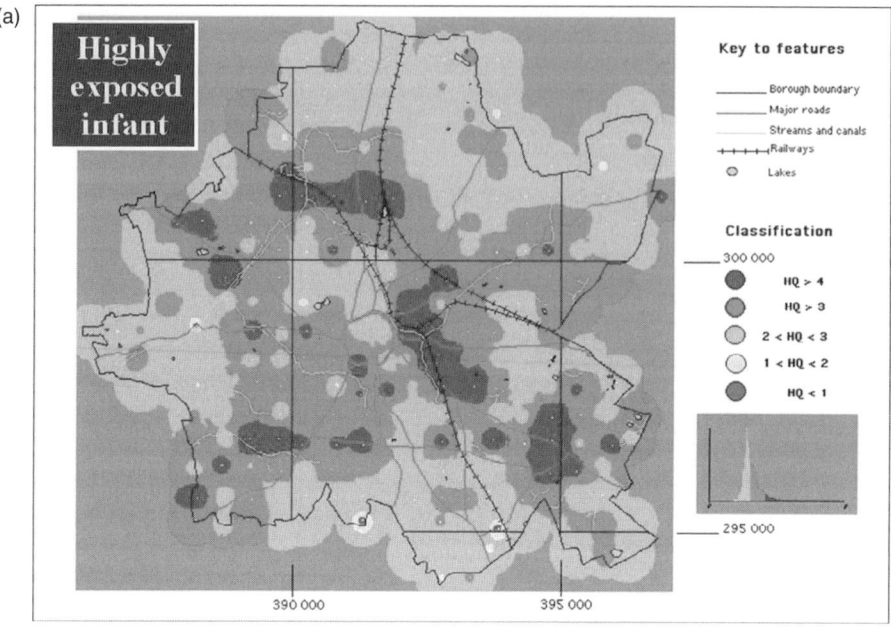

(b)
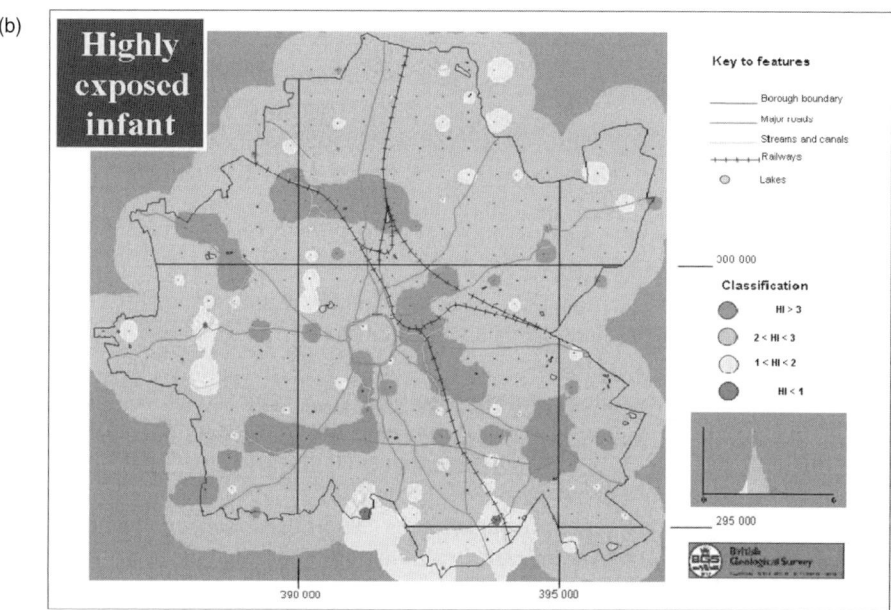

Figure 12.7 Maps of relative risk associated with land contamination in an UK urban area estimated using tolerable daily intakes derived from the principal source documents stipulated by (a) the UK Environment Agency [31]; and (b) the USEPA (Reproduced with permission from R.L. Hough et al., Assessing potential risk of heavy metal exposure from consumption of home-produced vegetables by urban populations, Environmental Health Perspectives, 2004, 112, 215–221 [31].)

Most countries require that reference doses (and, hence, tolerable daily intakes) are derived using a hierarchy of specific source documents. In the United Kingdom, principal source documents reflect toxicological interpretations derived from authoritative UK expert bodies. When further evidence or data are required, the hierarchy of information sources defers to European documentation, with non-EC source documents only used as a last resort. In the United States, a similar approach is adopted, except that information/data derived from authoritative US bodies are most valued. These and other stipulations become embedded into exposure assessment tools and health criteria values and inevitably lead to wide differences in risk characterizations adopted in different countries (Figure 12.7).

In the United States, the quantitative risk assessment (QRA) approach has provoked strong opposition, especially from campaigning groups who have claimed that decisions derived from QRA are too sensitive to the subjective judgements of 'experts'. The environmental justice movement has been critical, recognizing that by '[...] changing an assumption here, or tweaking an equation there, risk assessors can justify virtually any policy they desire' [13]. The inference is that decisions informed by QRA can be highly influenced by political, economic and scientific pressures. Although some decision-makers see a numerical presentation of a risk as an improvement over more ambiguous characterizations, others view these numerical graduations as deceptive, shrouding subjectivity with numerical certainty [13].

12.6 Assessment of Mixtures and Disparate Risks

The nature of toxicological and epidemiological study, and the risk assessment paradigm itself, means that for the majority of situations where trace elements may pose a risk to health or the environment, each trace element will be assessed in turn. This has been a long-standing limitation of the risk assessment approach, and in many ways is an extension of bioavailability. At one time, all risk assessments of trace elements would have been carried out using an estimation of the 'total' concentration, or some other operationally defined fraction that could be considered as a proxy measure of bioavailability. More recently, analysts have moved towards risk estimates that relate only to the bioavailable fraction of the trace element of interest. However, where exposure to more than one trace element was occurring (as is usually the case where receptors are being exposed to soil), risks from individual elements were usually combined additively [31]. This approach fails to take into account any interactions that may occur between different trace elements. For example, it has been shown that small, individually nontoxic amounts of certain environmental pollutants can cause significant changes in animal physiology when combined with other pollutants at nontoxic levels [91]. Investigations looking at physiological responses of sheep grazing on pastures amended with sewage sludge applied at levels below regulatory limits have also revealed toxic responses [92–94], although the statistical significance of these findings is marginal. Measured concentrations of potentially toxic elements in sheep tissues have also been nonintuitive. For example, concentrations of Zn in lamb tissues were shown to be higher in control rather than sludge-exposed animals. Differences were also observed between ewes and lambs [93]. These differences in pattern may reflect different patterns of accumulation in animal tissues resulting from different patterns of uptake. Specific evidence for this has been shown for Pb and Cd uptake by lactating rodents [95].

Similarly, it is difficult to assess overall risk to a receptor that is being exposed to trace elements from a variety of different pathways. Knowledge of which exposures have the greatest impacts for the same dose, and how different routes of exposure may interact is very limited. Under current methodologies, it is usual to combine all doses into a single dose that is then compared to some form of 'safe' dose, for example a reference value. Alternatively, risks from different exposure routes will be calculated separately, and these risk estimates then combined (usually additively) to provide a single prediction of risk. As with exposures to multiple trace elements, the uncertainties associated with applying such approaches are of unknown magnitude.

12.7 Conclusions

Assessing potential risks associated with trace elements in soils has become an integral part of environmental and sustainable development policy in many countries. However, it is important to be aware of the limitations and uncertainties built into the risk assessment paradigm. Quantitative risk assessment, because the output of the process is usually a definite quantified statement, can provide an illusion of accuracy and certainty. It is important to remember that most risk estimates for trace elements are based on dose–response data from artificial experimental situations that look at supra-environmental concentrations of single chemicals. There are huge uncertainties associated with extrapolation from high to low doses, from animal models to potential human impacts, from single studies focused on one specific route of exposure to a more general environmental situation where receptors are exposed through a variety of routes. Our understanding of how potentially toxic elements interact in the environment, the most appropriate ways of measuring or predicting bioavailability (both in terms of uptake by food crops, and once ingested), and how impacts vary across different routes of exposure, is extremely limited in toxicology terms.

Although risk assessment approaches provide a useful framework for the analysis of toxicological and epidemiological evidence, and the evaluation of potential ill effects associated with trace elements, the paradigm can only provide a guide. The incorporation of some aspects of bioavailability (both in terms of uptake by food plants, and in terms of adsorption across the gut post ingestion) into the risk assessment process was a first step to improving its relevance to real environmental situations. The next big steps will be to learn more about mixtures, interactions and combined doses at subtoxic levels.

References

[1] Samet, J.M., Risk assessment and childe health; Pediatrics 2004, 113, 952–956.
[2] El-Sayed, N.A.; Mahfouz, A.A.; Nofal, L.; Ismail, H.M.; Gad, A.; Abu Zeid, H., Iodine deficiency disorders among children in Upper Egypt: an epidemiologic study. J Trop. Pediatr 1998, 44, 270–274.

[3] Staessen, J.A.; Roels, H.A.; Emelianov, D. *et al.*, Environmental exposure to cadmium, forearm bone density, and risk of fractures: perspective population study; Lancet 1999, 353, 1140–1144.
[4] Hough, R.L., Soil and human health: an epidemiological review; Eur. J. Soil Sci. 2007, 58, 1200–1212.
[5] Hough, R.L.; Tye, A.M.; Crout, N.M.J.; McGrath, S.P.; Zhang, H.; Young, S.D., Evaluating a 'Free Ion Activity Model' applied to metal uptake by *Lolium perenne* L. grown in contaminated soils; Plant Soil 2005, 270, 1–12.
[6] NRC (National Research Council), Risk Assessment in the Federal Government: Managing the Process; National Academy Press, Washington, DC, USA, 1983.
[7] Hooda, P.S.; Henry, C.J.K.; Seyoum, T.A.; Armstrong, L.D.M.; Fowler, M.B., The potential impact of soil ingestion on human mineral nutrition; Sci. Total Environ. 2004, 333, 75–87.
[8] Paustenbach, D.J., Human and Ecological Risk Assessment: Theory and Practice; John Wiley & Sons, Inc., New York, 2002.
[9] USEPA, Toxic Substances Control Act. Discussion of premanufacture testing policies and technical issues: request for comment; Federal Register 1979, 44, 16240–16292.
[10] Mensink, B.J.W.G.; Montforts, M.; Wijkhuizen-Maslankiewicz, L.; Tibosch, H.; Linders, J.B.H.J., Manual for Summarising and Evaluating the Environmental Aspects of Pesticides, Report no. 67910022; Blithoven: National Institute of Public Health and the Environment, RIVM, 1995.
[11] Hart, J.W.; Hansen, B.G.; Karcher, W., Hazard assessment and risk assessment of chemical substances in the EU; in: Pugh, D.M.; Tarazona, J.V. (eds), Regulation for Chemical Safety in Europe: Analysis, Comment and Criticism. Kluwer Academic, 1998, pp. 113–125.
[12] Tarazona, J.V.; Fresno, A.; Aycard, S.; Ramos, C.; Vega, M.M.; Carbonell, G., Assessing the potential hazard of chemical substances for the terrestrial environment. Development of hazard classification criteria and quantitative environmental indicators; Sci. Total Environ. 2000 247, 151–164.
[13] Wartenberg, D.; Chess, C., The risk wars: assessing risk assessment; in: Levenstein, C.; Wooding, J. (eds.), Work, Health, and Environment: Old Problems, New Solutions; The Guildford Press, 1997, 258–274.
[14] DoE (Department of the Environment), A Guide to Risk Assessment and Risk Management for Environmental Protection; London, The Stationary Office, 1995.
[15] Pollard, S.J.T.; Hickman, G.A.W.; Irving, P. *et al.*, Exposure assessment of carcase disposal options in the event of a notifiable exotic animal disease – application to avian influenza virus; Environ. Sci. Technol. 2008. doi: 10.1021/es702918d.
[16] Stephens, C.; Hough, R.L.; Busby, A.; Edwards, D., Assessing the Health Risk and Impact Associated with Living on the Thames View Estate: a Rapid Literature Review of methods; Barking & Dagenham Primary Care Trust, Barking, London, January 2004.
[17] Rodricks, J.V., Calculated Risks: The Toxicity and Human Health Risks of Chemicals in Our Environment. Cambridge University Press, Cambridge, 1992.
[18] Reilly, C., Metal Contamination of Food; Elsevier Applied Science, London and New York, 1991.
[19] Sterrett, S.B.; Chaney, R.L.; Gifford, C.H.; Mielke, H.W., Influence of fertilizer and sewage sludge compost on yield and heavy metal accumulation by lettuce grown in urban soils; Environ. Geochem. Health 1996, 18, 135–142.
[20] Chronopoulos, J.; Haidouti, C.; Chronopoulou-Sereli, A.; Massas, I., Variations in plant and soil lead and cadmium content in urban parks in Athens, Greece; Sci. Total Environ. 1997, 196, 91–98.
[21] Moir, A.M.; Thornton, I., Lead and cadmium in urban allotment and garden soils and vegetables in the United Kingdom; Environ. Geochem. Health 1989, 11, 113–119.
[22] Sanchez-Camazano, M.; Sanchez-Martin, M.J.; Lorenzo, L.F., Lead and cadmium in soils and vegetables from urban gardens of Salamanca (Spain); Sci. Total Environ. 1994, 146/147, 163–168.
[23] Wong, J.W.C., Heavy metal contents of vegetables and market garden soils in Hong Kong; Environ. Technol. 1996, 17, 407–414.
[24] Culbard, E.B.; Thornton, I.; Watt, J.; Wheatley, M.; Moorcroft, S.; Thompson, M., Metal contamination in British urban dusts and soils; J. Environ. Qual. 1988, 17, 226–234.
[25] Crews, H.M.; Davies, B.E., Heavy metal uptake from contaminated soils by six varieties of lettuce (*Lactuca sativa* L.); J. Agric. Sci. 1985, 105, 591–595.

[26] Hamon, R.E.; Wundke, J.; McLaughlin, M.; Naidu, R., Availability of zinc and cadmium to different plant species; Aust. J. Soil Res. 1997, 35, 1267–1277.
[27] Bell, P.F.; Chaney, R.L.; Angle, J.S., Free metal activity and total metal concentrations of micronutrient availability to barley *Hordeum vulgare* (L.) cv 'Klages'; Plant Soil 1991, 130, 51–62.
[28] Davies, B.E., Inter-relationships between soil properties and the uptake of cadmium, copper, lead and zinc from contaminated soils by radish (*Raphanus sativus* L.); Water Air Soil Pollut. 1992, 63, 331–342.
[29] Andrewes, P.; Town, R.M.; Hedley, M.J.; Loganathan, P., Measurement of plant available cadmium in New Zealand soils; Aust. J. Soil Res. 1996, 54, 441–452.
[30] Cieslinski, G.; Nielsen, G.H.; Hogue, E.J., Effect of soil cadmium application and pH on growth and cadmium accumulation in roots, leaves and fruit of strawberry plants (*Fragaria* × *ananassa* Dutch.); Plant Soil 1996, 180, 267–276.
[31] Hough, R.L.; Breward, N.; Young, S.D. *et al.*, Assessing potential risk of heavy metal exposure from consumption of home-produced vegetables by urban populations; Environ. Health Perspect. 2004, 112, 215–221.
[32] Kacker, T.; Haupt, E.T.K.; Garms C.; Francke, W.; Steinhart, H., Structural characterisation of humic bound PAH residues in soil by 13c-CPMAS-NMR-Spectroscopy: evidence of covalent bonds; Chemosphere 2002, 48, 117–131.
[33] Keefer, R.F.; Singh, R.N.; Horvath, D.J., Chemical composition of vegetables grown on an agricultural soil amended with sewage sludges; J. Environ. Qual. 1986, 15, 146–152.
[34] Bidwell A.M.; Dowdy, R.H., Cadmium and zinc availability to corn following termination of sewage sludge applications; J. Environ. Qual. 1987, 16, 438–442.
[35] Jensen, H.; Mosbaek, K.H., Relative availability of 200 years old cadmium from soil to lettuce; Chemosphere 1990, 20, 693–702.
[36] Jackson, A.P.; Alloway, B.J., The bioavailability of cadmium to lettuce and cabbage in soils previously treated with sewage sludges; Plant Soil 1991, 132, 179–186.
[37] Pieri, L.A.D.; Buckley, W.T.; Kowalenko, C.G., Micronutrient content of commercially grown vegetables and of soils in the Lower Fraser Valley of British Columbia; Can. J. Soil Sci. 1996 76, 173–182.
[38] Jinadasa, K.B.P.N.; Milham, P.J.; Hawkins, C.A. *et al.*, Survey of cadmium levels in vegetables and soils of Greater Sydney, Australia; J. Environ. Qual. 1997, 26, 924–933.
[39] Sauve, S.; Hendershot, W.; Allen, H.E., Solid-solution partitioning of metals in contaminated soils: dependence on pH, total metal burden, and organic matter; Environ. Sci. Technol. 2000, 34, 1125–1131.
[40] Konz, J.; Lisi, K; Friebele, E., Exposure Factors Handbook, USEPA, Office of Health and Environmental Assessment, EPA/600/8–89/043, 1989.
[41] Stanek, E.J.; Calabrese, E.J.; Zheng, L., Soil ingestion estimates in children: Influence of sex and age; Trace Substances in Environmental Health 1990, XXIV, 33–43.
[42] Calabrese, E.J.; Stanek, E.J.; Pekow, P.; Barnes, R.M., Soil ingestion estimates for children residing on a superfund site. Ecotoxicol. Environ. Saf. 1997, 36, 258–268.
[43] Calabrese, E.J.; Stanek, E.J.; Barnes, R.M., Soil ingestion rates in children identified by parental observation as likely high soil ingesters; J. Soil Contam. 1997, 6, 271–279.
[44] Duggan, M.J.; Williams, S., Lead-in-dust in city streets; Sci. Total Environ. 1977, 7, 91–97.
[45] Davies, S., Quantitative estimates of soil ingestion in normal children between the ages of 2 and 7 years: population based estimates using aluminium, silicon and titanium as soil tracer elements; Arch. Environ. Health 1990, 45, 112–122.
[46] Ferguson, C.; Marsh, J., Assessing human health risks from ingestion of contaminated soil; Land Contam. Reclam. 1993, 1, 177–185.
[47] Lepow, M.L.; Bruckman, L.; Rubino, R.A.; Markowitz, S.; Gillette, M.; Kapish, J., Investigations into sources of lead in the environment of urban children; Environ. Res. 1975, 10, 415–426.
[48] Lepow, M.L.; Bruckman, L.; Rubino, R.A.; Markowitz, S.; Gillette, M.; Kapish, J., Role of airborne lead and increased body burden of lead in Hartford children; Environ. Health Perspect. 1974, 7, 99–102.
[49] Roels, H.A.; Buchet, J.P.; Lauwerys, R.R., Exposure to lead by the oral and pulmonary routes of children living in the vicinity of a primary lead smelter; Environ. Res. 1980, 22, 81–94.

[50] Sheppard, S.C.; Evenden, W.G., Concentration enrichment of sparingly soluble contaminants (U, Th and Pb) by erosion and by soil adhesion to plants and skin; Environ. Geochem. Health 1992, 14, 121–131.
[51] Stanek, E.J.; Calabrese, E.J., Daily estimates of soil ingestion in children; Environ. Health Perspect. 1995, 103, 276–285.
[52] Calabrese, E.J.; Stanek, E.J.; James, R.C.; Roberts, S.M., Soil ingestion: a concern for acute toxicity in children; J. Environ. Health 1999, 61, 18–23.
[53] Stanek, E.J.; Calabrese, E.J.; Zorn, M., Soil ingestion estimates for Monte Carlo risk assessment in children; Hum. Ecol. Risk Assess. 2001, 7, 357–368.
[54] Feldman, M.D., Pica: current perspectives; Psychosomatics 1986, 27, 519–523.
[55] Sheppard, S.C., Geophagy: who eats soils and where do the possible contaminants go? Environ. Geol. 1998, 33, 109–114.
[56] Calabrese, E.J.; Stanek, E.J.; Gilbert, C.E., Lead exposure in a soil pica child; J. Environ. Sci. Health 1993, A28, 353–362.
[57] Calabrese, E.J.; Stanek, E.J.; Gilbert, C.E., Evidence of soil-pica behaviour and quantification of soil ingestion; Hum. Exp. Toxicol. 1991, 10, 245–249.
[58] Edwards, R.D.; Yurkow, E.; Lioy, P.J., Seasonal deposition of housedusts onto household surfaces; Sci. Total Environ. 1998, 224, 69–80.
[59] Thatcher, T.L.; Layton, D.W., Deposition, re-suspension, and penetration of particles within a residence; Atmos. Environ. 1995, 29, 1487–1497.
[60] Trowbridge, P.R.; Burmaster, D.E., A parametric distribution for the fraction of outdoor soil in indoor dust; J. Soil Contam. 1997, 6, 161–168.
[61] Georgopoulos, P.G.; Roy, A.; Yonone-Lioy, M.J.; Opiekun, R.E.; Lioy, P.J., Environmental copper: its dynamics and human exposure issues; J. Toxicol. Environ. Health 2001, B4, 341–394.
[62] McArthur, B., Dermal measurement and wipe sampling methods: a review; Appl. Occupational Environ. Hyg. 1992, 7, 599–606.
[63] Lewis, R.G.; Fortmann, R.C.; Camann, D.E. Evaluation of methods for monitoring the potential exposure of small children to pesticides in the residential environment; Arch. Environ. Contam. Toxicol. 1994, 26, 37–46.
[64] Kenny, L.C.; Aitkin, R.; Chalmers, C. et al., A collaborative European study of personal inhalable aerosol sampler performance; Ann. Occup. Hyg. 1997, 41, 135–153.
[65] Charlesworth, S.; Everett, M.; McCarthy, R.; Ordonez, A.; Miguel, E.D., A comparative study of heavy metal concentration and distribution in deposited street dusts in a large and a small urban area: Birmingham and Coventry, West Midlands, UK; Environ. Int. 2003, 29, 563–573.
[66] Pakkanen, T.A.; Kerminen, V.; Loukkola, K. et al., Size distributions of mass and chemical components in street-level and rooftop PM_1 particles in Helsinki; Atmos. Environ. 2003, 37, 1673–1690.
[67] Hammel, S.C.; Buckley, B.; Lioy, P.J., Bioaccessibility of metals in soils for different liquid to solid ratios in synthetic gastric fluid; Environ. Sci. Technol. 1998, 32, 358–362.
[68] Gasser, U.G.; Walker, W.J.; Dahlgren, R.A.; Borch, R.S.; Burau, R.G., Lead release from smelter and mine waste impacted materials under simulated gastric conditions in relation to speciation; Environ. Sci. Technol. 1996, 32, 761–769.
[69] Ruby, M.V.; Davis, A.; Kempton, J.H.; Drexler, J.W.; Bergstrom, P.D., Lead bioavailability: Dissolution kinetics under simulated gastric conditions; Environ. Sci. Technol. 1992, 26, 1242–1248.
[70] Hack, A.; Selenka, K., Mobilization of PAH and PCB from contaminated soil using a digestive tract model; Toxicol. Lett. 1996, 88, 199–210.
[71] Diamond, G.L.; Goodrum, P.E.; Felter, S.P.; Ruoff, W.L., Gastrointestinal absorption of metals; Drug Chem. Toxicol. 1998, 21, 223–251.
[72] Ruoff, W.L.; Diamond, G.L.; Velazquez, S.F.; Stiteler, W.M.; Gefell, D.J., Bioavailability of cadmium in food and water: a case study on the derivation of relative bioavailability factors for inorganics and their relevance to the reference dose; Regul. Toxicol. Pharm. 1994, 20, S139–S160.
[73] Casteel, S.W.; Cowart, R.P.; Weis, C.P. et al., Bioavailability of lead to juvenile swine dosed with soil from the Smuggler Mountain NPL site of Aspen, Colorado; Fund. Appl. Toxicol. 1997, 36, 177–187.

[74] Fouchecourt, M.O.; Arnold, M.; Berny, P.; Videmann, B.; Rether, B.; Riviere, J.L., Assessment of the bioavailability of PAHs in rats exposed to a polluted soil by natural routes: reduction of EROD activity and DNA adducts and PAH burden in both liver and lung; Environ. Res. 1999, 80, 330–339.
[75] Linden, A.; Olsson, I.; Bensryd, I.; Lundh, T.; Skerfving, S.; Oskarsson, A., Monitoring of cadmium in the chain from soil via crops and feed to pig blood and kidney; Ecotoxicol. Environ. Saf. 2003, 55, 213–222.
[76] Blake, K., Absorption of ^{203}Pb from gastrointestinal tract of man; Environ. Res. 1976, 11, 1–4.
[77] Rabinowitz, M.; Koppel, J.; Wetherill, G., Effect of food intake on fasting gastrointestinal lead absorption in humans; Am. J. Clin. Nutr. 1980, 33, 1784–1788.
[78] Rabinowitz, M.; Wetherill, G.; Koppel, J., Kinetic analysis of lead metabolism in healthy humans; J. Clin. Invest. 1976, 58, 260–270.
[79] Heard, M.; Chamberlain, A., Effect of minerals and food on uptake of lead from the gastrointestinal tract of humans; Hum. Toxicol. 1982, 1, 411–445.
[80] Blake, K.; Barbezat, G.; Mann, M., Effect of dietary constraints on the gastrointestinal absorption of ^{203}Pb in man; Environ. Res. 1983, 30, 182–187.
[81] Blake, K.; Mann, M., Effect of calcium and phosphorus on the gastrointestinal absorption of ^{203}Pb in man; Environ. Res. 1983, 30, 188–198.
[82] James, H.M.; Milburn, M.E.; Blair, J.A., Effect of metals and meal times on uptake of lead from the gastrointestinal tract of humans; Hum. Toxicol. 1985, 4, 401–407.
[83] Lanphear, B.P.; Matte, T.D.; Rogers, J. et al., The contribution of lead contaminated house dust and residential soil to children's blood lead levels. A pooled analysis of 12 epidemiologic studies; Environ. Res. 1998, 79, 51–68.
[84] Meyer, I.; Heinrich, J.; Lippold, U., Factors affecting lead, cadmium, and arsenic levels in housedust in a smelter town in Eastern Germany; Environ. Res. 1999, 81, 32–44.
[85] Roberts, J.W.; Dickey, P., Exposure of children to pollutants in housedust and indoor air; Rev. Environ. Contam. T. 1995, 143, 59–78.
[86] USEPA., Guidelines for Carcinogen Risk Assessment. US Environmental Protection Agency, Risk Assessment Forum, Washington DC; EPA/630/P-03/001 B, 2005.
[87] Evans, J. S.; Rhomberg, L.R.; Williams, P.L. et al., Reproductive and developmental risks from ethylene oxide: a probabilistic characterization of possible regulatory thresholds; Risk Anal. 2001, 21, 697–717.
[88] Crump, K.S.; Howe, R. B., A review of methods for calculating statistical confidence limits in low dose extrapolation. In: Clayson, D.B. et al. (eds), Toxicological Risk Assessment, vol. 1: Biological and Statistical Criteria; CRC Press, Boca Raton, FL, 1985, pp. 187–203.
[89] Chiu, W.A.; Chen, C.; Hogan, K.; Lipscomb, J.C.; Scott, C.S.; Subramaniam, R., High-to-low dose extrapolation: issues and approaches; Hum. Ecol. Risk Assess. 2007, 13, 46–51.
[90] Armstrong, B.G., The effects of measurement errors on relative risk regressions; Am. J. Epidemiol. 1990, 132, 1176–1184.
[91] Rajapakse, N.; Silva, E.; Kortenkamp, A., Combining xenoestrogens at levels below individual no-observed-effect concentrations dramatically enhances steroid hormone action; Environ. Health Perspect. 2002, 110, 917–921.
[92] Erhard, H.W.; Rhind, S.M., Prenatal and postnatal exposure to environmental pollutants in sewage sludge alters emotional reactivity and exploratory behaviour in sheep; Sci. Total Environ. 2004, 332, 101–108.
[93] Rhind, S.M.; Kyle, C.E.; Owen, J., Accumulation of potentially toxic metals in the liver tissue of sheep grazed sewage sludge-treated pastures; Animal Sci. 2005, 81, 107–113.
[94] Paul, C.; Rhind, S.M.; Kyle, C.E.; Scott, H.; McKinnell, C.; Sharpe, R.M., Cellular and hormonal disruption of fetal testis development in sheep reared on pasture treated with sewage sludge; Environ. Health Perspect. 2005, 113, 1580–1587.
[95] Institute for Environmental Health, IEH Report on Factors Affecting the Absorption of Toxic Metals from the Diet (Report R8); MRC Institute for Environment and Health, Leicester, UK, 1998.

13

Regulatory Limits for Trace Elements in Soils

Graham Merrington,[1] Sohel Saikat[2] and Albania Grosso[1]

[1]*wca environment ltd, Brunel House, Volunteer Way, Faringdon, Oxfordshire, UK*
[2]*Health Protection Agency, London, UK*

13.1 Introduction

Derivation and use of limit values for metals in soils are one of several ways of addressing and controlling potential risks from trace elements to the environment and human health. The development of limit values for metals in soil is a relatively new regulatory activity compared with human health protection in the workplace, or ecosystem protection in the aquatic compartment.

Regulatory limit values for trace elements/metals in soils may be necessary for at least three key reasons: remediation and assessment, limiting industrial and domestic emissions, and voluntary initiatives. The selection and prioritization of metals for which development of a limit value is necessary may be influenced by several technical and political drivers underpinned by the need to ensure public safety and environmental protection. Technical considerations for environmental protection should include trace element toxicity, mobility/availability, and frequency of occurrence. The use of hazard criteria for prioritization, often based on physicochemical properties, is suitable for some organic chemicals, but is less relevant for metals which do not share relatively predictable environmental behaviour. For human health, technical considerations such as fate and transport are considered within an exposure scenario, but these are often based on several assumptions. For example, the arsenic limit value in the United Kingdom is based on a child's deliberate or accidental ingestion of soil. This has fostered a growing debate in the United Kingdom and many other Western European and North American countries about the need for a more 'proportionate approach' to risk assessment [1].

Whatever the drivers and decisions behind identification of the need to develop a soil regulatory limit for a metal, the process must be transparent and auditable if it is to be widely accepted. With the growing trend in legal challenge of limit values by the regulated community, it is imperative that the limit value is robust in terms of both science and process.

Regulatory limit values are normally numerical values related to soil that form part of a decision-making process and are ascribed to a particular ecological target/receptor (for example humans) or designated soil use. Many countries have regulatory limits for trace elements in soils that are legally enforceable numerical values. However, there are also other regulatory limits that are not mandatory but are contained in guidelines, codes of practice, advice on critical loads or sets of criteria for deciding individual cases. There is also increasing reliance on screening or trigger values, used as part of a tiered assessment scheme [2,3]. Some regulatory limits are not set by governments, but carry authority for other reasons, especially the scientific eminence or market power of those who set them (for example World Health Organization Guidelines). This chapter contains examples of soil regulatory limit values and their derivation from across jurisdictions in Europe, North America, Australia and New Zealand. However, we have deliberately avoided direct comparison of metal limit values across jurisdictions because it is fundamentally inappropriate to compare values derived and used for entirely different purposes, protection goals and policy [3], as we discuss later.

All types of limit value are set with the intention of meeting a specific protection goal or regulatory end point (the mandate). For example, for the recycling of sewage sludge to agricultural land in the United Kingdom the stated protection goal for soils is to 'prevent harmful effects on soil, vegetation, animals and man' [4]. The linkage of a limit value to a specific protection goal or target is a key characteristic of a robust and implementable standard for any chemical. Figure 13.1 shows what this may mean in practice, with different standards represented by horizontal lines for the same metal for different protection levels related to the sensitivity of the ecological receptor [5]. Importantly, there is a minimum protection level above which the resilience of the ecosystem is not guaranteed. Above this minimum level there is the opportunity for policy makers to influence the protection level by choice. Furthermore, such a scheme may also have implications for how the regulatory limit is derived (with biological effects data being used for less-sensitive sites and 'no effects' data being used for sensitive land uses). There are also implications for how the regulatory limit is used, with the burden of proof and implications of exceedance varying with land use.

It is therefore imperative that the methodologies and assumptions used in calculating limit values and their intended management purpose are understood before application to a soil investigation or assessment. There is also questionable technical merit in listing limit values for trace elements from a number of jurisdictions without information on protection goals and understanding of derivation methods [6]. Table 13.1 gives an example of threshold toxicity values for Cd from The Netherlands and United States Environmental Protection Agency (USEPA), including information on the derivation context. Direct

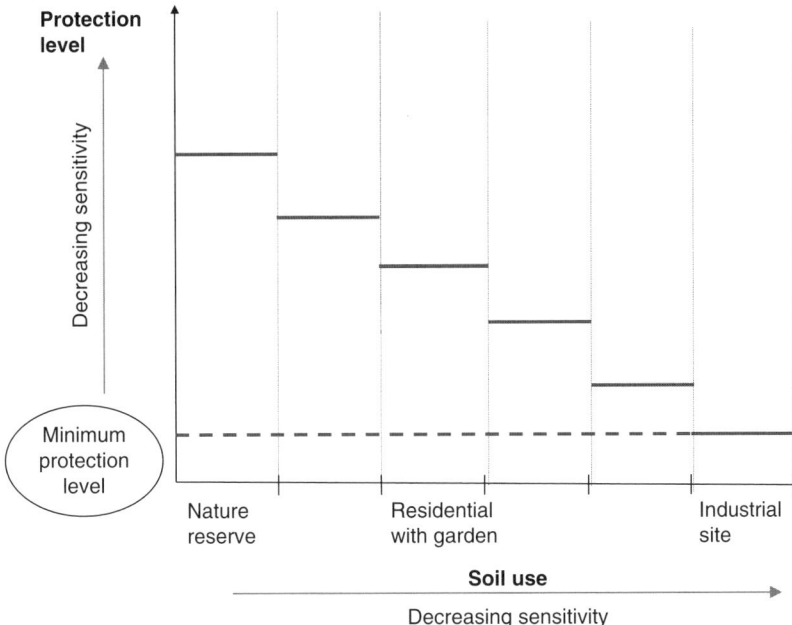

Figure 13.1 *Limit values for metals vary with protection level and sensitivity of soil use*

Table 13.1 *Threshold toxicity values for Cd from the Netherlands and US EPA with some derivation context*

Jurisdiction or organization	Name	What is considered?	Protection goal	Derivation basis	Values (mg kg^{-1})
The Netherlands [68]	Target Values	Background concentration and bioavailability	95% of all species	SSD of NOECs and EC$_{10}$s	0.8
	Intervention Values		50% of all species	As above	13
US EPA [58]	Ecological Soil Screening Levels	High metal availability	Plants	Geometric mean of the lowest NOEC	32
			Invertebrates		140

comparison of these limit values without this information would be a meaningless exercise. However, there are numerous examples where soil limit values for metals derived for the purpose of one environmental protection goal are inappropriately applied to another protection goal in another regulatory regime. In the United Kingdom the sewage sludge limit values are also used in at least six other regulatory regimes concerned with application of composts, sediment dredgings and food wastes to land, and aerial deposition from stack emissions. There is no scientific evidence to suggest that the different protection goals in these other regimes are met by the sludge limits, but no other metal limit values currently exist in the United Kingdom for environmental protection of soils. In addition to the potential scientific and technical mismatch arising from use of sludge limits as 'default' limits for all soils, this also creates confusion and inconsistency across the regulated community [7].

Regulatory soil limit values for metals (or indeed any chemical) should be considered as one tool within a process or regulatory framework, rather than as an end in themselves [8]. Unlike the aquatic compartment, soil limit values are rarely used as compliance measures (that is used as mandatory pass/fail criteria). Furthermore, a broader consideration of limit values should also include an assessment of the economic (for example cost-benefit), technical (for example Best Available Technique Not Entailing Excessive Cost, BATNEEC) and sociopolitical implications of setting a particular value [7,9].

For human health, a regulatory limit value for a metal is generally based on a relationship between the exposure and its effect. This relationship is ideally based on human data, but in practice such data are rarely available for soil-borne trace metals. Toxicological studies are usually conducted with animals (as surrogates of humans) in order to obtain an understanding of the dose and effect relationship. Typically these experiments are conducted with mice and rats, with single metal forms, at high doses and over short exposure periods. However, human exposure is usually chronic, low-level and from complex mixtures [10]. This means that generalization and extrapolation are usually required when deriving limit values, so such values may be conservative. However, the metabolism and deposition of metals varies considerably and effects may be clinically manifested, for example on the fetus.

13.2 Derivation of Regulatory Limits for Trace Elements

13.2.1 Environmental Protection Limit Values for Soils

The methods and processes employed to derive metal limit values for soils have not always been transparent or easy to follow [7,11,12]. Furthermore, methodologies used to set limit values have often included 'expert judgement' or default assumptions and extrapolations which were not always clearly detailed [13]. Limit values for metals in soils are commonly given as 'total' metal concentrations, or in some circumstances as concentrations in the soils as extracted using a range of dilute inorganic salt solutions [14]. Historically the data on which these metal limit values are based were primarily from agricultural field and pot-based trials. These studies targeted effects of metals on crop plants and grazing livestock, but have limited relevance to a broader range of ecoreceptors and ecosystems [13,15,16]. The increased societal desire in the developed world for long-term sustainable land use has led to the need to address the risks from metals in soils more broadly than just in the agro-ecosystem [17,18].

Current methodologies for the derivation of environment limit values for metals use laboratory-based ecotoxicity data [19]. These data are generally for a number of soil test

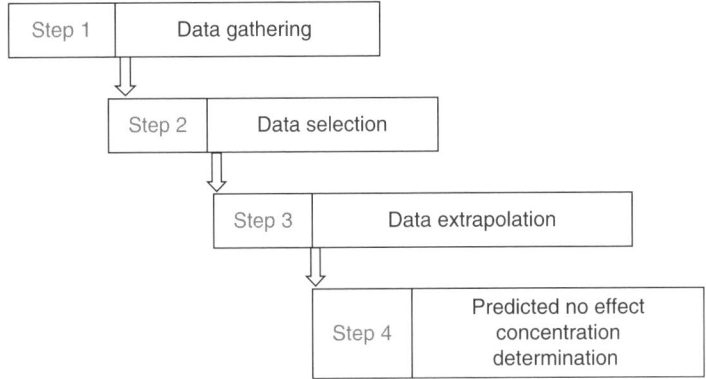

Figure 13.2 *Schematic of a general methodology used for the derivation of environment limit values [5,20]*

organisms exposed to a range of metal concentrations, from which a dose–response relationship can be established and Effect Concentrations (EC_x) or No Observed Effect Concentrations (NOECs) derived. Figure 13.2 shows a generalized schematic of a methodology used for the derivation of environment limit values in parts of the European Union (EU) [5,20]. The process starts with the collation of ecotoxicology data for the metal of interest, from both short- and long-term studies. These data are then reviewed and checked for quality (for example Were the test concentrations measured? Was the ecotoxicity test performed under standard conditions?). It is often the case that terrestrial ecotoxicity data are very limited in quantity relative to aquatic data for the same metal. For example, for vanadium there may be as few as three valid acute soil studies compared with 52 studies for water [5]. Extrapolation of these limited data to account for interspecies variability and interlaboratory and field differences is usually undertaken through the use of assessment factors. The origin of these assessment factors and their applicability to soils ecotoxicity data, and particularly to metals, is unclear.

Some metals are relatively rich in terrestrial ecotoxicity data (for example Cu has 251 reliable chronic NOEC/EC10 values, including 68 for plants, 105 for invertebrates and 78 for microbes). For metals which have large ecotoxicity datasets it is possible to use a probabilistic approach to construct a statistical frequency distribution or species sensitivity distribution (SSD) which describes the relationship between the concentration of the metal in the soil and the fraction of species that might be affected [21]. From these distributions the soil metal concentration at which 95% of the species are protected (the HC_5 or hazardous concentration to 5% of the population) can be determined. If the dataset used in the SSD is highly representative of the system to be protected and the mode of toxic action is well understood, then this may, in some circumstances, be considered as the limit value. Such a plot also enables the limit value to be understood in terms of the proportion of the species affected at different concentrations, for example by identifying the concentration of metal which may affect 50% of the population. The data demands of this methodology for setting limit values vary between jurisdictions, along with how the limit value is finally used. For some jurisdictions these data demands are relatively high ($n \geq 10$ NOECs in the EU Technical Guidance Document with species covering at least eight taxonomic groups [2]). In other jurisdictions (for example the Netherlands) the required base set of ecotoxicological data is relatively limited ($n \geq 4$ NOECs for soil organisms from different taxonomic groups) [22].

Whatever the extrapolation methodology, one drawback when deriving metal limit values is that there is often no consideration of how soil physicochemical and organism-specific factors influence potential toxicity. Importantly, the majority of limit values for metals, like the ecotoxicity data from which they are derived, are based on strong acid (often *aqua regia*) extractable metal. This is primarily because the methodology is standardized and relatively simple to perform. However, the concentrations of soil metals determined from this extraction approach are likely to be a significant overestimation of concentrations that are environmentally relevant (Chapter 11). An additional drawback in using laboratory ecotoxicity data to derive a metal limit value for soils in the field is that the studies are usually undertaken over a relatively short duration and the metals are added to the experimental system as soluble metal salts. Yet in the field metals are often added to soil in a number of different forms (such as in waste materials, as sludges or as aerially deposited particles) which are not as soluble as the salts used in laboratory toxicity tests. Furthermore, on addition to soils, metals age and with time the metal ion availability in the soil decreases [23]. Therefore, limit values for metals derived from laboratory ecotoxicity data can be very precautionary and are often of limited ecological relevance in the field if the methods are applied without common sense, although they are likely to protect even the most sensitive species.

Some jurisdictions acknowledge the importance of soil properties by deriving limit values that all vary generically with pH [4], pH and texture [8] and background metal levels and CEC [22]. However, the specific relationships and data on which these are based are not always easy to establish or scrutinize. The significant effect of soil physicochemical properties upon metal availability, and potentially on ecotoxicology, makes the use of total soil metal concentrations without consideration of 'soil type' a very blunt instrument with which to make decisions associated with metals risk. The problems of using total metal concentrations in a regulatory context have been widely acknowledged across the scientific and regulatory community and include problems in the interpretation of total metals data in the context of environmental behaviour and toxicological effect, lack of ecological relevance and no consideration of the huge natural variation in metal background concentrations in soils [24–26].

Most countries have limit values for metals in soils which are set for either the protection of human health or ecosystem receptors [27]. However, the Canadian Environmental Quality Guidelines and the Dutch intervention and target values integrate both human health and ecological receptors [6,8]. Both methodologies provide detailed and transparent protocols whereby separate limit values for both the environment and human health are derived, and the lower of the two values for a number of land use scenarios is considered as the final value.

Internationally the derivation of limit values for metals is an area of continual change and development in response to improvements in understanding of the ecotoxicity and risk posed by trace elements. This change also reflects different national, regional and state policies associated with soil protection and land management, which is an obvious reason for limited international consistency. However, the development of evidence-based limit values for metals in soils by environmental regulators which incorporate an understanding of metal-specific availability and bioavailability provide a counterpoint to policy makers who call for zero inputs of metals to ensure sustainable soil use and return to 'natural backgrounds' [28].

13.2.2 Human Health Protection Limit Values for Soils

The approach to deriving human health limit values for metals follows a similar principle to that used for environmental limit values. Similarly, values for the same metal vary between countries, reflecting national, state and regional environmental, socioeconomic, technical and legal priorities. As indicated earlier, the objective of setting limit values for metals is risk-based to prevent hazards in the population caused by metals.

Through data evaluation the key hazard is determined and the critical effect on human health is identified. The toxic effects of a metal observed at the lowest concentration or dose (the NOAEL or LOAEL) often provide the basis for setting the limit value. Where a NOAEL (no observed adverse effect level) cannot be identified a LOAEL (lowest observed adverse effect level) is used. Having identified the 'effect level', uncertainty factors (called assessment factors in environmental assessments) are used to extrapolate from an effect level to the tolerable daily intake (TDI). The TDI is an estimate of the amount of a chemical that can be ingested daily over a lifetime without appreciable health risks. It is expressed as mg kg^{-1} body weight per day.

The selection of uncertainty factors depends on chemical toxicity, population age group exposure scenario and the quality of the data. In extrapolating results from animal experiments to predict effects in humans, uncertainty factors of 10 to 1000 are generally used to allow for within-species and between-species differences in sensitivity and to account for differences in the adequacy of the study or database and for the nature and severity of the effect [29]. These factors largely follow the WHO conventions [30]. However, once an uncertainty level of 10 000 or more is reached, authorities generally do not undertake risk assessments and must use other methods to control exposure and determine risk.

For genotoxic and carcinogenic metalloids and metals (for example arsenic (As) and cadmium (Cd)), where a threshold value for the toxic effect cannot be identified (Figure 13.3), the concept of 'lifetime excess risk' can be based on quantitative risk assessment for the derivation of the toxicological value used in the risk assessment. Quantitative risk assessment involves the use of numerical estimates of cancer and risk-based models used to interpret toxicological data [31]. No distinction is made between different types of cancer. The acceptable lifetime excess risk for an individual person can range over orders of magnitude (for example 10^{-2} to 10^{-6}) between different organizations and jurisdictions. An acceptable risk level of 1×10^{-5} indicates one additional cancer in every 100 000 people in an exposed population. Risk levels of 1 in 10 000 (10^{-4}) (for example the Netherlands), 1 in 100 000 (10^{-5}) (for example New Zealand and the United States), and 1 in 1 000 000 (10^{-6}) (for example Canada and Denmark) are used by different countries and sometimes in different types of legislation in the same country. In the United Kingdom, lifetime risks of between 10^{-4} and 10^{-6} are the norm [30], and for As lifetime risk is currently set at approximately 10^{-4} [31]. However, currently the United Kingdom does not use quantitative risk assessment to derive the toxicological value used in the risk assessment of soil contaminants.

300   *Bioavailability, Risk Assessment and Remediation*

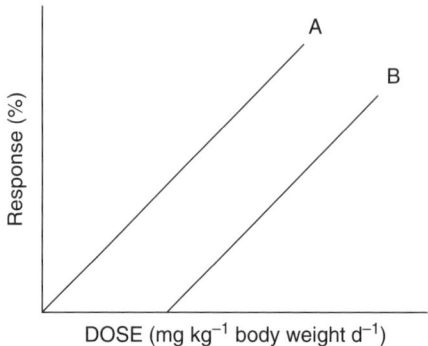

A – Metal has no threshold, there is some risk at any level of exposure
B – Metal has a threshold, there is a finite dose below which adverse effects are not discernible

Figure 13.3   *Threshold and nonthreshold concepts of toxic chemicals*

After deriving the TDI for a metal, the limit values for soils are calculated by taking account of daily exposure from the various exposure routes (oral intake, dermal contact, air inhalation, etc.) and pathways. Limit values could form the basis for setting quality criteria for soil. The nomenclature associated with the quality criteria vary between countries and may alternatively be termed screening values, screening levels, guideline values, soil quality guidelines or intervention levels. However, they all are designed to provide generic guidance to risk assessors on when there should be further evaluation of metal concentrations in soil. They are based on typical daily exposure of an individual given a specific exposure scenario. Exposure to contamination from other sources can be taken into account to ensure that total exposure does not exceed the ADI/TDI values (acceptable daily intake / tolerable daily intake), but this depends on the jurisdiction and legislation.

In the derivation of limit values, modelling of human exposure to metals in soil is a highly complex process associated with various scales of uncertainty [32]. Metal limit values can also vary on the basis of land use such that different values can be derived for the same metal for residential, allotments and commercial/industrial uses. For example, the arsenic guideline value in the United Kingdom for residential use with plant uptake is 20 mg kg^{-1}, whereas for commercial/industrial land use it is 500 mg kg^{-1} [33].

Two types of models are generally used to model human exposure to contaminants: deterministic and probabilistic. In a deterministic model, every parameter is defined by a single value. Many of these values are assigned on the basis of either average or conservative measurement and by professional judgement. In a probabilistic model parameters are not described by a single value but rather by probability distributions. Specifically, this type of model replaces some single value parameters in the exposure assessment with a family of values selected from a defined probability distribution. Each time the model estimates exposure it selects a value from this family. By repeating the assessment, a probabilistic model builds a range of predicted exposures rather than

Table 13.2 List of selected exposure models

Model[a]	Deterministic	Probabilistic	Chemicals
CLEA 2002		X	Range of chemical
CLEA 2007	X		Range of chemical
IEUBK	X		Lead only
ISE		X	Lead only
RBCA	X		Range of chemical
Risk*Assistant 1.1	X		Range of chemical
RISC – Human	X		Range of chemical

[a]CLEA, Contaminated Land Exposure Assessment; IEUBK, Integrated Exposure Uptake Biokinetic; ISE, Integrated Stochastic Exposure; RBCA, Risk Based Corrective Action; RISC, Risk Integrated Software for Clean-up.

providing a single outcome [32]. However, interpretation of the output of probabilistic modelling requires a greater understanding of the model and associated uncertainty, which is not as important for a deterministic model [34]. For this reason, most of the models used worldwide are deterministic (Table 13.2). In the United Kingdom, two types of exposure model are available: the SNIFFER (Scotland Northern Ireland Forum for Environmental Research) model (deterministic) and the CLEA (Contaminated Land Exposure Assessment) model (probabilistic, although a revised version of CLEA which is currently under development will be deterministic). These models are designed to assess exposure to different soil-borne metals via the food chain, ingestion of soil particles and dermal contact.

13.3 National and International Initiatives in Setting Limit Values

A recent EU-based review by Carlon [35] of Soil Screening Values across Member States has demonstrated the potential for harmonization, especially in relation to scientific information that underpins derivation methods. This review also highlights some of the key national differences across the EU in assessing risks to human and ecological health through the use of limit values.

For human health the establishment of limit values for international application can be scientifically as well as politically challenging because of the uncertainties surrounding the selection of data and the difficulty in defining a global exposure scenario. In addition, although the technical database for establishing numerical limits may be refined, inherent variability in diets, dietary habits, environmental exposure to pollutants, and soil and plant partition coefficients will always be problematic and difficult to generalize [36]. For example, for arsenic various types of values are available for soils: 8 mg kg^{-1} as a maximum permissible concentration in soil receiving sludge [36], 12 mg kg^{-1} as a soil quality guideline for residential land use in Canada [37], 0.8 mg kg^{-1} as a soil screening level for direct soil ingestion [38], 55 mg kg^{-1} as an integrated intervention value for all land uses [39], 20 mg kg^{-1} as a quality criterion for soil for very sensitive land use [40] and 20 mg kg^{-1} as a soil guideline value for residential land use [32]. However, as indicated earlier, differences in legislation, policy, background concentration, land use, contexts of derivation and so on. contribute to the wide variability in limit values.

Although the overall framework in deriving regulatory limit values may vary, in recent years there has been shared interest in addressing some of the default assumptions that underlie limit values. One such area is the estimation of bioavailability for metals on a site-specific basis. However, the use and regulatory definitions of bioavailability for metals differ from human health to environmental and ecosystem protection. In human health bioavailability is generally always used as an adjustment or correction factor that accounts for the ability of a metal to be absorbed by the body. Absolute bioavailability is the fraction or percentage of an external chemical dose which reaches the systemic circulation, that is, the ratio of internal dose to an applied dose [41].

In studies of absolute bioavailability, the internal/absorbed dose is often determined by measuring the concentration of the metal in blood, liver kidney and so on and/or by measuring the mass of the metal in excreta such as urine, faeces, or exhaled air. Relative bioavailability is the comparison of the absolute bioavailabilities of different forms of a chemical or for different exposure media containing the chemical [42]. Relative bioavailability is important because the 'matrix effect' may significantly reduce the bioavailability of a soil-borne metal when compared with the form of the metal and dosing medium used in a critical toxicity study utilized to derive a health criterion value. Therefore, incorporation of relative bioavailability into an exposure assessment can result in a better estimate of the external dose [43].

Direct measurement of the bioavailability of soil-borne metals in humans is often impractical. Various *in vivo* animal models have been employed to study the bioavailability of toxic chemicals as a surrogate for children [42,44–47]. The limitations of such approaches are that bioassays are specific to the test organism, metal and matrix; therefore, interpretation of data, and extrapolation to humans, needs careful consideration. Despite these limitations, the use of animal bioavailability assays has increased in the last two and half decades, particularly in the United States, in order to set appropriate clean-up goals for contaminated sites [45].

As stated previously limit values for metals for the environment are similar to those for human health in that there are few, if any, internationally recognized and broadly accepted numerical values for soils. This contrasts significantly with the aquatic compartment, where there are several EU-wide aquatic and drinking water limit values for metals [48]. This move to internationally applicable limit values for metals for the aquatic compartment may be followed in the EU for soils (beyond Member State sludge limits, which differ numerically) through developments under a Thematic Strategy for Soil Protection [49]. Limit values developed under one jurisdiction and adopted by another have occurred in the past, but questions often remain about the applicability and suitability of those values where they have been developed under climatic and soil conditions very different from those under which they are then proposed for use [15].

Incorporating consideration of metal availability and bioavailability into limit values is a laudable scientific aim for regulators and practitioners, yet the prosaic reality is often that this can be a contentious and complicated step [50]. The term 'bioavailability' is often incorrectly used in regard to metals and ecosystems, a situation that leads to considerable confusion and misunderstanding [51]. Recent developments in the understanding of metal behaviour and fate in soils provide regulators with an opportunity to consider availability and bioavailability in limit value setting and interpretation [27,52,53]. However, as discussed in the next section, this opportunity must be in a form that is pragmatic and implementable for both the regulator and regulated community.

13.4 Forward Look

Limit values are instrumental in risk assessment for metals in soils. The success of risk assessment and subsequent risk management can be significantly influenced by these values. If they lead to overestimation of risks, financial and social resources can be wasted in over-remediation of relatively unpolluted sites. On the other hand, if they lead to underestimation of risk, health and environmental protection can be compromised. In both cases the public can be mislead about the actual risks, and hence the communication of risk and the framing and contextualization of the limit value are critical.

One of the most regularly heard criticisms of limit values among the regulated community is that they are not proportionate or realistic, and that they are overly stringent [1]. For example, elevated arsenic concentrations in soils are now found in many parts of the world and there are examples where this land has been used for housing and development [54,55]. However no significant reports are available to link elevated arsenic deposits with any apparent human health effects, although scientific uncertainty still exists about the possible health effects of chronic low-level arsenic exposure. Arsenic is likely to be bound with soil as a result of sequestration, and may have restricted mobility within soil and therefore between soil and humans. Arsenic deposits in land in the United Kingdom, parts of the United States, Canada and Australia are likely to fall into this category. Lack of linkage between increased arsenic in soil and health effects often raises questions about the validity of limit values and current risk assessment approaches.

Values derived for human health protection do not often differentiate between natural and anthropogenic sources of metals. In theory, geogenic arsenic is likely to be less mobile (and therefore less toxic) than anthropogenic arsenic due to the possible effect of ageing [55]. There can also be difficulty in defining background concentrations from natural and anthropogenic inputs, although there are relatively simple tools available to do this [56,57]. In addition, limit values do not discriminate between different species (of a given metal) which could be unique in terms of fate and transport, bioavailability and toxicity [58]. Uncertainty also exists over whether a metal which is carcinogenic by one route of exposure may be carcinogenic by other routes. For example, cancers from drinking arsenic-contaminated water are well documented, but no adequate information is available on levels of cancer in people living in areas with elevated levels of soil arsenic who do not ingest arsenic-contaminated water.

Many of the uncertainties related to limit values for metals in soils could be addressed by undertaking appropriate and transparent validation exercises. It may seem strange, but very few limit values for metals used by jurisdictions across the world for environmental and ecosystem protection have been validated. Validation in this sense means an assessment in the field in order to establish whether the protection goal for which the limit value was set is achieved [59]. For example, if the limit value for a metal has been established in order to protect soil fertility then it is possible to assess the merit of the limit value in relation to the protection goal by using established biological metrics of soil fertility on a range of soils which have historically received metal inputs under the specific regime of interest [60]. The economic and social implications of the decisions that are based in part (or wholly) on these limit values suggests that validation might be supported by both industry and regulators. Recent evidence from experiments on sludge amended soils in Europe suggest that well-run trials can call into question whether long-established metal limit values are protective or robust [61,62].

For human health limit values such validation is not so easy, but a possible way to address validation would be to test laboratory-based toxicology data with real exposure scenarios. This could be undertaken through the use of biomonitoring, preferably coupled with systematic and focused epidemiological studies. These data then need to be integrated with improved geochemical understanding of chemical mobility within the soils and the effects that chemical mobility has on exposure and absorption in humans.

Research undertaken by regulators in the United States [42], Denmark [63] and the Netherlands [64] provides a way forward for potential use of bioavailability data in routine risk assessment when levels of contamination exceed limit values for a given metal. However, the use of *in vitro* models as an alternative to *in vivo* studies to produce bioavailability data for various metals has not yet been considered as a validated and robust approach by some regulators [42,65].

For a tiered approach to be effective in the implementation of metal limit values it must account for a number of potentially complex processes and strands of evidence. This needs to be done within individual tiers while also providing a straightforward, intuitive and interpretable flow of information leading to a clear decision. Furthermore, the tiers should be structured so that information requirements increase and conservatism decreases with the levels of identifiable risk. The incorporation of site-specific corrections for metal availability or bioavailability in the assessment of risk needs to be undertaken in a manner which fits with this approach, but as limit values for metals are often used within the early phases of a risk assessment [8], it is important that this is done in a pragmatic and straightforward way. Recent research undertaken internationally, but under the auspices of the European Existing Substances Regulations (793/93/EEC), has demonstrated the potential for incorporation of availability and bioavailability in setting and interpreting limit values for Cu, Ni and Zn. These approaches incorporate changes in metal availability between laboratory tests and the field, as well as making corrections for site-specific soil characteristics that govern bioavailability. The Environment Agency of England and Wales and the Scottish Environment Protection Agency are currently assessing the regulatory performance of these methods in collaboration with industrial partners.

13.5 Conclusions

Soil limit values for metals may come in many forms and be derived and used in a number of ways. The most valuable limits for environmental and human health protection are used as tools in a clear decision-making process, and are accompanied by contextual and process information on how the value should be used and interpreted.

It is appropriate to have a number of limit values for metals and a decision process based on several strands of evidence which are transparent and implementable, rather than a single metal limit value which is treated as a mandatory pass-or-fail criterion in isolation from other evidence. It is not possible to say with a great degree of certainty what scale of risk associated with metals in soils is tolerable to ecosystems or humans. Therefore it is not possible or appropriate for a single limit value to reflect this uncertainty [66]. If a metal limit value is part of a larger decision-making processes then the burden of proof of an effect can be weighted towards the protection goal, because if the limit is exceeded there are additional steps that can be taken and more information that could be gathered. The regulatory significance of failure of a limit value is not so great if that value is used within a tiered risk assessment instead of as a mandatory 'bright line' to assess compliance.

There are several other issues which need to be considered by regulators or policy makers before using a limit value. These include the need for consistency in limit values between regulatory regimes. 'Pollution swapping' can occur if an especially stringent limit value is set to apply to one regime without consideration of other regimes across all media. For example, setting an especially stringent limit value for a metal in the aquatic compartment may effectively mean that more of that metal is removed at water treatment plants, which subsequently results in higher levels in sewage sludge for disposal onto soils [8].

The communication of information between regulators, the regulated community and researchers needs to be a three-way discourse if it is to be effective in reducing the risks of metals in soils in a proportionate and technically sound manner. Too often views and requirements are not fully communicated and frustrations on all sides can lead to slow progress in developing this proportionate approach to metals regulation. If reliable evidence is presented to regulators to highlight mismatches between metal limit values and protection goals then it is the responsibility of the policy makers and government to act. However, this is often a slow and painstaking process. For example, the European Union Maximum Permissible Concentration of 0.235 mg Cd kg^{-1} in grain is routinely breached when using the current UK soil sludge Cd limit of 3 mg kg^{-1} [67]. Yet these limits values for Cd remain in place and, as outlined previously in the chapter, are widely used across several other UK regulatory regimes for managing Cd in soils.

References

[1] Department for Environment, Food and Rural Affairs, CLAN 6/06 Soil Guideline Values: the Way Forward; Department for Environment, Food and Rural Affairs, London, 2006.
[2] ECB, Technical Guidance Document on Risk Assessment in support of Commission Directive 93/67/EEC on Risk Assessment for new notified substances, Commission Regulation (EC) No 1488/94 on Risk Assessment of existing substances and Directive 98/8/EC of the European Parliament and of the Council concerning the placing of biocidal products on the market; European Commission, Joint Research Centre, EUR 20418 EN, 2003.
[3] Environment Agency, Soil screening values for use in UK ecological risk assessment. Report SP5-091-TR-E-E; Environment Agency, Bristol, UK, 2004.
[4] Ministry of Agriculture, Fisheries and Food, The Soil Code: Code of Good Agricultural Practice for the Protection of Soil; Ministry of Agriculture, Fisheries and Food Publications, London, UK, 1998.

[5] van Vlaardingen, P.L.A.; Posthumus, R.; Posthuma-Doodeman, C.J.A.M., Environmental risk limits for nine trace elements; RIVM Report 601501029; National Institute of Public Health and the Environment, Bilthoven, The Netherlands, 2005.
[6] Ministry for the Environment, Contaminated Land Management Guidelines No. 2; Ministry for the Environment, Wellington, New Zealand, 2003.
[7] Royal Commission on Environmental Pollution, 21st Report: Setting Environmental Standards; Royal Commission on Environmental Pollution, London, 1998.
[8] Canadian Council of Ministers of the Environment, A protocol for the derivation of environmental and human health soil quality guidelines; Canadian Council of Ministers of the Environment, Winnipeg, MB, 2006.
[9] Streffer, C.; Bücker, J.; Cansier, A. et al., Environmental Standards: Combined Exposures and Their Effects on Human Beings and Their Environment; Springer, Germany, 2003.
[10] Kolluru, R.; Barteli, S.; Pitbiado, R.; Stricoff, S., Risk Assessment and Management Handbook for Environmental, Health and Safety Professionals; McGraw-Hill, Inc., 1996.
[11] Interdepartmental Committee on the Redevelopment of Contaminated Land, Notes on the restoration and aftercare of metalliferous mining sites for pasture and grazing; ICRCL Guidance Note 70/90 1st edn; Department of Environment, HMSO, London, 1990.
[12] National Environmental Protection Council, Guideline on the investigation levels for soil and groundwater; National Environmental Protection Council, Canberra, Australia, 1999.
[13] Ministry of Agriculture, Fisheries and Food, The Analysis of Agricultural Materials; Reference Book RB 427, Ministry of Agriculture, Fisheries and Food, HMSO, London, 1981.
[14] Swiss Agency for the Environment, Forestry and Landscape, Commentary on the Ordinance 1 July 1998 Relating to Impacts on the Soil; Swiss Agency for the Environment, Forestry and landscape, Berne, 2001.
[15] Whatmuff, M., Biosolids application to agricultural land: considerations of contamination by heavy metals; in: Biosolids Research in NSW. Proceedings of the Biosolids Summit; Richmond, New South Wales, 1995, pp. 237–247.
[16] Smith, S.R., Agricultural Recycling of Sewage Sludge and the Environment; CAB International, Wallingford, 1996.
[17] Countryside Agency, A strategy for sustainable land management in England; Countryside Agency Publications, Wetherby, Yorkshire, 2001.
[18] Beeton, R.J.S.; Buckley, K. I.; Jones, G. J.; Morgan, D.; Reichelt, R.E.; Trewin, D., Australia State of the Environment 2006, Independent report to the Australian Government Minister for the Environment and Heritage, Department of the Environment and Heritage, Canberra, 2006.
[19] Alloway, B.J. Soil pollution and land contamination; in: R.M Harrison (ed.) Pollution, Causes, Effects and Control, 3rd edn; The Royal Society of Chemistry, London, 1996, pp. 318–339.
[20] Merrington, G.; Fishwick, S.; Brooke, D., The derivation and use of soil screening values for metals for the ecological risk assessment of contaminated land: a regulatory perspective; Land Contam. Reclam. 2006, 14, 673–684.
[21] Aldenberg, T.; Slob, W., Confidence limits for hazardous concentrations based on logistically distributed NOEC toxicity data; Ecotoxicol. Environ. Saf. 1993, 25, 48–63.
[22] Verbruggen, E.M.J.; Postumus, R.; Van Wezel, A.P., Ecotoxicological serious risk concentrations for soil, sediment and groundwater: updated proposals for first risk series of compounds; RIVM report 711701020; National Institute of Public Health and the Environment. Bilthoven, 2001.
[23] McLaughlin, M.J., Ageing of metals in soils changes bioavailability; ICME Fact Sheet on Environmental Risk Assessment, No. 4; ICME, Ottawa, ON, 2001.
[24] Impellitteri, C.A.; Saxe, K.J.; Cochran, M.; Janssen, G.M.C.M.; Allen, H.E., Predicting the bioavailability of Cu and Zn in soils: modelling the partitioning of potentially bioavailable Cu and Zn from soil solid to solution ratio; Environ. Toxicol. Chem. 2003, 22, 1380–1386.
[25] Basta, N.T.; Ryan, J.A.; Chaney, R.L., Trace element chemistry in residual-treated soil: key concepts and metal bioavailability; J. Environ. Qual. 2005, 34, 49–63.
[26] Meers, E.; Unamumo, V.; Vandeghuchte, M. et al., Soil-solution speciation of Cd as affected by soil characteristics in unpolluted and polluted soils; Environ. Toxicol. Chem. 2005, 24, 499–509.

[27] McLaughlin, M.J.; Whatmuff, M.; Warne, M. et al., A field investigation of solubility and food chain accumulation of biosolid-cadmium across diverse soil types; Environ. Chem. 2006, 3, 428–432.
[28] Witter, E., Towards zero accumulation of heavy metals in soils: an imperative or a fad?; Nutr. Cycl. Agroecosys. 1995, 43, 225–233.
[29] IGHRC. Uncertainty Factors: Their Use in Human Health Risk Assessment by UK Government; Interdepartmental Group on Health Risks from Chemicals, Institute for Environment and Health, Leicester, UK, 2003.
[30] Ferguson C.; Darmendrail D.; Freier K. et al. (ed), Risk Assessment for Contaminated Sites in Europe, Vol. 1, Scientific Basis; LQM Press Nottingham, 1998.
[31] Defra and Environment Agency, Contaminants in soil: collation of toxicological data and intake values for humans; R & D Publication CLR9. Bristol, UK, 2002.
[32] Defra and Environment Agency, Assessment of risks to human health from land contamination: an overview of the development of soil guideline values and related research; R&D Publication CLR7. Bristol, UK, 2002.
[33] Defra and Environment Agency, Soil guideline values for arsenic contamination; R&D Publication SGV1. Bristol, UK, 2002.
[34] US Environmental Protection Agency, General Principles for Performing Aggregate Exposure and Risk Assessments; US EPA, Office of Pesticide Programs, Washington DC, 2001.
[35] Carlon, C., Derivation methods of soil screening values in Europe: a review and evaluation of national procedures towards harmonisation; EU 22805-EN; European Commission, Joint Research Centre, Ispra, 2007.
[36] WHO. Developing Human Health-related Chemical Guidelines for Reclaimed Waster and Sewage Sludge Application in Agriculture; WHO Geneva, 2002.
[37] Canadian Council of Ministers of the Environment, Canadian Soil Quality Guidelines for the Protection of Environmental and Human health: Arsenic (inorganic); Canadian Council of Ministers of the Environment, Winnipeg, MB, 2001.
[38] US Environmental Protection Agency, Soil screening guidance: technical background document; US EPA Office of Solid Waste and Response, EPA/540/R95/128, Washington DC, 1996.
[39] RIVM, Human Exposure to Soil contamination: A Qualitative and Quantitative Analysis Towards Proposals for Human Toxicological Intervention Values; RIVM report 725201011; National Institute of Public Health and the Environment, Bilthoven, 1994.
[40] Danish Environmental Protection Agency, Environmental Guidelines: Guidelines on Remediation of Contaminated Sites, No. 7; Danish Environmental Protection Agency, 2002.
[41] Hrudy, S.E.; Chen, W.; Rousseaux, C.G., Bioavailability in Environmental Risk Assessment; CRC Press, Boca Raton, FL, 1996.
[42] US Environmental Protection Agency, Estimation of relative bioavailability of lead in soil and soil-like materials using in-vivo and in-vitro methods; Office of Solid Waste and Emergency Response, US EPA, Washington, DC, 2005.
[43] Naval Facilities Engineering Service Centre, Guide for incorporating bioavailability adjustments into human health and ecological risk assessments at US Navy and Marine Corps facilities (Part 2: Technical background document for assessing metals bioavailability); User's Guide UG-2041-ENV, Naval Facilities Engineering Service Centre, prepared by Battelle and Exponent, Washington, DC, 2000.
[44] Freeman, G.B.; Johnson, J.D.; Killinger, J.M. et al., Bioavailability of arsenic in soil impacted by smelter activities following oral administration in rabbits; Fundam. Appl. Toxicol. 1992, 21, 83–88.
[45] Casteel, S.W.; Cowart, R.P.; Weis, C.P. et al., Bioavailability of lead to juvenile swine dosed with soil from the Smuggler Mountain NPL site of Aspen, Colorado; Fundam. Appl. Toxicol. 1997, 36, 117–187.
[46] National Environmental Policy Institute, Assessing the bioavailability of metals in soils for use in human health risk Assessment; National Environmental Policy Institute, Washington, DC, 2000.
[47] Maddaloni, M.A.; Lolacono, N.; Manton, W.; Blum, C.; Drexler, J.; Graziano, J., Bioavailability of soil-borne lead in adults, by stable isotope dilution; Environ. Health Perspect. 1998, 106(S6), 1589–1594.

[48] European Commission, Proposal for a directive of the European Parliament and of the Council on Environmental Quality Standards in the Field of Water Policy and Amending Directive 2000/60/EC; COM(2006)397 Final, European Commission, Brussels, Belgium, 2006.

[49] European Commission, Proposal for a directive of the European Parliament and of the Council on establishing a framework for the protection of soil and amending Directive 2004/35/EC; COM(2006)232 Final, European Commission, Brussels, Belgium, 2006.

[50] Merrington, G.; Boekhold, S.; Haro, A. *et al.*, Derivation and use of environmental quality and human health standards for chemical substances in groundwater and soil; in: M. Crane, D. Maycock (eds), Environmental Quality and Human Health Standards for Chemical Substances in Water and Soil; SETAC, Pensacola, FL, 2008.

[51] Barber, S.A. Soil Nutrient Bioavailability: A Mechanistic Approach; John Wiley & Sons, New York, 1995.

[52] Van Gestel, G., Setting new soil standards for metals: improved relations with soil properties; Proc. SETAC Europe 17th Meeting, Multiple Stressors for the Environment and Human Health Present and Future Challenges and Perspectives, 20–24 May, Porto, 2007, P. 243.

[53] Broos, K.; Warne, M. St. J.; Heemsbergen, D.A. *et al.*, Soil factors controlling the toxicity of copper and zinc to microbial processes in Australian soils; Environ. Toxicol. Chem. 2007, 26, 583–590.

[54] Environment Agency, Questionnaire Survey on the Use of In-vitro Bioaccessibility in Human Health Risk Assessment; Science Report: SC040060/SR1, Environment Agency, Bristol, UK, 2006.

[55] Environment Agency, Inter-laboratory comparison of in vitro bioaccessibility measurements for arsenic, lead and nickel in soil. Science Report – SC040060/SR2; Environment Agency, Bristol, UK, 2008.

[56] Hamon, R.E.; McLaughlin, M.J.; Gilkes, R.J. *et al.*, Geochemical indices allow estimation of heavy metal background concentrations in soils; Global Biogeochem. Cycles 2004, 18, GB1014. doi:10.1029/2003GB002063.

[57] Zhao, F.J.; McGrath, S.P.; Merrington, G., Estimates of ambient background concentrations of trace metals in soils for risk assessment; Environ. Pollut. 2007, 148, 221–229.

[58] US Environmental Protection Agency. Ecological Soil Screening values for Cd. Interim Final, 2005; http://www.epa.gov/ecotox/ecossl/pdf/eco-ssl_cadmium.pdf (accessed 22 November 2009).

[59] Bright, D.A.; Sanborn, M.; Sawatsky, N., Relative sensitivity of different soil-associated flora and fauna to petroleum hydrocarbon releases: Current state of the knowledge and implications for environmental protection goals (Draft); Alberta Environment, Canada, 2006.

[60] De Jong, F.; Verbruggen, E.; Vos, J., The use of field studies in derivation of environmental quality standards; Proc. SETAC Europe 17th Meeting, Multiple Stressors for the Environment and Human Health Present and Future Challenges and Perspectives, 20–24 May, Porto, 2007, p. 352.

[61] De Brouwere, K.; Smolders, E., Yield response of crops amended with sewage sludge in the field is more affected by sludge properties than by final soil metal concentration; Eur. J. Soil Sci. 2006, 57, 858–867.

[62] Gibbs, P.A.; Chambers, B.J.; Chaudri, A.M. *et al.*, Long-term sludge experiments in Britain: the effects of heavy metals on soil micro-organisms; in: Y. Zhu, N. Lepp, R. Naidu (eds), Biogeochemistry of Trace Elements in the Environment: Environmental Protection, Remediation and Human Health; Proc. 9th International Conference on the Biogeochemistry of Trace Elements in the Environment; Tsinghua University Press, Beijing, 2007, pp 526–527.

[63] Grøn, C., Danish EPA report on test for bioaccessibility of heavy metals and PAH from soil; Environmental Agency sponsored English Translated version; Environment Agency, Bristol, UK, 2005.

[64] Oomen, A.G.; Brandon, E.F.A.; Swartjes, F.A.; Sips, A.J.A.M., How can information on oral bioavailability improve human health risk assessment for lead-contaminated soils?; RIVM report 711701042/2006; National Institute of Public Health and the Environment, Bilthoven, 2006.

[65] Environment Agency, Report on the international workshop on the potential use of bioaccessibility testing in risk assessment of land contamination; Report SC040054, Environment Agency, Bristol, UK, 2005.

[66] Faber, J.H.; van der Pol, J.; Rutgers, M., Soil quality and ecosystem services: a land use perspective; Proc. SETAC Europe 17th Meeting, Multiple Stressors for the Environment and Human Health Present and Future Challenges and Perspectives. 20–24 May, Porto, 2007, p. 243.
[67] Chaudri, A.M.; McGrath, S.; Gibbs, P. et al., Cadmium availability to wheat grain in soils treated with sewage sludge or metal salts; Chemosphere 2007, 66, 1415–1423.
[68] Smolders, E.; McGrath, S.; Fairbrother, A. et al., Hazard assessment of inorganic metals and metal substances in terrestrial systems; in: W.J. Adams, P.M. Chapman (eds), Assessing the Hazard of Metals and Inorganic Metal Substances in Aquatic and Terrestrial Systems; SETAC Publications, CRC Press, Boca Raton, FL, 2007, pp. 113–133.

14
Phytoremediation of Soil Trace Elements

Rufus L. Chaney[1], *C. Leigh Broadhurst*[2] *and Tiziana Centofanti*[2]

[1]*USDA-ARS-Environmental Management and Byproduct Utilization Laboratory, Beltsville, Maryland, USA*
[2]*University of Maryland, Department of Civil and Environmental Engineering, College Park, Maryland, USA*

14.1 Introduction

Phytoremediation includes many uses of plants to achieve remediation of soil risks. In relation to soil trace elements (hereafter, metals), the focus is phytoextraction, phytovolatilization, and phytostabilization. *Phytoextraction* uses growing and harvest of plants which accumulate high quantities of metals in shoots, allowing their removal from a contaminated site. If the plant biomass can be used as an alternative ore with monetary value, this can be labeled *phytomining*. *Phytovolatilization* uses plants and soil microbes to transform soil elements (Se, Hg) into volatile forms which leave the soil (($CH_3)_2Se$, Hg^0). *Phytostabilization* uses soil amendments which can cause formation of chemical forms of the metal with lower phytoavailability and bioavailability. The use of plants to prevent erosion and to support a sustainable ecosystem safe for wildlife is also part of phytostabilization. These approaches are considered 'green' technologies, utilizing low-cost agricultural practices rather than earth-moving equipment.

Trace Elements in Soils Edited by Peter S. Hooda
The contribution of Chaney has been written in the course of his official duties as a US government employee and is classified as a US government work, which is in the public domain in the United States of America.

Much progress has been made in understanding phytoremediation processes in the last twenty years. Because phytoextraction effectiveness depends on the amount of metals accumulated in the harvestable biomass per year, rare plants which hyperaccumulate metals in their shoots are especially valuable [1]. The word 'hyperaccumulator' was coined in 1977 [2], although Jaffré and Schmid [3] earlier used the French word *hypernickelophore*, while 'phytoextraction' was coined in 1983 [4]. The earliest review of the use of hyperaccumulator plants for phytoextraction was by Baker and Brooks [5]. These authors [5] suggested that species which accumulated Zn or Mn at 10 000 mg kg^{-1} dry weight (DW), Ni, Co, Pb or As at 1000 mg kg^{-1} DW, and Cd at 100 mg kg^{-1} DW could be called hyperaccumulator species. The levels noted are about 100 times higher than found in normal plants under most circumstances. Debate continues about what the hyperaccumulator concentration should be for each element, but the listed figures have received wide acceptance. Reeves [6] provided a fuller definition for Ni hyperaccumulators that should be applied to all elements and hyperaccumulator species: '... it is suggested that a hyperaccumulator of nickel be defined as "a plant in which a nickel concentration of at least 1000 mg kg^{-1} has been recorded in the dry matter of any aboveground tissue in at least one specimen growing in its natural habitat".' The implication of this definition is that tests which use artificial soils, spiked soils, or nutrient solutions with single elements added do not reflect the natural environment of hyperaccumulator plants, or of multielement contaminated soils. Many claims of hyperaccumulation based on Cd-only spiked soils are senseless because geogenic soil Zn concentrations are usually at least 100 times that of Cd. Therefore Zn inhibition of Cd uptake will easily prevent practical Cd phytoextraction by the claimed species (for example, [7]). If crop plants have $\geq$25% yield reduction at 500 mg Zn kg^{-1}, leaves will reach $\leq$5 mg Cd kg^{-1} even if they could attain >100 mg Cd kg^{-1} in a Cd spiked soil or nutrient solution. Relevant tests must be done with field multielement contaminated soils.

For phytoextraction to be accepted for remediation of contaminated soils, the process must be cost effective. If no one will pay for remediation of a contaminated site to reduce potential risks to humans and the environment of soil metals, no actions will be taken. Thus the costs of growing a crop to conduct phytoremediation must recover the costs of growing and harvesting and use or disposal of the crop. Producing value in the crop can offset these costs and support soil remediation.

Several authors have developed economic models that illustrate these concepts. In the case of soil selenium (Se), Banuelos and Mayland [8] showed that biomass of a Se-rich phytoremediation crop (that also did some phytovolatilization of Se) could be mixed in livestock feeds and replace Se salts normally added to feeds. Considering the difficulty of adding 0.2–1.0 mg Se kg^{-1} to feed, mixing ground Se-rich biomass would likely make better-mixed feeds than adding very small amount of Se salts. Analogously, Se-enriched vegetable crops might be marketed as phytonutrient-enhanced or naturally biofortified foods. Because of high sulfate levels in the same soils, crop Se is limited to levels which are safe in foods [9]. Robinson *et al.* [10] note the cost aspect for designing a phytotechnology to alleviate risk from boron in leachate from a treated wood waste pile, and discuss a model to estimate value of products and costs of phytoremediation.

Several groups have proposed growing willow as a biomass energy crop with an additional benefit of Cd removal if specific cultivars are utilized (for example, [11]). If enough Cd was removed from the soil, the ash would have to be placed in a landfill rather than recycled on forest land to return the nutrients removed in growing the crop (see discussion below). In this cost model, unless farmers considered the reduced soil Cd as a value to their land, they would not accept the loss in higher-value vegetable crop production due to growing willow to remove Cd [11]. Thus, unless the sale of food crops required soil remediation according to regulations, farmers would not consider remediation. Although the European Union (EU), United States, Japan, and China have much land which contains Cd contamination, governments have not ordered remediation or paid for remediation; therefore little has been undertaken other than research and development of Cd phytoextraction technology.

Phytoextraction using hyperaccumulators could be a cost-effective method to mine Ni from soils [12], or to remediate Ni phytotoxic soils. Ni metal sells for $15 kg^{-1} at this writing; it had reached more than $50 kg^{-1} in 2007. Ash from incineration of *Alyssum murale* biomass containing about 20% Ni is easily used as an ore in electric furnace refining of Ni. Plant ash is free of interfering factors (such as Fe and Mn oxides and silicates) present in lateritic Ni ores that require expensive processing to release Ni. An effective crop of 20 t ha^{-1} dry biomass containing 2.0% Ni yields 400 kg Ni ha^{-1}, and offers more profit than most agronomic crops, especially considering the infertile serpentine soils which are Ni phytomining substrates. Alternatively, the hyperaccumulator biomass can be used as an organic Ni fertilizer [13].

However, other elements in the plant biomass or biomass ash are seldom valuable enough to cover the costs of growing the crop. For example, phytoextracted Cd has negligible value; not even the sum of Cd and Zn has enough value to offset the costs of crop production and harvest. Similarly, crop Pb or As is unmarketable, hence plant biomass must be placed in a landfill – generating another cost as opposed to offsetting cost. Chaney *et al.* [12] discuss the disposal of plant biomass from phytoextraction. In contrast with some claims [14], phytoextraction biomass disposal is not a difficult problem, only a disposal cost rather than a source of profit.

Considerable research has been conducted on phytoextraction of soil metals, and several authors have noted problems encountered in development of commercial technologies. For example, Robinson *et al.* [15] list these potential limitations in using phytoextraction to remediate contaminated sites: (i) long period required for cleanup; (ii) limited number of target metals which can be phytoextracted; (iii) limited depth that can be accessed by roots; (iv) difficulty in producing a high biomass crop of the desired species; (v) potential of plant metals to enter environmental food chains; and (vi) inevitable leaching of metals if chelators are added to induce phytoextraction.

Ernst [16,17] has rightly stressed the 'hype' of hyperaccumulator use in phytoextraction. Many who have conducted research on hyperaccumulators did not appreciate the complex soil chemistry of trace elements, and the difficulty of moving polyvalent cations through plant membranes. It is nearly impossible to achieve useful phytoextraction of Pb, Cr, Cu, and some other important trace elements because they are too insoluble in soils or retained in plant roots. Adding chelating agents to 'induce' phytoextraction was never a good idea; it causes leaching of metals to groundwater, and is extremely expensive. Ernst [18] initially stressed the poor yield and low metal concentration in biomass harvested from sites where *Thlaspi caerulescens* occurred along with metal tolerant grasses. This criticism is valid in that the full yield of *T. caerulescens* is not high, but a reduced yield due to competition/shading from other species can be controlled by selective herbicides. Ernst also notes the potential difficulty of public acceptance of transgenic metal hyperaccumulator crops made from crop species. We should listen carefully to Dr. Ernst's comments because of his extensive experience with metallophytes across Europe [19]. Most of the early literature on *Thlaspi caerulescens* came from ecological and physiological studies by Ernst and cooperators. And Russian biogeochemists/botanical prospectors [20] reported Ni hyperaccumulators before the surge of research since 1977. Ernst, Brooks and Baker are the 'fathers of phytoextraction' who helped spread the idea which has led to development of important knowledge and technology.

On the other hand, Ernst [16] failed to consider the use of agronomic management practices to maximize yields of the hyperaccumulator crops. Appropriate use of herbicides and tillage, and optimal use of fertilizers and soil amendments to adjust pH and improve productivity can give useful harvestable yields of the wild plants. Grasses overwhelm the biomass of *T. caerulescens* if not controlled by herbicide, and may strongly reduce annual phytoextraction quantity. And any plant breeder would at least suggest that, given time, they could improve the harvestable yields of even rosette plants such as *T. caerulescens*.

Biotechnology is widely believed to offer the ability to combine critical activities of hyperaccumulator plants into plants with high harvestable biomass yields, which are easier to produce as an agricultural crop. How many proteins are required to transform a normal plant into a *T. caerulescens*, *Alyssum murale* or *Pteris vitatta* equivalent hyperaccumulator is simply not known, but recent research still supports the view that translocation from root to shoot, and storage in leaf cell vacuoles are the key functions needed [21]; selectivity of the root uptake transporter may also contribute, and may increase metal tolerance. But the 'hypertolerance' noted for hyperaccumulators is mostly dependent on effective storage in leaf cell vacuoles. Only continuing research and development will determine whether cost-effective bioengineered phytoextraction crops will become available for commercial use. Improved cultivars of natural hyperaccumulators have already been used in the field [22]. And it may take production of sterile hybrids of phytoextraction plants to allay fears of transferring metals into environmental food chains, especially for transgenics.

Although this review describes methods for phytoextraction which appear to be cost effective and ready to apply, one should always consider phytostabilization rather than phytoextraction for many elements. For Ni and Zn phytotoxic soils, a ready immediate remediation is available via making the soil calcareous and adding appropriate fertilizers to maintain soil fertility [23–28]. Only if the value of Ni or Zn in the plant biomass can support phytomining would phytoextraction be a better alternative than phytostabilization.

Some have suggested that yields of all hyperaccumulators are too low to give useful phytoextraction of soil metals [29]. This claim is contradicted by the effective phytomining of Ni by *Alyssum* species, which can provide more net income than growing normal crop plants on fertile soils [12]. Useful phytoextraction cannot be achieved by crop plants except in the case of Se [8] and that opportunity still requires development. Plants for phytoextraction must be highly metal tolerant, and accumulate high concentrations of the target elements in harvestable shoots so that the annual removal of metal from the site is economic.

Because of the unusual potential of these 'green' technologies to achieve soil remediation, much research has been conducted and reported. Substantial progress has been made even on the genetics of metal hyperaccumulator species, and the biology and biochemistry of phytoextraction. At the same time, much research has used addition of chelating agents to conduct 'induced phytoextraction,' but the added chelators cause leaching of target or other metals to groundwater. And many papers have reported studies of plant species with no promise for practical phytoextraction whatsoever because the researchers did not understand the practical side of soil contamination and hyperaccumulation. For example, species such as *Brassica juncea* never had promise for practical phytoextraction; it is not a metal tolerant species, and does not accumulate Pb or other elements from soils without addition of chelators. Even for Se phytoextraction, *B. juncea* has little value because it does not accumulate Se in the presence of high levels of sulfate normally found in Se contaminated or mineralized soils as do the Se hyperaccumulators.

14.2 The Nature of Soil Contamination where Phytoextraction may be Applied

Remediation of contaminated soils is conducted in response to government decisions about land use, or to provision of funding for remediation. Farmers may desire Cd removal from soils with mine or smelter contamination, where high Cd-phosphates were applied for decades, or high Cd biosolids were applied before regulations were developed, but the decision to proceed will be based on the economics of their farm operation. As noted below, the main impediment to development of phytoextraction technology is the failure of government to require and fund soil remediation. Usually the contamination must be severe enough to markedly limit crop selection, or to produce barren soils or unsafe food before government acts. The situation is complex because acidic soil pH increases soil metal phytoavailability and simple application of limestone may reduce metal phytotoxicity and metal uptake by crop plants and restore crop production freedom. When severe contamination occurs from mine wastes or smelter emissions, soils are often acidic, highly contaminated, and simple limestone application does not restore crop cover. In the United States, the Superfund program deals with such contaminated soils if humans or aquatic ecosystems are threatened by the soil contaminants, but simple soil ecosystem disruption is not usually enough to trigger Superfund status and action by government.

Soil metals must be present in chemical forms/solubilities which plants can absorb and translocate to shoots or the phytoextraction option is null. That is why Cr(III), Pb, Sn, and many other elements have such poor absorption, and Pb is also trapped within the fibrous root system as insoluble lead phosphate. Plants can only absorb metals from the volume of soils that root explore. High density of plant roots is generally limited to the surface 0–15 or 0–30 cm of soil depth, so metals in deeper soil layers are not likely to be phytoextracted effectively. Tap roots that obtain water from deeper soil depths may not contribute to metal accumulation. Deep-rooted hyperaccumulators, and trees grown with roots at depth within 10-m deep wells [30] to provide access to deeper soils can achieve removal of groundwater and contaminants from deeper soil than the surface 0–30 cm.

Tillage of contaminated soils may cause dispersal of the contamination by rainfall, especially on slopes. Phytoextraction can be applied only where the soil can be tilled, or competing plants controlled by selective herbicides, and the shoot biomass harvested mechanically. Fortunately, methods have been developed for surface application of amendments (organic matter plus limestone and fertilizers) on highly sloping contaminated soils, which can achieve effective phytostabilization of these soils [27].

14.3 Need for Metal-Tolerant Hyperaccumulators for Practical Phytoextraction

Soils which need phytoremediation for metals are often barren or have cover with only a few species of metal-tolerant plants [31]. Although some authors propose to grow crop plants because of potentially high yields of shoot biomass, if the species cannot tolerate the soil metals, one must modify the soil or choose another species for cover. The exceptional tolerance of shoot metals by hyperaccumulator plants has brought attention to these unusual species since at least 1960 [32]. Brooks [33] reviewed hyperaccumulator plants and covered many aspects of these unusual species. Table 14.1 lists identified element hyperaccumulator species with more than 1% of metal in shoots as sampled in the field on contaminated or mineralized soils.

In introducing the concept of phytoextraction, Chaney [4] provided an example of growing corn (*Zea mays* L.) or *Alyssum murale* on a Ni-mineralized or contaminated site (Table 14.2). Even at 50% yield reduction with 100 mg Ni kg^{-1} dry shoots, the high-yielding corn shoots contained only 1 kg Ni ha^{-1}. It should be clear from this example that crop plants cannot absorb high enough amounts of Ni to support phytoextraction. Annual removal of Ni is too small for crop plants even when they are suffering considerable yield reduction. However, even moderate yielding species such as *Alyssum murale* or *A. bertolonii* can accumulate 1–2% Ni [22,34,35], and improved agronomic management (fertilizers, herbicides, etc.) and cultivars produced by plant breeding can accumulate more than 3% Ni in dry shoots [12,22].

Although crop plants can accumulate higher levels of Zn than Ni before yield is substantially reduced, claims that crop plants may be useful for phytoextraction [36]

Table 14.1 Example plant species which hyperaccumulate elements to over 1% of their shoot dry matter; usually to at least 100-fold levels tolerated by crop species

Element	Plant species	Maximum metal concentration (mg kg^{-1} DW)	Location collected	Reference
Zn	*Thlaspi caerulescens*[a]	39 600	Germany	275
Cd	*Thlaspi caerulescens*	2908	France	77
Cu[‡]	*Aeolanthus biformifolius*	13 700	Zaire	149
Ni	*Phyllanthus serpentinus*	38 100	New Caledonia	276
Co[b]	*Haumaniastrum robertii*	10 200	Zaire	148
Se	*Astragalus racemosus*	14 900	Wyoming, USA	277
Mn	*Alyxia rubricaulis*	11 500	New Caledonia	278
As	*Pteris vittata*	22 300	Florida, USA	210
Tl	*Biscutella laevigata*	15 200	France	177

[a] Ingrouille and Smirnoff [279] summarize consideration of names for *Thlaspi* species; many species and subspecies were named by collectors over many years [275,280,281].
[b] Although Cu and Co hyperaccumulation were confirmed in field collected samples, similar concentrations have not been attained in controlled studies.
Reproduced from Chaney, R.L.; Angle, J.S.; Broadhurst, C.L.; Peters, C.A.; Tappero, R.V.; Sparks, D.L., Improved understanding of hyperaccumulation yields commercial phytoextraction and phytomining technologies; J. Environ. Qual. 2007, 36, 1429–1443 [12].

Table 14.2 Estimated Ni phytoextraction by corn (Zea mays L.) vs. Alyssum murale grown as a phytomining crop (adapted from [4,12]); assume control soil contains 25 mg Ni kg^{-1}, and the Ni-rich soil contains 2500 mg Ni kg^{-1} = 10 000 kg Ni (ha 30 cm)$^{-1}$; assume soil Ni is sufficiently phytoavailable that corn has 50% yield reduction compared to corn grown on similar soil without Ni mineralization. Research has shown that unimproved Alyssum murale can easily yield 10 Mg ha^{-1} with fertilizers, and selected cultivars can exceed 20 Mg ha^{-1} with appropriate soil and crop management on serpentine soils [42]. Most crop plant species suffer 25% yield reduction when the shoots contain 100 mg Ni kg^{-1} dry weight [24]. Ni concentration in ash is limited by formation of NiCO$_3$ with only 49% Ni

Species	Soil	Yield (dry Mg ha^{-1})	Ni in the crop			Ash-Ni (%)
			(mg kg^{-1})	(kg ha^{-1})	(% of soil)	
Corn	Control	20	1	0.02	0.01	0.002
Corn (50% YD)	Ni-rich	10	100	1.	0.01	0.20
Wild Alyssum murale	Ni-rich	10	20 000	200.	2.0	20–25
Alyssum murale cultivar	Ni-rich	20	25 000	500	5.0	25–30

Reproduced from Chaney, R.L.; Angle, J.S.; Broadhurst, C.L.; Peters, C.A.; Tappero, R.V.; Sparks, D.L., Improved understanding of hyperaccumulation yields commercial phytoextraction and phytomining technologies; J. Environ. Qual. 2007, 36, 1429–1443 [12].

ignore the evidence of Zn phytotoxicity at <1000 mg Zn kg^{-1}. As will be discussed below regarding phytoextraction of soil Cd, most soils have 200-fold higher concentration of Zn than Cd, so if the plant does not tolerate very high Zn levels, it cannot survive to phytoextract soil Cd.

Some researchers have predicted that phytochelatins would be important in metal tolerance and accumulation in plants, but research has shown these compounds have no identifiable role in Zn–Cd-hyperaccumulator plants [37,38]. Bioengineering of plants to express several enzymes which may increase the biosynthesis of phytochelatins caused only a small increase in plant metal tolerance or accumulation, far below those of natural hyperaccumulators. Clearly, phytochelatins are not a part of the hyperaccumulator phenotype.

14.4 Phytoremediation Strategies: Applications and Limitations

14.4.1 Phytomining Soil Nickel

Nickel (Ni) provides the best example for using natural hyperaccumulator plants to phytoextract soil Ni for profit [12]. The Ni phytomining technology [39] was developed by selecting for development promising high-yielding species with strong Ni hyperaccumulation for testing under controlled conditions. Seeds of diverse genotypes were collected where native plants grew in southern Europe. The composition of field-collected specimens is influenced by many factors which vary between sites and collectors; thus, only a field comparison of species or genotypes of a species can show the true potential of any species or genotype.

A growth test was conducted on serpentine soils in southwestern Oregon, USA. Greenhouse and growth chamber studies showed the seeding depth required for obtaining effective germination, and fertilizers and pH adjustment necessary to get maximum yield and Ni accumulation on infertile serpentine soils [22]. Greenhouse testing showed which herbicides did not harm *Alyssum* while controlling species which commonly grow on these serpentine soils (mostly grasses). Testing also showed that *Alyssum murale* and *Alyssum corsicum* were self-incompatible, so breeding to improve cultivars required use of recurrent selection rather than simpler methods. To minimize issues of variation in the field, reference genotypes were grown in each planting block with six other entries, and covariance correction for the reference genotype Ni concentration was used in genotype comparisons. Substantial genetic variation in shoot Ni concentration was found in both species evaluated (Figure 14.1).

Among other important observations was the genetic variation in leaf abscission during flowering. Because leaves contain much higher Ni concentration than stems, loss of leaves after flowering would reduce potential Ni yield. Thus, retention of leaves during early flowering was a selection factor during the recurrent selection program, along with shoot biomass yield, shoot Ni concentration, and plant form (multiple flowering stems, height, etc.).

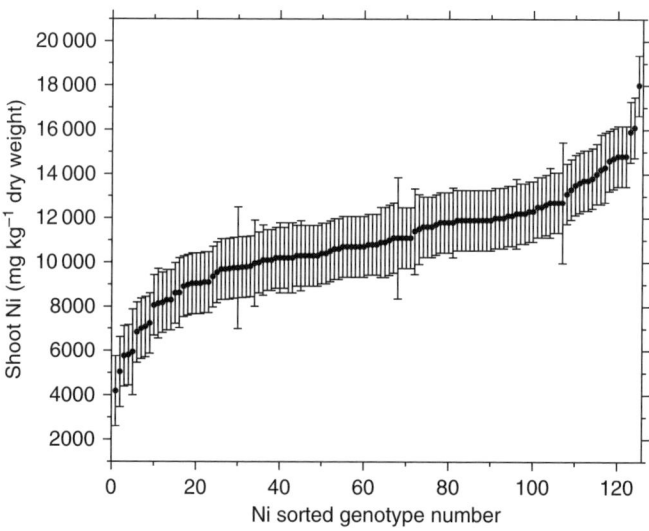

Figure 14.1 Genetic variation among Alyssum murale ecotypes in shoot Ni concentration when grown on a uniform test field of serpentine soils in Oregon, USA; points are means of four replications; bars are standard errors (Adapted from Li, Y.-M.; Chaney, R.L.; Brewer, E.; Roseberg, R.J. et al., Development of a technology for commercial phytoextraction of nickel: economic and technical considerations, Plant Soil 2003, 249, 107–115 [22].)

Further, it was shown that allowing the cut plants to dry in the sun for several days before baling the biomass did not cause loss of leaf biomass. Although full details have not been released, Viridian LLC (the company which is commercializing this technology) has reported substantial improvement in yield and Ni concentration in improved *Alyssum* cultivars grown on serpentine soils. *Berkheya coddii* was tested under the same field conditions and accumulated only 0.5–1.0% Ni in shoots compared with 2–3% Ni in shoots of improved *Alyssum* cultivars. Similar low shoot Ni was reported for *B. coddii* by Robinson *et al.* [40] even with fertilizer and pH adjustment. It is not evident why field samples of *B. coddii* grown in serpentine soils contain much lower Ni than reported for selected field collections [41]. Perhaps strong nutrient deficiency with very strongly reduced shoot yield caused higher Ni accumulation as reported for *A. murale* in the test by Li *et al.* [42] (Table 14.3). But note the high ability of *A. murale* to obtain adequate phosphate at low P-fertilizer rates. High-Fe serpentine soils require very high P-fertilization rates to support full yields of crop plants [43]. Fertilizer N has repeatedly been shown to increase shoot biomass yields but not substantially reduce shoot Ni concentration [44]. Split application of N-fertilizer will need to be considered to match the growth patterns of the hyperaccumulator species in different climates.

Table 14.3 Effect of soil amendments on trace element accumulation in shoots of *Alyssum murale* grown 120 days on a serpentine soil. For pH treatments, nitric acid was added and allowed to react, then salts were leached from the soil. Gypsum treatments are in Mg ha^{-1}; phosphate, treatments are in kg P ha^{-1}; all except control received 150 mg N as NH_4NO_3, 150 mg K ha^{-1} as KCl and 1 kg B ha^{-1} as H_3BO_3[a]

Treatment	Final pH	GM[b]-Yield (g/pot)	GM-Ni	GM-Co	GM-Mn (mg kg^{-1})	GM-Zn	GM-Fe	Cu
Control								
None	6.56 a	4.1 c	14740. a	34.3 c	56.5 e	63.4 bc	154. b	3.0 cd
Phosphate treatments								
0 P	5.82 e	1.6 d	6250. cd	19.4 ef	62.3 cde	118. a	273. a	2.8 d
100 P	6.24 b	24.5 a	6270. cd	19.9 ef	60.9 cde	59.9 bc	112. cd	3.6 bc
250 P	6.14 bcd	23.2 ab	6810. bc	22.6 def	65.2 cde	60.2 bc	104. d	4.2 ab
500 P	6.16 bc	26.5 a	5690. d	18.1 f	67.2 cde	55.1 cd	92. d	4.0 ab
pH treatments								
Lo	5.42 g	27.4 a	6150. cd	224. a	462. a	63.1 bc	144. bc	4.4 ab
Med-1	5.69 f	26.2 a	6800. bc	50.4 b	132. b	68.7 b	117. bcd	4.6 a
Med-2	5.89 e	27.0 a	5990. cd	28.8 cd	73.1 cd	58.2 bcd	96. d	3.6 bc
As is pH	6.24 b	24.5 a	6270. cd	19.9 ef	60.9 cde	59.9 bc	112. cd	3.6 bc
Gypsum treatments								
0.0 Ca	6.10 cd	19.3 b	7860. b	21.1 ef	55.6 e	49.4 d	87. d	3.1 cd
1.0 Ca	6.24 b	24.5 a	6270. cd	19.9 ef	60.9 cde	59.9 bc	112. cd	3.6 bc
2.5 Ca	6.04 cd	25.2 a	6050. cd	18.4 ef	58.2 de	59.6 bc	87. d	3.8 bc
5.0 Ca	6.03 d	24.2 a	5630. d	24.4 de	78.5 c	63.3 bc	93. d	3.6 bc

[a]Means followed by the same letter are not significantly different ($P < 0.05$ level) according to the Duncan–Waller K-ratio t-test.
[b]GM is geometric mean.
Source: adapted from [41], and unpublished data.

One of the most unexpected findings from this research was that although acidification of serpentine soils increased dissolved Ni in soil solution and that extracted by many extracting reagents, Ni accumulation by both *Alyssum murale* and *Alyssum corsicum* was reduced by acidification [42]. In contrast, raising pH of most soils increased Ni concentration and yield [45]. For especially Fe-oxide-rich serpentine soils (~20% Fe), liming above about pH 6.3, however, reduced Ni accumulation. Robinson *et al.* [40] found that *Berkheya coddii* followed the normal pattern of lower shoot Ni concentration with raised soil pH. Thus, use of ammonium acetate extraction to predict phytoextractable Ni from different soils is not appropriate for *Alyssum* species, although it may be relevant for *Berkheya*.

The potential use of chelators to increase Ni accumulation by Ni hyperaccumulators was tested for both *Alyssum murale* and *Berkheya coddii*. Robinson *et al.* [46] showed that added EDTA and NTA reduced Ni uptake significantly. Li, Chen, Chaney and Angle (unpublished results) also found that added EDTA significantly reduced Ni accumulation by *Alyssum murale*. Apparently it is less likely that the added chelator will injure the roots of these hardy plants enough to facilitate metal-chelate accumulation by the plants, in contrast with *B. juncea*.

For phytoextraction to be successful, the forms of Ni in soil must be phytoavailable. Several groups have reported that Ni in serpentine soils was associated with the Fe and Mn oxides or silicates [47,48]. In smelter contaminated soils, Ni was associated with organic matter, Fe oxides, as the Ni-Al layered double hydroxide, and as NiO [49]. NiO has very slow dissolution kinetics [50]; the half-time for NiO dissolution at pH 7 is 20 years. NiO was likely a form emitted from the nickel refinery. A basic test of whether *Alyssum* roots could dissolve this comparatively inert compound showed that *Alyssum* accumulated negligible Ni during a month of root exposure to 1 μm NiO particles [51]. Phytoextraction cannot deal with NiO or similar kinetically inert metal species.

An important characteristic of Ni hyperaccumulators is that they obtain soil Ni from the same pool of 'labile' Ni as other plant species [52,53]. No method is yet known to attack the nonlabile pool of soil Ni except allow time for equilibration to more labile forms. Interestingly, when soil microbes from the rhizosphere of *A. murale* growing on serpentine soils were cultured, several species/strains were found which could induce higher shoot Ni concentration and quantity when the microbe was inoculated into either sterile or non-sterile serpentine soil [54,55]. The mechanism by which soil microbes increase Ni hyperaccumulation remains unknown [56,57]. Several groups have studied changes in the chemistry of rhizosphere soil from hyperaccumulator vs. nonhyperaccumulator species [58,59]. There is little evidence that plants or soil microbes secrete amino or organic acids or other specific Ni chelators which might aid Ni dissolution or uptake.

The most profitable use of Ni-hyperaccumulator biomass is as Ni ore. Chaney *et al.* [12] reported that *Alyssum* biomass ash could enter a Ni refinery at the electric furnace stage rather than going through all the separation and purification steps to separate Ni from rock ores. Biomass energy might offset the costs of crop production, but at this time, no higher value can be obtained from this biomass.

An alternative valuable use of Ni-rich *Alyssum* biomass was recently demonstrated as a Ni fertilizer for Ni-deficient soils [13]. Ni deficiency of coastal plain soils in Georgia, USA, was demonstrated for pecan trees. Simple sprays of $NiSO_4$ or a water extract of *A. murale* cured Ni deficiency. Ground *Alyssum* biomass could be applied to the soil and prevent future deficiency. Other old coarse-textured soils managed at pH 6.5 or above may also suffer Ni deficiency, and the market for inexpensive organic Ni fertilizer could become significant. Other possible crop Ni deficiency is being examined by Wood et al. [60].

14.4.2 Soil Cadmium Contamination Requiring Remediation to Protect Food Chains

Extensive areas of paddy rice soils in Japan, China and Thailand have become so contaminated by Zn, Pb or Cu mine and smelter emissions that rice grown on the soils has caused human cadmium (Cd) disease [61,62]. Such locations require Cd remediation in order to continue growing rice for food. Paddy soil remediation in the Jinzu River Valley of Japan used soil removal and replacement, or soil inversion, at a cost of $2.5 million per hectare [63]. Soil inversion can be successful for rice because paddy rice absorbs Cd only from the surface layer [64]; inversion would not be adequate to alleviate food-chain risk for other crops. Rice readily accumulates Cd in excess of limits (0.4 mg kg^{-1} fresh weight of brown rice) when soils contain as low as 1.5 mg Cd kg^{-1} [65] due to the soil chemistry of flooded/drained soils, and metal uptake properties of rice (see Chapter 17).

Although many soils are Cd- and Zn-contaminated in western nations, there is little evidence that humans have been harmed by food-chain transfer of soil Cd. How this could occur is discussed in the chapter on Cd and Zn in this book (Chapter 17). Reeves and Chaney [66] and Chaney et al. [65] have described how rice transfers bioavailable Cd to humans much better than other crops due to both the Cd uptake relative to Zn and transport of Cd to the grain without Zn being increased, and the Zn and Fe malnutrition induced by subsistence rice diets which promotes Cd absorption in humans.

Other contaminated areas received Cd-rich phosphate fertilizers [67], or high-Cd biosolids [68], or were mineralized with high Cd:Zn ratio marine shale parent rocks (phosphorite source) [69,70]. Crops produced on these soils can contain Cd above guideline or permitted levels. Tobacco also accumulates soil Cd very effectively and like rice may exceed desired Cd levels on acidic soils with low-level Cd contamination [71,72], but few nations have limits on Cd in tobacco. Lack of Cd limits for tobacco is irrational; smoking tobacco contributes strongly to accumulation of Cd in kidney, and hence to Cd risks

Phytoextraction has been considered a promising method for removing Cd from soils that comprise risk to humans through crop accumulation of soil Cd [73,74]. The species first recognized to hyperaccumulate Cd were *Thlaspi caerulescens* and *Arabidopsis halleri* [5]. However, the 'Prayon' ecotype initially studied did not accumulate Cd well enough to support commercial phytoextraction [75]. Li et al. [76] and Chaney et al. [77] reported ecotypic variation in Cd accumulation, showing that strains from southern France accumulated about 10-times higher Cd from the same soils as did 'Prayon'. Reeves et al. [78] reported on the ecotypic variation in *T. caerulescens* Cd accumulation across Europe, showing more of the pattern that strains from southern France had remarkable ability to phytoextract soil Cd. The southern France ecotypes absorb and translocate Cd more effectively than 'Prayon' apparently due to the root Cd transporter [79]. Li et al. [80] eventually obtained a US patent for phytoextraction of soil Cd using such ecotypes and crop management practices to maximize annual removal of Cd.

Even though *T. caerulescens* can hyperaccumulate Cd to more than 1000 mg kg^{-1} dry shoot biomass when grown on contaminated soils [76,78–84], it has a rosette growth pattern and even at harvest after vernalization is seldom over 30 cm tall. It is not an optimal plant for Cd phytoextraction, and *A. halleri* is even smaller. Research has shown that good management can produce as much as 5 Mg dry biomass ha^{-1} in the field [75,84–87], although some authors found much lower yields. The 5 Mg ha^{-1} yield can result from a full year of growth after transplanting including the flowering period before the plants start to drop leaves as seeds are set. It appears that no group has obtained funding to breed improved strains of *T. caerulescens* for field phytoextraction of soil Cd. There appears to be considerable variation in Cd accumulation by plants grown from seeds of a single mother plant, and the source of this variation has not been clarified (Figure 14.2) [74,88]. These findings indicate that normal plant breeding should be able to produce cultivars with even higher Cd phytoextraction ability.

Because this review is focused on more practical aspects of phytoremediation of soil trace elements, it will not include a detailed review of the research on physiology and biochemistry of *T. caerulescens* hyperaccumulation. Several recent reviews cover these topics [89–91].

Acidic soil pH strongly increases Cd and Zn accumulation by *T. caerulescens*. When pH falls to levels which allow Al or Mn phytotoxicity, yields are reduced. Annual

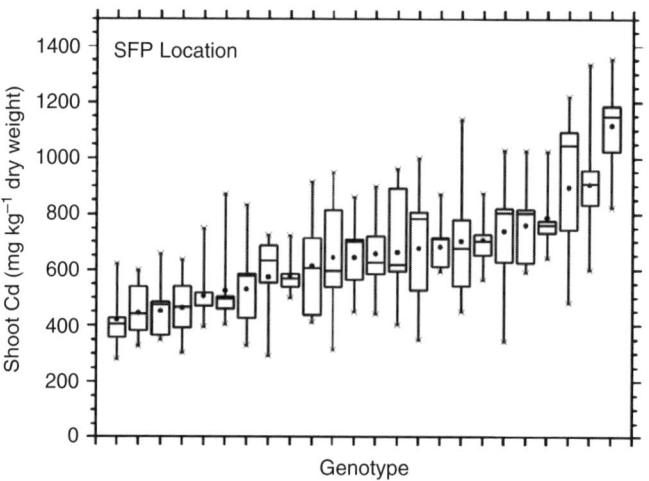

Figure 14.2 *Genetic diversity of Cd accumulation by Thlaspi caerulescens. Box plots of Cd in* T. caerulescens *genotypes from St. Félix de Pallières (SFP), France; each plot is of results for eight plants grown from seed of a single mother plant (the ends of the box are the 25th and 75th centiles of shoot Cd distribution for a genotype, and the whiskers are the 5th and 95 centiles of shoot Cd; the dot is the mean and the bar the median shoot Cd). Plants were grown for 62 days in pots of soil containing 47.6 mg Cd kg^{-1} and 821 mg Zn kg^{-1} due to application of high Cd:Zn biosolids for many years. 'Prayon' ecotype grown in the same experiment contained only 34 mg Cd kg^{-1} dry weight (Adapted from Perner, H.; Chaney, R.L.; Reeves, R.D. et al., Variation in Cd- and Zn-accumulation by French genotypes of Thlaspi caerulescens J. & C. Pres. grown on a Cd and Zn contaminated soil (forthcoming) [290].)*

phytoextraction maximum occurs at about pH 5.5–5.7 [81]. The yield responds to N fertilization, but seldom to P, K, or S unless the soil has very low levels of these nutrients [92]. Like *Alyssum* species, *Thlaspi* is adapted to low-phosphate soils and minimal P fertilization will be required for full yield potential. Although higher soil chloride usually causes increased Cd uptake by plants (for example, [67]), if soil pH is lowered to favor Cd uptake, addition of chloride caused little change in plant Cd concentrations [93].

It has been shown that *T. caerulescens* extracts soil Cd from the same 'labile pool' as other plant species rather than extracting nonlabile forms [94–97]. Several researchers have reported that roots of some strains of *T. caerulescens* tend to grow into spots of soil with higher Zn or Cd contamination [98–100]. Haines [100] compared 'Prayon' and 'Bradford Dale' ecotypes and found that only 'Prayon' had the zincophilic root growth behavior. This is another possible contributor to genotypic variation in Zn and Cd accumulation by *Thlaspi*. Others believed that ligands (organic and amino acids) were secreted/exuded by the roots of hyperaccumulator plants and that allowed them to dissolve soil Zn and Cd. But when *T. caerulescens*, wheat, and canola were compared in nutrient solutions, only wheat secreted ligands (phytosiderophores) which could bind significant amount of metals [101].

Several groups have attempted to model uptake of Cd and other elements by *T. caerulescens* and other species. They collected data on the kinetics of Cd uptake by roots, the equilibrium chemistry of Cd in test soils, and the growth of root length over time. These were important first attempts to understand the soil–plant relationships in Cd uptake by hyperaccumulator vs. normal plants [102,103]. There are difficulties in conducting such studies, including measuring Cd uptake at the activity of Cd^{2+} ions in the soil solution. All available evidence indicates that adsorption by the soil limits uptake, so that uptake is diffusion limited. Thus, the actual Cd^{2+} activity at the cell membrane is much lower than the displaced soil solution. The models used to date presume that the entire root length/area is involved in ion uptake, but earlier research indicated that polyvalent cations were absorbed–translocated only by young roots [104]. Thus, continuing root growth to maintain young roots would be very critical to high accumulation. Much remains to understand the mechanism of Cd hyperaccumulation from soils.

What are the alternatives to *T. caerulescens* for soil Cd phytoextraction? Although there have been many papers about other species that can accumulate high levels of Cd, nearly all of them came from study of Cd-salt spiking of test soils without the normal high soil Zn which inhibits Cd uptake by nearly all species. These studies will not be described because they have no application to Cd phytoextraction.

A few plants species today appear to have promise to outperform *T. caerulescens* in Cd phytoextraction, and some can be used in semitropical rice soils where the greatest need exists. A field test of *T. caerulescens* was conducted on the contaminated fields in Thailand with effective Cd removal [105], but climate was a significant limit on growing *T. caerulescens*; monsoon rains ended the first planting in the study. Ae and Arao [106] tested rice genotypes grown under upland conditions and found high enough Cd in shoots, and high enough yield potential to suggest that a technology could be developed. Compared with most other species, rice accumulates high levels of Cd relative to Zn, therefore Cd phytoextraction can be effective even on rice soils with Zn mine waste contamination. They stress that this species grows well in rice soils with tropical weather, and although the potential Cd concentration in shoots is far lower than *T. caerulescens*, in these soils and climate, *T. caerulescens* does not grow well if at all. The Japanese team tested more genotypes of rice, and conducted greenhouse and field tests of Cd phytoextraction from contaminated fields that needed remediation to lower Cd in rice grain [106–108]. The most complete demonstration to date was two years of using rice to remove soil Cd, followed by growing food rice on the phytoextracted and untreated soils [109]. The test was clearly successful; the soil contained 1.63 mg Cd kg^{-1} and 134 mg Zn kg^{-1}, and the initial pH was 5.6 (water); phytoextraction removed 883 g Cd ha^{-1}, lowering soil Cd by 38%; rice grain Cd was lowered 47%. Rice grain contained 1.02 mg Cd kg^{-1} on the control treatment, and 0.54 mg Cd kg^{-1} on the phytoextracted treatment with the best phytoextraction cultivar. Normal rice production practices were used. In their present method, they grew the 'Chokoukoku' rice with flooding until flowering and the shoots contained 0.42 mg Cd kg^{-1} on July 7, but the fields were drained at flowering and at harvest on October 19 the rice shoots contained 70 mg Cd kg^{-1}. This shows again the key role of field drainage in Cd accumulation by rice, and that the combination of high biomass yield and moderate Cd accumulation gave successful Cd phytoextraction. Murakami *et al.* [110] had earlier shown a similar reduction in soybean Cd after using rice to remove soil Cd. Because the allowable levels of Cd in rice in Japan have been lowered from 1.0 mg kg^{-1} to the new Codex Alimentarius standard of 0.4 mg Cd kg^{-1}, Murakami *et al.* [109] estimate that more than 40 000 ha of rice paddy soils in Japan will require Cd remediation.

In addition to the promising results for rice, others have examined plants growing on Cd+Zn mineralized or contaminated soils to see if other natural Cd hyperaccumulators with high yields could be found. Chen *et al.* [111] suggested that co-cropping *Athyrium yokoscense* and *Arabis flagellosa* might give useful Cd removals, but both species have low yield potential and lack high tolerance of Zn. Yang *et al.* [112] found that *Sedum alfredii* Hance collected at a metal-rich mine waste area in China hyperaccumulated Cd and Zn and did physiological research on this species. It is difficult to obtain seeds from *S. alfredii*, but the growth form is taller than *T. caerulescens*, and the plant tolerates semitropical climate. Deng *et al.* [113] found considerable genetic variation in metal accumulation and tolerance by *S. alfredii* (from 1.1 to 1051 mg Cd kg^{-1} shoot biomass), so there may be hope of developing a higher biomass Cd hyperaccumulator *S. alfredii*. Xu *et al.* [114] found that another *Sedum* species, *Sedum jinianum*, may also be able to phytoextract Cd from tropical soils. *Sedum* lacks the exceptional Cd accumulation of the southern France ecotypes of *T. caerulescens*, but they can grow in tropical environments. A tropical fruit, carambola (*Averrhoa carambola* L.) (star fruit), was also found to accumulate relatively high levels of Cd in shoots, and by growing plants at high density for shoot harvest (compared to normal orchard density), the authors believed that phytoextraction using carambola could be useful [115,116]. Phaenark *et al.* [117] examined native plants growing on Zn mine waste contaminated or Zn-Cd-mineralized soils in Thailand and reported four new Cd hyperaccumulator species based on field data. The species identified and shoot Cd concentrations were: *Chromolaena odoratum* (a perennial shrub), 166 mg kg^{-1}; *Gynura pseudochina* (an annual herb), 458 mg Cd kg^{-1}; *Justicia procumbens* (an annual herb), 548 mg Cd kg^{-1}; and *Impatien violaeflora* (an annual herb), 212 mg Cd kg^{-1} dry weight. These authors felt that *J. procumbens* had the greatest promise for practical phytoextraction because of its growth characteristics, but they have not yet characterized the farming of these species, effect of pH and fertilizer application, and so on, on Cd phytoextraction effectiveness. Presently, reproduction of this species is based on stem cuttings, which would make its use more difficult than that of a seed plant (P. Pokethitiyook, Mahidol University, Bangkok, Thailand; personal communication, June, 2009). They also noted that *C. odoratum* grew year-round as opposed to growing only during the rainy season, so it may also offer promise. Only controlled field testing over several years will show how useful these species might be.

The promising phytoextraction technologies for soil Cd in the most common contaminated soils with 100 times more Zn than Cd include: (1) improved domesticated *T. caerulescens* from southern France with pH management, for perennial culture; (2) domesticated *Sedum alfredii* or other *Sedum* species; (3) high Cd accumulating rice cultivars; (4) one of the newly identified Cd hyperaccumulators from Thailand; or (5) if biomass energy production is profitable, using willow to remove Cd while producing energy. Until government requires soil Cd remediation, it is likely that no practical technology will be developed.

14.4.3 Phytoextraction or Phytovolatilization of Soil Selenium

Soils can be geochemically enriched or contaminated anthropogenically in selenium (Se) such that the Se comprises a health risk to humans [118] or wildlife [119,120]. The Kesterson reservoir case in the Central Valley of California, USA, stimulated research because fish and birds suffered severe birth defects from excessive Se accumulation [119,121]. These drainage water evaporation pond aquatic ecosystems allowed Se biomagnification and much harm, though not directly caused by Se-mineralized cropland.

The existence of Se accumulator plants was well known [122] when studies testing different plant species for their ability to accumulate or exclude Se as possible solutions to Se enriched soils were initiated [122–126]. Banuelos [9] noted the economic limitations of using hyperaccumulator plants to achieve phytoremediation of Se-contaminated soils. He examined production of nonfood crops, and of foods with higher than normal Se levels to be sold as improved Se-rich crops, and production of forages with enough Se to replace Se supplements normally added to livestock diets. Banuelos and Mayland [8] tested the efficacy of Se supplementation using the Se-rich crop plants and found them to be a valuable Se supplement. Perhaps the best use of Se-rich biomass is to recycle the Se into livestock diets in place of virgin Se. By producing canola (*Brassica rapa* L.) for oil (food or biodiesel) and its deoiled seed press cake which could be used for Se-enriched livestock feed or organic crop Se-fertilizer, the grower could obtain value for producing the crop while reducing Se release to irrigation drainage waters [127].

An alternative phytoremediation approach, phytovolatilization, was also considered for remediation of Se-rich soils and water [128]. Both microbes in the rhizosphere and plants can produce dimethyl selenide which is volatilized from soils, roots and leaves. Plant shoots were not required to achieve useful phytovolatilization of soil Se [129]. Hansen *et al.* [130] attempted to convert this into a technology irrigating Se-rich wastewater on wetlands to remove Se from industrial wastewater, but found that more Se was accumulated in the soil than was phytovolatilized, so long-term risk from the treatment site had to be managed at higher cost.

An important principle of Se accumulation by plants is the normal competition between sulfate and selenate for root uptake. Bell *et al.* [131] found that although sulfate inhibited uptake of selenate by crop plants, it did not inhibit selenate uptake by the Se hyperaccumulator species *Astragalus bisulcatus*. This principle was extended by White *et al.* [132], who suggest that different selenate selectivity of the high affinity sulfate transporter in roots is the fundamental characteristic of Se accumulator plants. Inversely, when Banuelos [133] tested growing crop plants such as canola and broccoli on Se-rich soils, he found one could depend on the high sulfate level in the same soils to limit Se accumulation by the crops, allowing them to be marketed as Se-rich foods.

Thus, several approaches to phytoremediation of soil Se have been reported. In one, crop plants are grown which accumulate moderate levels of Se and might be sold as Se-enriched foods for extra value. In a second, crops are grown to support phytovolatilization and reduce Se in drainage waters. And in a third, phytoextraction using hyperaccumulator species may remove soil Se over time and produce soils which no longer threaten food safety or drainage water due to excessive Se. Phytoextraction of soil Se is complicated because part of the soil Se is in organic forms and in soil organisms, which are not immediately available for plant uptake.

Because Se hyperaccumulators were well known when problems arose in California [122], researchers tested growing *Astragalus* species to remove Se [123,131]. The concentration and biochemistry of Se in *Astragalus* vary with species, accession, tissue, and season [134–136]. Hyperaccumulators make a methyl derivative of selenocysteine which prevents it from entering proteins, protecting the plant from Se toxicity experienced by nearly all other species. Several researchers have noted the difficulty in growing *Astragalus* species (difficult germination of hard seed; one needs to treat seed with sulfuric acid to obtain germination) and looked at alternative Se hyperaccumulator species [137,138]. Feist and Parker [139] examined *Stanleya pinnata* collected from different sites and found considerable ecotypic variation in Se accumulation when grown in a uniform test. Improved phytoextraction cultivars might be available through breeding. Galeas *et al.* [140] reported seasonal variation in Se concentration in shoots of Se hyperaccumulators which could be used to determine the optimum time for harvest.

The alternative to development of hyperaccumulators was to produce transgenic plants which tolerate Se better than crop plants [141]. Eventually a field trial of transgenic *B. juncea* confirmed that they could achieve improved Se phytoremediation in the field [142]. The transgenic plants do not yet have the selective uptake characteristic of Se hyperaccumulators, and thus do not accumulate such high levels. On the other hand, they are easier to grow than the hyperaccumulators and offer the biodiesel plus oilseed feed supplement as economic products of cropping to alleviate Se risks.

Research on Se hyperaccumulator plants has shown that Se can be accumulated in specific tissues rather than dispersed evenly [143], and that Se reduces predation by leaf-eating insects [144]. This finding supports the 'defense' hypothesis regarding evolution of hyperaccumulators. Interestingly, Freeman *et al.* [145] found that mutant Diamondback moths (*Plutella xylostella*) had evolved on areas where *Stanleya pinnata* was dense and the moths tolerated high levels of dietary plant Se without injury; further, they found wasps which parasitized the moths which were also tolerant of high Se. Evolution generated hyperaccumulators, and then generated insects which tolerated the hyperaccumulators.

Plant Se levels as low as 38 mg kg^{-1} dry weight reduced predation by prairie dogs [146], showing evidence that hyperaccumulation can stop mammals as well as insects and microbes from harming plants with this property. Plants emitting $(CH_3)_2Se$

have a strong garlic odor, which may cause livestock to avoid consuming them unless little other forage is available [122]. Chronic diets with >3 mg Se kg^{-1} are considered a risk to livestock [147]. Thus, moderate Se accumulation gave the Se-hyperaccumulators protection against insect predators.

14.4.4 Phytoextraction of Soil Cobalt

Brooks *et al.* [148–150] reported unusual African plants (*Haumaniastrum, Aeolanthus*) growing on copper (Cu) and Co (cobalt) mineralized soils that hyperaccumulated Co, giving hope that Co phytoextraction could be developed. Ni hyperaccumulators they collected in that same time period had much poorer Co accumulation from serpentine soils and had low tolerance of absorbed Co [151]. Li *et al.* [41] reported that *Alyssum murale* could accumulate some Co from serpentine soils, and that plant Co concentration increased with soil acidification in contrast with Ni accumulation, which decreased with soil acidification. Keeling *et al.* [152] examined Ni and Co phytoextraction using *Berkheya coddii*, and found that Co phytotoxicity limits the use of Berkheya in Co phytomining. Malik *et al.* [153] tested the *Alyssum* Ni hyperaccumulators, *Brassica juncea*, and known Co accumulator species *Nyssa sylvatica*. When Ni was not present at the normal 10-fold ratio to Co of serpentine soils, *Alyssum* species accumulated and tolerated over 1000 mg Co kg^{-1} dry shoots, but Ni strongly inhibited Co accumulation by the tested species. *B. juncea* suffered strong Co phytotoxicity when *Alyssum* and *Nyssa* grew normally.

The initial controlled greenhouse test of Co uptake by these species used Co-salt-amended potting media and measured high Co accumulation in several species [154], so their ability to hyperaccumulate Co (>1000 mg kg DW (dry weight)) under this laboratory condition was established. However, subsequent research on the African Co accumulator species has raised questions about the analysis of these plants due to mineralized soil contamination of leaf samples [155]. Recent studies using mineralized soils showed weaker plant accumulation of Co and Cu [156].

For the *Alyssum* phytoextraction technology, raising soil pH to maximize Ni phytomining reduced Co accumulation [45], but at the end of Ni phytomining, the soil could be acidified without much risk of Ni phytotoxicity to allow more effective Co phytomining. Alternatively, *Alyssum* species could be grown on Co-mineralized or Co-contaminated soils. A striking finding was reported by Tappero *et al.* [157]: Co was not stored in leaf vacuoles similarly to Ni storage in *Alyssum murale*; Co remained within the xylem system or cytoplasm or exited leaves at the tips, forming a deposit on the leaf surface. Because Co is 3–4 times more valuable than Ni [158], Co phytomining could be cost-effective if plants and agronomic practices were developed to maximize annual biomass Co quantity.

14.4.5 Phytoextraction of Soil Boron

One example of effective boron (B) phytoextraction has been reported by Robinson *et al.* [159]. A site in New Zealand where wood was treated with boron chemicals had a large pile of waste materials from which B had leached to groundwater. Boron levels exceeding allowed limits were measured in nearby surface water. Robinson *et al.* tested growing hybrid poplar to evapotranspire water and limit off-site movement of the B-rich plume of groundwater, and irrigated groundwater onto the waste pile to supply water during dry periods of the year. The poplar leaves accumulated high B levels, more than 1000 mg kg^{-1} DW, much higher than the woody tissues. They proposed harvesting the trees on a three-year cycle, with the crop used for energy production or stock fodder, or even harvesting the leaves as a B-rich organic-fertilizer. Although sale of leaves as organic B-fertilizer has promise, the wood treatment company chose to simply prevent stream contamination.

High B levels in Se-contaminated irrigation drainage waters limits species which can tolerate soils managed to obtain phytoremediation of Se in the drainage waters. Thus Banuelos [133] looked at preventing Se from entering the drainage waters by growing crops which phytovolatilized Se. The high B, salinity, and Se in the drainage waters put severe limitations on treatment alternatives.

14.4.6 Phytovolatilization of Soil Mercury

Mercury (Hg) is a complicated element for phytoremediation. First, normal plants do not absorb and translocate important amounts of Hg from soils through the xylem to shoots; most shoot Hg arrives by volatilization from soils [160–165]. Photovolatilization is very important in releasing Hg vapor from soils, so heavy plant cover can strongly inhibit photovolatilization [166]. Roots and rhizosphere microbes can reduce Hg^{2+} to Hg^0 with volatilization from the soil surface, but hardly from below a few millimeters into a soil [167].

There are no land plants which hyperaccumulate Hg when growing in a Hg-rich soil except by capturing volatilized Hg^0 more effectively than other plant species growing on the soil, for example, mushrooms [168], but such fungi would not provide a convenient species for phytoextraction of Hg. Many studies of Hg uptake fail to control Hg^0 volatilization from the surface of the growth medium, which allows volatilized rooting medium Hg to reach the shoots and confound the interpretation of the uptake to shoots. Plant transpiration of root-absorbed Hg is a minor process [163], although some Hg does reach shoots by the xylem.

A team lead by Meagher developed transgenic plants expressing bacterial mercuric reductase (MerA), and organic-mercurial lyase (MerB) to see whether transgenic plants could be developed to achieve phytoremediation of soil Hg [169,170]. Effective expression of these genes in plants required changing some codons from those used in bacteria. Meagher *et al.* showed this approach allowed plants to phytovolatilize Hg from Hg^{2+} in soils when MerA was expressed in plants, and from CH$_3$Hg$^+$ when both MerA and MerB were expressed. They expressed the genes in several plant species to seek species for practical soil Hg phytoremediation [171–173]. Although the technology was scientifically successful, it was not accepted by regulatory agencies or commercial users because Hg would be released to the air from soil, and that Hg would eventually return to earth and possibly cause Hg risk where it was deposited. Subsequently, Meagher *et al.* attempted to develop plants which accumulated Hg inside the plant by expressing phytochelatin synthase [174]; although this increased As and Hg tolerance, it did not achieve accumulation of Hg in the plant biomass. In addition, their test did not prevent Hg volatilized by treated roots from reaching the shoots and cannot be considered definitive. The last advance in Hg phytoremediation was a demonstration that expression of the organic mercurial lyase secreted through root cell walls could increase plant tolerance of organic mercurials, and increase Hg volatilization better than general expression of the transgene in the plant [175].

Soil Hg comprises risk through soil ingestion and through aquatic ecosystem biomagnification of methyl-Hg from sediments. Large areas of soil are Hg-enriched where Au or Hg mining occurred historically. Such soils may be barren and eroding into streams where the Hg could cause much greater risk. Bare soils rich in Hg are highly subject to photovolatilization. Phytostabilization of such soils using combinations of organic matter-rich nutrient sources and required alkalinity could strongly reduce Hg release to the environment.

Moderately contaminated soil could be phytoremediated using the MerA-MerB transgenic plants described above. The likelihood of local deposition of the phytovolatilized Hg^0 is very small, and the benefit to society is large. The dig and haul alternative is much more expensive and could disperse the contaminants. The poplar, rice and other transgenic Hg remediation crops offer much promise.

14.4.7 Induced Phytoextraction of Soil Gold

Gold (Au) may be a cost-effective induced phytomining opportunity. Economic Au phytomining first requires standard ore mining and grinding, then placing the ground ore over plastic membranes to prevent leachate loss so that cyanide, thiocyanate or thiourea used to induce phytoextraction can be irrigated on the soil to promote Au uptake by growing plants [176–178]. Others have looked for formation of gold nanoparticles in the plant biomass because the nanoparticles may have higher value for use as catalysts, and other uses, than the gold content alone [179] and tested diverse plants to phytoextract soil Au [180,181]. Australian researchers estimated that Au phytomining could make more return per hecatre than Ni phytomining [182], so this research topic remains active.

14.4.8 Induced Phytoextraction of Soil Lead

In 1995, Kumar *et al.* [29] reported accumulation of more than 1% lead (Pb) in shoots of some genotypes of *Brassica juncea* grown in sand culture supplied soluble Pb with no phosphate or sulfate in the nutrient solution. These conditions allowed high uptake of Pb and severely harmed the plants, which were then harvested and analyzed. On the basis of these data the authors obtained a patent for phytoextraction which they believed covered all plants and soil management for phytoextraction. Fortunately, their patent excluded the 'low-yielding' natural hyperaccumulator plants and did not stop R&D on practical phytoextraction.

Subsequently the Rutgers University/Phytotech team tested Pb uptake from contaminated soils and found very low uptake by the same genotypes of *B. juncea* that worked so well in nutrient solution, and then developed the addition of chelating agents to induce phytoextraction of Pb from soils [183,184]. The patent was licensed to a Phytotech, Inc. which conducted Pb phytoextraction field tests. Further research showed the leaching of Pb and other metal chelates from treated soils, and Pb contamination of groundwater. The company eventually went bankrupt, and it is no longer possible to obtain permits to conduct field scale induced phytoextraction of soil Pb in the United States or European Union. It is very unfortunate that many researchers were misled by these studies to believe that Pb phytoextraction might be practical. Literally hundreds of papers have since been published on Pb phytoextraction with different plant species and different chelating agents. None have demonstrated a cost-effective and environmentally acceptable phytoextraction technology for soil Pb.

It had long been known that if plants grown in soil or nutrient solution were deficient in phosphate and no phosphate fertilizer was added, the plants could accumulate and translocate high amounts of Pb (for example, [185]). The presence of phosphate causes formation of an insoluble Pb-phosphate compound in roots. When adequate phosphate exists in the soil to produce a normal crop yield, even 3200 kg Pb ha^{-1} applied to a field soil caused little increase in Pb in *Zea mays* L. [186]. Cotter-Howells *et al.* [187] showed that plant roots/microbes caused formation of the nonphytoavailable Pb compound chloropyromorphite when growing in Pb-rich soils. Thus, normal soil and plant Pb chemistry prevents plant Pb uptake–translocation to any meaningful extent. Two minor accumulators (<1000 mg Pb kg^{-1}) of Pb (hemp dogbane, *Apocynum* sp. and common ragweed, *Ambrosia* sp.) were reported by Cunningham and Berti [188], who collected and analyzed plants growing on a Pb-contaminated field; other species contained <10 mg Pb kg^{-1} after washing to remove soil and dust from leaves – a normal level for most plants growing in such strongly contaminated soil.

Valuable studies conducted by Blaylock *et al.* [183], Huang *et al.* [184], and Cooper *et al.*, [189] clarified the need for adding chelating agents to soils in order for Pb to be accumulated by plants. With added chelators, several species could accumulate over 1% Pb in dry shoots [190]. In general, the plant is grown for some period to give biomass with effective evapotranspiration; then the chelating agent is applied in soluble form to the surface of the soil. Within a week or two, the plants are dead from multiple metal or chelator toxicity and ready to harvest. Research was conducted which showed that intact Pb-EDTA did reach the shoots of several species [191,192]. And basic studies showed that EDTA had to be in excess of Pb or other strongly chelated cations in the nutrient or soil solution for the EDTA to stimulate high uptake of Pb [191–193]. It is believed that free EDTA has to harm the root cell membrane enough to make it leaky, and then the solution containing Pb-EDTA (and other metal-EDTA chelates) enters the transpiration stream and rises to plant shoots [194]. Although Pb-EDTA may have low phytotoxicity to tested species, other metals and free EDTA strongly affect plant health when high levels are supplied to soils for induced-phytoextraction.

Many soils have mixed metal contamination such that uptake of Zn, Cu, and other ions are also greatly increased by chelation treatment because the chemistry of the specific chelating agent utilized controls which soil elements are dissolved [183,195,196]. Discussion of chelation equilibria can be found in Nowack [197], and Parker *et al.* [195]. Chelator selectivity among elements, surface area binding of the metal in soil, and activity of metals in the soil control which metals are dissolved and how rapidly they are dissolved.

In tests we conducted with *B. juncea* and Ni and Co accumulation, the growth habit of the plant seemed ill-suited to phytoextraction [153]. *B. juncea* is normally sown in the fall, grows over winter, and flowers when long days return in the spring. If the plant is started in the spring, it begins to flowers in about 4 weeks, greatly limiting biomass and potential transpiration needed to accumulate much Pb-EDTA from soil. Some *B. juncea* phytoextraction research used 10-hour days to prevent conversion to flowering growth (vernalization) [193]. Banuelos [126] initially studied *B. juncea* and collected local strains of this species in India because it might be useful in Se phytoextraction. Others chose to use it for Pb and many other elements for which there was never any evidence to suggest this species could be useful.

Many papers have since discussed concern about metal leaching when chelating agents and detergents were applied to induce phytoextraction [196,198–201], and some have shown strong experimental evidence of the leaching [202–208]. There can be little question that metal leaching precludes field application of EDTA-induced metal phytoextraction. It was noted above that cyanide-induced Au phytomining could be conducted commercially only because the 'soil' or ground ore was placed on a plastic membrane to conduct cyanide heap-leaching of Au before using plants to phytoextract some of the remaining Au. Open field cyanide leaching is not acceptable, similarly to open field EDTA-induced phytoextraction.

Other chelating agents were tested using different timings of chelator addition and split applications, combined with other treatments (for example, [183]). In particular, EDDS (ethylenediamine disuccinic acid) was tested because it is more rapidly biodegraded and hence less likely to cause unacceptable leaching. Testing showed that leaching was still a problem under any condition which allowed EDDS addition to stimulate Pb uptake [196].

Another important consideration of chelator-induced Pb phytoextraction is the cost of the added chelator. Chaney *et al.* [12] obtained the price of truckload quantities of EDTA ($4.30 kg^{-1} in 2000) and calculated the cost of applying 10 mmol EDTA kg^{-1} soil for 1 ha–15 cm deep (2· × 10^6 kg), the dose required for optimal Pb phytoextraction [183,184]. One application of EDTA would cost $30 000 ha^{-1}. When EDTA is added at these levels, one must wait before planting *B. juncea* again because of phytotoxicity until the metal chelates leach or are biodegraded. Thus this method is very expensive as well as comprising risk to groundwater contamination if liners are not used. With the highly effective *in situ* inactivation of soil Pb described below, it seems clear that inactivation of soil Pb is the more desirable approach for soil Pb remediation.

14.4.9 Phytoextraction of Soil Arsenic

Soil arsenic (As) is a risk to children through inadvertent soil ingestion, and to all ages of humans through dietary exposure from rice accumulation of As from contaminated soils, especially those irrigated with As-rich water. Current calculations by the US-EPA indicate that the limit for soil As should be 4.3 mg kg^{-1} to protect against 10^{-5} increased lifetime cancer risk based on soil ingestion. This concentration is within the normal range of soil As levels for background uncontaminated soils (5th and 95th centiles = 2 and 12 mg As kg^{-1}) [209], so it is not clear that the US-EPA limit is justified. In practice, soil cleanup to 20 mg kg^{-1} has been selected for several US Superfund sites in urban areas. Because Department of Defense activities during World War I caused soil As contamination in the Washington, DC, area, and US-EPA designated the site a Superfund site, soil remediation was conducted by Edenspace, Inc. under contract. Edenspace licensed use of the As hyperaccumulator fern *Pteris vitatta* from the University of Florida [210] and conducted field operations. Because soils contained in the order of 40 mg kg^{-1}, with localized higher concentrations, Edenspace used a combination of excavation of highly contaminated soil, and tillage prior to phytoextraction to reduce hot spots. This is either an advantage of phytoextraction compared to soil removal, or a problem with phytoextraction. Tillage is both normal and required for commercial phytoextraction. When the risk and clean-up goal is based on average surface soil As concentration, alleviating hot spots clearly reduces risk. For risk to occur, children must ingest the soil from the soil surface, so reducing As levels in the surface soil is the valid goal and tillage contributes to this goal. After a few years of As phytoextraction, Edenspace was able to reduce soil As to regulatory limits (M. Blaylock, Edenspace Inc., personal communication, June, 2009). In another field test of As phytoextraction, the point-to-point soil As variation of the 30.3-m^2 plot was so great that two crops of *P. vitatta* did not significantly reduce soil As concentration while they reduced average soil As from 190 to 140 mg kg^{-1} [211].

The As hyperaccumulator fern, *P. vitatta*, was discovered by Ma et al. [210] by analyzing many plant species growing at an As-contaminated site. They have done extensive research to develop this technology, including study of the chemical form of As in soil, and the effect of fertilizers and pH management on phytoextraction. Interestingly, although As enters most plants on the high-affinity phosphate transporter [212], and solution phosphate inhibits As uptake, phosphate did not inhibit As accumulation by *P. vitatta* [213]. Others have studied fern variation in As accumulation, finding other *Pteris* species, and other fern genera, that contain As hyperaccumulators [214–216]. Wang et al. [217] examined variation of As accumulation by ferns collected at different locations in South China and found genotypic variation within *P. vitatta* that could be useful in breeding improved cultivars.

The effect of rhizobacteria from the rhizosphere of field-grown ferns on As uptake, translocation, and fern yield in high As solutions and soils was tested [218]. Although rhizobacteria increased shoot As in nutrient solutions, they had no effect when used to inoculate ferns grown in sterilized soils. Mycorrhizae were found to significantly increase As and P phytoextraction by *P. vitatta* [219], but the benefit of inoculation in nonsterilized soils has not been reported.

The fern appears to tolerate high biomass As levels because it can store the As in vacuoles or make phytochelatins to bind part of the As within cells [220,221]. Only about 1–3% of total shoot As was found to be bound by phytochelatins [222], so the significance of As binding by phytochelatins is uncertain. Although the roots absorb arsenate, arsenite is the dominant form of As in the leaves [223]. Several papers suggested that reduction occurred in the shoots, but a careful test by Su et al. [224] measured As species in bleeding xylem exudate of *P. vitatta* and showed that arsenite was the dominant form present, confirming that reduction largely occurred in the roots. Previous tests used pressure expression of xylem sap and obtained very different and apparently incorrect results [225]. Arsenic has localized distribution within the fern plant (tissue concentration is much higher in young than old fronds) [226,227] and appears to be stored in epidermal cell vacuoles similar to other hyperaccumulators with other metals, and in trichomes [228].

Although *Pteris vittata* can give effective shoot biomass yields, it does not tolerate cold environments, and is difficult or time-consuming to establish; transplants are used after starting the fern propagules. Thus development of a transgenic seed plant with field As phytoextraction potential could be important for solving soil As contamination problems [174]. Dhanker et al. [229] found that if they transferred genes for arsenate reductase and γ-glutamyl-cysteine synthetase into plant species, plant tolerance of As was greatly increased [174]. These species were still not as able to hyperaccumulate As from soils as well as the natural fern hyperaccumulators, but if the arsenate reductase in *Arabidopsis thaliana* roots was silenced by RNA interference, arsenate was more freely translocated to shoots (a 10- to 16-fold rate of wild type) [230]. Apparently the root uptake or arsenate reduction characteristics and/or arsenite translocation characteristic of the ferns must also be transferred into transgenics in order to obtain seed plants with effective As hyperaccumulation.

Arsenic is also a potential risk through drinking water, and with recent lowering of drinking water As limits, many water producers must find a way to reduce As in their product. Chemical removal can be expensive. Elless *et al.* [231] tested use of *P. vitatta* to phytoextract As from raw drinking water and found they could reduce water As to $< 2 \, \mu g \, l^{-1}$, well below US limits (10–15 $\mu g \, l^{-1}$) for As in drinking water.

Yan *et al.* [232] demonstrated that incineration of As-rich biomass caused volatilization of the plant As, but that should have been expected. Copper smelters would emit large amount of As if not required to control all emissions, which is readily achieved using existing technology. If biomass energy production could reduce the cost of As phytoextraction, the incineration could be conducted using proper exhaust treatment. As noted above, to date As-rich phytoextraction biomass has been disposed in landfills.

In situ inactivation or phytostabilization of soil As has been demonstrated by several research groups. Reduced phytoavailability was shown for Fe additions to soils [233]. Reduction of *in vitro* bioavailable or bioaccessible soil As has been demonstrated by Subacz *et al.* [234], by Beak *et al.* [235], and by Smith *et al.* [236] by adding Fe to soils. By comparing the EXAFS As speciation results with bioavailability results from pig feeding studies, Beak *et al.* [235] showed that As associated with Fe was essentially nonbioavailable to young pigs. Thus *in situ* treatment may be an effective treatment for soil As risks. With the extensive contamination with As of irrigated rice soils in Bangladesh, some As remediation solution is needed to limit As uptake into rice grain and to prevent future risk to children from ingestion of these soils [237]. Similarly, the extensive area of As-rich soils in south-east England attributed to historic tin mining requires As remediation [238].

14.4.10 Phytoextraction of Other Soil Elements

Phytoextraction of a few other elements has been studied to some extent. Soil thallium (Tl) may comprise risk to humans at some locations where local industry emitted Tl, or major Pb smelters contaminated large areas, so studies were conducted comparing vegetable crop accumulation of Tl. Green cabbage accumulated much higher Tl levels than most other vegetable crops [239,240]. Fortunately, unusual hyperaccumulator species were discovered for Tl and they accumulate high plant Tl from soils with moderate to high Tl contamination [241,242]. The strong hyperaccumulator, *Iberis intermedia*, accumulated Tl in the leaf vascular system rather than in vacuoles or trichomes as seen for some other hyperaccumulated elements [243].

Cesium (Cs) risk is most important for the radionuclide (^{137}Cs) which is spread rapidly after nuclear fission releases. It was well known that Cs could be easily fixed within soil clay, and that high soil K inhibited Cs uptake by plants. Substantial species variation in Cs accumulation from soils has been demonstrated, but no special hyperaccumulator type was identified [244]. Researchers compared a number of species relative uptake of Cs and found that red root pigweed (*Amaranthus retroflexus* L.) accumulated much higher shoot Cs levels than other studied species [245,246]. Redroot pigweed bioaccumulated Cs from soils into shoots, while most other species accumulated Cs in roots. This species also has a relatively high potential shoot yield for phytoextraction. Depending on the extent of contamination, the clean-up period could be longer than desired [247]. Dushenkov *et al.* [248] tested using sunflower and *Brassica juncea* to phytoremediate ^{137}Cs contaminated soils near the Chernobyl nuclear site, with limited success. An alternative approach was tested in Belarus, growing canola on the ^{137}Cs-contaminated soils so that the canola oil could be used for biodiesel and the seed meal used for feed or fertilizer. Most elements do not follow the oil during crushing of oilseeds, and ^{137}Cs was hardly present in canola oil from seeds with lots of radioactivity. Here again the economic solution may not support phytoextraction, but the principles of soil–plant science for each contaminant are the basis for public decisions. The original plan was to use the canola oil as biodiesel, but when it was found to have such low ^{137}Cs activity, it was all used for human food.

Uranium (U) phytoremediation research has taken several directions. In one, plant roots were used to remove U from contaminated groundwater in a process called *rhizofiltration* [249]. After testing several species in the laboratory, an innovative field trial was conducted using alfalfa sprouts in plastic frames over which well water containing excessive U was irrigated; this rhizofiltration was quite effective in removing U, but not as inexpensive as the traditional removal using resins [249]. Natural uptake and translocation of U by plant roots are quite low, so Huang *et al.* [250] tested using chelator-induced phytoextraction for soil U; citrate increased U dissolution and uptake substantially. Whether this could be cost-effective has not been reported, but the technology has not been commercialized. Citrate addition can increase leaching of U, but because citrate is readily biodegradable, the potential for adverse impact is less than seen for addition of EDTA.

Molybdenum (Mo) and tungsten (W) can accumulate in soils to phytotoxic or zootoxic levels which require remediation. Certain legumes are able to accumulate much higher levels of Mo than most grasses [251], but no effective phytoextraction technology using legumes which accumulate Mo has been reported. Treatment of the soil with phosphate or vermiculite raised Mo uptake for phytoextraction, while other treatments lowered plant Mo and alleviated potential for Mo zootoxicity [252]. An overview of soil Mo risks and management strategies was reported by O'Connor *et al.* [253].

14.5 Phytostabilization of Zinc-Lead, Copper, or Nickel Mine Waste or Smelter-Contaminated Soils

It is evident from the previous discussion that phytoextraction is not practical for Zn and Pb, which are among the most common soil contaminants of public concern. Zinc, Cu, and Ni are the most common cause of soil metal phytotoxicity for most plant species and cause barren areas where acidic soils are contaminated. Often erosion of the contaminated soils, or the presence of low nutrient mine wastes causes severe infertility as well metal stress. Lead is more important as a risk to children and wildlife health through ingestion of soil than through plant uptake. Lead is included in this discussion because most Pb and Zn ores are mixed and include Cd as well, so remediation of Pb and Cd must also be achieved when Zn phytotoxicity is remediated. Research has increasingly shown that combinations of soil amendments to make the soil calcareous and fertile can remediate these risks and allow effective soil cover which protects wildlife. It is not certain that sensitive crop plants will succeed on such *in situ* remediated soils, but many common grasses and legumes do very well after soils have equilibrated.

In situ phytostabilization was first considered as an alternative to establishing metal tolerant ecotypes of grass [254] which had been studied and developed by Bradshaw *et al.* [255]. The 'Merlin' red fescue (*Festuca rubra*) they developed does very well on fertilized Zn toxic soils compared with other species (for example, [26]), but such plantings require regular fertilization to persist. In the absence of other legumes and plant species with similar genetic tolerance to soil metals, an ecosystem cannot develop. There are no legume ecotypes with high metal tolerance needed to supply N for other species on Zn phytotoxic acidic soils.

Gadgil [256] considered using biosolids and municipal solid waste compost coupled with metal-tolerant grass ecotypes to revegetate barren soil contaminated by Zn, Ni, and Cu. In some cases she included alkaline pulverized fuel ash from coal-fired power generators, which raised soil pH and aided in reducing metal phytoavailability.

Research has since shown that organic matter, Fe and Mn oxides, and phosphate in biosolids and other soil amendments can increase metal sorption or precipitation in contaminated soils, and aid in revegetation [27,257–259]. Combined with limestone to make the amended soil calcareous, biosolids or composts can provide effective revegetation and limit plant uptake of metals to protect wildlife food chains [260]. The effective control of Pb uptake and bioavailability from ingested soil was demonstrated by Ryan *et al.* [261]. Depending on the soil amendments, Pb is converted to chloropyromorphite or Pb sorbed to Fe oxides [262]. Field treatment of high-Pb soil with soluble triple superphosphate or phosphoric acid caused a 69% reduction in soil Pb bioavailability to humans within 1.5 years [261]; rock phosphate is less reactive with soil Pb than the soluble phosphates although it may be cost-effective for acidic soils. Uptake of Zn, Cd, and Pb by normal grasses did not threaten wildlife (for example, [28]).

Because the cost of *in situ* phytostabilization of soil metals is so much lower than that of alternatives, this approach is receiving increasing consideration (for example, [260,263]). Unfortunately, for rice soils which induce human Cd disease, phytostabilization is not adequate to protect rice consumers; farmers would have to stop growing rice to avoid soil Cd risks. Phytoextraction of soil Cd, or change in crops grown is needed to protect consumers [65]. Because crop Zn protects against nearly all soil Cd risks, it is not evident that most cropland contaminated with geogenic 1:200 Cd:Zn ratio Zn contamination will require Cd remediation – rather, only Zn phytotoxicity remediation will be required and can be easily obtained with limestone and other soil amendments. Stuczynski *et al.* [264] fed plant cover from a remediated zinc smelter slag site in Poland compared with the same plant species grown on control soils without or with added Cd and Zn salts to equal the levels found in the plants from the contaminated site. Salt metals added to control hay caused much higher Cd accumulation in kidney and liver of calves than did the hay from the contaminated site. Plants growing on natural contaminated sites may be a valuable resource for selection of cover crops or metal-tolerant excluder species for revegetation of contaminated sites [31].

14.6 Recovery of Elements from Phytoextraction Biomass

Several authors have raised issues about disposal of phytoremediation biomass, suggesting that disposal will be so expensive that it will prevent many opportunities for commercial phytoextraction [14]. In the case of Ni, Au, Co, and perhaps Tl phytomining, biomass would be incinerated to produce energy, and the metal recovered from the ash would provide considerable profit potential [12,158,182].

The value of the biomass energy may be high enough to cover the cost of crop production, but not enough to make the technology profitable if the metals have no value. Radionuclides in biomass are clearly a more difficult issue because few incinerators are permitted to burn radioactive materials, and shipping the biomass to a permitted site could add substantially to the cost [265]. Biomass from Pb and As commercial phytoextraction has been disposed in normal landfills because the biomass was not hazardous according to US-EPA testing protocols [207].

Several groups have proposed biomass energy production (incineration or pyrolysis) with a side benefit of Cd phytoextraction using willow [266–271]. As noted elsewhere, high enough Cd in this ash would require its disposal in landfill rather than application to return nutrients to the forest where the biomass was produced. Although there has been considerable discussion of pyrolysis of biomass, there is no evidence that pyrolysis is ready to be deployed for phytoextraction biomass conversion. Further, because metals and nutrients are higher in leaves than wood, and ash from leaves can foul turbine generators which are direct-fired, phytoextraction biomass with leaves is an even greater issue. Disposal of the ash in landfill would incidentally increase the cost of growing the willow because more fertilizer and limestone would be needed if the ash were not returned to the field. It may be possible to use the high temperature of biomass incineration to separate volatile elements into the fly ash fraction and retain nutrients and alkalinity in cyclone and bottom ash [272] rather than be concerned about loss because of volatilization. Effective exhaust gas treatment systems are available to prevent metal loss during biomass burn.

Use of ash from Ni phytomining biomass as an alternative Ni ore has been demonstrated [12]. *Alyssum* ash was processed in an electric arc furnace and was a very successful Ni ore. Harris *et al.* [182] and Brooks *et al.* [158] also discussed the value of Ni phytomining biomass. Ni phytomining on serpentine soils rich in Ni, or on contaminated soils, can provide more profit than most common agronomic crops. Ni is much less volatile than Cd, Zn, Pb, and As [272] so it is relatively easy to retain Ni during biomass processing.

14.7 Risks to Wildlife during Phytoextraction Operations

High levels of Se and other metals in ingested plant phytoextraction biomass may be toxic to wildlife (for example, [146]). It is important to keep in mind that the sites being remediated have been identified because they cause metal risk to humans or the environment (wildlife). Lack of remediation continues the harm to wildlife. During remediation operations, sites will exclude large wildlife, but small mammals which are resident in small areas might be harmed. On the other hand, dig and haul will destroy all local wildlife [273].

Animal feeding tests with hyperaccumulator plants have not been reported, so no data are available to estimate the risk from ingestion of these plants. Another consideration is needed for plant Cd risk to animals. Most researchers cite NRC [147] as the authoritative source for tolerable levels of diet Cd. Readers should recognize that these limits are based on feeding diets with added metal salts, not crops with intrinsic Cd; Zn was not increased in proportion to Cd as occurs with most plant species. As noted above, increased Cd in test feeds with intrinsic Cd as high as 2.5 mg Cd kg^{-1} caused no increase in kidney or liver Cd of test animals [274], so the 0.5 mg Cd kg^{-1} suggested limit lacks a valid technical basis. Further, that limit was based not on animal health, but on protecting against increased Cd in liver and kidney used as human food. Such limits should be based on the bioavailable Cd taking into account the Zn present in the same plant material, not total Cd in the crop [65]. Stuczynski *et al.* [264] found that adding Cd salt to forage diets for calves gave considerably higher kidney and liver Cd than feeding forage grown on a phytostabilized Zn smelter slag site, and that omitting Zn addition caused greater Cd accumulation in tissues than when both Cd and Zn were added equivalent to the forage from the phytostabilized site.

In the case of *Alyssum* species used in Ni phytoextraction, the dense trichomes on all leaves strongly inhibit feeding by livestock and wildlife. In the Mediterranean serpentine soils where these species occur naturally, pastures rich in *Alyssum* with high Ni levels (>1% DW) are grazed by cows, sheep, goats and wildlife which avoid consuming *Alyssum*. In field tests of *Alyssum* species in Oregon, USA, cattle, deer and rabbits did not graze on *Alyssum* [12]. The trichomes give the plants a wiry texture which appears to be unpalatable to animals, thus protecting the animals from the high *Alyssum* Ni levels. The seeds contain ~7000 mg Ni kg^{-1} but are so small that they are not useful for forage/feed.

Selenium-accumulator plants can accumulate toxic levels of Se with little difficulty, and the Se is bioavailable and toxic to livestock and wildlife according to many veterinary toxicologists; Se accumulator species are considered to be toxic weeds [122] but these plants are seldom ingested by livestock due to the garlic odor of Se rich plants. Even crop plants without the selenate/sulfate selectivity of the Se hyperaccumulators may accumulate high enough of Se (>3 mg kg^{-1} DW) [123] that if the crop were 100% of livestock diets, the Se would be toxic based on NRC [147] estimates of levels of minerals tolerated by livestock. But as noted above, the Se-rich biomass could be used as a Se supplement for livestock feeds to replace the usual chemical Se addition.

14.8 Conclusions

The promise of phytoextraction, phytomining, phytovolatilization, and phytostabilization as important tools for society to deal with contaminated soils remains important despite the conclusions of others [282–285]. Phytoextraction goals should be based on reduction of risk, not on arbitrary soil metal concentrations that are not specifically related to risk. Food Cd risk is real on contaminated rice soils as discussed above, and reduction in soil Cd to allow production of rice with acceptable Cd levels may be achieved using high Cd-accumulating rice cultivars grown using nonflooded culture in acidic soil [109]. Yet, *T. caerulescens* did not perform adequately in rice soils because of the climate. But this species is still promising for European and northern United States soils which comprise risk to humans. Risk levels to protect wildlife are estimated to be much lower than would be found if bioavailability were taken into account. And Cd does not magnify in terrestrial food chains [286].

Presumed risk from soil Zn and Ni based on additions of soluble metal salts to soils are woefully in error, as discussed in Chapter 17. When pH is adjusted to allow production of crops sensitive to excessive soluble soil Zn, lettuce can be grown on Zn smelter-contaminated soils and is safe for human consumption [287] even in soils with 100 mg Cd kg^{-1} and 10 000 mg Zn kg^{-1}. Nodulated white clover grows well in soils with high Zn if soil pH permits growth of the crop [288]. Thus, phytoextraction goals based on risk can be achieved by known hyperaccumulator crops. If the land involved has high value and taking it out of production in order to conduct phytoextraction would be unacceptable to land owners, they can use dig-and-haul methods with rapid remediation. For most arable soils, dig and haul will not be the rational choice for remediation.

Commercial phytoextraction practices continue to be developed and are being tested in the field. Because other soil remediation methods are so much more expensive, it seems likely that phytoextraction will continue to be developed. Improved crops will be bred for commercial application, and bioengineered plants will be developed with unique properties. Despite their technical value, bioengineered plants may not be accepted by the public even for phytoremediation [273,282], especially bioengineered strains of food plants.

Some phytoextraction can be profitable as a farming/phytomining business on contaminated or mineralized soils. Ni phytomining offers high profit potential [12]. Some ask why the technology has not been fully commercialized. The company (Viridian LLC) which licensed the patents obtained by Chaney, Angle, Li, and Baker [39,289] has chosen to not operate the technology by contracting with farmers to grow *Alyssum* crops on serpentine soils, but to attempt to do an Initial Public Offering of stock to recover their costs to support development of the technology and to obtain a profit. They have contracted with Vale-Inco to successfully test Ni phytomining on smelter-contaminated soils and mine waste deposits, and are considering phytomining on Vale-Inco properties. But this has not proceeded beyond planning, frustrating the scientists who considered the technology completely ready for commercial operations since 2001.

For most elements, phytoextraction can offer only lower cost of soil remediation. The biggest impediment to development and commercialization of many phytoextraction technology opportunities is the failure of environmental regulatory authorities to require remediation of highly contaminated soils. Until this market for soil metal phytoremediation service develops, only phytomining technologies will be practiced, along with basic research to understand hyperaccumulator plant biology and phytoextraction soil and plant chemistry.

Phytostabilization will remain a valid remediation technology for most contaminated sites. Mixed metal contamination can be handled by phytostabilization in most cases, except where food-chain transfer would continue risk to wildlife. This may be more important for soils with Zn and other metal phytotoxicity and with simultaneous Se or Mo contamination. Liming to alleviate Zn phytotoxicity would maximize Se or Mo accumulation in plants and threaten wildlife.

Public acceptance of *in situ* Pb inactivation will aid adoption of phytostabilization of mixed Zn-Pb-Cd contaminated sites such as the Joplin, MO, USA site [28,261] where field testing showed the forage was safe for livestock, and soil feeding tests showed strong reduction in soil Pb bioavailability. Such remediated sites may not be permitted to become housing areas or playgrounds, in order to avoid exposure of children even to apparently remediated Pb risks. Controlled land use after phytostabilization can protect humans and the environment from soil trace element risks. Continued development of phytotechnologies will provide more choices for remediation, and demonstrate in the field the value to society these technologies offer.

References

[1] Baker, A.J.M.; McGrath, S.P.; Reeves, R.D.; Smith, J.A.C., Metal hyperaccumulator plants: A review of the ecology and physiology of a biological resource for phytoremediation of metal-polluted soils; in: N. Terry, G.S. Banuelos (eds), Phytoremediation of Contaminants in Soil and Water; CRC Press, Boca Raton, FL, 1999, pp. 85–107.

[2] Brooks, R.R.; Lee, J.; Reeves, R.D.; Jaffré, T., Detection of nickeliferous rocks by analysis of herbarium specimens of indicator plants; J. Geochem. Explor. 1977, 7, 49–57.

[3] Jaffré, T.; Schmid, M., Accumulation du nickel par une Rubiacée de Nouvelle Calédonia: *Psychotria douarrei* (G. Beauvisage) Däniker (in French); Compt. Rendus Acad. Sci., Paris, Ser. 1974, D-278, 1727–1730.

[4] Chaney, R.L., Plant uptake of inorganic waste constituents; in: J.F. Parr *et al.* (ed.), Land Treatment of Hazardous Wastes; Noyes Data Corp., Park Ridge, NJ, 1983, pp. 50–76.

[5] Baker, A.J.M.; Brooks, R.R., Terrestrial higher plants which hyperaccumulate metal elements – a review of their distribution, ecology, and phytochemistry; Biorecovery 1989, 1, 81–126.

[6] Reeves, R.D., The hyperaccumulation of nickel by serpentine plants; in: A.J.M. Baker, J. Proctor, R.D. Reeves (eds), The Vegetation of Ultramafic (Serpentine) Soils; Intercept Ltd., Andover, 1992, pp. 253–277.

[7] De la Rosa, G.; Peralta-Videa, J.R.; Montes, M.; Parsons, J.G.; Cano-Aguilera, I.; Gardea-Torresdey, J.L., Cadmium uptake and translocation in tumbleweed (*Salsola kali*), a potential Cd-hyperaccumulator desert plant species; ICP/OES and XAS studies; Chemosphere 2004, 55, 1159–1168.

[8] Banuelos, G.S.; Mayland, H.F., Absorption and distribution of selenium in animals consuming canola grown for selenium phytoremediation; Ecotoxicol. Environ. Saf. 2000, 46, 322–328.

[9] Banuelos, G.S., Phyto-products may be essential for sustainability and implementation of phytoremediation; Environ. Pollut. 2006, 144, 19–23.

[10] Robinson, B.; Green, S.; Mills, T.; Clothier, B. *et al.*, Phytoremediation: Using plants as biopumps to improve degraded environments; Aust. J. Soil Res. 2003, 41, 599–611.

[11] Lewandowski, I.; Schmidt, U.; Londo, M.; Faaij, A., The economic value of the phytoremediation function – Assessed by the example of cadmium remediation by willow (*Salix* ssp); Agric. Syst. 2006, 89, 68–89.

[12] Chaney, R.L.; Angle, J.S.; Broadhurst, C.L.; Peters, C.A.; Tappero, R.V.; Sparks, D.L., Improved understanding of hyperaccumulation yields commercial phytoextraction and phytomining technologies; J. Environ. Qual. 2007, 36, 1429–1443.

[13] Wood, B.W.; Chaney, R.L.; Crawford, M. Correcting micronutrient deficiency using metal hyper accumulators: Alyssum biomass as a natural product for nickel deficiency correction; HortScience 2006, 41, 1231–1234.

[14] Sas-Nowosielska, A.; Kucharski, R.; Małkowski, E.; Pogrzeba, M.; Kuperberg, J.M.; Kryński, K., Phytoextraction crop disposal –an unsolved problem; Environ. Pollut. 2004, 128, 373–379.

[15] Robinson, B.; Schulin, R.; Nowack, B.; Roulier, S. *et al.*, Phytoremediation for the management of metal flux in contaminated sites; For. Snow Landsc. Res. 2006, 80, 221–234.

[16] Ernst, W.H.O., Commentary: Evolution of metal hyperaccumulation and phytoremediation hype; New Phytol. 2000, 146, 357–358.

[17] Ernst, W.H.O., Phytoextraction of mine wastes - Options and impossibilities; Chemie der Erde 2005, 65(S1), 29–42.

[18] Ernst, W.H.O., Bioavailability of heavy metals and decontamination of soils by plants; Appl. Geochem. 1996, 11, 163–167.
[19] Ernst, W.; Schwermetallvegetation der Erde; Fisher, Stuttgart, 1974.
[20] Malyuga, D.P.; Biogeochemical Methods of Prospecting; Acad. Sci. Press, Moscow, 1963; Translated Consultants Bureau, 1964, NY, 205 pp.
[21] Hanikenne, M.; Talke, I.N.; Haydon, M.J.; Lanz, C. et al., Evolution of metal hyperaccumulation required *cis*-regulatory changes and triplication of *HMA4*; Nature 2008, 453, 391–395.
[22] Li, Y.-M.; Chaney, R.L.; Brewer, E.; Roseberg, R.J. et al., Development of a technology for commercial phytoextraction of nickel: Economic and technical considerations; Plant Soil 2003, 249, 107–115.
[23] Ruttens, A,; Colpaert, J.V.; Mench, M.; Boisson, J.; Carleer, R.; Vangronsveld, J., Phytostabilization of a metal contaminated sandy soil. II: Influence of compost and/or inorganic metal immobilizing soil amendments on metal leaching; Environ. Pollut. 2006, 144, 533–539.
[24] Kukier, U.; Chaney, R.L., In situ remediation of Ni-phytotoxicity for different plant species; J. Plant Nutr. 2004, 27, 465–495.
[25] Siebielec, G.; Chaney, R.L.; Kukier, U., Liming to remediate Ni contaminated soils with diverse properties and a wide range of Ni concentration; Plant Soil 2007, 299, 117–130.
[26] Li, Y.-M.; Chaney, R.L.; Siebielec, G.; Kershner, B.A., Response of four turfgrass cultivars to limestone and biosolids compost amendment of a zinc and cadmium contaminated soil at Palmerton, PA; J. Environ. Qual. 2000, 29, 1440–1447.
[27] Brown, S.L.; Henry, C.L.; Chaney, R.L.; Compton, H.; DeVolder, P.S., Using municipal biosolids in combination with other residuals to restore metal-contaminated mining areas; Plant Soil 2003, 249, 203–215.
[28] Brown, S.; Chaney, R.; Hallfrisch, J.; Ryan, J.A.; Berti, W.R., In situ soil treatments to reduce the phyto- and bioavailability of lead, zinc, and cadmium; J. Environ. Qual. 2004, 33, 522–531.
[29] Kumar, P.B.A.N.; Dushenkov, V.; Motto, H.; Raskinj, I., Phytoextraction: The use of plants to remove heavy metals from soils; Environ. Sci. Technol. 1995, 29, 1232–1238.
[30] Quinn, J.J.; Negri, M.C.; Hinchman, R.R.; Moos, L.P.; Wozniak, J.B.; Gatliff, E.G., Predicting the effect of deep-rooted hybrid poplars on the groundwater flow system at a large-scale phytoremediation site; Int. J. Phytomed. 2001, 3, 41–60.
[31] Whiting, S.N.; Reeves, R.D.; Richards, D.; Johnson, M.S. et al., Research priorities for conservation of metallophyte biodiversity and their potential for restoration and site remediation; Restor. Ecol. 2004, 12, 106–116.
[32] Cannon, H.L., Botanical prospecting for ore deposits; Science 1960, 132, 591–598.
[33] Brooks, R.R. (ed.), Plants That Hyperaccumulate Heavy Metals: Their Role in Phytoremediation, Microbiology, Archaeology, Mineral Exploration and Phytomining; CAB International, Wallingford, 1998.
[34] Robinson, B.H.; Chiarucci, A.; Brooks, R.R.; Petit, D. et al., The nickel hyperaccumulator plant *Alyssum bertolonii* as a potential agent for phytoremediation and phytomining of nickel; J. Geochem. Explor. 1997, 59, 75–86.
[35] Bani, A.; Echevarria, F.; Sulçe, S.; Morel, J.L.; Mullai, A., In-situ phytoextraction of Ni by a native population of *Alyssum murale* on an ultramafic site (Albania); Plant Soil 2007, 293, 79–89.
[36] Ebbs, S.D.; Kochian, L.V., Phytoextraction of zinc by oat (*Avena sativa*), barley (*Hordeum vulgare*) and Indian mustard (*Brassica juncea*); Environ. Sci. Technol. 1998, 32, 802–806.
[37] Ebbs, S.; Lau, I.; Ahner, B.; Kochian, L., Phytochelatin synthesis is not responsible for Cd tolerance in the Zn/Cd hyperaccumulator *Thlaspi caerulescens* (J. & C. Presl); Planta 2002, 214, 635–640.
[38] Schat, H.; Llugany, M.; Vooijs, R.; Hartley-Whitaker, J.; Bleeker, P.M., The role of phytochelatins in constitutive and adaptive heavy metal tolerances in hyperaccumulator and non-hyperaccumulator metallophytes; J. Exp. Bot. 2002, 53, 2381–2392.
[39] Chaney, R.L.; Angle, J.S.; Baker, A.J.M.; Li, Y.-M., Method for phytomining of nickel, cobalt and other metals from soil; 1988, US Patent 5,711,784.
[40] Robinson, B.H.; Brooks, R.R.; Clothier, B.E., Soil amendments affecting nickel and cobalt uptake by *Berkheya coddii*: Potential use for phytomining and phytoremediation; Ann. Bot. 1999, 84, 689–694.

[41] Morrey, D.R.; Balkwill, K.; Balkwill, M.J., Studies on serpentine flora: Preliminary analyses of soils and vegetation associated with serpentinite rock formations in the south-eastern Transvaal; S. Afr. J. Bot. 1989, 55, 171–177.
[42] Li, Y.-M.; Chaney, R.L.; Brewer, E.P.; Angle, J.S.; Nelkin, J., Phytoextraction of nickel and cobalt by hyperaccumulator *Alyssum* species grown on nickel-contaminated soils; Environ. Sci. Technol. 2003, 37, 1463–1468.
[43] L'Huillier, L; Edighoffer, S., Extractability of nickel and its concentration in cultivated plants in Ni-rich ultramafic soils of New Caledonia; Plant Soil 1996, 186, 255–264.
[44] Bennett, F.A.; Tyler, E.K.; Brooks, R.R.; Greg, P.E.H.; Stewart, R.B., Fertilisation of hyperaccumulators to enhance their potential for phytoremediation and phytomining; in: Brooks, R.R. (ed.), Plants That Hyperaccumulate Heavy Metals; CAB International, Wallingford, 1998, pp. 249–259.
[45] Kukier, U.; Peters, C.A.; Chaney, R.L.; Angle, J.S.; Roseberg, R.J., The effect of pH on metal accumulation in two *Alyssum* species; J. Environ. Qual. 2004, 32, 2090–2102.
[46] Robinson, B.H.; Brooks, R.R.; Clothier, B.E., Soil amendments affecting nickel and cobalt uptake by *Berkheya coddii*: Potential use for phytomining and phytoremediation; Ann. Bot. 1999, 84, 689–694.
[47] Mizuno, N.; Nosaka, S., The distribution and extent of serpentinized areas in Japan; in: B.A. Roberts, J. Proctor (eds), The Ecology of Areas with Serpentinized Rocks: A World View. 1992; Kluwer Academic, Dordrecht, pp. 271–311.
[48] Massoura, S.T.; Echevarria, G.; Becquer, T.; Ghanbaja, J.; Leclerc-Cessac, E.; Morel, J.L., Control of nickel availability by nickel bearing minerals in natural and anthropogenic soils; Geoderma 2006, 136, 28–37.
[49] McNear, D.H., Jr.; Chaney, R.L.; Sparks, D.L., The effects of soil type and chemical treatment on nickel speciation in refinery enriched soils: A multi-technique investigation; Geochim. Cosmochim. Acta 2007, 71, 2190–2208.
[50] Ludwig, C.; Casey, W.H., On the mechanisms of dissolution of bunsenite [NiO(s)] and other simple oxide minerals; J. Colloid Interface Sci. 1996, 178, 176–185.
[51] Fellet, G.; Centofanti, T.; Chaney, R.L.; Green, C.E., NiO(s) (bunsenite) is not available to *Alyssum* species; Plant Soil. 2009, 319, 219–223,
[52] Shallari, S.; Echevarria, G.; Schwartz, C.; Morel, J.L., Availability of nickel in soils for the hyperaccumulator *Alyssum murale* Waldst. & Kit.; S. Afr. J. Sci. 2001, 97, 568–570
[53] Denys, S.; Echevarria, G.; Leclerc-Cessac, E.; Massoura, E.; Morel, J.L., Assessment of plant uptake of radioactive nickel from soils; J. Environ. Radioact. 2002, 62, 195–205.
[54] Abou-Shanab, R.A.; Angle, J.S.; Delorme, T.A. *et al.*, Rhizobacterial effects on nickel extraction from soil and uptake by *Alyssum murale*. New Phytol. 2003, 158, 219–224.
[55] Abou-Shanab, R.A.I.; Angle, J.S.; Chaney, R.L., Bacterial inoculants affecting nickel uptake by *Alyssum murale* from low, moderate and high Ni soils; Soil Biol. Biochem. 2006, 38, 2882–2889.
[56] Becerra-Castro, C.; Monterroso, C.; García-Lestón, M.; Prieto-Fernández, A.; Acea, M.J.; Kidd, P.S., Rhizosphere microbial densities and trace metal tolerance of the nickel hyperaccumulator *Alyssum serpyllifolium* Subsp. *lusitanicum*; Int. J. Phytoremed. 2009, 11, 525–541.
[57] Mengoni, A.; Grassi, E.; Barzanti, R.; Biondi, E. *et al.*, Genetic diversity of bacterial communities of serpentine soil and of rhizosphere of the nickel-hyperaccumulator plant *Alyssum bertolonii*; Microb. Ecol. 2004, 48, 209–217.
[58] Puschenreiter, M.; Wieczorek, S.; Horak, O.; Wenzel, W.W., Chemical changes in the rhizosphere of metal hyperaccumulator and excluder *Thlaspi* species; J. Plant Nutr. Soil Sci. 2003, 166, 579–584.
[59] Wenzel, W.W.; Bunkowski, M.; Puschenreiter, M.; Horak, O., Rhizosphere characteristics of indigenously growing nickel hyperaccumulator and excluder plants on serpentine soil; Environ. Pollut. 2003, 123, 131–138.
[60] Wood, B.W.; Reilly, C.C.; Nyczepir, A.P., Field deficiency of nickel in trees: Symptoms and causes; Acta Hortic. 2006, 721, 83–97.
[61] Nogawa, K.; Kobayashi, R.; Okubo, Y.; Suwazono, Y., Environmental cadmium exposure, adverse effects and preventive measures in Japan; BioMetals 2004, 17, 581–587.

[62] Jin, T.; Nordberg, G.; Ye, T. et al., Osteoporosis and renal dysfunction in a general population exposed to cadmium in China; Environ. Res. 2004, 96, 353–359.

[63] Iwamoto, A., Restoration of Cd-polluted paddy fields in the Jinzu River Basin – Progress and prospects of the restoration project; in: K. Nogawa, M. Kurachi, M. Kasuya (eds), Advances in the Prevention of Environmental Cadmium Pollution and Countermeasures; Eiko Laboratory, Kanazawa, 1999, pp. 179–183.

[64] Chino, M.; Baba, A., The effects of some environmental factors on the partitioning of zinc and cadmium between roots and tops of rice plants; J. Plant Nutr. 1981, 3, 203–214.

[65] Chaney, R.L.; Reeves, P.G.; Ryan, J.A.; Simmons, R.W.; Welch, R.M.; Angle, J.S., An improved understanding of soil Cd risk to humans and low cost methods to remediate soil Cd risks; BioMetals 2004, 17, 549–553.

[66] Reeves, P.G.; Chaney, R.L., Bioavailability as an issue in risk assessment and management of food cadmium: a review; Sci. Total Environ. 2008, 398, 13–19.

[67] McLaughlin, M.J.; Tiller, K.G.; Naidu, R.; Stevens, D.P., The behaviour and environmental impact of contaminants in fertilizers; Aust. J. Soil Res. 1996, 34, 1–54.

[68] Chaney, R.L.; Filcheva, E.; Green, C.E.; Brown, S.L., Zn deficiency promotes Cd accumulation by lettuce from biosolids amended soils with high Cd:Zn ratio; J. Residual Sci. Technol. 2006, 3, 68–75.

[69] Burau, R.G., National and local dietary impact of cadmium in south coastal California soils; Ecotoxicol. Environ. Saf. 1983, 7, 53–57.

[70] Garrett, R.G.; Porter, A.R.D.; Hunt, P.A.; Lalor, G.C., The presence of anomalous trace element levels in present day Jamaican soils and the geochemistry of Late-Miocene or Pliocene phosphorites; Appl. Geochem. 2008, 23, 822–834.

[71] Lugon-Moulin, N.M.; Zhang, F.; Gadani, L. et al., Critical review of the science and options for reducing cadmium in tobacco (*Nicotiana tabacum* L.) and other plants. Adv. Agron. 2004, 83, 111–180.

[72] Bell, P.F.; Mulchi, C.L.; Chaney, R.L., Microelement concentrations in Maryland air-cured tobacco; Commun. Soil Sci. Plant Anal. 1992, 23, 1617–1628.

[73] Schwartz, C.; Echevarria, G.; Morel, J.L., Phytoextraction of cadmium with *Thlaspi caerulescens*; Plant Soil 2003, 249, 27–35.

[74] Chaney, R.L.; Angle, J.S.; McIntosh, M.S. et al., Using hyperaccumulator plants to phytoextract soil Ni and Cd; Z. Naturforsch. 2005, 60C, 190–198.

[75] Brown, S.L.; Angle, J.S.; Chaney, R.L.; Baker, A.J.M., Zinc and cadmium uptake by *Thlaspi caerulescens* and *Silene cucubalis* grown on sludge-amended soils in relation to total soil metals and soil pH; Environ. Sci. Technol. 1994, 29, 1581–1585.

[76] Li, Y.-M.; Chaney, R.L.; Angle, J.S.; Chen, K.-Y.; Kerschner, B.A.; Baker, A.J.M., Genotypical differences in zinc and cadmium hyperaccumulation in *Thlaspi caerulescens*; Agron. Abstr. 1996, 1996, 27.

[77] Chaney, R.L.; Li, Y.-M.; Angle, J.S. et al., Improving metal hyperaccumulator wild plants to develop commercial phytoextraction systems: approaches and progress; in: N. Terry, G.S. Bañuelos (eds), Phytoremediation of Contaminated Soil and Water; CRC Press, Boca Raton, FL, 2000, pp. 131–160.

[78] Reeves, R.D.; Schwartz, C.; Morel, J.L.; Edmondson, J., Distribution and metal-accumulating behavior of *Thlaspi caerulescens* and associated metallophytes in France; Int. J. Phytorem. 2001, 3, 145–172.

[79] Zhao, F.J.; Hamon, R.E.; Lombi, E.; McLaughlin, M.J.; McGrath, S.P., Characteristics of cadmium uptake in two contrasting ecotypes of the hyperaccumulator *Thlaspi caerulescens*; J. Exp. Bot. 2002, 53, 535–543.

[80] Li, Y.-M.; Chaney, R.L.; Reeves, R.D.; Angle, J.S.; Baker, A.J.M., *Thlaspi caerulescens* subspecies for Cd and Zn recovery. 2006, US Patent 7,049,492, pp. 1–8.

[81] Wang, A.S.; Angle, J.S.; Chaney, R.L.; Delorme, T.A.; R.D. Reeves, R.D., Soil pH effects on uptake of Cd and Zn by *Thlaspi caerulescens*; Plant Soil 2006, 281, 325–337.

[82] Basic, N.; Keller, C.; Fontanillas, P.; Vittoz, P.; Besnard, G.; Galland, N., Cadmium hyperaccumulation and reproductive traits in natural *Thlaspi caerulescens* populations; Plant Biol. 2006, 8, 64–72.

[83] Basic, N.; Salamin, N.; Keller, C.; Galland, N.; Besnard, G., Cadmium hyperaccumulation and genetic differentiation of *Thlaspi caerulescens* populations; Biochem. Syst. Ecol. 2006, 34, 667–677.
[84] Robinson, B.H.; Leblanc, M.; Petit, D.; Brooks, R.R.; Kirkman, J.H.; Gregg, P.E,H., The potential of *Thlaspi caerulescens* for phytoremediation of contaminated soils; Plant Soil 1998, 203, 47–56.
[85] Hammer, D.; Keller, C., Phytoextraction of Cd and Zn with *Thlaspi caerulescens* in field trials; Soil Use Manage. 2003, 19, 144–149.
[86] Keller, C.; Hammer, D., Metal availability and soil toxicity after repeated croppings of *Thlaspi caerulescens* in metal contaminated soils; Environ. Pollut. 2004, 131, 243–254.
[87] McGrath, S.P.; Lombi, E.; Gray, C.W.; Caille, N.; Dunham, S.J.; Zhao, F.J., Field evaluation of Cd and Zn phytoextraction potential by the hyperaccumulators *Thlaspi caerulescens* and *Arabidopsis halleri*; Environ. Pollut. 2006, 141, 115–125.
[88] Schwartz, C.; Sirguey, C.; Peronny, S.; Reeves, R.D.; Bourgaud, F.; Morel, J.L., Testing of outstanding individuals of *Thlaspi caerulescens* for cadmium phytoextraction; Int. J. Phytoremed. 2006, 8, 339–357.
[89] Peer, W.A.; Baxter, I.R.; Richards, E.L.; Freeman, J.L.; Murphy, A.S., Phytoremediation and hyperaccumulator plants; Topics in Curr. Genet. 2006, 14, 299–340.
[90] Verbruggen, N.; Hermans, C.; Schat, H., Molecular mechanisms of metal hyperaccumulation in plants; New Phytol. 2009, 181, 759–776.
[91] Krämer, U.; Talke, I.N.; Hanikenne, M., Transition metal transport; FEBS Lett. 2007, 58, 2263–2272.
[92] Sirguey, C.; Schwartz, C.; Morel, J.-L., Response of *Thlaspi caerulescens* to nitrogen, phosphorus and sulfur fertilisation; Int. J. Phytoremed. 2006, 8, 149–161.
[93] Hattori, H.; Kuniyasu, K.; Chiba, K.; Chino, M., Effect of chloride application and low soil pH on cadmium uptake from soil by plants; Soil Sci. Plant Nutr. 2006, 52, 89–94.
[94] Gerard, E.; Echevarria, G.; Sterckeman, T.; Morel, J.L., Cadmium availability to three plant species varying in cadmium accumulation pattern; J. Environ. Qual. 2000, 29, 1117–1123.
[95] Hutchinson, J.J.; Young, S.D.; McGrath, S.P.; West, H.M.; Black, C.R.; Baker, A.J.M., Determining uptake of 'non-labile' soil cadmium by *Thlaspi caerulescens* using isotopic dilution techniques; New Phytol. 2000, 146, 453–460.
[96] Ayoub, A.S.; McGaw, B.A.; Shand, C.A.; Midwood, A.J., Phytoavailability of Cd and Zn in soil estimated by stable isotope exchange and chemical extraction; Plant Soil 2003, 252, 291–300.
[97] Hammer, D.; Keller, C.; McLaughlin, M.J.; Hamon, R.E., Fixation of metals in soil constituents and potential remobilization by hyperaccumulating and non-hyperaccumulating plants: Results from an isotopic dilution study; Environ. Pollut. 2006, 143, 407–415.
[98] Schwartz, C.; Morel, J.L.; Saumier, S.; Whiting, S.N.; Baker, A.J.M., Root development of the Zn-hyperaccumulator plant *Thlaspi caerulescens* as affected by metal origin, content and localization in soil; Plant Soil 1999, 208, 103–115.
[99] Whiting, S.N.; Leake, J.R.; McGrath, S.P.; Baker, A.J.M., Positive response to Zn and Cd by roots of the Zn and Cd hyperaccumulator *Thlaspi caerulescens*; New Phytol. 2000, 145, 199–210.
[100] Haines, B.J., Zincophilic root foraging in *Thlaspi* caerulescens; New Phytol. 2002, 155, 363–372.
[101] Zhao, F.J.; Hamon, R.E.; McLaughlin, M.J., Root exudates of the hyperaccumulator *Thlaspi caerulescens* do not enhance metal mobilization; New Phytol. 2001, 151, 613–620.
[102] Whiting, S.N.; Broadley, M.R.; White, P.J., Applying a solute transfer model to phytoextraction: Zinc acquisition by *Thlaspi caerulescens*; Plant Soil 2003, 249, 45–56.
[103] Sterckeman, T.; Perriguey, J.; Ca, M.; Schwartz, C.; Morel, J.L., Applying a mechanistic model to cadmium uptake by *Zea mays* and *Thlaspi caerulescens*: Consequences for the assessment of the soil quantity and capacity factors; Plant Soil 2004, 262, 289–302.
[104] Harrison-Murray, R.S.; Clarkson, D.T., Relationships between structural development and the absorption of ions by the root system of *Cucurbita pepo*; Planta 1973, 114, 1–16.
[105] Simmons, R.W.; Chaney, R.L.; Angle, J.S.; Kruatrachue, M.; Klinphoklap, S.; Reeves, R.D., Towards practical Cd phytoextraction with *Thlapsi caerulescens*; Abstracts Int. Phytotechnol. Soc., St. Louis, MO, 1–4 December, 2009, 38.

[106] Ae, N.; Arao, T., Utilization of rice plants for phytoremediation in heavy metal polluted soils; Farming Japan 2002, 36(6), 16–21.

[107] Ishikawa, S.; Ae, N.; Murakami, M.; Wagatsuma, T., Is *Brassica juncea* a suitable plant for phytoremediation of cadmium in soils with moderately low cadmium contamination? – Possibility of using other plant species for Cd-phytoextraction; Soil Sci. Plant Nutr. 2006, 52, 32–42.

[108] Murakami, M.; Ae, N.; Ishikawa, S., Phytoextraction of cadmium by rice (*Oryza sativa* L.), soybean (*Glycine max* (L.) Merr.), and maize (*Zea mays* L.); Environ. Pollut. 2007, 145, 96–103.

[109] Murakami, M.; Nakagawa, F.; Ae, N.; Ito, N.; Arao, T., Phytoextraction by rice capable of accumulating cd at high levels: Reduction of Cd content of rice grain; Environ. Sci. Technol. 2009, 43, 5878–5883.

[110] Murakami, M.; Ae, A.; Ishikawa, S.; Ibaraki, T.; Ito, M., Phytoextraction by a high-Cd-accumulating rice: reduction of Cd content of soybean seeds; Environ. Sci. Technol. 2008, 42, 6167–6172.

[111] Chen, Z.; Setagawa, M.; Kang, Y.; Sakurai, K.; Aikawa, Y.; Iwasaki, K., Zinc and cadmium uptake from a metalliferous soil by a mixed culture of *Athyrium yokoscense* and *Arabis flagellosa*; Soil Sci. Plant Nutr. 2009, 55, 315–324.

[112] Yang, X.E.; Long, X.X.; Ye, H.B.; He, Z.L.; Calvert, D.V.; Stoffella, P.J., Cadmium tolerance and hyperaccumulation in a new Zn-hyperaccumulating plant species (*Sedum alfredii* Hance); Plant Soil 2004, 259, 181–189.

[113] Deng, D.M.; Shu, W.S.; Zhang, J. *et al.*, Zinc and cadmium accumulation and tolerance in populations of *Sedum alfredii*; Environ. Pollut. 2007, 147, 381–386.

[114] Xu, L.; Zhou, S.; Wu, L.; Li, N.; Cui, L.; Luo, Y.; Christie, P., Cd and Zn tolerance and accumulation by *Sedum jinianum* in east China; Int. J. Phytoremed. 2009, 11, 283–295.

[115] Li, J.T.; Qiu, J.W.; Wang, X.W.; Zhong, Y.; Lan, C.Y.; Shu, W.S., Cadmium contamination in orchard soils and fruit trees and its potential health risk in Guangzhou, China; Environ. Pollut. 2006, 143, 159–165.

[116] Li, J.T.; Liao, B.; Lan, C.Y.; Qiu, J.W.; Shu, W.S., Zinc, nickel and cadmium in carambolas marketed in Guangzhou and Hong Kong, China: Implication for human health; Sci. Total Environ. 2007, 388, 405–412.

[117] Phaenark, C.; Pokethitiyook, P.; Kruatrachue, M.; Ngernsansaruay, C., Cd and Zn accumulation in plants from the Padaeng zinc mine area; Int. J. Phytoremed. 2009, 11, 479–495.

[118] Yang, G.; Wang, S.; Zhou, R.; Sun, S., Endemic selenium intoxication of humans in China; Am. J. Clin. Nutr. 1983, 37, 872–881.

[119] Ohlendorf, H.M., Bioaccumulation and effects of selenium in wildlife; in: L.W. Jacobs (ed.), Selenium in Agriculture and the Environment. SSSA Spec. Publ. 23; Soil Sci. Soc. Am., Madison, WI, 1989, pp. 133–177.

[120] Lemly, A.D.; Ohlendorf, H.M., Regulatory implications of using constructed wetlands to treat selenium-laden wastewater; Ecotoxicol. Environ. Saf. 2002, 52, 46–56.

[121] Ohlendorf, H.M.; Oldfield, J.E.; Sarka, M.K.; Aldrich, T.W., Embryonic mortality and abnormalities of aquatic birds: apparent impacts by selenium from irrigation drain water; Sci. Total Environ. 1986, 52, 49–63.

[122] Rosenfeld, I.; Beath, O.A., Selenium: Geobotany, Biochemistry, Toxicity and Nutrition; Academic Press, New York, 1964.

[123] Parker, D.R.; Page, A.L.; Thomason, D.N., Salinity and boron tolerances of candidate plants for the removal of selenium from soils; J. Environ. Qual. 1991, 20, 157–164.

[124] Banuelos, G.S.; Meek, D.W., Selenium uptake by different species in selenium enriched soils; J. Environ. Qual. 1990, 19. 772–777.

[125] Banuelos, G.S.; Ajwa, H.A.; Mackey, B. *et al.*, Evaluation of different plant species used for phytoremediation of high soil selenium; J. Environ. Qual. 1997, 26, 637–646.

[126] Banuelos, G.S.; Ajwa, H.A.; Wu, L.; Guo, X.; Akohoue, S.; Zambrzuski, S., Selenium-induced growth reduction in *Brassica* land races considered for phytoremediation; Ecotoxicol. Environ. Saf. 1997, 36, 282–287.

[127] Banuelos, G.S.; Mead, R.; Hoffman, G.J., Accumulation of selenium in wild mustard irrigated with agricultural effluent; Agric. Ecosyst. Environ. 1993, 43, 119–126.

[128] Terry, N.; Carlson, C.; Raab, T.K.; Zayed, A.M., Rates of selenium volatilization among crop species; J. Environ. Qual. 1992, 21, 341–344.
[129] Zayed, A.M.; Terry, N., Selenium volatilization in roots and shoots: Effects of shoot removal and sulfate level; J. Plant Physiol. 1994, 143, 8–14.
[130] Hansen, D.; Duda, P.; Zayed, A.M.; Terry, N., Selenium removal by constructed wetlands: Role of biological volatilization; Environ. Sci. Technol. 1998, 32, 591–597.
[131] Bell, P.F.; Parker, D.R.; Page, A.L., Contrasting selenate-sulfate interactions in selenium-accumulating and nonaccumulating plant species; Soil Sci. Soc. Am. J. 1992, 56, 1818–1824.
[132] White, P.J.; Bowen, H.C.; Marshall, B.; Broadley, M.R., Extraordinarily high leaf selenium to sulfur ratios define 'Se-accumulator' plants; Ann. Bot. 2007, 100, 111–118.
[133] Banuelos, G.S. Irrigation of broccoli and canola with boron- and selenium-laden effluent; J. Environ. Qual. 2002, 31, 1802–1808.
[134] Davis, A.M., Selenium uptake in *Astragalus* and *Lupinus* species; Agron. J. 1986, 78, 727–729.
[135] Pickering, I.J.; Wright, C.; Bubner, B. *et al.*, Chemical form and distribution of selenium and sulfur in the selenium hyperaccumulator *Astragalus bisulcatus*; Plant Physiol. 2003, 131, 1460–1467.
[136] Sors, T.G.; Ellis, D.R.; Na, G.N. *et al.*, Analysis of sulfur and selenium assimilation in *Astragalus* plants with varying capacities to accumulate selenium; Plant J. 2005, 42, 785–797.
[137] Parker, D.R.; Feist, L.J.; Varvel, T.W.; Thomason, D.N.; Zhang, Y.Q., Selenium phytoremediation potential of *Stanleya pinnata*; Plant Soil 2003, 249, 157–165.
[138] Goodson, C.C.; Parker, D.R.; Amrhein, C.; Zhang, Y., Soil selenium uptake and root system development in plant taxa differing in Se-accumulating capability; New Phytol. 2003, 159, 391–401.
[139] Feist, L.J.; Parker, D.R., Ecotypic variation in selenium accumulation among populations of *Stanleya pinnata*; New Phytol. 2001, 149, 61–69.
[140] Galeas, M.L.; Zhang, L.H.; Freeman, J.L.; Wegner, M.; Pilon-Smits, E.A.H., Seasonal fluctuations of selenium and sulfur accumulation in selenium hyperaccumulators and related nonaccumulators; New Phytol. 2006, 173, 517–525.
[141] Pilon-Smits, E.A.H.; Hwang, S.; Lytle, C.M. *et al.*, Overexpression of ATP sulfurylase in Indian mustard leads to increased selenate uptake, reduction, and tolerance; Plant Physiol. 1999, 119, 123–132.
[142] Banuelos, G.; Terry, N.; Leduc, D.L.; Pilon-Smits, E.A.H.; Mackey, B., Field trial of transgenic Indian mustard plants shows enhanced phytoremediation of selenium-contaminated sediment; Environ. Sci. Technol. 2005, 39, 1771–1777.
[143] Freeman, J.L.; Zhang, L.H.; Marcus, M.A.; Fakra, S.; McGrath, S.P.; Pilon-Smits, E.A.H., Spatial imaging, speciation, and quantification of selenium in the hyperaccumulator plants *Astragalus bisulcatus* and *Stanleya pinnata*; Plant Physiol. 2006, 142, 124–134.
[144] Hanson, B.; Lindblom, S.D.; Loeffler, M.L.; Pilon-Smits, E.A.H., Selenium protects plants from phloem-feeding aphids due to both deterrence and toxicity; New Phytol. 2004, 162, 655–662.
[145] Freeman, J.L.; Quinn, C.F.; Marcus, M.A.; Fakra S.; Pilon-Smits, E.A.H., Selenium-tolerant diamondback moth disarms hyperaccumulator plant defense; Curr. Biol. 2006, 16, 2181–2192.
[146] Quinn, C.F.; Freeman, J.L.; Galeas, M.L.; Klamper, E.M.; Pilon-Smits, E.A.H., The role of selenium in protecting plants against prairie dog herbivory: Implications for the evolution of selenium hyperaccumulation; Oecologia 2008, 155, 267–275.
[147] NRC (National Research Council). Mineral Tolerance of Domestic Animals; National Academy of Sciences, Washington, DC, 1980.
[148] Brooks, R.R., Copper and cobalt uptake by *Haumaniastrum* species; Plant Soil 1977, 48, 541–544.
[149] Brooks, R.R.; Morrison, R.S.; Reeves, R.D.; Malaisse, F., Copper and cobalt in African species of *Aeolanthus* Mart. (Plectranthinae, Labiatae); Plant Soil 1978, 50, 503–507.
[150] Brooks, R.R.; Naidu, S.M.; Malaisse, F.; Lee, J., The elemental content of metallophytes from the copper/cobalt deposits of Central Africa; Bull. Soc. Roy. Botan. Belg. 1987, 119, 179–191.

[151] Homer, F.A.; Morrison, R.S.; Brooks, R.R.; Clemens, J.; Reeves, R.D., Comparative studies of nickel, cobalt, and copper uptake by some nickel hyperaccumulators of the genus *Alyssum*; Plant Soil 1991, 138, 195–205.
[152] Keeling, S.M.; Stewart, R.B.; Anderson, C.W.N.; Robinson, B.H., Nickel and cobalt phytoextraction by the hyperaccumulator *Berkheya coddii*: Implications for polymetallic phytomining and phytoremediation; Int. J. Phytoremed. 2003, 5, 235–244.
[153] Malik, M.; Chaney, R.L.; Brewer, E.P.; Li, T.-M.; Angle, J.S., Phytoextraction of soil cobalt using hyperaccumulator plants; Int. J. Phytoremed. 2000, 2, 319–329.
[154] Morrison, R.S.; Brooks, R.R.; Reeves, R.D.; Malaisse, F., Copper and cobalt uptake by metallophytes from Zaïre; Plant Soil 1979, 53, 535–539.
[155] Faucon, M.-P.; Shutcha, M.N.; Meerts, P., Revisiting copper and cobalt concentrations in supposed hyperaccumulators from SC Africa: Influence of washing and metal concentrations in soil; Plant Soil 2007, 301, 29–36.
[156] Faucon, M.-P.; Colinet, G.; Mahy, G.; Luhembwe, M.N.; Verbruggen, N.; Meerts, P., Soil influence on Cu and Co uptake and plant size in the cuprophytes *Crepidorhopalon perennis* and *C. tenuis* (Scrophulariaceae) in SC Africa; Plant Soil 2009, 317, 201–212.
[157] Tappero, R.; Peltier, E.; Gräfe, M. *et al.*, Hyperaccumulator *Alyssum murale* relies on a different metal storage mechanism for cobalt than for nickel; New Phytologist 2007, 175, 641–654.
[158] Brooks, R.R.; Chambers, M.F.; Nicks, L.J.; Robinson, B.H., Phytomining; Trends Plant Sci. 1998, 3, 359–362.
[159] Robinson, B.H.; Green, S.R.; Chancerel, B.; Mills, T.M.; Clothier, B.E., Poplar for the phytomanagement of boron contaminated sites; Environ. Pollut. 2007, 150, 225–233.
[160] Frescholtz, T.F.; Gustin, M.S.; Schorran, D.E.; Fernandez, G.C.J., Assessing the source of mercury in foliar tissue of quaking aspen; Environ. Toxicol. Chem. 2003, 22, 2114–2119.
[161] Cocking, D.; Rohrer, M.; Thomas, R.; Walker, J.; Ward, D., Effects of root morphology and Hg concentration in the soil on uptake by terrestrial vascular plants; Water Air Soil Pollut. 1995, 80, 1113–1116.
[162] Mosbæk, H.; Tjell, J.C.; Sevel, T., Plant uptake of airborne mercury in background areas; Chemosphere 1988, 17, 1227–1236.
[163] Greger, M.; Wang, Y.; Neuschütz, C., Absence of Hg transpiration by shoot after Hg uptake by roots of six terrestrial plant species; Environ. Pollut. 2000, 134, 201–208
[164] Gustin, M.S.; Ericksen, J.A.; Schorran, D.E.; Johnson, D.W.; Lindberg, S.E.; Coleman, J.S., Application of controlled mesocosms for understanding mercury air-soil-plant exchange; Environ. Sci. Technol. 2004, 38, 6044–6050.
[165] Fay, L.; Gustin, M., Assessing the influence of different atmospheric and soil mercury concentrations on foliar mercury concentrations in a controlled environment; Water Air Soil Pollut. 2007, 181, 373–384.
[166] Gustin, M.S.; Biester, H.; Kim, C.S., Investigation of the light-enhanced emission of mercury from naturally enriched substrates; Atmos. Environ. 2002, 36, 3241–3254.
[167] Johnson, D.W.; Benesch, J.A.; Gustin, M.S.; Schorran, D.S.; Lindberg, S.E.; Coleman, J.S., Experimental evidence against diffusion control of Hg evasion from soils; Sci. Total. Environ. 2003, 304, 175–184.
[168] Lodenius, M., Herranen, M., Influence of a chlor-alkali plant on the mercury contents of fungi; Chemosphere 1981, 10, 313–318.
[169] Rugh, C.L.; Wilde, H.D.; Stack, N.M.; Thompson, D.N.; Summers, A.O.; Meagher, R.B., Mercuric ion reduction and resistance in transgenic *Arabidopsis thaliana* plants expressing a modified bacterial *merA* gene; Proc. Natl. Acad. Sci. USA 1996, 93, 3182–3187.
[170] Bizily, S.P.; Rugh, C.L.; Summers, A.O.; Meagher, R.B., Phytoremediation of methylmercury pollution: merB expression in *Arabidopsis thaliana* confers resistance to organomercurials; Proc. Natl. Acad. Sci. USA 1999, 96, 6808–6813.
[171] Che, D.; Meagher, R.B.; Rugh, C.L.; Kim, T.; Heaton, A.P.; Merkle, S.A., Expression of organomercurial lyase in eastern cottonwood enhances organomercury resistance; In Vitro Cell. Develop. Biol. Plant 2006, 42, 228–234.

[172] Heaton, A.C.P.; Rugh, C.L.; Kim, T.; Wang, N.J.; Meagher, R.B., Toward detoxifying mercury-polluted aquatic sediments with rice genetically engineered for mercury resistance; Environ. Toxicol. Chem. 2003, 22, 2940–2947.
[173] Heaton, A.C.P.; Rugh, C.L.; Wang, N.J.; Meagher, R.B., Physiological responses of transgenic merA-tobacco (*Nicotiana tabacum*) to foliar and root mercury exposure; Water Air Soil Pollut. 2005, 161, 137–155.
[174] Meagher, R.B.; Heaton, A.C.P., Strategies for the engineered phytoremediation of toxic element pollution: mercury and arsenic; J. Ind. Microbiol. Biotechnol. 2005, 32, 502–513.
[175] Bizily, S.P.; Kim, T.; Kandasamy, M.; Meagher, R.B., Subcellular targeting of methylmercury lyase enhances its specific activity for organic mercury detoxification in plants; Plant Physiol. 2003, 131, 463–471.
[176] Anderson, C.W.N.; Brooks, R.R.; Stewart, R.B.; Simcock, R., Harvesting a crop of gold in plants; Nature 1988, 395, 553–554.
[177] Anderson, C.W.N.; Brooks, R.R.; Chiarucci, A. *et al.*, Phytomining for nickel, thallium, and gold; J. Geochem. Explor. 1999, 67, 407–415.
[178] Anderson, C.; Moreno. F.; Meech, J., A field demonstration of gold phytoextraction technology; Minerals Eng. 2005, 18, 385–392.
[179] Marshall, A.T.; Haverkamp, R.G.; Davies, C.E.; Parsons, J.G.; Gardea-Torresdey, J.L.; van Agterveld, D., Accumulation of gold nanoparticles in *Brassic juncea*; Int. J. Phytoremed. 2007, 9, 197–206.
[180] Piccinin, R.C.R.; Ebbs, S.D.; Reichman, S.M.; Kolev, S.D.; Woodrow, I.E.; Baker, A.J.M., A screen of some native Australian flora and exotic agricultural species for their potential application in cyanide-induced phytoextraction of gold; Minerals Eng. 2007, 20, 1327–1330.
[181] Rodriguez, E.; Peralta-Videa, J.R.; Sanchez-Salcido, B. *et al.*, Improving gold phytoextraction in desert willow (*Chilopsis linearis*) using thiourea: a spectroscopic investigation; Environ. Chem. 2007, 4, 98–108.
[182] Harris, A.T.; Naidoo, K.; Nokes, J.; Walker, T.; Orton, F., Indicative assessment of the feasibility of Ni and Au phytomining in Australia; J. Cleaner Prod. 2009. 17. 194–200.
[183] Blaylock, M.J.; Salt, D.E.; Dushenkov, S.; Zakharova, O., Enhanced accumulation of Pb in Indian mustard by soil-applied chelating agents; Environ. Sci. Technol. 1997, 31, 860–865.
[184] Huang, J.W.; Chen, J.; Berti, W.R.; Cunningham, S.D., Phytoremediation of lead-contaminated soils: role of synthetic chelates in lead phytoextraction; Environ. Sci. Technol. 1997, 31, 800–805.
[185] Malone, C.; Koeppe, D.E.; Miller, R.J., Localization of lead accumulated by corn plants; Plant Physiol. 1974, 53, 388–394.
[186] Baumhardt, G.R.; Welch, L.F., Lead uptake and corn growth with soil-applied lead; J. Environ. Qual. 1972, 1, 92–94.
[187] Cotter-Howells, J.D.; Champness, P.E.; Charnock, J.M., Mineralogy of Pb-P grains in the roots of *Agrostis capillaris* L-by ATEM and EXAFS; Mineral. Mag. 1999, 63, 777–789.
[188] Cunningham, S.D.; Berti, W.R., Remediation of contaminated soils with green plants: an overview; In-Vitro Cell Dev. Biol. 1993, 29P, 207–212.
[189] Cooper, E.M.; Sims, J.T.; Cunningham, S.D.; Huang, J.W.; Berti, W.R., Chelate-assisted phytoextraction of lead from contaminated soils; J. Environ. Qual. 1999, 28, 1709–1719.
[190] Huang, J.W.; Cunningham, S.D., Lead phytoextraction: species variation in lead uptake and translocation; New Phytol. 1996, 134, 75–84.
[191] Epstein, A.L.; Gussman, C.D.; Blaylock, M.J. *et al.*, EDTA and Pb-EDTA accumulation in *Brassica juncea* grown in Pb-amended soil; Plant Soil 1999, 208, 87–94.
[192] Wu, J.; Hsu F.C.; Cunningham, S.D., Chelate-assisted Pb phytoextraction: Pb availability, uptake, and translocation constraints; Environ. Sci. Technol. 1999, 33. 1898–1904.
[193] Vassil, A.D.; Kapulnik, Y.; Raskin, I.; Salt, D.E., The role of EDTA in lead transport and accumulation by Indian mustard; Plant Physiol. 1998, 117, 447–453.
[194] Luo, C.; Shen, Z.; Li, X.; Baker, A.J.M., The role of root damage in the chelate-enhanced accumulation of lead by Indian mustard plants; Int. J. Phytoremed. 2006, 8, 323–337.
[195] Parker, D.R.; Chaney, R.L.; Norvell, W.A., Equilibrium computer models: Applications to plant nutrition research; in: R.H. Loeppert, A.P. Schwab, S., Goldberg (eds), Chemical

Equilibrium and Reaction Models. Soil Science Society of America Special Publ. No. 42; Soil Sci. Soc. Am./Am. Soc. Agron., Madison, WI, 1995, pp. 163–200.

[196] Nowack, B.; Schulin, R.; Robinson, B.H., Critical assessment of chelant-enhanced metal phytoextraction; Environ. Sci. Technol. 2006, 40, 5225–5232.

[197] Nowack, B., Environmental chemistry of aminopolycarboxylate chelating agents; Environ. Sci. Technol. 2002, 36, 4009–4016.

[198] Elless, M.P.; Blaylock, M.J., Amendment optimization to enhance lead extractability from contaminated soils for phytoremediation; Int. J. Phytoremed. 2000, 2, 75–89.

[199] Römkens, P.; Bouwman, L.; Japenga, J.; Draaisma, C., Potentials and drawbacks of chelate-enhanced phytoremediation of soils; Environ. Pollut. 2001, 116, 109–121.

[200] Schmidt, U., Enhancing phytoextraction: The effect of chemical soil manipulation on mobility, plant accumulation, and leaching of heavy metals; J. Environ. Qual. 2003, 32, 1939–1954.

[201] Tandy, S.; Schulin, R.; Nowack, B., Uptake of metals during chelant-assisted phytoextraction with EDDS related to the solubilized metal concentration; Environ. Sci. Technol. 2006, 40, 2753–2758.

[202] Jardine, P.M.; Taylor, D.L., Fate and transport of ethylene-diaminetetraacetate chelated contaminants in subsurface environments; Geoderma 1995, 67, 125–140.

[203] Madrid, F.; Liphadzi, M.S.; Kirkham, M.B., Heavy metal displacement in chelate-irrigated soil during phytoremediation; J. Hydrol. 2003, 272, 107–119.

[204] Liphadzi, M.S.; Kirkham, M.B., Heavy metal displacement in EDTA-assisted phytoremediation of biosolids soil; Water Sci. Technol. 2006, 54, 147–153.

[205] Evangelou, M.W.H.; Ebel, M.; Schaeffer, A., Chelate assisted phytoextraction of heavy metals from soil. Effect, mechanism, toxicity, and fate of chelating agents; Chemosphere 2007, 68, 989–1003.

[206] Hu, N.; Luo, T.; Wu, L.; Song, J., A field lysimeter study of heavy metal movement down the profile of soils with multiple metal pollution during chelate-enhanced phytoremediation; Int. J. Phytoremed. 2007, 9, 257–268.

[207] Parra, R.; Ulery, A.L.; Elless, M.O.; Blaylock, M.J., Transient phytoextraction agents: Establishing criteria for the use of chelants in phytoextraction of recalcitrant metals; Int. J. Phytoremed. 2008, 10, 415–429.

[208] Neugschwandtner, R.W.; Tlustoš, P.; Komárek, M.; Száková, J., Phytoextraction of Pb and Cd from a contaminated agricultural soil using different EDTA application regimes: Laboratory versus field scale measures of efficiency; Geoderma 2008, 144, 446–454.

[209] Smith, D.B.; Cannon, W.F.; Woodruff, L.G. et al., Major- and trace-element concentrations in soils from two continental-scale transects of the United States and Canada; US Geol. Surv. Open-File Report 2005-1253.

[210] Ma, L.Q.; Komar, K.M.; Tu, C.; Zhang, W.H.; Cai, T.; Kennelley, E.D., A fern that hyperaccumulates arsenic; Nature 2001, 409, 579.

[211] Kertulis-Tartar, G.M.; Ma, L.Q.; Tu, C.; Chirenje, T., Phytoremediation of an arsenic-contaminated site using *Pteris vittata* L.: a two-year study; Int. J. Phytoremed. 2006, 8, 311–322

[212] Zhao, F.-J.; Ma, J.F.; Meharg, A.A.; McGrath, S.P., Arsenic uptake and metabolism in plants; New Phytol. 2008, 181, 777–794.

[213] Chen, T.-B.; Fan, Z.; Lei, M.; Huang, Z.; Wei, C.-Y., Effect of phosphorus on arsenic accumulation in As-hyperaccumulator *Pteris vittata* L. and its implication; Chin. Sci. Bull. 2002, 47, 1876–1879.

[214] Mehang, A.A., Variation in arsenic accumulation hyperaccumulation in ferns and their allies; New Phytol. 2003, 157, 25–32.

[215] Zhao, F.-J.; Dunham, S.J.; McGrath, S.P., Arsenic hyperaccumulation by different fern species; New Phytol. 2002, 156, 27–31.

[216] Luongo, T.; Ma, L.Q., Characteristics of arsenic accumulation by *Pteris* and non-*Pteris* ferns; Plant Soil 2005, 277, 117–126.

[217] Wang, H.B.; Wong, M.H.; Lan, C.Y. et al., Uptake and accumulation of arsenic by 11 *Pteris* taxa from southern China; Environ. Pollut. 2007, 145, 225–233.

[218] Jankong, P.; Visoottiviseth, P.; Khokiattiwong, S., Enhanced phytoremediation of arsenic contaminated land; Chemosphere 2007, 68, 1906–1912.
[219] Liu, Y.; Zhu, Y.G.; Chen, B.D.; Christie, P.; Li, X.L., Influence of the arbuscular mycorrhizal fungus *Glomus mosseae* on uptake of arsenate by the As hyperaccumulator fern *Pteris vittata* L.; Mycorrhiza 2005, 15, 187–192.
[220] Zhao, F.J.; Wang, J.R.; Barker, J.M.A.; Schat, H.; Bleeker, P.M.; McGrath, S.P., The role of phytochelatins in arsenic tolerance in the hyperaccumulator *Pteris vittata*; New Phytol. 2003, 159, 403–410.
[221] Cai, Y.; Su, J.; Ma, L.Q., Low molecular weight thiols in arsenic hyperaccumulator *Pteris vittata* upon exposure to arsenic and other trace elements; Environ. Pollut. 2004, 129, 69–78.
[222] Raab, A.; Feldmann, J.; Meharg, A.A., The nature of arsenic-phytochelatin complexes in *Holcus lanatus* and *Pteris cretica*; Plant Physiol. 2004, 134, 1113–1122.
[223] Webb, S.M.; Gaillard, J.F.; Ma, L.Q.; Tu, C., XAS speciation of arsenic in a hyper-accumulating fern; Environ. Sci. Technol. 2003, 37, 754–760.
[224] Su, Y.H.; McGrath, S.P.; Zhu, Y.G.; Zhao, F.J., Highly efficient xylem transport of arsenite in the arsenic hyperaccumulator *Pteris vittata*; New Phytol. 2008, 180, 434–441.
[225] Kertulis, G.M.; Ma, L.Q.; MacDonald, G.E. *et al.*, Arsenic speciation and transport in *Pteris vittata* L. and the effects on phosphorus in the xylem sap; Environ. Exp. Bot. 2005, 54, 239–247.
[226] Hokura, A.; Omuma, R.; Terada, T. *et al.*, Arsenic distribution and speciation in an arsenic hyperaccumulator fern by X-ray spectrometry utilizing a synchrotron radiation source; J. Anal. Atomic Spectrom. 2006, 21, 321–328.
[227] Lombi, E.; Zhao, F.J.; Fuhrmann, M.; Ma, L.Q.; McGrath, S.P., Arsenic distribution and speciation in the fronds of the hyperaccumulator *Pteris vittata*; New Phytol. 2002, 156, 195–203.
[228] Li, W.; Chen, T.; Yang, C.; Mei, L., Role of trichome of *Pteris vittata* L. in arsenic hyperaccumulation; Sci. China 2005, C48, 148–154.
[229] Dhankher, O.P.; Li, Y.; Rosen, B.P. *et al.*, Engineering tolerance and hyperaccumulation of arsenic in plants by combining arsenate reductase and γ-glutamylcysteine synthetase expression; Nature Biotechnol. 2002, 20, 1140–1145.
[230] Dhankher, O.P.; Rosen, B.P.; McKinney, E.C.; Meagher, R.B., Hyperaccumulation of arsenic in the shoots of *Arabidopsis* silenced for arsenate reductase (ACR2); Proc. Natl. Acad. Sci. USA 2006, 103, 5413–5418.
[231] Elless, M.P.; Poynton, C.Y.; Willms, C.A. *et al.*, Pilot-scale demonstration of phytofiltration for treatment of arsenic in New Mexico drinking water; Water Res. 2005, 39, 3863–3872.
[232] Yan, X.-L.; Chen, T.-B.; Liao, X.-Y. *et al.*, Arsenic transformation and volatilization during incineration of the hyperaccumulator *Pteris vittata* L.; Environ. Sci. Technol. 2008, 42, 1479–1484.
[233] Warren, G.P.; Alloway, B.J., Reduction of arsenic uptake by lettuce with ferrous sulfate applied to contaminated soil; J. Environ. Qual. 2003, 32, 767–772.
[234] Subacz, J.L.; Barnett, M.O.; Jardine, P.M.; Stewart, M.A., Decreasing arsenic bioaccessibility/bioavailability in soils with iron amendments; J. Environ. Sci. Health 2007, A42, 1317–1329.
[235] Beak, D.G.; Basta, N.T.; Scheckel, K.G.; Traina, S.J., Bioaccessibility of lead sequestered to corundum and ferrihydrite in a simulated gastrointestinal system; J. Environ. Qual. 2006, 35, 2075–2083.
[236] Smith, E.; Naidu, R.; Weber, J.; Juhasz, A.L., The impact of sequestration on the bioaccessibility of arsenic in long-term contaminated soils; Chemosphere 2008, 71, 773–780.
[237] Meharg, A.A.; Rahman, M.M., Arsenic contamination of Bangladesh paddy field soils: implications for rice contribution to arsenic consumption; Environ. Sci. Technol. 2003, 37, 229–234.
[238] Abrahams, P.W.; Thornton, I., The contamination of agricultural land in the metalliferous province of southwest England: implications to livestock; Agric. Ecosys. Environ. 1994, 48, 125–137.
[239] Kurz, H.; Schulz, R.; Römheld, V., Selection of cultivars to reduce the concentration of cadmium and thallium in food and fodder plants; J. Plant Nutr. Soil Sci. 1999, 162, 323–328.

[240] LaCoste, C.; Robinson, B.; Brooks, R., Uptake of thallium by vegetables: Its significance for human health, phytoremediation, and phytomining; J. Plant Nutr. 2001, 24, 1205–1215.

[241] Al-Najar, H.; Schulz, R.; Römheld, V., Plant availability of thallium in the rhizosphere of hyperaccumulator plants: a key factor for assessment of phytoextraction; Plant Soil 2003, 249, 97–105.

[242] LaCoste, C.; Robinson, B.; Brooks, R.; Anderson, C.; Chiarucci, A.; Leblanc, M., The phytoremediation potential of thallium-contaminated soils using *Iberis* and *Biscutella* species; Int. J. Phytoremediat. 1999, 1, 327–338.

[243] Scheckel, K.G.; Hamon, R.; Jassogne, L.; Rivers, M.; Lombi, E., Synchrotron X-ray absorption-edge computed microtomography imaging of thallium compartmentalization in *Iberis intermedia*; Plant Soil 2007, 290, 51–60.

[244] Broadley, M.R.; Willey, N.J., Differences in root uptake of radio caesium by 30 plant taxa; Environ. Pollut. 1997, 97, 11–15.

[245] Fuhrmann, M.; Lasat, M.M.; Ebbs, S.D.; Kochian, L.V.; Cornish, J., Uptake of cesium-137 and strontium-90 from contaminated soil by three plant species: Application to phytoremediation; J. Environ. Qual. 2002, 31, 904–909.

[246] Fuhrmann, M.; Lasat, M.; Ebbs, S.; Cornish J.; Kochian, L., Uptake and release of cesium-137 by five plant species as influenced by soil amendments in field experiments; J. Environ. Qual. 2003, 32, 2272–2279.

[247] Vandenhove, H.; Van Hees, M., Phytoextraction for clean-up of low-level uranium contaminated soil evaluated; J. Environ. Rad. 2004, 72, 41–45.

[248] Dushenkov, S.; Mikheev, A.; Prokhnevsky, A.; Ruchko, M.; Sorochinsky, B., Phytoremediation of radiocesium-contaminated soil in the vicinity of Chernobyl, Ukraine; Environ. Sci. Technol. 1999, 33, 469–475.

[249] Dushenkov, S.; Vasudev, D.; Kapulnik, Y. et al., Removal of uranium from water using terrestrial plants; Environ. Sci. Technol. 1997, 31, 3468–3474.

[250] Huang, J.W.; Blaylock, M.J.; Kapulnik, Y.; Ensley, B.D., Phytoremediation of uranium-contaminated soils: Role of organic acids in triggering uranium hyperaccumulation in plants; Environ. Sci. Technol. 1998, 32, 2004–2008.

[251] Kubota, J.; Welch, R.M.; Van Campen, D., Soil related nutritional problem areas for grazing animals; Adv. Soil Sci. 1987, 6, 189–215.

[252] Neunhäuserer, C.; Berreck, M.; Insam, H, Remediation of soils contaminated with molybdenum using soil amendments and phytoremediation; Water Air Soil Pollut. 2001, 128, 85–96.

[253] O'Connor, G.A.; Brobst, R.B.; Chaney, R.L. et al., A modified risk assessment to establish molybdenum standards for land application of biosolids; J. Environ. Qual. 2001, 30, 1490–1507.

[254] Antonovics, J.; Bradshaw, A.D.; Turner, R.G., Heavy metal tolerance in plants; Adv. Ecol. Res. 1971, 7, 1–85.

[255] Smith, R.A.H.; Bradshaw, A.D., Reclamation of toxic metalliferous wastes using tolerant populations of grass; Nature 1970, 227, 376–377.

[256] Gadgil, R.L., Tolerance of heavy metals and the reclamation of industrial waste; J. Appl. Ecol. 1969, 6, 247–259.

[257] Corey, R.B.; King, L.D.; Lue-Hing, C.; Fanning, D.S.; Street, J.J.; Walker, J.M., Effects of sludge properties on accumulation of trace elements by crops; in: A.L. Page, T.J. Logan, J.A. Ryan (eds), Land Application of Sludge – Food Chain Implications; Lewis Publishers Inc., Chelsea, MI, 1987, pp. 25–51.

[258] Hettiarachchi, G.M.; Ryan, J.A.; Chaney, R.L.; La Fleur, C.M., Sorption and desorption of cadmium by different fractions of biosolids-amended soils; J. Environ. Qual. 2003, 32, 1684–1693.

[259] Kukier, U.; Chaney, R.L.; Ryan, J.A.; Daniels, W.L.; Dowdy, R.H.; Granato, T.C., Phytoavailability of cadmium in long-term biosolids amended soils; J. Environ. Qual. 2010, 39, 519–530.

[260] Allen, H.L.; Brown, S.L.; Chaney, R.L. et al., 2007. The use of soil amendments for remediation, revitalization and reuse; US Environmental Protection Agency, Washington, DC. EPA 542-R-07-013.

[261] Ryan, J.A.; Berti, W.R.; Brown, S.L. *et al.*, Reducing children's risk from soil lead: Summary of a field experiment; Environ. Sci. Technol. 2004, 38, 18A–24A.
[262] Scheckel, K.G.; Ryan, J.A., Spectroscopic speciation and quantification of lead in phosphate-amended soils; J. Environ. Qual. 2004, 33, 1288–1295.
[263] Dickinson, N.M.; Baker, A.J.M.; Doronila, A.; Laidlaw, S.; Reeves, R.D., Phytoremediation of inorganics: realism and synergies; Int. J. Phytoremed. 2009, 11, 97–114.
[264] Stuczynski, T.I.; Siebielec, G.; Daniels, W.L.; McCarty, G.C.; Chaney, R.L., Biological aspects of metal waste reclamation with sewage sludge; J. Environ. Qual. 2007, 36, 1154–1162.
[265] Negri, C.M.; Hinchman, R.R., The use of plants for the treatment of radionuclides; in: I. Raskin (ed.), Phytoremediation of Toxic Metals; Wiley-Interscience, New York, 2000, pp. 107–132.
[266] Granel, T.; Robinson, B.; Mills, T.; Clothier, B.; Green, S.; Fung, L., Cadmium accumulation by willow clones used for soil conservation, stock fodder, and phytoremediation; Austr. J. Soil Res. 2002, 40, 1331–1337.
[267] Vervaeke, P.; Luyssaert, S.; Mertens, J.; Meers, E.; Tack, F.M.G.; Lust, N., Phytoremediation prospects of willow stands on contaminated sediment: a field trial; Environ. Pollut. 2003, 126, 275–282
[268] Vervaeke, P.; Tack, F.M.G.; Navez, F.; Martin, J.; Verloo, M.G.; Lust, N., Fate of heavy metals during fixed bed downdraft gasification of willow wood harvested from contaminated sites; Biomass Bioenergy 2006, 30, 58–65.
[269] Keller, C.; Ludwig, C.; Davoli, F.; Wochele, J., Thermal treatment of metal-enriched biomass produced from heavy metal phytoextraction; Environ. Sci. Technol. 2005, 39, 3359–3367.
[270] Maxted, A.P.; Black, C.R.; West, H.M.; Crout, N.M.J.; Mcgrath, S.P.; Young, S.D., Phytoextraction of cadmium and zinc by *Salix* from soil historically amended with sewage sludge; Plant Soil 2007, 290, 157–172.
[271] Thewys, T.; Kuppens, T., Economics of willow pyrolysis after phytoextraction; Int. J. Phytoremed. 2008, 10, 561–583.
[272] Ljung, A.; Nordin, A., Theoretical feasibility for ecological biomass ash recirculation: Chemical equilibrium behavior of nutrient elements and heavy metals during combustion; Environ. Sci. Technol. 1997, 31, 2499–2503.
[273] Angle, J.S.; Linacre, N.A., Metal phytoextraction - A survey of potential risks; Int. J. Phytoremed. 2005, 7, 241–254.
[274] Chaney, R.L.; Stoewsand, G.S.; Furr, A.K.; Bache, C.A.; Lisk, D.J., Elemental content of tissues of Guinea pigs fed Swiss chard grown on municipal sewage sludge-amended soil; J. Agr. Food Chem. 1978, 26, 944–997.
[275] Reeves, R.D.; Brooks, R.R., Hyperaccumulation of lead and zinc by two metallophytes from mining areas of Central Europe; Environ. Pollut. 1983, A31, 277–285.
[276] Kersten, W.J.; Brooks, R.R.; Reeves, R.D.; Jaffré, T., Nickel uptake by New Caledonian species of *Phyllanthus*; Taxon 1979, 28, 529–534.
[277] Beath, O.A.; Eppsom, H.F.; Gilbert, C.S., Selenium distribution in and seasonal variation of type vegetation occurring on seleniferous soils; J. Am. Pharm. Assoc. 1937, 26, 394–405.
[278] Brooks, R.R.; Trow, J.M.; Veillon, J.-M.; Jaffré, T., Studies on manganese-accumulating *Alyxia* from New Caledonia; Taxon 1981, 30, 420–423.
[279] Ingrouille, M.J.; Smirnoff, N., *Thlaspi caerulescens* J. & C. Presl. (*T. alpestre* L.) in Britain; New Phytol. 1986, 102, 219–233.
[280] Reeves, R.D.; Brooks, R.R., European species of *Thlaspi* L. (Cruciferae) as indicators of nickel and zinc; J. Geochem. Explor. 1983, 18, 275–283.
[281] Reeves, R.D., Nickel and zinc accumulation by species of *Thlaspi* L., *Cochlearia* L., and other genera of the Brassicaceae; Taxon 1988, 37, 309–318.
[282] Linacre, N.A.; Whiting, S.N.; Baker, A.J.M.; Angle, J.S.; Ades, P.K., Transgenics and phytoremediation: the need for an integrated risk assessment, management, and communication strategy; Int. J. Phytoremed. 2003, 5, 181–185.
[283] Koopmans, G.F.; Römkens, P.F.A.M.; Fokkema, M.J. *et al.*, Feasibility of phytoextraction to remediate cadmium and zinc contaminated soils; Environ. Pollut. 2008, 156, 905–914.

[284] Maxted, A.P.; Black, C.R.; West, H.M.; Crout, N.M.J.; McGrath, S.P.; Young, S.D., Phytoextraction of cadmium and zinc from arable soils amended with sewage sludge using *Thlaspi caerulescens*: development of a predictive model; Environ. Pollut. 2007, 150. 363–372.

[285] Van Nevel, L.; Mertens, J.; Oorts, K.; Verheyen, K., Phytoextraction of metals from soils: How far from practice? Environ. Pollut. 2007, 150, 34–40.

[286] Beyer, W.N.; A reexamination of biomagnification of metals in terrestrial food chains; Environ. Toxicol. Chem. 1986, 5, 863–864.

[287] Chaney, R.L.; Ryan, J.A., Risk Based Standards for Arsenic, Lead and Cadmium in Urban Soils; DECHEMA, Frankfurt, 1994. (ISBN 3-926959-63-0).

[288] Broos, K.; Beyens, H.; Smolders, E., Survival of rhizobia in soil is sensitive to elevated zinc in the absence of the host plant; Soil Biol. Biochem. 2005, 37, 573–579.

[289] Chaney, R.L.; Angle, J.S.; Li, Y.-M.; Baker, A.J.M., Recovering metals from soil. 2007, US Patent 7,268,273. Issued 11 September 2007.

[290] Perner, H.; Chaney, R.L.; Reeves, R.D. *et al.*, Variation in Cd- and Zn-accumulation by French genotypes of *Thlaspi caerulescens* J.&C. Pres. grown on a Cd and Zn contaminated soil; Int. J. Phytoremed, Submitted. Forthcoming, in press.

15

Trace Element Immobilization in Soil Using Amendments

Jurate Kumpiene

Division of Waste Science and Technology, Luleå University of Technology, Luleå, Sweden

15.1 Introduction

The immobilization or stabilization of trace elements in soil is a remediation method that reduces contaminant mobility and bioavailability through the use of soil amendments. The aim of such a method is to immobilize/stabilize the labile trace element fraction in soil – that is, the fraction that can be easily released into soil solution, leached to groundwater and enter the biological cycle. Immobilization of trace elements can be achieved by changing their chemical state using soil amendments that are able to adsorb, complex or (co)precipitate trace elements in soil. The approach is therefore also termed chemical stabilization. It is a remediation technique based on the naturally occurring processes in soil; hence, it is also known as assisted natural remediation or attenuation [1]. The stabilization of trace element-contaminated soil has been considered as a possible alternative to conventional soil remediation methods (for example excavation and disposal at landfills), acknowledging that the risk to human health and the environment generally arises from the contaminant mobility and bioavailability, and not from its total concentration in soil.

Although fast and relatively simple, excavation and landfilling is not always possible due to the large size of some contaminated sites and associated prohibitive remediation costs. The stabilization of large contaminated areas (for example due to mining activities and agriculture) present a more realistic solution than digging out thousands of tonnes of contaminated masses, transporting them long distances (relocating the problem) and replacing the excavated material with clean soil. In such situations, stabilization techniques can offer a more sustainable and cost-effective solution by decreasing the mobility and bioavailability of trace elements. Soil stabilization can be further improved by vegetation. Tolerant plant species that are capable of growing in soil with high trace element concentrations and immobilize pollutants within the root zone have a potential to be used for phytostabilization [2–4]. Such plants present a reduced risk of translocation of the contaminant from soil through plant roots to shoots and further to other parts of ecosystem. Furthermore, dense plant root systems can physically stabilize surface soil and reduce soil erosion risk and spreading of contaminated particles. Reduced trace element solubility and toxicity, as well as improved soil nutrient status, following soil amendment and re-vegetation, can increase soil biological activity and help restore its functionality.

This chapter reviews present research on amendments that are used to chemically stabilize trace element-contaminated soil, their usefulness and shortcomings, as well as feasibility of the soil stabilization techniques.

15.2 Soil Amendments for Trace Element Immobilization

The main purpose of *in situ* trace element immobilization is to aid plant establishment, recover soil micro and mezzo fauna and restore soil functions (storing and recycling chemical elements and organic matter) through reduced trace element mobility and bioavailability.

Soil amelioration using amendments and fertilizers such as organic matter, lime, and phosphates for plant growth improvement has traditionally been used in agriculture for many years. The observed ability of the ameliorants to modify trace element uptake by plants prompted the development of research on the utilization of amendments for soil remediation purposes by deliberately reducing trace element bioavailability.

The choice of amendments has been extended to industrial by-products such as coal or wood fly ashes, sewage and paper mill sludge, by-products from the iron and aluminium industries (for example furnace slag, red mud), and so on. The most common soil amendments being tested for their ability to immobilize trace elements in soils include phosphates, natural and synthetic aluminosilicates, iron oxides, biosolids and their combinations.

15.2.1 Metal Oxides

15.2.1.1 Iron Oxides

Iron is one of the eight major elements composing up to 5% mass of the Earth's crust. In most soils iron occurs mainly as various poorly crystalline iron oxides and as noncrystalline coatings of soil particles rather than pure Fe crystals [5]. Oxidation and hydrolysis of primary minerals result in the formation of Fe oxides and hydroxides in soil that have extremely low solubility products, for example $K_{sp} = 10^{-37}$–10^{-39} for ferrihydrite and 10^{-40}–10^{-44} for goethite and hematite [5].

Goethite (alpha-FeOOH) and hematite (alpha-Fe_2O_3) are thermodynamically the most stable Fe oxides in aerobic environments and therefore are the most common forms of iron in soils [5]. Ferrihydrite ($Fe_5HO_8 \cdot 4H_2O$) is also widespread in natural environments and plays an important role as an active sorbent owing to its very high surface area (600 m^2 g^{-1}). Ferrihydrite is a metastable, poorly crystalline soil mineral that is easily transformed to hematite in warm regions and to goethite in humid temperate regions [6].

Iron oxides can reduce the mobility of As, Cd, Cu, Ni, Pb, and Zn by sorption on oxide surface exchange sites, coprecipitation, or formation of secondary oxidation minerals containing trace elements [7]. The surface of Fe oxide particles can be positively or negatively charged depending on pH, making them amphoteric, that is capable of sorbing both anions and cations [8].

Studies on trace element immobilization using natural and synthetic Fe oxides [9,10] have also considered the use of other iron-rich amendments, such as iron salts [11–13], metallic iron grit [13–17] and Fe-containing industrial by-products [18–21]. Iron oxides are particularly effective for immobilization of arsenic and can reduce dissolved As concentration in soil by more than 90% [13,15]. Arsenic is adsorbed by exchanging ligand of As species for OH_2 and OH^- groups on Fe oxyhydroxide surfaces [22]. Arsenic adsorption on goethite, lepidocrocite, hematite and ferrihydrite can occur through the formation of inner-sphere surface complexes resulting from bidentate corner-sharing between AsO_4 and FeO_6 polyhedra [23].

The pH of natural soils is rarely low enough for dissolution of iron oxides and bound trace elements to occur. Instead, iron remobilization and hence trace element release in soil occurs mainly due to the redox reactions and microbial activity of, for example, sulfate-reducing bacteria *Desulfovibrio desulfuricans* or iron-reducing bacteria *Shewanella putrefaciens* [24]. However, drastic soil acidification due to anthropogenic activities, together with formation of complexes with organic matter can also contribute to the dissolution of iron oxides [25]. *Thiobacillus ferrooxidans*, as well as availability of oxygen or Mn oxides, on the other hand, can oxidize Fe in soil [26] and cause readsorption of trace elements.

The transformation of amorphous and poorly crystalline Fe oxides to more crystalline forms of smaller active surface area over time may release sorbed trace elements. However, the transformation of Fe compounds to more crystalline forms, as well as the dissolution of Fe oxides and Fe corrosion, can be retarded by the presence of other sorbates, such as silicates, phosphates, a range of organics and co-precipitated Al [5]. The rate of dissolution of ferrihydrite precipitated in the presence of arsenate was strongly retarded compared with the dissolution of the pure ferrihydrite [27], showing that sorbed arsenic, similar to other sorbates, can reduce surface activity of Fe oxides. Melitas and co-workers [28] observed that the presence of 0.1 mg l^{-1} As^{5+} decreased the iron corrosion rate by up to a factor of 5 compared with a blank electrolyte solution. Hence, studies of pure single-component systems may provide different, but not necessarily better, results than complex soil systems. Several trace elements simultaneously occurring in soil solution, as is often the case in contaminated soils, can have a positive influence on each other's stability by forming coprecipitates on iron oxide surfaces. Arsenic anions for example were observed to increase Zn sorption to goethite by an order of magnitude, while occurrence of Zn increased sorption of As 5-fold [29]. The presence of Cr^{3+} was also demonstrated to enhance the adsorption of Zn and Ni on ferrihydrite [30]. In the presence of several cations and oxyanion, mixed hydroxide species can form and strongly adsorb on Fe oxide surface; for example, AsO_4 and Zn form an adamite-like [$Zn_2(AsO_4)OH$] surface precipitate [29]. Scheckel and Ryan [31] noted that co-application of Fe-rich materials to soil enhanced efficiency of phosphate amendments to immobilize Pb through its faster transformation to pyromorphite.

Ferrous sulfate and especially ferric sulfate successfully reduced As mobility in soil compared with other Fe amendments (that is metallic iron and goethite) [12,13] while metallic iron [13,15,17] and red mud [18] were highly efficient for the immobilization of metals, such as Cu, Cr and Zn. Metallic or zero-valent iron (Fe^0) has certain advantages over other Fe-containing compounds. It is a commonly available material containing several times more iron per unit weight than most iron salts, it has fewer impurities than Fe-rich by-products, and the oxidation of Fe^0 causes minor changes in soil pH. The added Fe^0 corrodes in soil, forming iron oxides of high specific surface area which are available for trace element sorption [5,32]. In contrast, precipitation of Fe oxides, followed by iron sulfate application, can considerably reduce soil pH, due to the release of sulfuric acid, triggering mobilization of cationic elements, for example Cu and Zn [13]. Although very effective for As sorption, the addition of Fe^{2+}/Fe^{3+} sulfates also requires some simultaneous acidity-counteracting measures to reduce metal solubility [8,11–13]. Lime is usually used to neutralize soil pH with a lime/Fe oxide ratio higher than 1:1 being necessary to compensate for soil acidification [33,34]. Furthermore, Fe^0 corrodes in soil slowly and continuously release Fe^{2+} ions, creating, for example, favourable conditions for oxidation of As^{3+} to As^{5+}, which is then adsorbed by freshly formed iron hydroxides [32]. In contrast, iron added to soil as easily soluble salts rapidly produces an excess of Fe^{2+} ions that can adsorb to newly formed Fe hydroxides and reduce their activity [35,36]. Humic acids can also inactivate the iron surface and decrease Fe^0 corrosion, but that is believed to actually prolong the lifetime of zero-valent iron [37].

The natural environment is rich in ions that are sorbed to Fe oxides or form solid solutions (that is when atoms or molecules of homogeneous crystalline structure are partly substituted by other atoms or molecules causing no structural changes) during the formation of ferrihydrite. Therefore, the application of Fe can also reduce the mobility of elements (for example nutrients) that were not intended to be immobilized. The use of Fe amendments can considerably reduce concentrations of Ca, K and Mg in soil solution and hinder plant development. Little or no improvement in vegetation growth was observed in trace element-contaminated soils treated with various Fe amendments (that is Fe^{2+}/Fe^{3+} sulfates + lime, metallic iron, magnetite), despite the significantly reduced concentrations of soluble trace elements – this, however, made no significant change in trace element plant uptake [10,38]. Iron rich industrial by-products often contain impurities, such as Ni, that are released to soil at increasing iron application rates [21], which may also contribute to the slow plant development in Fe-treated soils.

Although iron compounds are among amendments capable of reducing the mobility of multiple trace elements and having least negative effects on soil functions, co-application of nutrient-rich substrates, for example compost or/and ashes, is often necessary to further reduce soil phytotoxicity and metal accumulation in plants, as well as to improve vegetation growth [39].

15.2.1.2 Manganese and Aluminium Oxides

Beside Fe oxides, other common soil metal oxides, such as Mn and Al, are important scavengers of trace elements, although less studied as soil amendments for immobilization/stabilization of trace elements.

Manganese oxides are usually minor soil constituents, but have a broad range of surface areas (5–360 m^2 g^{-1}) and can strongly bind trace elements in soils [40]. Manganese oxides can immobilize trace elements through specific adsorption and by forming inner-sphere complexes [41,42], which would lead to a strong retention of trace elements. Arsenite, a reduced inorganic highly toxic As species, As^{3+}, can be oxidized by Mn^{2+} to a less toxic arsenate, As^{5+}, which then can form sparingly soluble bidentate corner-sharing complexes with Mn oxides [43] and precipitate as manganese arsenates (Mn$_3$(AsO$_4$)$_2$·8H$_2$O; MnHAsO$_4$·8H$_2$O) [44–46].

Manganese oxides, especially synthetic (birnessite and cryptomelane), were shown to be more efficient sorbents of Pb than Fe oxides [42]. The high efficiency was related to the presence of internal reactive sites of the Mn oxide structures that are absent in Fe oxides. Applied to a contaminated soil, 1% (w/w) of Mn oxides decreased Ca(NO$_3$)$_2$-extractable Pb concentration more efficiently than the equivalent amount of Fe oxides, but both amendments were equally effective in reducing metal uptake into wheat shoots (by 23–60% for Cd and by 32–64% for Pb) [47]. Yousfi and Bermond [48] observed in laboratory experiments that addition of 1% (w/w) synthetic Mn oxide to soil decreased Ba(ClO$_4$)$_2$-extractable Zn concentration by 90% and Cd by 60% and that it was more efficient than 1.25% (w/w) of synthetic ferrihydrite. Furthermore, the application of Mn oxides (1%, w/w) was found to be the most efficient means for reducing Cd concentration in plant tissues for a range of soil types and plant species compared to Fe oxides, aluminosilicates and alkaline materials [3].

Manganese is an essential micronutrient for plants and animals with a narrow tolerance range; therefore, high Mn oxide addition rates to contaminated soil might induce Mn toxicity. Also, the use of Mn amendments to immobilize As in chromated copper arsenate (CCA)-contaminated soil increases the risk of Cr remobilization. Manganese can readily oxidize Cr^{3+} to more toxic and mobile Cr^{6+} through the following reaction [49,50]:

$$Cr^{3+} + 3\,MnO_2(s) + 4\,H_2O \rightarrow HCrO_4^- + 3\,MnOOH(s) + 4\,H^+ \quad (15.1)$$

The use of Mn amendments therefore should be avoided in soils containing elevated concentrations of Cr. Amorphous aluminium oxide, on the other hand, was suggested to be a suitable sorbent for removal of hexavalent chromium, Cr^{6+}, from aqueous solutions [51].

Despite the common occurrence of Al in soils, pure Al oxides (for example gibbsite, boehmite, diaspore) are less common compared to Fe oxides, likely due to the tendency of Al to form aluminosilicates [40]. Aluminium oxides have high specific surface area (100–220 m^2 g^{-1}) and, owing to their variable surface charge, can chemisorb both cations and anions [40]. Aluminium oxides are usually stable between pH 4 and 10, but can dissolve outside this pH range [52] and release sorbed trace elements.

A polynuclear aluminium complex [AlO$_4$Al$_{12}$(OH)$_{24}$(H$_2$O)$_{12}$]$^{7+}$ or Al$_{13}$ polymer that partially transforms into Al(OH)$_3$ was shown to effectively sorb Ni, Zn and Cd, and to a lesser extent Cu and Pb from aqueous solution [53]. The Al$_{13}$ polymer was suggested as a binding agent suitable to immobilize trace elements in contaminated soils without affecting soil pH. Although amorphous Al oxide was shown to remove less Cu and Pb from solution than Fe oxides, Al oxide was capable of accumulating considerably higher metal concentrations than kaolinite [54]. Synthetic $Al(OH)_3$ was observed to decrease the water-extractable As fraction by 55–100% in amended soil having an efficiency similar to that of synthetic ferrihydrite, while natural Fe oxides (limonite) and clay minerals (bentonite) were significantly less effective [55]. Although Al oxides have potential as trace element immobilizers, more research demonstrating possible environmental effects are still needed.

15.2.1.3 Stability of Metal Oxides in Soil

Stability of trace elements immobilized by metal oxides depends to a large extent on the stability of the metal oxides themselves. Soil pH and redox conditions are the main factors affecting dissolution of metal oxides. Increasing soil acidity accelerates dissolution of Fe, Mn and Al oxides, while Al oxides also dissolve in alkaline pH.

Reductive dissolution of metal oxides is much faster than proton-promoted dissolution (that is due to acidification) [56]. Oxygen is necessary for metal oxides to exist; therefore, development of a low redox environment, for example due to extended period of soil saturation, intense microbial activity or presence of reductive agents, may interfere with metal oxide stability. Although all oxides are affected by a redox drop, the reductive dissolution of Mn oxides starts at higher redox potential (∼200 mV) than for Fe oxides (∼100 mV) [57,58]. Therefore, if reducing conditions begin to develop in soil, Mn oxides will release sorbed trace elements earlier than will Fe oxides, although some released metals that have a high affinity to solid surfaces, such as Pb, could be readsorbed onto the soil phases, like clays or organic matter [41]. Biologically generated ligands, for example citrate, malonate and oxalate, should also be considered, as these have a stronger effect on metal oxide dissolution than H^+ protons [56].

Iron oxides are the most studied among metal oxides for trace element immobilization in soil at various scales. A full-scale field project was recently implemented at La Combe du Saut industrial mining complex in the Salsigne area, south of France, where metallic iron was used as amendment for immobilization of As during restoration of the closed mine [59].

15.2.2 Natural and Synthetic Aluminosilicates

15.2.2.1 Clays

Clays are secondary aluminosilicate minerals formed in soils due to the weathering of the primary minerals (igneous and metamorphic) and are grouped together according to their particle size (<2 μm). Clays are reactive soil components due to the negatively charged surface area available for sorption of cationic elements (based on the diffuse double layer theory – see Chapter 2). Some clay minerals, for example kaolinite, can develop a net positive charge at low pH, therefore are able to sorb anions in acidic soil conditions [60]. Depending on the structural distribution of the building blocks (silicon tetrahedra and aluminium octahedra) and links between their sheets, specific surface area of clays can range from 15–20 m^2 g^{-1} for kaolinite to 280–500 m^2 g^{-1} for montmorillonite [61].

The ability of clays to retard contaminant migration to groundwater due to ion exchange and chemisorption is well established. Most of the recent studies on sorptive capacity of clays include batch tests with aqueous solutions of trace elements, for example As, Pb, Cd, Cu, Zn, Ni, Mn (for example [53,62–66]) and radionuclides, for example Cs, Sr, Eu, Th [67]. An in-depth review of trace element adsorption on natural and modified clays is given by Bhattacharyya and Gupta [68].

Recent studies on clay applications to contaminated soils in order to reduce trace element mobility are rather rare. Álvarez-Ayuso and García-Sánchez [69,70] studied metal retention in highly contaminated mining soils amended with magnesium aluminium silicate clays, palygorskite (attapulgite) (Mg,Al)$_2$Si$_4$O$_{10}$(OH)·4H$_2$O) and sepiolite (Mg$_4$Si$_6$O$_{15}$(OH)$_2$·6H$_2$O). The authors observed that water-soluble and relatively mobile (NH$_4$NO$_3$-extractable) fractions of Cd and Zn were reduced by 84–99% at 4% sepiolite application rate. The same amount of palygorskite reduced the leaching of Pb, Cu, Zn and Cd by 50–66%. Dissolved Cu concentration in soils amended with bentonite (montmorillonite-type clay) was shown to decrease with increasing amount of amendment and at 1.0 g kg^{-1} soil and above Cu was undetectable in solution [71]. The findings further demonstrated that bentonite addition decreased Cu desorption rates at low pH (2.5) nearly by half compared to unamended soil [71]. Garcia-Sanchez and co-authors [55] observed As immobilization in soils amended with bentonite to be pH-dependent. In one of the soils (pH 4.6), bentonite (10%) addition reduced As leaching by 50%, while in the more acidic soil (pH 3.8), bentonite was shown to become incapable of sorbing As.

Modified clay minerals, for example thiol-grafted synthetic beidellite (montmorillonite type clay), applied to contaminated soil reduced Cd uptake by rye grass 5-fold compared with unamended soil, but Pb uptake by plants was unaffected [72].

Other soil amendments might be preferred over clays due to the stated inability to immobilize high concentrations of multiple elements that often occur in contaminated soils [63] and expected high desorption rates, especially at lower pH, reaching up to 100% of sorbed amount of trace elements [71,73,74]. A combination of clays with other amendments, for example organic matter, may reduce the rate of trace element desorption. Clays (for example kaolinite) could outcompete dissolved humic substances for metal (for example Cu) sorption [75] and form ternary complexes with cations and organic matter (clay–cation–organic ligands), thereby increasing the strength of metal (for example Pb, Cu, Cd) sorption [76,77].

15.2.2.2 Zeolites

Zeolites are hydrated aluminosilicate minerals naturally formed when volcanic ash and rocks are in contact with surface water and groundwater. Zeolites can also form in nonvolcanic environments where salt-affected soils interact with strongly alkaline solutions [25]. Zeolites are similar to clay minerals in their predominantly Si-Al composition, but have different crystalline structure and properties. Porous zeolite minerals have a robust crystalline structure and, in contrast to most clays, do not exhibit swelling/shrinking characteristics. Zeolites are referred to as molecular sieves, that is they are able to selectively sort molecules based on their size [78].

Zeolites can be easily synthesized mainly from silica and alumina, but industrial by-products (for example fly ash) are also used for this purpose [79,80]. Under warm conditions (30–60 °C), formation of zeolites can occur directly in fly ash-amended soils [81,82]. Due to the microporous structure and distinctive adsorption, catalytic, cation-exchange, and dehydration–rehydration properties, zeolite applications cover many fields of human activities, spanning from construction, medicine and cosmetics to agriculture and environmental protection applications [78].

Natural and synthetic zeolites have also been tested for trace element immobilization in contaminated soils as they offer a high retention capacity for positively and negatively charged ions. Zeolites are reported to have a high selectivity for Cd [83] and Pb [84], but can also effectively sorb Cu, Zn, Tl, Mn, Co and As [47,62,85–87]. However, results of trace element immobilization in contaminated soils following zeolite application are highly variable and were demonstrated to be both positive and negative depending on the soils tested and zeolite types [85,86]. Such variation occurs both in field and laboratory tests. Müller and Marchner [85], for example, observed significantly lower dissolved concentrations of Cd (below detection limit) in a field experiment compared with a pot test (25% reduction in dissolved Cd), while zeolites had an opposite effect on Zn – 40% reduction in the pot test and no changes in soluble Zn in the field test. It is unclear whether this variability is due to the matrix effect or differences in experimental designs. Factors, such as heterogeneity of contaminant distribution in pot versus field soil and their contact with zeolite amendments can contribute to the large variation in experimental results.

Even when zeolites decrease trace element concentrations in soil solution, a reduction in plant uptake is not always apparent. For example, zeolites added to a contaminated garden soil (8.4 kg m^{-2}) reduced relatively mobile (1 M NH_4NO_3-extractable) concentrations of both cations (Cd and Pb) and oxyanions (As) below detection limit, but uptake of Cd and Pb by lettuce was unaffected, while uptake of As was noticeably higher compared with unamended soil [85]. In contrast, Chen and co-workers [47] reported a significant reduction in Cd and Zn accumulation in wheat shoots (by 54–60% and 86–95%, respectively) following a 1% application of synthetic zeolite to contaminated soils.

In addition to different types of zeolites and soil properties, variation in soil pH induced by zeolite addition is an important factor influencing trace element mobility. The pH of treated soil usually increases with increasing amount of added zeolites and so does the trace element immobilization in soil [88,89]. However, neutral pH is unfavourable for As immobilization, which makes zeolites less efficient for As sorption than for other trace metal contaminants. Therefore, a successful immobilization of As in soil by zeolites may not be achieved [62].

Attempts have been made to improve sorptive properties of natural zeolites by modifying them, for example by Fe oxide coatings [90]. Natural zeolite (clinoptilolite) added to a solution during goethite synthesis increased specific surface area of the material 5-fold compared to pure clinoptilolite and nearly tripled the maximum amount of sorbed Cu [90]. Such Fe-improved zeolites may be more effective also for As sorption due to the high affinity As has for Fe oxides.

Beside trace element immobilization, zeolites have other beneficial properties. Zeolites can modify soil texture and decrease hydraulic permeability of sandy soils, while increasing it in clayey soils [91]. Furthermore, natural and modified zeolites can gradually release major cations (Ca^{2+}, Mg^{2+}, Na^+ and K^+) and nitrate and improve soil nutrient status

[83,90,92]. However, the impact on soil microbial activity does not always respond to the reduced concentrations of soluble trace elements. Despite a significant reduction of dissolved Cd, Zn and Pb in treated soil, zeolite addition (10%) had no effect on the number of culturable fast-growing heterotrophic bacteria in soil, measured six months after the soil treatment [93].

Long-term studies are needed to evaluate changes in soil biological properties, as well as the stability of sorbed trace elements and zeolites themselves. Release of sorbed metal contaminants can occur after the saturation of sorptive sites. Also, dissolution of the Si-Al zeolite framework may occur upon soil acidification [90,94], which in turn would affect trace element mobility.

15.2.3 Ashes

Ashes might be considered as the very first soil ameliorant used. Woodland burning was practised as early as during interglacial times by primitive hunters and plant-gathering peoples to improve vegetation growth for their herds and wild animals [95]. Woodland clearance for arable farming started in the early Neolithic period, when the ash produced from the wood burning was mixed with the ploughed layer of the topsoil during cultivation [96]. In some countries (for example Sweden), fly ash from wood burning is widely used for forest fertilization. But in this case, the main reason is to beneficially dispose of ashes rather than to burn wood for land reclamation or fertilizer production.

Billions of tonnes of ashes are generated every year around the world from the combustion of coal and biofuel for energy recovery. The term *fly ash* is used as a generic name for all types of fine-sized ash and sorbent material collected in incineration air pollution control (APC) systems [97]. Chemical composition and physical properties (for example particle shape and size, see Figure 15.1) of fly ash differ enormously depending on the type of incinerator, the material burned and the APC systems used, which in turn affects their chemical properties. The noncombustible fraction that forms a residue on the hearth of a furnace is refereed to as *bottom ash* – sometimes also called slag or grate ash [97].

Coal and biofuel fly ashes are mainly composed of ferroaluminosilicate minerals, containing high amounts of Ca, Mg, Na, K and various concentrations of trace elements (for example As, B, Ba, Cd, Cr, Cu, Mo, Ni, Pb, Sb, Sn, V, Zn). The pH of fly ash can vary from acidic to highly alkaline depending on the coal composition. Although coal ashes are considered as waste and most often disposed of by landfilling, research on their utilization for soil amelioration purposes also has received considerable scientific attention [98].

Often the purpose of ash application to soil is to improve the biomass production by providing plant nutrients (S, Ca, K, Mg) and reducing soil acidification [99–101]. Ashes that have high alkalinity and acid neutralization capacity are considered suitable for reclamation of mine spoils and re-vegetation of barren sites [102–106]. Application of ash to sandy soils can improve soil texture and water-holding capacity [107,108] and reduce the swelling potential of clays [109]. In addition, ash can effectively immobilize trace elements in contaminated soils and reduce their toxicity [105,106,110–112].

Fly ashes are most efficient for immobilization of cationic trace elements, such as Cu, Pb, Zn, Ni, but might not be as effective for stabilization of oxyanions, for example arsenates, chromates. Mobility of oxyanions increases in the neutral to alkaline pH

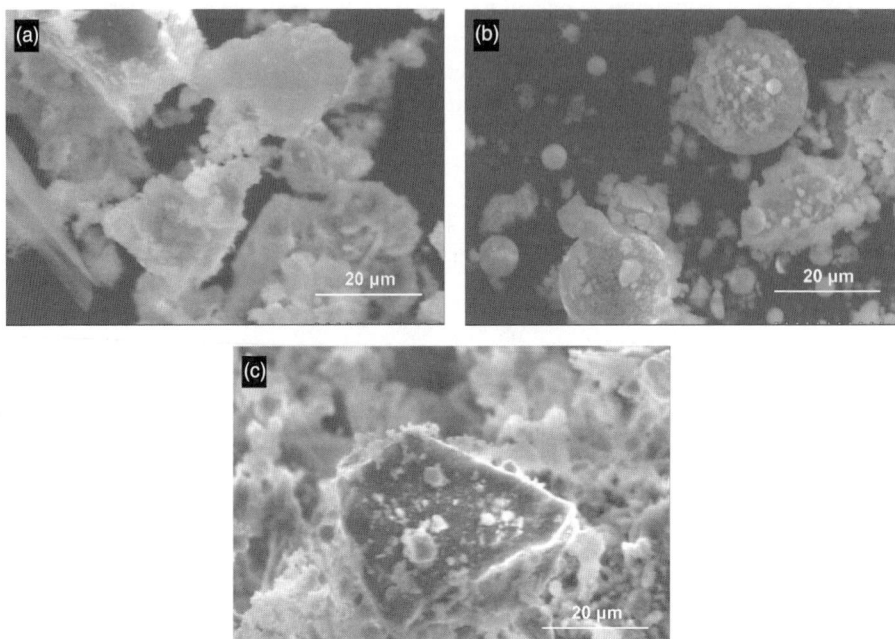

Figure 15.1 Scanning electron micrograms of fly ash from combustion of (a) paper mill sludge, (b) pulverized coal/peat, and (c) peat (Reproduced with permission from H. Sundblom, Fly ashes from co-combustion as a filler material in concrete production, Värmeforsk Report Q4-219, Vattenfall Research and Development, 2004 [112].)

range that is often obtained by ash addition. The effect of the same ash on oxyanions' mobility may vary substantially in different soils (Table 15.1; Figure 15.2), while the reasons for such variations are not well-understood.

Although an exact mechanism of trace element retention by fly ash is difficult to identify, the main factors seem to be (i) an increase in pH causing trace element precipitation and (ii) an increase in the specific surface area of the soil, promoting trace element sorption via a combination of surface complexation and cation exchange reactions.

Table 15.1 Some properties of As-contaminated soils collected from northern Sweden and treated using coal fly ash (CFA) (unpublished data, J. Kumpiene)

	Total As (mg kg^{-1})	Soil texture	Organic carbon (%)	Amount of added CFA (wt%)	pH after treatment
Soil A	270 ± 24	Sandy loam	0.84	5	9.6
Soil B	2714 ± 75	Not determined	0.77	5	10.3
Soil C	4589 ± 192	Sand	0.61	5	9.9
Soil D	4673 ± 194	Sand	0.15	3	8.9

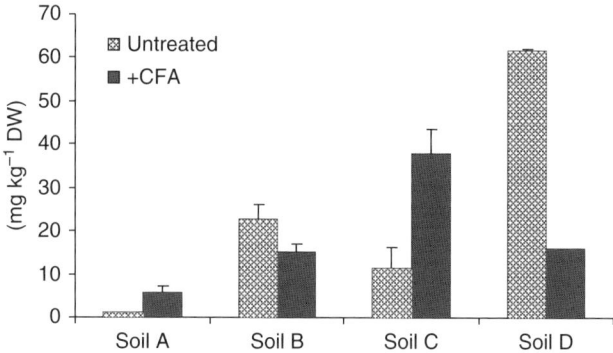

Figure 15.2 Impact of coal/wood fly ash (CFA) on As leaching in CCA contaminated soils (unpublished data, J. Kumpiene) as assessed by a batch leaching test using distilled water at liquid-to-solid ratio of 10 l kg^{-1} (n = 3; the error bars represent standard deviations)

The use of ash is shown to be effective for metal immobilization and is especially suitable for amelioration of acidic soils with low organic matter and clay contents, but data on the long-term effectiveness are scarce. Nachtegaal and co-workers [113] used extended X-ray absorption fine structure spectroscopy (EXAFS) to study changes in Zn speciation in a smelter-contaminated soil amended with fly ash and compost. No major differences in Zn speciation were found between the treated and nontreated soil samples 12 years after the treatment. Doubts were expressed regarding the long-term stability of Zn under soil acidification (pH 4), since 50–70% of soil Zn was identified in acid-soluble precipitates [113].

Application of fly ash as a soil ameliorant can also have a negative impact on soil quality due to possible increases in soil salinity, elevated concentrations of trace elements to toxic levels and their causing nutrient imbalance. Boron (B) and molybdenum (Mo), commonly present in ash, can be phytotoxic even at a moderate ash application rate, 20 Mg ha^{-1} [110,114]. Additionally, little is known about the possible effects of other rarely studied ash constituents, such as antimony (Sb) and tin (Sn) (concentrations can be up to grams per kilogram of ash), on plants and soil organisms.

Even when ashes contain rather high concentrations of K (up to several per cent), their application to soil was demonstrated to actually induce K, as well as P deficiency [115]. Similar to the iron amendments, excess of reactive ash constituents (Ca, Al, Fe) can sorb and precipitate nutrients, causing their imbalance in amended soils [114,115]. As a result of nutrient deficiency, application of ashes (80 Mg ha^{-1}) was observed to reduce maize biomass by half compared with a fertilized (using 20N–20P$_2$O$_5$–20K$_2$O kg ha^{-1}) control treatment [104]. Co-application of biosolids however has been suggested to counteract nutrient imbalance and maintain biomass production [99,115,116]. The composition of ash varies greatly and requires a comprehensive characterization prior to the utilization for *in situ* contaminant immobilization or soil amelioration purposes. Natural analogues, for example soils containing volcanic ash, can be used to study long-term processes occurring in soil–metal–ash systems.

15.2.4 Phosphates

The application of phosphorus-containing materials for contaminant immobilization in soil is based on the tendency of phosphate to precipitate, forming stable solid phases. Apatite $(Ca_{10}(PO_4)_6(F, Cl, OH)_2$ is the most common P mineral, which can dissolve in acidic soil conditions and reprecipitate together with metals (for example Pb, Cu, Zn, Cd) fully or partly substituting Ca [117–119].

Various sources of P have been tested for soil remediation purposes, including synthetic and natural phosphate minerals [120–125], phosphate-based salts (Na_2HPO_4, $MgHPO_4$, $(NH_4)_2HPO_4$, $9Ca(H_2PO_4)_2$ [17,85,123,125–127], phosphoric acid [122,127–132], industrial by-products [17,86], biogenic phosphate (bone meal, bone char and ashes) [133,134] and their combinations in both laboratory and field experiments.

The use of phosphorus-containing materials has been suggested to be especially successful for immobilization of Pb. In soils amended with hydroxyapatite (HA), the main Pb immobilization mechanism is dissolution of HA followed by precipitation of highly stable hydroxypyromorphite, $Ca_{10-x}Pb_x(PO_4)_6(OH)_2$ [117]. Similarly, addition of other apatite minerals or phosphate compounds can result in the formation of sparingly soluble pyromorphite-type minerals [$Pb_{10}(PO_4)_6(F, Cl, B)_2$], with precipitation dominating over surface sorption/complexation reactions [121]. The newly formed minerals have been demonstrated to significantly reduce Pb bioaccessibility and bioavailability [122,135–137]. A review of Pb immobilization by phosphate amendments is given by Chrysochoou and co-authors [138].

The solubility of Pb and P compounds determines the rate and effectiveness of Pb immobilization in soil. Lead can occur in soils in various forms (for example PbO, $PbCO_3$, $PbSO_4$) that dissolve at different pH values [119]. Dissolution of Pb and P amendments, which is enhanced by decreasing pH, is necessary for pyromorphite-type minerals to form. Phosphate-based salts and phosphoric acid are more soluble than phosphate rock and therefore more efficient for Pb immobilization [122]. The soil acidification, however, may cause dissolution of other trace elements and increase soil toxicity. Decrease in soil pH to <6 and elevated concentrations of P following triple superphosphate (TSP) application, for example, caused complete mortality of earworms in treated soils [123]. Addition of 1% of TSP decreased the soil pH from 6.2 to 5.6 and doubled the Cu concentration in the soil solution [17].

Co-application of P amendments with various solubilities, for example phosphate salts and phosphate rock, or following addition of lime, were suggested to counteract soil acidification [126,128]. Among a number of tested soil amendments (rock phosphate, TSP, H_3PO_4, limestone, various biosolids, cyclonic ashes, steel shots and red muds), application of TSP and phosphoric acid followed by liming were the most effective in reducing soil solution, bioaccessible and NH_4NO_3-extractable Pb and its concentration in plant tissues [122]. The phosphate amendments were shown to be similarly effective in reducing the dissolved, extractable and plant tissue Zn and Cd [122].

Phosphate rock, PR (1 wt%) application caused a significant reduction in 1 M CH_3COOH-extractable Pb in seven and $CaCl_2$-extractable Pb in eight out of 10 tested contaminated soils from four countries, but PR was less successful than lime or a mixture of ash and metallic iron in reducing $CaCl_2$-extractable Pb and mitigating plant phytotoxicity [120]. Phosphorus in soil can easily precipitate as apatite and manganese hydrogenphosphate, MnHPO$_4$ [50], which would reduce Ca and Mn availability and hinder plant development. Co-association of Pb, P and Ca was detected on the root surface in soil amended with hydroxyapatite [139], which was suggested to be the reason for increased nutrient deficiency in maize grown in HA-amended soil [140].

Phosphate amendments should be avoided when soil is contaminated with oxyanions, such as As, Se and W. Phosphate competition with arsenate for sorption sites may dramatically increase As mobility and bioavailability [120,140,141]. Friesl and co-workers [86] observed an increase in 1 M NH$_4$NO$_3$-extractable As by up to 74% following phosphatic slag addition and by an order of magnitude when the soil was amended with comparable amount of P using TSP. A significant increase in mobile As fractions observed following soil amendment with TSP increased As accumulation up to 9-fold for carrots and up to 10-fold for lettuce [142]. Nearly a 7-fold increase in mobile As concentration observed following 1% TSP application, however, had no effect on As uptake by lettuce compared with unamended soil [85], while As uptake by maize varied depending on the HA addition dose [140]. This illustrates that various plant species respond differently to increased concentrations of dissolved As.

The effectiveness of P amendments decreases in multielement-contaminated soils. The competitive metal sorption was observed to reduce the binding capacity of hydroxyapatite for, for example, Pb, Cu and Zn when simultaneously present in the solution [118,121]. Addition of 0.2% of ammonium phosphate as a P source to soil resulted in a 12- and 3-fold increase in the concentration of dissolved Cu and Zn, respectively, even though the soil pH remained unchanged [17]. Furthermore, P can easily react with major soil cations and, according to Porter *et al.* [46], stable pyromorphites can form in soil only after all soluble Ca and Mn are excluded from soil solution. Molecular-level studies using EXAFS showed that only less than 45% of the total Pb was converted to a stable pyromorphite 32 months after *in situ* soil treatment despite the satisfactory prerequisites, such as low pH, high concentration of dissolved Pb and P [31]. Higher treatment efficiencies have largely been reported based on the results of extraction tests (for example [127–129]). Reliable evidence (for example using sensitive non-destructive methods, such as EXAFS) of actual pyromorphite formation in amended soils, however, is generally lacking. Pyromorphite formation may occur during acidic extractions, which can give a significant overestimation of the treatment success [143].

A higher dose of added P can enhance contaminant immobilization, but that would require increasing P amounts to 20 times over the normal P concentration in soil [46], which would lead to undesirable changes in soil structure and the increased risk of the eutrophication of surface waters [144]. For successful Pb immobilization through pyromorphite formation, low soil pH, high solubility of Pb and P, and high P/Pb ratios are required, which all are high environmental risk factors. Therefore, Pb immobilization using phosphates in field applications should be considered very carefully and might have more disadvantages than benefits [138].

15.2.5 Organic Amendments

Organic matter (OM) is an important soil constituent that influences physical properties, nutrient status and biological activity of soil. Soil organic matter generally consist of a complex mixture of humic and fulvic acids. The high content of functional groups in the humic substances, primarily carboxyl COOH and phenolic-OH, can dissociate their H^+ ions and contribute to ion exchange, complex formation, adsorption, and/or chelation reactions with trace elements (see Chapter 2).

Loss of OM is not uncommon in highly contaminated soils, as toxic levels of trace elements hinder soil biological activity, vegetation development and generation of OM. Therefore, soil amendment with OM would help to restore soil properties and retain nutrients within soil. Trace elements strongly incorporated into OM, for example through the formation of a chelate (ring) structure, such as with a COO–phenolic-OH site combination [145], are expected to remain 'locked' or retained within the soil organic matter as long as the soil does not undergo acidification (below pH 3), the organic matter does not dissolve and its level does not decline. On the other hand, OM often appears among the factors influencing the trace element mobility in soil (for example [146,147]). Organic acids are strong ligands that can hold trace elements in solution by forming soluble metal-OM complexes (see Chapter 2).

The influence of OM on metal mobility depends greatly on its properties, for example degree of humification (decomposition), ratio between soluble low-molecular-weight organic acids (which act mainly as trace element carriers) and high-molecular-weight compounds (which tend to retain trace elements). Alkaline conditions can lead to dissolution of OM, thus increasing the mobility of trace elements. In an acidic environment, surface protonation of organic molecules leads to loss of negative surface charge and the ability to adsorb cationic trace elements. It is therefore important to keep soil pH within the neutral range to achieve the best effect of OM on trace element immobilization.

15.2.5.1 Peat

Adsorption and/or ion exchange by particulate media such as peat might be a suitable option for removal of colloidal and dissolved metals from soil and could provide long-term storage for contaminants [148]. Peat was shown to have the greatest affinity for Zn and Cd accumulation as compared to clay (kaolin), especially at low pH (~4) [149]. Peat application (5%, w/w) to contaminated soil can significantly reduce Cu and Zn concentration in soil solution at pH 5 [17]. Peat was believed to be responsible for a lower Cu and Pb release from an ash–peat mixture amended soil exposed to acidification [111]. Therefore, peat applied with ash can expand the effective pH range of trace element stability in amended soil. However, peat has a high affinity for water, poor chemical stability, a tendency to shrink and swell and is susceptible to decomposition with drying up [150], which makes it rather poor material in terms of physicochemical stability. Furthermore, the amount of nutrients and the pH of most peat types (2–4) can be too low to sustain the vegetation. The pH values for most plant growth processes are optimal in the circum-neutral range. Therefore, admixing high amounts of peat to contaminated soils may require pH neutralization and additional fertilizer application [151].

15.2.5.2 Biodegradable Waste

Recycling of organic or biodegradable waste (for example sewage sludge, animal manures, compost) through land applications has drawn a lot of attention among scientists, land users and regulators. Stopping the ocean dumping of all municipal sewage sludge and industrial waste in the United States as a result of the Ocean Dumping Ban Act of 1988 [152] and restrictions on landfilling of biodegradable waste introduced in European Union countries by the Council Directive 1999/31/EC [153], uphold sewage sludge land application as a potential option for material recycling. The European Community [154] and Environmental Protection Agencies in the United States [155] and Australia [156] and other countries around the world accept utilization of sewage sludge through land applications. Sewage sludge is rich in organic matter and generally contains high amounts of P and N and considerable concentrations of micronutrients and therefore is seen as a resource of potential agricultural value. Despite extensive studies on sewage sludge reuse by land applications (for example [147,157–159]), conclusions presented in the literature as to whether it is a beneficial or risky practice are still debatable [160,161]. Utilization of sewage sludge for agricultural purposes is often pursued by concerns of contaminant release, increased availability of trace elements in amended soils and overall risks for ecological and human health [161–163]. Organic amendments can alter oxidation state and chemical form of redox-sensitive elements, such as As, and increase their mobility. At neutral soil pH, reduction of As(V) to As(III) and consequent leaching was observed in a compost-amended soil [141].

Nevertheless, the utilization of biodegradable by-products alone or in combination with mineral soil amendments is being considered as a possible option for reclamation and re-vegetation of large areas degraded by, for example, smelter emissions or mine wastes [164,165]. Concentrations of trace elements in biosolids are rather low in comparison with those found in mining waste. Application of biosolids has successfully restored vegetative cover on several large contaminated sites, for example in Katowice (Poland), Bunker Hill (Idaho, USA), Leadville (Colorado, USA) and Jasper County (Missouri, USA) [166,167]. In most of the cases, neutralization of soil pH is necessary to increase trace element stability and reduce ecotoxicity [168], which is achieved by successive application of alkaline amendments (for example calcium carbonate, limestone, lime).

Similarly to ash, the potential for trace element immobilization and release from biosolid-amended soil varies greatly depending on the properties of biosolids (especially degree of their maturity) and should be comprehensively evaluated prior to their application to contaminated soil. For example, paper mill sludge, an industrial by-product mainly composed of carbonates, silicates and organic matter, has lower intrinsic ecotoxicity than sewage sludge and was observed to be very efficient in decreasing dissolved concentrations of trace elements (Cu, Pb, Zn, Ni and Cd) and their uptake by plants [169–171].

15.2.6 Liming Compounds

Lime is the most common material used to neutralize soil acidity and has traditionally been used in agriculture to improve plant growth and for reducing plant uptake of elevated concentrations of trace elements [172].

Liming compounds such as $CaCO_3$, $Ca(OH)_2$, CaO and MgO, and industrial by-products such as sugar foam (sugar beet lime) rich in $CaCO_3$, have been considered for soil remediation and compared with other soil amendments for their effectiveness [for example 17,122,173,174]. Increased soil pH by use of liming amendments leads to reduced mobility of trace elements through induced precipitation, increased number of pH-dependent positive surface charges of clay minerals and OM and consequent sorption of trace elements. This in turn reduces contaminant uptake by plants and increases plant biomass and growth rate [19,122,174]. Neutralization of soil acidity and reduction of soil toxicity can positively influence soil microbial and enzymatic activity [173] and accelerate the recovery of soil functions at degraded sites.

Alkaline materials have varying impacts on mobility of oxyanionic elements, such as arsenate and chromate, depending on the induced changes in soil pH. Soil liming and consequent increase in soil pH above neutral favours the oxidation of Cr^{3+} to the more mobile and toxic Cr^{6+} [175,176], which can increase Cr uptake by vegetation [177]. Increased As mobility is often observed in lime-amended soils [16,173], but alkaline materials can also have an opposite effect. Addition of sugar foam (containing 80% $CaCO_3$), for example, to a contaminated soil was observed to decrease dissolved As concentration below the detection limit (at pH 7.5) [173]. The presence of high concentrations of Ca in liming materials can cause As precipitation as calcium hydrogen arsenate ($CaHAsO_4$) and calcium arsenate ($Ca_3(AsO_4)_2$) at moderate pH conditions [46]. The possible formation of As–Ca complexes was also suggested to be the reason for slightly reduced As leaching in a lime-amended soil [13]. Lombi et al. [178] suggested that lime addition to soil can increase As sorption to carbonates, while later studies showed that lime had no significant effect on As immobilization in soils with different sources of contamination [10,179,180]. Based on a sequential extraction developed for As and scanning electron microscopy/X-ray spectrophotometry results, the significance of Ca for As retention is suggested to be negligible even in calcareous soils, since As is primarily bound to Fe oxides [181,182]. Even if Ca–As compounds are formed in soil, dissolved As concentration in soil solution in equilibrium with calcium arsenates in the pH range of 4.5–8.5 might be too high to mitigate ecotoxicological risks [183].

Although liming is conventionally a good measure to reduce the mobility of metal cations, it is not effective for oxyanions and in many cases can even increase their mobility and ecotoxicity. In addition, *in situ* soil treatment by altering pH might have only a short-term effect on metal immobilization. The reactions may be reversible, with eventual soil acidification, and annual measures to keep pH and trace elements stable would be required [104].

15.2.7 Gypsum

Industrial by-products containing gypsum, such as phosphogypsum (rock phosphate processing residue) and red gypsum (titanium processing residue) have been tested for their suitability for *in situ* remediation of contaminated soils. Gypsum amendments were observed to be especially efficient in reducing Pb mobility and to a lesser extent for Cd and Cu [122,184,185]. Garrido et al. [184] observed the formation of Al-hydroxy polymers in a gypsum-amended soil. This, together with ternary complexes formed between metals, Fe/Al oxy-hydroxide surfaces and SO_4/PO_4 anions were responsible for Cd, Cu and Pb retention in the soil. In addition, exchanged by Ca^{2+} and released to the soil solution, Pb^{2+} can precipitate as rather stable anglesite-type minerals in the presence of SO_4^{2-} [184].

Retention efficiency of oxyanions in gypsum-amended soil can be lower than that of cations due to increased competitive sorption between SO_4^{2-} and, for example, AsO_4^{3-}

anions [185]. Conversely, association of As, as well as Cd and Tl with Al-hydroxy polymers in phosphogypsum-amended soils, as demonstrated by SEM-EDX analysis, accounted for the increased trace element retention [185]. Red gypsum has elevated concentrations of soluble salts and this was believed to be the reason for lower plant yields grown in soil containing red gypsum compared to, for example, phosphorus-amended soil [122].

15.3 Method Acceptance

The common remediation policy of many environmental protection agencies is primarily to remove the contaminants or reduce their total amount in the soil. As a consequence, excavation and disposal of trace element-contaminated soil to landfills is a common soil remediation practice.

In countries that have established guideline values for trace elements in soils, a site is considered contaminated if the total concentration of trace elements exceeds the guideline values. Establishing the stabilization method for *in situ* soil treatment in such cases has a legal hindrance, because trace element concentration in soil does not change. While stabilization does not alter the contamination level of soil, it reduces the risk contaminants pose to human health and the environment by modifying the solubility and plant availability of trace elements. The absence of a common and validated procedure of stability evaluation is one of the reasons for limited acceptance of the *in situ* immobilization approaches. Site-specific solutions demand long-term monitoring programmes, the necessity of maintaining databases with thorough records and detailed information, and restricted future land use, which also hinders the general acceptance of stabilization methods.

Nevertheless, the soil stabilization approaches for large, moderately contaminated sites or mine-affected areas where excavation is practically unfeasible offer a considerably lower-cost alternative to conventional remediation techniques while delivering environmentally acceptable results. Several successful site restoration projects, where site re-vegetation after application of amendments (for example biosolids, iron oxides, lime and fertilizers) has been implemented, including Sudbury, Canada (restoration of about 30 km^2 of highly contaminated land around a mining and smelting region) [186], La Combe du Saut, France (reclamation of an As-mining site) [59], the Upper Silesia region, Poland (mining spoils and nonferrous smelter tailings) [166], the Bunker Hill site, USA (600 ha of land around a mining and smelting site) and several other Superfund sites [166].

One of the most important questions to answer is whether the efficiency of trace element stabilization is a long-term process, long enough to approach hundreds of years. It is a challenging task, if possible at all, to predict the trace element stability for a long time period in such a dynamic environment as soil. The treatment effectiveness is usually assessed within several weeks or months, and only a few treatments have been evaluated after several years (for example [14,16,113,187,188]). A combination of chemical (quantitative) and biological (qualitative) tests is necessary for assessing treatment efficiency and residual risks. The selection of a test battery should be relevant for a given environment and future use of a site. Establishment and continuous observations of large-scale long-lasting experiments, although difficult with short-term projects, are needed for the collection of reliable data on trace element fate in treated soils.

15.4 Concluding Remarks

The presence of multiple trace elements in contaminated soils is common and this is challenging for a successful *in situ* soil remediation. Contaminants that occur as cations (for example Cd, Zn, Cu, Ni, Pb) and oxyanions (for example As, Cr, Mb, V) often react differently when exposed to the same amendments. A combination of mineral additives (for example iron grit, ash) with organic matter (for example biosolids) usually gives better

Table 15.2 Commonly studied soil amendments and their impact on trace element immobilization

	Effects			Comments
	Positive (immobilization)	Varying[a]	Negative (solubilization)	
Metal oxides				
Fe oxides	As	Cd, Cu, Ni, Pb, Zn		Immobilization of cations depends on the final soil pH
Mn oxides	As, Zn, Cd		Cr	Too high Mn concentration are phytotoxic
Al oxides	Cr, Ni, Zn, Cd, Cu, Pb			Dissolves at alkaline pH
Natural and synthetic aluminosilicates				
Clays	Pb, Cu, Zn, Cd	As		Not effective at high contaminant concentrations
Zeolites		Cu, Zn, Mn, Co, Tl, As		Can dissolve at low pH, releases sorbed TE[b] after surface saturation
Ashes	Cu, Pb, Zn, Ni	As		Very high salinity
Phosphates	Pb, Cu, Zn, Cd		As	pH neutralizing amendments are often necessary
Organic amendments				
Peat	Cu, Pb, Zn, Cd			Poor physicochemical stability
Biodegradable waste		Cu, Pb, Zn, Ni, Cd, As		Formation of soluble OM-TE complexes can form and increase TE mobility
Liming compounds	Cu, Zn, Cd, Pb	As, Cr		Reversible with soil acidification
Gypsum	Pb, Cd, Cu	As		

[a] Impact varies from immobilization and no changes in mobility to increased solubility.
[b] TE, trace element(s).

results in terms of counteracting nutrient deficiency induced by mineral amendments and for recovering biodiversity. Utilization of industrial by-products (from the energy sector, the steel industry, etc.) for soil treatment can be economically and environmentally beneficial, especially in large areas affected by, for example, mining activities. Application of such materials, however, includes both risks and benefits and might be seen as an 'additional' pollutant source. Irresponsible application of sewage sludge, which is often considered as a resource of agricultural value, causing contamination of arable soils, is one such example.

Even though amendment application has been shown by numerous studies to stabilize trace elements in soil (Table 15.2), concerns are often expressed regarding the long-term efficiency of such soil stabilization approaches.

The soil characterization should cover a set of relevant chemical and biological tests applied to understand the behaviour of the treated material and its impact on the environment under specific field scenarios. This is a crucial point to determine the safety of the treated soil and, hence, acceptance of the method for *in situ* applications. Long-term experiments lasting decades rather than years and continuous studies of naturally attenuated sites are needed, which can serve as reference research sites and be highly valuable sources of information on the dynamics of trace elements in contaminated soils.

References

[1] Adriano, D.C.; Wenzel, W.W.; Vangronsveld, J., Bolan, N.S., Role of assisted natural remediation in environmental cleanup; Geoderma 2004, 122, 121–142.

[2] Mench, M.; Vangronsveld, J.; Lepp, N.W.; Edwards, R., Physico-chemical aspects and efficiency of trace element immobilisation by soil amendments; in: Vangronsveld, J., Cunningham, S.D. (eds), Metal-contaminated Soils: In Situ Inactivation and Phytorestoration; Springer, Heidelberg, Berlin, 1998, pp. 151–182.

[3] Mench, M.; Vangronsveld, J.; Clijsters, H.; Lepp, N.W.; Edwards, R., In situ metal immobilisation and phytostabilisation of contaminated soils; in: Terry, N., Banuelos, G. (eds.), Phytoremediation of Contaminated Soil and Water; Lewis Publishers, Boca Raton, FL, 2000, pp. 323–358.

[4] Knox, A.; Seaman, J.C.; Mench, M.J.; Vangronsveld, J., Remediation of metal- and radionuclides-contaminated soils by in situ stabilization techniques; in: Iskandar, I.K. (ed), Environmental Restoration of Metal-contaminated Soils; Lewis Publishers, Boca Raton, FL, 2001, pp. 21–60.

[5] Schwertmann, U.; Cornell R.M., Iron Oxides in the Laboratory: Preparation and Characterization, 2 edn; Wiley-VCH Verlag GmbH, Weinheim, 2000.

[6] Kabata-Pendias, A.; Pendias, H., Trace Elements in Soils and Plants 3rd edn; CRC Press, Boca Raton, FL, 2001.

[7] Berti, W.R.; Cunningham, S.D., Phytostabilization of metals; in: Raskin, I; Ensley, B.D. (eds), Phytoremediation of Toxic Metals, Using Plants to Clean Up the Environment; John Wiley & Sons, Inc., New York, 2000, pp. 71–88.

[8] Cornell, R.M.; Schwertmann, U., The Iron Oxides: Structure, Properties, Reactions, Occurences and Uses; Wiley-VCH Verlag GmbH, Weinheim, 2003.

[9] García-Sánchez, A.; Alvarez-Ayuso, E.; Rodriguez-Martin, F., Sorption of As(V) by some oxyhydroxides and clay minerals: application to its immobilisation in two polluted mining soils; Clay Miner. 2002, 37, 187–194.

[10] Hartley, W.; Lepp, N.W., Remediation of arsenic contaminated soils by iron-oxide application, evaluated in terms of plant productivity, arsenic and phytotoxic metal uptake; Sci. Total Environ. 2008, 390, 35–44.

[11] Moore, T.J.; Rightmire, C.M.; Vempati, R.K., Ferrous iron treatment of soils contaminated with arsenic-containing wood-preserving solution; Soil Sediment Contam. 2000, 9, 375–405.
[12] Kim, J.-Y.; Davis, A.P.; Kim, K.-W., Stabilization of available arsenic in highly contaminated mine tailings using iron; Environ. Sci. Technol. 2003, 37, 189–195.
[13] Hartley, W.; Edwards, R.; Lepp, N.W., Arsenic and heavy metal mobility in iron oxide-amended contaminated soils as evaluated by short- and long-term leaching tests; Environ. Pollut. 2004, 131, 495–504.
[14] Mench, M.; Bussière, S.; Boisson, J. et al., Progress in remediation and revegetation of the barren Jales Gold Mine spoil after in situ treatments; Plant. Soil 2003, 249, 187–202.
[15] Kumpiene, J.; Ore, S.; Renella, G.; Mench, M.; Lagerkvist, A., Maurice, C., Assessment of zerovalent iron for stabilization of chromium, copper, and arsenic in Soil; Environ. Pollut. 2006, 144, 62–69.
[16] Mench, M.; Vangronsveld, J.; Beckx, C.; Ruttens, A., Progress in assisted natural remediation of an arsenic contaminated agricultural soil; Environ. Pollut. 2006, 144, 51–61.
[17] Bes, C.; Mench, M. Remediation of copper-contaminated topsoils from a wood treatment facilities using in situ stabilisation; Environ. Pollut. 2008, 156, 1128–1138.
[18] Ciccu, R.; Ghiani M.; Serci, A.; Fadda, S.; Peretti, R.; Zucca, A., Heavy metal immobilization in the mining-contaminated soils using various industrial wastes; Miner. Eng. 2003, 16, 187–192.
[19] Gray, C.W.; Dunham, S.J.; Dennis, P.G.; Zhao, F.J.; McGrath, S.P., Field evaluation of in situ remediation of a heavy metal contaminated soil using lime and red-mud; Environ. Pollut. 2006, 142, 530–539.
[20] Al-Abed, S.R.; Jegadeesan, G.; Purandare, J.; Allen, D., Arsenic release from iron rich mineral processing waste: influence of ph and redox potential; Chemosphere 2007, 66, 775–782.
[21] Kumpiene, J.; Ragnvaldsson, A.; Lövgren, D.; Tesfalidet, S.; Gustavsson, B.; Lättström, L.; Leffler, P.; Maurice, C., Impact of water saturation level on arsenic and metal mobility in the Fe-amended soil. Chemosphere 2009, 74, 206–215.
[22] Jain, A.; Ravan, K.P.; Loeppert, R.H., Arsenite and arsenate adsorption on ferrihydrite: Surface charge reduction and net OH-release stoichiometry. Env. Sci. Technol. 1999, 33, 1179–1184.
[23] Sherman, D.M.; Randall, S.R., Surface complexation of arsenic(V) to iron(III) hydroxides: structural mechanism from ab initio molecular geometries and EXAFS spectroscopy. Geochim. Cosmochim. Acta 2003, 67, 4223–4230.
[24] Taylor, K.G.; Perry, C.T.; Greenaway, A.M.; Machent, P.G. Bacterial iron oxide reduction in a terrigenous sediment-impacted tropical shallow marine carbonate system, North Jamaica; Mar. Chem. 2007, 107, 449–463.
[25] Churchman, G.J. The alteration and formation of soil minerals by weathering; in: Sumner, M.E. (ed), Handbook of Soil Science; CRC Press; Boca Raton, FL, 2000, pp. F3–F76.
[26] Schrenk, M.O.; Edwards, K.J.; Goodman, R.M.; Hamers R.J.; Banfield, J.F., Distribution of *Thiobacillus ferrooxidans* and *Leptospirillum ferrooxidans*: implications for generation of acid mine drainage; Science 1998, 279, 1519–1521.
[27] Paige, C.R.; Snodgrass, W.J.; Nicholson, R.V.; Scharer, J.M., An arsenate effect on ferrihydrite dissolution kinetics under acidic oxic conditions; Water Res. 1997, 31, 2370–2382.
[28] Melitas, N.; Wang, J.; Conklin, M.; O'Day, P., Farrell, J., Understanding soluble arsenate removal kinetics by zerovalent iron media; Environ. Sci. Technol. 2002, 36, 2074–2081.
[29] Gräfe, M.; Nachtegaal, M.; Sparks, D.L., Formation of metal-arsenate precipitates at the goethite-water interface; Environ. Sci. Technol. 2004, 38, 6561–6570.
[30] Crawford, R.J.; Harding, I.H.; Mainwaring, D.E., Adsorption and coprecipitation of multiple heavy-metal ions onto the hydrated oxides of iron and chromium; Langmuir 1993, 9, 3057–3062.
[31] Scheckel, K.G.; Ryan, J.A., Spectroscopic speciation and quantification of lead in phosphate-amended soils; J. Environ. Qual. 2004, 33, 1288–1295.
[32] Leupin, O.X.; Hug, S.J., oxidation and removal of arsenic (III) from aerated groundwater by filtration through sand and zero-valent iron; Water Res. 2005, 39, 1729–1740.

[33] Warren, G.P.; Alloway, B.J.; Lepp, N.W.; Singh, B.; Bochereau, F.J.M.; Penny, C., Field trials to assess the uptake of arsenic by vegetables from contaminated soils and soil remediation with iron oxides; Sci. Total Environ. 2003, 311, 19–33.

[34] Warren, G.P.; Alloway, B.J. Reduction of arsenic uptake by lettuce with ferrous sulfate applied to contaminated soil; J. Environ. Qual. 2003, 32, 767–772.

[35] Tamura, H.; Kawamura, S.; Hagayama, M., Acceleration of the oxidation of Fe^{2+} ions by Fe(III) oxyhydroxides; Corros. Sci. 1980, 20, 963–971.

[36] Hacherl, E.L.; Kosson, D.S.; Young, L.Y.; Cowan, R.M., Measurement of iron(III) bioavailability in pure iron oxide minerals and soils using anthraquinone-2,6-disulfonate oxidation; Environ. Sci. Technol. 2001, 35, 4886–4893.

[37] Xie, L.; Shang, C., Role of humic acid and quinone model compounds in bromate reduction by zerovalent iron; Environ. Sci. Technol. 2005, 39, 1092–1100.

[38] Kumpiene, J.; Andreas, L.; Lagerkvist, A., Utilization of chemically stabilized soil in a landfill top cover. The 5th Intercontinental Landfill Research Symposium, Copper Mountain conference center, Colorado, USA. 10–12 September 2008.

[39] Ruttens, A.; Mench, M.; Colpaert, J.V.; Boisson, J.; Carleer, R.; Vangronsveld, J., Phytostabilization of a metal contaminated sandy soil. I: Influence of compost and/or inorganic metal immobilizing soil amendments on phytotoxicity and plant availability of metals; Environ. Pollut. 2006, 144, 524–532.

[40] Kämpf, N.; Scheinost, A.C.; Schulze, D.G., Oxide minerals; in: Sumner, M.E. (ed), Handbook of Soil Science; CRC Press, Boca Raton, FL, 2000, pp. F125–168.

[41] Davranche, M., Bollinger, J.-C., Heavy metals desorption from synthesized and natural iron and manganese oxyhydroxides: effect of reductive conditions; J. Colloid. Interf. Sci. 2000, 227, 531–539.

[42] O'Reilly, S.E. and Hochella, M.F. Jr., Lead sorption efficiencies of natural and synthetic Mn and Fe-oxides; Geochim. Cosmochim. Acta 2003, 67, 4471–4487.

[43] Manning, B.A.; Fendorf, S.E.; Bostick, B.; Suarez, D.L., arsenic(III) oxidation and arsenic(V) adsorption reactions on synthetic birnessite; Environ. Sci. Technol. 2002, 36, 976–981.

[44] Tournassat, C., Charlet, L., Bosbach, D., Manceau, A., arsenic(III) oxidation by birnessite and precipitation of manganese(II) arsenate; Environ. Sci. Technol. 2002, 36, 493–500.

[45] Lenoble, V.; Laclautre, C.; Serpaud, B.; Deluchat, V.; Bollinger, J.-C., As(V) retention and As(III) simultaneous oxidation and removal on a MnO_2-loaded polystyrene resin; Sci. Total Environ. 2004, 326, 197–207.

[46] Porter, S.K., Scheckel, K.G., Impellitteri, C.A., Ryan, J.A., Toxic metals in the environment: thermodynamic considerations for possible immobilisation strategies for Pb, Cd, As, and Hg; Crit. Rev. Env. Sci. Tec. 2004, 34, 495–604.

[47] Chen, Z.S.; Lee, G.J.; Liu, J.C., The effects of chemical remediation treatments on the extractability and speciation of cadmium and lead in contaminated soils; Chemosphere 2000, 41, 235–242.

[48] Youstı, I. and Bermond, A., Physical-chemical approach to asses the effectiveness of several amendments used for in situ remediation of trace metals-contaminated soils by adding solid phases; in: Iskandar, I.K. (ed), Environmental Restoration of Metals-Contaminated Soil; Lewis Publishers, Boca Raton, FL, 2000, pp. 3–20.

[49] Guha, H.; Saiers, J.E.; Brooks, S.; Jardine, P.; Jayachandran, K., Chromium transport, oxidation, and adsorption in manganese-coated sand; J. Contam. Hydrol. 2001, 49, 311–334.

[50] Kim, J.G.; Dixon, J.B., Oxidation and fate of chromium in soils; Soil Sci. Plant Nutr. 2002, 48, 483–490.

[51] Álvarez-Ayuso, E.; García-Sánchez, A.; Querol, X., Adsorption of Cr(VI) from synthetic solutions and electroplating wastewaters on amorphous aluminium oxide; J. Hazard. Mater. 2007, 142, 191–198.

[52] Chadwick, O.A., Graham, R.C., Pedogenic processes; in: Sumner, M.E. (ed), Handbook of Soil Science; CRC Press, Boca Raton, FL, 2000, pp. E41–E75.

[53] Lothenbach, B.; Furrer, G.; Schulin, R., Immobilization of heavy metals by polynuclear aluminium and montmorillonite compounds; Environ. Sci. Technol. 1997, 31, 1452–1462.

[54] Potter, H.A.; Yong, R.N., Influence of iron/aluminium ratio on the retention of lead and copper by amorphous iron–aluminium oxides; Appl. Clay Sci. 1999, 14, 1–26.
[55] Garcia-Sanchez, A.; Alvarez-Ayuso, E.; Rodriguez-Martin, F., Sorption of As(V) by some oxyhydroxides and clay minerals: Application to its immobilisation in two polluted mining soils; Clay Miner. 2002, 37, 187–194.
[56] Martin, S.T., Precipitation and dissolution of iron and manganese oxides, Environmental Catalysis, 2005, 61–81.
[57] Patrick, Jr. W.H.; Jugsujinda, A., Sequential reduction and oxidation of inorganic nitrogen, manganese, and iron in flooded soil. Soil Sci. Soc. Am. J. 1992, 56, 1071–1073.
[58] Stüben, D.; Berner, Z.; Chandrasekharam, D.; Karmakar, J., Arsenic enrichment in groundwater of West Bengal, India: geochemical evidence for mobilization of As under reducing conditions; Appl. Geochem. 2003, 18, 1417–1434.
[59] Programme Difpolmine, Diffuse pollution from mine activity. Project realisation. www.difpolmine.org (accessed 26 November 2009).
[60] Krebs, R.D.; Thomas, G.W.; Moore J.E., Anion influence on some soil physical properties. Clays and clay minerals, Vol. 9. Proceedings of the Ninth National Conference on Clays and Clay Minerals, Swineford, A. (ed.); Pergamon Press, New York, 1962, pp. 260–268.
[61] Skopp, J.M., Physical properties of primary particles; in: Sumner, M.E. (ed.), Handbook of Soil Science, CRC Press, New York, 2000, pp. A3–17.
[62] García-Sánchez, A.; Alastuey, A.; Querol, X., Heavy metal adsorption by different minerals: application to the remediation of polluted soils; Sci. Total Environ. 1999, 242, 179–188.
[63] Li, Z.; Beachner, R.; McManama, Z.; Hanlie, H., Sorption of arsenic by surfactant-modified zeolite and kaolinite; Micropor. Mesopor. Mater. 2007, 105, 291–297.
[64] Abollino, O.; Giacomino, A.; Malandrino, M.; Mentasti E., Interaction of metal ions with montmorillonite and vermiculite; Appl. Clay Sci. 2008, 38, 227–236.
[65] Gupta, S.S.; Bhattacharyya, K.G., Immobilization of Pb(II), Cd(II) and Ni(II) ions on kaolinite and montmorillonite surfaces from aqueous medium; J. Environ. Manage. 2008, 87, 46–58.
[66] Panuccio, M.R.; Sorgonà, A.; Rizzo, M.; Cacco, G., Cadmium adsorption on vermiculite, zeolite and pumice: batch experimental studies; J. Environ. Manage. 2009, 90, 364–374.
[67] Lauber, M.; Baeyens, B.; Bradbury, M.H., Physico-chemical characterization and sorption measurements of Cs, Sr, Ni, Eu, Th, Sn and Se on opalinus clay from Monti Terri; Nagra Technical Report 00-11, Nagra, Wettingen, Switzerland, 2000.
[68] Bhattacharyya, K.G.; Gupta, S.S., Adsorption of a few heavy metals on natural and modified kaolinite and montmorillonite: a review; Adv. Colloid. Interfac. 2008, 140, 114–131.
[69] Álvarez-Ayuso, E.; García-Sánchez, A., Palygorskite as a feasible amendment to stabilize heavy metal polluted soils; Environ. Pollut. 2003, 125, 337–344.
[70] Álvarez-Ayuso, E.; García-Sánchez, A., Sepiolite as a feasible soil additive for the immobilization of cadmium and zinc; Sci. Total Environ. 2003, 305, 1–12.
[71] Ling, W.; Shen, Q.; Gao, Y.; Gu, X.; Yang, Z., Use of bentonite to control the release of copper from contaminated soils; Aust. J. Soil Res. 2007, 45, 618–623.
[72] Diaz, M.; Cambier, P.; Brendlé, J.; Prost R., Functionalized clay heterostructures for reducing cadmium and lead uptake by plants in contaminated soils; Appl. Clay Sci. 2007, 37, 12–22.
[73] Atanassova, I., Competitive effect of copper, zinc, cadmium and nickel on ion adsorption and desorption by soil clays; Water Air Soil Pollut. 1999, 113, 115–125.
[74] Subramanian, B.; Gupta, G., adsorption of trace elements from poultry litter by montmorillonite clay; J. Hazard. Mater. 2006, 128, 80–83.
[75] Wu, J.; West, L.J.; Stewart D.I., effect of humic substances on Cu(II) solubility in kaolin-sand soil; J. Hazard. Mater. 2002, 94, 223–238.
[76] Arias, M.; Barral, M.T.; Mejuto, J.C., Enhancement of copper and cadmium adsorption on kaolin by the presence of humic acids; Chemosphere 2002, 48, 1081–1088.
[77] Hizal, J.; Apak, R., Modeling of copper(II) and lead(II) adsorption on kaolinite-based clay minerals individually and in the presence of humic acid; J. Colloid. Interf. Sci. 2006, 295, 1–13.

[78] Mumpton F.A., La roca magica. Uses of natural zeolites in agriculture and industry; Proc. Natl. Acad. Sci. USA 1999, 96, 3463–3470.

[79] Derkowski, A.; Franus, W.; Beran, E.; Czímerová, A., properties and potential applications of zeolitic materials produced from fly ash using simple method of synthesis; Powder Technol. 2006, 166, 47–54.

[80] Rayalu, S.S.; Udhoji, J.S.; Munshi, K.N.; Hasan M.Z., Highly crystalline Zeolite-A from flyash of bituminous and lignite coal combustion; J. Hazard. Mater. 2001, 88, 107–121.

[81] Terzano, R.; Spagnuolo, M.; Medici, L.; Tateo, F.; Ruggiero, P., Zeolite synthesis from pretreated coal fly ash in presence of soil as a tool for soil remediation; Appl. Clay Sci. 2005, 29, 99–110.

[82] Terzano, R.; Spagnuolo, M.; Medici, L.; Dorriné, W.; Janssens, K.; Ruggiero, P., Microscopic single particle characterization of zeolites synthesized in a soil polluted by copper or cadmium and treated with coal fly ash; Appl. Clay Sci. 2007, 35, 128–138.

[83] Panuccio, M.R.; Crea, F.; Sorgonà, A.; Cacco, G., adsorption of nutrients and cadmium by different minerals: experimental studies and modelling; J. Environ. Manage. 2007, in press.

[84] Ponizovsky, A.A. and Tsadilas, C.D., Lead(II) retention by alfisol and clinoptilolite: cation balance and ph effect; Geoderma 2003, 115, 303–312.

[85] Müller, I. and Marchner, B., Are immobilizing soil additives suitable to remediate heavy metal contaminated garden areas? Results from a field study; 9th International FZK/TNO Conference on Contaminated Soil: ConSoil 2005; 3–7 October 2005, Bordeaux, France, 2005.

[86] Friesl, W.; Friedl, J.; Platzer, K.; Horak, O.; Gerzabek, M.H., Remediation of contaminated agricultural soils near a former Pb/Zn smelter in Austria: Batch, pot and field experiments; Environ. Pollut. 2006, 144, 40–50.

[87] Li, L.Y.; Tazaki, K.; Lai, R. et al., Treatment of acid rock drainage by clinoptilolite – adsorptivity and structural stability for different pH environments; Appl. Clay Sci. 2008, 39, 1–9.

[88] Querol, X.; Alastuey, A.; Moreno, N. et al., Immobilization of heavy metals in polluted soils by the addition of zeolitic material synthesized from coal fly ash; Chemosphere 2006, 62, 171–180.

[89] Mahabadi, A.A.; Hajabbasi, M.A.; Khademi, H.; Kazemian, H., Soil Cadmium stabilization using an iranian natural zeolite; Geoderma 2007, 137, 388–393.

[90] Doula, M.K., Dimirkou A., Use of an iron-overexchanged clinoptilolite for the removal of Cu^{2+} ions from heavily contaminated drinking water samples; J. Hazard. Mater. 2008, 151, 738–745.

[91] Reháková, M.; Čuvanová, S.; Dzivák, M.; Rimár, J.; Gavalová Z., Agricultural and agrochemical uses of natural zeolite of the clinoptilolite type; Curr. Opin. Solid. State Mater. 2004, 8, 397–404.

[92] Li, Z., Use of surfactant-modified zeolite as fertilizer carriers to control nitrate release; Micropor. Mesopor. Mater. 2003, 61, 181–188.

[93] Garau, G.; Castaldi, P.; Santona, L.; Deiana, P.; Melis, P., Influence of red mud, zeolite and lime on heavy metal immobilization, culturable heterotrophic microbial populations and enzyme activities in a contaminated soil; Geoderma 2007, 142, 47–57.

[94] Ören, A.H. and Kaya, A., factors affecting adsorption characteristics of Zn^{2+} on two natural zeolites; J. Hazard. Mater. 2006, 131, 59–65.

[95] Steensberg, A., Some recent Danish experiments in neolithic agriculture; Agr. Hist. Rev. 1957, 5, 66–73.

[96] Huang, C.C.; Pang, J.; Chen, S. et al., Charcoal records of fire history in the holocene loess–soil sequences over the southern Loess Plateau of China; Palaeogeogr. Palaeocl. 2006, 239, 28–44.

[97] Chandler, A.J.; Eighmy, T.T.; Hartlen, J. et al., Municipal Solid Waste Incinerator Residues, Studies in Environmental Science 67; Elsevier Science B.V., Amsterdam, 1997.

[98] Jala, S.; Goyal, D., Fly ash as a soil ameliorant for improving crop production — a review; Bioresource Technol. 2006, 97, 1136–1147.

[99] Sajwan, K.S.; Paramasivam, S.; Alva, A.K.; Adriano, D.C.; Hooda P.S., Assessing the feasibility of land application of fly ash, sewage sludge and their mixtures; Adv. Environ. Res. 2003, 8, 77–91.

[100] Rautaray, S.K.; Ghosh, B.C.; Mittra, B.N. Effect of fly ash, organic wastes and chemical fertilizers on yield, nutrient uptake, heavy metal content and residual fertility in a rice–mustard cropping sequence under acid lateritic soils; Bioresource Technol. 2003, 90, 275–283.

[101] Mahmood, S.; Finlay, R.D.; Fransson, A.-M.; Wallander, H., Effects of hardened wood ash on microbial activity, plant growth and nutrient uptake by ectomycorrhizal spruce seedlings; FEMS Microbiol. Ecol. 2003, 43, 121–131.

[102] Vangronsveld, J.; Van Assche, F.; Clijsters, H., Reclamation of a bare industrial area contaminated by non-ferrous metals: in situ metal immobilization and revegetation; Environ Pollut 1995, 87, 51–59.

[103] Xenidis, A.; Mylona, E.; Paspaliaris, I., Potential use of lignite fly ash for the control of acid generation from sulphidic wastes; Waste Manage. 2002, 22, 631–641.

[104] Seoane, S.; Leiros, M.C., Acidification-neutralisation in a linite mine spoil amended with fly ash or limestone; J. Environ. Qual. 2001, 30, 1420–1431.

[105] Gorman, J.M.; Sencindiver, J.C.; Horvath, D.J.; Singh, R.N.; Keefer, R.F., Erodibility of fly ash used as a topsoil substitute in mineland reclamation; J. Environ. Qual. 2000, 29, 805–811.

[106] Ciccu, R.; Ghiani, M.; Serci, A.; Fadda, S.; Peretti, R.; Zucca, A., Heavy metal immobilisation in the mining-contaminated soils using variuos industrial wastes; Miner. Eng. 2002, 16, 187–192.

[107] Adriano, D.C.; Weber, J.T., Influence of fly ash on soil physical properties and turfgrass establishment; J. Environ. Qual. 2001, 30, 596–600.

[108] Pathan, S.M.; Aylmore, L.A.G.; Colmer, T.D., Properties of several fly ash materials in relation to use as soil amendments; J. Environ. Qual. 2003, 32, 687–693.

[109] Nalbantoğlu, Z., Effectiveness of class C fly ash as an expansive soil stabilizer; Constr. Build. Mater. 2004, 18, 377–381.

[110] Schumann, A.W. and Sumner, M.E., Chemical evaluation of nutrient supply from fly ash–biosolids mixtures; Soil Sci. Soc. Am. J. 2000, 64, 419–426.

[111] Kumpiene, J.; Ore, S.; Lagerkvist, A.; Maurice, C., Stabilization of Pb and Cu contaminated soil using coal fly ash and peat; Environ. Pollut. 2007, 145, 365–375.

[112] Sundblom, H., Fly ashes from co-combustion as a filler material in concrete production; Värmeforsk Report Q4-219 (in Swedish), 2004.

[113] Nachtegaal, M.; Marcus, M.A.; Sonke, J.E. et al., Effects of in situ remediation on the speciation and bioavailability of zinc in a smelter contaminated soil; Geochim. Cosmochim. Acta 2005, 69, 4649–4664.

[114] Doran J.W.; Martens D.C., Molybdenum availability as influenced by the application of fly ash to soil; J. Environ. Qual. 1972, 1, 186–189.

[115] Schumann, A.W.; Sumner, M.E. Plant nutrient availability from mixtures of fly ashes and biosolids; J. Environ. Qual. 1999, 28, 1651–1657.

[116] Mittra, B.N.; Karmakar, S.; Swain, D.K.; Ghosh, B.C., Fly ash—a potential source of soil amendment and a component of integrated plant nutrient supply system; Fuel 2005, 84, 1447–1451.

[117] Ma, Q.Y.; Tralna, S.J.; Logan, T.J., In situ lead immobilizaiton by apatite; Environ. Sci. Technol. 1993, 27, 1803–1810.

[118] Corami, A.; Mignardi, S.; Ferrini, V., Copper and zinc decontamination from single- and binary-metal solutions using hydroxyapatite; J. Hazard. Mater. 2007, 146, 164–170.

[119] Cao, X.; Ma, L.Q.; Singh, S.P.; Zhou, Q., Phosphate-induced lead immobilization from different lead minerals in soils under varying Ph conditions; Environ. Pollut. 2008, 152, 184–192.

[120] Geebelen, W.; Adriano, D.C.; van der Lelie, D. et al., Selected bioavailability assays to test the efficacy of amendment-induced immobilisation of lead in soils; Plant. Soil. 2002, 249, 217–228.

[121] Cao, X.; Ma, L.Q.; Rhue, D.R.; Appel, C.S., mechanisms of lead, copper, and zinc retention by phosphate rock; Environ. Pollut. 2004, 131, 435–444.

[122] Brown, S.; Christensen, B.; Lombi, E. et al., An inter-laboratory study to test the ability of amendments to reduce the availability of Cd, Pb, and Zn in situ; Environ. Pollut 2005, 138, 34–45.

[123] Ownby, D.R.; Galvan, K.A.; Lydy, M.J., Lead and zinc bioavailability to Eisenia Fetida after phosphorus amendment to repository soils; Environ. Pollut. 2005, 136, 315–321.
[124] Shi, Z.; Erickson, L.E., Mathematical model development and simulation of in situ stabilization in lead-contaminated soils; J. Hazard. Mater. 2001, 87, 99–116.
[125] Raicevic, S.; Kaludjerovic-Radoicic, T.; Zouboulis, A.I., In situ stabilization of toxic metals in polluted soils using phosphates: theoretical prediction and experimental verification; J. Hazard. Mater. 2005, 117, 41–53.
[126] McGowen, S.L.; Basta, N.T.; Brown, G.O., Use of diammonium phosphate to reduce heavy metal solubility and transport in smelter-contaminated soil; J. Environ. Qual. 2001, 30, 493–500.
[127] Cao, R.X.; Ma, L.Q.; Chen, M.; Singh, S.P.; Harris, W.G., Phosphate-induced metal immobilization in a contaminated site; Environ. Pollut. 2003, 122, 19–28.
[128] Melamed, R.; Cao, X.; Chen, M.; Ma, L.Q., Field assessment of lead immobilization in a contaminated soil after phosphate application; Sci. Total Environ. 2003, 305, 117–127.
[129] Chen, M.; Ma, L.Q.; Singh, S.P.; Cao, R.X.; Melamed, R., Field demonstration of in situ immobilization of soil Pb using P amendments; Adv. Environ. Res. 2003, 8, 93–102.
[130] Impellitteri, C.A., Effects of ph and phosphate on metal distribution with emphasis on as speciation and mobilization in soils from a lead smelting site; Sci. Total Environ. 2005, 345, 175–190.
[131] Scheckel, K.G.; Ryan, J.A.; Allen, D.; Lescano, N.V., Determining speciation of Pb in phosphate-amended soils: method limitations; Sci. Total Environ. 2005, 350, 261–272.
[132] Yang, J.; Mosby, D., Field assessment of treatment efficacy by three methods of phosphoric acid application in lead-contaminated urban soil; Sci. Total Environ. 2006, 366, 136–142.
[133] Hodson, M.E.; Valsami-Jones E.; Cotter-Howells, J.D., Bonemeal additions as a remediation treatment for metal contaminated soil; Environ. Sci. Technol. 2000, 34, 3501–3507.
[134] Deydier, E.; Guilet, R.; Cren, S.; Pereas, V.; Mouchet, F.; Gauthier, L., Evaluation of meat and bone meal combustion residue as lead immobilizing material for in situ remediation of polluted aqueous solutions and soils: Chemical and ecotoxicological studies; J. Hazard. Mater. 2007, 146, 227–236.
[135] Hettiarachchi, G.M.; Pierzynski, G.M.; Ransom, M.D., In situ stabilization of soil lead using phosphorus; J. Environ. Qual. 2001, 30, 1214–1221.
[136] Scheckel, K.G.; Ryan, J.A., In vitro formation of pyromorphite via reaction of Pb sources with soft-drink phosphoric acid; Sci. Total Environ. 2003, 302, 253–265.
[137] Arnich, N.; Lanhers, M.-C.; Laurensot, F.; Podor, R.; Montiel, A.; Burnel, D. In vitro and in vivo studies of lead immobilization by synthetic hydroxyapatite; Environ. Pollut. 2003, 124, 139–149.
[138] Chrysochoou, M.; Dermatas, D.; Grubb, D.G., Phosphate application to firing range soils for Pb immobilization: the unclear role of phosphate; J. Hazard. Mater. 2007, 144, 1–14.
[139] Laperche, V.; Logan, T.J.; Gaddam, P.; Traina, S.J., Effect of apatite amendments on plant uptake of lead from contaminated soil; Environ. Sci. Technol. 1997, 31, 2745–2753.
[140] Boisson, J.; Ruttens, A.; Mench, M.; Vangronsveld, J., Evaluation of hydroxyapatite as a metal immobilizing soil additive for the remediation of polluted soils. Part 1. influence of hydroxyapatite on metal exchangeability in soil, plant growth and plant metal accumulation; Environ. Pollut. 1999, 104, 225–233.
[141] Shiralipour, A.; Ma, L.Q.; Cao, R.X., Effects of compost on arsenic leachability in soils and arsenic uptake by a fern; Report #02–04, University of Florida, Gainesville, FL, 2002.
[142] Cao, X.; Ma, L.Q., Effects of compost and phosphate on plant arsenic accumulation from soils near pressure-treated wood; Environ. Pollut. 2004, 132, 435–442.
[143] Scheckel, K.G.; Impellitteri, C.A.; Ryan, J.A., McEvoy, T., Assessment of a sequential extraction procedure for perturbed lead-contaminated samples with and without phosphorus amendments; Environ. Sci. Technol. 2003, 37, 1892–1898.
[144] Brown, S.; Chaney, R.; Hallfrisch, J.; Ryan, J.A.; Berti, W.R., In situ soil treatments to reduce the phyto- and bioavailability of lead, zinc and cadmium; J. Environ. Qual. 2004, 33, 522–531.

[145] Stevenson, F.J., Humus Chemistry: Genesis, Composition, Reactions, 2nd edn; John Wiley & Sons, Inc. New York, 1994.
[146] McBride, M.; Sauve, S.; Hendershot, W., Solubility control of Cu, Zn, Cd and Pb in contaminated soils; Eur. J. Soil Sci. 1997, 48, 337–346.
[147] Antoniadis, V.; Alloway, B.J., The role of dissolved organic carbon in the mobility of Cd, Ni and Zn in sewage sludge-amended soils; Environ. Pollut. 2002, 117, 515–521.
[148] Jones, D.R.; Chapman, B.M. Wetlands to treat AMD – facts and fallacies; in: Grundon, N.J., Bell, L.C. (eds), Proceedings of the 2nd Australian Acid Mine Drainage Workshop; Australian Centre for Minesite Rehabilitation Research, Brisbane, 1995, pp. 127–145.
[149] Ledin, M.; Krantz-Rulcker, C.; Allard, B., Zn, Cd and Hg accumulation by microorganisms, organic and inorganic soil components in multi-compartment systems; Soil Biol. Biochem. 1996, 28, 791–799.
[150] Brown, P.A.; Gill, S.A.; Allen, S.J., Metal removal from wastewater using peat; Water Res. 2000, 34, 3907–3916.
[151] Järvan, M., Available plant nutrients in growth substrate depending on various lime materials used for neutralizing bog peat; Agron. Res. 2004, 2, 29–37.
[152] US Environmental Protection Agency; Ocean Dumping Ban Act of 1988.
[153] Commission of European Communities; Council Directive 1999/31/EC of 26 April 1999 on the Landfill of Waste.
[154] Commission of European Communities; Council Directive 86/278/EEC of 4 July 1986 on the Protection of the Environment, and in Particular of the Soil, When Sewage Sludge Is Used in Agriculture.
[155] US Environmental Protection Agency; Technical Support Document for Land Application of Sewage Sludge; Vol. 1; Washington, DC; publ. 822/R-93–00a, 1992.
[156] Environmental Protection Agency Victoria; Guidelines for Environmental Management. Biosolids Land Application; Australia; publ. 943, ISBN 0 7306 7641 2, 2004.
[157] Albiach, R.; Canet, R.; Pomares, F.; Ingelmo, F., Organic matter components, aggregate stability and biological activity in a horticultural soil fertilized with different rates of two sewage sludges during ten years; Bioresource Technol. 2001, 77, 109–114.
[158] Debosz, K.; Petersen, S.O.; Kure, L.K.; Ambus, P., Evaluating effects of sewage sludge and household compost on soil physical, chemical and microbiological properties; Appl. Soil Ecol. 2002, 19, 237–248.
[159] Kizilkaya, R., Cu and Zn accumulation in earthworm *Lumbricus terrestris* L. in sewage sludge amended soil and fractions of Cu and Zn in casts and surrounding soil; Ecol. Eng. 2004, 22, 141–151.
[160] Merrington, G.; Oliver, I.; Smernik, R.J.; McLaughlin, M.J., The influence of sewage sludge properties on sludge-borne metal availability; Adv. Environ. Res. 2003, 8, 21–36.
[161] McBride, M.B., Toxic metals in sewage sludge-amended soils: has promotion of beneficial use discounted the risks?; Adv. Environ. Res. 2003, 8, 5–19.
[162] Harrison, E.Z.; McBride, M.B.; Bouldin, D.R., Land application of sewage sludges: an appraisal of the US regulations; Int. J. Environ. Pollut. 1999, 11, 1–36.
[163] Swanson, R.L.; Bortman, M.L.; O'Connor, T.P.; Stanford, H.M., Science, policy and the management of sewage materials. The New York City experience; Mar. Pollut. Bull. 2004, 49, 679–687.
[164] Caravaca, F.; Alguacil, M.M.; Figueroa, D.; Barea, J.M.; Roldán, A., Re-establishment of Retama Sphaerocarpa as a target species for reclamation of soil physical and biological properties in a semi-arid Mediterranean area; Forest Ecol. Manage. 2003, 182, 49–58.
[165] Maddocks, G.; Lin, C.; McConchie, D., Effects of Bauxsol^{TM} and biosolids on soil conditions of acid-generating mine spoil for plant growth; Environ. Pollut. 2004, 127, 157–167.
[166] Brown, S.; Chaney, R.L.; Sprenger, M.; Compton, H., Soil–plant–animal pathway. Soil remediation using biosolids. Part I. BioCycle 2002, 43, 41–44.
[167] Brown, S.; Chaney, R.L.; Sprenger, M.; Compton, H., Soil–animal pathway. Assessing impact to wildlife at biosolids remediated sites. Part II. BioCycle 2002, 43, 50–58.
[168] Conder, J.M.; Lanno, R.P.; Basta, N.T., Assessment of metal availability in smelter soil using earthworms and chemical extractions; J. Environ. Qual. 2001, 30, 1231–1237.

[169] Calace, N.; Campisi, T.; Iacondini, A.; Leoni, M.; Petronio, B.M.; Pietroletti, M., Metal-contaminated soil remediation by means of paper mill sludges addition: chemical and ecotoxicological evaluation; Environ. Pollut. 2005, 136, 485–492.

[170] Battaglia, A.; Calace, N.; Nardi, E.; Petronio, B.M.; Pietroletti, M., Paper mill sludge–soil mixture: kinetic and thermodynamic tests of cadmium and lead sorption capability; Microchem. J. 2003, 75, 97–102.

[171] Battaglia, A.; Calace, N.; Nardi, E.; Petronio, B.M.; Pietroletti, M., Reduction of Pb and Zn bioavailable forms in metal polluted soils due to paper mill sludge addition effects on Pb and Zn transferability to barley; Bioresource Technol. 2007, 98, 2993–2999.

[172] Bolan, N.S.; Adriano, D.C.; Curtin, D. Soil acidification and liming interactions with nutrient and heavy metal transformation and bioavailability; Adv. Agron. 2003, 78, 215–272.

[173] Pérez-de-Mora, A.; Madejón, E.; Burgos, P.; Cabrera, F., Trace element availability and plant growth in a mine-spill contaminated soil under assisted natural remediation; I. Soils. Sci. Total Environ. 2006, 363, 28–37.

[174] Geebelen, W.; Sappin-Didier, V.; Ruttens, A. *et al.*, Evaluation of cyclonic ash, commercial Na-silicates, lime and phosphoric acid for metal immobilisation purposes in contaminated soils in Flanders (Belgium); Environ. Pollut. 2006, 144, 32–39.

[175] Pantsar-Kallio, M.; Reinikainen, S.-P.; Oksanen, M., Interactions of soil components and their effects on speciation of chromium in soils; Anal. Chim. Acta 2001, 439, 9–17.

[176] Seaman, J.C.; Arey, J.S.; Bertsch, P.M., Immobilization of nickel and other metals in contaminated sediments by hydroxyapatite addition; J. Environ. Qual. 2001, 30, 460–469.

[177] Rai, U.N.; Pandey, S.; Sinha, S.; Singh, A.; Saxena, R.; Gupta, D.K., Revegetating fly ash landfills with *Prosopis Juliflora* L.: impact of different amendments and *Rhizobium* inoculation; Environ. Int. 2004, 30, 293–300.

[178] Lombi, E.; Hamon, R.E.; Wieshammer, G.; McLaughlin, M.J.; McGrath, S.P., Assessment of the use of industrial by-products to remediate a copper- and arsenic-contaminated soil; J. Environ. Qual. 2004, 33, 902–910.

[179] Madejón, E.; Pérez de Mora, A.; Felipe, E.; Burgos, P.; Cabrera, F., Soil amendments reduce trace element solubility in a contaminated soil and allow regrowth of natural vegetation; Environ. Pollut. 2006, 139, 40–52.

[180] Hartley, W.; Lepp, N.W., Effect of in situ soil amendments on arsenic uptake in successive harvest of ryegrass (*Lolium Perenne* cv Elka) grown in amended As-polluted soils; Environ. Pollut. 2008, 157, 1030–1040.

[181] Wenzel, W.W.; Kirchbaumer, N.; Prohaska, T.; Stingeder, G.; Lombi, E.; Adriano, D.C., Arsenic fractionation in soils using an improved sequential extraction procedure; Anal. Chim. Acta 2001, 436, 309–323.

[182] Lombi, E.; Sletten, R.S.; Wenzel, W.W., Sequentially extracted arsenic from different size fractions of contaminated soils; Water, Air Soil Pollut. 2000, 124, 319–332.

[183] Magalhães, M.C.F., Arsenic – an environmental problem limited by solubility; Pure Appl. Chem. 2002, 74, 1843–1850.

[184] Garrido, F.; Illera V.; Garcia-Gonzalez, M.T., Effect of the addition of gypsum- and lime-rich industrial by-products on Cd, Cu and Pb availability and leachability in metal-spiked acid soils; Appl. Geochem. 2005, 20, 397–408.

[185] Aguilar-Carrillo, J.; Barrios, L.; Garrido, F.; García-González, M.T., Effects of industrial by-product amendments on As, Cd and Tl retention/release in an element-spiked acidic soil; Appl. Geochem. 2007, 22, 1515–1529.

[186] Winterhalder, K., Environmental degradation and rehabilitation of landscape around Sudbury, a major mining and smelting area. Environ. Rev. 1996, 4, 185–224.

[187] Mench, M.; Renella, G.; Gelsomino, A.; Landi, L.; Nannipieri, P., Biochemical parameters and bacterial species richness in soils contaminated by sludge-borne metals and remediated with inorganic soil amendments; Environ. Pollut. 2006, 144, 24–31.

[188] Renella, G.; Landi, L.; Ascher, J. *et al.*, Long-term effects of aided phytostabilisation of trace elements on microbial biomass and activity, enzyme activities, and composition of microbial community in the Jales contaminated mine spoils; Environ. Pollut. 2008, 152, 702–712.

Section 4

Characteristics and Behaviour of Individual Elements

16

Arsenic and Antimony

Yuji Arai

*Department of Entomology, Soils and Plant Sciences, Clemson University,
Clemson, South Carolina, USA*

16.1 Introduction

Arsenic (As) and antimony (Sb) are chalcophilic metalloids that share numerous similarities in biogeochemical properties. This chapter reviews the chemical properties, environmental/geochemical reactions, phytoaccumulation and toxicology of the two elements.

Arsenic belongs to Group 15 in the periodic table. The electron configuration is [Ar]$3d^{10}$ $4s^2$ $4p^3$, and it has four major oxidation states (+5, +3, 0, and −3). Arsenic has multiple isotopes, and most of them have very short half-life ($t_{1/2}$) of microseconds to milliseconds. Of these, ^{71}As, ^{72}As, ^{73}As, ^{74}As and ^{74}As have $t_{1/2}$ of approximately 65 h to 80 days, and ^{75}As is the only stable isotope. In the soil and water environment, inorganic As is mainly present in two oxidation states (+3 and +5). Arsenite, As(III), commonly exists as arsenious acid, As(OH)$_3$, in reduced environments. Conversely, an oxidized environment contains more arsenate, As(V), as arsenic acid (for example, HAsO$_4^{2-}$). Antimony also belongs to Group 15 in the periodic table. The electron configuration is [Kr] $5s^2$ $4d^{10}$ $5p^3$, and it has four major oxidation states (+5, +3, 0 and −3). The most common oxidation states in low temperature environments are +5 and +3. Although Sb has multiple isotopes, most of them have short $t_{1/2}$ of <1 day. Only a few of its isotopes have $t_{1/2}$ of more than a few days: ^{122}Sb (2.7 days), ^{124}Sb (60.2 days), ^{126}Sb (12.3 days), and ^{127}Sb (3.85 days). Antimony has two stable isotopes, ^{121}Sb and ^{123}Sb. Antimony-125 is a fission product of ^{235}U, and has a half life of ~2.76 years. While antimonite, Sb(III), commonly exists as antimonous acid,

$Sb(OH)_3$, in reduced environments, antimonate, $Sb(V)$, as $Sb(OH)_6^-$ is commonly present in oxidized environments.

These elements have commonly been used in industrial, medical, and agricultural applications. One of the common industrial uses of As is in tanneries for preserving animal hides. Arsenic has also been used in agriculture as a component of insecticides, herbicides, fungicides, algaecides, sheepdips, wood preservatives, deworming agents for livestock, and vaccinations for poultry and swine [1,2].

Antimony was historically used in the gold extraction processes. The Sb sulfide stibnite (Sb_2S_3) is the principal ore of antimony, and has a brilliant metallic luster. The color was used for cosmetic purposes (for example, eyeliner) in ancient times. In more recent years, stibnite is used in the semiconductor industry to produce diodes and infrared detectors. It is also commonly used as a mixing compound with lead to strengthen the hardness of batteries, alloys, bullets, cable sheathing, matches, medicines, plumbing, and soldering [3,4]. It can be found in flame-retardants in plastics and textiles, and medical drugs to treat some tropical diseases (for example, visceral leishmaniasis).

Arsenic and Sb in soil and water environments originate from indigenous sources (for example, mineral weathering) and anthropogenic inputs (for example, mining, industrial processes, and pesticide application). Due to their undesirable toxicological effects, these metalloids pose threats to human health and the wider environment. In particular, As contamination occurs worldwide, with hotspots in Thailand, Taiwan, mainland China, Argentina, and Chile. There are also many watersheds and drinking water sources in the United States affected by As contamination. In some parts of South East Asia, Bangladesh and West Bengal, the total As level can be as high as 150 $\mu g \, l^{-1}$ in drinking water [5], while the maximum concentration level (MCL) set by the US Environmental Protection Agency (USEPA) and the World Health Organization (WHO) is 10 $\mu g \, l^{-1}$ [6,7].

The maximum drinking water limit for Sb was set to 5 $\mu g \, l^{-1}$, though this is considered provisional until a greater understanding of its toxicity is achieved [7]. While the MCL of Sb is low, typical Sb concentrations in uncontaminated waters are less than 1 $\mu g \, l^{-1}$ [7]. However, there are some reports of elevated Sb levels in natural geothermal systems ranging from 500 to 100, 000 $mg \, kg^{-1}$ [8–11].

Arsenic can have negative impacts on both human and ecological health, because of the carcinogenicity (for example, skin, lung, and bladder), phytotoxicity and biotoxicity [12,13]. Arsenate, As(V) is known to replace phosphate in substituted monosaccharides, along with inhibiting ATP synthesis by uncoupling oxidative phosphorylation, leading to the breakdown of energy metabolism [14]. Arsenite, As(III), is generally more toxic than As(V) due to its preferential reaction with sulfhydryl groups in mammalian enzymes [15], resulting in inhibition of the pyruvate and succinate oxidation pathways and the tricarboxylic acid cycle, impaired gluconeogenesis, and reduced oxidative phosphorylation. Labile As(III) as arsenious oxide (As_2O_3) is absorbed through the lungs and intestines, and biochemically acts to coagulate proteins, form complexes with coenzymes, and inhibit the production of the essential enzyme andenosine triphosphate (ATP) in metabolic processes [16]. Arsine gas is known to cause a dose-dependent intravascular hemolysis and multisystem cytotoxicity in rodents (for example, [17]).

The sudden infant death syndrome drew public attention to Sb toxicity [18,19]. It was hypothesized that stibine (SbH_3) gas formed by fungal transformation of fire retardants containing Sb in cot mattresses might be a cause of sudden death syndrome [19]. However, there is insufficient epidemiological evidence to support the theory [20]. In adult humans, inhalation of stibine gas is known to cause hemolytic anemia, hemolysis, myoglobinuria, hematuria, renal failure, nausea, vomiting, and headache. Pneumoconiosis, other respiratory effects, and cutaneous effects (for example, a transient and pustules known as 'antimony spots') have been reported in workers occupationally exposed to dusts and fumes containing Sb [21,22]. Cardiovascular mortality and morbidity have been attributed to inhalation of the Sb trisulfide [23]. Patients treated with antileishmaniasis agents, sodium stibogluconate, showed side-effects of renal tubular acidosis, thrombocytopenia, and pancreatitis [24–26]. However, the mechanism of toxicity of Sb(V) compounds is not clearly understood.

16.2 Geogenic Occurrence

16.2.1 Arsenic

The average total As content in uncontaminated soils is approximately 5 mg As kg^{-1} [27,28]. Volcanic soils may contain up to 20 mg As kg^{-1} [15]. Due to the application of As-containing pesticides and defoliants, the As content of contaminated soils can range up to 2553 mg As kg^{-1} [29]. The average As level in agricultural soils in 12 US states was approximately 165 mg As kg^{-1} [30].

Naturally occurring As is found in about 245 mineral species including arsenides, sulfides, sulfosalts, and oxidation products such as oxides, arsenites, and arsenates. These are generally associated with basin-filled deposits of alluvial-lacustrine origin, and volcanic deposits [2,31]. Naturally occurring As-S minerals are arsenopyrite (FeAsS), enargite (Cu_3AsS_4), orpiment (As_2S_3), and realgar (AsS). In terrestrial environments, inorganic forms are generally predominant over organic forms. The speciation of the inorganic forms (arsenite and arsenate) is highly affected by pH and redox conditions. Most of the organic fractions are associated with methyl groups (–CH_3) such as methanearsonic acid (MAA) and dimethylarsinic acid (DMAA) [32]. In agricultural drainage/evaporation ponds, approximately 40 % and 15 % of total As are found as DMAA and MAA, respectively [33]. Eventually organic As species are converted into CO_2 and inorganic As by oxidative degradation, or into volatile As compounds or arsine (AsH_3) gas by reduction and further methylation. Due to the low solubility of the volatile compounds, they readily complex with atmospheric particulates before being deposited on the ground [2].

16.2.2 Antimony

In geothermal systems, antimony is commonly associated with gold and sulfur [34]. The average total Sb content in uncontaminated soils is in the order of a few micrograms per kilogram [35,36]. Martin and Whitefield [37] reported the average Sb concentration in soils of approximately 1 μg kg^{-1}. The majority of natural Sb sources are mineral deposits from hot springs, volcanic ore deposits, boreholes, and gold mines in schist [9,11].

Elevated concentrations (8–61 mg Sb kg^{-1}) of Sb have been observed near smelting plants and outfalls of sewage and fertilizer facilities [38–41]. There is a report about Sb-enriched sandstone from Zimbabwe containing total Sb of up to $\sim$5000 mg kg^{-1} [42]. Areas near old mining and Cu and Pb smelting can have residual Sb levels of 103–260 mg kg^{-1} [39,43,44].

Naturally occurring antimony is commonly combined with S, Pb, and As to form over 100 different minerals. In hydrothermally altered rocks, Sb is often found in stibnite [Sb$_2$S$_3$] and/or kermisite [Sb$_2$S$_2$O], guettardite, [Pb(Sb,As)$_2$S$_4$], cervantite [Sb(III)Sb(V)O$_4$] and para-docrasite [Sb(III)$_2$(Sb(III),As)$_2$]. Of these minerals, only stibnite (SbS$_3$) is commercially mined as a source for metallic antimony. A trace quantity of Sb is generally present in Ag, Cu, and Pb ores. Like As, Sb can be volatilized as SbH$_3$ or methylated species in the environment [45].

16.3 Sources of Soil Contamination

Arsenic contamination sources originate from both indigenous and anthropogenic inputs, including atmospheric deposition. Weathering of As-containing primary minerals, the main natural source of contamination, could yield an annual global input of 45 000 Mg As y^{-1} [14]. While shales and coal contain mean As concentration of 13 mg kg^{-1} and 25 mg kg^{-1}, respectively, sedimentary rocks could contain As concentrations as high as 2000 mg kg^{-1}. The major anthropogenic sources of As originate from industrial processes such as smelting of As-containing ores and by-products of fossil fuel combustion (for example, fly coal ash), and agricultural uses. Approximately 60 % of the anthropogenic As emissions are largely associated with two sources: Cu-smelting and coal combustion [46]. The total As input into the atmosphere including volcanic activity, wood preservatives grassland fires marine aerosols and biogenic liberation is estimated to be approximately 2.8×10^7 kg As y^{-1} [46,47].

Agricultural applications of As include insecticides, herbicides, fungicides, algaecides, sheepdips, wood preservatives, deworming and antibacterial agents for livestock, vaccination for poultry and swine, and recycled poultry litter [1,48]. Poultry litter containing an organic As compound (Roxasone) as an antibacterial agent has been used in agriculture at the rate of 8.96–20.16 Mg ha^{-1}, with total As inputs in three US states (Delaware, Maryland, and Virginia) estimated at between 20 and 50 Mg As y^{-1}[49].

Most Sb soil pollution originates from mining and smelting industries and from the outfalls of sewage, shooting ranges and fertilizer facilities [1,38–41]. In the United States alone, approximately 5.5×10^6 kg of Sb-associated compounds were released into the aquatic and terrestrial environments between 1993 and 1997, with more than 97 % released onto the land [50]. In addition to the terrestrial inputs, significant amounts of air-borne Sb have also contributed to overall Sb inputs. Global atmospheric input of Sb can be as high as 6.1×10^3 Mg y^{-1}. While $\sim$60 % of the atmospheric inputs are from anthropogenic sources including the incineration of waste, fossil fuel combustion, and the smelting of metal, the rest are from natural sources including wind-blown dust, sea salt spray, volcanic activity, and forest fires [51–54].

16.4 Chemical Behavior in Soils

16.4.1 Arsenic Speciation and Solubility

In general, the solubility of inorganic As increases in reduced environments. Adsorbed As can be released through reductive dissolution of the adsorbent, for example conversion from the ferric ion to the ferrous ion. A direct reduction of As(V) to As(III) also increases the solubility of total As due to weak As(III) sorption on soil components. Arsenic can form solubility products with calcium, aluminum, sulfur and iron in the soil/water environment. The solubility constant values (10^{-11}) for iron and aluminum arsenates are smaller compared to that of calcium arsenate (10^{-5}), indicating that iron and aluminum control the availability of As in soils [29]. In oxic environments, the predominant chemical species is arsenate, which readily binds to clay minerals. pH also affects the solubility of arsenate. While arsenate ions precipitate out with trivalent metals, for example Al(III) and Fe(III)) to form scorodite and amorphous aluminum arsenate at acidic pH, calcium arsenate precipitates (for example, $Ca_3(AsO_4)_2$) at alkaline pH conditions [55]. Barium arsenate ($Ba_3(AsO_4)_2$) is known to be most insoluble of the As precipitates. In reduced systems, arsenite forms solubility products with sulfides (for example, realgar, AsS, and orpiment, As_2S_3).

In low-temperature geochemical environments, inorganic As is mainly present in two oxidation states (+3 and +5). Figure 16.1 shows the effects of redox potential

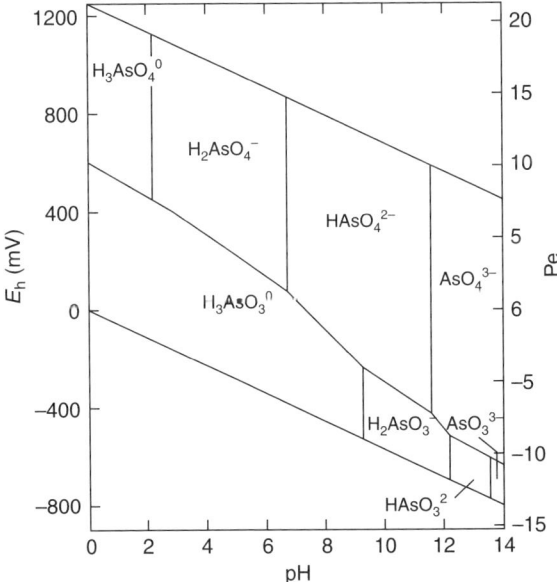

Figure 16.1 E_h–pH diagram for aqueous As species in the system As–O_2–H_2O at 25 °C and 1 bar total pressure (Reprinted from Applied Geochemistry, 17, P.L. Smedley and D.G. Kinniburgh, A review of the source, behavior and distribution of arsenic in natural waters, 517–568. Copyright 2002, with permission from Elsevier [56].)

(E_h) and pH on aqueous As speciation. Arsenite, As(III), commonly exists as arsenious acid, As(OH)$_3$, in reduced environments (for example, E_h -200 mV $< +300$ mV over a pH range of 4 to 8) [2]. It has weak acid characteristics similar to boric acid (Equation 16.1):

$$\text{As(OH)}_3 + \text{H}_2\text{O} \longleftrightarrow \text{As(OH)}_4^- + \text{H}^+, \quad \log K = -9.29 \quad (16.1)$$

A fully protonated form is expected to be predominant in reduced soil environments due to the high pK values (p$K_1 = 9.22$, and p$K_2 = 12.13$). Conversely, oxidized environments contain more arsenate, As(V), as arsenic acid. Dissociation constants of arsenic acid (p$K_1 = 2.20$, p$K_2 = 6.97$, and p$K_3 = 13.4$) predict deprotonated forms of arsenic (H$_2$AsO$_4^-$ and HAsO$_4^{2-}$) in acidic to neutral environments.

In neutral oxygenated aquatic systems, As(III) oxidation has been reported to have a half-life of 1 year [57], and no oxidation occurred over a 37-day period in distilled, deionized water [58]. Arsenite oxidation in a 0.0005 M NaCl solution was stable below pH 9 after 72 h [59].

In soil and water environments, As solubility is controlled not only by the redox potential but also by the types of sorbent available, that is Fe(III) and Mn(IV) hydroxides, and the As minerals themselves. In general, reducing conditions and/or the presence of reductants readily promote sorbent dissolution, causing the release of sorbed As. Increased As solubility in reduced soils has been reported by many researchers [60–63]; however, decreased As solubility in long-term flooded soils has also been observed [63,64]. Resorption of As on solids [65] and co-precipitation of Mn$_3$(AsO$_4$)$_2$-like phases [61] were suggested to explain the decreased As solubility under long-term and moderately reduced conditions (0–100 mV).

The influence of redox potential on the solubility of As from several metal-arsenate minerals was investigated in a Santa silt loam soil from northern Idaho using an equilibrium thermodynamic study coupled with XANES (X-ray absorption near edge structure) analysis of arsenic oxidation [66]. Arsenic solubility decreased under oxidized conditions as: Ca$_3$(AsO$_4$)$_2$ = Na$_2$AsO$_4$ > AlAsO$_4$ > Mn$_3$(AsO$_4$)$_2$ > Fe(AsO$_4$)$_2$. In contrast, under anoxic conditions (<0 mV), the relative solubilities were: Fe(AsO4)$_2$ > Ca$_3$(AsO$_4$)$_2$ = Na$_2$AsO$_4$ > AlAsO$_4$ > Mn$_3$(AsO$_4$)$_2$. XANES analysis showed that aluminum arsenate is rapidly converted to solid-phase arsenite, indicating the most susceptible metal-arsenate phase occurs under reducing conditions [66].

16.4.2 Arsenic Retention in Soils

The retention of As on soils is highly dependent on the physicochemical properties of the soils, especially the nature and abundance of crystalline and amorphous iron and aluminum oxides, and clay and calcium contents. Several researchers have reported that As(III) and As(V) retention in soils are highly associated with ammonium oxalate-extractable iron (Fe$_{ox}$) and/or aluminum (Al$_{ox}$) and dithionite–citrate–bicarbonate (DCB)-extractable iron and/or aluminum [67–74]. High As(V) retention in calcium rich soils with pH > 7 has also been reported [75]. The sorption of As(III) and As(V) on three California soils was studied at varying As concentration, pH, and ionic strength [69]. The soils with the highest DCB-extractable Fe and the highest clay content had the highest affinity for As(III) and

As(V). Sorption isotherms showed that As(V) species sorbed more strongly than As(III) under most conditions.

Ionic strength (I) can have an influence on both the rate of the elementary reaction and the type of surface complexation (inner- and/or outer-sphere complexation) [76,77]. Gupta and co-workers [78] investigated the effects of ionic strength on As(III) and As(V) sorption kinetics on alumina. As(V) sorption kinetic experiments (pH $\sim$6.6) showed that an initial sorption rate in seawater was much slower than that in (nonsaline) water. The sorption of As(III) at pH $\sim$8 in seawater was also slower than that in water. Arai et al. [79] also reported two different I dependencies on As reactivity on alumina; while As(V) was insensitive to changes in I at pH 4–8, As(III) sorption decreased with increasing I and pH.

Arsenate sorption on soil minerals is influenced by the surface charge density of solids and the speciation of As(V). In general, As(V) sorption on inorganic minerals increases with decreasing pH of bulk fluid (pH_b) due to (i) the negatively charged chemical species, that is $HAsO_4^{2-}$ and (ii) the positively charged mineral surfaces, when pH_b < PZC (point of zero charge) of the solids. The first dissociation constant of As(V) is approximately 2.2, followed by constants of $\sim$7 and $\sim$12.8. At most environmental pH values (4–8), the arsenic species are predominantly in deprotonated forms (negatively charged species), and the charge properties of metal oxides are positive due to the PZC of the solids (that is, 6.5–8.5 for iron oxides, 8.2–9.1 for aluminum oxides, an exception is manganese oxides, for example birnessite $\sim$2.8). Therefore, As(V) is expected to sorb onto metal oxide surfaces strongly via electrostatic interaction when pH_b − PZC is less then zero, and to predominantly sorb via ligand exchange when pH_b − PZC is greater than zero.

pH-dependent sorption behavior of As(V) has been observed for ferrihydrite, hematite and aluminum oxyhydroxides [80–85]. In general, As(V) sorption is maximized at acidic pH, and gradually decreases with increasing pH. Figure 16.2 shows the pH-dependent As(V) sorption on hematite surfaces.

As metal oxyhydroxides (for example, ferrihydrite, hematite, and schwertmannite) exhibit a strong affinity for As(V) at acidic pH values [86–88], various phyllosilicate minerals also show a similar pH-dependent As(V) sorption behavior. Arsenate sorption on alumina, kaolinite, illite, and montmorillonite gradually increases from pH 3 to 5, and then decreases with increasing pH up to 10 [75,82,85,89]. Conversely, the As(V) sorption envelope on quartz showed no significant adsorption (less than 15 % of net adsorption) between pH 2.8 and 9.5 [85]. Arai and co-workers [79] investigated As(V) sorption complexes forming at the γ-Al_2O_3–water interface using X-ray absorption spectroscopy (XAS). The XAS data indicates that As(V) forms inner-sphere complexes with a bidentate binuclear configuration, as evidenced by an As(V)–Al bond distance of $\sim$ 3.11–3.15 Å. A similar molecular configuration was reported on allophone surfaces. Several other XAS studies indicated that arsenate inner-sphere sorption mechanisms on metal hydroxide surfaces [91,93–96]. The results of inner-sphere surface species are consistent with the findings of Arai and co-workers' study [79]. The results of spectroscopic investigations are summarized in Table 16.1.

While the sorption of As(V) on metal oxides and phyllosilicate minerals is highly dependent on the relation between pH of the bulk solution and the PZC of solids, As(III) sorption is more dependent on its speciation. At most environmental pH values (4–8),

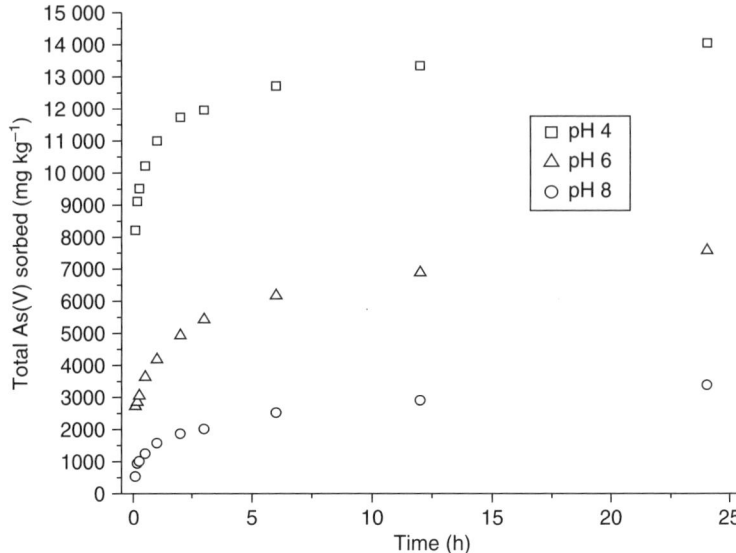

Figure 16.2 Arsenate sorption kinetics at the hematite–water interface, with suspension density 4g l^{-1}, ionic strength = 0.01 M NaCl, and [As(V)]$_{total}$ = 1.35 mM.

arsenite (as As(OH)$_3$)(aq) does not have any charge, and the negative charge increases with increasing pH due to its weak acidity (pK_1 = 9.22, and pK_2 = 12.13). As(III) sorption on metal oxides and phyllosilicate minerals increases with increasing pH, whereas As(V) sorption decreases with increasing pH. Figure 16.3 shows the contrasting pH dependent sorption behavior of As(III) and As(V) on γ-Al$_2$O$_3$ surfaces.

The As(III) sorption envelope and isotherms on goethite, kaolinite, illite, montmorilloinite, and aluminum hydroxide were investigated earlier [59,83,109]. Although there were slight differences in As(III) uptake in different sorbents, As(III) sorption generally increased with increasing pH from 4 to 6, and then gradually decreased. A similar pH-dependent As(III) sorption was also observed in goethite and in arid soils that had high (> 4000 mg kg^{-1}) DCB-extractable iron [97]. Raven and co-workers [86] reported the biphasic As(III) sorption reactions on ferrihydrite; the initial fast sorption was nearly completed within the first few hours, and followed by slow uptake. At initial As concentrations of 0.27–13.3 mol As mol^{-1} Fe ferrihydrite, the As(III) sorption was greater at higher pH. At the highest initial arsenite concentration of 13.3 mol As kg^{-1} of ferrihydrite, a distinct sorption maximum was observed for arsenite adsorption at approximately pH 9.0, which corresponds closely to the first pK_a (~9.2) of H$_3$AsO$_3$.

The presence of As(III) inner-sphere surface species on aluminum oxide, goethite, and birnessite has been reported in several spectroscopic studies (see Table 16.2). While most mineral surfaces retain As(III) via ligand exchange mechanisms, there are a few reports about the formation of As(III) outer-sphere surface complexes on aluminum oxide and ferrihydrite surfaces [79,101].

The oxidation of As(III) in geomedia containing oxidants (for example, manganese oxides) has been reported by several researchers [99,110–113]. Arsenite can be readily

Table 16.1 A summary of sorption mechanisms of arsenate, arsenite, antimonite, and antimonate in geomedia

Adsorbate	Adsorbent	Suggested surface species	References
As(III)	γ-Al$_2$O$_3$	Inner-sphere bidentate and outer-sphere	[79]
As(III)	Goethite	Inner-sphere bidentate	[97]
		Inner-sphere bidentate	[98]
As(III)	Birnessite	Inner-sphere bidentate	[99]
As(III)	Hydrous Mn oxides	Inner-sphere bidentate	[100]
As(III)	Ferrihydrite	Inner-sphere bidentate and outer-sphere	[101]
		Inner-sphere bidentate	[102]
As(III)	Amorphous Al oxides	Outer-sphere	[101]
As(V)	γ-Al$_2$O$_3$	Inner-sphere bidentate	[79]
As(V)	Gibbsite	Inner-sphere bidentate	[100]
		Inner-sphere bidentate	[103]
As(V)	HFO	Inner-sphere bidentate	[104]
	Ferrihydrite		[93,105]
As(V)	Goethite	Inner-sphere bidentate	[93,105]
	Lepidocrocite, and Akagenite		
	Goethite	Inner-sphere bidentate and Inner-sphere monodentate	[94]
	Goethite	Inner-sphere bidentate	[92]
	Goethite	Inner-sphere bidentate	[96]
	Goethite	Inner-sphere bidentate	[98]
As(V)	Amorphous Fe oxides		[101]
As(V)	Hematite	Inner-sphere bidentate	[86]
As(V)	Green rust and Lepidocrocite	Inner-sphere bidentate	[106]
As(V)	Birnessite	Inner-sphere bidentate	[95]
As(V)	Birnessite	Inner-sphere bidentate	[91]
As(V)	Allophane	Inner-sphere bidentate	[90]
Sb(III)	Goethite	Inner-sphere bidentate	[107]
Sb(V)	Goethite	Inner-sphere bidentate	[107]
Sb(V)	Amorphous Fe oxyhydroxide	Inner-sphere	[108]

oxidized to As(V) at the mineral surface and the mineral–water interface either by surface catalysis or direct oxidation. For example, Scott and Morgan [110] reported that the uptake of As(III) by synthetic birnessite (δ-MnO$_2$) increased with decreasing pH, and oxidation products [110]. The reaction, which was independent of oxygen concentration, indicated that birnessite acted as a direct oxidant for As(III). The reaction rate was positively correlated to the molar ratio of MnO$_2$/As(III), and the respective reaction order was 1.5 [111].

Surface-catalyzed As(III) oxidation on birnessite was shown in molecular scale analysis, using XANES [99,112,114]. To explain the reaction mechanisms, several researchers suggest the importance of Mn(IV) and/or Mn(III) in birnessite contributing to the initial reactivity of manganese oxides in the experimental systems [115–118].

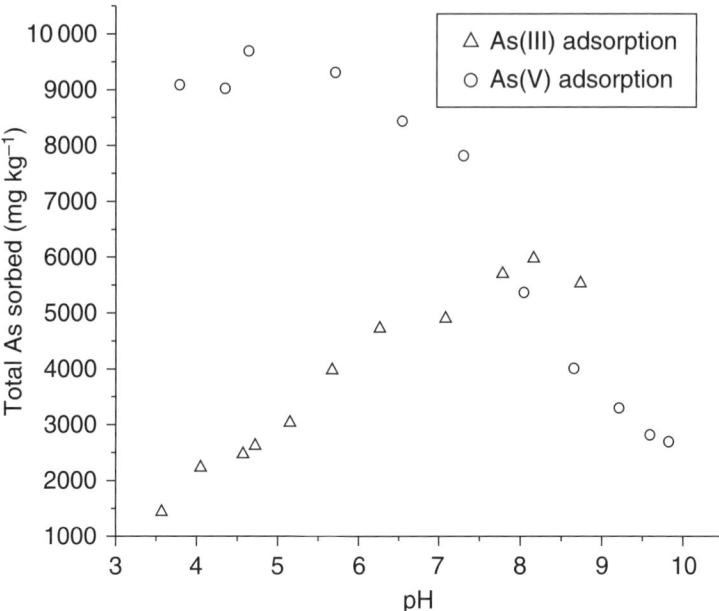

Figure 16.3 *Arsenate and arsenite sorption envelopes on γ-Al_2O_3 (As(III) and As(V))$_{total}$ = 1 mM, suspension density = 5g l^{-1}, and ionic strength = 0.01 M NaCl)*

16.4.3 Arsenic Desorption in Soils

The extent of contaminant release greatly affects the bioavailability and transport processes in aquatic and terrestrial environments. Slow desorption and irreversible/hysteretic reactions have been reported in several studies. Slow As(V) desorption was observed in goethite and aluminum oxide [92,119,120]. The desorption rate was significantly slower than the rate of adsorption under similar reaction conditions. Some have found that incubation times strongly influence the reversibility of adsorbed As(V) from soils and soil components (aluminum oxide, goethite, halloysite, kaolinite, illite, montmorillonite, and chlorite) [113,119–121]. The As(V) recovery decreased with increasing aging time. Arai and Sparks [119] reported that the As(V) release from aluminum oxide decreased with increasing aging time from 3 days to 1 year at pH 4.5 and 7.8. The longer the residence time (3 days–1 year), the greater the decrease in As(V) desorption at both pH values, suggesting irreversible reactions. Lin and Puls [122] also reported the irreversible As(V) sorption on clay mineral surfaces. The As(V) recovery decreased with increasing aging time from 1 to 75 days, and the effect was most pronounced in halloysite and kaolinite.

Several researchers tested the effects of ligands on As(V) release from soil components. O'Reilly and co-workers [92] reported that phosphate was more effective than sulfate in removing As(V). Jackson and Miller compared the ability of phosphate and hydroxyl ions to remove various As compounds (that is, As(III), As(V), dimethylarsenic acid, monomethylarsonic acid, *p*-arsanilic acid, and roxasone) from goethite and amorphous iron oxyhydroxide [123]. Although phosphate released As(III) from goethite surfaces, it was not as effective as hydroxyl ions. They reported that hydroxyl ions removed >70 % of all As compounds on these mineral surfaces. The effect of alkaline pH on As release was also documented by Masscheleyn et al. [61].

16.4.4 Antimony Speciation and Solubility

Like arsenic, the valence state of Sb greatly affects the solubility in the low-temperature geochemical environments. This means the solution speciation of Sb(III) and Sb(V) influences its retention and transport processes in soils. As discussed above, the speciation of oxyanions is influenced by pH. Oxyanion sorption on soil components is a function of both the net surface charge density of the sorbent and the chemical speciation of the sorbate, which in turn are dependent on the pH of the bulk fluid (pH_b). Antimonite (Sb(III)) commonly exists as antimonous acid, $Sb(OH)_3$, in the reduced environment [35,124,125]. It has weak acid characteristics similar to boric acid as shown in Equation (16.2).

$$Sb(OH)_3 + H_2O = Sb(OH)_4^- + H^+, \log K = -11.82 \quad (16.2)$$

A fully protonated form is expected to be predominant under reduced conditions due to the high pK value. In the presence of aqueous sulfide under reduced conditions, thioantimony complexes (for example, $Sb(III)S_2^-$ and $Sb(III)_2S_4^{2-}$) are known to readily form at neutral to alkaline conditions [10,126]. Under reducing conditions, solubility of Sb is limited by the solubility of Sb(III) sulfides (for example, stibnite) and oxides (for example, $Sb(OH)_3$, Sb_2O_3 (valentinite, senarmontite), Sb_2O_4 (cervantite)). Krupka and Serne [127] demonstrated the solubility calculation of these species under the total Sb concentration (Sb_t) of $10^{-14.6}$ mol l^{-1}. The system was undersaturated with respect to these solids (Figure 16.4). When Sb_{total} increases to 10^{-7} mol l^{-1}, the solubility products (Sb_2O_4, Sb_2S_3, Sb_2O_4, and/or $Sb(OH)_3$) readily form at acidic to alkaline pH values under reducing conditions (−500 mV to 750S mV) (Figure 16.5).

In addition to the sulfide complexes, Sb(III) is also known to make complexes with chlorine at high chloride concentrations. Under reducing conditions, the aqueous species of Sb(III) readily complex with dissolved Cl^- to form $SbCl^{2+}$, $SbCl_2^+$, $SbCl_3^0$(aq), and $SbCl_4^-$ in acidic aqueous solutions [128].

Conversely, in oxidized environments antimonite, Sb(V) is the dominant Sb species [35,129]. Due to its larger atomic size, the coordination of Sb(V) is different from that of arsenate. It is octahedrally coordinated with six oxygen atoms. A protonation constant of antimonite is shown in Equation (16.3).

$$Sb(OH)_5 + H_2O = Sb(OH)_6^- + H^+, \log K = -2.47 \quad (16.3)$$

This equation predicts negatively charged species ($Sb(OH)_6^-$) in acidic to neutral environments. Antimonate is very soluble in oxic environments [130]. Although the general thermodynamic prediction of Sb(III) species can be applied to reduced environments, Sb(III) species can also be detected in oxic conditions [35]. Researchers have suggested that the metastability of Sb(III) under oxic conditions may have been linked to biotic processes and/or a slow kinetic effect of Sb(III) oxidation. Similarly, Sb(V) is frequently found in anoxic systems [35]. Metastability of Sb(V) might be attributed to the formation

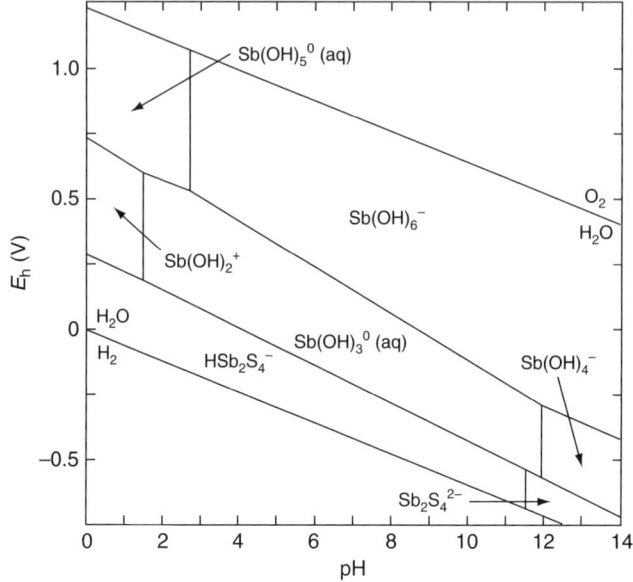

Figure 16.4 E_h–pH diagram of aqueous speciation of antimony (calculated at 25 °C and a concentration of $10^{-14.6}$ mol l^{-1} total dissolved antimony (Reproduced with permission from K.M. Krupka and R.J. Serne, Geochemical factors affecting the behavior of antimony, cobalt, europium, technetium, and uranium in vadose sediments, Pacific Northwest National Laboratory, 2002 [127].)

of Sb(V) thiocomplexes (for example, SbS_4^{3-}) and/or a slow Sb(V) reduction process in anoxic environments. The presence of Sb(V) sulfide complexes under reducing conditions have been reported by several researchers [131–133].

While the aqueous speciation at low concentrations of total dissolved Sb ($<100\ \mu M$) can be explained by the studies mentioned above [131–133], Sb(V) may undergo polymerization when dissolved antimony concentration is >0.001 M under acidic conditions, resulting in polynuclear species such as $Sb_{12}(OH)_{64}^{4-}$ and $Sb_{12}(OH)_{66}^{6-}$ [129]. No polymerized Sb(III) aqueous species have been reported.

16.4.5 Antimony Adsorption and Desorption in Soils

Unlike arsenic, the sorption and desorption processes of Sb in geomedia have not been extensively investigated. However, several researchers have studied the macroscopic Sb(III) and Sb(V) sorption in metal oxyhydroxides, phyllosilicate minerals and soils [108,134–137].

In an earlier investigation, Crecelius et al. [135] reported strong sorption of anionic Sb(V) hydrolytic species in uncontaminated Puget Sound sediments from Washington, USA. The sorption increased with decreasing pH, and the sorbed Sb(V) fraction was associated with extractable iron and aluminum components in the sediments. Ambe [134] investigated the sorption kinetics of Sb(V) on α-Fe_2O_3 surfaces at pH 4. The rate of sorption was proportional to the square of both the concentration of Sb(V)$_t$ and the

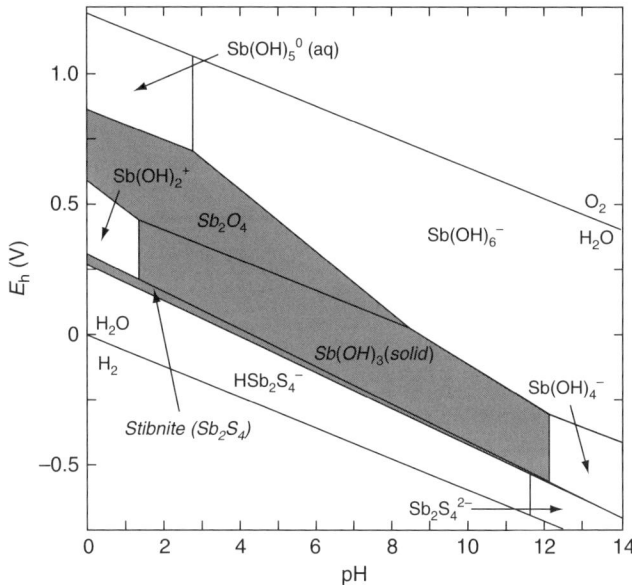

Figure 16.5 E_h–pH diagram of aqueous speciation and solubility products of antimony. (The diagram was calculated at 25 °C and a concentration of 10^{-7} mol/L total dissolved antimony.) Solubility products are shown shaded (Reproduced with permission from K.M. Krupka and R.J. Serne, Geochemical factors affecting the behavior of antimony, cobalt, europium, technetium, and uranium in vadose sediments, *Pacific Northwest National Laboratory*, 2002 [127].)

surface area of α-Fe$_2$O$_3$. The sorption capacity of the α-Fe$_2$O$_3$ sample for Sb(V) ions was ~ 7 mg g^{-1}, which is much less than the calculated value of monolayer coverage (45 mg g^{-1}). Subsequent desorption studies indicated the slow desorption of Sb(V) from the α-Fe$_2$O$_3$ surfaces at pH 4–10.

Tighe *et al.* [136] studied Sb(V) sorption in two organic rich soils, and two model phases mimicking those dominant in these soils, namely a solid humic acid and an amorphous Fe(OH)$_3$, at pH 2.5–7 [136]. Antimonate sorption increased with decreasing pH in these soils and in the humic acid. In contrast, the amorphous ferric oxyhydroxide showed less pH dependency of Sb(V) sorption within the pH range examined.

Effects of ionic strength and pH on Sb sorption were investigated on goethite surfaces [137]. Antimonate sorption on goethite decreases with increasing pH from 4 to 10. However, the uptake of Sb(V) was reduced at pH > 6 when ionic strength was increased from 0.01 M to 0.1 M KClO$_4$. The authors suggested that the sorption phenomena in the higher ionic strength solutions was caused by the formation of KSb(OH)$_6^0$ ion pair [137]. On the contrary, antimonite, Sb(III) sorption was not strongly affected by changes in the pH or ionic strength [137]. Antimonite adsorption was insensitive to changes in I at pH 2–6. A modified triple-layer model was successfully used to describe the inner-sphere Sb(III) and Sb(V) sorption processes in goethite under the reaction conditions studied [137].

More recently, a few spectroscopic studies were conducted to investigate the sorption mechanisms of Sb(III) and Sb(V) on mineral surfaces [106,107]. Scheinost *et al.* [107]

reported that the presence of specific Sb–Fe interatomic distances corresponded to the edge sharing inner-sphere surface species of Sb(III) and Sb(V) on goethite surfaces (Table 16.1). McComb and co-workers [108] used Attenuated Total Reflectance infrared (IR) spectroscopy to understand Sb(V) retention at the amorphous iron oxyhydroxide–water interface. The chemisorption of Sb(V) increased with decreasing pH from 8 to 3. They suggested a ligand exchange reaction (that is, inner-sphere sorption mechanism) based on the Fourier transform infrared (FTIR) spectra observation; the loss of an OH stretching mode upon the Sb(V) reaction on the mineral surfaces (Table 16.1). They also observed an increase in the extent of Sb(V) release at alkaline pH values.

Belzile et al. [138] studied the Sb(III) sorption in amorphous iron oxyhydroxides and birnessite. They reported rapid surface catalyzed Sb(III) oxidation on these surfaces at different pH values, determined via differential pulse adsorptive stripping voltammetry analysis. Leuz and co-workers [137] also reported the surface catalyzed oxidation of Sb(III) on goethite surfaces.

16.5 Risks from Arsenic and Antimony in Soils

Due to various anthropogenic inputs to (for example, mining, smelting industries) and indigenous sources in (for example, mineral weathering) soils, As and Sb are often transported to surface waters and groundwaters, resulting in aqueous concentrations of these elements greater than the current MCL in many parts of the world [2,8,9,11,139]. High concentrations of As and Sb in drinking water supplies raise serious concerns regarding protection of human and ecological health. In recent years many researchers have assessed the environmental risks of As and Sb accumulation in terrestrial biota, such as plant species at contaminated sites.

As described above in Section 16.4, solubility and mobility of As and Sb are highly influenced by pH, redox status and types of adsorbents in soils. Antimony and As are considered relatively immobile under oxidizing conditions due to the strong fixation mechanisms of their pentavalent species in soil matrices. However, some studies have shown that a fraction of these elements may be bioavailable in oxic environments, depending on soil contamination levels and the specific plant species growing at contaminated sites. Arsenic and Sb in soils can be readily taken up by a wide variety of plant species [140,141]. Background As and Sb content in terrestrial vascular plants ranges from 0.2–50 μg g^{-1} and 0.009–1.5 μg g^{-1}, respectively [140,142–144]. In terrestrial vascular plants, the background Sb content was found to range from 0.2 to 50 μg g^{-1} [140,142,145]. Tyler and co-workers [146] reported the amounts of As and Sb in developing, maturing, and wilting leaves. The amounts of As and Sb were very low in leaves (<1 μg per 100 leaves). However, total element concentrations gradually increased throughout the growing season and usually in the subsequent winter.

While the above studies show a wide range of the background As and Sb concentrations in plants growing at uncontaminated sites, bioaccumulation of As and Sb can be elevated in some plant species at or near mining sites [38,147,148]. Some plant species (for example, *Agrostis canina*, *Achillea tenuis*, *Pseudosuga taxifolia*, and *Pityrogramma calomelanos*, *Pteris vittata*, and moss species (for example, *Pohlia wahlenbergii* and

Table 16.2 A summary of arsenic and antimony accumulation in plant species at metalloid-contaminated sites

Study sites	Total metalloid conc. in soils (mg kg^{-1})	Plant species	Metalloid uptake in plant tissues (mg kg^{-1})	References
Abandoned Portuguese mines	As: 76 Sb: 663	*Digitalis purpurea* *Erica umbellate* *Calluna vulgaris* *Cistus ladanifer* Old needles of *Pinus pinaster* *Calluna vulgaris* *Chamaespartinum tridentatum* leaves of *Cistus ladanifer* *Erica umbellate* *Quercus ilex* subsp. *ballota*	Sb: 139 Sb: 1.74 Sb: 1.25 Sb: 3.65 As: 30 As: 0.62 As: 0.54 As: 2.77 As: 0.64 As: 3.6	[158]
An abandoned Sb-mining area in southern Italy	Sb: 139	*Achillea ageratum* basal leaves *Achillea orescences* English Plantain (*Plantago lanceolata*) in roots Maidenstears (*Silence vulgaris*) in shoots	Sb: 1367 Sb: 1105 Sb: 1150 Sb: 1164	[141]
Arsenic-contaminated soils in southwest England	As: 8500–26,500	Old leaves of velvet bentgrass (*Agrostis canina*) and bentgrass (*Agrostis tenuis*)	As: 6,640	[150]
Chromated copper arsenate-contaminated soils in Central Florida	As: 18.8–1,603	Brake fern (*Pteris vittata*)	As: 3,280–4,980	[151]
Arsenic-contaminated sites in Thailand that are affected by mine tailings	As: 21–16,000	Dixie silverback fern (*Pityrogramma calomelanos*) and ferns (*Pteris vittata*) Herb (*Mimosa pudica*) Shrub(*Melastoma malabrathricum*) Tree (*Bauhinia glauca* Wall.)	As: up to 8,350 As: 24–76 As: 18–37 As: 15–32	[155]
Former Sb-mining areas	As: 5.3–2035	Roots and leaves of (*Mentha aquatica*) Roots of (*Phragmites australis*)	540 and 216 688	[141]
Historic gold-mining sites in northern Westland, New Zealand	As: 2000–40,000	Manuka (*Leptospermum scoparium*) Gorse (*Ulex europaeus*) Tree fuchsia (*Fuchsia excorticate*) Broadleaf (*Griselinia littoralis*) Reeds (*Juncus spp.*) Mosses (*Pohlia wahlenbergii*)	As: <10 As: <10 As: <10 As: <10 As: <100 As: Up to 30,000	[156]

Table 16.3 A summary of arsenic and antimony chemical speciation in different plant species in metalloid-contaminated environments

Study sites	Total metalloid conc. in soils	Biota/plant species	Chemical species	References
Abandoned As, Sb and Zn mines	As: 51.9–21 200 mg kg^{-1}	Scots pine needles, *Pinus sylvestris* Hard rush shoots (*Juncus inflexus* L.), boxtree leaves (*Buxus sem-perviren*), and Ferns (*Dryopteris filix-max*)	Arsenite and/or arsenate Methylarsonate (0.012–0.585 mg kg^{-1}), dimethylarsinate (0.013–0.569 mg kg^{-1}), trimethylarsine oxide (0.02–0.5 mg kg^{-1}) and tetramethylarsonium ion (<0.19 mg kg^{-1})	[157]
Laboratory study	Switzerland soils (Sb >1000 mg kg^{-1})	Microbes and filamentous fungi	Volatile trimethylantimony [(CH$_3$)$_3$S] species	[159]
	0.1 mg kg^{-1} of Sb in liquid culture			[160]
Keg Lake and Kam Lake, near Yellowknife, Northwest Territories, Canada	Sb in pondweeds: 48–68 mg kg^{-1}	Pondweed (*Potamogetan pectinatus*)	Sb as Sb(III), methylstibine, CH$_3$SbH$_2$, dimethylstibine (CH$_3$)$_2$SbH, and trimethylantimony (CH$_3$)$_3$Sb	[161]
NaH$_2$AsO$_4$-spiked uncontaminated soils	As: 50 mg kg^{-1}	Fern (*Pteris vittata* L.)	As(III)/As(III)-S in the leaves	[162]
Arsenic-contaminated mine waste, and arsenic-amended liquid cultures	Six hydroponic treatments: 0, 1, 2, 5, 15, 30 and 60 mg kg^{-1}, and As contaminated soil samples(< 1,100 mg kg^{-1})	Radish (*Rhaphanus sativus*)	Arsenite, arsenate, and As(III)– sulfur compounds in leaf, stem and root sections of plants	[163]

Streams and puddles receiving mine effluent	Sb: 3.7–380 µg kg^{-1}	Moss (*Funuria hygmmerrica* and *Drepanocladus sp.*), cattail (*Typha lutifoliu*) and burmarigold (*Bidens cernuu*), duckweed (*Lemna minor*), water milfoil (*Myriophyllum sp.*) Pondweed (*Potomogetan richanisonii*), and bur-reed (*Spurgunium augustifolium*) and fungi lichen, pixie cup lichen (*Cladoniu sp.*), puffball mushroom (*Lycopenion sp.*) and shaggy mane mushroom (*Coprinus comatus*).	Sb (V), Sb(III) and methylated (Me) antimony species in all samples Me$_3$Sb ranged from 5 to 170 µg kg^{-1} Sb(III) and Sb(V) concentrations were from 5 to 12 and 28 to 5000 µg kg^{-1}, respectively	[164]
Former mining sites in Catalonia, Spain	Not available	Moss (*Hypnum cupressiforme* Hedw), fern (*Dryopteris filix-max* (L.) Schott.), stitchwort (*Stellaria halostea*), *Dryopteris filix-max* (L.) Schott. and figwort (*Chaenorhinum asarina*)	A sum of Sb(III), Sb(V) and trimethyl Sb species in these plant species yielded <28 % of total Sb. Sb(V) was a major Sb species in moss Sb(III) was a dominant Sb species (1.75 mg kg^{-1}) in stitchwort	[165]
Laboratory study	40 µg kg^{-1} of Sb(III) in river and sea water	*Spirulina platensis* (cyanobacterium) and *Phaseolus*	Sb(III) and Sb(V) species	[166]

Brachythecium cf. reflexum) can accumulate high levels (>1000 mg kg^{-1}) of As and Sb from soils [149–157]. The results of metalloid phytoaccumulation studies are summarized in Table 16.2. While many focused on the measurements of operationally defined chemical extractable and/or total As and Sb concentrations in plant tissues, some researchers pointed out the importance of chemical speciation with respect to toxicity and phytoaccumulation processes. Recent research has shown the presence of As(III), As(V), Sb(III), Sb(V), and methylated As and Sb species in plant tissues (Table 16.3). It is apparent that phytoaccumulation of As and Sb differs across plant species growing at contaminated sites. However, the relationship between phytoaccumulation mechanisms and the chemical speciation found in plants and soils is not well understood.

16.6 Conclusions and Future Research Needs

This chapter has discussed the chemical properties of As and Sb, toxicity, and reactivity in soil–water–plant systems. Although these elements share numerous similarities in biogeochemical properties, their reactivity is highly influenced by changes in environmental conditions. Whereas the trivalent states of As and Sb are mobile in the reduced soil–water systems, the pentavalent species are relatively immobile under oxidizing conditions due to the strong fixation mechanisms in soil matrices.

The metalloid accumulation in plant species readily occurs at As and Sb naturally contaminated sites. Some shrubs, grasses, reeds, and mosses can tolerate As and Sb and largely remove these metalloids from surface soils. Populations of a variety of plant species colonized at contaminated sites are responding to these metalloids by exclusion or accumulation.

The extent/rate of bioaccumulation and sorption/desorption processes are highly influenced by the chemical species specific in the plant–water–soil systems. Unfortunately, the chemical speciation of As and Sb in soils and plant tissues are rarely correlated with reaction processes (for example, plant uptake and desorption processes) at the field scales and soil physicochemical properties (for example, redox status, pH, and hydrologic properties). Using modern microscopic and spectroscopic techniques (for example, transmission electron microscope spectroscopy, synchrotron based X-ray techniques), solid-state speciation in soils and biological-media can be better characterized, and these research findings will further lead to better toxicological/risk assessment of As and Sb in field-scale settings. These comprehensive research results will be helpful in making better regulatory decisions, and in designing effective *in situ* remediation technologies and environmental management programs to enhance environmental quality and ecological health.

References

[1] Adriano, D.C., Trace Elements in the Terrestrial Environment; Springer-Verlag, New York, 1986.
[2] Gao, S.; Tanji, K.K.; Goldberg, S., Session 1: Potentially toxic trace elements in soils and sediments Paper 4: Reactivity and transformation of arsenic; in: Dudley LM, Guitjens J (eds), Agroecosystems: Sources, Control, and Remediation of Oxyanions. Symposium on Sources,

Control, and Remediation of Oxyanions in Agroecosystems, 19–22 June, 1994; San Francisco: Proc. Symp., Pacific Div., Am. Assoc. Adv. Sci., 1997.
[3] Carlin, J.F. Jr., Antimony; U.S. Geological Survey Mineral Commodity Summaries. 2000. http://minerals.usgs.gov/minerals/pubs/commodity/antimony/(accessed 7 December 2009).
[4] Onish, H., Antimony; Springer Verlag, Heidelberg, 1969.
[5] Berg, M.; Tran, H.C.; Nguyen, T.C.; Pham, H.V. Schertenleib, R; Giger, W., Arsenic contamination of groundwater and drinking water in Vietnam: A human health threat; Environ Sci Technol. 2001, 35, 2621–2626.
[6] United States Environmental Protection Agency. EPA to implement 10ppb standard for arsenic in drinking water; Report No. EPA 815-F-01-010, 2001.
[7] WHO. Guidelines for Drinking-water Quality, vol. 2: Health Criteria and Other Supporting Information; World Health Organization, Geneva, 1996.
[8] Ritchie, J.A., Arsenic and antimony in some New Zealand thermal waters. NZJ Sci. 1961, 4, 218–29.
[9] Weissberg, B.G.; Browne, P.R.L.; Seward, T.M., Ore metals in active geothermal systems; in: Barnes H.L. (ed.), Geochemistry of Hydrothermal Ore Deposits; John Wiley & Sons, Inc., New York, 1979, pp. 738–7380.
[10] Kolpakova, N.N., Laboratory and field studies of ionic equilibria in the Sb_2S_3–H_2O–H_2S system; Geochem Int. 1982, 19, 46–54.
[11] Stauffer, E.; Thompson, J.M., Arsenic and antimony in geothermal waters of Yellowstone National Park, Wyoming, USA; Geochim. Cosmochim. Acta 1984, 48, 2547–2561.
[12] DaCosta, E.W.B., Variation in the toxicity of arsenic compounds to microorganisms and the suppression of the inhibitory effects by phosphate; Appl. Microbiol. 1972, 23, 46–53.
[13] Sheppard, S.C., Summary of phytotoxic levels of soil arsenic; Water Air Soil Pollut. 1991, 64, 539–550.
[14] Tamaki, S.; Frankenberger, Jr. W.T., Environmental biochemistry of arsenic; in: Whitacre, D.M. (ed.), Reviews of Environmental Contamination and Toxicology; Springer-Verlag, New York, 1992, vol. 124, pp. 79–110.
[15] Faust, S.D.; Aly, O.M., Chemistry of Natural Waters; Science Publishers Inc., Ann Arbor, MI, 1981.
[16] Manahan, S.E., Toxicological chemistry of chemical substances; Environmental Chemistry, 6th edn; Lewis Publishers, Ann Arbor, 1994, pp. 675–704.
[17] Blair, P.C.; Thompson, M.B.; Morrissey, R.E.; Moorman, M.P.; Sloane, R.A.; Fowler, B.A., Comparative toxicity of arsine gas in B6C3F1 mice, Fischer 344 rats, and Syrian Golden Hamsters: system organ studies and comparison of clinical indices of exposure; Toxicol. Sci. 1990, 14, 776–787.
[18] Flemming, P.J.; Cooke, M.; Chantlet, S.M.; Golding, J., Fire retardants, biocides, plasticisers, and sudden infant deaths; Brit. Med. J. 1994, 309, 1594–1596.
[19] Richardson, B.A., Sudden infant death syndrome; a possible primary cause; J. Forensic Sci. Soc., 1994, 34, 199–204.
[20] Blair, P.; Fleming, P.; Bensley, D.; Smith, I.; Bacon, C.; Taylor, E., Plastic mattresses and sudden infant death syndrome; Lancet 1995, 345, 720.
[21] Apostoli, P.; Porru, S.; Alessio, L., Antimony, biological indicators for the assessment of human exposure to industrial chemicals; European Commission, Luxembourg, 1994.
[22] McCallum, R.I., Occupational exposure to antimony compounds; J. Environ. Monit. 2005, 7, 1245–50.
[23] ATSDR AfTSaDR, Toxicological Profile for Antimony; US Public Health Service, US Department of Health and Human Services, Altanta, 1992.
[24] Horber, F.F.; Lerut, J.; Jaeger, P., Renal tubular acidosis, a side effect of treatment with pentavalent antimony; Clin Nephrol. 1991, 36:213.
[25] Braconier, J.H.; Miorner, H., Recurrent episodes of thrombocytopenia during treatment with sodium stibogluconate; J Antimicrob. Chemother. 1993, 31, 187–8.
[26] Gasser, R.A.; Magill, A.J.; Oster, C.N.; Franke, E.D. Grogel, M; Berman, J.D., Pancreatitis induced by pentavalent antimonial agents during treatment of leishmaniasis. Clin. infect. Dis. 1994 18, 83–90.

[27] Colburn, P.; Alloway, B.J.; Tompson, I., Arsenic and heavy metals in soil associated with regional geochemical anomalies in Southwest England; Sci. Total Environ. 1975, 4, 359–63.

[28] Voigt, D.E.; Brantley, S.L., Chemical fixation of arsenic in contaminated soils. Appl Geochem. 1996, 11, 633–43.

[29] Walsh, L.M.; Keeny, D.R., Behavior and phytotoxicity of inorganic arsenicals in soils; in: Woolson, E.A. (ed.), Arsenical Pesticides; ACS, Washington, DC, 1975, pp. 35–52.

[30] Woolson, E.A.; Axley, J.H.; Kearney, P.C., Correlation between available soil arsenic, estimated by six methods, and response of corn (*Zea Mays L.*); Soil Sci. Soc. Am. Proc. 1971, 35, 101–5.

[31] Committee on Medical and Biological Effects of Environmental Pollutants NRC. Chemistry and distribution of arsenic in the environment. Arsenic, Medical and Biological Effects of Environmental Pollutants, National Academy of Science, Washington, DC, 1977, pp. 4–79.

[32] Braman, R.S., Environmental reaction and analysis method; in: Fowler, B.A. (ed.), Biological and Environmental Effects of Arsenic; Topics in Environmental Health; Elsevier Science Publishers, New York, 1983, pp. 141–154.

[33] Tanji, K.K.; Dalgren, R.A., Acidification of evaporation ponds to reduce contaminant hazards to wildlife-phase I (Chemical aspects). DWR Agreement #: B58194; 1993.

[34] Bagby, W.C.; Berger, B.R., Geologic characteristics of sediment-hosted, disseminate precious metal deposits in the western United States; in: Berger, B.R., Bethke, P.M. (eds), Geology and Geochemistry of Epithermal Systems; Rev. Econ., 1985, 165–202.

[35] Filella, M.; Belzile, N.; Chen, Y.W., Antimony in the environment: a review focused on natural waters I. Occurrence; Earth Sci. Rev. 2002, 57, 125–176.

[36] Fink, C.R., A perspective on metals in soils; J. Soil Contam. 1996, 5, 329–359.

[37] Martin, J.-M.; Whitefield, M., The significance of the river input of chemical elements to the ocean; in: Wong, C.S., Boye, E., Bruland, K.W., Burton, J.D., Goldberg, E.D. (eds), Trace Metals in Sea Water; NATO Adv. Res. Inst., Plenum, New York, 1983, pp. 265–296.

[38] Ainsworth, N.; Cooke, J.A.; Johnson, M.S., Biological significance of antimony in contaminated grassland; Water, Air Soil Pollut. 1991, 57–58, 1993–1999.

[39] Crecelius, EA, Johnston, CJ, Hofer, GC., Contamination of soils near a copper smelting by arsenic, antimony and lead. Water, Air, Soil Pollut. 1974, 3, 337–342.

[40] Asami, T, Kubota, M, Saito, S., Simultaneous determination of anitimony and bismuth in soils by continuous hydride generation-atomic absorption spectrometry. Water Air Soil Pollut. 1992, 62, 349–355.

[41] Papakostidis, G.; Grimanis, A.P.; Zafiropoulos, D., Heavy metals in sediments from the Athens sewage outfall area; Mar Pollut Bull. 1975, 6, 136–139.

[42] Wild, H., Geobotanical anomalies in Rhodesia: 4. The vegetation of arsenical soils. Kirkia 1974, 9, 243–264.

[43] Li, X.; Thornton, I., Arsenic, antimony and bismuth in soil and pasture herbage in some old metalliferous mining area in England. Environ; Geochem. Health. 1993, 15, 135–144.

[44] Ragaini, R.C.; Ralston, H.R.; Roberts, N., Environmental trace metal contamination in Kellogg, Idaho, near a lead smelting complex; Environ. Sci. Technol. 1977, 11, 773–781.

[45] Andreae, M.O.; Asmodi, J.-F.; Foster, P.; Van't dack, l., Determination of antimony(III), antimony(V), and methylantimony species in natural waters by atomic absorption spectrometry with hydride generation; Anal. Chem. 1981, 53, 1766–1771.

[46] Matschullat, J., Arsenic in the geosphere-a review; Sci. Total Environ. 2000, 249, 297–312.

[47] Chilvers, D.C.; Peterson, P.J., Lead, mercury, cadmium and arsenic in the environment; in: Hutchinson, T.C., Meema, K.M. (eds), Global Cycling of Arsenic; John Wiley & Sons, Inc., New York, 1987, pp. 279–301.

[48] Anderson, C.E., Arsenicals as feed additives for poultry and swine; in: Lederer, W.H., Fensterheim, R.J. (eds), Arsenic: Industrial, Biomedical, Environmental Perspectives; Van Nostrand Reinhold, New York, 1983, pp. 89–97.

[49] Christen, K., Chickens, manure, and arsenic; Environ. Sci. Technol. 2001, 35(5), 184A–5A.

[50] United States Environmental Protection Agency. Toxics Release Inventory, USEPA, Washington, DC, 1998.

[51] Nriagu, J.O.; Pacyna, J.M., Quantitative assessment of worldwide contamination of air, water and soils by trace metals; Nature 1988, 333, 134–139.

[52] Nriagu, J.O., A global assessment of natural sources of atmospheric trace metals; Nature 1989, 338, 47–49.
[53] Nriagu, J.O. Global metal pollution. Poisoning the biosphere? Environment 1990, 32, 28–33.
[54] Maeda, S., Safety and Environmental Effects, John Wiley & Sons, Inc., New York, 1994.
[55] Fergusson, J.E., The Heavy Metals: Chemistry, Environmental Impact and Health Effects, Pergamon Press, Oxford, 1990.
[56] Smedley, P.L. and Kinniburgh, D.G., A review of the source, behavior and distribution of arsenic in natural waters. Appl. Geochem. 2002, 17, 517–568.
[57] Tallman, D.E.; Shaikh, A.U., Redox stability of inorganic arsenic(III) and arsenic(V) in aqueous solution; Anal. Chem. 1980, 52, 196–199.
[58] Eary, L.E.; Schranke, J.A., Chemical modeling of aqueous systems II, American Chemical Society, Washington, DC, 1990.
[59] Manning, B.A.; Goldberg, S., Adsorption and stability of arsenic (III) at the clay mineral–water interface; Environ. Sci. Technol. 1997, 31, 2005–2011.
[60] Deuel, L.E.; Swoboda, A.R., Arsenic solubility in a reduced environment; Soil Sci. Soc. Am. Proc. 1972, 36, 276–278.
[61] Masscheleyn, P.H.; Delaune, R.D.W.H.; Patrick, J., Effect of redox potential and pH on arsenic speciation and solubility in a contaminated soil; Environ. Sci. Technol. 1991, 25(8), 1414–1418.
[62] Masscheleyn, P.H.; Delaune, R.D.W.H.; Patrick, J., Arsenic and selenium chemistry as affected by sediment redox potential and pH; J. Environ. Qual. 1991, 20, 522–527.
[63] McGeeham, S.L.; Naylor, D.V., Sorption and redox transformation of arsenite and arsenate in two flooded soils; Soil Sci. Soc. Am. J. 1994, 58, 337–342.
[64] Onken, B.M.; Hossner, L.R., Plant uptake and determination of arsenic species in soil solution under flooded conditions; J. Environ. Qual. 1995, 24, 373–381.
[65] Onken, B.M.; Adriano, D.C., Arsenic availability in soil with time under saturated and subsaturated conditions; Soil Sci. Soc. Am. J. 1997, 61, 746–752.
[66] Rochette, E.A.; Li, G.C.; Fendorf, S.E., Stability of arsenate minerals in soil under biotically generated reducing conditions; Soil Sci. Soc. Am. J. 1998, 62, 1530–1537.
[67] Arai, Y.; Lanzirotti, A.; Sutton, S.; Newville, M; Dyer, J.; Sparks, D.L., Spatial and temporal variability of arsenic solid-state speciation in historically lead arsenate contaminated soils; Environ. Sci. Technol. 2006, 40, 673–679.
[68] Jacobs, L.W.; Syers, J.K.; Keeney, D.R., Arsenic sorption by soils; Soil Sci. Soc. Am. Proc. 1970, 34, 750–754.
[69] Manning, B.A.; Goldberg, S., Arsenic(III) and arsenic(V) adsorption on three California soils; Soil Sci. 1997, 162, 886–895.
[70] Livesey, N.T.; Huang, P.M., Adsorption of arsenate by soils and its relation to selected chemical properties and anions; Soil Sci. 1981, 131, 88–94.
[71] Sakata, M., Relationship between adsorption of As(III) and boron by soil and soil properties. Environ. Sci. Technol. 1987, 21, 1126–1130.
[72] Violante, A.; Colombo, C.; Buondonno, A., Competitive adsorption of phosphate and oxalate by aluminum oxides; Soil Sci. Soc. Am. J. 1991, 55, 65–70.
[73] Fordham, A.W.; Norrish, K., Arsenate-73 uptake by components of several acidic soils and its implications for phosphate retention, Aust. J. Soil Res. 1979, 17, 307–316.
[74] Norrish, K.; Rosser, H., Mineral Phosphate; Academic Press, Melbourne, 1983.
[75] Goldberg, S.; Glaubig, R.A., Anion sorption on a calcareous, montmorillonitic soil-arsenic; Soil Sci. Soc.Am. J. 1988, 52, 1297–1300.
[76] Hayes, K.F.; Papelis, C.; Leckie, J.O., Modeling ionic strength effects of anion adsorption at hydrous oxide/solution interface; J Colloid Interface Sci. 1988, 125, 717–726.
[77] Stumm, W.; Morgan, J.J., Theory of elementary processes. Aquatic Chemistry, Chemical Equilibria and Rates in Natural Water, John Wiley & Sons, Inc., New York, 1995, pp. 69–76.
[78] Gupta, S.K.; Chen, K.Y., Arsenic removal by adsorption; J. Water Pollut. Cont. Feder. 1978, 50, 493–506.

[79] Arai, Y.; Elzinga, E.J.; Sparks, D.L., X-ray absorption spectroscopic investigation of arsenite and arsenate adsorption at the aluminum oxide–water interface; J. Colloid Interface Sci. 2001, 235, 80–88.

[80] Anderson, M.A.; Ferguson, J.F.; Gavis, J., Arsenate adsorption on amorphous aluminum hydroxide; J. Colloid Interface Sci. 1976, 54, 391–399.

[81] Bleam, W.F.; Pfeffer, P.E.; Goldberg, S.; Taylor, R.W.; Dudley, R., A ^{31}P solid-state nuclear Magnetic resonance study of phosphate adsorption at the boehmite/aqueous solution; Langmuir 1991, 7, 1702–1712.

[82] Chen, Y.R.; Butler, J.N.; Stumm, W., Adsorption of phosphate on alumina and kaolinite from dilute aqueous solutions; J. Colloid Interface Sci. 1973, 43, 421–436.

[83] Pierce, M.L.; Moore, C.B., Adsorption of arsenite and arsenate on amorphous iron hydroxide; Water Res. 1982, 16, 1247–1253.

[84] Shang, C.; Stewart, J.W.B.; Huang, P.M., pH effect on kinetics of adsorption of organic inorganic phosphates by short-range ordered aluminum and iron precipitates; Geoderma 1992, 53, 1–14.

[85] Xu, H.; Allard, B.; Grimvall, A., Influence of pH and organic substance on the adsorption of As(V) on geologic materials; Water Air Soil Pollut. 1988, 40, 293–305.

[86] Arai, Y.; Sparks, D.L.; Davis, J.A., Effects of dissolved carbonate on arsenate adsorption and surface speciation at the hematite-water interface; Environ. Sci. Technol. 2004, 38, 817–824.

[87] Raven, K.P.; Jain, A.; Loeppert, R.H., Arsenite and arsenate adsorption on ferrihydrite: Kinetics, equilibrium, and adsorption envelopes; Environ. Sci. Technol. 1998, 32, 344–349.

[88] Fukushi, K.; Sato, T.; Yanase, N., Solid-solution reactions in As(V) sorption by schwertmannite; Environ. Sci. Technol. 2003, 37, 3581–3586.

[89] Edzwald, J.K;, Toensing, D.C.; Leung, M.C. Phosphate adsorption reaction with clay minerals; Environ. Sci. Technol. 1976, 10, 485–490.

[90] Arai, Y.; Sparks, D.L.; Davis, J.A., Arsenate adsorption mechanisms at the allophane-water interface; Environ. Sci. Technol. 2005. 39, 2537–2544.

[91] Manning, B.A.; Fendorf, S.E.; Bostick, B.; Suarez, D.L., Arsenic(III) oxidation and arsenic(V) adsorption reactions on synthetic birnessite; Environ. Sci. Technol. 2002, 36, 976–981.

[92] O'Reilly, S.E.; Strawn, D.G.; Sparks, D.L., Residence time effects on arsenate adsorption/ desorption mechanisms on goethite; Soil Sci. Soc. Am. J. 2001, 65, 67–77.

[93] Waychunas, G.A.; Rea, B.A.; Fuller, C.C.; Davis, J.A., Surface chemistry of ferrihydrite: Part 1. EXAFS studies of the geometry of coprecipitated and adsorbed arsenate; Geochim. Cosmochim. Acta 1993, 57, 2251–2264.

[94] Fendorf, S.E.; Eick, M.J.; Grossl, P.; Sparks, D.L., Arsenate and chromate retention mechanisms on goethite. 1. Surface structure; Environ. Sci. Technol. 1997, 31, 315–320.

[95] Foster, A.L.; G.E.; Brown, J.; Parks, G.A., X-ray absorption fine structure study of As(V) and Se(IV) sorption complexes on hydrous Mn oxides; Geochim. Cosmochim. Acta 2003, 67, 1937–1953.

[96] Lumsdon, D.G.; Fraser, A.R.; Russell, J.D.; Livesey, N.T., New infrared band assignments for the arsenate ion adsorbed on synthetic goethite (α-FeOOH); J. Soil Sci. 1984, 35, 381–386.

[97] Manning, B.A.; Fendorf, S.E.; Goldberg, S., Surface structure and stability of arsenic(III) on goethite: spectroscopic evidence for Inner-sphere complexes; Environ Sci Technol. 1998, 32, 2383–2388.

[98] Sun, X.; Doner, H.E., An investigation of arsenate and arsenite bonding structures on goethite by FTIR; Soil Sci. 1996, 161, 865–872.

[99] Tournassat, C.; Charlet, L.; Bosbach, D.; Manceau, A., Arsenic(III) oxadation by birnessite and precipitation of manganese(II) arsenate; Environ. Sci. Technol. 2002, 36, 493–500.

[100] Foster, A.L.; Brown, Jr. G.E., Tingle, T.N.; Parks, G.A., Quantitative arsenic speciation in mine tailing using X-ray absorption spectroscopy; Am. Mineral. 1998, 83, 553–568.

[101] Goldberg, S.; Johnston, C.T., Mechanisms of arsenic adsorption on amorphous oxides evaluated using macroscopic measurements, vibrational spectroscopy, and surface complexation modeling; J. Colloid. Interface Sci. 2001, 234, 204–216.

[102] Suarez, D.L.; Goldberg, S.; Su, C., Evaluation of oxyanion adsorption mechanisms in oxides using FTIR spectroscopy and electrophoretic mobility; in: Sparks D.L., Grundl T.J. (eds), Mineral–Water Interfacial Reactions Kinetics and Mechanisms, ACS, Washington, DC, 1998, pp. 136–178.

[103] Ladeira, A.C.Q.; Ciminelli, V.S.T.; Duarte, H.A.; Alves, M.C.M.; Ramos, A.Y., Mechanism of anion retention from EXAFS and density functional calculation: arsenic (V) adsorbed on gibbsite; Geochim. Cosmochim. Acta 2001, 65, 1211–1217.

[104] Manceau, A., The mechanism of anion adsorption on iron oxides: Evidence for the bonding of arsenate tetrahedra on free $Fe(O, OH)_6$ edges; Geochim. Cosmochim. Acta 1995, 59, 3647–3653.

[105] Waychunas, G.A.; Davis, J.A.; Fuller, C.C., Geometry of sorbed arsenate on ferrihydrite and crystalline FeOOH: Re-evaluation of EXAFS results and topological factors in predicting sorbate geometry, and evidence for monodentate complexes; Geochim Cosmochim Acta 1995, 59, 3655–3661.

[106] Randall, S.; Sherman, D.M.; Ragnarsdottir, K.V., Sorption of As(V) on green rust $(Fe_4(II)Fe_2(III)(OH)12SO_4\ 3H_2O)$ and lepidocrocite (γ-FeOOH): Surface complexes from EXAFS spectroscopy; Geochim. Cosmochim. Acta 2001, 65, 1015–1023.

[107] Scheinost, A.C.; Rossberg, A.; Vantelon, D. et al., Quantitative antimony speciation in shooting-range soils by EXAFS spectroscopy; Geochim Cosmochim Acta 2006, 70, 3299–3312.

[108] McComb, K.A.; Craw, D.; McQuillan, A.J., ATR-IR spectroscopic study of antimonate adsorption to iron oxide; Langmuir 2007, 23(24), 12125–12130.

[109] Frost, R.R.; Griffin, R.A., Effect of pH on adsorption of arsenic and selenium from landfill leachate by clay minerals; Soil Sci. Soc. Am. J. 1977, 41, 53–57.

[110] Scott, M.J.; Morgan, J.J., Reactions at oxide surface. 1. oxidation of As(III) by synthetic birnessite; Environ Sci Technol. 1995, 29, 1898–1905.

[111] Oscarson, D.W.; Huang, P.M.; Liaw, W.K., The oxidation of arsenite by aquatic sediments; J. Environ. Qual. 1980, 9, 700–703.

[112] Sun, X.; Doner, H.E.; Zavarin, M., Spectroscopy study of arsenite [As(III)] oxidation on Mn-substituted goethite; Clays Clay Miner. 1999, 47(4), 474–480.

[113] Driehaus, W.; Seith, R.; Jekel, M., Oxidation of arsenate(III) with manganese oxides in water treatment;Water Res. 1995, 29, 297–305.

[114] Foster, A.L.; G.E.; Brown, J.; Parks, G.A., X-ray absorption fine structure spectroscopy study of photocatalyzed, heterogeneous As(III) oxidation on kaoline and anatase; Environ. Sci. Technol. 1998, 32, 1444–1452.

[115] Kostka, J.E.; George, W.; Luther, I.; Nealson, K.H., Chemical and biological reduction of Mn (III)-pyrophosphate complexes: potential importance of dissolved Mn (III) as an environmental oxidant; Geochim. Cosmochim. Acta 1995, 59, 885–894.

[116] Nico, P.S.; Zasoski, R.J., Importance of Mn(III) availability on the rate of Cr(III) oxidation on d-MnO_2; Environ. Sci.Technol. 2000, 34, 3363–3367.

[117] Nesbitt, H.W.; Canning, G.W.; Bancroft, G.M., XPS study of reductive dissolution of 7Å-birnessite by H_3AsO_3, with constraints on reaction mechanism; Geochim. Cosmochm. Acta 1998, 62, 2097–2110.

[118] Chiu, V.Q.; Hering, J.G., Arsenic adsorption and oxidation at manganite surfaces. 1. Method for simultaneous determination of adsorbed and dissolved arsenic species; Environ. Sci Technol. 2000, 34, 2029–2034.

[119] Arai, Y.; Sparks, D.L., Residence time effects on arsenate surface speciation at the aluminum oxide–water interface; Soil Sci. 2002, 167, 303–314.

[120] Grossl, P.R.; Sparks, D.L., Evaluation of contaminant ion adsorption/desorption on goethite using pressure-jump relaxation kinetics; Geoderma 1995, 67, 87–101.

[121] Woolson, E.A.; Axley, J.H.; Kearney, P.C., The chemistry and phytotoxicity of arsenic in soils: II. Effects of time and phosphorus; Soil Sci. Soc. Am. Proc. 1973, 37, 254–259.

[122] Lin, Z.; Puls, R.W., Adsorption, desorption and oxidation of arsenic affected by clay minerals and aging process; Environ. Geol. 2000, 39, 753–759.

[123] Jackson, B.P.; Mille, W.P., Effectiveness of phosphate and hydroxide for desorption of arsenic and selenium species from iron oxides; Soil Sci. Soc. Am. J. 2000, 64, 1616–1622.

[124] Gayer, K.H.; Gerrett, A.B., The equilibria of antimonous oxide(rhombic) in dilute solutions of hydrochloric acid and sodium hydroxide at 258 °C; J. Am. Chem. Soc. 1952, 72, 2353–2354.

[125] Ahrland, S.; Bovin, J., The complex formation of antimony(III). in perchloric and nitric acid solutions – a solubility study; Acta Chem. Scand A 1974, 28, 1089–1100.

[126] Wood, S.A., Raman spectroscopic determination of the speciation of ore metals in hydrothermal solutions: I. Speciation of antimony in alkaline sulfide solutions at 25 °C; Geochim Cosmochim Acta 1989, 53, 237–244.

[127] Krupka, K.M.; Serne, R.J., Geochemical factors affecting the behavior of antimony, cobalt, europium, technetium, and uranium in vadose sediments; Pacific Northwest National Laboratory, Richland, Washington, 2002.

[128] Oelkers, E.H.; Sherman, D.M.; Vala Ragnarsdottir, K.; Collins, C., An EXAFS spectroscopic study of aqueous antimony(III)-chloride complexation at temperatures from 25 to 250 °C; Chem. Geo. 1998, 151, 21–27.

[129] Baes, C.F.R.E.M., The Hydrolysis of Cations. John Wiley & Sons, Inc., New York, 1976.

[130] Rai, D.; Zachara, J.M.; Schwab, A.P.; Schmidt, R.L.; Girvin, D.C.; Rogers, J.E., Chemical Attenuation Rates, Coefficients, and Constants in Leachate Migration. Volume 1: A Critical Review; Electric Power Research Institute; Palo Alto, 1984.

[131] Helz, G.R.; Valerio, M.S.; Capps, N.E., Antimony speciation in alkaline sulfide solutions: Role of zerovalent sulfur; Environ. Sci. Technol. 2002, 36, 943–948.

[132] Mosselmans, J.F.W.; Helz, G.R.; Pattrick, R.A.D.; Charnock, J.M.; Vaughan, D.J., A study of speciation of Sb of bisulfide solutions by X-ray absorption spectroscopy; Appl. Geochem. 2000, 15, 879–989.

[133] Sherman, D.M.; Ragnarsdottir, K.V.; Oelkers, E.H., Antimony transport in hydrothermal solutions: an EXAFS study of antimony(V) complexation in alkaline sulfide and sulfide-chloride brines at temperatures from 25 °C to 300 °C at Psat; Chem. Geo. 2000, 167, 161–167.

[134] Ambe, S., Adsorption-kinetics of antimony(V) ions onto alpha-Fe_2O_3 surfaces from an aqueous solution; Langmuir 1987, 3, 489–493.

[135] Crecelius, E.A.; Bothner, M.H.; Carpenter, R., Geochemistries of arsenic, antimony, mercury, and related elements in sediments of Puget Sound; Environ. Sci. Technol. 1975, 9, 325–333.

[136] Tighe, M.; Lockwood, P.; Wilson, S., Adsorption of antimony(V) by floodplain soils, amorphous iron(III) hydroxide and humic acid; J. Environ. Monit. 2005, 7, 1177–1185.

[137] Leuz, A.K.; Monch, H.; Johnson, C.A., Sorption of Sb(III) and Sb(V) to goethite: influence on Sb(III) oxidation and mobilization; Environ. Sci. Technol. 2006, 40, 7277–7282.

[138] Belzile, N.; Chen, Y-W.; Wang, Z., Oxidation ofatimony(III) by amorphous iron and manganese oxyhydroxides; Chem. Geo. 2001, 174, 379–387.

[139] Meharg, A., Venomous Earth. How Arsenic Caused the World's Worst Mass Poisoning; Macmillan Science, New York, 2005.

[140] Coughtrey, P.J.; Jackson, P.J.; Thorne MC. Radionuclide distribution and transport in terrestrial and aquatic ecosystems; AA Balkema, Rotterdam, 1983.

[141] Baroni, F.; Boscagli, A.; Protano, G.; Riccobono, F., Antimony accumulation in *Achillea ageratum, Plantago lanceolata* and *Silene vulgaris* growing in an old Sb-mining area; Environ. Pollut. 2000, 109, 347–352.

[142] Brooks, R.R., Geobotany and biogeochemistry in mineral exploration; Harper and Row, New York, 1972.

[143] Bowen, H.J.M., Environmental Chemistry of the Elements; Academic Press, London, 1979.

[144] Kabata-Pendias, A.; Pendias, H., Trace Elements in Soils and Plants; CRC Press, Boca Raton, FL, 1984.

[145] Bowden, J.W.; Nagarajah, S.; Barrow, N.J.; Posner, A.M.; Quirk, J.P., Describing the adsorption of phosphate, citrate and selenite on the variable charge mineral surface; Aust. J. Soil Res. 1980, 18, 49–60.

[146] Tyler, G.; Olsson, T., The importance of atmospheric deposition, charge and atomic mass to the dynamics of minor and rare elements in developing, ageing, and wilted leaves of beech (*Fagus sylvatica* L.); Chemosphere 2006, 65, 250–260.

[147] Shacklette, H.T.; Erdman, J.A.; Harms, T.F., Trace elements in plant foodstuffs; in: Oehme F.H. (ed.), Toxicity of Heavy Metals in the Environment, Part 1; Marcel Dekker, New York, 1978, 25–43.
[148] Istas, J.; De Temmeran, L.; Dupire, S.; Hoenig, M., Luchtverontreiniging door een metallurgisch bedrijf: een systematisch onderzoek (Final Report). Belgian National Research and Development Programme on Environment – Air; Brussels, Belgium, 1982.
[149] Peterson, P.J., Unusual accumulation of elements by plants and animals; Sci. Prog. 1971, 59, 505–526.
[150] Porter, E.K.; Peterson, P.J., Arsenic accumulation by plants on mine waste (United Kingdom); Sci. Total Environ. 1975, 4, 365–371.
[151] de Koe, T., Agrostis castellana and Agrostis delicatula on heavy metal and arsenic enriched sites in NE Portugal; Sci. Total Environ. 1994, 145, 103–109.
[152] de Koe, T.; Beek, M.A.; Haarsma, M.S.; Ernst, W.H.O., Heavy metals and arsenic in grasses and soils of mine spoils in North East Portugal, with particular reference to some Portuguese goldmines; in: Nath B. (ed.), Environmental Pollution, Proc. Int. Conf, ICEP-1, 1991, pp. 373–380.
[153] Ma, L.Q.; Komar, K.M.; Tu, C.; Zhang, W.; Cai, Y.; Kennelley, E.D., A fern that hyperaccumulates arsenic; Nature 2001, 409, 579.
[154] Warren, H.V.; Delavault, R.E.; Barasko, J., The arsenic content of Douglas Fir as a guide to some gold, silver and base metal deposits; Can. Mining Metall. Bull. 1968, 7, 1–9.
[155] Visoottiviseth, P.; Francesconi, K.; Sridokchan, W., The potential of Thai indigenous plant species for the phytoremediation of arsenic contaminated land; Environ. Pollut. (Series B) 2002, 118, 453–461.
[156] Craw, D.; Rufaut, C.; Haffert, L.; Paterson, L., Plant colonization and arsenic uptake on high arsenic mine wastes, New Zealand; Water Air Soil Pollut. 2007, 179, 351–364.
[157] Ruiz-Chancho, M.J.; Lopez-Sanchez, J.F.; Schmeisser, E.; Goessler, W.; Francesconi, K.A.; Rubio, R., Arsenic speciation in plants growing in arsenic-contaminated sites; Chemosphere. 2008, 71, 1522–1530.
[158] Pratas, J.; Prasad, M.N.V.; Freitas, H.; Conde, L., Plants growing in abandoned mines of Portugal are useful for biogeochemical exploration of arsenic, antimony, tungsten and mine reclamation; J Geochem. Explor. 2005, 85, 99–107.
[159] Gürleyük, H.; Van Fleet Stalder, V.; Chasteen, T.G., Confirmation of the biomethylation of antimony compounds; Appl. Organomet. Chem. 1997, 11, 471–483.
[160] Jenkins, R.O.; Craig, P.J.; Miller, D.P.; Stoop, L.C.A.M.; Ostah, N.; Morris, T.A., Antimony biomethylation by mixed cultures of micro-organisms under anaerobic conditions; Appl. Organomet. Chem. 1998, 12, 449–455.
[161] Dodd, M.; Pergantis, S.A.; Cullen, W.R.; Li, H.; Eigendorf, G.K.; Reimer, K.J., Antimony speciation in freshwater plant extracts by using hydride generation–gas chromatography–mass spectrometry; Analyst 1996, 121, 223–228.
[162] Webb, S.M.; Gaillard, J.F.; Ma, L.Q.; Tu, C., XAS speciation of arsenic in a hyperaccumulating fern; Environ. Sci. Technol. 2003, 37, 754–760.
[163] Smith, P.G.; Koch, I.; Reimer, K.J., Uptake, transport and transformation of arsenate in radishes (*Raphanus sativus*); Sci. Total Environ. 2008, 390, 188–197.
[164] Koch, I.L.W.; Feldmann, J.; Andrewes, P.; Reimer, K.J.; Cullen, W.R. Antimony species in environmental samples; Int. J. Environ. Anal. Chem. 2000, 77, 111–131.
[165] Miravet, R.; Bonilla, E.; López-Sánchez, J.F.; Rubio, R., Antimony speciation in terrestrial plants. Comparative studies on extraction methods; J Environ. Monit. 2005, 7, 1207–1213.
[166] Madrid, Y.; Barrio-Cordoba, M.E.; Cámara, C., Biosorption of antimony and chromium species by *Spirulina platensis* and *Phaseolus*. Applications to bioextract antimony and chromium from natural and industrial waters; Analyst 1998, 123, 1593–1598.

17

Cadmium and Zinc

Rufus L. Chaney

USDA-ARS-Environmental Management and By-product Utilization Laboratory, Beltsville, Maryland, USA

17.1 Introduction

Cadmium (Cd) is well known for causing adverse health effects in subsistence rice farmers in Asia, and as a subject of food-chain concern, but is seldom important as a cause of phytotoxicity in the field. On the other hand, zinc (Zn) is commonly both a deficient and phytotoxic element in soils, the latter due to industrial contamination. And Zn is associated with significant food-chain deficiency concerns, but not food-chain toxicity risk. In nearly all cases, soil Cd contamination occurs with 200-fold higher Zn contamination. Because these elements are usually co-contaminants, have similar properties in soils and plants, and are readily absorbed and translocated to plant shoots, Cd and Zn should be considered together. Further, their uptake and transport to plant shoots are competitive, therefore both elements need to be considered together to understand either Cd or Zn in detail. Thus the focus of this chapter is potential food-chain transfer of soil Cd risks, and remediation of Zn phytotoxicity. The interaction of soil Zn limiting Cd risk to food-chains is very important in understanding why rice has caused essentially all Cd disease attributed to soil Cd.

17.2 Geogenic Occurrence and Sources of Soil Contamination

What are normal levels of Cd and Zn in soils? Surveys of topsoil trace elements have been conducted in many countries. When agricultural soils are sampled, the slow accumulations

---

Trace Elements in Soils Edited by Peter S. Hooda
The contribution of Chaney has been written in the course of his official duties as a US government employee and is classified as a US government work, which is in the public domain in the United States of America.

Table 17.1 Concentrations of Cd and Zn in agricultural soils of the United kingdom and United States

Cd (mg kg^{-1})		Zn (mg kg^{-1})		Number	Reference
Median	Range	Median	Range		
0.7	<0.2–40.9	82	5–3648	5665	McGrath and Loveland [1]
0.20	<0.005–2.0	53.	1.5–264	3045	Holmgren et al. [2]
0.20	<0.1–5.2	56.	8–377	254	Smith et al. [3]

of Cd and Zn from agricultural amendments are included in the result. And soil contamination may be included in the set of samples, so the high end concentrations are due to contamination. But the median soil concentrations are representative of background concentrations of metals in surface soils. Table 17.1 shows the results of surveys in the United States and United Kingdom [1–3].

The most important sources of soil Zn and Cd contamination are anthropogenic with ore-like mixtures of Zn and Cd (mine wastes, smelter emissions). Most soils with ore-like Cd+Zn contamination have about 200-fold higher Zn than Cd (Cd:Zn = 0.005); this is the average ratio in Zn ores around the globe, with some variation (0.002–0.04). Huge land areas have become Zn/Cd co-contaminated from Zn, Pb, Cu, and Ag ores from dispersal of mine wastes or smelter emissions in many countries, and the contamination is persistent. Examples include the many locations in Japan, China, Korea, and Thailand where mine waste-contaminated rice soils have caused human Cd disease [4–7]. Zinc mine and smelter waste contamination of soils have been reported in many countries; large areas with historic mine waste contamination in Somerset [8] and Wales [9] were reported in the United Kingdom, and the Palmerton, Pennsylvania [10–12] and Bunker Hill, Idaho [13,14] Zn smelters are famous for denuded mountains where vegetation was killed and could not regrow due to Zn phytotoxicity. At the most contaminated locations, surface soils near smelters exceed 4 % Zn, and deposits of pyrite–rich Zn-Pb mine wastes are over 1 % Zn and pH < 4 with no living plants [15].

A few sources of geogenic Cd enrichment without 200-fold Zn contamination of soils have been identified. The high Cd:Zn ratio of these soils makes them much higher potential Cd risk to humans. In California, USA, marine shale parent soils caused Cd enrichment without much Zn enrichment, which allows the soil Cd to be much more mobile in crops [16–18]. Alum shale in Norway was also geochemically enriched in Cd [19]. Another unusual geogenic Cd enrichment case occurs in Jamaica where several 'Parishes' have soils with up to 200 mg Cd kg^{-1} in calcareous soils with bauxitic parent materials [20]. Garrett et al. [21] did an extensive analysis of the source of the Jamaica Cd enrichment and were able to trace it back to marine phosphorite, similar to other marine shale Cd enrichment cases. Many phosphate ore deposits have surface soils with substantial Cd and Cd:Zn enrichment.

Besides the ore-like Zn + Cd and phosphorite-Cd enrichment of soils, other important anthropogenic sources of high Cd or Zn contamination include Cd plating wastes (and indirectly through biosolids) (Cd:Zn ~ 0.05–0.10); Cd-Ni battery production wastes (and indirectly through biosolids) (Cd:Zn > 1.0) [22]; Cd pigment production and use (and indirectly through biosolids) (Cd:Zn > 1.0) [23]; Cd plastic stabilizer production

(Cd:Zn > 1.0) [24]); Cu-Cd alloy production emissions (Cd:Zn ~ 0.3) [25,26]; steel production emissions (Cd:Zn < 0.01); phosphate fertilizers (Cd:Zn ~ 0.13) [27]; phosphogypsum (Cd:Zn ~ 0.5) [28]; galvanized steel use (Cd:Zn << 0.01) [29]; by-product limestone from Zn-Pb mine wastes (Cd:Zn ~ 0.01) [30]; Zn in hazardous steel mill fume wastes applied as limestone [31]; by-product Zn fertilizers (Cd:Zn << 0.01) [32,33]; orchard Zn fungicide sprays [34]; and rubber products (Cd:Zn << 0.01) (contain 1–2 % Zn but only about 0.2–5 mg Cd kg^{-1} because purified ZnO is used in rubber manufacturing) used in potting media or soils contaminated by burned rubber [35,36]. In previous years, biosolids from industrialized countries contained as high as 3400 mg Cd kg^{-1}, and as high as 49 000 mg Zn kg^{-1} [37]. But in more recent decades, industrial pretreatment and regulation of biosolids use on land yielded extensive improvement in biosolids quality. Modern biosolids had a median Cd of 2 mg kg^{-1} dry weight (DW), and the 90th centile was only 4 mg Cd mg^{-1} DW, and Zn in the order of 500–1500 mg kg^{-1} [38].

As noted above, when biosolids came under intense research starting about 1970, it was found that many biosolids were highly contaminated by industrial discharges in some cities (for example, [39]). Cd and Zn plating wastes, Cd-Ni battery manufacturing wastes, Cd-pigment production or use wastes, among others, caused biosolids to exceed 1000 mg Cd kg^{-1} and 10 000 mg Zn kg^{-1} at some cities. These levels are far above the levels achieved by the year 2000 after enforcement of industrial pretreatment programs. As noted in Stehouwer et al. [38], pretreatment allowed production of biosolids with low levels of Cd and Zn in essentially all cities if they wanted to achieve agricultural utilization of their biosolids. With the establishment of regulations for use of biosolids on land, cities learned of the need to produce biosolids with low levels of metals such as Cd and Zn.

Unfortunately, the legacy of already applied contaminated biosolids has not been dealt with in the United States or other countries except in research. In the early 1970s, we found more than 1000 mg Cd kg^{-1} DW and high Cd:Zn ratio in some biosolids from three Pennsylvania cities where metal plating and Cd pigment users discharged Cd into wastewater [39]. In crop uptake tests grown on farmers fields and matched control soils, Chaney and Hornick (in [40]) found remarkable Cd accumulation by representative crops when they were grown on strongly acidic soils that had been amended with the highly Cd-contaminated biosolids (Table 17.2). Thus, the Cd in the historically applied biosolids remains plant available long after application and some remedy is needed for these legacy contaminated sites. In the case of the city with Cd-pigment waste contamination, the city was able to sue the source of the Cd and obtain funds to raise all the treated fields to pH 7 or above [41], lowering risks from the applied Cd. But because of the high Cd:Zn of these biosolids, that cannot be considered a final remediation. Development of these soils for home sites has been prohibited because of the high soil Cd and Cd:Zn levels.

By-product Zn fertilizers were listed above as a source of Cd contamination. Several experimental studies of Cd uptake by indicator crops on soils amended with the steel mill fume waste-derived Zn fertilizers showed that Cd accumulation was not important [32,33]. This appears to be a consequence of the Cd:Zn ratio of the fume wastes, much lower than 0.01 because recycled steel had relatively higher Zn content than Cd. In a recent study by Kuo et al. [42], varied rates were applied over time and the additional Zn with each increment caused lower slopes for Cd uptake by lettuce.

Another class of deliberate contamination of ZnSO$_4$ fertilizer and feed ingredients occurred starting in 1998. In this case, a commercial ZnSO$_4$ seller certified that their

Table 17.2 Composition of Swiss chard (Beta vulgaris var. cicla) and oat (Avena sativa L.) grown in 1977 on long-term biosolids utilization farms and matched control soils at six cities in the north-eastern United States (Chaney and Hornick in [40]). Cities are identified by numbers; C = control field; S = biosolids amended field; U = unlimed; L = limed; H = at near neutral pH as managed by farmer. Soil pH as measured in May 1977. Chard leaves harvested by July 1977. Soil pH is 1:1 soil:water mixture, by volume, after 1 hour. Biosolids were applied to Hagerstown silt loam soil from 1962 to 1975 at city 4; to Lansdale sandy loam from 1967 to 1973 at city 9; to Readington silt loam from 1967 to 1974 at city 13; to Lansdale loam from 1967 to 1975 at city 1; on Genesee silt loam from 1960 to 1976 at city 19; and on Hagerstown silt loam from 1960 to 1971 at city 39. Biosolids sampled during the experiment contained 22.1 mg Cd kg^{-1} DW and 1720 mg Zn kg^{-1} DW at city 4; 169 mg Cd kg^{-1} DW and 5050 mg Zn kg^{-1} DW at city 9; 683 mg Cd kg^{-1} DW and 6430 mg Zn kg^{-1} DW at city 13; 100 mg Cd kg^{-1} DW and 3450 mg Zn kg^{-1} DW at city 1; 6.6 mg Cd kg^{-1} DW and 1430 mg Zn kg^{-1} DW at city 19; and 95 mg Cd kg^{-1} DW and 2460 mg Zn kg^{-1} DW at city 39

City	Soil total metals (mg kg^{-1} dry soil)		Soil pH	Metals in chard leaves (mg kg^{-1} dry weight)		Metals in oat grain (mg kg^{-1} dry weight)	
	Zn	Cd		Zn	Cd	Zn	Cd
4C	73.1	0.22	5.4	179.	1.48	30.8	0.034
4C-L	63.0	0.16	6.4	49.3	0.63	29.2	0.025
4S	156.	0.98	4.9	800.	3.24	50.4	0.209
4S-L	154.	0.94	6.0	105.	0.82	33.7	0.065
9C	52.9	0.18	4.9	189.	3.34	34.5	0.104
9C-L	50.6	0.15	6.4	51.7	0.94	32.2	0.052
9S	82.4	1.66	4.9	1230.	94.8	62.6	1.96
9S-L	91.0	2.10	6.3	65.6	2.20	35.7	0.259
13C	58.6	0.10	5.3	183.	2.65	29.7	0.076
13C-L	60.7	0.10	6.1	50.3	0.37	27.1	0.064
13S	146.	9.10	5.5	823.	54.3	46.6	2.24
13S-L	129.	7.02	6.2	90.	5.34	30.7	0.277
1C	53.3	0.07	5.9	63.4	0.44	23.9	0.051
1C-L	52.5	0.07	6.3	52.0	0.33	22.2	0.044
1S	146.	3.26	5.5	896.	9.51	38.2	0.299
1S-L	212.	4.50	6.2	212.	3.68	34.1	0.193
1H	150.	2.54	6.6	109.	1.89	32.0	0.125
19CH	51.8	0.09	6.1	99.4	0.41	27.6	0.065
19SH	156.	0.41	5.9	40.6	0.64	33.0	0.060
39CH	56.3	0.05	5.6	31.6	0.30	23.9	0.022
39SH	602.	12.7	6.7	162.	3.32	59.8	1.22

Adapted from Chaney R.L.; Hornick, S.B.; Simon, P.W., Heavy metal relationships during land utilization of sewage sludge in the Northeast, Land as a Waste Management Alternative, Ann Arbor Science Publishers, Inc., 1977 [40].

product contained low Cd, but they had mixed very high levels of Cd wastes into the product. Table 17.3 shows the analysis of five samples of the product received in Seattle, WA, USA and analyzed by the Washington Department of Agriculture, compared with normal ZnSO$_4$ fertilizer products. Because Washington State had established limits on Cd

Table 17.3 Cd and Zn concentration and Cd:Zn ratio of Cd-contaminated Zn-fertilizer product delivered to north-western United States/Canada

Sample	Cd (mg kg^{-1} DW)	Zn (mg kg^{-1} DW)	Cd:Zn (g g^{-1})
China-Zn-1	46 400	345 000	0.135
China-Zn-2	72 800	313 000	0.233
China-Zn-4	215 000	216 000	0.995
China-Zn-5	199 000	230 000	0.865
Cenes[a] ZnSO$_4$	7.1	320 000	0.000 022
Blue-Min[b]	49.0	420 000	0.000 127

[a] Zn fertilizer grade product.
[b] By-product Zn-fertilizer.
Data provided by Washington State Department of Agriculture

levels in Zn fertilizer products, they could prohibit importation or sale of the product. In subsequent years, similar episodes of sale of highly Cd contaminated ZnSO$_4$ were identified in the European Union (EU), Australia, South Africa, and Kenya. The product had been used as Zn fertilizer, or as Zn feed additive and the livestock manure was spread on farmland before the contamination was identified. These Cd contamination episodes are also legacy situations where the high Cd:Zn ratio of the applied materials promotes uptake by plants, and the high uptake and potential risk will persist.

Zinc-fungicide orchard spray caused accumulation of Zn in acidic light-textured soils, which caused Zn phytotoxicity to subsequent crops [34]. Farmers limed soils to prevent this adverse effect of the accumulated Zn. Later, partly because of the higher soil pH, and partly because of Zn inhibition of Ni uptake by plants, Ni deficiency was observed in pecan trees growing on old orchard soils in Georgia, USA [43]. This Zn-contaminated soil-induced Ni deficiency is a new mechanism for adverse effects of accumulated soil Zn.

Livestock manures also contain Cd and Zn (Table 17.4) [44] and, because high cumulative quantities of manure are applied on agricultural land, it can be the most important source of soil enrichment over time. Several research groups have attempted to model

Table 17.4 Zn and Cd levels in livestock manures

Species	Zn (mg kg^{-1} dry matter)		Cd (mg kg^{-1} dry matter)		Cd:Zn (g g^{-1})
	Mean	Range	Mean	Range	
Dairy cattle FYM[a]	153	99–238	0.38	<0.10–0.53	0.0025
Dairy cattle Slurry	209	<5–727	0.33	<0.10–1.74	0.0016
Beef cattle FYM	81	41–274	0.13	<0.10–0.24	0.0016
Beef cattle slurry	133	68–235	0.26	0.11–0.53	0.0020
Pig FYM	431	206–716	0.37	0.19–0.53	0.000 86
Pig slurry	575	<5–2500	0.30	<0.10–0.84	0.000 52
Broiler/turkey litter	378	208–473	0.42	0.20–1.16	0.0011
Layer manure	459	350–632	1.06	0.44–2.04	0.0023

[a] FYM, farmyard manure.
Adapted from Nicholson, F.A.; Chambers, B.J.; Williams, J.R.; Unwin, R.J., Heavy metal contents of livestock feeds and animal manures in England and Wales; Bioresource Technol 1999, 70, 23–31 [44].

Table 17.5 Metal accumulation in soils of England and Wales from different sources

	Zn (Mg y^{-1})	Cd (Mg y^{-1})	Cu (Mg y^{-1})	Cd:Zn (g g^{-1})
Aerosol deposition	2457	21.	631	0.0085
Livestock manure	1858	4.2	643	0.0022
Biosolids	385	1.6	271	0.0042
Phosphate fertilizers	213	10.0	30	0.047

Adapted from Nicholson, F.A.; Smith, S.R.; Alloway, B.J.; Carlton-Smith, C.; Chambers, B.J., An inventory of heavy metals inputs to agricultural soils in England and Wales; Sci. Total Environ. 2003, 311, 205–219 [45].

long-term additions of Cd, Zn, Cu, and P in soils, considering atmospheric inputs, fertilizer inputs, biosolids/compost inputs, and manure inputs. Table 17.5 shows an inventory of average Cd and Zn additions to the soils of England and Wales based on studies by Nicholson *et al.* [45]. Livestock manure was the most important agricultural source of added Zn, while phosphate fertilizers were more important than manure, which was more important than biosolids as a source of added Cd. Atmospheric deposition was larger than the usual soil amendments. It is not clear how much of the measured atmospheric deposition resulted from suspended soil/dust, which should be considered recycled soil rather than input unless the loss by wind erosion is also measured. Manures are important sources because of both the importation of feedstuffs from off farm and the supplementation of diets with Zn and Cu salts. Both swine and poultry diets are often greatly enriched in Zn and Cu to promote growth, and local utilization of the manure enriches soil Zn and Cu levels. Many nations regulate the levels of Zn and Cu in pre-mixed feeds, but not in locally mixed feeds.

Manure is a significant source of Cd addition but because Zn is usually added at much higher rates, the Cd:Zn of manures are usually quite low, <0.002 (Table 17.4). The plant uptake of manure-applied Cd was examined by Jones and Johnston [46] on long-term plots at Rothamsted Experimental Station in the United Kingdom. In that case, plant uptake from aerosol-deposited plus fertilizer Cd and Zn was compared with plots amended annually with livestock manure, which added considerably more Cd and Zn to the soils. Interestingly, over time plant uptake of Cd declined on the manure plots but increased on the aerosol deposition plus fertilizer plots. Repeated application of manure raised soil pH, while the pH of the control plot soils fell over time.

Another kind of inventory considers the multiple sources of fertilizer nutrients and likelihood of application on particular farms. For example, livestock farms produce manure rich in N and P, so they would not need to purchase P-fertilizers or accept biosolids as a P-fertilizer source. Table 17.6 shows this assessment of Cd application from significant sources to soils of Sundgau Canton of Switzerland [47]. Again, manure was the most important Cd source among normal amendments, but atmospheric deposition was still larger. And among farm types, livestock farms accumulated Cd faster than others.

An assessment of potential additions of various potentially toxic elements including Cd and Zn was recently conducted for Canada. The approach was different from the two discussed above, but considered the same sources. Both manure (Zn, Cu) and P-fertilizers (Cd) were important sources of Zn and Cd enrichment, and livestock farms received the most important additions of Zn and Cd [48].

Table 17.6 Estimated Cd accumulation in soils of Sundgau Canton from different sources, and for different farm types

Fluxes	Arable farming	Dairy and mixed	Animal husbandry	Whole region
Input fluxes (g ha^{-1} y^{-1})				
Manure	<0.1	0.7	8.2	0.6
Biosolids	0.4	0.1	<0.1	0.1
Commercial fertilizers	0.6	0.2	<0.1	0.4
Deposition	2.1	2.1	2.1	2.1
Output fluxes (g ha^{-1} y^{-1})				
Crop removal	0.6	1.9	0.6	1.4
Leaching	0.4	0.4	0.4	0.4
Net flux (g ha^{-1} y^{-1})	2.3	0.8	9.1	1.4

Adapted from Keller, A.; Schulin, R., Modelling regional-scale mass balances of phosphorus, cadmium and zinc fluxes on arable and dairy farms; Eur. J. Agron. 2003, 20, 181–198 [47].

17.3 Chemical Behavior in Soils

The chemical form of Cd and Zn in soils is important because added soluble metals react over time to less phytoavailable forms in soils. Further, because rice has comprised such an important source of human Cd risk, the chemistry of Cd in flooded (anaerobic; reduced) soils is especially important. In general, both elements co-occur in sphalerite as a solid solution of ZnS. When sphalerite is mixed in aerobic soil, the sulfide is oxidized by microbes and soluble Zn and Cd are released to react with the soil [49]. At least temporarily some ZnS and CdS are present in the mine waste-contaminated soil [50]. In flooded soils, several changes occur sequentially that affect the phytoavailability of Zn and Cd [51]. As reduction of soil chemicals increases over time, soil pH rises and CdS has been shown to be formed [52] by using X-ray absorption spectroscopy (XAS). And upon drainage of the flooded soil, pH falls and CdS is oxidized rapidly. The oxidation of CdS and drop in pH greatly increase soil Cd phytoavailability [51,53], allowing little adverse effect of Zn on rice yields but for grain to contain high Cd levels at harvest. In the same reduced soils, Zn does not appear to form sphalerite [54]. Fields are normally drained at rice flowering to maximize yields and prepare the fields for harvest. An important management method to reduce rice Cd accumulation is to keep the field flooded until grain maturity, but growers seldom want to use this method because of reduced yield and inconvenience at harvest. Rainfall during the grain filling period can slow soil oxidation and reduce Cd accumulation in rice [55], causing variation in grain Cd from year to year. Because of the flooding management, most rice roots are restricted to the near surface soil [51], which promotes Cd uptake after drainage.

In aerobic soils, it is clear that added soluble Zn reacts to form adsorbed and occluded species with lower phytoavailability. When Zn fertilizers are applied, even to acidic soils, the Zn becomes less phytoavailable over time [56–58]. When the labile pool of Zn in soils was assessed, the labile fraction varied among soils with generally lower lability in alkaline soils than in acidic soils [59]. More recent studies using XAS have shown that Zn–Al layered double hydroxides (LDH) can be formed from Zn added to soils, and that this form becomes more stable over time by additional reactions with aluminosilicates [60–62]. Roberts et al. [63] and Scheinost et al. [64] used EXAFS (extended X-ray absorption fine structure) spectroscopy to examine forms of Zn in a highly contaminated soil from Palmerton, PA, USA that contained 6200 mg Zn kg^{-1}, had a pH value of 3.2 and was rich in organic matter. They found 2/3 franklinite ($ZnFe_2O_4$) and 1/3 sphalerite (ZnS), along with some Zn adsorbed on Fe and Mn oxides. These minerals could have been present in wind-blown ore particles which contaminated the area. More samples of highly Zn-contaminated soils need to be characterized by XAS to determine the kinds of Zn species which can form after Zn reacts with soils, especially in contaminated soils made calcareous to prevent adverse Zn effects.

Cadmium's reactions in soils are substantially different from those of Zn. Although several groups have tested for formation of LDH compounds from Cd, none have found this reaction to occur [62]. Instead, most Cd remains in the labile pool over time. This is especially evident in soils amended with biosolids, where more than 90 % of added Cd remained labile over 25 years after the biosolids had been land-applied [65]. Cadmium in phosphate-amended soils may be slowly transformed to nonlabile forms, but the process is slow and incomplete [66]. Mine waste-contaminated soils may have a lower fraction of labile Cd because CdS persisted in the soil, or because the Cd was occluded in other mineral [50].

17.4 Plant Accumulation of Soil Cadmium and Zinc

Plant uptakes of Cd and Zn are affected by soil properties, plant species and cultivar, fertilizers, agronomic management, and properties of the metal source. So many factors can affect crop Cd and Zn accumulation in particular cases that only the key factors are discussed here. More information can be found in reviews of soil and plant factors that may affect accumulation of Cd in edible crop tissues [67–70]. Because soil Cd remains in the labile pool in Cd-enriched soils, the Cd remains phytoavailable; no method to transform soil Cd into low phytoavailable forms has been identified although soil management can reduce uptake such as by raising soil pH.

Membrane transporters are now understood to absorb Cd and Zn from soil solution. The primary uptake uses transporters of the ZIP family cloned from *Arabidopsis* [71], and other types of transporters are believed to move Cd and Zn into the xylem, into leaf cells, and into vacuoles. Two membrane transporters for uptake of soil solution Zn have been cloned from *Arabidopsis thaliana* L. [72]. The OsZIP1 protein is a high-affinity Zn transporter with $K_m = 16.3$ μM, which is much higher than found in soil solutions of fertile soils [73]. The HMT4 protein was shown to control Zn transfer into the xylem of *Arabidopsis halleri*, a Zn hyperaccumulator species [74].

In other studies, Cohen et al. [75] reported that the ferrous transporter of dicot roots also transported Zn, Cd, Cu, and Mn. Subsequently, they reported studies at activities of these other ions more like those in soil solutions and noted that the IRT1 ferrous transporter was likely only important for Fe uptake, not for Cd or Zn [76]. Thus, Cd^{2+} and Zn^{2+} ions are believed to be accumulated across the root plasma membrane by a specific permease or transport protein. The study of Hart et al. [77] showed that Zn and Cd uptake were mutually inhibitory when practical concentrations were examined; Zn^{2+} is usually at much higher activity in soil solution than is Cd^{2+}. Transport from roots to shoots varies widely among plant species [78], and also among cultivars within species (see below).

Major soil and plant factors which influence Cd and Zn accumulation include (i) plant species and cultivar; (ii) soil Cd and Zn phytoavailability and Cd:Zn ratio; (iii) soil pH; (iv) soil sorption surfaces including organic matter, and amorphous Fe and Mn oxides; (v) chloride; (vi) fertilizer N and P; and (vii) crop rotation or preceding crop. The role of plant species and cultivar variation in Cd accumulation is discussed by Grant *et al.* [79] and will not be addressed in detail here. All species which have been tested have been found to have substantial genetic variation in Cd accumulation. Breeding programs have been conducted for durum wheat and sunflower to obtain lower Cd levels in grain/kernels [80,81] with good success. In durum wheat, the key difference between lower- and higher-Cd genotypes was translocation from root to shoots [82]. But breeding for lower Cd is expensive and competes with normal breeding goals such as yield, disease resistance, and so on. Many groups are attempting to breed lower-Cd rice cultivars (for example, [53,83]), but this is unlikely to be fully successful unless contaminated soils are kept flooded to lower soil Cd phytoavailability to rice [79].

The soil Cd and Zn concentrations and Cd:Zn ratio of the Cd source is one key factor. The lower the Cd and Cd:Zn ratio, the less likely that increased bioavailable Cd will accumulate in crops. Zn and Cd are affected similarly by soil properties in many ways, and Zn inhibits Cd uptake by roots and transport to shoots [77,84]. For some crops, Zn inhibits Cd transfer to grain or other storage tissues [85]. Chaney *et al.* [85] also illustrated the strong effect of Cd:Zn ratio on maximum Cd concentrations which can be reached in lettuce shoots (Table 17.7); when shoot Zn exceeded 400–500 mg kg^{-1} dry leaves, plants became chlorotic and had reduced yield. Studies of lettuce grown on Zn-smelter-contaminated garden soils showed the very strong reductions in yield if highly Cd+Zn contaminated soils were acidic [37]. When no edible crop yield is produced, humans are not at risk from the soil Cd. The strong impact of Cd:Zn ratio of land-applied biosolids in subsequent Cd uptake if the soil becomes acidified is illustrated in Table 17.2.

Soil pH is usually considered a central factor in Cd accumulation; more acidic soils bind Cd less strongly, allowing more Cd uptake [86,67]. A few rare cases have been reported where increasing pH did not reduce crop Cd concentration [87]. Soils on one farm where a high-Cd and high-Cd:Zn biosolids had been applied before regulations were developed caused lettuce to accumulate very high Cd concentrations whether acidic or calcareous (Table 17.8) [23]. The high Cd:Zn ratio apparently induced a Zn deficiency stress that caused the lettuce to keep trying to take up Zn, but the plant accumulated more Cd because Zn activity was so low in the calcareous soil. Actual Zn deficiency has been repeatedly observed to cause higher Cd accumulation by plants [68,88].

Chloride was identified as very important in Cd uptake from neutral pH soils [89]. High soil chloride causes formation of $CdCl^+$ complex that is soluble in soil solution. The mechanism by which this complex is absorbed has not been determined, but high chloride caused high crop Cd accumulation even from alkaline soils [89–91]. A study by Smolders and McLaughlin [92] tested the role of Cd^{2+} activity versus chloride in Cd uptake by using a chelating resin to buffer Cd^{2+} activity; added chloride caused more Cd to be dissolved in the resin-buffered nutrient solution, and more Cd to be absorbed by the plant even though Cd^{2+} activity was tightly controlled. For soils irrigated with

Table 17.7 Interactions of Cd and Zn in uptake by lettuce grown on acidic soil: Cd and Zn in lettuce shoots in relation to Cd:Zn ratio and Zn phytotoxicity response to the treatments [86]. A shows the metal additions and Cd:Zn ratio of the added Cd and Zn; B shows the leaf Cd, and C, shows the leaf Zn [85]. Bold type denotes nonphytotoxic Zn; low Cd:Zn ratio = edible crop

Added soil Cd (mg kg^{-1})	Added soil Zn (mg kg^{-1})					
	0	30	60	90	120	150
A	Cd:Zn of added metals					
0.00	<**0.001**	<**0.001**	<**0.001**	^{a}	^{a}	^{a}
0.75	>0.10^b	0.025^b	**0.012**	**0.008**	0.006^a	0.005^a
1.50	>0.10^b	0.050^b	0.025^b	0.017^b	0.012^a	0.010^a
3.00	>0.10^b	0.100^b	0.050^b	0.033^b	0.025^b	0.020^{ab}
B	Cd (mg Cd kg^{-1} dry leaves)					
0.00	**1.15**	**0.60**	**1.16**	1.16^a	0.37^a	1.07^a
0.75	8.14^b	**4.21**^{b}	**4.21**	4.55	4.63^a	4.95^a
1.50	18.5^b	7.75^b	7.60^b	9.41^b	7.88^a	8.69^a
3.00	25.2^b	15.2^b	15.0^b	19.3^b	17.2^b	21.4^{ab}
C	Zn (mg Zn kg^{-1} dry leaves)					
0.00	**30.**	**165.**	**262.**	401.^{a}	508.^{a}	615.^{a}
0.75	31.^{b}	154.^{b}	**240.**	312.^{b}	425.^{a}	842.^{a}
1.50	31.^{b}	128.^{b}	205.^{b}	320.^{b}	478.^{a}	672.^{a}
3.00	37.^{b}	102.^{b}	185.^{b}	293.^{b}	406.^{b}	522.^{ab}

^{a}Zn phytotoxicity caused yield reduction >35%.
^{b}Cd:Zn ≥ 0.015 limit of recommended biosolids.
Adapted from R.L. Chaney, J.A. Ryan, Y.-M. Li and J.S. Angle, Transfer of cadmium through plants to the food chain, Proceedings Workshop 'Environmental Cadmium in the Food Chain: Sources, Pathways, and Risks', Scientific Committee on Problems of the Environment, 2001 [85].

Table 17.8 Effect of adding limestone, Zn, or peat plus Zn to pH 6.2 soil which contained 6.8 mg Cd kg^{-1} and 89 mg Zn kg^{-1}; a Cd-rich, high-Cd:Zn biosolids was applied for many years. Zn deficiency was induced by liming in the presence of high soil Cd levels, causing higher shoot Cd concentration than the unlimed treatment

Treatments	pH	Yield (g per pot)	Lettuce CD (mg kg^{-1})	Lettuce Zn (mg kg^{-1})
Control	5.6	5.9 c^a	14.3 b	12.8 e
CaCO$_3$	7.5	1.6 d	19.5 a	9.9 f
CaCO$_3$ + Zn	7.4	6.8 abc	1.9 g	53.7 c
CaCO$_3$ + peat	7.3	1.2 e	12.0 bc	9.3 f
CaCO$_3$ + peat + Zn	7.3	7.9 ab	2.2 f	62.8 b

^{a}Means followed by the same letter are not significantly different ($P < 0.05$) according to the Waller–Duncan K-ratio t test.
Adapted from R.L. Chaney et al., Zn deficiency promotes Cd accumulation by lettuce from biosolids amended soils with high Cd:Zn ratio, J. Residual Sci. Technol, 2006, 68–75 [23].

high-chloride waters, or soils with geogenic chloride, Cd is increased in crops as much as 10-fold compared with similar soils with basal chloride. Interestingly, increasing soil chloride does not increase crop Zn, so the increased Cd in crops may have higher bioavailability to animals that ingest the crops. One possible mechanism adding to the effect of chloride on Cd uptake may be induced Zn-deficiency; Khoshgoftarmanesh *et al.* [93] found that chloride reduced Zn activity and uptake by wheat, and that adding Zn fertilizer in this case significantly reduced wheat Cd. Because chloride is an essential element, and applied to reduce disease incidence, tests have been conducted with fertilizer levels of chloride application [94]; fertilizer chloride rates had little effect on crop Cd.

The role of Cd in P-fertilizers in soil and crop Cd concentrations is a complicated picture. As noted above, regular small additions of Cd in P-fertilizers may accumulate in soil, but uptake is usually little affected except by the quite high Cd phosphates such as those from Narau [27]. In many studies the form of N-fertilizer applied has greater effect on crop Cd than the amount of Cd in the P-fertilizer applied [96]. The recent work of Huang *et al.* [95] evaluated repeated applications of P-fertilizers over five crops of lettuce and found low additivity of the applied Cd on crop Cd. It appears that P reaction with Fe oxides may increase the selective adsorption of Cd, thereby reducing Cd phytoavailability. As noted above, studies of long-term fertilizer plots with low-Cd phosphates has shown little or no affect of long-term repeated applications of P-fertilizers on Cd accumulated in staple grains (for example, [97]). European Union regulations have encouraged production of phosphates with lower Cd levels [98].

One aspect of Cd in crops has been difficult to explain, the effect of previous crops (crop rotation) on Cd uptake by the current crop. Oliver *et al.* [99] found that previous legume crops caused higher Cd uptake by wheat than did previous cereal crops. Grant *et al.* [68] noted some evidence of a similar effect of a previous Cd-accumulating crop such as flax or sunflower on Cd in wheat grain. Khoshgoftarmanesh and Chaney [100] found that when sunflower stover was removed from the field, the following crop of wheat had lower grain Cd than following wheat. None of the expected factors (pH, organic matter, etc.) seems to explain these results.

Another complication is soil moisture in the tillage depth where applied Cd accumulates. If active roots are not present in the topsoil during grain filling, properties of the topsoil (pH, Zn) may have little effect on grain Cd accumulation [68,88]. Further, chloride in the subsurface soil can move up during dry conditions and have large impact on plant Cd accumulation.

17.5 Risk Implications for Cadmium in Soil Amendments

Biosolids, Composts, and Manures
As noted above, rice and tobacco soils are unique in their potential to transfer soil Cd to humans when soils contain ore-like Zn+Cd enrichment. Today most soil amendments (manure, compost, biosolids) in the European Union, North America, and Australasia have favorable Cd:Zn ratios to limit Cd transfer and bioavailability. Biosolids have received much attention regarding Cd risk, but readers should recognize that the apparent risks of Cd in biosolids came from experiments that used biosolids that are no longer produced or

permitted under national regulatory systems. Industrial pretreatment and reduced use of Cd in products had greatly improved biosolids quality at nearly every treatment works [38].

Subsequent research on Cd applied in biosolids has shown that when high-quality (low Cd and low Cd:Zn ratio) biosolids were applied, crop Cd was only marginally increased and Zn came along with the Cd so that bioavailable Cd in leafy vegetables was not increased [101,102]. Over 20 years after field plots were established, Cd uptake by lettuce was not increased over time for biosolids with lower Cd levels even though added organic matter was largely biodegraded [103]. Research by Hettiarchchi *et al.* [104] and Kukier *et al.* [65] showed that the presence of Fe and Mn oxides and phosphates in applied biosolids had increased the specific Cd adsorption capacity of the inorganic fraction of biosolids-amended soils, which substantially lowered the phytoavailability of soil Cd (see Figure 17.1) (see also Basta *et al.* [105]).

Many sources of error have been identified in testing potential Cd uptake by crops, from using Cd salts without the matrix of a soil amendment, to using greenhouse pots which restrain all roots to the contaminated soil depth (see [27]). Because of the strong adsorption of Cd and other metals by the matrix of biosolids, especially those which contain Fe oxides from Fe added to remove phosphate during wastewater treatment, and the low bioavailability of Cd in plants when Cd:Zn ratio is similar to geogenic, there is no longer any basis to predict increase in human Cd risk due to land-application of biosolids, composts, or

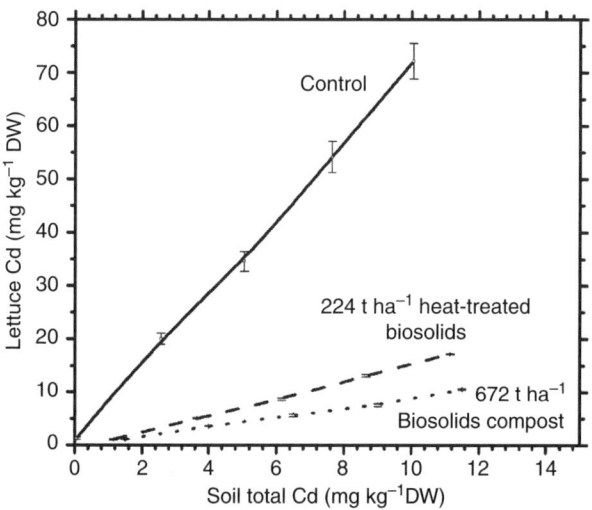

Figure 17.1 *Effect of historic biosolids application on the phytoavailability of applied Cd salts to Romaine lettuce (Lactuca sativa var. longifolia). Soils were collected from long term biosolids amended and control soils from plots established in 1976 on Christiana fine sandy loam. Soils with all treatments were amended with Cd salt and adjusted to pH 6.5 in 0.01 M Ca(NO$_3$)$_2$ before growing 'Parris Island' Romaine lettuce to maturity; Cd in the heat-treated biosolids was 13.4 mg Cd kg^{-1}, and in the composted biosolids, 7.2 mg Cd kg^{-1} (Modified from Kukier, U.; Chaney, R.L.; Ryan, J.L.; Daniels, W.L.; Dowdy, R.H.; Granato, T.C., Phytoavailability of cadmium in long-term biosolids amended soils' J. Environ. Qual. 2010 [65].)*

livestock manure except for rice or tobacco soils. An examination of the literature shows that biosolids caused high crop Cd only when they had high Cd and Cd:Zn ratio, and soils were strongly acidic [37,106].

Crops with Unusually High Normal Cadmium Concentrations
We have described above how rice and tobacco can transfer soil Cd into humans to cause health risks. No other crops have been implicated in human Cd disease. Some foods naturally contain comparatively high Cd levels (for example, sunflower kernels, flax, poppy seeds, durum wheat, chocolate [107–110]) (Figure 17.2), but only small amounts of these are consumed daily. The Cd in these crops appear to have low bioavailability as evidenced by a one-year sunflower kernel consumption trial with human volunteers who showed no increase in blood Cd despite doubling of Cd intake [111]. This result was similar to that of Vahter *et al.* [112] who sampled nonsmoking young Swedish women, some of whom ingested shellfish, which tripled their Cd intakes, compared with women not consuming shellfish in their diets. Shellfish significantly increased daily Cd intakes, but this did not cause increased blood Cd. The shellfish also supplied high levels of bioavailable Zn and Fe that lowered the absorption of diet Cd compared to non-shellfish-containing diets. Therefore, the mere fact that a diet contains elevated levels of Cd does not mean that humans will absorb higher amounts of Cd with concomitant adverse effects on health. Fox *et al.* [113] stressed that estimates of food-Cd risks should be based on intakes of Cd similar to those of human diets, not the massive levels often studied. Fox *et al.* [113] illustrated trace element deficiency interactions in Cd absorption and used

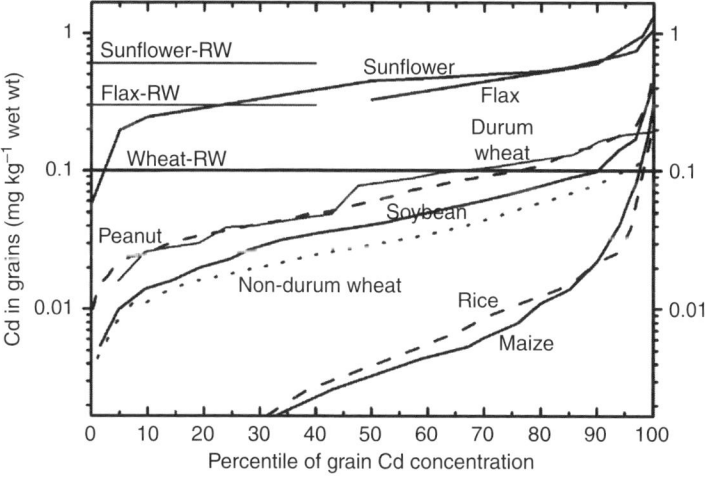

Figure 17.2 *Variation among crop in statistical distribution of Cd concentration [106]. Data for US wheat, rice, corn, soybean, peanut from Wolnik et al. [107,108]; data for flax from Klein and Weigert [109]; data for sunflower from Ocker et al. [110]. German Richtwert for crops shown with horizontal lines (Modified from Chaney, R.L.; Ryan, J.A.; Li, Y.-M. et al., Phytoavailability and bio-availability in risk assessment of Cd in agricultural environments, Proceedings of the OECD Workshop on Sources of Cadmium in the Environment, 1996 [106].)*

'nutritional' rather than 'toxicological' approaches to the study of dietary Cd risk. The nutritional perspective seems even more important after our recent research showing that Fe-Zn-Ca 'marginal' diets can cause a large increase in diet Cd bioavailability [114]. Indeed, despite many papers showing a possible role of metallothionein in intestinal Cd metabolism, when Reeves *et al.* [115] used metallothionein-null mice and the nutritional approach for testing, they found no effect of the metallothionein genes on uptake or retention of Cd, or on the large effect of Fe-Zn-Ca marginal deficiency on Cd retention. A fuller discussion of the complexity of dietary Cd bioavailability is reported by Reeves and Chaney [114].

17.6 Plant Uptake of Cadmium and Zinc in Relation to Food-Chain Cadmium Risk

When soils are strongly acidic, both Cd and Zn are more soluble and are taken up more readily by plants. Both elements are relatively easily translocated to plant shoots, so food-chain risks and benefits need to be considered. At some combination of acidic pH and nearly all Zn+Cd co-contamination, Zn becomes phytotoxic but Cd remains far below phytotoxic concentration because of the Cd:Zn ratio present in the soil. This is illustrated by Table 17.7 which reports the concentration of Cd and Zn in Romaine lettuce (*Lactuca sativa* L. var. *rotundifolium*) shoots when the plants were grown on pH 5.5 soil with factorial additions of salt Zn and Cd at different Cd:Zn ratios [85]. At some rate of Zn application, lettuce accumulated more than 400 mg Zn kg^{-1} DW and had 35 % yield reduction or greater. And at some rate of Cd application, the Cd:Zn ratio was higher than geogenic ratios such that Cd uptake to shoots was high enough that the lettuce should not be used as food (a CODEX limit for Cd in lettuce was proposed at 0.05 mg Cd kg^{-1} fresh weight which would be equal to 4 mg Cd kg^{-1} dry weight; see [134]). But when the Cd:Zn ratio was in the range of geogenic contamination, Zn phytotoxicity occurred before excessive Cd accumulation occurred in lettuce shoots.

In addition, when a food accumulates high levels of Zn along with increased levels of Cd in plant shoots such as lettuce or spinach (*Spinacia oleracea* L.), the much higher Zn inhibits absorption of Cd in the animal that ingests the crop [101,102,116,117]. Thus at the usual geogenic ratio of Cd and Zn, even crops (except rice and tobacco) with relatively high ability to accumulate Cd do not comprise risk to animals because Zn is also present. In addition, tests of Cd accumulation by lettuce grown on smelter-contaminated soils showed that whenever Zn reached levels near 500 mg kg^{-1} DW, yield was strongly reduced; and that this phytotoxicity of Zn limited the potential yield of garden crops and hence risk from home garden soils. The raw data of Baker and Bowers [118] were presented graphically by Chaney and Ryan [37]; the strong Zn phytotoxicity protection against any food-chain Cd risk is evident.

Potential phytotoxicity from Cd or Zn raises the issue of Cd and Zn speciation in plant tissues. Ordinarily Zn^{2+} in cytoplasm is mostly complexed with organic and amino acids. Grill *et al.* [119] discovered that when cell cultures were fed toxic levels of Cd, they accumulated a new form of Cd chelated with 'phytochelatins'. Phytochelatins are derivatives of glutathione with additional γ-glutamyl-cysteine residues attached to make a more selective Cd chelator. After much research on phytochelatins, it became evident that they were only formed under conditions of severe Cd phytotoxicity and have little relevance to normal plant metabolism of accumulated Cd or Zn [120,121]. Even when phytochelatins are produced in roots, only a small fraction of total Cd is present as phytochelatins [121]. Other research indicates that the presence of Cd drains reaction products from the glutathione metabolism system rather than inducing biosynthesis at the genetic level [122]. When Romaine lettuce was grown to contain basal or food-type levels of Cd, the Cd was not present as phytochelatins [123].

On the other hand, rice soils have been contaminated by geogenic Zn+Cd and caused human Cd disease without Zn phytotoxicity at many locations. From the locations in Japan where environmental Cd disease was discovered [4,124], it is clear that rice has a very different ability to accumulate soil Cd and Zn than found in other crops. It was found that rice grown at Tohyama, Jinzu Valley, Japan, had a large increase in grain Cd but no increase in grain Zn even though the soil was contaminated by Zn mine wastes which gave about 10 mg Cd kg^{-1} and 1000 mg Zn kg^{-1} soil [125]. The cause of environmental Cd disease in Japan was discovered by a soil scientist, Dr. Jun Kobayashi, who had a spectrograph and could measure Cd in water, soil, rice and human tissue samples, and who had experience with mine contamination of rice soils [4]. Dr. Kobayashi worked with the local medical doctor, Dr. Hagino, to identify the cause of bone disease and an epidemiologist at the Prefecture Medical School worked with them to test for specific effects of Cd and confirmed the Cd disease [126–128]. About 80 % of farm families more than 50 years old experienced the proximal renal tubular dysfunction that is the first adverse effect of excessive dietary Cd intakes. At Tohyama, about 300 women and one man also suffered *itai-itai* disease (osteomalacia) with repeated bone fractures. Cadmium-induced bone disease is rarely induced by dietary Cd, but was also found at one location in China [129].

Since the original finding at Tohyama, Japan, more than 50 other locations in Japan have been found to have been contaminated by mine or smelter emissions and produced rice with excessive grain Cd [130]. Then locations were discovered in China [6] where rice and tobacco contributed to human Cd disease. Locations with excessive Cd in rice grain have also been discovered in Korea and Thailand [7] and are suspected in other locations. The case in Thailand illustrates the relationships between normal Zn mining and human risk when rice is the main local crop. Subsistence farmers consume the rice grain they produce on local soils. If those soils become contaminated by dispersal of mine or smelter emissions or industrially contaminated sewage, excessive Cd in rice grain is unavoidable.

J. Scott Angle and Rufus L. Chaney (unpublished) sought a location in Asia that had not been examined previously where the need to remediate risks from rice Cd may support development of commercial Cd phytoextraction. We had developed and patented Cd phytoextraction using southern France populations of *Thlaspi caerulescens* J.&C. Presl. (alpine pennycress) and wanted to test its use in rice fields [131–133]. We looked at other nations where rice was a common crop and where Zn ores were mined and/or smelted; and where western patents were respected. Mae Sot, Thailand, was identified and we cooperated with the Thai Ministry of Agriculture to design an evaluation of soil and crop contamination. Those results were first reported in detail by Simmons *et al.* [7], who show the same relationship found in Japan. In this case, soil Zn was as high as 7000 mg kg^{-1} and Cd as high as 200 mg kg^{-1}, remarkable levels. As shown in Figure 17.3, even with the extremely high soil Zn levels, rice grain Zn was not increased, while rice grain Cd

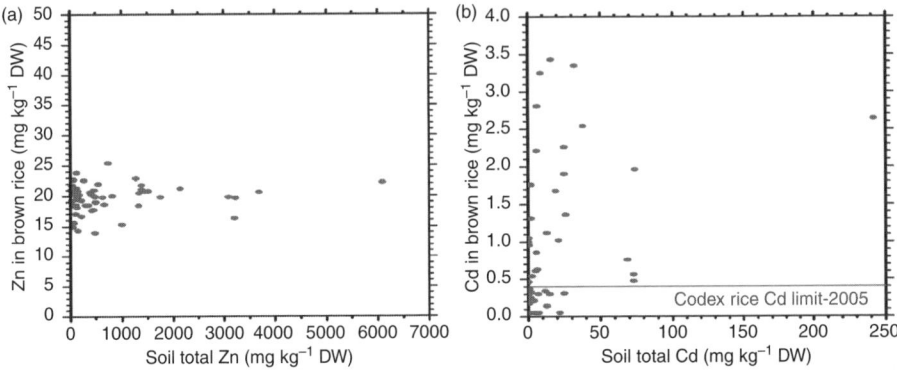

Figure 17.3 a) Zinc and (b) cadmium concentrations in brown rice grain samples in different paddy fields near Mae Sot, Thailand, Zn and Cd-contaminated by mine wastes (based on data in Simmons et al., [7]). The line across the Cd figure is the Codex limit for Cd in rice grain [134] (Modified from Simmons, R.W.; Pongsakul, P.; Chaney, R.L.; Saiyasitpanich, D.; Klinphoklap, S.; Nobuntou, W., The relative exclusion of zinc and iron from rice grain in relation to rice grain cadmium as compared to soybean: Implications for human health, *Plant Soil 2003, 257, 163–170 [7].)*

was increased to extreme levels in all fields irrigated with the water supply contaminated by mine discharge. Simmons *et al.* [134] reported more details on the patterns of contamination across irrigation networks; fields were arranged on terraced hillsides with water entering the highest field and moving sequentially through lower fields until discharged. Because much of the contamination was from suspended solids, the highest contamination was deposited near the entrance of the highest field; soil Cd and Zn concentrations fell strongly in fields lower in the landscape because the suspended solids were deposited in fields above. Unfortunately, grain Cd is not a simple linear function of soil Cd, so even fields with only moderate Cd contamination produce rice with Cd level higher than allowed to be used as food (0.4 mg Cd kg^{-1} rice grain set by the Codex Alimentarius Commission in 2006 [135]. Finally, examination of the older residents who consumed the rice grown at Mae Sot showed renal tubular dysfunction was prevalent in the population [136].

Why Does Cadmium Comprise Such High Risk to Subsistence Rice Consumers?

Rice is the only staple grain normally produced in flooded soils. Upon flooding, soil pH rises toward 7 and insoluble CdS is formed in flooded soils; both changes reduce Cd phytoavailability. But the growers usually drain their fields at flowering to improve yields. When the fields are drained, sulfides can be rapidly oxidized, as can the ammonium, ferrous, and manganous ions produced during flooding. This can cause a sharp decline in soil pH and increase in solubility of Cd and other metals. Because of the unusual properties of drained rice soils, and rice plants, little or no increase in rice grain Zn occurs (although Zn is increased in rice leaves), but Cd can be highly increased in rice grain [7,55]. This relative exclusion of soil Zn from rice grain was apparent in the first reports on the Jinzu Valley rice Cd poisoning case, and in research on rice grown in soils contaminated by mining wastes [55,125]. But the exclusion of Zn from grain on even highly Zn+Cd-contaminated rice soils was not appreciated until later research on contaminated Thai soils by Simmons et al. [7] (Figure 17.3). Interestingly, most soil testing is done on air-dried soils, and the extractable Cd and Zn levels do not give useful estimates of rice accumulation of Cd or Zn. However, when field-moist soil collected at grain maturity is tested immediately, extractable Cd is highly correlated with grain Cd [137].

In addition to the flooded-soil effect on phytoavailable Cd, the composition of rice grain promotes Cd absorption in animals. Rice grain is inherently low in Fe, Zn, and Ca concentrations, and milling of the grain [138] removes much of these elements in the brown rice. Interestingly, a much smaller fraction of Cd is removed during milling than of Zn and Fe [139,140]. Because Zn and Fe deficiency stress increase the absorption of dietary Cd, the natural composition of milled rice promotes Cd absorption by humans [114,141]. It has recently been discovered that most Cd enters the intestinal cells on the ferrous transporter (DMT1) in the duodenum [142,143]. Reeves and Chaney [144] found that a high fraction of diet Cd entered the duodenum and turned over in that tissue for many days; the longer retention in animals fed the Fe-Zn-marginal diets apparently caused 10-fold higher net absorption of diet Cd than found for animals with adequate Fe and Zn nutrition In addition, higher dietary Zn can also inhibit Cd absorption by animals (for example, [116,145]).

These findings should influence decisions about tolerable Cd in diets and crops. Presently, Cd is considered to be equally bioavailable in all crops and at all crop Zn levels. But the marginal supply of Fe and Zn in polished rice grain is now recognized to comprise an international health problem due to practical Fe and Zn deficiency in subsistence rice consumers [146]. Further, higher plant Zn is known to reduce absorption of dietary Cd [116]. In addition, Cd in spinach was found to be half as bioavailable as Cd in lettuce or other foods [116,147]. This likely results from Cd co-precipitation with Ca-oxalate in the intestine. Other studies of diet Cd absorption by humans have shown that increased Cd ingestion in shellfish [112] or in sunflower kernels [111] did not cause higher blood or urine Cd. This is in strong contrast with rice diets [148].

Several northern European populations have been reported to possibly suffer adverse effect of dietary Cd at much lower dietary Cd, blood Cd, and urinary Cd [149,150] than other populations without identified adverse Cd effects [148]. These reports are not explicable in terms of known aspects of Cd metabolism, and remain debated among scientists. Ezaki et al. [151] reported tests on over 10 000 middle-aged nonsmoking Japanese women which strongly contradict the European results. Because of high rice consumption and somewhat high Cd levels in Japanese rice, these women consumed considerably more dietary Cd than the European subjects. Blood and urine Cd were higher for the Japanese women, but the renal dysfunction indicators revealed no case of adverse effects at least to 3 μg Cd l^{-1} urine. Compare this with claims of adverse effects at lower than 2.5 μg Cd l^{-1} in the Europeans. The difference in interpretation appears to be due to the 'normal' values used for comparisons of urine indicator proteins, and the Ikeda et al. team has provided a detailed examination of the control population normal values and how to express them [152].

Other crops

Although it is clear that geogenic Zn+Cd in rice soils can readily cause human Cd disease in subsistence farmers, other crop production systems with even higher geogenic Zn+Cd contamination have not been shown to cause such Cd disease. Smelters and mine wastes were also dispersed in western countries, and villages developed on the land such that home gardens and some farm crops were produced on soils with very high Zn+Cd contamination. Examples include Shipham, UK [8,153–155], Palmerton, PA, USA [12,156], Stolberg, Germany [157,158], and France [159]. Kreis *et al.* [160] estimated increase in Cd ingestion from garden crops in the Kempenland, a smelter contaminated area of the Netherlands and Belgium, finding insufficient increase to cause individuals to exceed the allowable Cd intake. Full epidemiologic studies were conducted on persons who lived in the contaminated soil areas and similar villages without soil contamination. Thus garden foods did not cause human Cd disease even when grown on highly Zn+Cd-contaminated soils. One location in Belgium has been reported to have adverse Cd effects on human health, and crop uptake of Cd was measured [161], but as was discussed above, the Cd risk from crop exposure does not agree with the other studies [148,152].

Besides the studies of populations exposed to highly contaminated soils and crops, a population of oyster fishers in New Zealand was shown to consume large amount of oyster Cd, but no adverse effects on kidney or bone health were observed [162,163]. A similar outcome was observed for Swedish young women who consumed shellfish compared with those who did not consume shellfish; despite their consuming three times more Cd, blood and urine Cd levels were not different between the groups [112]. Oysters also accumulate high levels of Zn and Fe, which are important in inhibition of Cd absorption in humans. Thus the co-accumulated Zn was able to prevent adverse effects of high accumulation of Cd in oysters from Bluff, New Zealand. And consuming sunflower kernels which doubled daily Cd intake for one year had no effect on blood or urine Cd [111].

Cadmium-Accumulating Crops

As noted above, a few crops accumulate relatively more Cd in their edible tissues than do the major staple grains such as wheat. Sunflower kernels, flaxseed, poppy seed, and chocolate have substantially higher Cd than wheat or rice when grown on normal soils, and accumulate Cd effectively when soil Cd or chloride are high. Figure 17.2 shows the statistical distribution of Cd concentrations for major grains and some of these accumulating species. Figure 17.2 also shows the 'richtwert' limits for importing sunflower kernels and flax into Germany. Because the levels in these crops are not a result of growing them on contaminated soils, and the daily consumption of these crops is small, limits are higher than those for wheat.

Tobacco has exceptional ability to move soil Cd into human kidneys. Tobacco accumulates Cd from soils about as well as lettuce and spinach. Tobacco is a potentially adverse source of Cd when soils are contaminated [164]. Smoking transfers about 10 % of tobacco Cd into mainstream smoke, and that Cd is effectively absorbed in the lung (about 40 %; [165]) regardless of Fe-Zn-Ca nutritional status. In the study of Cai *et al.* [6], high tobacco Cd caused about half of the Cd absorption to result from tobacco and half from rice. The commercial market for tobacco products controls Cd concentration by not purchasing tobacco containing high Cd levels. Home-grown tobacco, produced on Cd-contaminated soils, may comprise as high a risk as home-grown rice in Asia. A recent paper reported that 'illicit' cigarettes are on average considerably higher in Cd than authentic major brand cigarettes, although likely not enough higher to comprise likelihood of causing tubular dysfunction in a lifetime [166]. But production of tobacco on contaminated soil, and smoking such home-grown tobacco can clearly comprise Cd risk to humans.

17.7 Food-Chain Zinc Issues

Zinc Deficiency

Many soils are deficient in phytoavailable Zn to support crop production, and Zn fertilizers are commonly applied for crops with high Zn demand or poor Zn uptake ability. Rice (*Oryza sativa* L.) and maize (*Zea mays* L.) are well known for being susceptible to Zn deficiency [167]. Most deficient soils are alkaline; high pH reduces Zn phytoavailability, and may hasten transformation of added soluble Zn into nonphytoavailable forms in the soil [57,58]. Even some strongly acidic soils can become Zn deficient over time as cropping removes Zn.

A more important Zn deficiency issue is the low concentration and low bioavailability of Zn in some staple food grains [168]. Because many grains contain high levels of phytate, and phytate reduces the bioavailability of dietary Zn, rice, maize, and some other food grains have been found to accumulate insufficient Zn to support human health. Perhaps one billion persons on Earth are Zn deficient [169]. Zn deficiency retards neurologic development and causes birth defects. The same grains are often Fe deficient for human health. The major international 'Harvest-Plus' project aims to improve the density of bioavailable Fe and Zn in staple grains [146]. These researchers have also reported that sowing seeds with higher Zn concentration can promote early growth of crops in low-Zn soils, and high enough Zn can substantially reduce yield reduction on deficient soils. One method to increase seed Zn to obtain this 'primer' effect is to spray the plants with soluble Zn-including detergent at flowering [170].

Milling grains is known to remove much of seed Zn, Fe, and phytate from cereal grains [138,171]. Whole grains which are not leavened carry high levels of phytate into diets, reducing dietary Zn bioavailability. Part of the goal of the 'increase grain Zn bioavailability' research program is to simultaneously reduce phytate, or to move more of the Zn into the embryo rather than the pericarp of cereal grains. Some success has been achieved in finding genetic sources of higher grain Zn [172,173]. 'Biofortification' was coined to express the concept of increasing the concentration and/or bioavailability of trace elements in staple foods.

Zinc Phytotoxicity Issues

Plant leaves accumulate about 400–500 mg Zn kg^{-1} at the soil Zn addition which reduces yield significantly [174]. Several papers covered large numbers of species, showing about 500 mg Zn kg^{-1} at 25 % yield reduction [175,176]. Other papers suggest that 290 mg Zn kg^{-1} dry leaves is associated with reduced yields, but base that on the theoretical first point where yield would be decreased from control yields [177,178]. The method used by Beckett and Davis [177] was sand culture with complete nutrient solutions including FeEDTA. However, under the conditions of their experiments, Zn displaces Fe from the FeEDTA chelate [179], strongly reducing the phytoavailability of added Fe to grasses. Grasses use phytosiderophores to dissolve soil Fe, and actively accumulate Fe-phytosiderophore chelates into the young roots [180], and Zn can also inhibit Fe chelation by phytosiderophores [181]. Zn salt but not Zn-mugineic acid chelate inhibited Fe-mugineic acid uptake by barley [182]. It seems likely that the lower Zn phytotoxicity thresholds reported by Beckett and Davis [177] may have resulted from use of FeEDTA in their sand culture testing. In fundamental tests of Zn toxicity to plants, Pedler *et al.* [183] tested the effect of various cations on the toxicity, finding that toxicity was substantially lowered by low levels of added Mg ion.

Most studies of Zn phytotoxicity have used addition of soluble Zn salts so that the added Zn could be presumed to be 100 % phytoavailable at the time of addition. This method can cause significant artifacts in estimation of Zn phytotoxicity in soils because the addition causes a reduction in soil pH as the added Zn displaces sorbed protons from the soil solid phases [184,185]. The displaced protons cause soil pH to fall by 1–2 pH units depending on the exchange capacity of the soil and the initial soil pH. Thus many of the tests of Zn phytotoxicity are so confounded by pH lowering due to the Zn salt addition that the results should not be considered useful. In addition, the presence of the anion added with the Zn salt increases Ca^{2+} in soil solution, and increased Ca^{2+} can displace more Zn from adsorption sites. Only leaching and aging after Zn salt addition can allow the added salt-Zn to approach phytoavailability levels found in field contaminated soils [186].

Estimation of toxicity to soil microbial process and soil organisms of Zn salts added to soils is increasingly recognized to be strongly affected by artifacts of pH shift, anion solubility, leaching, and aging. The organisms being studied are already present in the soil when the Zn salts are added to the soil, allowing exposures that are quite unrepresentative of soil equilibrium. And soil organisms can adapt to metals over time. McLaughlin and Smolders [187] found that soils with higher Zn levels tolerated higher Zn additions before toxicity was observed than soils with lower Zn levels and suggested a 'metalloregion' approach to setting soil Zn limits. Smolders *et al.* [188] studied field contaminated soils compared with Zn salt-amended paired control soils without and with leaching and aging and 'concluded that there is a large discrepancy in microbial responses to elevated Zn between spiked soils and field-contaminated soils'. Smolders *et al.* [189] found the spiked soils were up to 100-fold more toxic (median factor 3.4) than corresponding field contaminated soils and confirmed that leaching and aging strongly reduced toxicity. Smolders *et al.* [189] extended this finding and suggest that 'soil microbial processes are sensitively affected in soils that are freshly spiked with Zn salt but these effects are undetectable in the field'. It appears likely that Zn-sensitive plant species may be the most sensitive organism for excessive soil Zn (rather than soil microbial processes or soil fauna), and the same artifacts of spiking soils strongly affect Zn toxicity to higher plants.

Study of cultivar variation in tolerance of soil Zn has been reported for soybean [184,190]. Interestingly, variation in tolerance was a property of the scion, while uptake and translocation was a property of the rootstock [191]. Tolerance variation was great enough that one could breed for greater Zn tolerance, but this property has little value in normal agriculture. Another interpretation of these results is that the concentration of Zn which would be diagnostic of phytotoxicity varies among cultivars.

Remediation of Zinc Phytotoxicity
As noted by Chaney [174], Zn phytotoxicity is the most common practical metal toxicity resulting from soil contamination. This results partly from the frequency of Zn emissions and dispersal of Zn mine wastes, and partly because Zn is less strongly adsorbed by soils than are Pb and Cu. Especially when soils are acidic, Zn can achieve soluble Zn^{2+} levels which induce toxicity to sensitive plants. In acidic soils, Zn usually reduces root length and induces Fe deficiency chlorosis, but direct toxicity to leaf processes also limits growth.

Thus there is need for methods to remediate Zn phytotoxicity. In the author's experience, every Zn phytotoxicity in the field has been alleviated by making soils calcareous if attention is paid to soil fertility needs of the remediated soils. Often the contaminated soils are barren and eroded, removing much of the stored soil organic matter, P, and N. Because P reacts with Fe oxides to increase specific Zn and Cd adsorption capacity of such soils, additions of P and Fe along with limestone and organic-N help achieve persistent remediation [192]. Organic matter improves soil structure and fertility so that application of organic amendments with limestone is beneficial. Even when soils exceed 10 000 mg Zn kg^{-1}, remediation has been very effective [14,15,193–196].

Potential Zinc Toxicity to Animals Consuming High-Zinc Crops
Zinc is essential for animals, and active homeostasis of body Zn is observed [197]. When soluble Zn salts are mixed with practical diets, Zn toxicity to domestic livestock occurs at 300–1000 mg kg^{-1} dry diet [199]. Zinc toxicity appears to result from Zn-induced Cu deficiency [198,199–202]. Sheep are more sensitive to excessive dietary Zn if Cu in the feed is marginal to deficient [200]; as low as 300 mg Zn kg^{-1} diet induced Cu deficiency and lower gain rates in sheep fed diets low in bioavailable Cu. Unfortunately, the toxicity of intrinsic plant Zn to livestock has not been reported.

Forage crops comprise the worst case for evaluation of excess soil Zn. Under conditions of high soil Zn supply, grain and fruit contain lower Zn concentrations than leaves, and Zn toxicity from feed grains is very unlikely to occur. Research by Ott *et al.* [203], Campbell and Mills [200], L'Estrange [204], Bremner [201], and Bremner *et al.* [205] indicates that even crop uptake of Zn to leaves is very unlikely to cause Zn poisoning of cattle or sheep.

Although humans could consume appreciable amounts of high Zn leafy vegetables (acid soils; mild phytotoxicity), it is very unlikely that Zn toxicity would occur in humans in even this worst case. Many humans presently consume low or deficient amounts of Zn [169]; increased food Zn would be beneficial in many cases. Assuming lettuce accumulated 500 mg kg^{-1} dry weight, that is only 25 mg kg^{-1} fresh weight; consuming 50 g fresh lettuce would provide only 1.25 mg of the daily required 15 mg of Zn. Thus garden crops are very unlikely to comprise Zn risk to consumers.

Wildlife comprises a greater risk case than do domestic animals because wildlife could chronically consume Zn-phytotoxic foliage as most of their diet. It seems likely that high-Zn sludges and industrial wastes could cause adverse health effects (due to Zn-induced Cu deficiency) in wildlife such as that resulting from Zn smelters [202].

In summary, land application/treatment of sewage sludges and industrial wastes is very unlikely to cause Zn toxicity in domestic livestock or humans even under worst-case conditions. Wastes unusually high in Zn but low in Cu may cause Zn toxicity in animals if mismanaged.

References

[1] McGrath, S.P.; Loveland, P.J., The Soil Geochemical Atlas of England and Wales. Blackie Academic and Professional, London, 1992.
[2] Holmgren, G.G.S.; Meyer, M.W.; Chaney, R.L.; Daniels, R.B., Cadmium, lead, zinc, copper, and in agricultural soils of the United States of America; J. Environ. Qual. 1993, 22, 335–348.

[3] Smith, D.B.; Cannon, W.F.; Woodruff, L.G. *et al.*, Major- and trace-element concentrations in soils from two continental-scale transects of the United States and Canada; USGS Open-File Report 2005–1253, 2005.
[4] Kobayashi, J., Pollution by cadmium and the *itai-itai* disease in Japan; in: F.W. Oehme (ed.), Toxicity of Heavy Metals in the Environment. Marcel Dekker, New York, 1978, pp. 199–260.
[5] Asami, T., Pollution of soils by cadmium; in: J.O. Nriagu (ed.), Changing Metal Cycles and Human Health. Springer-Verlag, Berlin, 1984, pp. 95–111.
[6] Cai, S.; Yue, L.; Hu, Z. *et al.*, Cadmium exposure and health effects among residents in an irrigation area with ore dressing wastewater; Sci. Total Environ. 1990, 90, 67–73.
[7] Simmons, R.W.; Pongsakul, P.; Chaney, R.L.; Saiyasitpanich, D.; Klinphoklap, S.; Nobuntou, W., The relative exclusion of zinc and iron from rice grain in relation to rice grain cadmium as compared to soybean: Implications for human health; Plant Soil 2003, 257, 163–170.
[8] Morgan, H., The Shipham Report – An investigation into cadmium contamination and its implications for human health: 2. Metal contamination at Shipham; Sci. Total Environ. 1988, 75, 11–20.
[9] Davies, B.E.; Roberts, L.J., The distribution of heavy metal contaminated soils in Northeast Clwyd, Wales; Water Air Soil Pollut. 1978, 9, 507–518.
[10] Buchauer, M.J., Contamination of soil and vegetation near a zinc smelter by zinc, cadmium, copper and lead; Environ. Sci. Technol. 1973, 7, 131–135.
[11] Beyer, W.N., Damage to the forest ecosystem on Blue Mountain from zinc smelting; in: D.D. Hemphill (ed.), Trace Substances in Environmental Health-22; University of Missouri, Columbia, 1988, pp. 249–262.
[12] Chaney, R.L.; Beyer, W.N.; Gifford, C.H.; Sileo, L., Effects of zinc smelter emissions on farms and gardens at Palmerton, PA; in: D.D. Hemphill (ed.), Trace Substances in Environmental Health-22; University of Missouri, Columbia, 1988, pp. 263–280.
[13] Hansen, J.E.; Mitchell, J.E., The role of terraces and soil amendments in revegetating steep, smelter-affected land; Reclam. Rev. 1978, 1, 103–112.
[14] Brown, S.L.; Henry, C.L.; Chaney, R.L.; Compton, H.; DeVolder, P.S., Using municipal biosolids in combination with other residuals to restore metal-contaminated mining areas; Plant Soil 2003, 249, 203–215.
[15] Brown, S.; Sprenger, M.; Maxemchuk, A.; Compton, H., Ecosystem function in alluvial tailings after biosolids and lime addition; J. Environ. Qual. 2005, 34, 139–148.
[16] Burau, R.G., National and local dietary impact of cadmium in south coastal California soils. Ecotoxicol; Environ. Saf. 1983, 7, 53–57.
[17] Lund, L.J.; Betty, E.E.; Page, A.L.; Elliott, R.A., Occurrence of naturally high cadmium levels in soils and its accumulation by vegetation; J. Environ. Qual. 1981, 10, 551–556.
[18] Chaney, R.L.; Green, C.E.; Ajwa, H.; Smith, R., Zinc fertilization plus liming to reduce cadmium uptake by Romaine lettuce on Cd-mineralized Lockwood soil; Proc. Int. Plant Nutr Colloq. 25–28 August, 2009, Sacramento, CA, 2009.
[19] Mellum, H.K.; Arnesen, A.K.M.; Singh, B.R., Extractability and plant uptake of heavy metals in alum shale soils; Commun. Soil Sci. Plant Anal. 1998, 29, 1183–1198.
[20] Lalor, G.C.; Rattray, R.; Simpson, P.; Vutchkov, M., Heavy metals in Jamaica. Part 3: The distribution of cadmium in Jamaican soils; Rev. Int. Contam. Ambien. 1998, 14, 7–12.
[21] Garrett, R.G.; Porter, A.R.D.; Hunt, P.A.; Lalor, G.C., The presence of anomalous trace element levels in present day Jamaican soils and the geochemistry of Late-Miocene or Pliocene phosphorites; Appl. Geochem. 2008, 23, 822–834.
[22] Legret, M.; Divet, L.; Juste, C., Movement and speciation of heavy metals in a soil amended with sewage sludge containing large amounts of Cd and Ni; Water Res. 1988, 22, 953–959.
[23] Chaney, R.L.; Filcheva, E.; Green, C.E; Brown, S.L., Zn deficiency promotes Cd accumulation by lettuce from biosolids amended soils with high Cd:Zn ratio; J. Residual Sci. Technol. 2006, 3, 68–75.
[24] Chen, Z.-S., Cadmium and lead contamination of soils near plastic stabilizing materials producing plants in northern Taiwan; Water Air Soil Pollut. 1991, 57–58, 745–754.

[25] Hunter, B.A.; Johnson, M.S.; Thompson, D.J., Ecotoxicology of copper and cadmium in a contaminated grassland ecosystem. I. Soils and vegetation contamination; J. Appl. Ecol. 1987, 24, 573–586.
[26] Clemente, R.; Dickinson, N.M.; Lepp, N.W., Mobility of metals and metalloids in a multi-element contaminated soil 20 years after cessation of the pollution source activity; Environ. Pollut. 2008, 155, 254–261.
[27] McLaughlin, M.J.; Tiller, K.G.; Naidu, R.; Stevens, D.P., The behaviour and environmental impact of contaminants in fertilizers; Aust. J. Soil Res. 1996, 34, 1–54.
[28] Rutherford, P.M.; Dudas, M.J.; Samek, R.A., Environmental impacts of phosphogypsum; Sci. Total Environ. 1994, 149, 1–38.
[29] Jones, R., Zinc and cadmium in lettuce and radish grown in soils collected near electrical transmission (hydro) towers; Water Air Soil Pollut. 1983, 19, 389–395.
[30] Davies, B.E.; Paveley, C.F.; Wixson, B.G., Use of limestone wastes from metal mining as agricultural lime: potential heavy metal limitations; Soil Use Manage. 1993, 9, 47–52.
[31] Davis, J.G.; Weeks, G.; Parker, M.B., Use of deep tillage and liming to reduce zinc toxicity in peanuts grown on flue dust contaminated land; Soil Technol. 1995, 8, 85–89.
[32] Mortvedt, J.J., Plant uptake of heavy metals in zinc fertilizers made from industrial by-products; J. Environ. Qual. 1985, 14, 424–427.
[33] Westfall, D.G.; Mortvedt, J.J.; Peterson, G.A.; Gangloff, W.J., Efficient and environmentally safe use of micronutrients in agriculture; Commun. Soil Sci. Plant Anal. 2005, 36, 169–182.
[34] Lee, C.R.; Page, N.R., Soil factors influencing the growth of cotton following peach orchards; Agron. J. 1967, 59, 237–240.
[35] Handreck, K.A., Zinc toxicity from tire rubber in soilless potting media; Commun. Soil Sci. Plant Anal. 1996, 27, 2615–2623.
[36] Borkert, C.M.; Cox. F.R., Effects of acidity at high soil zinc, copper, and manganese on peanut, rice, and soybean; Commun. Soil Sci. Plant Anal. 1999, 30, 1371–1384.
[37] Chaney, R.L.; Ryan, J.A., Risk Based Standards for Arsenic, Lead and Cadmium in Urban Soils; DECHEMA, Frankfurt, 1994. ISBN 3-926959-63-0.
[38] Stehouwer, R.C.; Wolf, A.M.; Doty, W.T., Chemical monitoring of sewage sludge in Pennsylvania: variability and application uncertainty; J. Environ. Qual. 2000, 29, 1686–1695.
[39] Chaney R.L.; Hornick, S.B.; Simon, P.W., Heavy metal relationships during land utilization of sewage sludge in the Northeast; in:R.C. Loehr (ed.), Land as a Waste Management Alternative. Ann Arbor Science Publishers, Inc., Ann Arbor, MI, 1977, pp. 283–314.
[40] Sommers, L.E. (ed.), Effects of sewage sludge on the cadmium and zinc content of crops. CAST Report No. 83; Council for Agricultural Science and Technology, Ames, IA, 1980.
[41] Logan, T.J.; Cassler, D.E., Case study for correction of a widespread cadmium soil contamination problem using local funds; in: D.D. Hemphill (ed., Trace Substances in Environmental Health-22; University of Missouri, Columbia, 1988, pp. 337–345.
[42] Kuo, S.; Huang, B.; Bembenek, R., The availability to lettuce of zinc and cadmium in a zinc fertilizer; Soil Sci. 2004, 169, 363–373.
[43] Wood, B.W.; Reilly, C.C.; Nyczepir, A.P., Mouse-ear of pecan: a nickel deficiency; HortSci. 2005, 39, 1238–1242.
[44] Nicholson, F.A.; Chambers, B.J.; Williams, J.R.; Unwin, R.J., Heavy metal contents of livestock feeds and animal manures in England and Wales; Bioresource Technol 1999, 70, 23–31.
[45] Nicholson, F.A.; Smith, S.R.; Alloway, B.J.; Carlton-Smith, C.; Chambers, B.J., An inventory of heavy metals inputs to agricultural soils in England and Wales; Sci. Total Environ. 2003, 311, 205–219.
[46] Jones, K.C.; Johnston, A.E., Cadmium in cereal grain and herbage from long-term experimental plots at Rothamsted; Environ. Pollut. 1989, 57, 199–216.
[47] Keller, A.; Schulin, R., Modelling regional-scale mass balances of phosphorus, cadmium and zinc fluxes on arable and dairy farms; Eur. J. Agron. 2003, 20, 181–198.
[48] Sheppard, S.C.; Grant, C.A.; Sheppard, M.I.; de Jong, R.; Long, J., Risk indicator for agricultural inputs of trace elements to Canadian soils; J. Environ. Qual. 2009, 38, 919–932.

[49] Qureshi, S.; Richards, B.K.; McBride, M.B.; Baveye, P.; Steenhuis, T.S., Temperature and microbial activity effects on trace element leaching from metalliferous peats; J. Environ. Qual. 2003, 32, 2067–2075.

[50] Ahnstrom, Z.S.; Parker, D.R., Cadmium reactivity in metal-contaminated soils using a coupled stable isotope dilution-sequential extraction procedure; Environ. Sci. Technol. 2001, 35, 121–126.

[51] Chino, M.; Baba A., The effects of some environmental factors on the partitioning of zinc and cadmium between roots and tops of rice plants; J. Plant Nutr. 1981, 3, 203–214.

[52] Khaokaew, S., M. Ginder-Vogel, R.L. Chaney and D.L. Sparks. 2009. Speciation and release kinetics of cadmium and zinc in Thai paddy soil. Abstract 741-7. American Society of Agronomy (Nov. 1–5, 2009; Pittsburgh, PA).

[53] Arao, T.; Ae, N., Genotypic variations in cadmium levels of rice grain; Soil Sci. Plant Nutr. 2003, 49, 473–479.

[54] Duxbury, J.M.; Bodruzzaman, M.; Johnson, S. *et al.*, Impacts of increased mineral micronutrient content of rice and wheat seed/grain on crop productivity and human nutrition in Bangladesh; in: C.J. Li *et al.* (eds.), Plant Nutrition for Food Security, Human Health and Environmental Protection; Tsinghua University Press, Beijing, 2005, pp. 30–31.

[55] Takijima, Y.; Katsumi, F., Cadmium contamination of soils and rice plants caused by zinc mining. 1. Production of high-cadmium rice on the paddy fields in lower reaches of the mine station; Soil Sci. Plant Nutr. 1973, 19, 29–38.

[56] Boawn, L.C., Residual availability of fertilizer zinc; Soil Sci. Soc. Am. Proc. 1974, 38, 800–803.

[57] Boawn, L.C., Sequel to 'residual availability of fertilizer zinc'; Soil Sci. Soc. Am. J. 1976, 40, 467–468.

[58] Brennan, R.F.; Bolland, M.D.A., Estimating the long-term residual value of zinc oxide for growing wheat in a sandy duplex soil; Aust. J. Agr. Res. 2007, 58, 57–65.

[59] Tiller, K.G.; Honeysett, J.L.; deVries, M.P.C., Soil zinc and its uptake by plants. II. Soil chemistry in relation to prediction of availability; Aust. J. Soil Res. 1972, 10, 165–182.

[60] Ford, R.G, Sparks, D.L., The nature of Zn precipitates formed in the presence of pyrophyllite; Environ. Sci. Technol. 2000. 34, 2479–2483.

[61] Voegelin, A.; Scheinost, A.C.; Bühlmann, K.; Barmettler, K.; Kretzschmar, R., Slow formation and dissolution of Zn precipitates in soil: A combined column-transport and XAFS study; Environ. Sci. Technol. 2002, 36, 3749–3754.

[62] Roberts, D.R.; Ford, R.G.; Sparks, D.L., Kinetics and mechanisms of Zn complexation on metal oxides using EXAFS spectroscopy; J. Coll. Interfac. Sci. 2003, 263, 364–376.

[63] Roberts, D.R.; Scheinost, A.C.; Sparks, D.L., Zinc speciation in a smelter-contaminated soil profile using bulk and microspectroscopic techniques; Environ. Sci. Technol. 2002, 36, 1742–1750.

[64] Scheinost, A.C.; Kretzschmar, R.; Pfister, S.; Roberts, D.R., Combining selective sequential extractions, X-ray absorption spectroscopy, and principal component analysis for quantitative zinc speciation in soil; Environ. Sci. Technol. 2002, 36, 5021–5028.

[65] Kukier, U.; Chaney, R.L.; Ryan, J.L.; Daniels, W.L.; Dowdy, R.H.; Granato, T.C., Phytoavailability of cadmium in long-term biosolids amended soils; J. Environ. Qual. 2010, 39, 519–530.

[66] Hamon, R.E.; McLaughlin, M.J.; Naidu, R.; Correll, R, Long-term changes in cadmium bioavailability in soil; Environ. Sci. Technol. 1998, 32, 3699–3703.

[67] Chaney, R.L.; Ryan, J.A., Li, Y.-M., Brown, S.L., Soil cadmium as a threat to human health; in: M.J. McLaughlin, B.R. Singh (eds), Cadmium in Soils and Plants; Kluwer Academic, Dordrecht, 1999, 219–256.

[68] Grant, C.A.; Bailey, L.D.; McLaughlin, M.J.; Singh, B.R., Management factors which influence cadmium concentrations in crops: a review; in: M.J. McLaughlin, B.R. Singh (eds.), Cadmium in Soils and Plants; Kluwer Academic, Dordrecht, 1999, pp. 151–198.

[69] Mench, M.J., Cadmium availability to plants in relation to major long-term changes in agronomy systems; Agric. Ecosyst. Environ. 1998, 67, 175–187.

[70] McLaughlin, M.J.; Hamon, R.E.; McLaren, R.G.; Speir, T.W.; Rogers, S.L., Review: A bioavailability-based rationale for controlling metal and metalloid contamination of agricultural land in Australia and New Zealand; Aust. J. Soil Res. 2000, 38, 1037–1086.

[71] Grotz, N.; Guerinot, M.L., Molecular aspects of Cu, Fe and Zn homeostasis in plants; Biochim. Biophys. Acta 2006, 1763, 595–608.

[72] Grotz, N.; Fox, T.; Connolly, E.; Park, W.; Guerinot, M.L.; Eide, D., Identification of a family of zinc transporter genes from *Arabidopsis* that respond to zinc deficiency; Proc. Natl. Acad. Sci. USA 1998, 95, 7220–7224.

[73] Ramesh, S.A.; Shin, R.; Eide, D.J.; Schachtman, D.P., Differential metal selectivity and gene expression of two zinc transporters from rice; Plant Physiol. 2003, 133, 126–134.

[74] Hanikenne, M.; Talke, I.N.; Haydon, M.J. *et al.*, Evolution of metal hyperaccumulation required *cis*-regulatory changes and triplication of *HMA4*; Nature 2008, 453, 391–395.

[75] Cohen, C.K.; Fox, T.C.; Garvin, D.F.; Kochian, L.V., The role of iron-deficiency stress responses in stimulating heavy-metal transport in plants; Plant Physiol. 1998, 116, 1063–1072.

[76] Cohen, C.K.; Garvin, D.F.; Kochian, L.V., Kinetic properties of a micronutrient transporter from *Pisum sativum* indicate a primary function in Fe uptake from the soil; Planta 2004, 218, 784–792.

[77] Hart, J.J.; Welch, R.M.; Norvell, W.A.; Kochian, L.V., Transport interactions between cadmium and zinc in roots of bread and durum wheat seedlings; Physiol. Plant. 2002, 116, 73–78.

[78] Jarvis, S.C.; Jones, L.H.P.; Hopper, M.J., Cadmium uptake from solution by plants and its transport from roots to shoots; Plant Soil 1976, 44, 179–191.

[79] Grant, C.A.; Clarke, J.M.; Duguid S.; Chaney, R.L., Selection and breeding of plant cultivars to minimize cadmium accumulation; Sci. Total Environ. 2008, 390, 301–310.

[80] Clarke, J.M.; Norvell, W.A.; Clarke, F.R.; Buckley, W.T., Concentration of cadmium and other elements in the grain of near-isogenic durum lines; Can. J. Plant Sci. 2002, 82, 27–33.

[81] Li, Y.-M.; Chaney, R.L.; Schneiter, A.A.; Miller, J.F., Genotypic variation in kernel cadmium concentration in sunflower germplasm under varying soil conditions; Crop Sci. 1995. 35, 137–141.

[82] Harris, N.S.; Taylor, G.J., Cadmium uptake and translocation in seedlings of near isogenic lines of durum wheat that differ in grain cadmium accumulation. BMC Plant Biol. 2004 (14 April), 4, 4. doi: 10.1186/1471-2229-4-4.

[83] Ishikawa, S.; Ae, N.; Yano, M., Chromosomal regions with quantitative trait loci controlling cadmium concentration in brown rice (*Oryza sativa*); New Phytol 2005, 168, 345–350.

[84] McKenna, I.M.; Chaney, R.L., Characterization of a cadmium-zinc complex in lettuce leaves; Biol. Trace Element Res. 1995, 48, 13–29.

[85] Chaney, R.L.; Ryan, J.A.; Li, Y.-M.; Angle, J.S., Transfer of cadmium through plants to the food chain; in: J.K. Syers, M. Gochfeld (eds.), Proceedings of the Workshop 'Environmental Cadmium in the Food Chain: Sources, Pathways, and Risks', 13–16 September, 2000, Brussels, Belgium; Scientific Committee on Problems of the Environment, Paris, 2001, pp. 76–81.

[86] Hooda, P.S.; Alloway, B.J., The effects of liming on heavy metal concentrations in wheat, carrots and spinach grown on previously sludge-amended soils; J. Agric. Sci. 1996, 127, 289–294.

[87] Pepper, I.L.; Bezdicek, D.E.; Baker, A.S.; Sims, J.M., Silage corn uptake of sludge-applied zinc and cadmium as affected by soil pH; J. Environ. Qual. 1983, 12, 270–275.

[88] Oliver, D.P.; Hannam, R.; Tiller, K.G.; Wilhelm, N.S.; Merry, R.H., The effects of zinc fertilization on cadmium concentration in wheat grain; J. Environ. Qual. 1994, 23, 705–711.

[89] McLaughlin, M.J.; Palmer, L.T.; Tiller, K.G.; Beech, T.W.; Smart, M.K., Increasing soil salinity causes elevated cadmium concentrations in field-grown potato tubers; J. Environ. Qual. 1994, 23, 1013–1018.

[90] Li, Y.-M.; Chaney, R.L.; Schneiter, A.A., Effect of soil chloride level on cadmium concentration in sunflower kernels; Plant Soil 1995, 167, 275–280.

[91] Norvell, W.A.; Wu, J.; Hopkins, D.G.; Welch, R.M., Association of cadmium in durum wheat grain with soil chloride and chelate-extractable soil cadmium; Soil Sci. Soc. Am. J. 2000, 64, 2162–2168.

[92] Smolders, E.; McLaughlin, M.J., Chloride increases Cd uptake in Swiss chard in a resin-buffered nutrient solution; Soil Sci. Soc. Am. J. 1996, 60, 1443–1447.
[93] Khoshgoftarmanesh, A.H.; Shariatmadari, H.; Karimian, N.; Kalbasi, M.; van der Zee, S.E.A.T.M., Cadmium and zinc in saline soil solutions and their concentrations in wheat; Soil Sci. Soc. Am. J. 2006, 70, 582–589
[94] Pelletier, S.; Simpson, R.; Randall, P. et al., Dietary cation–anion difference and cadmium concentration in grasses fertilized with chloride; Grass Forage Sci. 2007, 62, 416–428.
[95] Huang, B.; Kuo, S.; Bembenek, R., Availability of cadmium in some phosphorus fertilizers to field-grown lettuce; Water Air Soil Pollut 2004, 158, 37–51.
[96] Grant, C.A.; Sheppard, S.C., Fertilizer impacts on cadmium availability in agricultural soils and crops; Human Ecol. Risk Assess. 2008, 14, 210–228.
[97] Mortvedt, J.J., Cadmium levels in soils and plants from some long-term soil fertility experiments in the United States of America; J Environ Qual 1987, 16, 137–142.
[98] Nziguheba, G.; Smolders, E., Inputs of trace elements in agricultural soils via phosphate fertilizers in European countries; Sci Total Environ. 2008, 390, 53–57.
[99] Oliver, D.P.; Schultz, J.E.; Tiller, K.G.; Merry, R.H., The effect of crop rotations and tillage practices on cadmium concentrations in wheat grain; Aust. J. Agric. Res. 1993, 44, 1221–1234.
[100] Khoshgoftarmanesh, A.H.; Chaney, R.L., Preceding crop affects grain cadmium and zinc of wheat grown in saline soils of central Iran; J. Environ. Qual. 2007, 36, 1132–1136.
[101] Chaney, R.L.; Stoewsand, G.S.; Bache, C.A.; Lisk, D.J., Cadmium deposition and hepatic microsomal induction in mice fed lettuce grown on municipal sludge-amended soil; J. Agr. Food Chem. 1978, 26, 992–994.
[102] Chaney, R.L.; Stoewsand, G.S.; Furr, A.K.; Bache, C.A.; Lisk, D.J., Elemental content of tissues of Guinea pigs fed Swiss chard grown on municipal sewage sludge-amended soil; J. Agr. Food Chem. 1978, 26, 944–997.
[103] Brown, S.L.; Chaney, R.L.; Angle, J.S.; Ryan, J.A., The phytoavailability of cadmium to lettuce in long-term biosolids-amended soils; J. Environ. Qual. 1998, 27, 1071–1078.
[104] Hettiarachchi, G.M.; Ryan, J.A.; Chaney, R.L.; La Fleur, C.M., Sorption and desorption of cadmium by different fractions of biosolids-amended soils; J. Environ. Qual. 2003, 32, 1684–1693.
[105] Basta. N.T.; Ryan, J.A.; Chaney, R.L., Trace element chemistry in residual-treated soil: key concepts and metal bioavailability; J. Environ. Qual. 2005, 34, 49–63.
[106] Chaney, R.L.; Ryan, J.A.; Li, Y.-M. et al., Phytoavailability and bio-availability in risk assessment of Cd in agricultural environments. OECD Proceedings Sources of Cadmium in the Environment 15–22 October, Stockholm, Sweden; OECD, Paris, 1996, pp. 49–78.
[107] Wolnik, K.A.; Fricke, F.L.; Capar, S.G. et al., Elements in major raw agricultural crops in the United States. 1. Cadmium and lead in lettuce, peanuts, potatoes, soybeans, sweet corn, and wheat. J. Agr. Food Chem. 1983, 31, 1240–1244.
[108] Wolnik, K.A.; Fricke, F.L.; Capar, S.G. et al., Elements in raw agricultural crops in the United States. 3. Cadmium, lead, and eleven other elements in carrots, field corn, onions, rice, spinach, and tomatoes; J. Agr. Food Chem. 1985, 33, 807–811.
[109] Klein, H.; Weigert, P., Schwermetalle in Leinsamen [Heavy metals in linseed]; Bundesgesundheitsblatt 1987, 30, 391–395.
[110] Ocker, H.-D.; Brüggemann, J.; Rühl, C.S.; Klein, H., Cadmium in sunflower kernels, poppy and sesame [in German]; Bundesgesundheitsblatt 1991, 34, 556.
[111] Reeves, P.G.; Nielsen, E.G.; O'Brien-Nimens, C.; Vanderpool, R.A., Cadmium bioavailability from edible sunflower kernels: a long-term study with men and women volunteers; Environ Res 2001, A87, 81–91.
[112] Vahter, M.; Berglund, M.; Nermell, B.; Åkesson, A., Bioavailability of cadmium from shellfish and mixed diet in women; Toxicol. Appl. Pharmacol. 1996, 136, 332–341.
[113] Fox, M.R.S.; Jacobs, R.M.; Jones, A.O.L.; Fry, B.E., Jr., Effects of nutritional factors on metabolism of dietary cadmium at levels similar to those of man; Environ. Health Perspect. 1979, 28, 107–114.

[114] Reeves, P.G.; Chaney, R.L., Bioavailability as an issue in risk assessment and management of food cadmium: A review; Sci. Total Environ. 2008, 398, 13–19.
[115] Reeves, P.G.; Chaney, R.L.; Simmons, R.W.; Cherian, M.G., Metallothionein induction is not involved in cadmium accumulation in the duodenum of mice and rats fed diets containing high-cadmium-rice or sunflower kernels and a marginal supply of zinc, iron, and calcium; J. Nutr. 2005, 135, 99–108.
[116] McKenna, I.M.; Chaney, R.L.; Tao, S.H.; Leach, R.H., Jr.; Williams, F.M., Interactions of plant zinc and plant species on the bioavailability of plant cadmium to Japanese quail fed lettuce and spinach; Environ. Res. 1992, 57, 73–87.
[117] Stuczynski, T.I.; Siebielec, G.; Daniels, W.L.; McCarty, G.C.; Chaney, R.L., Biological aspects of metal waste reclamation with sewage sludge; J. Environ. Qual. 2007, 36, 1154–1162.
[118] Baker, D.E.; Bowers, M.E., Health effects of cadmium predicted from growth and composition of lettuce grown in gardens contaminated by emissions from zinc smelters; in: D.D. Hemphill, D.D. (ed.), Trace Substances in Environmental Health-22; University of Missouri, Columbia, 1988, pp. 281–295.
[119] Grill, E.; Winnacker, E.-L.; Zenk, M.H., Phytochelatins: The principle heavy-metal complexing peptides of higher plants; Science 1985, 230, 674–676.
[120] Schat, H.; Kalff, M.M.A., Are phytochelatins involved in differential metal tolerance or do they merely reflect metal-imposed strain? Plant Physiol. 1992, 99, 1475–1480.
[121] Rauser, W.E., Phytochelatins and related peptides: structure, biosynthesis and function; Plant Physiol. 1995, 109, 1141–1149.
[122] Vatamaniuk, O.K.; Mari, S.; Lu, Y.-P.; Rea, P.A., Mechanism of heavy metal ion activation of phytochelatin (PC) synthase: blocked thiols are sufficient for PC synthase-catalyzed transpeptidation of glutathione and related thiol peptides; J. Biol. Chem. 2000, 275, 31451–31459.
[123] McKenna, I.M.; Chaney, R.L., Characterization of a cadmium-zinc complex in lettuce leaves; Biol. Trace Element Res. 1995, 48, 13–29.
[124] Ishizaki, M.; Kido, T.; Honda, R. et al., Dose-response relationship between urinary cadmium and β_2-microglobulin in a Japanese cadmium exposed population; Toxicol. 1989, 58, 121–131.
[125] Fukushima, M.; Ishizaki, A.; Sakamoto, M.; Kobayashi, E., Cadmium concentration in rice eaten by farmers in the Jinzu River Basin [in Japanese]; Japan J. Hyg. 1973, 28, 406–415.
[126] Ishizaki, A.; Fukushima, M.; Sakamoto, M., Contents of cadmium and zinc in organs of Itai-itai disease patients and residents of Hokuriku District [in Japanese]; Japan. J. Hyg. 1971, 26, 268–73.
[127] Nogawa, K.; Ishizaki, A., A comparison between cadmium in rice and renal effects among inhabitants of the Jinzu River basin; Environ. Res. 1979, 18, 410–420.
[128] Nogawa, K.; Kobayashi, E.; Honda, R.; Ishizaki, A.; Kawano, S.; Matsuda, H., Renal dysfunctions of inhabitants in a cadmium-polluted area; Environ. Res. 1980, 23, 13–23.
[129] Jin, T.; Nordberg, G.; Ye, T. et al., Osteoporosis and renal dysfunction in a general population exposed to cadmium in China; Environ. Res. 2004, 96, 353–359.
[130] Tsuchiya, K. (ed.), Cadmium Studies in Japan: A Review. Elsevier/North-Holland Biomedical Press, New York, 1978.
[131] Chaney, R.L.; Reeves, P.G.; Ryan, J.A.; Simmons, R.W.; Welch, R.M.; Angle, J.S., An improved understanding of soil Cd risk to humans and low cost methods to remediate soil Cd risks; BioMetals 2004, 17, 549–553.
[132] Chaney, R.L.; Angle, J.S.; Broadhurst, C.L.; Peters, C.A.; Tappero, R.V.; Sparks, D.L. Improved understanding of hyperaccumulation yields commercial phytoextraction and phytomining technologies; J. Environ. Qual. 2007, 36, 1429–1443.
[133] Li, Y.-M.; Chaney, R.L.; Reeves, R.D.; Angle, J.S.; Baker, A.J.M., *Thlaspi caerulescens* subspecies for Cd and Zn recovery; US Patent 7,049,492, 2006, pp. 1–8.
[134] Simmons, R.W.; Pongsakul, P.; Saiyasitpanich, D.; Klinphoklap, S., Elevated levels of cadmium and zinc in paddy soils and elevated levels of cadmium in rice grain downstream of a zinc mineralized area in Thailand: implications for public health; Environ. Geochem. Health 2005, 27, 501–511.

[135] Codex Alimentarius Commission. Report of the 29th session of the Codex Alimentarius Commission; ALINORM 06/29/41; Codex Alimentarius Commission, Rome, 2006.
[136] Swaddiwudhipong, W.; Limpatanachote, P.; Mahasakpan, P.; Krintratun, S.; Padungtod, C., Cadmium-exposed population in Mae Sot District, Tak Province: 1. Prevalence of high urinary cadmium levels in the adults; J. Med. Assoc. Thai. 2007, 90, 143–148.
[137] Simmons, R.W.; Noble, A.D.; Pongsakul, P.; Sukreeyapongse, O.; Chinabut, N., Analysis of field-moist Cd contaminated paddy soils during rice grain fill allows reliable prediction of grain Cd levels; Plant Soil 2008, 302, 125–137.
[138] Pedersen, B.; Eggum, B.O., The influence of milling on the nutritive value of flour from cereal grains. 4. Rice.; Plant Foods Hum. Nutr. 1983, 33, 267–278.
[139] Yoshikawa, T.; Kusaka, S.; Zikihara, T.; Yoshida, T., Distribution of heavy metals in rice plants. I. Distribution of heavy metal elements in rice grains using an electron probe x-ray microanalyser (EPMA) [In Japanese]; J. Soc. Soil Manure, Japan 1977, 48, 523–528.
[140] Meharg, A.A.; Lombi, E., Williams, P.N. et al., Speciation and localization of arsenic in white and brown rice grains; Environ. Sci. Technol. 2008, 42,1051–1057.
[141] Reeves, P.G.; Chaney, R.L., Nutritional status affects the absorption and whole-body and organ retention of cadmium in rats fed rice-based diets; Environ. Sci. Technol. 2002, 36, 2684–2692.
[142] Park, J.D.; Cherrington, N.J.; Klaassen, C.D., Intestinal absorption of cadmium is associated with divalent metal transporter 1 in rats; Toxicol. Sci. 2002, 68, 288–294.
[143] Ryu, D.-Y.; Lee S.-J.; Park, D.W.; Choi, B.-S.; Klaassen, C.D.; Park, J.-D., Dietary iron regulates intestinal cadmium absorption through iron transporters in rats; Toxicol. Lett. 2004, 152, 19–25.
[144] Reeves, P.G.; Chaney, R.L., Marginal nutritional status of zinc, iron, and calcium increases cadmium retention in the duodenum and other organs of rats fed a rice-based diet; Environ. Res. 2004, 96, 311–322.
[145] Jacobs, R.M.; Jones, A.O.L.; Fry, B.E., Jr.; Fox, M.R.S., Decreased long-term retention of ^{115m}cadmium in Japanese quail produced by a combined supplement of zinc copper and manganese; J. Nutr. 1978, 108, 901–910.
[146] Graham, R.D.; Welch, R.M.; Saunders, D.A. et al., Nutritious subsistence food systems; Adv. Agron. 2007, 92, 1–74.
[147] Buhler, D.R., Availability of cadmium from foods and water; in: E.J. Calabrese, R.W. Tuthill, L. Condie (eds.), Inorganics in Drinking Water and Cardiovascular Disease; Princeton Scientific Publ. Co., Princton, NJ, 1985, pp. 271–287.
[148] Ikeda, M.; Ezaki, T.; Tsukahara, T. et al., Threshold levels of urinary cadmium in relation to increases in urinary β_2-microglobulin among general Japanese populations; Toxicol. Lett. 2003, 137, 135–141.
[149] Buchet, J.P.; Lauwerys, R.; Roels, H. et al., Renal effects of cadmium body burden of the general population; Lancet 1990, 336, 699–702.
[150] Järup, L.; Hellstrom, L.; Alfven, T. et al., Low level exposure to cadmium and early kidney damage: The OSCAR study; Occup Environ. Med. 2000, 57, 668–672.
[151] Ezaki, T.; Tsukahara, T.; Moriguchi, J. et al., No clear-cut evidence for cadmium-induced renal tubular dysfunction among over 10,000 women in the Japanese general population: a nationwide large-scale survey; Int. Arch. Occup. Environ. Health 2003 76, 186–196.
[152] Ikeda, M,; Ezaki, T.; Moriguchi, J. et al., No meaningful increase in urinary tubular dysfunction markers in a population with 3 μg cadmium/g creatinine in urine; Biol. Trace Elem. Res. 2006, 113, 35–44.
[153] Strehlow, C.D.; Barltrop, D., The Shipham Report – An investigation into cadmium concentrations and its implications for human health: 6. Health studies; Sci. Total Environ. 1988, 75, 101–133.
[154] Elliott, P.; Arnold, R.; Cockings, S. et al., Risk of mortality, cancer incidence, and stroke in a population potentially exposed to cadmium; Occup. Environ. Med. 2000, 57, 94–97.
[155] Philipp, R.; Hughes, A.; Elliott, P.; Jarup, L.; Quinn, M.; Thornton, I., Health risks from exposure to cadmium in soil; Occup. Environ. Med. 2000, 57, 647–648.

[156] Sarasua, S.M.; McGeehin, M.A.; Stallings, F.L. *et al.*, Final Report. Technical Assistance to the Pennsylvania Department of Health. Biologic indicators of exposure to cadmium and lead. Palmerton, PA. Part II. (May 1995); Agency for Toxic Substances and Disease Registry, US-DHHS, Atlanta, GA, 1995.
[157] Ewers, U.; Freier, I.; Turfeld, M. *et al.*, Heavy metals in garden soil and vegetables from private gardens located in lead/zinc smelter area and exposure of gardeners to lead and cadmium (in German); Gesundheitswesen 1993, 55, 318–325.
[158] Ewers, U.; Brockhaus, A.; Dolgner, R. *et al.*, Environmental exposure to cadmium and renal function of elderly women living in cadmium-polluted areas of the Federal Republic of Germany; Int. Arch. Occup. Environ. Health 1985, 55, 217–239.
[159] de Burbure, C.; Buchet, J.P.; Bernard, A. *et al.*, Biomarkers of renal effects in children and adults with low environmental exposure to heavy metals; J. Toxicol. Environ. Health 2003, A66, 783–798.
[160] Kreis, I.A.; Wijga, A.; van Wijnen, J.H., Assessment of the lifetime accumulated cadmium intake from food in Kempenland; Sci. Total Environ. 1992, 127, 281–292.
[161] Staessen, J.A.; Vyncke, G.; Lauwerys, R.R. *et al.*, Transfer of cadmium from a sandy acidic soil to man: a population study; Environ Res 1992, 58, 25–34.
[162] Sharma, R.P.; Kjellström, T.; McKenzie, J.M., Cadmium in blood and urine among smokers and non-smokers with high cadmium intake via food; Toxicology 1983, 29, 163–171.
[163] McKenzie-Parnell, J.M.; Kjellström, T.E.; Sharma, R.P.; Robinson, M.F., Unusually high intake and fecal output of cadmium, and fecal output of other trace elements in New Zealand adults consuming dredge oysters; Environ. Res. 1988, 46, 1–14.
[164] Bell, P.F.; Mulchi, C.L.; Chaney, R.L., Residual effects of land applied municipal sludge on tobacco. III. Agronomic, chemical, and physical properties vs. multiple sludge sources; Tobacco Sci. 1988, 32, 71–76.
[165] Friberg, L.; Elinder, C.G.; Kjellstrom, T.; Nordberg, G.F. (eds), Cadmium and Health: A Toxicological and Epidemiological Appraisal, Vol. 1,. Exposure, Dose, and Metabolism; CRC Press, Boca Raton, FL, 1995.
[166] Stephens, W.E.; Calder, A.; Newton, J., Source and health implications of high toxic metal concentrations in illicit tobacco products; Environ. Sci. Technol. 2005, 39, 479–488.
[167] Graham, R.D; Rengel, Z., Genotypic variation in zinc uptake and utilization by plants; in: A.D. Robson (ed.), Zinc in Soils and Plants. Kluwer Academic, Dordrecht, 1993, pp. 107–118.
[168] Graham, R.D.; Welch, R.M.; Bouis, H.E., Addressing micronutrient malnutrition through enhancing the nutritional quality of staple foods: principles, perspectives and knowledge gaps; Adv. Agron. 2001, 70, 77–142.
[169] Black, R.E.; Allen, L.H.; Bhutta, Z.A. *et al.*, The Maternal and Child Undernutrition Study Group. Maternal and child undernutrition: global and regional exposures and health consequences; Lancet 2008, 371, 243–260.
[170] Rengel, Z.; Graham, R.D., Importance of seed Zn content for wheat growth on Zn-deficient soil: I. Vegetative growth; Plant Soil 1995, 173, 259–266.
[171] Pedersen, B.; Eggum, B.O., The influence of milling on the nutritive value of flour from cereal grains. 2. Wheat.; Plant Foods Hum. Nutr. 1983, 33, 51–61.
[172] Welch, R.M.; House, W.A.; Ortiz-Monasterio, I.; Cheng, Z., Potential for improving bioavailable zinc in wheat grain (*Triticum* species) through plant breeding; J. Agric. Food Chem. 2005, 53, 2176–2180.
[173] White, P.J.; Broadley, M.R., Biofortification of crops with seven mineral elements often lacking in human diets – iron, zinc, copper, calcium, magnesium, selenium and iodine; New Phytol. 2009, 182, 49–84.
[174] Chaney, R.L., Zinc phytotoxicity; in: A.D. Robson (ed.), Zinc in Soils and Plants. Kluwer Academic, Dordrecht, 1993, pp. 135–150.
[175] Boawn, L.C., Zinc accumulation characteristics of some leafy vegetables; Commun. Soil Sci. Plant Anal. 1971, 2, 31–36.
[176] Boawn, L.C.; Rasmussen, P.E., Crop response to excessive zinc fertilization of alkaline soil; Agron. J. 1971, 63, 874–876.

[177] Beckett, P.H.T.; Davis, R.D., Upper critical levels of toxic elements in plants; New Phytol. 1977, 79, 95–106.
[178] Davis, R.D.; Beckett, P.H.T., Upper critical levels of toxic elements in plants. II. Critical levels of copper in young barley, wheat, rape, lettuce and ryegrass, and of nickel and zinc in young barley and ryegrass; New Phytol. 1978, 80, 23–32.
[179] Parker, D.R.; Chaney, R.L.; Norvell, W.A., Equilibrium computer models: applications to plant nutrition research; in: R.H. Loeppert, A.P. Schwab, S. Goldberg (eds.), Chemical Equilibrium and Reaction Models; Soil Science Society of America Special Publ. No. 42; Soil Sci. Soc. Am./Am. Soc. Agron., Madison, WI, 1995, pp. 163–200.
[180] Römheld, V.; The role of phytosiderophores in acquisition of iron and other micronutrients in graminaceous species: an ecological approach; Plant Soil 1991, 130, 127–134.
[181] Zhang, F.S.; Treeby, M.; Römheld, V.; Marschner, H., Mobilization of iron by phytosiderophores as affected by other micronutrients; Plant Soil 1991, 130, 173–178.
[182] Ma, J.F. and K. Nomoto., Inhibition of mugineic acid–ferric complex uptake in barley by copper, zinc and cobalt; Physiol. Plant. 1993, 89, 331–334.
[183] Pedler, J.F.; Kinraide, T.B.; Parker, D.R., Zinc rhizotoxicity in wheat and radish is alleviated by micromolar levels of magnesium and potassium in solution culture; Plant Soil 2004, 259, 191–199.
[184] White, M.C.; Decker, A.M.; Chaney, R.L., Differential cultivar tolerance of soybean to phytotoxic levels of soil Zn. I. Range of cultivar response; Agron. J. 1979, 71, 121–126.
[185] Speir, T.W.; Kettles, H.A.; Percival, H.J.; Parshotam, A., Is soil acidification the cause of biochemical responses when soils are amended with heavy metal salts? Soil Biol. Biochem. 1999, 31, 1953–1961.
[186] Smit, C.E.; Van Gestel, C.A.M., Effects of soil type, prepercolation, and ageing on bioaccumulation and toxicity of zinc for the springtail *Folsomia candida*; Environ. Toxicol Chem. 1998, 17, 1132–1141.
[187] McLaughlin, M.J.; Smolders, E., Background zinc concentrations in soil affect the zinc sensitivity of soil microbial processes – a rationale for a metalloregion approach to risk assessments; Environ. Toxicol. Chem. 2001, 20, 2639–2643.
[188] Smolders, E.; McGrath, S.P.; Lombi, E. *et al.*, Comparison of toxicity of zinc for soil microbial processes between laboratory-contaminated and polluted field soils; Environ. Toxicol. Chem. 2003, 22, 2592–2598.
[189] Smolders, E.; Oorts, K.; van Sprang, V. *et al.*, Toxicity of trace metals in soils as affected by soil type and aging after contamination: using calibrated bioavailability models to set ecological soil standards. Environ. Toxicol. Chem. 2009, 28, 1633–1642.
[190] White, M.C.; Chaney, R.L.; Decker, A.M., Differential cultivar tolerance of soybean to phytotoxic levels of soil Zn. II. Range of soil Zn additions and the uptake and translocation of Zn, Mn, Fe, and P; Agron. J. 1979, 71, 126–131.
[191] White, M.C.; Chaney, R.L.; Decker, A.M., Role of roots and shoots of soybean in tolerance to excess soil zinc; Crop Sci. 1979, 19, 126–128.
[192] Allen, H.L.; Brown, S.L.; Chaney, R.L. *et al.*, The Use of Soil Amendments for Remediation, Revitalization and Reuse; EPA 542-R-07-013, US Environmental Protection Agency, Washington, DC, 2007.
[193] Vangronsveld, J.; Van Assche, F; Clijsters, H., Reclamation of a bare industrial area contaminated by non-ferrous metals: *in situ* metal immobilization and revegetation.Environ. Pollut. 1995, 87:51–59.
[194] Nachtegaal, M.; Marcus, M.A.; Sonke, J.E., Effects of *in situ* remediation on the speciation and bioavailability of zinc in a smelter contaminated soil; Geochim. Cosmochim. Acta 2005, 69:4649–4664.
[195] Brown, S.; Christensen, B.; Lombi, E. *et al.*, An inter-laboratory study to test the ability of amendments to reduce the availability of Cd, Pb, and Zn in situ. Environ. Pollut. 2005, 138, 34–45.
[196] Gadgil, R.L., Tolerance of heavy metals and reclamation of industrial waste; J. Appl. Ecol. 1969, 6, 247–259.
[197] Underwood, E.J. Trace Elements in Human and Animal Nutrition, 4th edn; Academic Press, New York, 1977.

[198] National Research Council, Mineral Tolerance of Domestic Animals; National Academy of Sciences, Washington, DC, 1980.
[199] Grant-Frost, D.B.; Underwood, E.J., Zn toxicity in the rat and its interrelation with Cu; Aust. J. Exp. Biol. Med. Sci. 1958, 36, 339–346.
[200] Campbell, J.K.; Mills, C.F., The toxicity of zinc to pregnant sheep; Environ. Res. 1979, 20, 1–13.
[201] Bremner, I., The toxicity of cadmium, zinc, and molybdenum and their effects on copper metabolism; Proc. Nutr. Soc. 1979, 38, 235–242.
[202] Gunson, D.E.; Kowalczyk, D.F.; Shoop, C.R.; Ramberg, C.F., Jr., Environmental zinc and cadmium pollution associated with generalized osteochrondrosis, osteoporosis, and nephrocalcinosis in horses; J. Am. Vet. Med. Assoc. 1982, 180, 295–299.
[203] Ott, E.A.; Smith, W.H.; Harrington, R.B.; Beeson, W.M., Zinc toxicity to ruminants. II. Effect of high levels of dietary zinc on gains, feed consumption and feed efficiency of beef cattle; J. Anim. Sci. 1966, 25, 419–423.
[204] L'Estrange, J.L., The performance and carcass fat characteristics of lambs fattened on concentrate diets. 4. Effects of barley fed whole, ground or pelleted and of a high level of zinc supplementation; Ir. J. Agr. Res. 1979, 18, 173–182.
[205] Bremner, I.; Young, B.W.; Mills, C.F., Protective effect of zinc supplementation against copper toxicosis in sheep; Brit. J. Nutr. 1976, 36, 551–561.

18
Copper and Lead

Rupert L. Hough

The Macaulay Land Use Research Institute, Aberdeen, Scotland, UK

18.1 Introduction

Copper (Cu) is essential to human life and health. Following zinc and iron, copper is the third most abundant trace element in the human body [1]. It also has the potential to be toxic. Therefore, biochemical mechanisms have evolved that control the absorption and excretion of Cu that offset the effects of temporary deficiency or excess in the diet.

Copper is a Group 11 element of the periodic table. It consists of two natural isotopes: ^{63}Cu and ^{65}Cu, which constitute 69.09 % and 31.91 %, respectively. It has two oxidation states (+I and +II; cuprous and cupric ions, respectively). It is reddish-brown in colour, takes on a bright metallic lustre, is malleable, ductile, and an excellent conductor of heat and electricity [2]. In nature, the divalent cation is most abundant, but Cu can also occur in the form of monovalent cations and complex anions. For example, Cu readily forms sulfates, sulfides, carbonates and sulfosalts. Copper is an essential micronutrient to nearly all higher plants and animals. In animals and humans it is primarily found in the bloodstream, as a cofactor in various enzymes, and in copper-based pigments, for example turacoverdin and turacin [3]. At least thirty Cu-containing enzymes (human and animal) are known, all of which function as redox catalysts (for example cytochrome oxidase, nitrate reductase) or dioxygen carriers (for example haemocyanin) [4–6]. However, in excess amounts, copper can be poisonous and even fatal to organisms [7].

Copper has played a significant role in the development of human civilization. Near the time of Christ, Assyrian, Egyptian and Hindu alchemists prepared copper compounds for

Trace Elements in Soils Edited by Peter S. Hooda
© 2010 Blackwell Publishing, Ltd

a variety of human ailments [8]. During the period of the Roman Empire, Cu was principally mined in Cyprus, hence the origin of the name of the metal as *cyprium*, 'Metal of Cyprus', later shortened to *cuprum*.

Copper production has increased significantly in recent times. From 1955 to 1958, annual US production of recoverable copper was about 9.0×10^6 Mg; by 1975 this had increased to 12.6×10^6 Mg. Since then, production in the United States has levelled off, with 1.31×10^6 Mg of recoverable Cu reported for 2008 [9]. Globally, an estimated 1.80×10^6 Mg of anthropogenic copper has entered the soil annually since 1975 [9]. A number of countries, such as Chile and the United States, still have sizeable reserves of the metal.

Lead (Pb) is a member of Group 14 of the periodic table. Two oxidation states Pb(II) and Pb(IV) are stable, but the environmental chemistry of the element is dominated by the divalent ion, Pb^{2+}. Lead rarely occurs uncombined with other elements and is primarily found in the sulfide form in its chief ore, galena (PbS). Other common minerals of Pb (lead carbonate, $PbCO_3$ (cerussite); lead sulfate, $PbSO_4$ (anglesite); and lead chlorophosphate, $Pb_5(PO_4)_3Cl$ (pyromorphite)) are highly insoluble. In its elemental state, Pb is a dense (11.35 g cm^{-3}) blue-grey metal. Its density is in part due to its high atomic number (relative atomic mass of 207). The low melting point (327 °C) has allowed Pb to be worked, smelted and melted even in primitive societies. The metal is very soft and tends to creep or flow under sustained pressure. It is readily cut and shaped. There is a long history of its use on roofs or as pipes. The term plumbing (that is water pipes) is derived from the Latin for lead, *plumbum*. The highest demand for architectural Pb is currently in the United Kingdom due to the style of architecture, climatic conditions, and tradition. A life-cycle analysis of Pb and two alternative polymeric materials for flashings, found that Pb compared favourably [10].

Metallic Pb has a high mass attenuation coefficient, particularly for higher-energy X-rays, and makes a valuable shield material in X-ray and radioisotope work [11]. Lead also has a very small neutron capture cross-section and thus does not become radioactive or unstable.

Lead readily alloys with other metals. The Pb/Sb alloy is primarily used to make battery plates and Pb/Sn alloys are often used as solder. Lead metal, in combination with PbO_2, is used to fabricate lead–acid accumulator batteries. Many paints still contain Pb oxides or Pb soaps in order to promote polymerization. The industrial use of lead in the United States doubled between 1940 and 1970. By 2005, annual consumption of lead was 13 241 700 Mg, with the storage battery industry as the principal consumer [9]. Substantial amounts of lead are used in pigments, ceramics, pesticides and plumbing. A comprehensive review of the various applications of lead can be found in Thornton *et al.* [12]. There is an extensive organic chemistry of Pb(IV) compounds, especially tetraalkyl and tetraaryl compounds [13]. The tetraalkyl compounds, such as $Pb(CH_3)_4$, are colourless, heavy liquids, insoluble in water and miscible with organic solvents. An excellent review of the organic compounds of lead has been published by Calingaert [14].

Lead is considered to be one of the major environmental pollutants and has been incriminated as a cause of accidental poisoning in domestic animals more than any other substance [15]. Pre-1985, one of the primary sources of lead contamination in the air, soil, and water was combustion of fuel containing lead additives. Underwood [16] provides an excellent review from the leaded-fuel era.

18.2 Copper

18.2.1 Sources and Content of Copper in Soils

With an average concentration of 70 mg kg^{-1}, Cu is estimated to rank 26th in abundance in the lithosphere. Measured values reported for the earth's crust range from 24 to 55 mg kg^{-1} [17], while background values for soils range from 13 to 24 mg kg^{-1} [18]. Highest concentrations tend to be associated with Steppe soils such as the kastanozems and chernozems, while lower concentrations are associated with podzols and histosols. Ferrasols and fluvisols are also widely reported to contain relatively high concentrations of Cu [19].

Baker [17], McBride [20] and Aubert and Pinta [21] all discuss the role of soil parent materials in the origin of copper in soils. Aubert and Pinta [21] published an extensive catalogue that included the abundance of copper in parent material and the total copper content of the soil types developed on this material. Typical copper contents of major rock types are summarized in Table 18.1.

Copper deposits are related to intermediate-felsic (dioritic-granitic) igneous rocks and also mafic-ultramafic (basaltic-peridotitic) igneous rocks. A simplified view is the following: Cu is not accommodated in the crystal lattices of common silicate minerals except in trace proportions. Therefore, when magmas crystallize, Cu becomes concentrated in residual liquids together with other incompatible and large-ion lithophile elements plus volatiles, which are mainly H_2O, CO_2, F and Cl. The result is large volumes of hydrothermal solutions containing small quantities of valuable metals. These pass through fractured rock within and around the igneous bodies where the metals are deposited.

This model applies particularly to instances where Cu deposits are related to porphyritic (some minerals forming larger crystals, that is phenocrysts, in medium-fine grained rocks) intermediate-felsic igneous rocks (diorite-quartz monzonite (a form of granite)). The main

Table 18.1 Typical copper contents (mg kg^{-1}) of major rock types [17,21,31–33]

Major rock type	Cu content (mg kg^{-1})	
	Range	Average
Basic igneous (basalts)	30–160	100
Acid igneous (granites)	4–30	10
Ultramafic (pyroxenites)	10–40	15
Shale and clay	30–150	45
Black shales	20–200	70
Volcanic rocks	5–20	—
Agrillaceous sediments	40–60	—
Limestones	5–20	—
Sanstones	2–40	—
Lithosphere	55–70	60
Soils	2–100	20–30

Cu mineral, chalcopyrite (CuFeS$_2$ plus minor–trace amounts of Co, Ni, Zn, As, Se, Ag, Au, Pt, Pb, V and Cr), is formed by precipitation from late-stage saline fluids in hydrothermal vein deposits of Cu-Pb-Zn-Ag, or Cu-Zn-As-bearing assemblages. Molybdenum and Au may also be associated. For example, the porphyry copper deposits of Bisbee (Arizona, USA), Butte (Montana, USA), and Bingham (Utah, USA) [22,23].

In some cases isotope data show an interaction between hydrothermal fluids and meteoric or sea water [24]. Some deposits indicate a contribution to the ores of bacteriogenic sulfide produced during sedimentation or early diagenesis, for example Kupferschiefer in Poland [25].

Porphyry copper deposits are the largest source of Cu, and are found in North and South America, Europe, Asia and Pacific islands. None are documented in Africa. The largest examples are found in the Andes in South America. This distribution has led to the theory that Cu deposits are related to island arc volcanism and plate tectonics [26–28].

Copper deposits (the main mineral is again chalcopyrite) occurring in veins in mafic and ultramafic igneous rocks are typically associated with Ni and platinum group elements (PGE). The metals are considered to have exsolved from a Cu-Fe-Ni-S solid solution, which was probably immiscible with the silicate liquid. Much of the S in magmatic Ni-Cu sulfide deposits is considered to have been derived from sulfidic wall rocks, commonly pyritic sediments. Once a liquid sulfide is formed, it will tend to equilibrate with the silicate magma, and this means acquiring the Ni, Cu and PGE from the magma according to the partition coefficients for those elements [29]. Well-known examples are the sulfide–nickel deposits of Sudbury, Canada and Norilsk, Russia.

Strong covalent bonds are formed between Cu^{2+} and sulfide (S^{2-}) anions. The greater Pauling electronegativity [30] of Cu^{2+} (2.0) compared with Fe^{2+} (1.8) and Mg^{2+} (1.3) limits isomorphous replacement of these elements by Cu in soil parent materials [20]. However, in silicate clays and mafic (high Mg^{2+} and Fe^{2+}) rocks, Cu can replace metals with sixfold coordination (Mg^{2+}, Fe^{2+}, Zn^{2+}, Ni^{2+} and Mn^{2+}). Typical concentrations of Cu found in soils derived from different parent materials are presented in Table 18.2.

Table 18.2 Typical copper contents 'total' (aqua regia-extractable) (mg kg^{-1}) in soils on various parent materials

Soil type	Typical Cu content (mg kg^{-1})
Peat (histosols)	15–40
Sandy soils on drift (arenosols, podzols)	2–10
Sandy soils on granite (arenosols, podzols)	av. 10
Silty clay loams on shales (gleysols, cambosols)	av. 40
Clays developed on clay rocks (gleysols)	10–27
Loams developed on basalt rock (cambisols, etc.)	40–150
Humic loams on chalk (rendzinas)	7–28
Organic-rich loams or loess (chernozems)	1–100 (av. 30)
Soils developed on pumice (lithosols/arenosols)	3–25
Tropical soils (ferralsols)	8–128

Reproduced with permission from V.M. Shorrocks and B.J. Alloway, Copper in Plant, Animal and Human Nutrition, Copper Development Association, Potters Bar, 1988 [31].

Atmospheric inputs of Cu to soils from rain and dry deposition varies considerably according to the proximity of point sources, air currents, and the nature and quantities of windblown dust. The total quantity of Cu that has been emitted to the atmosphere since 3800 BC has been estimated at 3.2×10^6 Mg (about 1 % of the 307×10^6 Mg Cu produced) [34]. This amount is about three orders of magnitude greater than the estimated present-day atmospheric Cu burden [34]. Due to the relatively short residence time for airborne Cu aerosols, it is unlikely that a substantial build-up of Cu in the atmosphere could occur. However, the atmosphere is the primary medium for the transportation of pollutant Cu to the most remote areas of the planet. For example, analysis of moss samples [35], Antarctic ice and snow [36] and polar ice cores [37] has inferred substantial increases in airborne Cu at locations remote from any source of Cu emissions [17]. About 80 % of the total world production of copper has been made in the 20th century, and it has been estimated that 30 % of total Cu production took place during the 1970s [34]. Total global Cu production is estimated as 307×10^6 Mg. This is about twice the Cu content of the top 2 cm layer of soils of the world, and is about an order of magnitude greater than the annual Cu requirements of all living land biota [34].

Inputs of Cu to soil also occur as a result of agricultural practice. Copper is present in some agrochemicals, animal feeds, as well as coproducts such as animal manures (in some cases as a result of Cu-containing feed), composts and sewage sludge (for example [38,39]). During wastewater treatment, only ~25 % of the mass of Cu present in sewage is removed [40]. While there have been substantial decreases in the inputs of Cu to sewage over the last 30 years [41,42], studies have shown elevated levels of Cu in crops from farmland previously treated with sewage sludge [38,43–46].

18.2.2 Chemical Behaviour in Soils

Copper in soil adsorbs strongly to clay minerals, iron and manganese oxides, and organic material. Because of this, the Cu tends to remain in horizons that have greater organic content. Sandy soils with low pH possess the greatest potential for Cu leaching. However, the effect of pH tends to be relatively less than on other divalent metals [47]. After $CuOH^+$, the divalent cation (Cu^{2+}) is considered to be the most mobile form of copper in the soil environment [48]. However, a number of other ionic species may occur (Cu^+, $CuOH^+$, $Cu(OH)_2^{2+}$, $Cu(CO_3)_2^{2-}$, $Cu(OH)_4^{2-}$, $Cu(OH)_3^-$, CuO_2^{2-}, $HCuO_2^-$). The majority of copper ions are held tightly on both organic and inorganic exchange sites. The processes controlling fixation by soil constituents are related to the following phenomena: adsorption, occlusion and coprecipitation, chelation and complexing with humic substances, and microbial fixation. Sorption phenomena of copper have been extensively studied, and a number of comprehensive reviews are available [20,49–53]. All soil minerals are capable of adsorbing copper ions from solution, depending on the surface charge carried by the absorbants. Surface charge is partly dependent on the pH, thus solubility can be described as a function of soil pH [49–53]. Adsorption of copper by minerals takes place between 0.001 and 1 μmol l^{-1} or from 30 to 1000 μmol g^{-1} [18]. Levels of adsorption towards the top end of this range are associated with Fe- and Mn-oxides (haematite, goethite, birnessite), amorphous Fe- and Al-hydroxides, and clays (montmorillonite, vermiculite, imogolite).

Apart from those soils with little or no soil organic matter, chelation and complexing are the principal reactions governing specific adsorption of Cu. Binding of copper by soils is related to the formation of organic complexes and is strongly associated with pH [54]. Maximum sorption capacity of different soils varies significantly with the physical and chemical properties of organic substances. Peat and humic acids have been shown to strongly immobilize the Cu^{2+} ion in direct coordination with functional oxygen of the organic substances [55,56]. Sapek [57] reported the maximum sorption capacity of peat soil to range from 130 to 190 meq per 100 g peat. Ovchatenko et al. [58] calculated sorption of Cu as 3.3 g kg^{-1} humic acid. Ponizovsky et al. [59] found that the retention of Cu^{2+} by organic-rich soil differs from mechanisms of exchange of alkali and alkali earth metal cations and could be regarded as triple Cu^{2+}, Ca^{2+} and H^+ cation exchange. Stevenson and Fitch [60] stated that the maximum amount of Cu^{2+} that can be bound to humic and fulvic acids is approximately equal to the content of acidic functional groups, corresponding to a sorption capacity of 48–160 mg Cu g^{-1} humic acid.

Occlusion and coprecipitation are involved in nonspecific adsorption of Cu. Aluminium- and Fe-hydroxides, carbonates, and phosphates, and to some extent silicate clays, have a great affinity to bind to soil Cu in a nondiffusible form.

The affinity of Cu for organic complexing has implications for its migration and bioavailability in soils. Bioavailability of soluble complexes of Cu will depend on their molecular weight. Compounds of low molecular weight (for example those liberated during decomposition of plant and animal residues), as well as those applied with coproducts (for example sewage sludge and compost), are likely to be relatively available for uptake by plants.

In many surface soils, microbial fixation is a significant mechanism of Cu binding. The amount of Cu fixed by microbial biomass in a given soil is highly dependent on a variety of factors, including the concentration of Cu in the soil; the physicochemical properties of the soil; and the time of year [61].

While Cu is one of the least mobile heavy metals in soils, it is abundant as free and complexed ions in soil solutions across all soil types. The most common forms of Cu in soil solutions are soluble organic chelates [62,63]. Reported concentrations of Cu in soil solutions, obtained by various techniques, range from 3 to 135 μg l^{-1} (0.047–2.125 μmol l^{-1}) [62,64–69]. The solubility of both cationic and anionic species decreases at soil pH values between 7 and 8. It has been estimated that hydrolysis products of Cu ($CuOH^+$, $Cu_2(OH)_2^{2+}$) are the most abundant Cu-species in the soil solution below pH 7. Above pH 8, anionic hydroxyl complexes ($Cu(OH)_3^-$, $Cu(OH)_4^{2-}$) are the dominant Cu species in the soil solution. The solubility of copper carbonate ($CuCO_3$) shows much lower pH-dependency, and it is the dominant inorganic soluble form of Cu in neutral and alkaline soil solutions.

18.3 Lead

18.3.1 Sources and Content of Lead in Soils

Lead is a trace element in rocks and soils. Natural sources of Pb in the surface environment arise from the weathering of geological materials, emissions to the atmosphere from

Table 18.3 Mean concentrations of Pb (mg kg^{-1}) in the main rock types (adapted from [17])

Rock type	Average Pb (mg kg^{-1})
Ultramafic	1.0
Basaltic	6.0
Granitic	18
Shales and clays	20
Black shales	30
Limestones	9.0
Sandstones	12

volcanoes, wind-blown dusts, sea spray, biogenic material, and forest fires [12]. Its ionic radius is 124 picometres (pm) and it isomorphously replaces K (133 pm) in silicate lattices. As a result, there is a general concentration gradient from ultrabasic (lower Pb contents) through to acid igneous rocks (higher Pb contents). Lead is chalcophilic and therefore has a strong affinity for S and concentrates in the S phases in rocks with the major ore mineral being the sulfide, galena (PbS).

The average concentration of Pb in the earth's crust has been estimated as approximately 16 mg kg^{-1} [70]. Table 18.3. details the mean concentrations of Pb reported in various rock types. Nriagu [34] calculated the mean Pb content of gabbro as 1.9 mg kg^{-1}, andesite as 8.3 mg kg^{-1} and granite as 22.7 mg kg^{-1}, demonstrating the tendency for Pb concentration to increase with increasing silica content. For the sedimentary rocks, shales and mudstones have been shown to contain on average 23 mg Pb kg^{-1}. Of the sedimentary rocks, black shales (which tend to be composed of very fine particles and rich in organic matter and sulfide minerals) tend to have relatively higher concentrations of Pb. Limestones, basalts and igneous ultramafic rocks contain only traces of Pb, with mean concentrations reported from 1 to 9 mg kg^{-1} [71]. Lead also has a strong affinity for S and therefore concentrates in the S phases in rocks, with the major ore mineral being galena (PbS).

Estimates of Pb in uncontaminated soils are compounded by widespread low-level contamination. Ure and Berrow [72] reported a mean value of 29 mg kg^{-1}, while Nriagu [34] has reported a mean concentration of 17 mg kg^{-1}. Reaves and Berrow [73] examined the distribution of Pb in 3944 samples from 896 Scottish soil profiles. The geometric mean Pb content of all mineral soil samples was 13 mg kg^{-1}, while organic and organomineral soils were found to have an average concentration of 30 mg kg^{-1}. In 1983, an analysis of soils data for England and Wales indicated that the (then) normal surface (0–5 cm) concentration of Pb was between 50 and 106 mg kg^{-1} (geometric mean 42 mg kg^{-1}) [74]. The distribution of Pb in 5000 agricultural soils (England and Wales) was reported as 10.9–145 mg kg^{-1} [75]. Reported concentrations of Pb for over 3000 agricultural soils (Ap horizons) in the United States were lower than those in the United Kingdom, with a median concentration of 11 mg kg^{-1} [76]. This may be due to differences in bedrock composition, or lower intensity/shorter history of industrial development. It is also worth noting that Reaves and Berrow [73] and Holmgren *et al.* [76] performed their sampling using pedological principles (that is sampling by horizon) rather than by fixed-depth sampling. The former may result in a more appropriate estimate of soil Pb.

Atmospheric deposition is a major source of Pb in the surface environment. Dusts and vapours of Pb can enter the atmosphere from natural as well as anthropogenic processes and sources. It has been estimated that natural emissions of Pb globally to the atmosphere are in the range 18.6×10^6 to 29.5×10^6 kg y^{-1}[77]. Of this, 60–85 % is thought to be windblown material, 5–10 % from vegetation, and the remainder from volcanoes, sea spray and meteorites [70,77]. Nriagu [78] estimated that 23 000 Mg of trace metals were emitted annually from volcanoes (about 5 % of which is Pb [79]). More recent estimates (for example [80]) put this figure at <10 000 Mg. These estimates are significantly smaller than the historical anthropogenic flux of lead in the atmosphere. Lead from natural emissions is increasing proportionately as the use of leaded petrol is discontinued (for example [81,82]).

18.3.2 Chemical Behaviour in Soils

A number of early studies indicate that Pb appears to accumulate naturally in the surface horizons of uncontaminated soil [83–90]. Lounamaa [83] reported this phenomenon for soils collected from remote, seminatural sites in Finland. Accumulations of Pb tend to be associated with organic-rich layers [91]; thus organic matter can be considered an important sink for Pb in soils [18]. For example, Wright *et al.* [84] inferred that Pb showed greatest proportional accumulation in surface horizons of podzols, brown podzolics and grey-brown podzols in Eastern Canada. However, it is impossible to consider the presence of Pb and its compounds in the environment and its potential toxicity to the ecosystem or to human health without considering its chemical and mineral form [12].

Lead exists in a number of different forms in the environment. Naturally occurring Pb may be as minerals such as galena (PbS), residues from Pb mining also tend to be in the form of the ore (PbS with some $PbSO_4$ (anglesite) or $PbCO_3$ (cerussite)), emissions from smelting are typically a mixture of Pb oxides, sulfates, and metallic Pb. All these forms have very low solubility in water.

As with Cu, only a small proportion of the total Pb present in soil is bioavailable. The main 'pools' of Pb in soil are the soil solution, the adsorption surfaces of the clay–humus exchange complex, precipitated forms, secondary Fe and Mn oxides and alkaline earth carbonates, the soil humus and silicate lattices [92–94].The majority of evidence suggests that soil solution Pb is the primary source for plant uptake, with a dynamic equilibrium occurring between the soil solution and the more 'exchangeable' or 'labile' pools of Pb within the soil matrix. In some soils, Pb may be concentrated with $CaCO_3$ or with phosphate. In soils with highly elevated concentrations of Pb, the formation of pyromorphite ($Pb_5Cl(PO_4)_3$) can occur, especially close to plant roots [95]. The solubility of Pb in soil is strongly pH-dependent. A high soil pH may precipitate Pb as hydroxide, phosphate, or carbonate, as well as promote the formation of relatively stable Pb–organic complexes [96]. While increasing acidity will increase the solubility of Pb, mobilization is usually at a slower rate than the rate of accumulation of Pb in the organic-rich layer [97].

There is a large body of experimental evidence describing the sorption of Pb by various clay minerals (for example [98–100]). It has been shown that adsorption is highly dependent on the kinds of ligands involved in the formation of hydroxyl complexes of Pb (for example $PbOH^+$ and $Pb_4(OH)_4^{4+}$) [100]. It has been suggested that adsorption of these species on montmorillonite can be interpreted as a cation exchange process, while on kaolinite and illite the process is competitive adsorption. Abd-Elfattah and Wada [101] found a higher selective sorption of Pb by iron oxides, halloysite, and imogolite than by humus, kaolinite, and montmorillonite. Other studies have reported manganese oxides as having the greatest affinity to sorb Pb [98,102].

18.4 Risks from Copper and Lead

18.4.1 Essentiality and Metabolism

The essentiality of Cu was suggested by McHargue as early as the 1920s [103]. The first conclusive evidence of the biological requirement for Cu was provided towards the end of the same decade [104]. This work demonstrated that anaemia in rats was not corrected by feeding iron supplementation alone, or by feeding liver extract alone. However the combined feeding of iron and liver extract together resulted in a significant elevation in both haemoglobin levels and packed cell volume. Assays on the liver revealed significant Cu content. Further anaemic rats were fed both Fe and Cu supplementation and the same response was recorded. Since then, the roles of Cu in ovine enzootic ataxia (swayback) [105], bovine falling disease [106,107], aortic rupture in swine and turkeys [108], wool and hair depigmentation [109], and anaemia [110] have been elucidated. Human hearts of people who have died from ischaemic heart disease have been found to be relatively low in Cu [111–115]. Kinsman *et al.* [116] found a significant correlation ($r = 0.67$) between Cu in leukocytes and the degree of patency of the coronary arteries of men. Numerous Cu-dependent enzymes, including lysyl oxidase, cytochrome-*c* oxidase, ferroxidase and tyrosinase, have all been recognized [117,118].

The level of dietary Cu required for good health is species dependent and is usually positively correlated with dietary levels of molybdenum and inorganic sulfur. Various data suggest that the Cu requirements for specific biological processes increase in the order: haemoglobin formation, growth, hair pigmentation, lactation [119]. When dietary conditions are optimal for utilization of copper, a diet comprising 4–5 mg Cu kg^{-1} for swine and poultry and 8–10 mg Cu kg^{-1} for ruminants appears adequate [16,110].

Copper absorption, for the majority of higher mammals and humans, tends to take place in the duodenum and jejunum [120,121]. The human gastrointestinal system can absorb 30–40 % of ingested Cu from the diets typically consumed in industrialized countries [120]. However, absorption can be as low as 12 % [122,123]. In domestic animal species, absorption rate may be a low as 10 % [124]. Absorption is significantly affected by the chemical form of the ingested copper. Generally, water-soluble forms including copper carbonate, sulfate, nitrate, and chloride, are absorbed to a greater extent than copper oxide [120]. Metallic Cu is poorly absorbed [120]. Absorption of Cu is also highly modified by the presence of other mineral elements. The antagonism between Zn and Cu has been known for decades [for example 125,126]. In fact, a ternary relation between Zn, Cu, and dietary protein can result in marginal Cu deficiency, especially when Zn intake is high and protein consumption low [127]. If dietary Zn is increased from 5 to 20 mg d^{-1}, the intake of Cu needs to be 60 % higher in order to maintain nutritional balance [128]. Other divalent cations may also act competitively on intestinal absorption sites. This has been shown for ferrous iron [129,130] and stannous tin [130,131].

Absorbed Cu appears first in blood plasma as the cupric ion bound loosely to albumin. During hepatic synthesis of ceruloplasmin, Cu is tightly bound to this metalloprotein, which is then released to the general circulation [120,121,132]. Ultimately, cuproprotein is present in brain, erythrocytes and liver as cerebrocuprein, erythrocuprein and hepatocuprein, respectively. The biliary system is the major excretory pathway for absorbed Cu in most species for which data are available [16,110,133]. Copper is also excreted during perspiration and lactation. Large quantities of Cu are excreted by the urinary system in cases of biliary obstruction or Wilson's disease [134]. Similarly to the role of molybdenum in Cu deficiency in sheep, Wilson's disease can be treated using molybdenum, which is introduced into the diet as a thiomolybdate complex. When ingested with foodstuffs, thiomolybdate effectively scavenges dietary Cu [135].

Lead is generally not considered to be an essential element for humans or animals. However, Schwarz [136] showed that the addition of 1 mg kg^{-1} dietary Pb as Pb subacetate increased the growth of rats by 16 % (1.79 vs 2.08 g d^{-1}) over controls receiving no supplementary Pb. Increasing the Pb level from 1.0 to 2.5 mg kg^{-1} dietary Pb decreased this growth response by 33 % from 0.29 to 0.19 g d^{-1} over controls. Lead oxide and lead nitrate have produced similar responses [110]. These studies suggest some level of essentiality, at least for rats, at very low doses.

Many early reviews are available concerning the metabolism of Pb [15,137–139]. The bioavailability of Pb has been shown to be affected by diet, growth rate, and physiological stresses, such as malnutrition, pregnancy, and lactation [140–144]. Lead tends to accumulate in the bones [145]; consequently, the majority of body Pb ($\sim$90 %) can be accounted for in the skeleton [146] and appears to be relatively immobile. Nonruminant animals absorb approximately 10 % of dietary Pb, while ruminants absorb roughly 3 % [147]. Mass balance studies on humans ingesting environmental quantities of Pb have shown that only 5–10 % of ingested lead is absorbed [148–150]. During chronic exposure a steady state appears to be reached in which metabolic excretion, by way of urinary and faecal excretion, approximately equals absorption. This occurs after initial tissue saturation level is reached; therefore, levels of Pb in many tissues and body fluids have been shown to increase with increasing exposure to Pb. Absorption of Pb in the human infant has been shown to occur at significantly higher rates than in adults.

18.4.2 Exposure and Toxicology

Risks from Cu can be associated with either deficiency or toxicity. The average daily intake of Cu in the United States is about 1 mg, with the primary source being the diet [1]. Copper ends up in food and beverages either through contact with processing equipment, through agricultural practices (for example use of Cu-based agrichemicals), or via uptake of Cu into food crops from soil. The bioavailability of Cu from the diet is about 65–70 % depending on a variety of factors including chemical form, interaction with other metals, and dietary components. The biological half-life of Cu from the diet is 13–33 days, with biliary excretion being the major route of elimination [150].

In order to understand risks from Cu, it is important to know who may be most susceptible and attempt to protect the most vulnerable. Apart from specific rare disorders, for example Wilson's disease [134,135,151], a number of population subgroups are 'unusually' susceptible to Cu toxicity [152]. Infants and children under one year old are at high risk because they have not yet developed mechanisms for clearing Cu and preventing its entry via the intestine [153–156]. Individuals with liver damage are also susceptible [157]. This is due to the liver's key role in Cu storage and excretion into bile; metabolic and pathological dysfunction is likely to disrupt Cu homeostasis. At high Cu intake, urinary excretion increases. Individuals with chronic renal disease have difficulty handling high Cu intake. Individuals with an inherited deficiency of the enzyme glucose-6-phosphate dehydrogenase are more likely to be susceptible to the toxic effects of oxidative stressors such as Cu [158].

The amount of oral Cu required to produce toxic effects is not well established; there are no human data (beyond individual case reports), and animal data for carcinogenicity are considered inadequate by the USEPA for the formulation of 'safe' doses (further explanation of this concept may be found in Chapter 12 on risk assessment). The USEPA report that Bionetics Research Labs (1968, unpublished) studied the carcinogenicity of copper hydroxyquinoline in two strains of mice (B6C3F1 and B6AKF1). Groups of 18 male and 18 female seven-day-old mice were administered 1000 mg Cu hydroxyquinoline kg^{-1} body weight until they were 28 days old, after which they were administered 2800 mg kg^{-1} Cu hydroxyquinoline in the feed for 50 additional weeks. No statistically significant increases in tumour incidence were observed in the treated 78-week-old animals. Gilman [159] administered intramuscular injections containing 20 mg of cupric oxide (16 mg Cu), cupric sulfide (13.3 mg Cu) and cuprous sulfide (16 mg Cu) into the left and right thighs of two- to three-month-old Wistar rats. After 20 months of observations, no injection-site tumours were observed in any animals, but other tumours were observed at very low incidence in the animals receiving cupric sulfide (2 out of 30 rats) and cuprous sulfide (1 out of 30 rats).

Relatively few cases of Cu deficiency have been reported (for example [160–162]), and these are often accompanied by confounding factors such as malnutrition, malabsorption, excessive gastrointestinal losses [152], and excess dietary zinc [163]. Clinically evident Cu deficiency is a relatively infrequent condition that can result in anaemia, compromised immune response and bone abnormalities (including osteoporosis and fractures) [164–166]. Acquired deficiencies, which occur mainly in infants, can be the consequence of decreased maternal Cu stores at birth [167]. It is suggested that the minimum Cu requirement for men is somewhere between 0.4 and 0.8 mg d^{-1} [168]. An adult basal Cu requirement of 0.6 mg d^{-1} for women and 0.7 mg d^{-1} for men was suggested by the World Health Organization [169]. A recommendation of 1.25 mg d^{-1} was derived by the WHO after adding margins of safety to the individual requirement, including variation in dietary conditions, variations in usual intakes, and individual variability. Most multi-vitamin dietary supplements on the market today contain 2 mg Cu to be taken daily. The daily dietary Cu intake recommended by the US National Research Council (NRC) in 1989 was 1.5–3.0 mg d^{-1} for adults [170]. The recommended daily range for children is from 0.4–0.6 mg d^{-1} for infants 0–6 months, to 1.5–2.5 mg d^{-1} for children over 11 years [171].

Environmental Pb is largely air-borne but returns to soil, water and plants as dust and can become a hazard, especially to grazing livestock. Clinical toxicosis in animals exposed chronically to Pb is indirect and probably results through interference in normal metal-dependent enzyme functions at specific cellular sites characterized by apparent clinical abnormalities in haematological, neural, renal, or skeletal systems. Lead poisoning in livestock is well documented, with a number of seminal reviews published during the 1970s (for example [172–177]). Lead toxicosis is characterized by one or more of several clinical signs and underlying pathophysiological effects [178]. The main clinical signs have been described as microcytic hypochromic anaemia; anorexia, fatigue, depression; intestinal colic (constipation, diarrhoea, abdominal pain); vomiting, increased salivation, oesophageal paralysis; nephropathy; irritability, peripheral neuropathy, encephalopathy, blindness in cattle; laryngeal paralysis in horses; weight loss; abortion; and maniacal excitement in young calves. The main pathological effects include derangement of porphyrin and haem synthesis; interference in protein and haem synthesis; increased mechanical fragility of cell membranes resulting in shortened life of red blood cells; enzyme changes where small concentrations of Pb ($\leq 10^{-6}$ M) may inhibit or enhance activities; renal tubular intracellular inclusion bodies, containing protein-bound Pb, Ca and P; basophilic stippling of erythrocytes and inhibitors of haemoglobin synthesis; and altered endocrine function.

Relatively large amounts of absorbed Pb can be sequestered preferentially in the skeleton with subsequent gradual release into the bloodstream. Chronic Pb toxicosis is rarely seen in ruminants, but is more common in the nonruminant. It is usually recognized, however, only when distinct signs of poisoning are apparent.

There are a number of early experiments exploring the dose–response to environmental levels of Pb. A comprehensive review of studies can be found in USEPA 2005 [179]. A no observed adverse effect level (NOAEL, see Chapter 12) was determined as early as 1950, when cattle fed 2 g Pb d^{-1} as lead acetate showed no effects [141]. Lead poisoning was produced in cattle within six to eight weeks when fed lead acetate at 7 mg Pb kg^{-1} body weight d^{-1} [180,181]. Dinius *et al.* [182] fed calves a concentrate diet containing 0, 10, and 100 mg Pb kg^{-1} as lead chromate for 100 days and saw no effect on feed consumption or weight gain. There was increased accumulation of Pb in the liver and kidney with 100 mg Pb kg^{-1}. Kelliher *et al.* [183] observed reduced growth and feed utilization when calves were fed 15 mg Pb kg^{-1} body weight d^{-1} for 283 days. There was no adverse effect on performance when lead acetate [Pb(C$_2$H$_3$O$_2$)$_2$·3H$_2$O] was fed to lambs at added dietary levels of 10, 100, 500, or 1000 mgPb kg^{-1} [184]. However, the authors do not indicate whether the doses administered were milligrams per kilogram body weight or milligrams per kilogram in the feed.

There is evidence that horses may be more susceptible to chronic Pb toxicosis than cattle. Horses were poisoned on pastures adjacent to a smelter where they experienced a Pb intake of 2.4 mg Pb kg^{-1} body weight d^{-1} [181]. Horses exposed to a daily intake as low as 1.7 mg Pb kg^{-1} body weight d^{-1} (approximately 80 mg Pb kg^{-1} in forage dry matter) were poisoned [185].

18.5 Concluding Remarks

Both Cu and Pb have played prominent roles in the development of humankind and are now included in a vast range of manufactured components. Both elements have also played important, but highly contrasting, roles in terms of health, diet and exposure. For example, Cu is an essential trace nutrient, with the majority of human/animal health issues associated with deficiency rather than toxicity. Lead, however, is considered one of the most widely spread potentially toxic elements and has been associated with a significant number of deleterious human and animal health outcomes. For these reasons, the coupling of Cu and Pb into the same chapter may seem unusual. However, these two elements are part of the same story. They are two of the most widely used metals for everyday purposes. The impacts of these metals on ordinary people's lives have been great. Our understanding of their impacts has been implicit in developments that have improved public health such as the conversion from lead to copper water pipes.

References

[1] Barceloux, D.G., Copper; Clin. Toxicol. 1999, 37, 217–230.
[2] Forster, U.; Wittmann, G.T.W., Metal Pollution in the Aquatic Environment; Springer-Verlag, Berlin, 1979.
[3] Fowler, M.E.; Miller, R.E., Zoo and Wild Animal Medicine, 5th edn; W.B. Saunders, Philadelphia, 2003.
[4] Pena, M.M.O.; Lee, J.; Thiele, D.J., A delicate balance: homeostatic control of copper uptake and distribution; J. Nutr. 1999, 129, 1251–1260.
[5] Linder, M.C., Biogeochemistry of Copper. Plenum Press, New York, 1991.
[6] Weser, U.; Schubotz, L.M.; Younes, M., Chemistry of copper proteins and enzymes; in: J.O. Nriagu (ed.), Copper in the Environment. Part II: Health Effects; John Wiley & Sons Canada Ltd, 1979, pp. 197–240.
[7] Spitalny, K.C.; Brondum, J.; Vogt, R.L.; Sargent, H.E.; Kappel, S., Drinking-water-induced copper intoxication in a Vermont family; Pediatrics 1984, 74, 1103–1106.
[8] Abdel-Mageed, A.B.; Oehme, F.W., A review of the biochemical roles, toxicity and interactions of zinc, copper and iron: II copper; Vet. Hum. Toxicol. 1990, 32, 230–234.
[9] United States Geological Survey. Mineral Resources Program, Mineral Commodity Summaries, 2009; USGS. Washington DC, 2009.
[10] Roorda, A.A.H.; van der Ven, B.L., Lead Sheet and the Environment. TNO Report TNO-MEO – R 98/503; TNO, Amsterdam, 1999.
[11] Lide, D.R., CRC Handbook of Chemistry and Physics, 79th edn; CRC Press, Boca Raton, FL, 1998.
[12] Thornton, I.; Rauti, R.; Brush, S., Lead: The Facts; IC Consultants Ltd. London, 2001.
[13] Davies, B.E., Lead; in: B.J. Alloway (ed.), Heavy Metals in Soils, 2nd ed;. Blackie Academic & Professional, London, 1995, Chapter 9, pp. 177–196.
[14] Calingaert, G., The organic compounds of lead; Chem. Rev. 1925, 2, 43–83.
[15] NRC. Lead: Airborne lead in perspective; National Academy of Sciences, Washington, DC, 1972.
[16] Underwood, E.J., Trace Elements in Human and Animal Nutrition, 4th edn;. Academic Press, New York, 1977.
[17] Baker, D.E., Copper; in: B.J. Alloway (ed.), Heavy Metals in Soils, 2nd edn; Blackie Academic & Professional, London, 1995, Chapter 7, pp. 152–178.
[18] Kabata-Pendias, A.; Pendias, H., Trace Elements in Soils and Plants; CRC Press, Boca Raton, FL, 2001.
[19] Shacklette, H.T.; Boerngen, J.G., Element concentrations in soils and other surficial materials of the conterminous United States; United States Geological Survey Circular, 1984, 1270, 105.
[20] McBride, M.B., Forms and distribution of copper in solid and solution phases of soil; in: J.F. Longeragan, A.D. Robson, R.D. Graham (eds), Copper in Soils and Plants; Academic Press, New York, 1981.

[21] Aubert, H., Pinta, M., Trace Elements in Soils; Elsevier, Amsterdam, 1977.
[22] Reed, B.L., Descriptive model of porphyry Sn, Mineral Deposit Models; US Geol. Surv. Bull. 1986, 1693, 108.
[23] Andrew R.L., Porphyry copper-gold deposits of the southwest Pacific; Min. Eng.-Littleton 1995, 1, 33–38.
[24] Mao, J.; Bierlein, F.P., The Dapingzhang polymetallic copper deposit; in: Mineral Deposit Research: Meeting the Global Challenge. Proceedings of the Eighth Biennial SGA Meeting Beijing, China, 18–21 August 2005.
[25] Sawlowicz, Z. Primary copper sulphides from the Kupferschiefer, Poland; Miner. Deposita 1990, 25, 262–271.
[26] Sillitoe, R.H., A plate tectonic model for the origin of porphyry copper deposits; Econ. Geol. 1972, 67, 184–197.
[27] Sillitoe, R.H., Characteristics and controls of the largest porphyry copper-gold and epithermal gold deposits in the circum-pacific region; Aust. J. Earth Sci. 1997, 44, 373–388.
[28] Zengqian, H.; Hongwen, M.; Zaw, K.; Yuquan, Z.; Mingjie, W.; Zeng, W. *et al.*, The Himalayan Yulong porphyry copper belt: product of large-scale strike-slip faulting in Eastern Tibet; Econ. Geol. 2003, 98, 125–145.
[29] Eckstrand O. R.; Hulbert L.J., Mineral Deposits of Canada: Magmatic Nickel-Copper-PGE deposits; Geological Survey of Canada, Ottawa, ON, 2008.
[30] Pauling, L., The nature of the chemical bond. IV. The energy of single bonds and the relative electronegativity of atoms; J. Am. Chem. Soc. 1932, 54, 3570–3582.
[31] Shorrocks, V.M.; Alloway, B.J., Copper in Plant, Animal and Human Nutrition; Copper Development Association, Potters Bar, 1988.
[32] Thornton, I., Copper in soils and sediments; in: J.O. Nriagu (ed.), Copper in the Environment, Part I: Ecological Cycling., John Wiley & Sons, Inc., New York, 1979.
[33] Baker, D.E.; Chesin, L., Chemical monitoring of soils for environmental quality and animal and human health; Adv. Agron. 1975, 27, 305–374.
[34] Nriagu, J.O., Copper in the atmosphere and precipitation; in: J.O. Nriagu (ed.), Copper in the Environment, Part I: Ecological Cycling., John Wiley & Sons, Inc., New York, 1979.
[35] Berg, T.; Røyset, O.; Steinnes, E.; Vadet, M., Atmospheric trace element deposition: principal component analysis of ICP-MS data from moss samples; Environ. Pollut. 1995, 88, 67–77.
[36] Wolff, E.W.; Suttie, E.D.; Peel, D.A., Antarctic snow record of cadmium, copper, and zinc content during the twentieth century; Atmos. Environ. 1999, 33, 1535–1541.
[37] Wolff, E.W., Signals of atmosphere pollution in polar snow and ice; Antarct. Sci. 1990, 2, 189–205.
[38] Keller, C.; Kayser, A.; Keller, A.; Schulin, R., Heavy-metal uptake by agricultural crops from sewage-sludge treated soils of the Upper Swiss Rhine Valley and the effect of time; in: I.K. Iskandar (ed.), Environmental Restoration of Metals – Contaminated Soils; Lewis Publishers, Boca Raton, FL, 2001, pp. 273–291.
[39] Soffe, R.E., The Agricultural Notebook, 18th edn; Blackwell Science, Oxford, UK, 1995.
[40] Netzer, A.; Beszedits, S., Removal of copper from wastewaters; in: J.O. Nriagu (ed.), Copper in the Environment, Part I: Ecological Cycling; John Wiley & Sons, Inc., New York, 1979.
[41] Smith, S.R., Agricultural Recycling of Sewage Sludge and the Environment; CAB International, Wallingford, 1996.
[42] Rowlands, C.L., Sewage sludge in agriculture; a UK perspective. Proceedings of the Water Environment Federation 65th Annual Conference and Exposition, New Orleans, 1992, pp. 303–315.
[43] Gardiner, D.T.; Miller, R.W.; Badamchain, B.; Azzari, A.S.; Sisson, D.R., Effects of repeated sewage sludge applications on plant accumulation of heavy metals; Agr. Ecosyst. Environ. 1995, 55, 1–6.
[44] Miner, G.S.; Guiterrez, R.; King, L.D., Soil factors affecting plant concentrations of cadmium, copper, and zinc on sludge-amended soils; J. Environ. Qual. 1997, 26, 989–994.
[45] Hooda, P.S.; McNulty, D.; Alloway, B.J.; Aitken, M.N., Plant availability of heavy metals in soils previously amended with heavy applications of sewage sludge; J. Sci. Food Agr. 1997, 73, 446–454.

[46] Hough, R.L.; Young, S.D.; Crout, N.M.J., Modelling of Cd, Cu, Ni, Pb and Zn uptake, by winter wheat and forage maize, from a sewage disposal farm; Soil Use Manage. 2003, 19, 19–27.
[47] Gerritse, R.G.; van Driel, W., The relationship between adsorption of trace metals, organic matter, and pH in temperate soils; J. Environ. Qual. 1984, 13, 197–204.
[48] Bodek, I.; Lyman, W.; Reehl, W.F.; Rosenblatt, D.H., Environmental Inorganic Chemistry; Pergamon Press, New York, 1988.
[49] McLaren, R.G.; Crawford, D.V., Studies on soil copper I. Fractionation of copper in soils; J. Soil Sci. 1973, 24, 172–181.
[50] McLaren, R.G.; Crawford, D.V., Studies on soil copper II. Specific adsorption of copper by soils; J. Soil Sci. 1973b 24, 443–452.
[51] McLaren, R.G.; Crawford, D.V., Studies on soil copper III. Isotopically exchangeable copper in soils; J. Soil Sci. 1974, 25, 111–119.
[52] James, R.O.; Barrow, N.J., Copper reactions with inorganic components of soils including uptake by oxide and silicate minerals; in: J.F. Longeragan, A.D. Robson, R.D. Graham (eds), Copper in Soils and Plants; Academic Press, New York, 1981.
[53] Kitagishi, K.; Yamane, I., Heavy Metal Pollution in Soils of Japan; Japan Science Society Press, Tokyo, 1981.
[54] Sparks, D.L., Kinetics of reactions in pure and mixed systems; in: D.L. Sparks (ed.). Soil Physical Chemistry; CRC Press, Boca Raton, FL, 1986.
[55] Bloom, P.R.; McBride, M.B., Metal ion binding and exchange with hydrogen ions in acid-washed peat; Soil Sci. Soc. Am. J. 1979, 43, 687–692.
[56] Bloomfield, C., The translocation of metals in soils; in: D.J. Greenland, M.H.B. Hayes (eds), The Chemistry of Soil Processess; John Wiley & Sons, Inc., New York, 1981.
[57] Sapek, B., Copper behaviour in reclaimed peat soil of grassland; Roczniki Nauk Roln 1980, 80f, 13 and 65.
[58] Ovcharenko, F.D.; Gordienko, S.A.; Glushchenko, T.F.; Gavrish, I.N., Methods and results of the complex formation of humic acids from peat; Transactions of the 6th International Symposium 'Humus et Planta', Prague, 1975, pp. 137.
[59] Ponizovsky, A.A.; Studenikina, T.A.; Mironenko, E.V., Regularities of copper (II) retention by chernozems, dernovo-podsolic and grey forest soils; Proceedings of the 5th International Conference on the Biogeochemistry of Trace Elements, Vienna, Austria, 1999.
[60] Stevenson, F.J.; Fitch, A., Reactions with organic matter; in: J.F. Longeragan, A.D. Robson, R.D. Graham (eds), Copper in Soils and Plants; Academic Press, New York, 1981.
[61] Kovalskiy, V.V.; Letunova, S.V., Geochemical ecology of microorganisms; Trans. Biogeochim. Lab. 1974, 13, 3.
[62] Hodgson, J.F.; Geering, H.R.; Norvell, W.A., Micronutrient cation complexes in soil solution I. Partition between complexed and uncomplexed forms by solvent extraction; Soil Sci. Soc. Am. J. 1965, 29, 665–669.
[63] Hodgson, J.F.; Geering, H.R.; Norvell, W.A., Micronutrient cation complexes in soil solution II. Complexing of zinc and copper in displaced solution from calcareous soils; Soil Sci. Soc. Am. J. 1966, 30, 723–726.
[64] Bradford, G.R.; Bair, F.L.; Hunsaker, V., Trace and major element contents of soil saturation extracts; Soil Sci. 1971, 112, 225–230.
[65] Yamasaki, S.; Yoshino, A.; Kishita, A., The determination of sub-microgram amounts of elements in soil solution by flameless atomic absorption specrophotometry with a heated graphite atomizer; Soil Sci. Plant Nutr. 1975, 21, 63–72.
[66] Heinrichs, H.; Mayer, R., The role of forest vegetation in the biogeochemical cycle of heavy metals; J. Environ. Qual. 1980, 9, 111–118.
[67] Itoh, S.; Yumura, Y., Studies on the contamination of vegetable crops by excessive absorption of heavy metals; Bulletin of the Vegetable and Ornamental Crops Research Station 1979, 6a, 145.
[68] Tiller, K.G., The availability of micronutrients in paddy soils and its assessment by soil analysis including radioisotopic techniques; Proceedings of the Symposium on Paddy Soils; Science Press, Beijing and Springer-Verlag, Berlin, 1981.
[69] Bergkvist, B.; Folkeson, L.; Berggren, D., Fluxes of Cu, Zn, Pb, Cr, and Ni in temperate forest ecosystems – a literature review; Water Air Soil Pollut. 1989, 47, 217–286.

[70] Thornton, I., Metals in the Global Environment: Facts and Misconceptions; ICME, Ottawa, ON, 1995.
[71] Cannon, H.L.; Connally, G.G.; Epstein, G.B.; Parker, J.G.; Thornton, I.; Wixson, B.G., Rocks: geological sources of most trace elements. Report to the workshop at South Seas Plantation, Captiva Islands, Florida, USA; Geochemistry and Environment 1978, 3, 17–31.
[72] Ure, A.M.; Berrow, M.L., The Elemental Constituents of Soils, Environmental Chemistry, Vol. 2; Royal Society of Chemistry, London, 1982.
[73] Reaves, G.A.; Berrow, M.L., Total lead concentrations in Scottish soils; Geoderma 1984, 32, 1–8.
[74] Davies, B.E., A Graphical estimation of the normal lead content of some British soils; Geoderma 1983, 29, 67–75.
[75] Archer, F.C.; Hodgson, H., Total and extractable trace-element contents of soils in England and Wales; J. Soil Sci. 1987, 38, 421–431.
[76] Holmgren, G.G.; Meyer, M.W.; Daniels, R.B.; Kubota, J.; Chaney, R.L., Cadmium, lead, zinc, copper and nickel in agricultural soils of the United States; Agronomy Abstracts 1983, 33
[77] Nriagu, J.O.; Pacyna, J.M., Quanititative assessment of worldwide contamination of air, water and soils by trace metals; Nature 1988, 333, 134–139.
[78] Nriagu, J.O., A global assessment of natural sources of atmospheric trace metals; Nature 1989, 338, 47–49.
[79] Patterson, C.C.; Settle, D.M., Magnitude of lead flux to the atmosphere from volcanoes; Geochim. Cosmochim. Acta 1987, 51, 675–681.
[80] Hinkley, T.K.; Lamothe, P.J.; Wilson, S.A.; Finnegan, D.L.; Gerlach, T.M., Metal emissions from Kilauea, and a suggested revision of the estimated worldwide metal output by quiescent degassing of volcanoes; Earth Planet. Sci. Lett. 1999, 170, 315–325.
[81] Boutron, C.F., Evidence in ice-core research; in: Proceedings of Workshop on the Atmospheric Transport and Fate of Metals in the Environment; International Council on Metals in the Environment, Ottawa, Canada, 1998, pp. 147–165.
[82] Boutron, C.F.; Görlach, U.; Candelone, J-P.; Bolshov, M.A.; Delmas, R.J., Decrease in anthropogenic lead, cadmium and zinc in Greenland snows since the late 1960s; Nature 1991, 353, 153–156.
[83] Lounamaa, J., Trace elements in plants; Ann. Bot. Soc. Vanamo 1955, 29, 1–96.
[84] Wright, J.R.; Levick, R.; Atkinson, H.J., Trace element distribution in virgin profiles representing four great soil groups; Soil Sci. Soc. Am. J. 1955, 19, 340–344.
[85] Merry, R.H.; Tiller, K.G.; Alston, A.M., Accumulation of copper, lead, and arsenic in Australian orchard soils; Aust. J. Soil Res. 1983, 21, 549–561.
[86] Swaine, D.J.; Mitchell, R.L., Trace element distribution in soil profiles; J. Soil Sci. 1960, 11, 347–368.
[87] Archer, F.C., Trace elements in some Welsh upland soils; J. Soil Sci. 1963, 14, 144–148.
[88] Presant, E.W.; Tupper, W.M., Trace elements in some New Brunswick soils; Can. J. Soil Sci. 1965, 45, 305–310.
[89] Bradley, R.I.; Rudeforth, C.C.; Wilkins, C., Distribution of some chemical elements in the soils of north west Pembrokeshire; J. Soil Sci. 1978, 29, 258–270.
[90] Friedland, A.J.; Johnson, A.H.; Siccama, T.G.; Mader, D.L., Trace metal profiles in the forest floor of New England; Soil Sci. Soc. Am. J. 1984, 48, 422–425.
[91] Fleming, G.A.; Walsh, T.; Ryan, P., Some factors influencing the content and profile distribution of trace elements in Irish soils; Proceedings of the 9th International Congress of Soil Science, Vol. 2, Adelaide, Australia, 1968, p. 341.
[92] Riffaldi, R.; Levi-Minzi, R.; Solfatini, G.E., Pb adsorption by soils. 2. Specific adsorption; Water Air Soil Pollut. 1976, 6, 119–128.
[93] Tidball, R.R., Lead in soils; In: T.G. Lovering (ed.), Lead in the Environment; US Geological Survey Professional Paper 957, 1976.
[94] Schnitzer, M.; Kerndorff, H., Reactions of fulvic acid with metal ions; Water Air Soil Pollut. 1981, 15, 97–108.
[95] Cotter-Howells, J.; Caporn, S., Remediation of contaminated land by formation of heavy metal phosphates; Appl. Geochem. 1996, 11, 335–342.

[96] Martínez, C.E.; Motto, H.L., Solubility of lead, zinc and copper added to mineral soils; Environ. Pollut. 2000, 107, 153–158.
[97] Levonmäki, M.; Hartikaimen, H.; Kairesalo, T., Effect of organic amendment and plant roots on the solubility and mobilization of lead in soils at a shooting range; J. Environ. Qual. 2006, 35, 1026–1031.
[98] Kabata-Pendias, A., Heavy metal sorption by clay minerals and oxides of iron and manganese; Mineral. Pollut. 1980, 11, 3–13.
[99] Hildebra, E.E.; Blum, W.E., Lead fixation by iron oxides; Naturwissenschaften 1974, 61, 169–170.
[100] Farrah, H.; Pickering, W.F., The sorption of mercury species by clay minerals; Water Air Soil Pollut. 1978, 9, 23–31.
[101] Abd-Elfattah, A.; Wada, K., Adsorption of lead, copper, zinc, cobalt and cadmium by soils that differ in cation exchange materials; J. Soil Sci. 1981, 32, 271–283.
[102] McKenzie, R.M., The adsorption of lead and other heavy metals on oxides of manganese and iron; Aust. J. Soil Res. 1980, 18, 61–73.
[103] McHargue, J.S., The association of copper with substances containing the fat-soluble A vitamin; Am. J. Physiol. 1925, 72, 583–594.
[104] Hart, E.B.; Steenbock, H.; Waddell, J.; Elvehjem, C.A., Iron in nutrition. VII. Copper as a supplement to iron for haemoglobin building in the rat; J. Biol. Chem. 1930, 77, 797–812.
[105] Harvey, J.M., Copper deficiency in ruminants in Queensland; Aust. Vet. J. 1952, 28, 209–216.
[106] Klevay, L.M., Cardiovascular disease from copper deficiency – a history; J. Nutr. 2000, 130, 489S–492S.
[107] Bennetts, H.W.; Hall, H.T., 'Falling disease' of cattle in the south-west of western Australia; Aust. Vet. J. 1939, 15, 152–159.
[108] Shields, G.S.; Coulson, W.F.; Kimball, D.A.; Cartwright, G.E.; Winthrobe, M., Studies on copper metabolism XXXII. Cardiovascular lesions in copper deficient swine; Am. J. Pathol. 1962, 41, 603–621.
[109] Danks, D.M., Copper deficiency in humans; Annu. Rev. Nutr. 1988, 8, 235–257.
[110] National Research Council, Mineral Tolerance of Domestic Animals. National Research Council Subcommittee on Toxicity in Animals, Committee on Animal Nutrition, Board on Agriculture and Renewable Resources, Commission on Natural Resources; National Academy of Sciences, Washington DC, 1980.
[111] Anderson, T.W.; Neri, L.C.; Schreiber, G.B.; Talbot, F.D.; Zdrojewski, A., Ischemic heart disease, water hardness and myocardial magnesium; Can. Med. Assoc. J. 1975, 113, 199–203.
[112] Chipperfield, B.; Chipperfield, J.R., Differences in metal content of the heart muscle in death from ischemic heart disease; Am. Heart J. 1978, 95, 732–737.
[113] Penttilä, O.; Neuvonen, P.J.; Himberg, J.J.; Siltanen, P.; Järvinen, A.; Merikallio, E., Auricular myocardial cation concentrations in certain heart diseases in man; Trace Elem. Med. 1986, 3, 47–51.
[114] Wester, P.O., Trace elements in human myocardial infarction determined by neutron activation analysis; Acta Med. Scand. 1965, 178, 765–788.
[115] Zama, N.; Towns, R.L., Cardiac copper, madnesium, and zinc in recent and old myocardial infarction; Biol. Trace Elem. Res. 1986, 10, 201–208.
[116] Kinsman, G.D., Howard, A.N., Stone, D.L., Mullins, P.A., Studies in copper status and atherosclerosis; Biochem. Soc. Trans. 1990, 18, 1186–1188.
[117] Prohaska, J.R., Biochemical changes in copper deficiency; J. Nutr. Biochem. 1990, 1, 452–461.
[118] O'Dell, B.L., Biochemistry of copper; Med. Clin. North Am. 1976, 60, 687–703.
[119] Lee, J.; Pena, M.M.O.; Nose, Y.; Thiele, D.J., Biochemicl characterization of the human copper transporter Ctrl1; J. Biol. Chem. 2002, 277, 4380–4387.
[120] Wapnir, R.A., Copper absorption and bioavailability; Am. J. Clin. Nutr. 1998, 67, 1054S–1060S.
[121] Cousins, R.J., Absorption, transport, and hepatic metabolism of copper and zinc: special reference to metallothionein and ceruloplasmin; Physiol. Rev. 1985, 65, 238–309.

[122] Turnlund, J.R.; Keyes, W.R.; Anderson, H.L.; Acord, L.L., Copper absorption and retention in young men at three levels of dietary copper by use of stable isotope ^{65}Cu, Am. J. Clin. Nutr. 1989, 49, 870–878.

[123] Turnlund, J.R.; King, J.C.; Gong, B.; Keyes, W.R.; Michel, M.C., A stable isotope study of copper absorption in young men: effect of phytate and α-cellulose; Am. J. Clin. Nutr. 1985, 42, 18–23.

[124] Comar, C.L., The use of radioisotopes of copper and molybdenum in nutritional studies; in: W.D. McElroy, B. Glass (eds), Symposium on Copper Metabolism; John Hopkins Press, Baltimore, MD, 1950.

[125] Van Campen, D.R.; Scaife, P.V., Zinc interference with copper absorption in rats.; J. Nutr. 1967, 91, 473–476.

[126] Hall, A.C.; Young, B.W.; Brenner, I., Intestinal metallothionein and the mutual antagonism between copper and zinc in the rat; J. Inorg. Biochem. 1979, 11, 57–66.

[127] Sandsted, H.H., Copper bioavailability and requirements; Am. J. Clin. Nutr. 1982, 35, 809–814.

[128] Greger, J.L.; Snedeker, S.M., Effect of dietary protein and phosphorus levels on the utilization of zinc, copper and manganese by adult males; J. Nutr. 1980, 110, 2243–2253.

[129] Yu, S.; West, C.E.; Beynen, A.C., Increasing intakes of iron reduces status, absorption and biliary excretion of copper in rats; Brit. J. Nutr. 1994, 71, 887–895.

[130] Wapnir, R.A.; Devas, G.; Solans, C.V., Inhabition of intestinal copper absorption by divalent cations and low-molecular-weight ligands in the rat; Biol. Trace Elem. Res. 1993, 36, 291–305.

[131] Pekelharing, H.L.M.; Lemmens, A.G.; Beynen, A.C., Iron, copper and zinc status in rats fed on diets containing various concentrations of tin; Brit. J. Nutr. 1994, 71, 103–109.

[132] Scheinberg, I.H.; Sternlieb, I., Copper metabolism; Pharmacol. Rev. 1960, 12, 355–381.

[133] Cartwright, G.E.; Wintrobe, M.M., Copper metabolism in normal subjects; Am. J. Clin. Nutr. 1964, 14, 224–232.

[134] Madden, J.W.; Ironside, J.W.; Triger, D.R.; Bradshaw, J.P.P., An unusual case of Wilson's disease; Q. J. Med. 1985, 55, 63–73.

[135] Brewer, G.J.; Dick, R.D.; Yuzbasiyan-Gurkan, V.; Tankanow, R.; Young, A.B.; Kluin, K.J., Initial therapy of patients with Wilson's disease with tetrathiomolybdate; Arch. Neurol-Chicargo 1991, 48, 42–47.

[136] Schwarz, K., New essential trace elements (Sn, V, F, Si): progress report and outlook; in: W.G. Hoekstra, J.W. Suttie, H.E. Ganther, W. Mertz (eds), Trace Element Metabolism in Animals – 2; University Park Press, Baltimore, MD, 1974.

[137] Vallee, B.L.; Ullmer, D.D., Biochemical effect of mercury, cadmium and lead; Annu. Rev. Biochem. 1972, 41, 91–128.

[138] Hammond, P.B., Metabolism and metabolic action of lead and other heavy metals; Clin. Toxicol. 1973, 6, 353–365.

[139] Neathery, M.W.; Miller, W.J., Metabolism and toxicity of cadmium, mercury and lead in animals: a review; J. Dairy Sci. 1975, 58, 1767–1781.

[140] White, W.B.; Clifford, P.A.; Calvey, H.O., Lethal dose of lead for the cow: the elimination of ingested lead through milk; J. Am. Vet. Med. Assoc. 1943, 102, 292–293.

[141] Allcroft, R., Lead as a nutritional hazard to farm livestock. IV. Distribution of lead in the tissues of bovines after ingestion of various lead compounds; J. Comp. Pathol. 1950, 60, 190–208.

[142] Blaxter, K.L., Lead as a nutritional hazard to farm livestock. II. The absorption and excretion of lead by sheep and rabbits; J. Comp. Pathol. 1950, 60, 140–159.

[143] Blaxter, K.L., Lead as a nutritional hazard to farm livestock. III. Factors influencing the distribution of lead in tissues; J. Comp. Pathol. 1950b, 60, 177–189.

[144] Jones, L.M., Veterinary Pharmacology and Therapeutics, 3rd edn; Iowa State University Press, Ames, IA, 1965.

[145] Rabinowitz, M.B., Toxicokinetics of bone lead; Environ. Health Persp. 1991, 91, 33–37.

[146] Schroeder, H.A.; Tipton, I.H., The human body burden of lead; Arch. Environ. Health 1968, 17, 965–978.

[147] Karacic, V.; Prpic-Majic, D.; Skender, L. Lead absorption in cows: biological indicators of ambient lead exposure; Bull. Environ. Contam. Toxicol. 1984, 32, 290–294.
[148] Kehoe, R.A., Normal metabolism of lead; Arch. Environ. Health 1964, 8, 232–243.
[149] Thompson, J.A., Balance between intake and output of lead in normal individuals; Brit. J. Ind. Med. 1971, 28, 189–194.
[150] Diamond, G.L.; Goodrum, P.E.; Felter, S.P.; Ruoff, W.L., Gastrointestinal absorption of metals; Drug Chem. Toxicol. 1998, 21, 223–251.
[151] Schroeder, H.A.; Nason, A.P.; Tipton, I.H.; Balassa, J.J., Essential trace metals in man: copper; J. Chron. Dis. 1966, 19, 1007–1034.
[152] Georgopoulos, P.G.; Roy, A.; Yonone-Lioy, M.J.; Opiekun, R.E.; Lioy, P.J., Environmental copper: its dynamics and human exposure issues; J. Toxicol. Environ. Health B 2002, 4, 341–394.
[153] Araya, M.; Koletzko, B.; Uauy, R., Copper deficiency and excess in infancy: developing a research agenda; J. Pediatr. Gastr. Nutr. 2003, 37, 422–429.
[154] Eife, R.; Weiss, M.; Barros, V. *et al.*, Chronic poisoning by copper in tap water: I. Copper intoxication with predominantly gastrointestinal symptoms; Eur. J. Med. Res. 1999, 4, 219–223.
[155] Eife, R.; Weiss, M.; Muller-Hocker, M. *et al.*, Chronic poisoning by copper in tap water. II. Copper intoxication with predominantly systemic symptoms; Eur. J. Med. Res. 1999, 4, 224–228.
[156] Muller-Hocker, J.; Meyer, U.; Wiebecke, B., *et al.*, Copper storage disease of the liver and chronic dietary copper intoxication in two further German infants mimicking Indian childhood cirrhosis; Pathol. Res. Pract. 1988, 183, 39–45.
[157] Gaetke, L.M.; Chow, C.K., Copper toxicity, oxidative stress, and antioxidant nutrients; Toxicology 2003, 189, 147–163.
[158] Chugh, K.S.; Sakhuja, V., Acute copper intoxication; Int. J. Artif. Organs 1979, 2, 181–182.
[159] Gilman, J.P.W., Metal carcinogenesis. II. A study on the carcinogenic activity of cobalt, copper, iron and nickel compounds; Cancer Res. 1962, 22, 158–166.
[160] Wu, J.; Ricker, M.; Muench, J., Copper deficiency as cause of unexplained hematologic and neurologic deficits in patient with prior gastrointestinal surgery; J. Am. Board Fam. Med. 2006, 19, 191–194.
[161] Iyoda, K.; Tsuda, M.; Nogami, H., Periosteal hyperostosis associated with copper deficiency – Report of a case; J. Bone Miner. Metab. 1989, 7, 42–45.
[162] Ruocco, L.; Baldi, A.; Cecconi, N., *et al.*, Severe pancytopenia due to copper deficiency. Case report; Acta Haematol.-Basel 1986, 76, 224–226.
[163] Hein, M.S., Copper deficiency anemia and nephrosis in zinc-toxicity: a case report; S. Dak. J. Med. 2003, 56, 143–147.
[164] Shaw, J.C.L., Copper deficiency in term and preterm infants; in: S.J. Fomon, S. Zlotkin (eds), Nutritional Anaemias Workshop; Raven Press, New York, 1992, pp. 109–119.
[165] Olivares, M.; Uauy, R., Copper as an essential nutrient; Am. J. Clin. Nutr. 1996, 63, 791S–796S.
[166] Olivares, M.; Uauy, R., Limits of metabolic tolerance to copper and biological basis for present recommendations and regulations; Am. J. Clin. Nutr. 1996, 63, 846S–852S.
[167] Amsden, M.P.; Sweetin, R.M.; Treilhard, D.G., Selection and design of Texas-Gulf Canada's copper smelter and refinery; JOM-J Min. Met. Mat. S. 1978, 30, 16–26.
[168] Turnlund, J.R., Copper; in: M.E. Shils, S.A. Olsen, M. Shike, A.C. Ross (eds), Modern Nutrition in Health and Disease, 9th edn; Williams & Wilkins, Baltimore, MD, 1999, pp. 870–878.
[169] World Health Organisation, Trace Elements in Human Nutrition and Health; WHO, Geneva, 1996.
[170] National Research Council, Biological Markers of Reproductive Toxicology; National Academy Press, Washington DC, 1989.
[171] Aalbers, T.G.; Houtman, J.P.W.; Makkink, B., Trace-element concentrations in human autopsy tissue; Clin. Chem. 1987, 33, 2057–5064.

[172] Lillie, R.J., Air Pollutants Affecting the Performance of Domestic Animals. A Literature Review; Agricultural Handbook No. 380; US Department of Agriculture, Agricultural Research Service, Washington, DC, 1970, pp. 870–878.
[173] Ammerman, C.B.; Fick, K.R.; Hansard II, S.L.; Miller, S.M., Toxicity of certain minerals to domestic animals. A review; FLA Agric. Exp. Stat. Bull. 1973, AL, 73–76.
[174] Clarke, E.G.C., Lead poisoning in small animals; J. Small Anim. Pract. 1973, 14, 183–193.
[175] Bremmer, I., Heavy metal toxicities; Q. Rev. Biophys. 1974, 7, 75–124.
[176] MacLeavey, B.J., Lead poisoning in dogs; New Zeal. Vet. J. 1977, 25, 395.
[177] Forbes, R.M.; Sanderson, G.C., Lead toxicity in domestic animals and wildlife; Topics Environ. Health 1978, 225.
[178] Ammerman, C.B.; Miller, S.M.; Fick, K.R.; Hansard II, S.L., Contaminating elements in mineral supplements and their potential toxicity: a review; J. Anim. Sci. 1977, 44, 485–508.
[179] USEPA, Ecological Soil Screening Levels for Lead. OSWER Directive 9285.7-70; US Environmental Protection Agency, Office of Solid Waste and Emergency Response, Washington, DC, 2005.
[180] Buck, W.B.; James, L.F.; Binns, W., Changes in serum transaminase activities associated with plant and mineral toxicity in sheep and cattle; Cornell Vet. 1961, 51, 568–585.
[181] Hammond, P.B.; Aronson, A.L., Lead poisoning in cattle and horses in the vicinity of a smelter; Ann. NY Acad. Sci. 1964, 111, 595–611.
[182] Dinius, D.A.; Brinsfield, T.H.; Williams, E.E., Effect of subclinical lead intake on calves; J. Anim. Sci. 1973, 37, 169–173.
[183] Kelliher, D.J.; Hilliard, E.P.; Poole, D.B.R.; Collins, J.D., Chronic lead intoxication in cattle: preliminary observations on its effect on the erythrocyte and on porphyrin metabolism; Irish J. Agr. Res. 1973, 12, 61–69.
[184] Fick, K.R.; Ammerman, C.B.; Miller, S.M.; Simpson, C.F.; Loggins, P.E., Effect of dietary lead on performance, tissue mineral composition and lead adsorption in sheep; J. Anim. Sci. 1976, 42, 515–523.
[185] Aronson, A.L., Lead poisoning in cattle and horses following long term exposure to lead; Am. J. Vet. Res. 1972, 33, 627–629.

19
Chromium, Nickel and Cobalt

Yibing Ma[1] and Peter S. Hooda[2]

[1]*Institute of Agricultural Resources and Regional Planning, Chinese Academy of Agricultural Sciences, Haidian, Beijing, China*
[2]*School of Geography, Geology and the Environment, Kingston University London, Kingston upon Thames, Surrey, UK*

19.1 Introduction

Chromium (Cr) is a d-block transition metal of Group 6 of the periodic table. It is a silver-grey lustrous, hard metal that can be highly polished. Naturally occurring Cr is composed of three stable isotopes, ^{52}Cr, ^{53}Cr and ^{54}Cr, with natural abundances of 83.789 %, 9.501 % and 2.366 %, respectively. Of the 19 radioisotopes, ^{50}Cr (4.345 % natural abundance) is the most stable with a half-life $> 1.8 \times 10^{17}$ years, and ^{51}Cr has a half-life of 27.7 days; all of the remaining Cr radioactive isotopes have short half-lives.

Chromium exhibits a wide range of possible oxidation states. The most common oxidation states of Cr in the soil and water environment are +3 and +6, with +3 being the most stable.

Chromium is produced from the chromite ore ($FeCr_2O_4$), which is the only Cr mineral of commercial importance. South Africa, Kazakhstan, India, Zimbabwe, Turkey, Russia, Finland and Brazil are the main producers of chromite ores. The world production of chromium (gross weight as Cr_2O_3) increased from 12×10^6 Mg in 1995 to 19.3×10^6 Mg in 2005 [1], reflecting a significant increase in its use.

The main uses of Cr are in steel fabrication, manufacturing of stainless steel and alloys, paint and pigment manufacturing and leather tanning. Much is used in metallurgy to impart

---

Trace Elements in Soils Edited by Peter S. Hooda
© 2010 Blackwell Publishing, Ltd

hardness, corrosion resistance and shiny finish. Chromium(IV) oxide, CrO_2, is mainly used in manufacturing magnetic tapes. Estimated annual world consumption of chromium (as metal) is about 12.5×10^6 Mg, of which 85 % is used in metallurgy, while its usages in chemicals and refractories account for 8 % and 7 %, respectively [2].

Chromium is a nonessential trace element for plants but its trivalent chromium form, Cr(III), is considered a beneficial nutrient in trace amounts for humans and animals. Although Cr(III) does not strictly meet the essential nutrient criteria, it is known to enhance the action of insulin [3,4], and is also considered to be directly involved in carbohydrate, fat and protein metabolism [3,5,6]. As a result it is generally accepted that Cr(III) is an 'essential element' for humans; the mechanisms of its action in the body and the dietary requirements, however, are not well defined. Chromium(VI) is a strong oxidizing agent and is extremely toxic to animals and humans [7]. In epidemiological studies, an association has been found between exposure to Cr(VI) by the inhalation route and increased incidence of lung cancer. The International Agency for Research on Cancer has classified Cr(VI) in group-1 human carcinogens [8], although the risk of cancer from Cr(VI) via the food chain has not been established.

Nickel (Ni) is a d-block transition metal of Group 10 of the periodic table. It is a silvery-white, hard, malleable, ductile metal, and like Cr it takes on a high polish. It is a fairly good conductor of heat and electricity and is one of the five ferromagnetic elements (iron, nickel, cobalt, gadolinium and dysprosium). Among the five naturally occurring stable isotopes of nickel (^{58}Ni, ^{60}Ni, ^{61}Ni, ^{62}Ni and ^{64}Ni), ^{58}Ni is the most abundant, with its natural abundance of 68.077 %. Of the 18 radioisotopes that have been characterized, ^{59}Ni is the most stable with a half-life of 76 000 years, while ^{63}Ni has a half-life of 100.1 years. The remaining radioactive isotopes have short half-lives.

Nickel can occur in a number of oxidation states: +1, +2, +3 and +4, but only Ni(II) is stable over the wide range of pH and redox conditions found in the soil environment [9]. Nickel is extracted from two types of ore deposits. The first type of deposits are laterites where the main ore minerals are nickeliferous limonite, (Fe, Ni)O(OH) and garnierite, a hydrous nickel silicate, $(Ni, Mg)_3Si_2O_5(OH)$. The second are magmatic sulfide deposits where the principal ore mineral is pentlandite, $(Ni, Fe)_9S_8$, which is the most commercially important Ni-bearing mineral [9]. The major producers of nickel ores include Russia, Australia, Canada, Indonesia and New Caledonia. The world Ni mining production increased from 1.04×10^6 Mg in 1995 to 1.48×10^6 Mg in 2005 [1]. Most of the Ni produced is used in manufacturing of stainless steel (65 %) and alloys (21 %). Nickel alloys are characterized by high strength, ductility and resistance to corrosion and heat. Other uses are in rechargeable batteries, foundry, plating, minting coins, catalysts and other chemicals [10].

Nickel is an essential nutrient for animals and a beneficial element for plants [11,12]. As with other trace metals, elevated Ni concentrations in soils have potential negative impact on plants, microorganisms and animals [13,14].

Cobalt (Co) is a d-block transition metal of Group 9 of the periodic table. It is a hard, lustrous, silver-grey and ferromagnetic metal. Naturally occurring cobalt is 'monoisotopic', that is only one isotope, ^{59}Co is stable. Of the 22 radioisotopes of cobalt, ^{60}Co is the most stable with a half-life of 5.2714 years. Other relatively stable radioisotopes are ^{57}Co, ^{56}Co and ^{58}Co, with a half-lives of 271.79, 77.27 and 70.86 days, respectively. ^{60}Co, as a gamma ray source, is often used for sterilization of medical supplies and waste, food radiation treatment (cold pasteurization) and industrial radiography. The radioisotopes, ^{57}Co, ^{58}Co and ^{60}Co, have been used in soil scientific research as tracers.

Cobalt can exhibit +1, +2, +3 and +4 oxidation states, with +3 and +2 being the most common. Cobalt was historically produced as a by-product of nickel and copper mining activities. Recently, cobalt has been the primary product of mining operations in Morocco and in Congo (Kinshasa). The main ores of cobalt are cobaltite, CoAsS, erythrite, $(Co,Ni)_3(AsO_4)_2·8H_2O$, glaucodot, (Co,Fe)AsS, and skutterudite, $(Co,Ni,Fe)As_3$.

Congo is the top producer of cobalt with about 40 % of the total world production, followed by Zambia, Australia, Canada, Russia and Cuba [1]. The world production of cobalt increased from 22 100 Mg in 1995 to 57 900 Mg in 2005. The main uses of cobalt are alloys, especially superalloys, chemicals and ceramic production, cemented carbides and steels. About one-fourth of global cobalt production is consumed in the forms of chemical compounds (organic and inorganic).

Cobalt is considered an essential nutrient in trace amounts for ruminant animals, largely due its requirement for rumen bacteria. It is also required for atmospheric N_2-fixation by microorganisms and for plants [15].

19.2 Geogenic Occurrences

The average concentration of Cr in rocks is 100 mg kg^{-1}. The mean concentrations of Cr in various types of rock (mg kg^{-1}) are 1800 in ultramafic igneous, 200 in basaltic igneous, 120 in shales and clays, 100 in black shales, 35 in sandstone, 20 in granitic igneous, and 10 in limestone [7]. Chromium (3$^+$) readily substitutes for Fe(2$^+$) which has a similar radius (0.067 nm) and geochemical properties to Cr^{3+}, so that chromite is mainly associated with mafic and ultramafic rocks and the higher concentration of Cr is in the soils derived from these rocks. The grand mean of Cr concentrations in surface soils worldwide is estimated to be 54 mg kg^{-1} [16]. Ure and Berrow [17], who reviewed published global data on Cr in soils, however, quoted an average concentration of 84 mg Cr kg^{-1}. The concentration of Cr in soils varies widely from 1 to 1500 mg kg^{-1} [10,16]. In some serpentine-derived soils, the Cr concentration can be as high as 125 000 mg kg^{-1} [18].

The average concentration of Ni in rocks is 75 mg kg^{-1} [9]. The mean concentrations of Ni in various types of rock (mg kg^{-1}) are 2000 in ultramafic igneous, 140 in basaltic igneous, 68 in shales and clays, 50 in black shales, 20 in limestone, 8 in granitic igneous, and 2 in sandstone [9]. Nickel, like Cr, is also found in association with mafic and ultramafic rock formations [9]. The concentrations of indigenous Ni in soils depend very much on the soil parent materials. The grand mean of Ni concentrations in surface soils worldwide is estimated between 20 and 40 mg kg^{-1} [9]. The Ni concentration in soils, however, varies widely, between 0.2 and 450 mg kg^{-1} [16]. In some serpentine-derived soils, Ni concentration as high as 7000 mg kg^{-1} has been measured [9].

The average concentration of Co in rocks is considered as 25 mg kg^{-1}, which varies widely between rocks type, for example 1–15 mg kg^{-1} (acid rocks), 0.1–20 mg kg^{-1} (sedimentary rocks) and 100–200 mg kg^{-1} (ultramafic rocks), as Co(2$^+$) is often associated with rocks containing Fe and Mg by their isomorphous substitution [16]. The concentration of geogenic Co in soils, like other trace metals, depends on the parent materials the soil is derived from. The total concentration of Co in soils varies widely from 0.05 to 300 mg kg^{-1} with an average in the range of 10–15 mg kg^{-1} [15].

19.3 Sources of Soil Contamination

The major sources of Cr inputs to agricultural soils are phosphate fertilizers, sewage sludge, atmospheric deposition, as well as poultry manures [19,20]. Large amounts of Cr are released to the atmosphere by human activity, especially from ferrochrome production [9]. The reported deposition fluxes of Cr (mg m^{-2} y^{-1}) range from 0.05 to 6.46 [19,20]. The average atmospheric Cr deposition fluxes (mg m^{-2} y^{-1}) for China, New Zealand and Europe are estimated to be 4.39, 2.78 and 0.93, respectively [19–21]. This seems to be a reflection of the differences in the industrial production and the implementation of pollution emission control strategies. In England and Wales, the most important sources of Cr to agricultural soils are inorganic fertilizers, atmospheric deposition and sewage sludge; while in China, they are livestock manures, atmospheric deposition and inorganic fertilizers (Table 19.1). Most of the chromium in inorganic fertilizers is from phosphate fertilizers or compound fertilizers. The use of phosphate fertilizers, sewage sludge and livestock manures is the major source of Cr in agricultural soils [19,20].

The major sources of Ni to agricultural soils are atmospheric deposition, livestock manures, and phosphate fertilizers [19,20]. Large amounts of anthropogenic Ni are released to the atmosphere by human activities, especially from burning of fossil fuel and residual oils, followed by Ni mining and smelting [9]. The atmospheric deposition fluxes of Ni (mg m^{-2} y^{-1}) range from 0.05 to 8.35 in the literature ([19,20] and references therein) in agricultural areas. The average atmospheric deposition fluxes of Ni (mg m^{-2} y^{-1}) in China, New Zealand and Europe are estimated to be 5.09, 0.95 and 1.06, respectively [19–21]. However, the deposition fluxes of metals depend on the proximity to point sources of pollution, such as heavy industry and major roads. The inventories of Ni inputs to agricultural soils in England and Wales and China [19,20] indicate that the most important sources of Ni are atmospheric deposition and livestock manures, followed by phosphate or compound fertilizers and sewage sludge (Table 19.1)

Table 19.1 Estimated annual input rates (g ha^{-1}) of Cr and Ni to agricultural land in China in 2005, and England and Wales in 2000 [19,20]

Source	China		England and Wales	
	Cr	Ni	Cr	Ni
Atmospheric deposition	60.6	58.1	7.48	16.0
Livestock manures	50.1	21.7	3.24	4.77
Inorganic fertilizers	28.1	4.13	11.4	3.33
Sewage sludges	0.70	0.30	7.03	2.52
Others	0.43	1.94	0.35	0.27
Total	140	86.2	29.5	26.9

Cobalt can enter into soils by a number of pathways. Apart from mining and smelting activities, the most important anthropogenic sources of Co in soils include phosphorus-containing fertilizers, sewage sludges and other wastes used as soil amendments, and atmospheric deposition [22]. The Co concentration in different soil amendments ranges between 8–21 mg kg^{-1} for P fertilizers, 0.6–1565 mg kg^{-1} for sewage sludges, 12–15 mg kg^{-1} for composted municipal refuses and 0.3–24 mg kg^{-1} for organic manures [22]. The atmospheric fallout of Co onto soils in remote regions (South Pole, North Norway, North West Canada), rural, and metropolitan and industrialized regions in Europe and North America are estimated as <0.001–0.04, 0.3–58 and 0.2–200 g ha^{-1} y^{-1}, respectively [22,23]. Apart from high Co concentration in the soils developed over serpentine rocks and ore deposits, significant sources of Co pollution are related to nonferrous smelters [16].

19.4 Chemical Behaviour in Soils

19.4.1 Chromium

The most common oxidation states of Cr in soils are Cr(III) and Cr(VI). There is a great contrast between Cr(III) and Cr(VI) in terms of the fate and chemical behaviour of Cr in soils. Figure 19.1 presents the valence states and hydrolysis speciation of Cr over a range of pH and E_h values. In the range of pH between 4 and 8 and E_h between 0.3 and 0.7 V, Cr(VI) exists mainly as HCrO$_4^-$ at pH < 6.5 and as CrO$_4^{2-}$ at pH > 6.5 in dilute aqueous systems. However, Cr(III) is the most stable form when E_h < 0.4 V (such as in paddy soils) and pH values between 4 and 8, and is present as a cationic species with the first or second hydrolysis products [24]. The aqueous inorganic Cr(III) concentration at pH >5 is limited by Cr precipitation. The low solubility of Cr(III) hydroxides (Cr(OH)$_3$, pK = 9.35) and strong retention on soil particle surfaces limits its mobility and bioavailability in soils with pH >5. Chromium(VI) as chromate, on the other hand, is highly water soluble [25] and is a strong oxidizing agent. This makes Cr(VI) a highly mobile and toxic form of Cr [9,24,26]. Under oxidized conditions, Cr(VI) is often thermodynamically the most stable species; however, Cr(VI) is readily reduced to Cr(III) in the presence of reducing agents, and thereafter Cr(III) can form highly soluble organic complexes under acidic conditions [26,27].

Most indigenous Cr exists generally as occluded lattice forms in soil minerals and partly associated with Fe(Al) oxides [26]. For example, the majority of Cr was found in the residual fraction (85.1 %) and in the crystalline Fe oxide fraction (9.8 %) in 25 typical soils from China [28]. However, in sludge-amended soils a large fraction of the total Cr can be associated with soil organic matter [26]. In extremely high Cr-containing ultramafic soils (>5000 mg Cr kg^{-1}), most of the Cr was found associated with Fe oxides [29].

The sorption and precipitation of Cr(III) and oxidation of Cr(III) to Cr(VI) are important processes in soils. The sorption of Cr(III) is strong in soils, and decreases in the presence of other inorganic cations or dissolved organic ligands in the soil solution [26]. Studies of Cr(III) sorption mechanism on silica show that at surface coverage of less than 20 % Cr(III) forms a surface-complexed monodentate while at a greater surface coverage discrete chromium hydroxide surface clusters are discerned [30,31]. Under aerobic conditions,

Cr(III) may be oxidized to Cr(VI), probably by Mn oxides [32]. For example, the Cr(VI) concentrations were greater in well-aerated colluvial soils with high levels of Mn oxides than in piedmont soil with a lower Mn-oxide content, or in lowland alluvial soil, where waterlogging occurs [33]. However, a large portion of Cr in soil will not be oxidized to Cr(VI), even in the presence of Mn oxides and favourable pH conditions, due to the poor availability of mobile Cr(III), and generally the oxidation of Cr(III) to Cr(VI) is a very slow process at pH values above 5 [7]. In anaerobic soils (such as $E_h < 0.4$ V), Cr(VI) could be reduced to Cr(III) by redox reactions with aqueous inorganic species, electron transfers at mineral surfaces, reaction with nonhumic organic substances such as carbohydrates and proteins, or reduction by soil humic substances [7]. This reduction of Cr(VI) to Cr(III) increases at lower pH values (Figure 19.1). In aerobic soils, the reduction of Cr(VI) to Cr(III) is also possible, particularly in acidic soils with sufficient labile organic carbon [7]. In most soils and sludge samples Cr is present predominantly in the Cr(III) state [7,34].

The factors that influence transformations between Cr(III) and Cr(VI) in soils have been extensively reviewed [9,24,26]. Avudainayagam et al. [26] summarized the most important reactions of Cr in the soil and water environments (Figure 19.2). The inorganic form of Cr, that is chromate, is mainly present as $HCrO_4^-$ or CrO_4^{2-} species in soil solution, but is subject to adsorption or precipitation reactions, uptake by plants, and leaching to subsurface layers under favourable soil pH and moisture conditions. The Cr(VI) can be reduced to Cr(III) in soils or in plants during photosynthesis [26]. The reduced Cr(III) in soils can

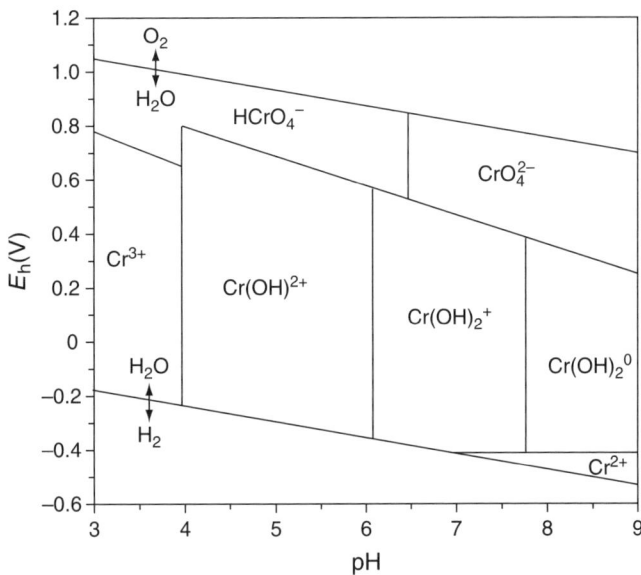

Figure 19.1 E_h–pH diagram depicting the aqueous inorganic Cr species in soil–water environments (Reprinted from Geoderma, 67, S.E. Fendorf, Surface reactions of chromium in soil and waters, 55–71. Copyright 1995, with permission from Elsevier [24].)

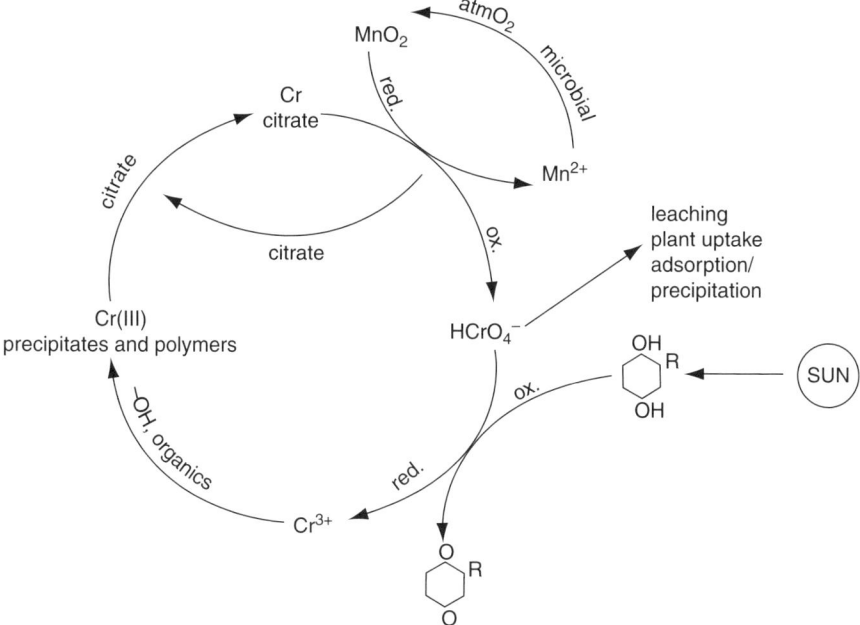

Figure 19.2 A general representation of chromium biogeochemical processes (adsorption, precipitation, reduction–oxidation, complexation, plant uptake and leaching), in soil–water environments (Reproduced from Reviews of Environmental Contamination & Toxicology, 178, 2003, Chemistry of chromium in soils with emphasis on tannery waste sites, S. Avudainayagam, M. Megharaj, G. Owens, R.S. Kookana, D. Chittleborough and R. Naidu, 53–91, with kind permission of Springer Science and Business Media [26].)

easily form highly soluble organic complexes, and when bound to soil solids by adsorption or precipitation, it (Cr-III) becomes immobile/insoluble [26]. However, organic ligands, such as citrate, can mobilize Cr(III). Soil pH and E_h values are the most important factors controlling the speciation, mobility, reaction and bioavailability of Cr in soils.

19.4.2 Nickel

Unlike Cr, Ni is relatively stable in aqueous solution. In soil solution, Ni may exist as aqueous Ni^{2+}, complexed with inorganic and organic ligands, and/or associated with suspended mineral colloids [35]. The organic complexes may be the dominant species in soil solution, instead of various inorganic species, for example Ni^{2+}, $NiCO_3^0$, $NiHCO_3^+$ and $NiOH^+$ [35]. The extent of Ni–organic complex formation, however, depends on the amount of dissolved organic carbon [35]. The concentration of Ni in solution is generally related to its total concentration in the soil and is often <100 μg l^{-1} ([35] and references therein) in uncontaminated nonserpentine soils. The serpentine-derived Ni-rich soils may contain soil solution concentrations ranging from 130 to 3250 μg Ni l^{-1} [36]. Apart from the total soil Ni content, the source of Ni in soils (geogenic or anthropogenic) and soil physicochemical characteristics also influence the partitioning of Ni between the solution and soil solid phases, which, in turn, controls its mobility and bioavailability [37–39]. In addition to the Ni in primary minerals, it is also associated mainly with Fe and Mn oxides, and layered silicates, in particular garnierite, chlorite and vermiculite [35]. In serpentine soils with high concentration of Ni and alkaline pH, Ni may be associated partly with carbonates [40].

Nickel in the solution phase could become rapidly adsorbed to any of the surfaces proffered by the soil solid phases, including layered silicates, Fe(Al) and Mn oxides and organic substances [35,41]. The sorption of Ni on soil solid surfaces involves the formation of either outer-sphere surface complexes or inner-sphere surface complexes. The Ni adsorbed though the formation of outer-sphere complexes is relatively readily exchangeable, because it exists as hydrated Ni species on the surfaces of solid phases, loosely attached to the surface by electrostatic forces and through H-bonding between the hydration shell and surface oxygen [35]. The specifically adsorbed Ni, on the other hand, is thought to exist as inner-sphere complexes on the surface of colloids as partially dehydrated or dehydrated species [42]. The Ni adsorbed on the soil surfaces, however, distributes from outer-sphere to inner-sphere complexes though without clear demarcation. Compared with other metals, the affinity of Ni^{2+} ions for soil mineral materials is generally lower than that of Pb^{2+}, Cu^{2+} and Zn^{2+} [35].

Stoichiometric precipitates such as nickel ferrite ($NiFe_2O_4$) and nickel phosphate ($Ni_3(PO_4)_2$) are considered rarely responsible for controlling the concentration of Ni in soil solution, even in heavily fertilized and very polluted soils [35]. Nevertheless, Ni precipitates have been detected on the surface of clay minerals and Al oxides at sorption densities below a theoretical monolayer of sorbed Ni [43]. Furthermore, these precipitates have formed at circumneutral pH values and Ni solution concentrations undersaturated with respect to the solid phases of Ni [44]. For example, the formation of Ni and Al layered double hydroxide (LDH) precipitates, Ni–Al LDH phases, have been observed in contaminated mineral field soils (3516–4902 mg Ni kg^{-1} soil) at circumneutral pH, using synchrotron micro-X-ray absorption fine structure and X-ray fluorescence spectroscopy. The formation of Ni–Al LDH stable surface phases/precipitates can lead to a reduction in Ni mobility [45]. Such Ni precipitates (Ni-Al LDH), however, have only been observed at pH values > 6.8, at very high solution Ni concentrations ($\sim$3 mM) and at high sorbed Ni concentrations (>2700 mg kg^{-1}). The extent to which these surface precipitates control the partitioning of Ni between the solution and solid phases in soils with normal range of Ni concentrations is still a matter of further research [46–50]. Nevertheless, the incorporation of Ni into stable surface precipitates, for example by liming, could help achieve long-term Ni sequestration in highly contaminated soils.

The most important factor affecting the behaviour of Ni in soils appears to be pH, while attributes such as clay content and Fe and Mn oxides in soils are of secondary importance. For instance, the mobility of Ni in soils increases as the pH and cation exchange capacity (CEC) decrease [9]. Other factors, such as waterlogging (redox), and addition of organic manures and sewage sludge to soil also influence the fate and behaviour of Ni in soils [35].

19.4.3 Cobalt

The divalent Co(II) and trivalent Co(III) are the most important oxidation states of Co in aqueous solution, and depend on the pH and E_h values (Figure 19.3). Also, the stability of

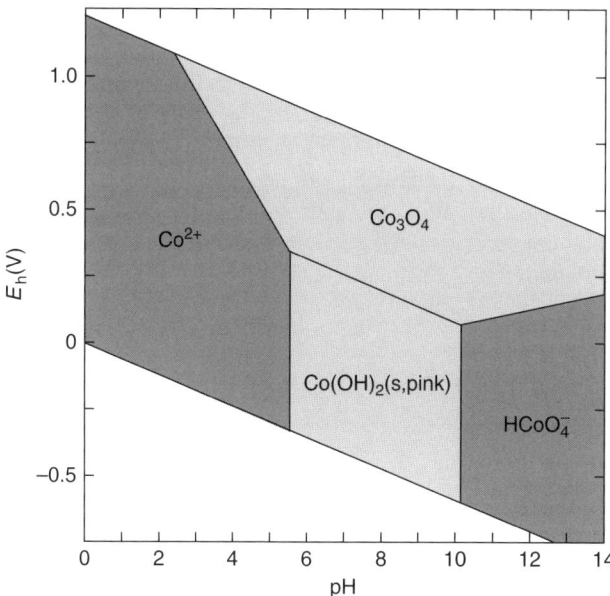

Figure 19.3 E_h–pH diagram showing the dominant aqueous Co species and solid phases (diagram is calculated at 25 °C and a concentration of $10^{-14.8}$ M total dissolved Co, using Geochemists Workbench Software and Thermodynamic Libraries) (Reprinted from Environment International, 34, J. Gál, A. Hursthouse, P. Tatner, F. Stewart and R. Welton, Cobalt and secondary poisoning in the terrestrial food chain: Data review and research gaps to support risk assessment, 821–838. Copyright 2008, with permission from Elsevier [66].)

these oxidation states is largely dependent upon the complexation of Co with ligands in solution. In soil solution, the cation Co^{2+} and its aquo-complexes are stable [51]. For example, only Co(II) species were identified in 0.01 M $CaCl_2$ soil extracts by HPLC-ICP-MS; the absence of Co(III) in soil extracts suggests that little or no oxidation of Co(II) to Co(III) occurs in soil extracts [51].

The behaviour of Co in soils is influenced to a large degree by the presence of Mn and Fe oxides which are known to have a great affinity for Co [52], as most of the Co (up to 79 %) has been found strongly associated with Fe and Mn oxyhydroxides in soils [52–55]. The strong relationship between Co and oxyhydroxide minerals occurs through surface oxidation of Co(II) to Co(III), with subsequent incorporation into structural vacancies, or surface precipitation as hydroxyl solids [56–60]. Because Co(III) species are less soluble, the bioavailability of Co is largely related to the reaction of Co(II) and its compounds (organic and inorganic species) in soils, and the extent of Co(II) oxidation to Co(III) on Mn oxide mineral surfaces, which is an important process to reduce its mobility and bioavailability in soils [51].

In general, except for soil Mn oxide content, the fate and behaviour of Co are greatly influenced by soil pH and E_h values. The sorption of Co(II) to the soil solid phase has been shown to increase with increasing pH [61,62]. The increased sorption and possible fixation of Co with increasing pH suggest it may be a major factor in controlling its bioavailability in soils. Soil moisture status has a significant influence on Co bioavailability as the soil–water regime mediates oxidation–reduction reactions [15]. For instance, under anaerobic conditions generated by waterlogging, Co may be released from the solid phases into solution either by changes in soil pH due to altered redox conditions, due to direct reduction of Co(III), or via reductive dissolution of Fe and Mn oxides [63,64]. The factors contributing to Co deficiency for grazing animals are mainly associated with alkaline or calcareous soils, light leached soils, and soils with high organic matter content [16]. The amount Co available for plant uptake is also dependent upon the total soil Co content and is inversely related to both the soil pH and the ability of the soil to adsorb Co [65].

19.5 Environmental and Human Health Risks

19.5.1 Chromium

Chromium is a nonessential element for plants, but trivalent chromium, Cr(III), is an essential/beneficial nutrient in trace amounts for sugar and cholesterol metabolism in humans and animals, whereas its hexavalent form Cr(VI) is a potent carcinogen and extremely toxic to animals and humans [7]. Plant uptake of Cr has received much attention recently. The low solubility of Cr(III) in soils leads generally to relatively small concentrations of Cr in plants. The concentration of Cr in plants grown on uncontaminated soils is generally between 0.02 and 1.0 mg kg^{-1}; plants grown on contaminated or serpentine-derived soils, however, can accumulate Cr within the 10–190 mg kg^{-1} range, depending upon the soil concentration and plant species [9]. The concentrations in foliar parts of plants show little correlation with total Cr in the soils, and have been found to correlate with easily extractable Cr in the soils [9]. The concentrations of Cr in plants depend on plant species and soil properties (for example pH, E_h, organic matter).

In general, Cr(VI) occurs in soils with higher pH, aerobic conditions, low amounts of organic matter and the presence of Mn and Fe oxides which oxidize chromium (III). Transformation of chromium (VI) to the trivalent form tends to occur in acidic, anoxic soils with high organic content [7]. Also, it has been shown that greater amount of Cr accumulates in plants when they are supplied with Cr(VI) than when supplied with Cr(III), which confirms the greater solubility and plant uptake of Cr(IV) compared with Cr(III) [67]. Usually, the distribution of Cr accumulation in different plant parts is in the order: root > leave or shoot > grain, which shows restricted translocation of Cr. The reason for high Cr accumulation in plant roots is likely due to its immobilization in the vacuoles of root cells [68].

Toxicity of Cr to plants depends on its oxidation state, with Cr(VI) being much more toxic than Cr(III). The chromium toxicity in plants has been reviewed by Shanker *et al.* [68]. It is reported that the toxic properties of Cr(VI) originate from the action of this form itself being an oxidizing agent, as well as from the formation of free radicals during the reduction of Cr(VI) to Cr(III) within the plant cells [68]. Chromium(III), on the other hand, apart from generating reactive oxygen species, if present in high concentrations, can cause toxicity due to its ability to coordinate various organic compounds, resulting in the inhibition of some metalloenzyme systems [68]. The differential toxicity of these two

Cr species may also be due to their different mechanisms of translocation and partitioning. Chromium(VI) is actively taken up by a metabolically driven process, whereas Cr(III) is probably passively taken up and retained by ion-exchange sites on the root cell wall, restricting its translocation to shoots and other above-ground plant parts [68]. In addition, Cr(VI) competes with various elements of similar electronic structure; hence, it seems that Cr(VI) has an advantage over Cr (III) at the entry level into the plant system [68].

Toxic effects of Cr on physiological processes such as photosynthesis, water movement and transpiration, and mineral nutrition and on enzymes or other metabolites affect plant growth and development [68]. Plant Cr toxicity symptoms appear as stunted plant growth, root injury, and chlorosis in young leaves, chlorotic band on cereals and brownish-red leaves on some plants [7].

Chromium(III) is largely immobile in the soil, which limits its accumulation in the food chain, in contrast with the highly mobile and toxic Cr(VI). Environmental contamination with Cr, especially Cr(VI), has therefore become a major area of concern. Among the different anthropogenic Cr sources, land-disposal of wastes from chromium industry and tanneries is a particularly large contributor of Cr pollution to water and soil resources [26,69].

Concerns have been increasing about the potential adverse environmental effects of elevated Cr especially in tannery industrial districts, such as Tamil Nadu in India and Veneto in Northern Italy [70]. For example, the Cr accumulation in soils ranges from 50 to 10 000 mg kg^{-1} (average 210 mg kg^{-1}), with Cr in plants as high as 222 mg kg^{-1} (average content 29 mg kg^{-1}) at a tannery industrial district in Italy [70]. In addition to Cr entry into the food chain via plants, high Cr in soils may represent a direct health hazard due ingestion or inhalation of soil. Recently, in the United States the dietary guidelines for daily Cr intake have been lowered from 50–200 μg to 35 μg and 25 μg for male and female adults, respectively [71].

Elevated Cr in soils also represents a terrestrial ecological risk. The toxicity of Cr (as $K_2Cr_2O_7$) to barley (*Hordeum vulgare* L.) evaluated using an artificial soil (pH 7.8, organic matter 0.27 %) and a natural forest soil (pH 7.6, organic matter 3.8 %), for example, indicated that root biomass was the most sensitive endpoint [72]. The EC_{50} value (the effective concentrations of added metals causing 50 % inhibition) for barley root elongation in the artificial soil was much lower (6.6 mg Cr kg^{-1}) than that in the forest soil (61.8 mg Cr kg^{-1}). The greater amount of organic matter in the forest soil mitigated the toxicity effect, possibly by immobilizing the soluble Cr added to the soils [72]. Lee et al. [73] evaluated the toxicity of Cr (as $K_2Cr_2O_7$) in terms of its effect on the growth of rice seedlings on paddy soils; the toxicity occurred at 300 mg Cr(VI) kg^{-1} for the soil with pH 3.8 (organic matter 2.5 %) and at 75 mg Cr(VI) kg^{-1} for the soils with pH 6.9 (organic matter 2.2 %) and 8.1 (organic matter 2.6 %), showing clear evidence of increased Cr uptake and hence toxicity with increasing soil pH. This is at least partly due to the decrease in Cr(VI) sorption caused by expected increasing competition between the oxyanions ($HCrO_4^-$ and CrO_4^{2-}) and OH^- anions for the positive charged sites on soil colloids as the pH increases On the other hand, at low pH values the dominant Cr(VI) species, $HCrO_4^-$ can be easily adsorbed at positively charged sites. In acidic environments Cr(VI) generally reduces to Cr(III), which is strongly retained onto soil colloids due to the availability of additional positive charges generated by the protonation of (organic and inorganic) colloidal surfaces. This strong retention considerably decreases the plant uptake and hence phytotoxicity of Cr(III) under acidic soil conditions [73].

Plant species differ in their Cr toxicity or tolerance thresholds. A study by Adema and Henzen [74] which assessed the effect of Cr (as $K_2Cr_2O_7$) on lettuce (*Lactuca sativa*), tomato (*Lycopersicon esculentum*) and oats (*Avena sativa*) grown in a loamy soil found lettuce more sensitive ($EC_{50} = 1.8$ mg kg^{-1}) to Cr toxicity than tomato ($EC_{50} = 6.8$ mg kg^{-1}) and oats ($EC_{50} = 7.4$ mg kg^{-1}). Comparing Cr(VI) with Cr(III), Sivakumar and Subbhuraam [75] measured their toxicity to an earthworm species, *Eisenia fetida* in 10 different soils and reported LC_{50} (lethal concentration which causes death of 50 % of the exposed organisms) values which varied across the soils within 1656–1902 mg Cr(III) kg^{-1} and 222–257 mg Cr(VI) kg^{-1}; the LC_{50} of Cr(III) was about 8 times higher than that of Cr(VI). However, compared with the results from Sivakumar and Subbhuraam [75], the relatively high toxicity of Cr(VI) to earthworm species *Octochaetus pattoni* (LC_{50} 15.1 mg kg^{-1}) was reported by Abbasi and Soni [76] in a soil mixed with animal manure. Clearly for Cr(VI or III) the toxicity threshold to earthworms can be remarkably variable across soil types, and also possibly between different species of earthworms. Generally, the toxicity of Cr in soils is affected mainly by soil properties (pH and organic matter), plant varieties and Cr species, that is Cr(VI) or Cr(III).

Critical loads or criteria of Cr in soils have been developed for environmental protection in many countries. The values for Cr vary widely; most of them are in the range 50–600 mg Cr kg^{-1} soil. The wide variation of soil criteria in different countries is at least partly due to insufficient Cr ecotoxicity data. Considering the difference in toxicity of Cr(III) and Cr(VI), the species of Cr, as well as soil properties should be considered for Cr critical limits or criteria in soils.

19.5.2 Nickel

Although Ni is thought to play a significant role in enzyme-catalyzed metabolic processes of higher plants as a cofactor of urease and hydrogenase, excessive Ni concentrations in higher plants can decrease photosynthesis and cause chlorosis of the plants when phytotoxicity occurs [11]. Nickel is actively transported into plant cells and is distributed differentially within individual plant species. Nickel is also present in substantial concentrations in the xylem and phloem sap and in the former it exists both as free ionic species (Ni^{2+}) and in a chelated form (Ni–citrate) [12]. Nickel has high mobility in the phloem, which explains its translocation and accumulation in leaves, fruits and seeds [12]. Hyperaccumulator plant species can tolerate tissue Ni concentrations up to 10 000-fold higher than in other plants [11]. However, concentrations of Ni in plants commonly range within 0.1–5 mg kg^{-1} for noncontaminated and nonserpentinic soils [9].

Soil contamination with Ni has a potential negative impact on plants, microorganisms and animals. A study in Port Colborne (Ontario, Canada), where Ni emissions from a local refinery contaminated 29 km^2 of adjacent land, with soil-Ni concentrations between 600 and 10 000 mg kg^{-1}, demonstrates the effect of Ni contamination on plants [45]. Plants in the impacted area showed typical toxicity symptoms. Further field and greenhouse pot experiments, conducted using the Ni-contaminated muck soils, displayed distinct

Ni-toxicity foliar symptoms with stunting, cupping and foliar necrosis in celery, beets, cabbage, lettuce and radish, reducing the yield of the tested plants by 28–100 % compared with their counterparts grown in comparable uncontaminated soils [77–79]. Liming these Ni-contaminated soils, which raised their pH from 5.1–5.9 to at least 6.9, showed that soil-extractable Ni could be reduced and Ni toxicity to celery, wheat, oat and redbeet crops mitigated [77,80].

Like other potentially toxic trace elements, elevated soil Ni concentrations can also affect the soil microbial community structure and function. For instance, Åkerblom et al. [81] studied the negative effects of trace elements including Ni on soil respiration and microbial community structure of forest mor layers using phospholipid fatty acid (PLFA) analysis. The work showed that the addition of Ni to the soils induced changes in the microbial community structure and affected microbial respiration negatively and that there was a direct relationship between the microbial community structure and the soil cumulative respiration [81]. There is, however, some evidence which suggests that soil microbes can become increasingly tolerant to metals, including Ni [82–84]. For example, in a soil with pH 6.4 from Woburn (UK), the pre-exposure of nitrifying bacteria to Ni was measured by an increased tolerance to Ni; the EC_{50} for pre-exposure treatment increased from 502 mg Ni kg^{-1} to 1064 mg Ni kg^{-1}, as compared with the treatment without pre-exposure [84].

Key soil properties are known to moderate the toxic effects of Ni on soil microbial activities, largely through their influence on metal retention–release processes. For instance, the toxicity of added Ni on three soil microbial activities (nitrification potential, glucose-induced respiration, and C-mineralization of a plant residue) varied considerably across a wide range of soils [85]. The toxicity decreased with increasing CEC and clay content for all three assays. Further work by the same research team showed that ageing (time after spiking with Ni) decreased Ni toxicity on soil microbial processes; the EC_{50} values of Ni in soils aged for 15 months were larger by a factor of 1 to 23 (median, 4.6) than those of freshly spiked soils [86]. Changes in the Ni toxicity due to ageing were generally largest in a soil with the highest pH, consistent with the largest relative decrease in soluble Ni concentrations [86]. Although above a certain threshold the ecotoxicity of Ni is unambiguous, establishing its toxicity relative to other trace elements is not that straightforward. This is largely due to the influence of soil properties in the manifestation or mitigation of Ni toxicity to soil microbial populations. For example, based on soil total concentration, the EC_{50} for Cu (35 mg kg^{-1}) was more than for Ni (90 mg kg^{-1}) when their inhibitory effects on the dehydrogenase activity were measured in a loess soil with pH 7.0 [87]. In a comparison across a wide range of soils, however, Ni was generally more toxic than Cu when their inhibitory effects on three soil microbial activities, namely nitrification potential, glucose-induced respiration, and C-mineralization were investigated [85].

Knowledge of Ni critical loads, as for other potentially toxic trace elements, is important in order to preserve soil ecosystem functioning. For this reason the toxicity thresholds of Ni have been extensively assessed in different soils for plants, microbes and invertebrates [13,14,88–90]. The toxicity thresholds in the literature vary widely between different soils/studies. For example, the EC_{50} values for barley root elongation and tomato shoot growth when grown in 16 European soils were in the ranges 52–1929 mg kg^{-1} for barley root elongation and 17–920 mg kg^{-1} for tomato shoot growth [90]. These rather wide differences in the EC_{50} values for barley and tomato were mainly due to the differences in soil types [90]. Generally, Ni availability/toxicity in soils depends on the geochemical origin of the metal and on soil characteristics such as pH, organic matter, clay content, CEC and E_h [90–92]. The critical loads or criteria of Ni in soils have been developed for environmental protection in many countries, although they are highly variable, 35–200 mg Ni kg^{-1} soil [93].

19.5.3 Cobalt

Cobalt is an essential trace nutrient for ruminant animals as their rumen bacteria require Co [15]. Cobalt may have some beneficial effect but is not essential as such for humans. It is also essential for atmospheric symbiotic and nonsymbiotic N_2-fixation [12,15,66]. Although its essentiality in higher, nonleguminous plants is not clearly proven, there is some evidence which suggests that Co has favourable effects on plant growth [16].

The concentration of Co in plants varies widely from 0.004 to 0.6 mg kg^{-1} depending upon plant species or soil Co content, and, in general, legume plants are known to accumulate more Co than grasses and grain crops [16]. Critical Co concentrations for ruminant diets are around 0.08–0.1 mg kg^{-1} dry matter of herbage. Cobalt in the xylem has been found mainly in cationic forms in tomatoes [94] and in the *Ricinus communis* phloem as organic complexes with molecular weight between 1000 and 5000 Da [95]. It is reported that Co mobility in plants is lower than that of Ni [66,96]. The limited translocation of Co in plants means it tends to accumulate in roots, with shoot concentrations generally being smaller than those in roots [97,98].

The primary interest in Co in soils is related to its essential role and potential deficiency in ruminant animals. However, high Co concentration in soils as found in the soils over serpentine rocks or ore deposits or near smelters [16,66] can result in far too large amounts of Co in the plants, which may result in phytotoxicity. The phytotoxicity of Co can be severe when plants accumulate Co above 50–100 mg kg^{-1} dry weight [99]. The symptoms of Co toxicity in plants often appear as chlorosis of young leaves with stunted plants [100,101]. Soils with naturally high Co concentrations usually also have high arsenic and nickel concentrations and, these elements are generally more toxic to plants and humans [102,103].

The Canadian Council of Ministers of the Environment [104] recommended an interim soil remediation criterion of 40 mg Co kg^{-1} to protect agricultural soils. However, there is little information about the ecological risk assessment of Co in soils, including plants, microorganisms and invertebrates, which is still an issue for further research and adoption of regulation for Co levels in soils.

19.6 Concluding Remarks

The fate and chemical behaviour of Cr, Co and Ni in soils have been studied extensively, including chemical processes, plant uptake and environmental impacts. Soil pH and the contents of Fe/Mn oxides, organic matter and clay minerals are important factors that control the portioning of these trace elements between the soil solid and solution phases, and ultimately their bioavailability and ecotoxicity.

The solubility of Co and Ni, like other cationic trace elements, decreases with increasing soil pH. Cr(III) is largely immobile in soil; this generally limits its entry into the food chain. On the other hand, Cr(VI) is highly mobile in neutral to alkaline soil pH conditions. This is problematic, as raising soil pH by liming is one of the most common and practical measures to mitigate the risk of excess plant uptake of trace elements. Liming will have quite the opposite effect in terms of Cr(VI) bioavailability and ecotoxicity. For Cr and Co, the influence of soil redox potential on their oxidation states plays a crucially important role in determining their retention–release processes and overall behaviour in soils.

References

[1] USDG, 2005 Minerals Yearbook; US Department of the Interior, US Geological Survey, Washington DC, 2007; http://minerals.usgs.gov/minerals/pubs/commodity (accessed 26 November 2009).
[2] Darrie, G., Commercial extraction technology and process waste disposal in the manufacture of chromium chemicals from ore; Environ. Geochem. Health 2001, 23, 187–193.
[3] Mertz, W., Chromium in human nutrition: a review; J. Nutr. 1993, 123, 626–633.
[4] Mertz, W., Interaction of chromium with insulin: a progress report; Nutr. Rev. 1998, 56, 174–177.
[5] Hopkins, L.L., Jr.; Ransome-Kuti, O.; Majaj, A.S., Improvement of impaired carbohydrate metabolism by chromium(III) in malnourished infants; Am. J. Clin. Nutr. 1968, 21, 203
[6] Vincent, J.B., The potential value and toxicity of chromium picolinate as a nutritional supplement, weight loss agent and muscle development agent; Sports Med. 2003, 33:213–230.
[7] Zayed, A.M.; Terry, N., Chromium in the environment: factors affecting biological remediation; Plant Soil 2003, 249, 139–156.
[8] International Agency for Research on Cancer (IARC), Chromium, Nickel and Welding, IARC Monographs on the Evaluation of Carcinogenic Risks to Humans, Vol. 49, Lyon 1990.
[9] McGrath, S.P., Chromium and nickel; in: B.J. Alloway (ed.), Heavy Metals in Soil, 2nd edn; Blackie Academic & Professional, London, 1995, Chapter 7, pp. 152–178.
[10] Mukherjee, A.B., Nickel: a review of occurrence, uses, emissions, and concentration in the environment in Finland; Environ. Rev. 1998, 6, 1–15.
[11] Phipps, T.; Tank, S.L.; Wirtz, T. et al., Essentiality of nickel and homeostatic mechanisms for its regulation in terrestrial organisms; Environ. Rev. 2002, 10, 209–261.
[12] Sharma, C.P., Plant Micronutrients: Roles, Responses, and Amelioration of Deficiencies; Science Publishers, Enfield, NH, 2006.
[13] Thakali, S.; Herbert, E.A.; Di Toro, D.M. et al., A terrestrial biotic ligand model. 1. Development and application to Cu and Ni toxicities to barley root elongation in soils; Environ. Sci. Technol. 2006, 40, 7085–7093.
[14] Thakali, S.; Herbert, E.A.; Di Toro, D.M. et al., A terrestrial biotic ligand model. 2. Application to Ni and Cu toxicities to plants, invertebrates, and microbes in soil; Environ. Sci. Technol. 2006, 40, 7094–7100.
[15] Smith, K.A.; Paterson, J.E., Manganese and cobalt; in: B.J. Alloway (ed.), Heavy Metals in Soils, 2nd edn; Blackie Academic & Professional, London, 1995, Chapter 10, pp. 225–244.
[16] Kabata-Pendias, A.; Pendias, H., Trace Elements in Soils and Plants; CRC Press; Boca Raton, FL, 2001.
[17] Ure, A.M.; Berrow, M.L., The elemental constituents of soils; in H.J.M. Bowen (ed.), Environmental Chemistry, Vol. 2; Royal Society of Chemistry, London, 1982, pp. 94–204.

[18] Adriano, D.C., Chromium; Trace Elements in the Environment; Springer-Verlag, New York, 1986, Chapter 5.
[19] Luo, L.; Ma, Y.B.; Zhang, S.Z.; Wei, D.P.; Zhu, Y.G., An inventory of heavy metal inputs to agricultural soils in China; J. Environ. Manage. 2009. doi:10.1016/j.jenvman.2009.01.011
[20] Nicholson, F.A.; Smith, S.R.; Alloway, B.J.; Carlton-Smith, C.; Chambers, B.J., An inventory of heavy metals inputs to agricultural soils in England and Wales; Sci. Total Environ. 2003, 311, 205–219.
[21] Gray, C.W.; McLaren, R.G.; Roberts, A.H.C., Atmospheric accessions of heavy metals to some New Zealand pastoral soils; Sci. Total Environ. 2003, 305, 105–115.
[22] Senesi, G.S.; Baldassarre, G.; Senesi, N.; Radina, B., Trace element input into soils by anthropogenic activities and implications for human health; Chemosphere 1999, 39, 343–377.
[23] Hamilton, E.I., The geobiochemistry of cobalt; Sci. Total Environ. 1994, 150, 7–39.
[24] Fendorf, S.E., Surface reactions of chromium in soil and waters; Geoderma 1995, 67, 55–71.
[25] Eiichiro, N.; Hiroyuki, T.; Tooru, K.; Taitiro, F., Dissolved state of chromium in seawater; Nature 1981, 290, 768–770.
[26] Avudainayagam, S.; Megharaj, M.; Owens, G.; Kookana, R.S.; Chittleborough, D.; Naidu, R., Chemistry of chromium in soils with emphasis on tannery waste sites; Rev. Environ. Contam. Toxicol. 2003, 178, 53–91.
[27] Calder, L., Chromium contamination of groundwater; in: J.O. Nriagu, E Nieboer (eds), Chromium in Natural and Human Environments; Advances in Envronmental Sciences and Technology; John Wiley & Sons, Inc., New York, 1988, pp. 215–229.
[28] Shao, X.H.; Xing, G.X.; Yang, W.X., Distribution of chemical forms for Co, Cr, Ni and V in typical soils of China; Pedosphere 1993, 3, 289–298.
[29] Garnier, J.; Quantin, C.; Martins, E.S.; Becquer, T., Solid speciation and availability of chromium in ultramafic soils from Niquelândia, Brazil; J. Geochem. Expl. 2006, 88, 206–209.
[30] Fendorf, S.E.; Sparks, D.L., Mechanisms of chromium (III) sorption on silica. 2. Effect of reaction conditions; Environ. Sci. Technol. 1994, 28, 290–297.
[31] Fendorf, S.E.; Lamble, G.M.; Stapleton, M.G.; Kelley, M.J.; Sparks, D.L., Mechanisms of chromium (III) sorption on silica. 1. Chromium (III) surface structure derived by extended x-ray absorption fine structure spectroscopy; Environ. Sci. Technol. 1994, 28, 284–289.
[32] Bartlett, R.J.; James, B., Behavior of Cr in soils: III. Oxidation; J. Environ. Qual. 1979, 8, 31–35.
[33] Becquer, T.; Quantin, C.; Sicot, M.; Boudot, J.P., Chromium availability in ultramafic soils from New Caledonia; Sci. Total Environ. 2003, 301, 251–261.
[34] Martin, R.R.; Naftel, S.J.; Sham, T.K.; Hart, B.; Powell, M.A., XANES of chromium in sludges used as soil ameliorants; Can. J. Chem. 2003, 81, 193–196.
[35] Uren, N.C., Forms, reaction, and availability of nickel in soils; Adv. Agron. 1992, 48, 141–203.
[36] Johnston, W.R.; Proctor, J., Growth of serpentine and non-serpentine races of *Festuca rubra* in solutions simulating the chemical conditions in a toxic serpentine soil; J. Ecol. 1981, 69, 855–869.
[37] Rieuwerts, J.S., The mobility and bioavailability of trace metals in tropical soils: a review; Chem. Spec. Bioavailab. 2007, 19, 75–85.
[38] Covelo, E.F.; Matias, J.M.; Vega, F.A.; Reigosa, M.J.; Andrade, M.L., A tree regression analysis of factors determining the sorption and retention of heavy metals by soil; Geoderma 2008, 147, 75–85.
[39] Buekers, J.; Amery, F.; Maes, A.; Smolders, E., Long-term reactions of Ni, Zn and Cd with iron oxyhydroxides depend on crystallinity and structure and on metal concentrations; Eur. J. Soil Sci. 2008, 59, 706–715.
[40] Doner, H.E.; Zavarin, M., The role of carbonates in trace and minor element chemistry; Adv. Geoecol. 1997, 30, 407–422.
[41] Arai, Y., Spectroscopic evidence for Ni(II) surface speciation at the iron oxyhydroxides-water interface; Environ. Sci. Technol. 2008, 42, 1151–1156.
[42] Businelli, D.; Casciari, F.; Gigliotti, G., Sorption mechanisms determining Ni(II) retention by a calcareous soil; Soil Sci. 2004, 169, 355–362.

[43] Nachtegaal, M.; Sparks, D.L., Nickel sequestration in a kaolinite–humic acid complex; Environ. Sci. Technol. 2003, 37: 529–534.
[44] Scheidegger, A.M.; Fendorf, M.; Sparks, D.L., 1996. Mechanisms of nickel sorption on pyrophyllite: macroscopic and microscopic approaches; Soil Sci. Soc. Am. J. 1996, 60, 1763–1772.
[45] McNear, Jr D.H.; Chaney, R.L.; Sparks, D.L., The effects of soil type and chemical treatment on nickel speciation in refinery enriched soils: a multi-technique investigation; Geochim. Cosmochim. Acta 2007, 71, 2190–2208.
[46] Scheidegger, A.M.; Lamble, G.M.; Sparks, D.L., Investigation of Ni sorption on pyrophyllite: an XAFS study; Environ. Sci. Technol. 1996, 30, 548–554.
[47] Scheidegger, A.M.; Lamble, G.M.; Sparks, D.L., Spectroscopic evidence for the formation of mixed-cation hydroxide phases upon metal sorption on clays and aluminum oxides; J. Colloid Interf. Sci. 1997, 186, 118–128.
[48] Scheckel, K.G.; Sparks, D.L., Kinetics of the formation and dissolution of Ni precipitates in a gibbsite/amorphous silica mixture; J. Colloid Interf. Sci. 2000, 229, 222–229.
[49] Scheckel, K.G.; Scheinost, A.C.; Ford, R.G.; Sparks, D.L., Stability of layered Ni hydroxide surface precipitates – a dissolution kinetics study; Geochim. Cosmochim. Acta 2000, 64, 2727–2735.
[50] Scheckel, K.G.; Sparks, D.L., Dissolution kinetics of nickel surface precipitates on clay mineral and oxide surfaces; Soil Sci. Soc. Am. J. 2001, 65, 685–694.
[51] Wendling, L.A.; Kirby, J.K.; McLaughlin, M.J., A novel technique to determine cobalt exchangeability in soils using isotope dilution; Environ. Sci. Technol. 2008, 42, 140–146.
[52] Taylor, R.M.; McKenzie, R.M., The association of trace elements with manganese minerals in Australian soils; Aust. J. Soil Res. 1966, 4, 29–39.
[53] McLaren, R.G.; Lawson, D.M.; Swift, R.S., The forms of cobalt in some Scottish soils as determined by extraction and isotopic exchange; J. Soil Sci. 1986, 37, 223–234.
[54] Li, Z.; McLaren, R.G.; Metherell, A.K., Fractionation of cobalt and manganese in New Zealand soils; Aust. J. Soil Res. 2001, 39, 951–967.
[55] Neaman, A.; Mouele, F.; Trolard, F.; Bourrie, G., Improved methods for selective dissolution of Mn oxides: applications for studying trace element associations; Appl. Geochem. 2004, 19, 973–979.
[56] Burns, R.G., The uptake of cobalt into ferromanganese nodules, soils, and synthetic manganese (IV) oxides; Geochim. Cosmochim. Acta 1976, 40, 95–102.
[57] Crowther, D.L.; Dillard, J.G.; Murray, J., The mechanism of Co(II) oxidation on synthetic birnessite; Geochim. Cosmochim. Acta 1983, 47, 1399–1403.
[58] Murray, J.W.; Dillard, J.G., The oxidation of cobalt (II) adsorbed on manganese dioxide; Geochim. Cosmochim. Acta 1979, 43, 781–787.
[59] Manceau, A.; Drits, V.A.; Silvester, E.; Bartoli, C.; Lanson, B., Structural mechanism of Co^{2+} oxidation by the phyllomanganite buserite; Am. Miner. 1997, 82, 1150–1175.
[60] Kay, J.T.; Conklin, M.H.; Fuller, C.C.; O'Day, P.A., Processes of nickel and cobalt uptake by a manganese oxide forming sediment in Pinal Creek, Globe Mining, Arizona; Environ. Sci. Technol. 2001, 35, 4719–25.
[61] Anderson, P.R.; Christensen, T.H., Distribution coefficients of Cd, Co, Ni, and Zn in soils; J. Soil Sci. 1988, 39, 15–22.
[62] Spark, K.M.; Johnson, B.B.; Wells, J.D., Characterizing heavy-metal adsorption on oxides and oxyhydroxides; Eur. J. Soil Sci. 1995, 46, 621–631.
[63] Han, F.X.; Banin, A., Long-term transformations of cadmium, cobalt, nickel, zinc, vanadium, manganese, and iron in arid-zone soils under saturated condition; Commun. Soil Sci. Plant Anal. 2000, 31, 943–957.
[64] Plekhanova, I.O.; Savel'eva, V.A., The transformation of cobalt compounds in soils upon moistening; Eurasian Soil Sci. 1999, 32, 514–520.
[65] McLaren, R.G.; Lawson, D.M.; Swift, R.S., The availability to pasture plants of native and applied soil cobalt in relation to extractable soil cobalt and other soil properties; J. Sci. Food Agric. 1987, 39, 101–112.
[66] Gál, J.; Hursthouse, A.; Tatner, P.; Stewart, F.; Welton, R., Cobalt and secondary poisoning in the terrestrial food chain: data review and research gaps to support risk assessment; Environ. Int. 2008, 34, 821–838.

[67] Zayed, A.; Lytle, C.M.; Qian, J.H.; Terry, N., Chromium accumulation, translocation and chemical speciation in vegetable crops; Planta 1998, 206, 293–299.
[68] Shanker, A.K.; Cervantes, C.; Loza-Tavera, H.; Avudainayagam, S., Chromium toxicity in plants; Environ. Int. 2005, 31, 739–753.
[69] Chandra, P.; Sinha, S.; Rai, U.N., Bioremediation of Cr from water and soil by vascular aquatic plants; in: E.L. Kruger, T.A. Anderson, J.R. Coats (eds), Phytoremediation of Soil and Water Contaminants; ACS Symposium Series 664; American Chemical Society, Washington, DC, 1997, pp. 274–282.
[70] Bini, C.; Maleci, L.; Romanin, A., The chromium issue in soils of the leather tannery district in Italy; J. Geochem. Expl. 2008, 96, 194–202.
[71] Vincent, J.B., Recent advances in the nutritional biochemistry of trivalent chromium; Proc. Nutr. Soc. 2004, 63, 41–47.
[72] Ali, N.A.; Ater, M.; Sunahara, G.I.; Robidoux, P.Y., Phytotoxicity and bioaccumulation of copper and chromium using barley (*Hordeum vulgare* L.) in spiked artificial and natural forest soils; Ecotoxicol. Environ. Saf. 2003, 57, 363–374.
[73] Lee, D.Y.; Huang, J.C.; Juang, K.W.; Tsui, L., Assessment of phytotoxicity of chromium in flooded soils using embedded selective ion exchange resin method; Plant Soil, 2005, 277, 97–105.
[74] Adema, D.M.M.; Henzen, L., A comparison of plant toxicity of some industrial chemicals in soil culture and soilless culture; Ecotoxicol. Environ. Saf. 1989, 18, 219–229.
[75] Sivakumar, S.; Subbhuraam, C.V., Toxicity of chromium(III) and chromium(VI) to the earthworm *Eisenia fetida*; Ecotoxicol. Environ. Saf. 2005, 62, 93–98.
[76] Abbasi, S.A.; Soni, R., Stress-induced enhancement of reproduction in earthworm *Octochaetus pattoni* exposed to chromium (VI) and mercury (II) – implications in environmental management; Int. J. Environ. Stud. 1983, 2, 43–47.
[77] Bisessar, S., Effects of lime on nickel uptake and toxicity in cereal grown on muck soil contaminated by a nickel refinery; Sci. Total Environ. 1989, 84, 83–90.
[78] Frank., R.; Stonefield, K. I.; Suda, P.; Potter, J. W., Impact of nickel contamination on the production of vegetables on an organic soil, Ontario, Canada, 1980–1981; Sci. Total Environ. 1982, 26, 41–65.
[79] Temple, P. J.; Besessar, S., Uptake and toxicity of nickel and other metals in crops grown on soil contaminated by a nickel refinery; J. Plant Nutr. 1981, 3, 473–482.
[80] Kukier, U.; Chaney, R.L., Remediating Ni-phytotoxicity of contaminated quarry muck soil using limestone and hydrous iron oxide; Can. J. Soil Sci. 2000, 80, 581–593.
[81] Åkerblom, S.; Bååth, E.; Bringmark, L.; Bringmark, E., Experimentally induced effects of heavy metal on microbial activity and community structure of forest mor layers; Biol. Fertil. Soils 2007, 44, 79–91.
[82] Doelman, P.; Haanstra, L., Short-term and long-term effects of cadmium, chromium, copper, nickel, lead and zinc on soil microbial respiration in relation to abiotic soil factors; Plant Soil 1984, 79, 317–327.
[83] Héry, M.; Nazaret, S.; Jaffré, T.; Normand, P.; Navarra, E., Adaptation to nickel spliking of bacterial communities in noecaledonian soils; Environ. Microbiol. 2003, 5, 3–12.
[84] Fait, G.; Broos, K.; Zrna, S.; Lombi, E.; Hamon, R., Tolerance of nitrifying bacteria to copper and nickel; Environ. Toxicol. Chem. 2006, 25, 2000–2005.
[85] Oorts, K.; Ghesquiere, U.; Swinnen, K.; Smolers, E., Soil properties affecting the toxicity of $CuCl_2$ and $NiCl_2$ for soil microbial processes in freshly spiked soils; Environ. Toxicol. Chem. 2006, 25, 838–844.
[86] Oorts, K.; Ghesquiere, U.; Smolers, E., Leaching and aging decrease nickel toxicity to soil microbial processes in soils freshly spiked with nickel chloride; Environ. Toxicol. Chem. 2007, 26, 1130–1138.
[87] Welp, G., Inhibitory effects of the total and water-soluble concentrations of nine different metals on the dehydrogenase activity of a loess soil; Biol. Fert. Soils 1999, 30, 132–139.
[88] Oort, K.; Ghesquiere, U.; Swinnen, K.; Smolders, E., Soil properties affecting the toxicity of $CuCl_2$ and $NiCl_2$ for soil microbial processes in freshly spiked soils; Environ. Toxicol. Chem. 2006, 25, 836–844.

[89] Lock, K., Janssen, C.L., Ecotoxicity of nickel to *Eisenia fetida, Enchytraeus albidus* and *Folsomia candia*; Chemosphere, 2002, 26, 197–200.

[90] Rooney, C.P.; Zhao, F.J.; McGrath, S.P., Phytotoxicity of nickel in a range of European soils: Influence of soil properties, Ni solubility and speciation; Environ. Pollut. 2007, 145, 596–605.

[91] Echevarria, G.; Morel, J.L.; Fardeau, J.C.; Leclerc-Cessac, E., Assessment of phytoavailability of nickel in soils; J Environ. Qual. 1998, 27, 1064–1070.

[92] Everhart, J.L.; McNear, Jr D.; Peltier, E. van der Lelie, D.; Chaney, R.L.; Sparks, D.L., Assessing nickel bioavailability in smelter-contaminated soils; Sci. Total Environ. 2006, 367, 732–744.

[93] Provost, J.; Cornelis, C.; Swartjes, J., Comparison of soil clean-up standards for trace elements between countries: why do they differ?; J. Soils Sediments 2006, 6, 173–181.

[94] Tiffin, L.O., Translocation of manganese, iron, cobalt and zinc in tomato; Plant Physiol. 1967, 42, 1427–1432.

[95] Wiersma, D.; van Goor, G.J., Chemical forms of nickel and cobalt in phloem of *Ricinus communis*; Physiol. Plant. 1979, 45, 440–442.

[96] Zeller, S.; Feller, U., Long-distance transport of cobalt and nickel in maturing wheat; Eur. J. Agron. 1999, 10, 91–98.

[97] Bakkaus, E.; Gouget, B.; Gallien, J-P.; Khodja, H.; Carrot, F.; Morel, J.L. *et al.*, Concentration and distribution of cobalt in higher plants: the use of micro-PIXE spectroscopy; Nucl. Instrum. Methods Phys. Res., Sect. A, 2005, 231, 350–356.

[98] Sasmaz, A.; Yaman, M., Distribution of chromium, nickel, and cobalt in different parts of plant species and soil in mining area of Keban, Turkey; Commun. Soil Sci. Plant Anal. 2006, 37, 1845–1857.

[99] Chaney, R.L., Potential Effects of Waste Constituents on the Food Chain; Park Ridge, NJ: Noyes Data Corp.; 1983.

[100] Chatterjee, J.; Chatterjee, C., Management of phytotoxicity of cobalt in tomato by chemical measures; Plant Sci. 2003, 164, 793–801.

[101] Kapustka, L.A.; Eskew, D.; Yocum, J.M., Plant toxicity testing to drive ecological soil screening levels for cobalt and nickel; Environ. Toxicol. Chem. 2006, 25, 865–874.

[102] Robinson, B.H.; Brooks, R.R.; Hedley, M.J., Cobalt and nickel accumulation in *Nyssa* (tupelo) species and its significance for New Zealand agriculture; NZ J. Agric. Res. 1999, 42, 235–40.

[103] Reeves, R.D.; Baker, A.J.M.; Becquer, T.; Echevarria, G.; Miranda, Z.J.G., The flora and biogeochemistry of the ultramafic soils of Goias state, Brazil; Plant Soil 2007, 293, 107–119.

[104] Canadian Council of Ministers of the Environment (CCME), Canadian Environmental Quality Guidelines, Chapter 7; Canadian Soil Quality Guidelines for the Protection of Environmental and Human Health, 1999.

20

Manganese and Selenium

Zhenli L. He[1,3], Jiali Shentu[2] and Xiao E. Yang[3]

[1]University of Florida, Institute of Food and Agricultural Sciences, Indian River Research and Education Center, Fort Pierce, Florida, USA
[2]College of Environmental and Resource Sciences, Zhejiang Gongshang University, Hangzhou, China
[3]Key Laboratory of Environmental Remediation and Ecosystem Health, College of Environmental and Resources Sciences, Zhejiang University, Hangzhou, China

20.1 Introduction

Manganese (Mn) is found as a free element in nature, often in combination with iron and as a constituent of many minerals. As a free element, Mn has many industrial metal alloy uses and is used in pigments and oxidation chemicals. Naturally occurring Mn is composed of one stable isotope, ^{55}Mn. Eighteen radioisotopes of Mn have been characterized, with the most stable being ^{53}Mn with a half-life of 3.7 million years, while ^{54}Mn and ^{52}Mn have half-lives of 312.3 and 5.591 days, respectively.

The most common oxidation states of Mn are +2, +3, +4, and +7, although it can occur in oxidation states from +1 to +7. However, +2 is the most stable oxidation state, which is common in the rock-forming minerals. The cation Mn^{2+} is known to substitute Fe^{2+} and Mg^{2+} in silicates and oxides. Manganese is essential to organisms, including microorganisms, plants, animals, and human beings. It is involved in many physiological and biochemical processes in plants as a cofactor of enzymes or as a catalyst and is thus a required trace mineral for all known living organisms.

Trace Elements in Soils   Edited by Peter S. Hooda
© 2010 Blackwell Publishing, Ltd

Selenium (Se) belongs to the Group 7 of the periodic table, is a nonmetallic element chemically related to sulfur and tellurium, and rarely occurs in its elemental state in nature. Selenium has six naturally occurring isotopes, five of which are stable: ^{74}Se, ^{76}Se, ^{77}Se, ^{78}Se, and ^{80}Se. Like Mn, Se exhibits multiple oxidation states: −2, 0, +2, +4, and +6 in nature.

Much of the selenium in the environment comes from selenium dioxide produced by burning of coal and other fossil fuels [1]. Selenium is used in the glass industry for either ruby red coloring (as cadmium sulfoselenide pigments) or for counteracting green tint caused by iron oxide impurities [2]. Its photoelectric and semiconducting properties are widely exploited in electronics [3]. Selenium fertilizers and dietary supplements are used to counteract selenium deficiency [2], and selenium sulfide is used in fungicides and in antidandruff shampoos [4]. An emerging application of selenium is the 'amorphous selenium detector' in mammographic instruments [5].

Selenium is not required by higher plants, but is essential to animals and human beings. It is required for the synthesis of glutathione peroxidase enzyme, which protects cells and erythrocytes from peroxidation. Deficiency of Se can cause a number of disorders such as white-muscle disease and cardiovascular disease in humans and animals, whereas an excess supply of Se produces toxicity to plants and animals [6,7]. Fresh interest in Se has arisen in the last few decades due to frequent reports of both deficiency and toxicity of Se occurring at a significant geographical scale such as the Se-deficiency belt in China [8] and Se toxicity to wildlife in California [9].

20.2 Concentrations and Sources of Manganese and Selenium in Soils

20.2.1 Manganese

Manganese is one of the most abundant trace elements in the lithosphere, and its common range in rocks is 350–2000 mg kg^{-1}, with the highest values related to mafic rocks [10]. The content of total Mn in global surface soils varies substantially, ranging from <7 to >9000 mg kg^{-1} with mean values of 270–530 mg kg^{-1}, affected by parent material, soil texture, and anthropogenic inputs [10–12]. Most soil Mn is inherited from parent materials. A small portion of Mn may be lost through mineral weathering processes, particularly under reducing conditions, and consequently Mn abundance in soil is generally lower than in the lithosphere. Other sources of soil Mn include atmospheric deposition, irrigation water, and the use of sewage sludge, fertilizers, liming materials, and manure (Table 20.1).

During soil development, Mn tends to accumulate as oxides and hydroxides and therefore, total concentration of Mn is higher than that of other trace elements such as Cu, Zn, B, Mo in most agricultural soils. The soil available Mn pool consists of water-soluble, exchangeable and readily reducible Mn and generally accounts for a small portion of total Mn, except under strongly reducing conditions. Consequently, extractable Mn is often a better indicator of Mn availability than total Mn in soil.

The commonly found Mn-containing primary minerals are mainly ferromagnesian silicate minerals, such as biotite, chlorite, and hornblende. During weathering, Mn compounds are oxidized under atmospheric conditions and the released Mn is rapidly transformed into oxides and/or hydroxides (secondary minerals) and accumulated in soils.

Table 20.1 Concentrations (mg kg^{-1}) of Mn and Se in source materials (data from [12])

Sources	Mn	Se
Rocks	350–2000	<0.01–600
Soils	200–1000	<0.1–5000
Atmosphere (ng m^{-3})^a	3–900	60–30 000
Sewage sludge	60–3900	2–10
Phosphate fertilizers	40–2000	0.5–25
Limestone	40–1200	0.08–0.1
Manure	30–550	2.4

^{a}Data (in ng m^{-3}) were pooled from European and American regions.

These oxides/hydroxides can be either noncrystalline material coated on soil particles or crystalline minerals of discrete solid phases such as nodules and concretions. The reported common crystalline Mn oxides include birnessite [(Na,Ca) (Mn^{3+}, Mn^{4+})$_7$O$_{14}$·2.8H$_2$O], pyrolusite (β-MnO$_2$), and manganite (MnOOH) [13]. Manganese oxides can coprecipitate with Fe oxides. Manganese oxides may, therefore, accumulate in horizons rich in Fe oxides and hydroxides. A small portion of total Mn in soil occurs in dissolved forms, adsorbed on mineral surfaces and chelated with organic matter, and these forms contribute most to the Mn pool that is available to plants [12,14].

20.2.2 Selenium

Total concentration of Se in soils varies widely, from <0.005 to >20 mg kg^{-1}; however, for most agricultural soils, the range is generally from 0.1 to 2.0 mg kg^{-1}, with a mean value of 0.2 mg kg^{-1}. Total Se content in surface soils varies widely, from <0.1 to 4 mg kg^{-1}, with a mean value of 0.33 mg kg^{-1} on a worldwide scale. The abundance of Se in lithosphere is reported to be 0.05–0.8 mg kg^{-1}. In China, total Se concentration in surface soil was found to vary from 0.006 to 9.13 mg kg^{-1}, with an algebraic mean of 0.29 mg kg^{-1} [8]. Sedimentary rocks generally contain more Se than magmatic rocks. In sedimentary rocks, Se is associated with clay fractions and thus sandstone and limestone contain less Se than shale and argillaceous sediments [10,12,15,16]. There was a trend of soil Se geographically increasing from north-west to south-east of China, highest in the oxisols and lowest in the inceptisols [8].

Soil is the major source of Se supply to humans and animals through food chains. The flux of Se from soil to humans is, to a large extent, determined by the content and availability of Se in the soils [6,17]. Keshan and Kashin–Beck diseases caused by Se deficiency mainly occurred in the regions with total soil Se below 0.127 mg kg^{-1} (the critical level of soil Se for human health). It was estimated that 25% of the soils in China contained total Se concentrations below or close to the critical level of Se deficiency [18]. However, in some extreme cases, total Se content in soil can be as high as over 1000 mg kg^{-1}, as in seleniferous soils, which causes toxicity to plants and animals that feed on Se-enriched plants and induces chronic conditions of 'alkali disease' and 'blind stagger' of cattle and sheep [8,10,12,16,19]. Soil Se is mainly of geogenic nature, that is inherited from parent materials, but is also affected by anthropogenic inputs, including atmospheric deposition and the use of Se-containing fertilizers and sewage sludge (Table 20.1). In the natural environment, elevated concentrations of Se in soils are associated primarily with volcanic materials, sulfide ore bodies, black shales and carbonaceous sandstones [8,19]. Soils derived from sedimentary rocks such as shales and schists generally contain higher concentration of Se. Selenium toxicity occurs in alkali soils of arid and semiarid areas with limited rainfall, whereas highly weathered acidic soils derived from igneous rocks are poor in Se, and the herbage or food grains grown on these soils do not contain sufficient Se to meet the requirement of grazing animals or humans dependent on local agricultural products for food supply.

20.3 Chemical Behaviour of Manganese and Selenium in Soils

The distribution of Mn and Se between the solution and solid phases and their speciation in soil solution are closely related to the pH, redox potential, solution composition, and the characteristics of soil surfaces.

20.3.1 Solution and Solid Forms

Manganese(II) is the predominant form in solution in equilibrium with exchange sites on soil surfaces, while Mn(III) and Mn(IV) are mainly associated with solid phase as oxides, carbonate or phosphate. The plant-available form of Mn is mainly Mn^{2+} and its complexes with organic and inorganic ligands. Its concentration in soil solution ranges from 10^{-3} to 10^{-9} M in well-aerated soils, decreasing as pH increases [12,14]. Under reducing conditions, Mn^{2+} concentration in solutions of highly weathered soils such as Oxisols and Ultisols can be so extremely high as to cause toxicity to crop plants such as rice [12–14].

Major inorganic complexes and ion pairs of Mn in soil solution include $MnCO_3$, $MnSO_4$, $MnCl_2$, $Mn(OH)^+$, and $MnHPO_4$. These forms of Mn are water soluble and readily bioavailable. Manganese(II) forms complexes in soil solution with organic substances such as organic acids, amino acids, and humic acids. Nuclear magnetic resonance spectroscopy (NMRS) analysis showed that Mn^{2+} bound to fulvic acid remains fully hydrated, indicating that the complex formed between humates and Mn^{2+} can be described as an outer-sphere complex wherein the ligand does not interact directly with the Mn electrons [20]. Therefore, Mn–organic ligand complexes may be still quite labile and available to plants and microorganisms.

Selenium in soils occurs in various organic and inorganic forms (Table 20.2). Organic Se may account for 20–50% of the total Se in soil and the commonly found organic Se includes dimethyl selenide and dimethyl diselenide, the former being produced through microbial activity and the latter being emitted from plants [21,22]. A significant portion of soil Se is also bound in humic substances, with 2/3 of the humic-Se being associated with fulvic acid fraction; however, in low -Se soils, more Se tends to bind in the humic acid fraction [22]. Inorganic Se includes selenide (Se^{2-}), elemental Se (Se^0), selenite ($HSeO_3^-$, SeO_3^{2-}) and selenate ($HSeO_4^-$, SeO_4^{2-}), of which selenite and selenate are water soluble and directly

Table 20.2 Major selenium forms in soils as affected by biotic and abiotic factors

Compounds	Chemical formula	Conditions
Selenides	Se^{2-}	Formation of metal selenides under strongly reducing conditions, stable
Hydrogen selenide	H_2Se	Gas, unstable in moist air and transformed to elemental Se in water
Dimethyl selenide	$(CH_3)Se$	Volatile gas, produced through the activity of bacteria and fungi
Dimethyl diselenide	$(CH_3)_2Se_2$	Volatile gas, emitted from plant
Elemental Se	Se^0	Stable under reducing conditions
Selenite	$HSeO_3^-$, SeO_3^{2-}	Water-soluble and bioavailable, stable under moderately aerated conditions and strongly adsorbed by soil
Selenate	$HSeO_4^-$, SeO_4^{2-}	Water-soluble and bioavailable, stable under well-oxidized conditions, readily subjected to leaching

available to plants. The speciation of inorganic Se in soils is largely determined by oxidation–reduction conditions, and so is its bioavailability. Generally speaking, water soluble and exchangeable Se is considered as plant-available [23]. In addition to ions of selenite and selenate, there may also be some complexes of metals with these anions, such as $CoSeO_3$, $MnSeO_4$, $NiSeO_4$, and $ZnSeO_4$ [24]. However, these soluble forms account for only a small portion of the total Se in soil solution and therefore are of minor importance to soil Se availability.

20.3.2 Ion-Exchange and Sorption–Desorption Reactions

Manganese(II) participates in ion exchange similarly to Ca^{2+}. The cation Mn^{2+} is attracted to negatively charged soil surfaces by means of electrostatic forces and the retained Mn^{2+} can be replaced by other divalent or monovalent cations such as Ca^{2+} and Na^+. The exchange reaction occurs on both internal and external surfaces of layer silicates and external surfaces of calcite and oxides at relatively high pH values (greater than the PZC (point of zero charge) of the oxides), and the dissociated functional groups of organic substances. The exchangeable Mn^{2+} is generally considered to be bioavailable.

The retention of Mn^{2+} on the surfaces of Fe, Al, Mn, and Ti oxides involves both nonspecific and specific adsorption [13]. In the former, Mn^{2+} is electrostatically held onto negatively charged exchange sites on the surface and this reaction occurs when soil pH > PZC of the dominant oxides, whereas in the latter, Mn^{2+} replaces H^+ and directly forms chemical bond with surface O^- to become a part of inner sphere surface complex and the specific adsorption can occur at the whole pH range normally found in soils and the adsorbed Mn^{2+} is more difficult to desorb. Adsorption of Mn^{2+} often results in reduced pH because of proton release [12].

Oxides and hydroxides of Mn such as birnessite and manganite have greater adsorption affinity and capacity for Mn^{2+} than Fe and Al oxides and hydroxides because of their larger specific surface area and relatively lower PZC, for instance, birnessites have a PZC between 1.5 and 2.6 and specific surface area of 50–300 m^2 g^{-1}. Adsorption of Mn^{2+} on birnessite and related hydrous oxides involves the formation of both inner-sphere and outer-sphere surface complexes.

pH plays a significant role in adsorption–desorption of Mn^{2+} in soils. Decreasing pH leads to a rapid decrease in net negative surface charge, particularly in soils with a significantly variable (pH-dependent) charge density and soil affinity for metal ions, which subsequently causes an increase in the metal desorption [25]. Desorption of Mn^{2+} has been reported to increase with decreasing pH [27–29], especially in variable-charge soils [30]. Organic constituents and temperature also influenced Mn^{2+} desorption. The organic constituents have high adsorptive capacity for Mn^{2+} [27]. Elevated temperature was observed to increase Mn^{2+} adsorption onto oxide minerals [31]), but decrease Mn^{2+} desorption onto lignite [29]. Rates of Mn(II) oxidation in soils have been reported to increase with temperature [32].

Both selenite (HSeO$_3^-$, SeO$_3^{2-}$) and selenate (HSeO$_4^-$, SeO$_4^{2-}$) participate in anion-exchange process. Selenate on exchange sites is readily replaced by sulfate (SO$_4^{2-}$) or phosphate (H$_2$PO$_4^-$, HPO$_4^{2-}$, PO$_4^{3-}$), whereas only a small portion of the sorbed selenite is exchangeable. The exchangeable Se usually accounts for less than 0.5–7% of the total Se in soils [12].

Similar to phosphate, selenite (HSeO$_3^-$, SeO$_3^{2-}$) is strongly sorbed by Fe, Mn and Al oxides and some layer silicates in soils [12]. The adsorption can be either nonspecific or specific [33]. In the former, selenite is adsorbed through electrostatic attraction (Equation 20.1), whereas the latter involves the formation of inner-sphere surface complex with chemical bond between the surface and selenite (Equation 20.2):

$$O{<}^{M\text{-}OH}_{M\text{-}OH_2^+} + HSeO_3^- \rightleftharpoons O{<}^{M\text{-}OH}_{M\text{-}OH_2^+ \ldots SeO_3H^-} + OH^- \quad (20.1)$$

$$O{<}^{M\text{-}OH}_{M\text{-}OH} + HSeO_3^- \rightleftharpoons O{<}^{M\text{-}OH}_{M\text{-}HSeO_3} + OH^- \quad (20.2)$$

In soils most of the selenite is retained through specific adsorption, whereas the majority of selenate is retained by electrostatic forces, and therefore selenate is readily leached out, particularly in well-oxidized neutral and alkaline soils. Major factors that influence Se sorption in soils include oxides/clay minerals, pH, organic matter, and competitive anions [34]. Based on the Langmuir model, monolayer adsorption capacity of various clay minerals follows the sequence: oxides > 1:1 layer silicates > 2:1 layer silicates, and among the oxides the order was Fe$_2$O$_3$ (1773 mg kg^{-1}) > MnO$_2$ (1318 mg kg^{-1}) > Al$_2$O$_3$ (1144 mg kg^{-1}) [35]. The 2:1 layer silicates such as montmorillonite and vermiculite showed a relatively small capacity of Se sorption, likely due to some oxide coating or through cation bridging to the surface, as was found with phosphate sorption [36]. The sorption of Se in soil generally decreases with increasing pH, because an increase in pH increases negative charge to both the surface and Se anions, thus enhancing the repulsion between them. The effect of organic matter on Se adsorption was related to the blockage of adsorbing sites by humic substances and competitive adsorption of organic anions with selenate/selenite [37].

Phosphate is the most important competitive anion for selenite/selenate adsorption in soils. Addition of phosphate at 5 mg l^{-1} was found to reduce selenite adsorption by 30–70% [36]. This is important in agricultural management as P fertilization potentially enhances both plant Se uptake and Se leaching loss from soils [38,39].

The sorbed selenate and selenite can be desorbed by other anions such as phosphate and sulfate [40–42]. Nonspecifically sorbed selenate can even be replaced by chloride ion through anion exchange [40]. Phosphate is an effective ligand to replace selenite specifically sorbed on oxide minerals [41] and clay minerals [42]. Dillon *et al.* [39] revealed that more Se was desorbed by KH_2PO_4 than by KCl from acidic soils, and that acidic soils retained mainly selenite on specific sites through ligand exchange mechanisms, whereas alkaline soils generally retained selenate in water-soluble nonspecifically sorbed forms.

The concentration of phosphate in soil solution affects the sorption–desorption reaction of Se in soil [43]. The extracted Se was lower in 0.1 M Na_2HPO_4 than in 1 M Na_2HPO_4 solution and most of the sorbed Se was desorbed in the 1 M Na_2HPO_4 solution. The sorption–desorption of Se was also influenced by pH [44]. Lowering soil pH generally enhances selenite sorption and subsequently reduce selenite desorption [39]. In acidic soils, a larger amount of sorbed Se was desorbed in KH_2PO_4 than in KCl solution, but the reverse was true for alkaline soils [41].

20.3.3 Precipitation–Dissolution and Oxidation–Reduction Reactions

The precipitation–dissolution of Mn in soils is closely related to the oxidation–reduction processes. Some common oxidation–reduction reactions in soils are as shown in Table 20.3.

The reduction of MnO_2 occurs after the depletion of O_2 and NO_3 with pe^0 (pH 7) at about 6.8. However, at this stage Mn^{2+} concentration in soil solution may not increase as fast as expected, because the released Mn^{2+} will be partially captured by Fe oxides and hydroxides through sorption processes. With further development of reducing conditions with pe^0 (pH 7) down to -3.13 or below, Fe oxides and hydroxides start to dissolve and release Fe^{2+} into soil solution. At this point Mn^{2+} concentration in soil solution will increase rapidly due to the dissolution of Fe oxides. pH is also critical to these reactions. Low pH favors the production of Mn^{2+} and vice versa. Table 20.4 presents major chemical equilibrium reactions of Mn compounds in soils. Oxides of Mn (III, IV) are generally

Table 20.3 Some common oxidation–reduction reactions in soils

Reactions	pe (pH 7)[a]
$O_2 + 4H^+ + 4e^- = 2H_2O$	13.80
$2NO_3 + 12H^+ + 10e^- = N_2 + 6H_2O$	12.66
$MnO_2 + 4H^+ + 2e^- = Mn^{2+} + 2H_2O$	6.80
$Fe(OH)_3 + 3H^+ + e^- = Fe^{2+} + 3H_2O$	−3.13
$SO_4^{2-} + 10H^+ + 8e^- = H_2S + 4H_2O$	−3.63
$N_2 + 8H^+ + 6e^- = 2NH_4^+$	−4.69
$2H^+ + 2e^- = H_2$	−7.00

[a]pe is the negative log of electron concentration.

Table 20.4 Major reactions of manganese compounds in soils

No.	Reaction formula	log K^0
1	$\beta\text{-MnO}_2 + 4\,H^+ + 2e^- = Mn^{2+} + 2\,H_2O$	41.89
2	$\gamma\text{-MnO}_{1.9} + 3.8\,H^+ + 1.8e^- = Mn^{2+} + 1.9\,H_2O$	38.89
3	$\delta\text{-MnO}_{1.8} + 3.6\,H^+ + 2e^- = Mn^{2+} + 1.8\,H_2O$	35.38
4	$\gamma\text{-MnOOH} + 3\,H^+ + e^- = Mn^{2+} + 2\,H_2O$	25.27
5	$Mn_2O_3 + 6\,H^+ + 2e^- = 2\,Mn^{2+} + 3\,H_2O$	51.46
6	$Mn_3O_4 + 8\,H^+ + 2e^- = 3\,Mn^{2+} + 4\,H_2O$	63.03
7	$MnCO_3 + 4\,H^+ = Mn^{2+} + CO_2(gas) + 2\,H_2O$	8.08
8	$Mn_2(SO_4)_3(s) + 2e = 2\,Mn^{2+} + 3\,SO_4^{2-}$	39.54
9	$Mn_3(PO_4)_2(s) + 4\,H^+ = 3\,Mn^{2+} + 2\,H_2PO_4^-$	11.78

found in aerobic conditions; soluble and exchangeable Mn^{2+} tends to be dominant under strongly acidic conditions; and Mn^{2+} is released when aerobic soils become reduced as a result of flooding or mineralization of organic matter [45]. Within the oxidation–reduction limits of natural soils, (pe + pH between 20.61 and 0), MnO_2, $MnOOH$, $MnCO_3$, and $Mn_3(PO_4)_2$ are the major compounds that control the concentration of Mn^{2+} in soil solution [45]. The reduction of $\beta\text{-MnO}_2$ can be expressed as

$$\beta\text{-MnO}_2 + 4\,H^+ + 2e^- \rightarrow Mn^{2+} + 2H_2O \quad \log K = 41.89 \tag{20.3}$$

$$\log(Mn^{2+} + 2pH) = 41.89 - 2(pe + pH)$$

Under well-oxidized conditions in equilibrium with atmospheric O_2 (pe + pH = 20.61), the solubility of Mn oxides increases in the order $MnO_2 < MnOOH < Mn_3O_4$. However, under moderately reducing conditions (pe + pH < 12.5), the order of expected solubility of these Mn oxides reverses completely. Under strongly reducing conditions, the solubility of all these oxides is very high and their persistence in soils or natural waters is unlikely at equilibrium. Carbonates and phosphates of Mn are often the solid phases that control the concentration of Mn^{2+}, with pe + pH at 12 or less [13].

A number of Se compounds were proposed to control Se concentration in soil solution based on laboratory studies. They include $BaSeO_4$, $MnSeO_3$, $Fe_3(SeO_3)_2$, Se^0, and a series of selenides [12]. However, based on their solubility and the measured Se concentration in soil solution, neither selenates nor selenites can persist in soils [12]. Se^0 is more soluble than selenides and may not be stable in soils. Most selenides in soils have very low solubility and under strongly reducing conditions, they may be the pool of Se fixed in soils. For instance, Cu_2Se is stable in acidic soils, and $PbSe$ and $SnSe$ are stable in neutral and alkaline soils.

Selenium is readily oxidized during weathering and becomes more mobile with increasing oxidation states. Like Mn, Se is subjected to transformation when oxidation–reduction conditions change, from Se (VI) $\rightarrow$ Se (IV) $\rightarrow$ Se(0) $\rightarrow$ Se^{2-}, depending upon the oxidation–reduction potential. The oxidation–reduction reaction of $SeO_4^{2-}/HSeO_3^{2-}$ and SeO_3^{2-}/Se^0 can be expressed as follows:

$$SeO_4^{2-} + 2e^- + 3\,H^+ = HSeO_3^{2-} + H_2O \quad pe^0(pH7) = 7.60 \tag{20.4}$$

$$SeO_3^{2-} + 4e^- + 6H^+ = Se^0 + 3H_2O \quad pe^0(pH7) = 4.56 \quad (20.5)$$

The redox potential of $SeO_4^{2-}/HSeO_3^{2-}$ is between that of NO_3/N_2 ($pe^0 = 12.66$ at pH 7) and MnO_2/Mn^{2+} ($pe^0 = 6.80$ at pH 7). This indicates that SeO_4^{2-} will be reduced after NO_3, but before MnO_2. Selenate is stable and becomes the predominant available Se form only in well-oxidized soils with pe^0 above 6.8 at pH 7 [13]. Similarly, SeO_3^{2-} is reduced to elemental Se after MnO_2, but before $Fe(OH)_3$ ($pe^0 = -3.13$ at pH 7). Therefore, SeO_3^{2-} is stable and becomes the predominant available Se form in moderately oxidized soils with pe^0 (pH 7) between 4.56 and 7.6, whereas Se^0 may exist in soils with pe^0 (pH 7) below 4.56 and HSe^- may appear at $pe^0 < 0$.

Precipitation–dissolution and oxidation–reduction equilibrium is affected by pH, and therefore, pe+pH may be a better indicator of Se speciation. At pe+pH < 7.5, selenides ($HSe^-/Se^{2-}/H_2Se/MSe$) may control Se concentration in soil solution; at pe+pH < 9.6, elemental Se is the controlling solid phase; at pe+pH = 7.5 to 15, $H_2SeO_3/HSeO_3^-/SeO_3^{2-}$ are the predominant Se species in solution; and $H_2SeO_4/HSeO_4^-/SeO_4^{2-}$ become predominant Se species at pe+pH>15 [12,16].

20.3.4 Availability of Manganese and Selenium in Soils

A series of soil extraction methods have been proposed for estimating available Mn in soils [14]. The most commonly used include water, 0.01 M $CaCl_2$, 1 M ammonium acetate (pH7), 0.1 N H_3PO_4, Mehlich-1 (double acids), DTPA, and 0.001M EDTA. In general, 0.1 N H_3PO_4 and Mehlich-1 appear to be better than other methods based on the relationship between extractable Mn and that taken up by plants [12]. However, many factors affect the determination of soil available Mn, such as soil moisture status of the samples, soil/solution ratio, and force and duration of agitation. Therefore, caution needs to be taken in the interpretation of the data while comparing results from different sources. The issues relating to trace element bioavailability can be found in Chapter 11.

Hou et al. [46] determined soil Se fractions based on their association with other soil components, that is Al-Se extracted by 0.5 N NH_4F, Fe-Se by 0.1 N NaOH, Ca-Se by 0.5 N H_2SO_4, occluded Se by dithionite–citrate–bicarbonate buffering solution, and residual Se (the difference between the total Se determined by digestion method and the sum of the above extractable fractions). This fractionation gives a good consideration of Se transformations in soil but fails to relate the fraction(s) to soil Se availability to plants. He et al. proposed a method for soil Se fractionation based on plant availability [23]. In this procedure, soil Se is fractionated into readily available Se (extracted by 0.5 M $NaHCO_3$ pH 8.5), potentially available Se (extracted by 0.1 N NaOH + 0.1 N $Na_4P_2O_7$, followed by 0.2 M $(NH_4)_2C_2O_4$, pH 3.2), occluded Se (extracted by dithionite–citrate–bicarbonate buffering solution), and residual Se. The advantages of this fractionation include: (i) it relates Se fractions to its plant availability, and (ii) it gives consideration to organic Se (extracted by 0.1 N $Na_4P_2O_7$), which may contribute to available Se pools in soils. Recently, a modified procedure has been applied for Se fractionation in seleniferous or Se-contaminated soils, in which NaCl, Na_2SO_4, and Na_2CO_3 were used to extract exchangeable Se, Na_3PO_4 to extract exchangeable and strongly bound fractions, and H_2O_2 and $KMnO_4$ to estimate organic Se [47]. An overview of sequential fractionation schemes, their applications, relevance, and limitations can be found in Chapter 11.

As discussed above, for both Mn and Se, the major soil factors that control their bioavailability and mobility are redox potential, pH, clay minerals and oxides, and organic matter. Redox potential controls the transformation and speciation of Mn and Se and subsequently the concentrations of Mn^{2+} and selenate/selenite in soil solution, thus determining their availability and mobility. Soil pH affects oxidation–reduction of Mn or Se and solubility of Mn or Se compounds in soils, as H^+ is a reactant in the redox reactions. In addition, pH affects surface charge characteristics (particularly for variable-charge soils) and speciation of Mn (Mn^{2+}, $MnOH^+$) or Se ($H_2SeO_4/HSeO_4^-/SeO_4^{2-}$, $H_2SeO_3/HSeO_3^-/SeO_3^{2-}$) ions, thus influencing adsorption-desorption of the involved ions of Mn or Se. Clay minerals and oxides control the distribution of water soluble Mn or Se between liquid and solid phase in soils through adsorption–desorption and surface precipitation processes, directly influencing plant uptake. Oxides of Fe, Al, Mn, and Ti can form inner-sphere surface complexes with Mn^{2+} or selenite anions, and the adsorbed species are very difficult to desorb, resulting in lower plant availability of Mn or Se. On the other hand, layer silicates, particularly 2:1 types, retain Mn^{2+} on exchange sites by electrostatic force, which are readily replaced by other cations, such as Ca^{2+} and Mg^{2+}. The effects of organic matter on the availability and mobility of Mn and Se are more complicated. Firstly, organic matter, particularly freshly added, can enhance the development of reducing conditions under flooding, as in paddy soils, thus affecting the transformation of Mn or Se. Secondly, humic substances, especially humate, can retain Mn^{2+} on their negatively charged functional groups such as R–COO-, R–O–, thus reducing its movement, whereas humic acids and fulvic acids can compete for adsorbing sites with selenite or selenate, thus increasing the availability of Se to plant. In addition, climate (temperature and rainfall) and hydrological conditions (soil moisture, water table) play an important role in the transformation, availability, and mobility of Se and Mn. They affect all the chemical and biological reactions, especially oxidation–reduction, adsorption-desorption, plant uptake, and methylation processes [12,42].

20.4 Effects on Plant, Animal and Human Health

The availability and mobility of both Mn and Se vary spatially and temporarily, as affected by soil type, hydrological conditions, pH, climate, and plant species. Manganese is essential to plants, but is required in small amounts (20–40 mg kg^{-1} dry weight (DW)). In well-oxidized neutral and alkaline soils (including calcareous soils), the availability and mobility of Mn are limited, and crop plants may suffer from Mn deficiency and leaching of Mn is minimal. In acidic soils, particularly those highly weathered soils such as Oxisols and Ultisols, Mn relatively accumulates as oxides and hydroxides under aerated conditions but quickly converts to Mn^{2+} up to excessive levels when subjected to flooding. For instance, rice plants on paddy soils derived from Ultisols or Oxisols frequently suffer from Mn^{2+} toxicity, particularly when a strongly reducing conditions develop after green manure or farm yard manure is applied [12]. In these paddy soils, vertical movement of Mn can be clearly seen in the soil profile, with Mn leached out from the submerged layers and deposited in the aerated layers due to water table fluctuation [12]. In contrast, the leaching of Se occurs mainly in well-oxidized neutral and alkaline soils where selenate is the predominant species and subjected to leaching loss due to its weak binding to soil surfaces, whereas under flooding and reducing conditions, elemental Se and selenide are the predominant Se forms, which are stable and subjected to minimal leaching.

Selenium is not required by plants, but is essential or beneficial to human beings and animals. Soil tests are mainly intended for indicating Se flux through the food chain and subsequent effects on human health. For instance, in China, Se deficiency-related diseases of humans such as Keshan and Kashin–Beck disease mainly occurred in places with plant Se concentration less than 0.1–0.15 mg kg^{-1}DW. Soil tests have been used for predicting Se plant availability on the basis of soil test index and Se uptake by plants. Similar to other trace elements, total Se in soils is not a good indicator of Se availability to plants. Efforts have been made to develop a reliable soil Se test, including single extraction and Se fractionation [23,46,48–50]. The use of 0.5 M NaHCO$_3$ (pH 8.5) was proposed by Hou et al. [49] for extracting soil available Se in soils. The extractable Se was reported to correlate highly with plant Se uptake by the authors [49]. This method is based on anion/ligand exchange of selenate and selenite by HCO$_3^-$, and most of selenate and part of selenite adsorbed on soil surfaces can be recovered by this method. However, this method may not be adequate for weathered acidic soils with high Fe and Mn oxides such as Oxisols due to its low extraction efficiency. For these soils, 0.1 M NaH$_2$PO$_4$ may be better because of its effectiveness in releasing the adsorbed selenite, which is potentially bioavailable [50]. Other proposed single-extraction procedures include 0.25 M KCl, AB-DTPA (ammonium bicarbonate–diethylenetriamine pentaacetic acid), and hot water. For seleniferous soils, hot-water-soluble Se was observed to correlate well with plant Se uptake [51].

Manganese is involved in many biochemical and physiological processes in plants, and its deficiency can severely inhibit plant growth and health, whereas excessive Mn is toxic to plants. Manganese deficiency is relatively common in certain crops (oats, peas, sugar beet) grown on neutral and calcareous soils [10,12], but diagnosis and correction of the deficiency is not well defined. For sensitive crops like oats, peas, sugar cane, sugar beet, adequate Mn concentrations in plant range from 20 to 100 mg kg^{-1}, and deficiency can occur when Mn concentrations in plants are below 15–20 mg kg^{-1} [12]. Manganese deficiency can be corrected by soil and foliar application. Excessive Mn in plant (>1000 mg kg^{-1}) [12] causes toxicity problems, which mainly occur on acidic soils with pH around 5.5 or lower in combination with high soil Mn level. Manganese toxicity was also reported on poorly drained soils at higher pH [12]. Toxicity of Mn to rice seedlings occurs in acidic paddy soils derived from Oxisols and Ultisols in the tropical and subtropical regions [12].

Selenium is unevenly distributed on the surface of the earth, resulting in both seleniferous and Se-deficient geo-ecosystems. Selenium deficiency or toxicity is mainly related to parent material or geological background although in certain circumstances atmospheric, and more recently, anthropogenic inputs may influence their composition [7,10,12]. In natural environments, elevated concentrations of Se in soils are associated primarily with volcanic materials, sulfide ore bodies, black shale, and carbonaceous sandstones, and low-Se soils are often related to sandstones, limestone, loess, and some low-Se igneous rocks. Good examples include the low-Se belt from north-east to south-west of China (Figure 20.1), where the selenosis occurred in Enshi County, Hubei province, China and the San Joaquin Valley in California, USA. The low Se-belt in China covered 16 provinces with over 350 counties (Figure 20.1). The mean values of total Se concentrations in top soils

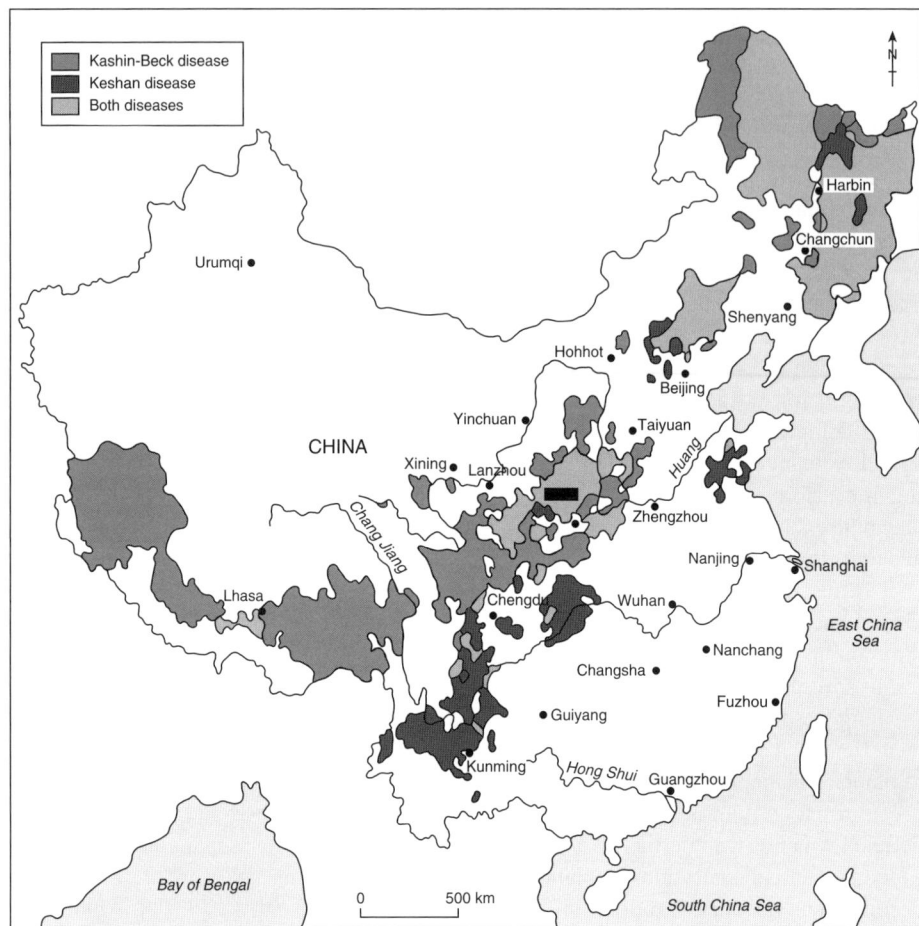

Figure 20.1 Distribution of Keshan disease and Kashin–Beck disease in China (by disease-affected counties) (Reprinted from Science of the Total Environment, 28, J.A. Tan et al., Selenium in soil and endemic diseases in China, 227–235. Copyright 2002, with permission from Elsevier.[8])

within this belt were below 0.125 mg kg^{-1}, compared with 0.239 mg kg^{-1}, the mean value of soils country-wide [8]. This belt is located in the temperate humid/subhumid regions in China.

Two major factors contributed to the formation of the low-Se belt: (i) low Se concentrations inherited from parent materials; and (ii) the biological and climatic conditions that promote the loss of Se from the soils [13]. The impact of Se deficiency on human health is evidenced by the occurrence of two endemic diseases in this belt during the 1960s and 1970s, one called Keshan disease, that is cardiomyopathy; and the other called Kashin–Beck disease resulting in deformed joints (becoming thickened, shortened). After the discovery of the association of these diseases with Se deficiency, patients were treated with Se-enriched tablets or injections, Se-enriched foods or other Se supplements.

This practice has proved a great success in the prevention and treatment of these two diseases for people living in this low-Se belt in China [8]. In addition, agricultural practices such as soil and foliar application of Se, importation of food from outside this belt, and Se fortification of food products has also helped improve Se nutrition of the habitants [12,18].

For Se, the gap between deficiency and toxicity is narrow. Selenium toxicity was reported to occur in Enshi of Hubei province, China where the soil total Se is up to 3.8 mg kg^{-1}, with a mean value of 2.3 mg kg^{-1}. The symptoms of Se toxicity include the loss of hair and nails in humans and 'blind stagger' in animals. The solution to the Se toxicity problems includes remediation of the seleniferous soils by soil amendment or phytoremediation [52].

The high Se concentration in soils of the western San Joaquin Valley in California, USA was also inherited from parent materials, the Jurassic and Cretaceous sediments comprising sandstones, shales, and conglomerates that contain selenosulfides of Fe. The exposure of the sediments due to geological movement and the development of soils enhanced the oxidation of selenides to selenite and selenate, which were then leached into drainage or released into surface run-off or lost with soil erosion. Eventually the Se-enriched drainage water from the valley caused the contamination of a receiving water body, the Kesterson Reservoir. The death of Se-sensitive wildlife such as water fowls caused public concern over Se pollution of the environment [9].

In the last decade, many efforts have been made to remediate Se contaminated soils and water, including phytoremediation and microbial remediation [52–55]. Plants have the ability to absorb and sequester Se and to convert inorganic Se to volatile forms of organic compounds that are harmlessly released into the atmosphere. Some plants such as ferns can accumulate Se over 1000 mg kg^{-1} in the root and fronds without visible toxicity symptoms [55].

References

[1] Rosen, B.P.; Liu, Z.J., Transport pathways for arsenic and selenium: a mini review; Environ. Int. 2009, 35, 512–515.

[2] USGS, Minerals Yearbook Selenium and Tellurium (2005, 2006); United States Geological Survey, Washington DC, 2007; http://minerals.usgs.gov/ds/2005/140/#selenium (accessed 26 November 2009).

[3] Haygarth, P.M., Global importance and global cycling of selenium; in: W.T. Frankenberger, S. Benson (eds), Selenium in the Environment; Marcel Dekker, New York, 1994, pp. 1–28.

[4] Selenium(IV) sulfide. Pharmacy Codes. http://pharmacycode.com/Selenium(IV) sulfide.html. Retrieved 2009-01-06. Retrieved 2009-01-06.

[5] Kabir, M.Z.; Emelianova, E.V.; Arkhipov, V.I. et al., The effects of large signals on charge collection in radiation detectors: application to amorphous selenium detectors; J. Appl. Phys. 2006, 99.

[6] Hartikainen, H., Biogeochemistry of selenium and its impact on food chain quality and human health; J. Trace Elem. Med. Biol. 2005, 18, 309–318.

[7] Tinggi, U., Selenium toxicity and its adverse health effects; Rev. Food Nutr. Tox. 2005, 4, 29–55.

[8] Tan, J.A.; Zhu, W.Y; Wang, W.Y. et al., Selenium in soil and endemic diseases in China; Sci. Total Environ. 2002, 28, 227–235.

[9] Presser, T.S., Geologic origin and pathways of selenium from the California Coast Ranges of the West-Central San Joaquin Valley; in: W.T. Frankenberger , S. Benson (eds), Selenium in the Environment; Marcel Dekker, New York, 1994.

[10] Kabata-Pendias, A. (ed.), Trace Elements in Soils and Plants, 3rd edn; CRC Press; Boca Raton, FL, 2001.
[11] Smith, K.A; Paterson, J.E., Manganese and cobalt; in: B. J. Alloway (ed.), Heavy Metals in Soils; Blackie Academic & Professional, London, 1995, 224–244.
[12] He, Z.L. (ed.), Chemical Equilibrium of Pollution and Beneficial Elements in Soil; China Environmental Science Press, Beijing, 1998 [In Chinese].
[13] Norvell, W.A., Inorganic reactions of manganese in soils; in: R.D. Graham. R.J. Hannam, N.C. Uren (eds); Manganese in Soils and Plants; Kluwer Academic Publishers; Dordrecht, 1988, 37–58.
[14] Reisenauer, H.M., Determination of plant-available manganese; in: R.D. Graham. R.J. Hannam, N.C. Uren (eds); Manganese in Soils and Plants; Kluwer Academic Publishers; Dordrecht, 1988, 87–100.
[15] Xia, W.P.; Tan, J.A., A comparative study of selenium content in Chinese rocks; Acta Sci. Circum. 1990, 10, 125–131 [in Chinese with English abstract].
[16] Neal, R.H., Selenium; in: B. J. Alloway (ed.), Heavy Metals in Soils; Blackie Academic & Professional, London, 1995, 260–283.
[17] Gissel-Nielsen, G.; Gupta, U.C.; Lamand, M. et al., Selenium in soils and plants and its importance in livestock and human nutrition; Adv. Agron. 1984, 37, 397–459.
[18] Tan, J.A.; Zhang, D.X.; Hou, S.F. et al., Selenium ecological geochemicogeography and endemic Keshan disease and Kashin–Beck disease in China; in: G. F. Combs, J.E. Spallholz, O.A. Levander, et al. (eds), Selenium in Biology and Medicine; Van Nostrrand Reinhold, New York, 1987, pp. 839–876.
[19] Luo, K.L.; Xu, L.R.; Tan, J.A. et al., Selenium source in the selenosis area of the Daba region, South Qinling Mountain, China; Environ. Geol. 2004, 45, 426–432.
[20] Gamble, D.S.; Langford, C.H.; Tong. J.P.K., The structure and equilibrium of a manganese II. Complex of fulvic acid studies by ion exchange and nuclear magnetic resonance; Can. J. Chem. 1976, 54, 1239–1245.
[21] Abrams, M.M.; Burau, R.G.; Zasoski, R.J., Organic selenium distribution in selected California soils; Soil Sci. Soc. Am. J. 1990, 54, 979–982.
[22] He, Z.L.; Yang, X.E.; Zhu, J. et al., Organic selenium and its distribution in soils; Acta Sci. Circum. 1993, 13, 281–287 [in Chinese with English abstract].
[23] He, Z.L; Yang, X.E; Zhu, Z.X. et al., Fractionation of soil selenium with relation to Se availability to plants; Pedosphere 1994, 4, 209–216.
[24] Séby, F.; Potin-Gautier, M.; Giffaut, E. et al., A critical review of thermodynamic data for selenium species at 25 °C; Chem. Geol. 2001, 171, 173–194.
[25] Wu, Z.H.; Gu, Z.M.; Wang, X.Y. et al., Effects of organic acids on desorption of lead onto montmorillonite, goethite and humic acid; Environ. Pollut. 2003, 121, 469–475.
[26] Willett, I.R.; Bond, W.J., Sorption of manganese, uranium, and radium by highly weathered soils; J. Environ. Qual. 1995, 24, 834–845.
[27] Saeki, K., Adsorption of Fe^{2+} and Mn^{2+} on silica, gibbsite, and humic acids; Soil Sci. 2004, 169, 832–840.
[28] Basta, N.T.; Ryan, J.A.; Chaney, R.L., Trace element chemistry in residual treated soil: key concepts and metal bioavailability; J. Environ. Qual. 2005, 34, 49–63.
[29] Mohan, D.; Chander, S., Single, binary, and multicomponent sorption of iron and manganese on lignite; J. Colloid Interface Sci. 2006, 299, 76–87.
[30] Bradl, H.B., Adsorption of heavy metal ions on soils and soils constituents; J. Colloid Interface Sci. 2004, 277, 1–18.
[31] Davies, S.H.R.; Morgan, J.J., Manganese (II) oxidation kinetics on metal oxide surfaces; J. Colloid Interface Sci. 1989, 129, 63–77.
[32] Sparrow, L.A.; Uren, N.C., Oxidation and reduction of Mn in acidic soils: effect of temperature and soil pH; Soil Biol. Biochem. 1987, 9, 143–148.
[33] Peak, D.; Sparks, D.L., Mechanisms of selenate adsorption on iron oxides and hydroxides; Environ. Sci. Technol. 2002, 36, 1460–1466.
[34] He, Z.L.; Zhang, M.K.; Huang, C.Y. et al., Studies on the status and availability of selenium in soils in Shenxan, Zhejiang Province; Acta Ecol. Sin. 1994, 14, 51–56 [in Chinese with English abstract].

[35] Tan, J.A.; Wang, W.Y.; Wang, D.C. et al., Adsorption, volatilization, and speciation of selenium in different types of soils in China; in: W.T. Frankenberger , S. Benson (eds), Selenium in the Environment; Marcel Dekker, New York, 1994, pp. 47–68.
[36] He, Z.L.; Yang, X.E.; Zhu, Z.X. et al., Effect of phosphate on the sorption, desorption and plant-availability of selenium in soil; Fert. Res. 1994, 39, 189–197.
[37] Ogaard, A.F.; Sogn, T.A.; Eich-Greatorex. S., Effect of cattle manure on selenate and selenite retention in soil; Nutr. Cycling Agroecosyt. 2006, 76, 39–48.
[38] Liu, Q.; Wang, D.J.; Jiang, X.J., Effects of the interactions between selenium and phosphorus on the growth and selenium accumulation in rice (*Oryza sativa*); Environ. Geochem. Health 2004, 26, 325–330.
[39] Nakamaru, Y.; Tagami, K.; Uchida, S., Effect of phosphate addition on the sorption-desorption reaction of selenium in Japanese agricultural soils; Chemosphere 2006, 63, 109–115.
[40] Dhillon K.S.; Dhillon, S.K., Adsorption–desorption reactions of selenium in some soils of India; Geoderma 1999, 93, 19–31.
[41] Ryden, J.C.; Syers, J.K.; Tielman, R.W., Inorganic anion sorption and interaction with phosphate sorption by hydrous ferric oxide gel; J. Soil Sci. 1987, 38, 211–217.
[42] Bar-Yosef, B.; Meek, D., Selenium sorption by kaolinite and montmorillonite. Soil Sci. 1987, 144, 11–19.
[43] Yasuo N.; Keiko, T.; Shigeo, U., Effect of phosphate addition on the sorption–desorption reaction of selenium in Japanese agricultural soils; Chemosphere 2006, 63, 109–115.
[44] Neal, R.H.; Sposito, G.; Holtzclaw, K.M. et al., Selenite adsorption on alluvial soils: I. Soil composition and pH effects; Soil Sci. Soc. Am. J., 1987, 51, 1161–1165.
[45] Bartlett, R.J., Manganese redox reactions and organic interactions in soils; in: R.D. Graham. R.J. Hannam, N.C. Uren (eds); Manganese in Soils and Plants; Kluwer Academic Publishers; Dordrecht, 1988.
[46] Hou, S.F.; Li, D.Z.; Wang, L.Z., Selenium fractionation and its distribution in Chinese soil; Geog. Res. 1990, 9, 17–24 [in Chinese].
[47] Lim, T.T.; Goh, K.H., Selenium extractability from a contaminated fine soil fraction: implication on soil cleanup; Chemosphere 2005, 58, 91–101.
[48] Chao, T.T.; Sanzolone, R.F., Fractionation of soil selenium by sequential partial dissolution; Soil Sci. Soc. Am. J. 1989, 53, 389–392.
[49] Hou, J.N.; Li, J.Y., Study on the available Se in soils and its relation to the soil properties; in: Proceedings of International Symposium on Environmental Life Elements and Health; China Environmental Science Press, Beijing, 1988, pp. 104–105.
[50] Alam, M.G.M.; Tokunaga, S.; Maekawa, T., Extraction of selenium from a contaminated forest soil using phosphate; Environ. Technol. 2000, 21, 1371–1378.
[51] Dhillon, K.S.; Rani, N.; Dhillon, S.K., Evaluation of different extractants for the estimation of bioavailable selenium in seleniferous soils of Northwest India; Aust. J. Soil Res. 2005, 43, 639–645.
[52] Banuelos, G.S.; Lin, Z.Q.; Wu, L. et al., Phytoremediation of selenium-contaminated soils and waters: fundamentals and future prospects; Rev. Environ. Health 2002, 17, 291–306.
[53] Azaizeh, H.A.; Salhani, N.; Sebesvari, Z. et al., The potential of rhizosphere microbes isolated from a constructed wetland to biomethylate selenium; J. Environ. Qual. 2003, 32, 55–62.
[54] Dungan, R.S.; Yates, S.R.; Frankenberger, W.T., Transformations of selenate and selenite by *Stenotrophomonas maltophilia* isolated from a seleniferous agricultural drainage pond sediment; Environ. Microbiol. 2003, 5, 287–295.
[55] Srivastava, M.; Ma, L.Q.; Cotruvo, J.A., Uptake and distribution of selenium in different fern species; Int. J. Phytoremed. 2005, 7, 33–42.

21
Tin and Mercury

Martin J. Clifford[1], Gavin M. Hilson[1] and Mark E. Hodson[2]

[1]*School of Agriculture, Policy and Development, University of Reading, Reading, UK*
[2]*Department of Soil Science, University of Reading, Reading, UK*

21.1 Introduction

Tin (Sn) occurs in Group 14 of the periodic table, together with carbon, silicon, germanium, and lead. Typically it occurs in two oxidation states in the natural environment, +2 and +4, both of which are stable. There are 10 stable isotopes of tin, the most of any element (Table 21.1) plus another 28 unstable isotopes with half-lives ranging from nanoseconds to years.

Mercury (Hg) occurs in Group 12 of the periodic table, together with zinc and cadmium. Typically it occurs in three states in the natural environment: native metallic mercury, and mercuric (Hg^{2+}) and mercurous (Hg_2^{2+}) ions. It has seven stable isotopes (Table 21.1) plus another 56 unstable isotopes with half-lives ranging from nanoseconds to years. Mercury is the only metallic element that occurs as a liquid at standard temperatures and pressures–hence the derivation of its chemical symbol from the Latinized Greek *hydrargyrum* meaning 'liquid silver'. In addition it has a high vapour pressure.

Tin has been used in metalwork for at least 5000 years, initially as a component of bronze due to its hardening effect on copper. In recent times, tin was used predominantly to provide a thin rust-proof coating to steel and iron goods. Tin is still used for this purpose, though synthetic organic compounds are more often used instead. Tin hit the environmental pollution headlines through the use of organotins as antifouling paints in the

Table 21.1 Stable isotopes of Sn and Hg

Tin (Sn)									
Mass number	112	114	115	116	117	118	119	120	122	124
Relative molar abundance	0.97	0.66	0.34	14.54	7.68	24.22	8.59	32.58	4.63	5.79

Mercury (Hg)						
Mass number	196	198	199	200	201	202	204
Relative molar abundance	0.15	9.97	16.87	23.10	13.18	29.86	6.87

Data from [1].

aquatic environment. There is limited use of organotins in the terrestrial environment in PVC as a stabilizer and as a biocide [2].

Mercury has a long history of usage [3,4], being used by the Egyptians around 600 BC for amalgamation and by Chinese and Indian cultures medicinally. In the Roman era it was used to produce the pigment vermillion. It is alleged that mercury usage in the hat manufacture trade and its toxicological affects gave rise to the phrase 'mad as a hatter', though this is the subject of some debate [5]. In the modern era, the principle uses of mercury remain its use in the recovery of gold and silver from their respective ores [6–11]; its use in devices for measuring temperature and pressure, which rely upon the constant rate of expansion of mercury in response to changes in temperature and pressure [12]; its use in dentistry to form amalgams for fillings [4,13]; its use as a pesticide and fungicide [3,14]; the production of disposable batteries of various sizes; and within chloro-akali cells used in the production of caustic soda [10,15,16]. In the developing world, supply and usage of mercury have decreased as environmental concerns have increased (for example Figure 21.1, Table 21.2). In the present-day context, the largest proportion of current and projected global demand can be attributed to use in areas of the globe that are undergoing rapid economic and industrial development, and/or where regulation governing applications, contamination and emissions are less stringent [15–17].

Within soils there is a relatively small dataset concerning inorganic tin and it appears to be an element that has limited environmental significance. A small amount of data is available on the occurrence of organotin compounds in soil. At present there is no suggestion that levels are sufficiently high to cause toxicity.

There is a large literature on the occurrence and impact of mercury in the environment and, more specifically, in soils. Use of mercury by humans is primarily responsible for the movement of mercury through the ecosystem. Mercury is naturally present in trace amounts in many rocks, but the dominant natural form of mercury, cinnabar (HgS) is chemically resistant. However, following extraction and use of mercury, it escapes to the atmosphere due to its volatile nature, resulting in subsequent pollution via wet and dry deposition. The widespread distribution of mercury due to anthropogenic activities has had a significant impact on the global biogeochemical cycle of mercury [18].

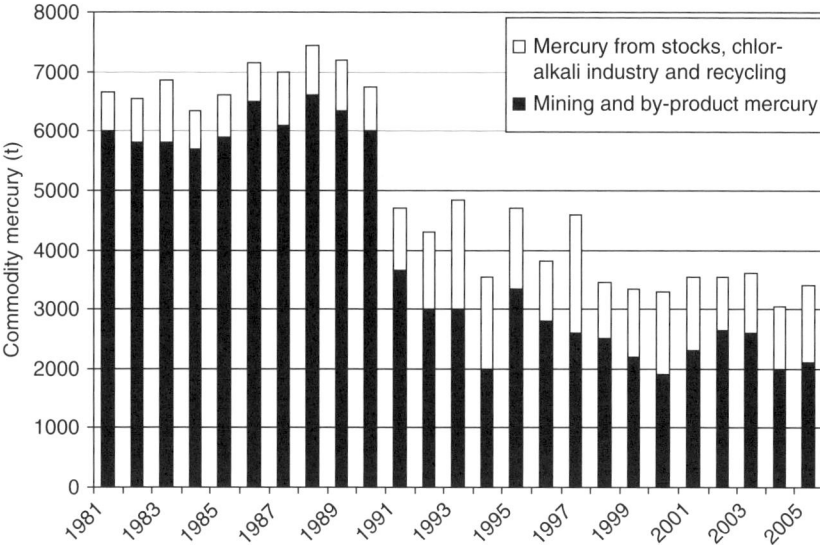

Figure 21.1 Global mercury supply 1981–2005.

Table 21.2 Global Hg demand in metric tonnes, calculated for 2005, and predicted for 2015 under two scenarios: a 'status quo' relating to current conditions, and secondly to demand under a 'focused reduction' in Hg use

Global mercury demand, by sector	2005	2015	
		'Status quo' scenario	'Focused Hg reduction' scenario
Small-scale/artisanal mining	650–1000	650	400
Vinyl chloride monomer (VCM) production	600–800	1000	1000
Chloro-akali production	450–550	350	250
Batteries	300–600	200	100
Dental use	240–300	270	230
Measuring and control devices	150–350	125	100
Lighting	100–150	125	100
Electrical and electronic devices	150–350	110	90
Other (paints, laboratory, pharmaceutical, etc.)	30–60	40	30
Total	3000–3900	2870	2300

Data from [15].

Tin is a nonessential element to both animals and humans. Tin and its inorganic compounds are thought to be nontoxic or show low toxicity but organic tin compounds are highly toxic [19].

Mercury is again a nonessential element and is reported to be one of the most toxic elements to humans and many higher animals. Inorganic and organic compounds of mercury are particularly toxic compared with elemental mercury. The degree of uptake and accumulation of mercury by plants is debated in the literature (for example [20–22]); variations are undoubtedly due to issues relating to bioavailability versus bulk soil concentration. Transfer of mercury up the food chain seems to occur predominantly through the concentration of organic mercury compounds in aquatic animals.

21.2 Geogenic Occurrence

21.2.1 Tin

Tin occurs at trace levels in a variety of minerals and in higher concentrations in certain ore minerals. The most common tin-bearing mineral is cassiterite (SnO_2). A variety of tin sulfide minerals also exist, for example stannite (Cu_2SnFeS_4) and montesite ($PbSn_4S_5$) [23]. The lack of tin-bearing silicate phases, and the low elemental abundance of tin results in an average crustal composition of about $\sim$2 mg kg^{-1}. Average concentrations of tin in a variety of rock types are given in Table 21.3. The relatively high concentrations of tin in argillaceous sediments reflect the siderophilic nature of the element.

The low concentration of tin in rocks is also reflected in the low concentrations in uncontaminated soils. There are not a great number of reported soil tin analyses in the literature and these are summarized in (Table 21.4).

Table 21.4 comprises data for mineral soils. In peat soils tin tends to have a higher concentration due to the strong association between tin and humic substances; tin concentrations of 50–300 mg kg^{-1} are reported in peats [29] that are not obviously 'contaminated' but rather appear to have accumulated tin deposited on them via atmospheric deposition that, in mineral soils, would not have been retained.

Table 21.3 Concentrations of tin in various rock types [24]

Rock type	Tin concentration (mg kg^{-1})
Magmatic rocks	
Dunites, periodotites, pyroxenites	0.35–0.50
Basalts, gabbros	0.9–1.5
Diorites, syenites	1.3–1.5
Granites, gneisses	1.5–3.6
Rhyolites, trachytes, dacites	2–3
Sedimentary rocks	
Argillaceous sediments	6–10
Shales	6
Sandstones	0.5
Limestones, dolomites	0.5

Data from [24].

Table 21.4 Tin content of surface soils

Range of tin concentration (mg kg^{-1})	Sample information	Reference
<0.1–7.9	A wide range of US soil types	[25]
1.1–4.6	Podzolic soils developed on sulfide deposits and granites	[26]
1.5–7.1	Four Scottish soils developed on sandstone, granite and basic igneous rocks	[27]
1.7–7.7	A range of soil certified reference materials	[28]

21.2.2 Mercury

Mercury is typically found in very small concentrations in any given rock type–no more than 0.05 mg kg^{-1} on average (Table 21.5), with the main mercury-bearing mineral being cinnabar (HgS) though over 20 mercury minerals are known [23]. Sizeable deposits do exist in tectonic fault regions that have permitted large-scale extraction of the element at sites in the United Sates, Italy, Slovenia, Spain, Kyrgyzstan, China and Algeria [12,15], but many of these sites are now inactive due to decreased demand and usage. Mercury vapour is naturally released into the environment in four different scenarios [30]: (i) areas of large heat outputs from the crust, that is surface volcanic and geothermal activity such as deep-sea trenches; (ii) places where faults and fractures permit the infiltration of elemental Hg vapour from deep within the crust to the surface rock, soil or air; (iii) seismically active regions, where the Hg vapour may be released from underlying rocks by the large mechanical forces in action; and (iv) zones rich in mineral and/or fossil fuel deposits.

Table 21.5 Concentrations of mercury (mg kg^{-1}) in a range of rocks and soils

Rock or soil type	Range of concentrations (mg kg^{-1})	Mean concentration (mg kg^{-1})
Crust/lithosphere	0.05–0.08	
Granite		0.08
Basalt	0.01–0.12	
Shale	0.18–0.50	
Limestone	0.05–0.16	
Sandstone	0.03–0.29	
Coal	0.01–0.021	
Podzols and sandy soils	0.01–0.95	0.06
Desert sands	0.008–0.32	0.06
Loamy soils	0.01–0.78	0.13
Soils developed on glacial tills	0.02–0.36	0.06
Forest soils	0.02–0.58	0.14
Paddy soils	0.02–1.00	0.30
Volcanic soils	0.03–0.10	0.08

Data from [3,4,24].

Links between high ambient Hg levels and mineralization have been made in numerous studies, and the association has formed the basis of geoprospecting techniques to locate subsurface deposits of precious resources such as Ag, Au, Pb and Zn, as well as oil and gas [30]. Predictions of the amount of Hg released from natural degassing are in the region of 18×10^6 kg per year [31,32].

As a result of extraction and use of mercury, establishing ambient or background levels in various soils is extremely problematic if not impossible. Human activity has altered levels of the element throughout global soil profiles to the extent that it is unrealistic to make reliable claims about its levels in 'uncontaminated' soils [33]. Work attempting to assess natural levels within soils relies upon data from proxy sources such as readings from extremely remote areas [34]. The variation between different soil types is much larger than in rock types, which is to be expected considering the greater number of environmental and chemical interactions occurring at this level of the ground. *Upper* limits to background Hg concentrations in most world soils are suggested to be in the range of 0.15–0.2 mg kg^{-1} [34]. These levels are dependent upon the tectonic and mineralogical features of the area, but also on the physicochemical characteristics of the soil [24,35,36]. Mercury shows a strong affinity for organic material in soils; this goes in some way to explain the higher levels of the element found in organic-rich, forest and peaty soils, as well as in areas used for paddy cultivation, which can be between 1 and 2 mg kg^{-1} (Table 21.5).

With the exception of very specific localities, it can be concluded that Hg and minerals bearing it are not found in great natural abundance in rock or soil types [35,36]. Elevated levels of the element in soil can therefore in most cases be identified with a certain degree of reliability to be of anthropogenic origin.

21.3 Sources of Soil Contamination

21.3.1 Tin

Within the restricted literature on tin contamination in soils the dominant source of recorded contamination is from smelters. Although it is not explicitly stated, it seems likely that such contamination occurs both in the form of both dry deposition (for example dust) and also wet deposition (dissolved material in rainfall). There are also reports of waste sludges that are spread on land as sources of Sn. While bronze statuary may represent local point sources for Sn contamination, a study indicated little release of Sn from the weathering of bronze in the urban environment [37]. Tin release over 2 years was below the detection limit (0.001 mg Sn l^{-1}). For organotin compounds, leakage from landfill sites, atmospheric deposition and biological activity are identified as sources. Fluxes of tin into the atmosphere from metal production, organotin formation and combustion processes were estimated to lie in the range 10–60×10^6 kg y^{-1} in 1982, assuming volatilization efficiency to be $\sim 10\,\%$ [38].

The majority of studies that specifically mention Sn concentrations in soil due to contamination focus on smelter-related pollution. Sterckman *et al.* [39] report concentrations of 277–884 mg kg^{-1} tin in smelter dust and 793 mg kg^{-1} in slag from the smelter. Tin concentrations in alluvial and loessic agricultural soils local to two smelters are recorded in the range 2.02–7.26 mg kg^{-1} compared with 1.36–3.20 mg kg^{-1} at control sites. The Sn concentrations decreased with distance from the smelter. Chatterjee and Banerjee [40] record Sn levels of 19–34 mg kg^{-1} in the vicinity of a lead factory compared to 0.64 mg kg^{-1} at control sites. Variation in Sn concentration was related to proximity to the factory and the direction of prevailing winds. Rawlins *et al.* [41] carried out a regional survey across soils developed on coal-rich bedrock and recorded Sn levels in topsoils from rural and peri-urban areas in the range 1–161 mg kg^{-1}. Higher Sn concentrations were suggested to be due to a combination of the local custom of spreading coal ash on soil either as a soil amendment or a waste disposal method, atmospheric deposition following coal combustion, the presence of tin in sewage waste spread on land and vehicular exhaust emissions. Cecchi *et al.* [42] measured Sn concentrations down the profile of a silt-rich, neutral-pH soil in the vicinity of a lead-recycling plant that had been in operation for 40 years. Tin concentrations of up to 37 mg kg^{-1} were recorded at the soil surface, but these had dropped to back ground levels of 1–2 mg kg^{-1} by a depth of 40 cm. Douay *et al.* [43] report Sn concentrations of 2.6–18.6 mg kg^{-1} in urban loamy, circum-neutral topsoils in the vicinity of two lead and zinc smelters. Concentrations were elevated by up to a factor of 3 compared with control sites. Rawlins *et al.* [44] recorded Sn concentration of 7.2 ± 16.6 mg kg^{-1} in topsoils and 4.8 ± 4.8 mg kg^{-1} in subsoils in the vicinity of a smelter, with concentrations being related to the prevailing wind direction.

McBride *et al.* [45] report concentrations of Sn in a sewage sludge of 95 mg kg^{-1}. Fifteen years after application of the sludge to a near neutral silty clay loam, the top soil contained ∼56 mg kg^{-1} Sn compared with 1.9 mg kg^{-1} at a control site. Average soil water concentrations, sampled 16–18 years after the sludge amendment were 0.09 μg Sn l^{-1}.

Huang *et al.* [46] recorded organotin compounds in forest soils, thought to have been derived from volatilization from marine sources and deposited via precipitation and fog. The compounds were concentrated in the upper layers of the soil horizon and generally concentrations were <10 μg Sn kg^{-1}. The exception to this was monobutyltin which reached concentrations of up to 30 μg Sn kg^{-1} in organic horizons. They went on to determine degradation rate constants for the organotin compounds of 0.05–1.54 pg Sn g^{-1} y^{-1} and determined that the rates decreased in the order mono ≥ di- ≥ tri-substituted organotin [47]. Pinel-Raffaitin *et al.* [48] recorded concentrations of 0.3–7.8 μg Sn l^{-1} due to organotin compounds in leachates from landfills, with the organotin compounds accounting for up to 38 % of the total Sn concentration. Biogas emitted from the landfills also contained organotin compounds, up to 25 μg Sn m^{-3}. Mersiowsky *et al.* [2] determined organotin concentrations of < 0.1–4 μg compound l^{-1} in landfill leachates and specifically identified methyl-, butyl- and octyltin. Duester *et al.* [49] recorded organotin compounds in soils from agricultural, garden and former industrial sites. Concentrations of up to 65 μg compound kg^{-1} were recorded with concentrations being higher in top soils, particularly if organic rich, and in the agricultural and garden soils due, it was suggested, to the higher levels of biological activity compared to the former industrial site soils.

21.3.2 Mercury

Mercury emitted into the atmosphere in vaporous or more solid forms becomes incorporated into soil via dry and wet deposition predominantly from fly ash and rainwater [13,30,31,34,50]. It is not solely direct uses of Hg in industrial activities such as gold extraction and caustic soda production that have contributed to elevated soil and atmospheric levels; just as important is Hg emitted as a by-product from other sources associated with human, industrial and economic development [10,13,14,34]. As Table 21.6 shows, the primary origins of air-borne Hg are combustion and industrial processes, with the manufacture of cement and metals, and the incineration of domestic waste being notable contributors alongside gold extraction and caustic soda production. Pacyna *et al.* [10] suggest that as much as two-thirds of anthropogenic Hg emissions can be attributed to combustion of fossil fuels for domestic and industrial energy. Total emissions for the year 2000 were estimated to be around 2190 Mg.

Table 21.6 shows that output from the Asian subcontinent is estimated to be over ten times larger than for Europe. It is probable that this trend can be attributed to the rapid industrial expansion occurring in many Asian countries, in particular China, which was identified as the worst polluter, emitting around 605 Mg of Hg [10]. It is already clear that the exponential development that has characterized this region of the globe in the last two decades has had significant implications for local and global emissions of atmospheric Hg [51]. The disparity is no doubt accentuated by the fact that the introduction of much stricter environmental regulations and cleaner forms of technology in potentially polluting industries, and an engagement with alternative energy sources in Europe [10,15,16] has

Table 21.6 Atmospheric Hg emissions (in Mg) attributed to anthropogenic activity in 2000

	Africa	Asia (excluding Russia)	Australasia	Europe (excluding Russia)	Russia	South America	North America	Total
Stationary combustion	205.2	878.7	112.6	88.8	26.5	31.0	79.6	1422.4
Cement production	5.3	89.9	0.8	26.5	3.7	6.5	7.7	148.6
Nonferrous metal production	7.9	87.6	4.4	10.0	6.9	25.4	6.4	148.6
Pig iron and steel production	0.4	11.6	0.3	10.6	2.7	1.4	4.3	31.3
Caustic soda production	0.3	30.7	0.7	12.4	8.0	5.0	8.0	65.1
Mercury production	0.1	0.1				22.8	0.	12.
Gold production	177.8	47.2	7.7			3.1	12.2	66.4
Waste disposal		21.6	0.1	11.5	3.5		18.7	66.4
Other	1.4	0.9		15.3	18.2		8.8	44.6
Total	398.4	1179.3	126.6	175.1	72.6	92.1	145.8	2189.9

Reprinted from Atmospheric Environment, 40, E.G. Pacyna, J.M. Pacyna, F. Steenhuisen and S. Wilson, Global anthropogenic mercury emission inventory for 2000, 4048–4063. Copyright 2006, with permission from Elsevier [10].

Table 21.7 Concentration range for Hg in materials added to land

	Sewage sludge	Composted refuse	Farmyard manure	Phosphate fertilizers	Nitrate fertilizers	Lime
Concentration (mg kg^{-1})	0.1–55	0.09–21	0.01–0.36	0.01–2.0	0.3–2.9	0.05

Data from [34].

coincided with a relocation of many forms of 'heavier' industry into these emerging economies.

Weathering, fluvial dispersion, and leaching from (often very extensive) metalliferous mining spoil heaps and tailings can be sources of localized Hg contamination [34,52,53]. Although current and disused sites of larger-scale mining have received a greater amount of attention with regard to potential pollution [54], the widespread smaller-scale artisanal gold mining activities in Africa, Asia and Latin America have more recently come to light as causes of great concern [7–9,11]. Practices vary between localities, but artisanal gold mining is noted for both a heavy reliance upon Hg in extraction of gold and the hazardous way in which it is used. In some cases, elemental Hg is poured directly onto filtering sluice boxes or into working pits [55]. The extremely unregulated nature of the sector makes estimation of the levels involved extremely hard, but one author puts the amount at as much as 900 Mg in 2003 [56], from which it has been suggested that around 66–75 % of the used Hg enters the soil [8], with the rest entering the atmosphere with of course the potential to be deposited locally or farther afield.

The application of fertilizing and pest-controlling chemicals, sewage sludge and manure composts onto agricultural land is another significant Hg input [34]. Some typical Hg ranges in agricultural substances are shown in Table 21.7.

21.4 Chemical Behaviour in Soils

21.4.1 Tin

The most important tin mineral, cassiterite is strongly resistant to mineral weathering and accumulates in soils developed from cassiterite-bearing parent material. Inorganic tin is similarly relatively immobile and tends to accumulate in soils, often being found at high concentrations in Fe- and Al- rich heavily weathered soils. Indeed, the mobility of tin in soils is rather similar to that of Al and Fe and is governed primarily by pH and E_h but also organic matter content as tin is strongly sorbed by organic matter. The tendency for tin to complex with organics, as well as enrichment due to aerial deposition, explains why tin is enriched in top soils relative to subsoils (for example [42,46]) and shows high concentrations in peats (for example [29]). In solution Sn^{2+}, which is a strong reducing agent, is found only under low-pH, low-E_h conditions [23].

21.4.2 Mercury

The main mercury mineral, cinnabar, is highly insoluble; together with the strong tendency to sorb to soil components, this results in mercury being relatively immobile in soils [57,58]. The chemical behaviour of mercury in soils is dominated by pH, E_h and organic matter and is well described by a number of authors (for example [59–61]). At acidic pH, and E_h greater than 0.4 V, Hg^{2+} is the stable form and, when in solution, the inorganic form typically occurs as the $HgCl_2^0$ complex. Under alkaline conditions $Hg(OH)^0$ is the dominant complex. Under reducing conditions Hg^0 is the stable oxidation state and, in the presence of sulfide ions precipitates as HgS. Mercury also forms strong complexes with dissolved organic matter [62,63], for example $(CH_3)_2Hg$, CH_3Hg^+ and CH_3HgS^- as well as poorly characterized humate and fluvate complexes and these tend to dominate in soils of intermediate E_h. Despite the theoretical stability of these complexes, in reality very little mercury occurs in solution due to the strong sorption of mercury to soil components. Mercury–chloride complexes are only weakly sorbed by soil components [64] whereas Hg^{2+} is strongly sorbed by organic matter [65]. Under acid conditions, sorption of Hg^{2+} by organic components is the dominant sorption mechanism. Under neutral and slightly alkaline conditions, clay minerals and iron oxides become more effective sorbents and mercury–hydroxide complexes are the dominant sorbed species [59]. The conversion of inorganic mercury to organic forms and vice versa appears to occur both through the action of organisms (particularly sulfate-reducing bacteria) and also as an inorganic process but is still not fully understood (for example [66–70]). It is an important process since organic forms of mercury may persist for longer in the aquatic and terrestrial environment than elemental Hg, which is lost by volatilization; also the organic forms are generally more mobile in the aquatic and terrestrial environment and more toxic than the inorganic forms.

Native mercury has a very high vapour pressure and this causes the metal to volatilize at temperatures and pressures typically found at the earth's surface. Additionally, some mercury–organic complexes such as dimethylmercury are also volatile. Mercury loss from soils due to volatilization increases with increasing concentrations of free mercury and mercury–organic complexes in solution, increased bacterial activity and rising temperature [50].

21.5 Risks from Tin and Mercury in Soils

21.5.1 Tin

Due to its low concentrations in soils and plants and low mobility, risks from inorganic tin are not considered significant in the literature. Typically contamination due to inorganic tin is due to atmospheric deposition and is accompanied by far higher concentrations of metals such as zinc and lead, which are more likely to have a toxic effect. In the studies referenced above citing levels of tin due to pollution from smelters, no reports of plant toxicity were included.

Uptake of tin from soils by plants is relatively low. Tin concentrations reported in plants lie in the range $<0.04–30$ mg Sn kg^{-1} [40,71–76] for plants growing in soils with background tin concentrations. Elevated plant concentrations, due to natural mineralization or contamination, as high as 2000 mg Sn kg^{-1} are recorded [24,77], with mosses and sedges being particularly high accumulators.

Organotins are far more toxic than inorganic tin and there is evidence of enrichment of these compounds in soils due to atmospheric deposition [46]. However, to date there have been no studies investigating transfer of these compounds through the food chain due to uptake by plants or low trophic level animals exposed to organotin compounds via the soil. Aquatic toxicity data [19] record toxic effects on aquatic organisms at tributyltin concentrations of a few micrograms per litre and less. Thus, the organotins in landfill leachate [2,48] could potentially have an adverse impact on soil fauna, as could organotins recorded in forest soils [46] depending on partitioning between the solid and solution phase. However, detecting such effects in the complex mixed system that is soil poses significant challenges.

21.5.2 Mercury

Free Hg is extremely mobile within the soil–plant system; soluble and vaporous forms of Hg are readily taken in from surrounding air and soil by plants, and translocated from the roots and stomata to various organs [24,33,78]. Much work has drawn direct correlations between soil and plant Hg content, although levels of uptake and tolerance to toxicity are very variable between different species [24]. Seedling plants and lower-order species are more sensitive to elevated levels of Hg than mature and higher-order individuals, respectively, with organic and methylated mercury compounds universally found to be most damaging. Sublethal damage to vital respiratory and photosynthetic functions may start from exposure as low as 1 μg kg^{-1}, but more critical problems are noted to occur between 0.5–8.0 μg kg^{-1}, again depending upon species. Universally displayed symptoms of toxicity are the stunting of seedling and root development, and the inhibition of photosynthesis resulting in reduced biodiversity or yields in agricultural crops. Kabata-Pendias [24] gives the mean ranges of Hg values found in agricultural food crops growing in uncontaminated soils as 2.6–86 μg kg^{-1} for vegetables, 0.6–70 μg kg^{-1} for fruit, and 0.9–21 μg kg^{-1} for the grains of cereal. Certain crops growing in contaminated soil conditions have displayed considerably elevated levels of Hg; apples (0.04–1.32 mg kg^{-1}), carrots (0.5–0.8 mg kg^{-1}), lettuce (0.09–0.36 mg kg^{-1}) and especially edible mushrooms (33.6–200 mg kg^{-1}) are cited as being particularly efficient at Hg uptake.

Given its use as a fungicide, it is unsurprising that microbial functions are affected by both organic and inorganic Hg at very low levels of around 5 μg l^{-1} for inorganic compounds, and much lower for organic ones [79]. Mercury binds tightly to proteins in the cell wall of microorganisms, inhibiting their functioning; this same feature also means that the metal easily attaches to tissues in higher-order organisms, and allows its transfer through the food chain.

The predominant source of intake of Hg in animals, including humans, comes from diet [13,14,80], but Hg compounds may also be inhaled in vapour form, with uptake rates of inhaled compounds in the lungs being at least 80 % [13,17,30,31]. Ingested elemental Hg has an extremely low level of uptake by the body, with <0.01 % being taken up through the gastrointestinal tracts in humans, but this figure is much higher for mercuric salts and organic forms, particularly methylmercury; about 7–15 % and 95 % respectively [13,17,30,31]. Dermal absorption is only a significant entry route for unusually large exposures [17].

Once it has passed across the cellular membranes lining the lungs or gastrointestinal tract, all forms of Hg are quickly distributed through the body owing to the lipophilic nature of the element [13,17]. The kidneys are an important target for ingested Hg, with 50–90 % of methylmercury being deposited here, as well as a large percentage of elemental Hg and mercuric salts. Of more primary concern is the ease with which elemental mercury and methylmercury are able to cross the blood–brain barrier and find their way into placental and fetal tissue [11,14,17,80]. Evidence suggests that both of these forms are gradually converted by oxidation (most probably in the kidneys) to mercuric salts, where they are excreted mainly in faeces and urine. These may alternatively leave the body through sweat, saliva, and exhaled air. Complete expulsion of Hg compounds is suggested to take around 1–3 months depending on dose, species of animal and sex [13,17,30,31].

Exposure to acute levels of organic and inorganic Hg can result in the disturbance and inhibition of numerous essential biological functions; symptoms include ataxia and extensive damage to many internal organs including the brain, heart, intestines and kidneys; heavy poisoning can ultimately lead to death [17,30,31,79–82]. Several tragic episodes serve to demonstrate mercury's potential harm [11,17,80,83]. Thousands were affected and hundreds died as a direct or indirect result of poisoning from metallic mercury and methylmercury discharged by industries in the Japanese cities of Niigata and Minamata during the 1950s and 1960s. In Iraq in 1972, over six and half thousand people were hospitalized and over 450 died from consuming bread produced from grain that had been treated with seed dressing containing Hg. Treatment of crops with such chemicals in countries like Sweden from the post-war period up until the 1960s had severe effects on wildlife, especially birds [80,81,84,85].

However, persistent low-dosage exposure either directly due to contaminated land and water systems or via food grown in such environments is more common than high-dosage exposure [13]. A principal target for both organic mercury and methylmercury is the central nervous system [80,81], where long-term exposure manifests itself, in humans and (as far as can be established) in animals as erethism – that is, heightened and irrational emotional behaviours such as delirium, depressive sometimes suicidal tendencies, memory loss, irritability and similar psychotic reactions. Other trait characteristics of the neurological disturbance caused by Hg are the development of gingivitis and a general loss of coordination and tremors, most noticeably in the hands [17,80,81]. It has been speculated that inorganic Hg and its salts can create an autoimmune response in individuals genetically predisposed to it [13,14].

Health concerns posed by Hg are most severe in relation to young individuals of human and animal kind, both alive and *in utero* (as described by [14,17,80,82]). Fetal development is shown to be at risk due to the passing of Hg to the placenta and other fetal tissues via the maternal blood. Animal experiments and human studies both indicate interference of fetal development and healthy delivery of babies. Finally, absorption of the element is reported to be higher in newborn animals tested, indicating a further factor of vulnerability even outside the womb. In male animals, decreased fertility is a well-observed consequence of Hg exposure, but the evidence is unclear with regards to male humans. In fact, many studies on both humans and animals have been inconclusive, and due to the ethical implications of planned exposure, often represent opportunistic research based around unfortunate one-off incidents. Furthermore, transferability of early research conducted on rats and similar small mammals is limited. What can be stated is that susceptibility to the detrimental effects of Hg appears to be highly variable between individuals within samples, the indication being that some individuals are very sensitive, while others are very tolerant.

Conversion of elemental Hg by environmental or biological processes into mercuric salts or organic forms such as methylmercury is one of the primary concerns surrounding mercury contamination [80]. Methylmercury has a strong bioaccumulative effect through the food chain, and as a result is found in the highest intensities and represents the greatest risks in higher-order animals, including humans [30,31,79,81,82]. The lipophilic nature of this form of mercury, coupled with the much higher fraction typically absorbed within the body via ingestion, means that it is very effectively stored in various tissues upon consumption of contaminated meats by carnivorous animals. Aquatic environments are particularly susceptible to containing large numbers of creatures with elevated methylmercury content due to the naturally higher ambient levels of dissolved Hg [86–88] being added to, sometimes extensively, by deposits coming from movement from the surrounding land and soil [89]. Predatory fish in particular store very large concentrations of methylmercury [14,81,90]. As such, humans and animals with large amounts of fish in their diets are shown to be most at risk of mercury poisoning [79–81,83,87,89].

References

[1] De Laetar, J.R.; Bohlke, J.K.; de Bievre, P. *et al*., Atomic weights of the elements. Review; Pure Appl. Chem. 2003, 75 683–799.
[2] Mersiowsky, I.; Brandsch, R.; Ejlertsson, J., Screening for organotin compounds in European landfill leachate; J. Environ. Qual. 2001, 30 1604–1611.
[3] Ferguson, J.E., The Heavy elements: Chemistry, Environmental Impact and Health Effects; Pergamon, London, 1990.
[4] Stiennes, E., Mercury; in: B.J. Alloway (ed.) Heavy Metal in Soils, 2nd edn; Blackie Academic and Professional, London, 1995, Chapter 11, pp. 245–259.
[5] Waldron, H.A., Did the Mad Hatter have mercury poisoning?; Brit. Med. J. (Clin. Res. Ed.) 1983, 287, 1961.
[6] Nriagu, J.O., (1994) Mercury pollution from the past mining of gold and silver in the Americas; Sci. Total Environ. 1994, 149, 167–181.
[7] Hilson, G., Abatement of mercury pollution in the small-scale gold mining industry: restructuring the policy and research agendas; Sci. Total Environ. 2006, 362, 1–14.
[8] Veiga, M.M.; Maxson, P.A.; Hylander, L.D., Origin and consumption of mercury in small-scale gold mining; J. Cleaner Prod. 2006, 14, 436–447.
[9] de Lacerda, L.D., Updating global Hg emissions from small-scale gold mining and assessing its environmental impacts; Environ. Geol. 2003, 43, 308–314.
[10] Pacyna, E.G.; Pacyna, J.M.; Steenhuisen, F.; Wilson, S., Global anthropogenic mercury emission inventory for 2000; Atmos. Environ. 2006, 40, 4048–4063.
[11] Hylander, L.D., Global mercury pollution and its expected decrease after a mercury trade ban; Water Air Soil Pollut. 2001, 125, 331–344.
[12] Schroeder, W.H.; Munthe, J., Atmospheric mercury – an overview; Atmos. Environ. 1998, 32, 809–822.

[13] World Health Organization (WHO), Concise International Chemical Assessment Document 50: Elemental Mercury and Inorganic Mercury Compounds: Human Health Aspects; WHO, Geneva, 2003.
[14] Sweet, L.I.; Zelikoff, J.T., Toxicology and immunotoxicology of mercury: a comparative review in fish and humans; J. Toxicol. Environ. Health Part B, 2001, 4, 161–205.
[15] United Nations Environment Programme (UNEP), Summary of supply, trade and demand information on mercury; Requested by UNEP Governing Council decision 23/9 IV. Geneva, November 2006.
[16] United Nations Environment Programme (UNEP), Meeting projected mercury demand without primary mercury mining; Ad-Hoc Open-ended Working Group on Mercury, Second meeting, Nairobi, Kenya 6–10 October 2008.
[17] Counter, S.A.; Buchanan, L.H., Mercury exposure in children: a review; Toxicol. Appl. Pharmacol. 2004, 198, 209–230.
[18] Mason, R.P.; Fitzgerald, W.F.; Morel, F.M.M., The biogeochemical cycling of elemental mercury: anthropogenic influences; Geochim. Cosmochim. Acta 1994, 58 3191–3198.
[19] World Health Organization (WHO), Environmental Health Criteria 116. Tributyltin compounds; WHO, Geneva, 1990.
[20] Hogg, T.J.; Stewart, J.W.B.; Bettany, J.R., Influence of the chemical form of mercury on its adsorption and ability to leach through soils; J. Environ. Qual. 1978, 7, 440–445.
[21] Lipsey, R.L., Accumulation and physiological effects of methyl mercury hydroxide on maize seedlings; Environ. Pollut. 1975, 8, 149–155.
[22] Anon, Environmental mercury and man. Report of the Department of the Environment; Central unit on Environmental Pollution, London, 1976.
[23] Wedepohl, K.H., Handbook of Geochemistry; Springer-Verlag, Berlin, 1978.
[24] Kabata-Pendias, A., Trace Elements in Soils and Plants; CRC Press, Boca Raton, FL, 2001.
[25] Shacklette, H.T.; Boerngen, J.G., Element concentrations in soils and other superficial material of the conterminous United States; US Geological Survey Professional Paper 1270, 1984.
[26] Presant, E.W., Geochemistry of iron, manganese, lead, copper, zinc, antimony, silver, tin and cadmium in the soils of the Bathurst area, New Brunswick; Geological Survey of Canada Bulletin 174; Geological Survey of Canada, Ottawa, ON, 1971.
[27] Ure, A.M.; Bacon, J.R., Comprehensive analysis of soils and rocks by spark-source massspectrometry; Analyst 1978, 103, 807–822.
[28] Govindaraju, K., Compilation of working values and sample description for 383 Geostandards; Geostandards Newsletter 18, 1994.
[29] Griffitts, W.R.; Milne, D.B., Tin; in: K.C. Beeson (ed.), Geochemistry and the Environment, Vol. 2. National Academy of Science, Washington DC, 1977.
[30] Schlüter, K., Review: evaporation of mercury from soils. An integration and synthesis of current knowledge; Environ. Geol. 2000, 39, 249–271.
[31] World Health Organization (WHO), Environmental Health Criteria 118: Inorganic mercury; WHO, Geneva, 1990.
[32] World Health Organization (WHO), Environmental Health Criteria 101: Methylmercury; WHO, Geneva, 1991.
[33] Dias, G.M., Edwards, G.C., Differentiating natural and anthropogenic sources of metals to the environment; Human Ecol. Risk Assess. 2003, 9(4), 699–721.
[34] Pais, I.; Benton,J. Jnr., The Handbook of Trace Elements; St. Lucie Press, Boca Raton, FL, 1997.
[35] Ross, S.M., Sources and forms of potentially toxic metals in soil-plant systems; in: S.M. Ross (ed.), Toxic Metals in Soil–Plant Systems; John Wiley and Sons, Ltd, Chichester, 1994, Chapter 1, pp. 4–25.
[36] Stiennes, E.; Friedland, A.J., Metal contamination of natural surface soils from long-range atmospheric transport: existing and missing knowledge; Environ. Rev. 2006, 14,169–186.
[37] Herting, G.; Goidanich, S.; Wallinder, I.O.; Leygraf, C., Corrosion-induced release of Cu and Zn into rainwater from brass, bronze and their pure metals. A 2-year field study; Environ. Monit. Assess. 2008, 144, 455–461.
[38] GESAMP (IMO, FAO, UNESCO, WMO, WHO, IAEA, UN, UNEP), Joint group of experts on the scientific aspects of marine pollution; Review of potentially harmful substances: cadmium, lead and tin, Reports and Studies 22; GESAMP, 1984.

[39] Sterckeman, T.; Douay, F.; Proix, N.; Fourrier, H.; Perdrix, E., Assessment of the contamination of cultivated soils by eighteen trace elements around smelters in the north of France; Water Air Soil Pollut. 2002, 135, 173–194.
[40] Chatterjee, A.; Banerjee, R.N., Determination of lead and other metals in a residential area of greater Calcutta; Sci. Total Environ. 1999, 227, 175–185.
[41] Rawlins, B.G.; Lister, T.R.; Mackenzie, A.C., Trace-metal pollution of soils in northern England; Environ. Geol. 2002, 42, 612–620.
[42] Cecchi, M.; Dumat, C.; Alric, A.; Felix-Faure, B.; Pradere, P.; Cuiresse, M., Multi-metal contamination of a calcic cambisol from a lead-recycling plant; Geoderma 2008, 144, 287–298.
[43] Douay, F.; Pruvot, C.; Roussel, H. *et al.*, Contamination of urban soils in an area of northern France polluted by dust emissions of two smelters; Water Air Soil Pollut. 2008, 188, 247–260.
[44] Rawlins, B.G.; Lark, R.M.; Webster, R.; O'Donnell, K.E., The use of soil survey data to determine the magnitude and extent of historic metal deposition related to atmospheric smelter emissions across Humberside, UK; Environ. Pollut. 2006, 143 416–426.
[45] McBride, M.B.; Richards, B.K.; Steenhius, T.; Spiers, G., Long-term leaching of trace elements in a heavily sludge-amended silty clay loam; Soil Science 1999, 164, 613–623.
[46] Huang, J.-H.; Schwesig, D.; Matzner, E., Organotin compounds in precipitation, fog and soils of a forested ecosystem in Germany; Environ. Pollut. 2004, 130, 177–186.
[47] Huang J.-H.; Matzner, E. (2004) Degradation of organotin compounds in organic and mineral forest soils; J. Plant Nutr. Soil Sci. 2004, 167, 33–38.
[48] Pinel-Raffaitin, P.; Amourouz, D.; LeHécho, I.; Rodrìguez-Gonzalez, P.; Potin-Gautier, M., Occurrence and distribution of organotin compounds in leachates and biogases from municipal landfills; Water Res. 2008, 42, 987–996.
[49] Duester, L.; Diaz-Bone, R.A.; Kösters, J.; Hirner, A.V., Methylated arsenic, antimony and tin species in soils; J. Environ. Monit. 2005, 7, 1186–1193.
[50] Schlüter, K., The fate of mercury in soil: a review of current knowledge; Commission of the European Communities, Directorate-General XIII, Information Technologies and Industries and Telecommunications, Brussels, 1993.
[51] Zhang, L.; Wong, M.H., Environmental mercury contamination in China: Sources and impacts; Environ. Int. 2007, 33, 108–121.
[52] de Lacerda, L.D.; Salomons, W. (eds), Mercury from Gold and Silver Mining: A Chemical Timebomb? Springer, New York, 1997.
[53] Veiga, M.M.; Hinton, J.J., Abandoned artisanal gold mines in the Brazilian Amazon: a legacy of mercury pollution; Nat. Resour. Forum 2002, 26, 15–26.
[54] McGowen, S.L.; Basta, N.T., Heavy metal solubility and transport in soil contaminated by mining and smelting; in: H.M. Selim, D.L. Sparks (eds), Heavy Metals Release in Soils; Lewis Publishers, London, 2001, Chapter 4, pp. 89–107.
[55] Hilson, G.; Vieira, R., Challenges with minimising mercury pollution in the small-scale gold mining sector: Experiences from the Guianas; Int. J. Environ. Health Res. 2007, 17, 429–441.
[56] Maxson, P., Mercury flows in Europe and the world: the impact of decommissioned chlor-akali plants; in: N. Pirone, K.R. Mahaffey (eds), Dynamics of Mercury Pollution on Regional and Global Scales. Atmospheric Processes and Human Exposures around the World; Springer, New York, 2005, pp. 25–50.
[57] Jonasson, I.R., Mercury in the natural environment: a review of recent work; Geological Survey of Canada, Ottawa, ON, 1970.
[58] Landa, E.R., The retention of metallic mercury vapour by soils; Geochim. Cosmochim. Acta 1978, 42, 1407–1411.
[59] Andersson, A., Mercury in soils; in J.O. Nriagu (ed.), The Biogeochemistry of Mercury in the Environment; Elsevier Biomedical Press, Amsterdam, 1979, Chapter 4, pp. 79–112.
[60] Brookins, D.G., Eh-pH Diagrams for Geochemistry; Springer-Verlag, Berlin, 1988.
[61] Schuster, E., The behaviour of mercury in the soil with special emphasis on complexation and adsorption processes – a review of the literature; Water Air Soil Pollut. 1991, 56, 667–680.
[62] Ravichandran, M., Interactions between mercury and dissolved organic matter – a review; Chemosphere 2004, 55, 319–331.

[63] Akerblom, S.; Meill, M.; Bringmark, L.; Johansson, K.; Kleja, D.B.; Bergkvist, B., Partitioning of Hg between solid and dissolved organic matter in the humus layer of boreal forests; Water Air Soil Pollut. 2008, 189, 239–252.
[64] Bodek, I.; Lyman, W.; Reehl, W.F.; Rosenblatt, D.H., Environmental Inorganic Chemistry; Pergamon Press, New York, 1998.
[65] Yin, Y.; Allen, H.E.; Huang, C.P.; Sanders, P.F., Adsorption/desorption isotherms of Hg (II) by soil; Soil Sci. 1997, 162, 35–45.
[66] Beckert, W.F.; Moghissi, A.A.; Au, F.H.E.; Bretthauer, F.W.; McFarlane, M., Formation of methyl mercury in a terrestrial environment; Nature 1974, 249, 674–675.
[67] Rogers, R.D., Abiological methylation of mercury in soil; J. Environ. Qual. 1997, 6, 463–467.
[68] Hempel, M.; Wilken, R.D.; Miess, R.; Hertwich, J.; Beyer, K., Mercury contaminated sites – behaviour of mercury and its species in lysimeter experiments; Water Air Soil Pollut. 1995, 80, 1089–1098.
[69] Benoit, J.M.; Gilmour, C.C.; Heyes, A.; Mason, R.P.; Miller, C.L., Geochemical and biological controls over methylmercury production and degradation in aquatic ecosystems; in: Biogeochemistry of Environmentally Important Trace Elements, ACS Symposium Series 835; American Chemical Society, Washington, DC, 2003, pp. 262–297.
[70] Skyllberg, U.; Qian, J.; Frech, W.; Xia, K.; Bleam, W.F., Distribution of mercury, methyl mercury and organic sulphur species in soil, soil solution and stream of a boreal forest catchment; Biogeochemistry 2003, 64, 53–76.
[71] Gough, L.P.; Shacklette, H.T.; Case, A.A., Element concentrations toxic to plants, animals and man; US Geological Survey Bulletin 1466; US Geological Survey, Washington DC, 1979.
[72] Zook, E.G.; Greene, F.E.; Morris, E.R., Nutrient composition of selected wheats and wheat products. VI. Distribution of manganese, copper, nickel, zinc, magnesium, lead, tin, cadmium, chromium and selenium as determined by atomic absorption spectroscopy and colorimetry; Cereal Chem. 1970, 47, 720–731.
[73] Duke, J.A., Ethnobotanical observations on the Chocó Indians; Econ. Bot. 1970, 24, 344–366.
[74] Connor, J.J.; Schacklette, H.T., Background geochemistry of some rocks, soils, plants and vegetables in the conterminous United States; United States Geological Survey Profession Paper 574f; United States Geological Survey, Washington DC, 1975.
[75] Chapman, H.D., Diagnostic Criteria for Plants and Soils; University of California. Riverside, 1972.
[76] Bowen, H.J.M., Environmental Chemistry of the Elements; Academic Press, London, 1979.
[77] Agoramoorthy, G.; Chen, F.A.; Hsu, M.J., Threat of heavy metal pollution in halophytic and mangrove plants of Tamil Nadu, India; Environ. Pollut. 2008, 155, 320–326.
[78] Golow, A.A.; Adzei, E.A., Mercury in surface soil and cassava crop near an alluvial goldmine at Dunkwa-on-Offin, Ghana; Bull. Environ. Contam. Toxicol. 2002, 69, 228–235.
[79] Clarkson, T.W., Mercury: major issues in environmental health; Environ. Health Perspect. 1993, 100, 31–38.
[80] Clarkson, T.W., The three modern faces of mercury; Environ. Health Perspect. 2002, 110, 11–23.
[81] Wolfe, M.F.; Schwarzbach, S.; Sulaiman, R.A., Effects of mercury on wildlife: a comprehensive review; Environ. Toxicol. Chem. 1997, 17, 146–160.
[82] Castoldi, A.F.; Coccini, T.; Ceccatelli, S.; Manzo, L., Neurotoxicity and molecular effects of methylmercury; Brain Res. Bull. 2001, 55, 197–203.
[83] Myers, G.J.; Davidson, P.W.; Cox, C.; Shamlaye, C.; Cernichiari, E.; Clarkson, T.W. Twenty-seven years studying the human neurotoxicity of methylmercury exposure; Environ. Res. 2000, 83, 275–285.
[84] Wallin, K., Decrease and recovery patterns of some raptors in relation to the introduction and ban of akyl-mercury and DDT in Sweden; Ambio 1984, 13, 263–265.
[85] Westermark, T.; Odsjö, T.; Johnels, A.G., Mercury content of bird feathers before and after Swedish ban on alkyl mercury in agriculture; Ambio 1975, 4, 87–92.
[86] Klein, D.H.; Goldberg, E.D., Mercury in the marine environment; Environ. Sci. Technol. 1970, 4, 765–768.
[87] Chan, H.M.; Scheuhammer, A.M.; Ferran, A.; Loupelle, C.; Holloway, J.; Weech, S., Impacts of mercury on freshwater fish-eating wildlife and humans; Human Ecol. Risk Assess. 2003, 9, 867–883.

[88] Golow, A.A.; Mingle, L.C., Mercury in river water and sediments in some rivers near Dunkwa-on-Offin, an alluvial goldmine, Ghana; Bull. Environ. Contam. Toxicol. 2003, 70, 379–384.
[89] Fitzgerald, W.F.; Clarkson, T.W., Mercury and monomethylmercury: present and future concerns; Environ. Health Perspect. 1991, 96, 159–166.
[90] Johnels, A.G.; Westermark, T.; Berg, W.; Persson, P.I.; Sjostrand, B., Pike (*Esox lucius* L.) and some other aquatic organisms in Sweden as indicators of mercury contamination in the environment; Oikos 1967, 18, 323–333.

22

Molybdenum, Silver, Thallium and Vanadium

Les J. Evans and Sarah J. Barabash

School of Environmental Sciences, University of Guelph, Guelph, Ontario, Canada

22.1 Introduction

Molybdenum (Mo), silver (Ag), thallium, (Tl) and vanadium (V) are all potentially toxic metals whose soluble salts in sufficient amounts can adversely affect the health of both plants and animals. They vary widely in their concentrations in the upper continental crust through 97 mg kg^{-1} for V, 1.1 mg kg^{-1} for Mo, 0.9 mg kg^{-1} for Tl, and 0.053 mg kg^{-1} for Ag [1]. Their distribution also differs in various rock types [2] (Table 22.1).

Molybdenum is an essential element for the growth and development of both plants and animals. It has even been suggested that because a considerable amount of Mo was sequestered in black shales, a deficiency of Mo in ancient oceans may have delayed the evolution of life on earth by 2 billion years [3]. In plants, the presence of Mo in certain enzymes allows the plant to absorb nitrogen. Most plant species, except for legumes, do not exhibit symptoms of Mo-deficiency because Mo requirements are low. Molybdenum is an essential micronutrient for animals because of the formation of Mo-containing enzymes which are necessary for the health of all animals. In ruminants, the dietary intake of excessive Mo causes hypocuprosis – Mo-induced Cu deficiency.

The major use for Mo is in steel making where its addition to steel makes it stronger and more highly resistant to heat. The iron and steel industries account for more than 75 % of the consumption of Mo. Stainless steels are used for water distribution systems, food

Table 22.1 Molybdenum, silver, thallium and vanadium in major rock types

Rock type	Mo (mg kg^{-1})	Ag (μg kg^{-1})	Tl (mg kg^{-1})	V (mg kg^{-1})
Magmatic rocks				
Ultramafic rocks	0.2–0.3	50–60	0.05–0.2	40–100
Mafic rocks	1.0–1.5	100	0.1–0.4	200–250
Intermediate rocks	0.6–1.0	50–70	0.5–1.4	30–100
Felsic rocks (plutonic)	1–2	40	0.6–2.3	40–90
Felsic rocks (volcanic)	2	50	0.5–1.8	70
Sedimentary rocks				
Argillaceous sediments	2.0–2.6	70	0.5–1.5	80–130
Shales	0.7–2.6	70–100	0.5–2.0	100–130
Sandstones	0.2–0.8	50–250	0.4–1.0	10–60
Limestones, dolomites	0.16–0.4	100–150	0.01–0.14	10–45

handling equipment, and chemical processing equipment. Molybdenum also is an important material for the chemical and lubricant industries. Molybdenum has uses as catalysts, paint pigments, corrosion inhibitors, smoke and flame retardants, and dry lubricants and is resistant to high loads and temperatures. As a pure metal, Mo is used because of its high melting temperature as filament supports in light bulbs, in metal-working dies, and in furnace parts.

Molybdenum occurs predominantly as Mo(VI), but can also exist in oxidation state Mo(IV) as its sulfide, molybdenite, $MoS_2(s)$. Its aqueous chemistry is dominated by the presence of anionic species and their conjugate acids, with polynuclear complexes becoming important at higher concentrations. Molybdenum is retained by soils under acidic conditions, principally on the surfaces of variable-charge minerals, such as the Fe oxides.

Silver has not been shown to be an essential element for any plant or animal species, but can be toxic to aquatic organisms, particularly aquatic invertebrates [4]. Plant accumulation of Ag is low, even in soils amended with sewage sludges. The antimicrobial properties of Ag are well known and nanoparticles of Ag are increasingly being used in both medical and clothing applications.

Emissions from smelting operations, disposal of certain photographic and electrical supplies, coal combustion, and cloud seeding are some of the important anthropogenic sources of Ag. Silver can exist in two oxidation states – Ag(I) and Ag(0). Because Ag(I) can form relatively insoluble precipitates with chloride ions and, under anaerobic conditions, with sulfide ions, it is not very mobile in the environment.

Thallium is a highly toxic element to humans and other mammals and also inhibits the growth of many plants and aquatic organisms. Its deleterious effects on human health and to the environment have been the subject of a number of recent reviews [5–7] and its toxicity is related to the ability of Tl^+ ions to replace K^+ ions in biological systems. It was once widely used as a rodenticide as $Tl_2SO_4^0$, but this chemical has been banned as a rat poison in the United States since the 1970s. Thallium and its compounds can enter a body through the skin, by inhaling dust or fumes, and by direct ingestion. As a result, strict rules about the use of Tl compounds have been created by the US Environmental Protection Agency (EPA).

Thallium is used in a number of electronic devices. It is used in selenium rectifiers, gamma radiation detection equipment, and infrared radiation detection and transmission equipment. It also has nonelectrical uses. For example, Tl is added to glass to increase its density and refractive index, that is, its ability to break light into its component colors. It is also used as a catalyst to create certain organic compounds. Radioactive Tl compounds are used in medical applications.

Thallium can exist in two oxidation states, Tl(III) and Tl(I), with the latter being the most predominant. Most studies on soils indicate that it is not strongly bound by soil components and has been shown to quickly de-sorb, and hence poses an environmental and health hazard [8].

Vanadium can be both essential and toxic to humans [9,10]. Vanadium compounds have been implicated in the pathogenesis of some human diseases and also in maintaining normal body functions. Vanadium interferes with an essential array of enzymatic systems, while V deficiency accounts for several physiological malfunctions. As early as 1953, vanadium was suggested as an essential element for plants [11], although this essentiality has been questioned. Vanadium toxicities in plants are uncommon because V is believed to precipitate in the roots as an insoluble calcium vanadate [12].

Vanadium is used extensively to make steel alloys for tools and for construction purposes. These hard, strong ferrovanadium alloys are used to make the frames of high-rise buildings and oil drilling platforms and also used in armor plating for military vehicles and other protective vehicles. Some V is used in other industrial applications. For example, vanadium pentoxide (V_2O_5) is used in the production of glass and ceramics and as a chemical catalyst.

Vanadium can exist in three oxidation states – V(V), V(IV), and V(III). V(V) species predominate in aerobic environments and at alkaline pH values. V(IV) species predominate in acidic and mildly reducing conditions, while V(III) exists only in anaerobic conditions. Vanadium has been shown to be relatively immobile in soils and hence has a reduced environmental risk.

22.2 Molybdenum

22.2.1 Geochemical Occurrences and Soil Concentrations

Molybdenite, MoS_2, a Mo sulfide, is the major ore mineral for Mo, with lesser occurrences of Mo found in the mineral wulfenite, $Pb(MoO_4)$, lead molybdate. Molybdenite generally forms in high-temperature environments, such as those associated with igneous rocks and contact metamorphism, but there is evidence that it can form at lower temperatures in soil and sediment under anaerobic conditions.

The rather uniform distribution of Mo among the common igneous rocks (Table 22.1) is largely due to its ability to replace a number of elements in rock-forming minerals [13]. Molybdenum is found in the feldspars and in the ferromagnesium minerals (biotite, amphibole, pyroxene). Substitution of Mo for Fe^{3+}, Ti^{4+}, Al^{3+} and possibly Si^{4+} is suggested. These substitutions are in general accord with the ionic radii of the elements involved. Shales poor in organic matter contain about as much Mo as igneous rocks; however, black shales are likely to be much richer in Mo with contents in excess of 100 mg kg^{-1}. Limestones and dolomites usually contain less than 0.5 mg kg^{-1} Mo.

The Mo concentration of soils ranges from 0.013 to 17.0 mg kg^{-1}, with a mean value of 1.8 mg kg^{-1} [2]. Ure and Berrow [14] had calculated a mean of 1.9 mg kg^{-1} from earlier data. A mean value of 5.8 mg kg^{-1} was calculated by Kubota [15] in soils in areas of the United States where Mo toxicity of cattle had been noted. Contents of up to 180 mg kg^{-1} have been noted in soils contaminated by industrial pollution [16].

22.2.2 Sources of Soil Contamination

The iron and steel industries account for more than 75 % of the worldwide consumption of Mo. Molybdenum is alloyed with steel, making the steel stronger and more highly resistant to heat. Molybdenum-containing stainless steels have the strength and corrosion-resistant requirements for use in water distribution systems, food handling equipment, chemical processing equipment, automotive parts, construction equipment, and gas transmission pipes. Molybdenum is also an important material for the chemicals and lubricant industries. Molybdenum is also used in catalysts, paint pigments, corrosion inhibitors, smoke and flame retardants, and dry lubricants. As a pure metal, molybdenum is used because of its high melting temperature as filament supports in light bulbs, metal-working dies and furnace parts. Molybdenum cathodes are used in special electrical applications. The overall uses of molybdenum have been estimated to be in machinery (35 %), for electrical applications (15 %), in transportation (15 %), in chemicals (10 %), in the oil and gas industry (10 %), and assorted others (15 %) [17].

Because of its widespread industrial uses, atmospheric deposition is an important anthropogenic source of soil Mo near major cities. It is a common constituent of biosolids, such as sewage sludges, and is added to soils when these wastes are land applied. Because of its possible deleterious effects on grazing cattle, the content of Mo in biosolids applied to agricultural land is regulated in most countries. In the United States, there is an upper limit of 75 mg kg^{-1} Mo in land-applied biosolids. McBride and Spiers [18] reported a range of Mo contents in P-containing fertilizers of 2. to 7.7 mg kg^{-1} and a mean value of 2.5 mg kg^{-1} in 20 dairy manures.

22.2.3 Chemical Behavior in Soils

22.2.3.1 Solution and Solid Phases

Molybdenum can exist in soil solutions and other natural waters as soluble species, sparingly soluble precipitates or adsorbed species. The major soluble Mo species in most waters are molybdic acid, $H_2MoO_4^0$, and its oxyanions, $HMoO_4^-$ and MoO_4^{2-}. At high Mo concentrations, greater than about 10^{-4} M, polynuclear complexes have been shown to form. These include a polymolybdic acid, $H_3Mo_7O_{24}^{3-}$, as well as three conjugate anions, $H_3Mo_7O_{24}^{4-}$, $HMo_7O_{24}^{5-}$ and $Mo_7O_{24}^{6-}$. In addition to these molydbic acid species, the molybdate ion, MoO_4^{2-}, has been shown to form complexes with Ca^{2+}, Mg^{2+} [19], K^+ and Na^+ ions [20], and with Al^{3+} ions [21]. Formation constants for these species are shown in Table 22.2.

Table 22.2 Formation constants for soluble molybdenum species

Reaction	log K	Reaction	log K
$MoO_4^{2-} + H^+ \rightleftharpoons HMoO_4^-$	4.21	$19MoO_4^{2-} + 34H^+ \rightleftharpoons Mo_{19}C_{59}^{4-} + 17H_2O$	183.60
$MoO_4^{2-} + 2H^+ \rightleftharpoons H_2MoO_4^0$	8.19	$MoO_4^{2-} + Ca^{2+} \rightleftharpoons CaMoO_4^0$	2.57[a]
$7MoO_4^{2-} + 8H^+ \rightleftharpoons Mo_7O_{24}^{6-} + 4H_2O$	52.99	$MoO_4^{2-} + Mg^{2+} \rightleftharpoons MgMoO_4^0$	3.03[a]
$7MoO_4^{2-} + 9H^+ \rightleftharpoons HMo_7O_{24}^{5-} + 4H_2O$	59.37	$MoO_4^{2-} + K^+ \rightleftharpoons KMoO_4^-$	1.29[b]
$7MoO_4^{2-} + 10H^+ \rightleftharpoons H_3Mo_7O_{24}^{3-} + 4H_2O$	64.15	$MoO_4^{2-} + Na^+ \rightleftharpoons NaMoO_4^-$	1.66[b]
$7MoO_4^{2-} + 11H^+ \rightleftharpoons H_3Mo_7O_{24}^{3-} + 4H_2O$	67.39	$6MoO_4^{2-} + 6H^+ + Al^{3+} \rightleftharpoons Al(OH)_6Mo_6O_{18}^{3-}$	54.98

[a,b] Values taken from Verweij [22], except [a] Essington [19] and [b] Essington [20].

The relative proportions of molybdic acid, both mononuclear and polynuclear, and their conjugate bases at two different concentrations are shown in Figure 22.1. Polynuclear complexes are not significant contributors to Mo speciation at concentrations below 2×10^{-4} M.

In the presence of soluble Ca and Mg at concentrations that might occur in soil solutions, a proportion of the soluble Mo in soil solutions occurs as the Ca– and Mg–molybdate complexes $CaMoO_4^0$ and $MgMoO_4^0$ (Figure 22.2).

The maximum concentration of soluble Mo in waters is controlled by the relatively insoluble mineral powellite, $CaMoO_4(s)$ [20]. For soil solutions in equilibrium with the mineral powellite, the actual amount of Mo in solution will depend on the soluble Ca concentration in the soil solution (Figure 22.3).

$$CaMoO_4(s) \rightleftharpoons Ca^{2+} + MoO_4^{2-} \quad \log K_{sp} = -8.05 \quad (22.1)$$

An insoluble oxide of Mo(VII), molybdenum trioxide, $MoO_3(s)$, exists but forms only at very high Mo concentrations and at very acidic pH values.

$$MoO_3(s) + H_2O \rightleftharpoons MoO_4^{2-} \; 2H^+ \quad \log K_{sp} = -8.0 \quad (22.2)$$

Other insoluble molybdate precipitates include those with Mg, Ag, and Pb (Table 22.3).

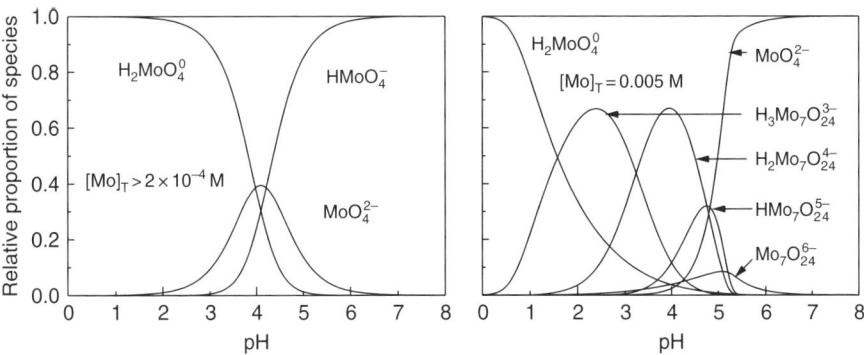

Figure 22.1 Speciation of molybdic acid, $H_2MoO_4^0$, as a function of pH and Mo concentration

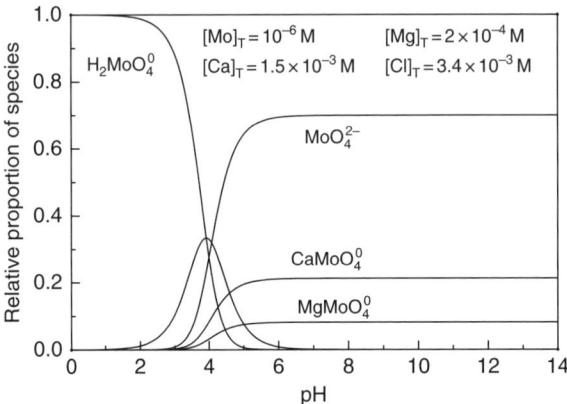

Figure 22.2 Speciation of molybdenum in the presence of Ca^{2+} and Mg^{2+} ions

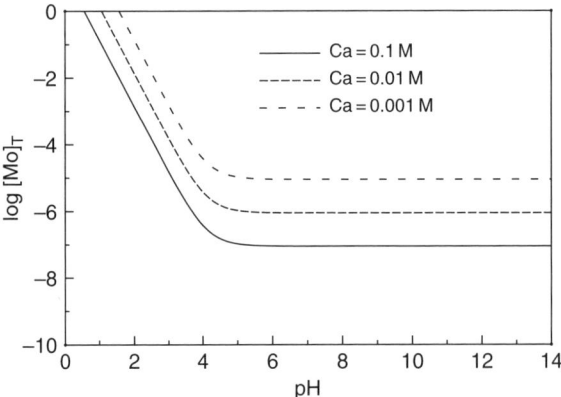

Figure 22.3 Molybdenum concentrations in equilibrium with the mineral powellite, $CaMoO_4(s)$

Table 22.3 Solubility products for molybdate precipitates

Reaction	log K_{sp}	Reaction	log K_{sp}
$MoO_3(s) + H_2O \rightleftharpoons MoO_4^{2-} + 2H^+$	−8.00	$Ag_2MoO_4^0(s) \rightleftharpoons MoO_4^{2-} + 2Ag^+$	−11.55
$CaMoO_4^0(s) \rightleftharpoons MoO_4^{2-} + Ca^{2+}$	−7.95	$PbMoO_4^0(s) \rightleftharpoons MoO_4^{2-} + Pb^{2+}$	−15.62
$MgMoO_4^0(s) \rightleftharpoons MoO_4^{2-} + Mg^{2+}$	−1.85		

Values taken from Verweij [22].

22.2.3.2 Sorption–Desorption Reactions

The adsorption of molybdate anions to soils occurs under acidic conditions, decreasing rapidly after pH 5, and is essentially complete at pH 8 [23,24]. Molybdate anions have been shown to be specifically adsorbed onto the surfaces of a number of naturally occurring minerals, which include goethite [25], hydrous ferric oxide [26], hematite [26], ferrihydrite [27], and the edges of kaolinite, montmorillonite, and illite [28] through the formation of surface inner-sphere complexes. A ligand exchange mechanism has been proposed for the adsorption of molybdate ions onto variable-charge surfaces to form surface complexes and results to date suggest that inner-sphere complexes are formed between the molybdate ions and the variable-charged surfaces [25,28]. Both monodentate and bidentate surface complexes have been considered when modelling Mo retention by soils [23].

Two surface complexes have been assumed to form at the edges of variable-charged minerals $\equiv$S-OMoO$_3^-$ and $\equiv$S-O-MoOHO$_2^0$. The surface complexation reactions for the adsorption of molybdate ions onto edge sites of hematite and kaolinite are shown in Table 22.4, together with some measured complexation constants.

These values, in addition to the intrinsic acidity constants for each variable-charge surface, have been used to model Mo retention by tropical soil containing only hematite and kaolinite as its variable-charge minerals (Evans, L.J., unpublished data, 2008). The results are shown in Figure 22.4 using the constant capacitance model.

Table 22.4 Complexation constants, $\log {}^c\beta^{int}$, for Mo on two variable-charge surfaces

Reaction	$\log {}^c\beta^{int}$	
	Hematite	Kaolinite
$\equiv S - OH^0 + H_2MoO_4^0 \rightleftharpoons \equiv S - OMoOHO_2^0 + H_2O$	3.15	4.95
$\equiv S - OH^0 + H_2MoO_4^0 \rightleftharpoons \equiv S - OMoO_3^- + H^+ + H_2O$	9.02	0.95

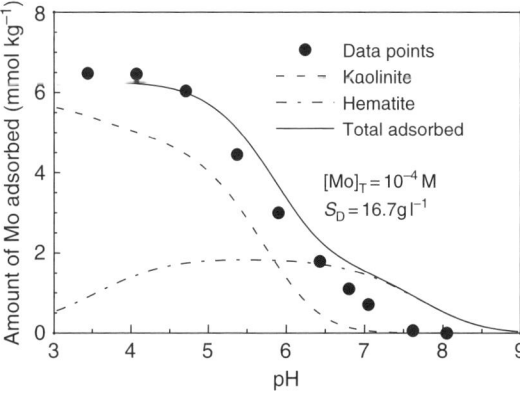

Figure 22.4 Actual and modelled Mo adsorption onto a tropical soil

Soil organic matter has been suggested as an important adsorbing constituent of soils [23,26] and Akiyoshi and Hisashi [24] reported that soil Mo was incorporated into the humic acid fraction of soils after extraction with alkaline solutions and precipitation with acidic solutions. The majority of anions would not be expected to adsorb to soil organic matter because of the repulsion of the negative charges on both adsorbent and adsorbing species. Molybdate ions have been shown to be adsorbed by soil humic acid [29], although the shape of the adsorption envelope was different from that obtained for oxide minerals. Molybdate ions can complex with carboxylic and phenol groups and have been shown to complex with oxalate, malate, citrate, and several catecholate derivatives [30].

22.2.3.3 Oxidation–Reduction Chemistry

The oxidation–reduction chemistry of Mo has not been extensively studied but it is known that Mo(VI), Mo(V), and Mo(IV) oxidation states can exist under certain conditions in aqueous systems. This has led to the suggestion that the formation of insoluble molybdenum disulfide, $MoS_2(s)$, may be a possible mechanism for the removal of Mo in anaerobic wetlands [31].

$$MoO_4^{2-} + 2SO_4^{2-} + 24H^+ + 18e^- \rightleftharpoons MoS_2(s) + 12H_2O \quad E^0 = 0.395V \quad (22.3)$$

The calculated E^0 value for this reaction, using the thermodynamic data given in Wagman et al. [32], was found to be 0.395 V and therefore the reaction is possible in aqueous systems (Figure 22.5).

A number of sulfate-reducing bacteria have been found which can reduce Mo(VI) to Mo(IV) [33]. Although there is at present no direct evidence for the reduction of Mo(VI) to Mo(IV) in soils or freshwater sediments, evidence from the geological literature suggests that $MoS_2(s)$ can be formed in oceanic sediments in the presence of sufficient amounts of HS^- ions and thiol groups from humic materials [34]. Reduction of Mo(VI) to molybdenum blue by bacteria has recently been reported [35]. Molybdenum blue is a macromolecule which contains 154 Mo atoms in both Mo(VI) and Mo(V) oxidation states.

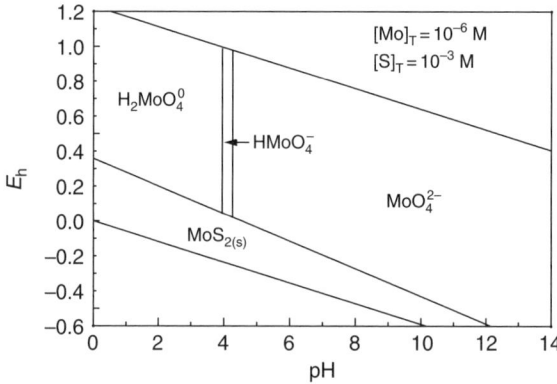

Figure 22.5 Oxidation–reduction reactions of molybdenum at 25 °C

22.3 Silver

22.3.1 Geochemical Occurrences and Soil Concentrations

Silver is a rare but naturally occurring metal and is often found in compounds with S, its homologues Se and Te, and their neighbors in the periodic table, As, Sb, and Bi [36,37]. These compounds have a metallic luster and are semiconductors. The most common Ag minerals are native silver (Ag), argentite (Ag_2S), pyrargyrite (Ag_3SbS_3), prousite $(Ag_3AsS)_3$, tetrahedriretennanite $[(Cu,Fe,Ag)_{12}(SbAs)_4S_{13}]$, cerargyrite (AgCl), and argentejarosite $[(AgFe_3(SO)_2(OH)_6]$ [38]. In addition, there are many other well-defined Ag minerals; examples of which are shown in Table 22.5) [36].

Silver is a minor component of the earth's crust, with common soil concentrations ranging from < 0.01 to 5 mg kg^{-1}, with most reported values below 0.1 mg kg^{-1} [39–42]. However, Ag concentrations in soils impacted by industry or by smelting activities may be elevated substantially with respect to background concentrations.

Soils from Wales, which were contaminated by past mining activities, had average concentrations of 9 mg kg^{-1} [43]. In a recent study by Figeurora *et al.* [44], soils from a silver mine in Mexico had concentrations ranging from 3 to 42 mg kg^{-1}. Sterckemen [45] found concentrations up to 2.3 mg kg^{-1} around lead and zinc smelters in France, which were 3–20 times the mean background concentrations. Tailings sampled from Cobalt, Ontario, Canada, had Ag concentrations that were elevated 100–1000 times with respect to average Ag concentrations in igneous rocks and ranged from 9 to 240 mg kg^{-1} [46].

22.3.2 Sources of Soil Contamination

The amount of Ag that was actively mined in 2006 was estimated by the World Silver Survey to be 646 million troy ounces (M oz) [47]. The top Ag-producing mines in 2006 were in Peru, China, Chile and Mexico. Approximately one-fifth of Ag production comes from old scrap (Ag recovered from used goods such as silverware and jewelry, photographic chemicals, and recycled computers); however, much is lost in various emissions to the environment [47]. The total global losses of Ag to the environment were estimated by Eckelmen and Gracdel [48] to be more than 1.3×10^7 kg annually, with tailings and landfills accounting for three-quarters of the total amount (Table 22.6).

High concentrations of Ag, other than that in natural mineral deposits, are found on contaminated sites where it has been added because of human activity and/or deposited from

Table 22.5 Some silver-containing minerals

Name	Formula	Name	Formula
Hessite	Ag_2Te	Schapbachite	$AgBiS_2$
Polybasite	Ag_9SbS_6	Argentopyrite	$AgFe_3S_5$
Andorite	$AgPbSb_3S_6$	Argyropyrite	$Ag_2Fe_7S_{11}$
Stromeyerite	$Ag_2Cu_2S_2$	Iodite	AgI

Table 22.6 Annual global emissions of silver to the environment [48]

Emission category	Mass flow (Mg)	% Total mass	Emission category	Mass flow (Mg)	% Total mass
Tailings	4 041	30.1	Dissipation to water	457	3.4
Slag	13	0.1	Dissipation to land	1 379	10.3
Leachate	1 428	10.6	Landfill	5 950	44.3
Particulate matter	151	1.1			
Total	13 420	100			

the atmosphere as a result of smelting and coal burning activities [40]. High concentrations also occur in aquatic systems because of erosion of natural sources, as a result of mining waste and industrial discharges, or as inputs from sewage treatment plants [40].

Inputs of Ag to the environment come from a variety of different sources, such as photographic materials and processing, mining and smelting activities, jewelry, silverware, solder, bearings, medical and dental applications, electrical contacts, circuit board manufacture, cloud seeding, catalysts, and sewage sludge [40,49,50]. Wastes arising from such commercial uses of Ag as photographic processing, printing, and plating are often discharged to municipal treatment facilities [51]. Soils that have been amended with sewage sludge can have up to ten times as much Ag as background levels [52].

Silver nanoparticles are becoming an increasingly popular antimicrobial agent. Silver therefore constitutes a new, and often overlooked, heavy metal contaminant in sewage sludge-amended soil [53]. Silver nanoparticles are incorporated into clothing, paints, food containers, interiors of automobiles, building materials, and medical equipment [50,54]. A recent study estimated that in the year 2010, biocidal plastics and textiles are predicted to account for 15 % of the total Ag released into water in the European Union [50].

22.3.3 Chemical Behavior in Soils

22.3.3.1 Soluble and Solid Phases

Silver is found within the environment in four oxidation states: 0, 1+, 2+ and 3+, with 0 and 1 being the most common. As a 'soft' metal cation, Ag(I) tends to coordinate and complex with soft bases and consequently exhibits a high affinity for S [55]. Silver weakly hydrolyzes in water to form two hydroxo complexes, $AgOH^0$, and $Ag(OH)_2^-$, and a carbonate complex, $AgCO_3^-$, (Table 22.7). The speciation of Ag in water in equilibrium with atmospheric $CO_2(g)$ is shown in Figure 22.6.

In the absence of reduced sulfur species, the speciation of silver is strongly influenced by ligands such as chloride and sulfate. Silver forms strong multihalide complexes and even at relatively low Cl^- concentrations these chloride complexes predominate in the speciation of Ag(I) over a large range of pH values (Figure 22.7).

Table 22.7 Formation constants for soluble Ag(I) species and solubility products [22]

Reaction	log K	Reaction	log K
Soluble			
$Ag^+ + H_2O \rightleftharpoons AgOH^0 + H^+$	−12.00	$Ag^+ + SO_4^{2-} \rightleftharpoons AgSO_4^-$	1.30
$Ag^+ + 2H_2O \rightleftharpoons Ag(OH)_2^- + 2H^+$	−24.00	$Ag^+ + 2SO_4^{2-} \rightleftharpoons Ag(SO_4)_2^{3-}$	0.57
$Ag^+ + CO_3^{2-} \rightleftharpoons AgCO_3^-$	3.40	$Ag^+ + 3SO_4^{2-} \rightleftharpoons Ag(SO_4)_3^{5-}$	−1.51
$Ag^+ + Cl^- \rightleftharpoons AgCl^0$	3.31	$Ag^+ + HS^- \rightleftharpoons AgHS^0$	13.85
$Ag^+ + 2Cl^- \rightleftharpoons AgCl_2^-$	5.25	$Ag^+ + 2HS^- \rightleftharpoons Ag(HS)_2^-$	17.91
$Ag^+ + 3Cl^- \rightleftharpoons AgCl_3^{2-}$	5.20	$Ag^+ + HS^- \rightleftharpoons AgS^-$	5.30
$Ag^+ + 4Cl^- \rightleftharpoons AgCl_4^{3-}$	6.96	$Ag^+ + 2HS^- \rightleftharpoons Ag(HS)S^{2-} + H^+$	9.99
Precipitates			
$Ag_2O(s) + 2H^+ \rightleftharpoons 2Ag^+ + H_2O$	12.57	$Ag_2CO_3(s) \rightleftharpoons 2Ag^+ + CO_3^{2-}$	−11.09
$AgCl(s) \rightleftharpoons Ag^+ + Cl^-$	−9.75	$Ag_2SO_4(s) \rightleftharpoons 2Ag^+ + SO_4^{2-}$	−4.82
$Ag_2S(s) + H^+ \rightleftharpoons 2Ag^+ + HS^-$	−36.22		

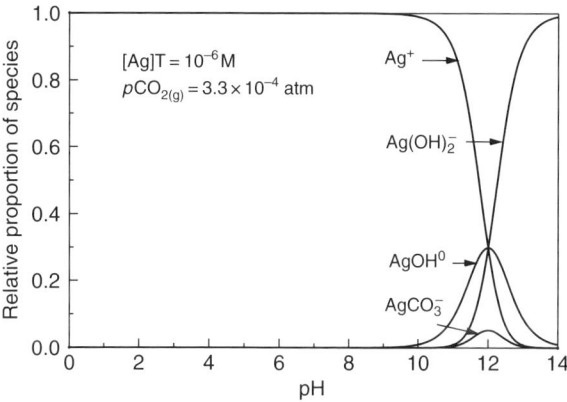

Figure 22.6 Speciation of Ag(I) in water in the presence of $CO_2(g)$

Silver is strongly bound to both inorganic and organic sulfur, S^{2-}, groups. The formation constants, log K, values for inorganic sulfides range from 14 to 21, and for organic sulfides range from 12 to 15 [41]. Compounds containing oxygen and nitrogen groups, however, display much weaker Ag binding; that is, EDTA has a log $K \sim 6$ [56].

Because of the ability of Ag to form strong sulfide complexes, Ag should out-compete most other metals for the available sulfide ions ($Hg^{2+} > Ag^+ > Cu^+ > Pb^{2+} > Zn^{2+}$) [57]. Due to this displacement, with the exception of Hg^{2+} and Cu^+, silver will rapidly sorb onto and into colloidal or particulate sulfides even at low nanomolar concentrations [58]. If sufficient Ag(I) is present, it will eventually replace all the other transition metals and form $Ag_2S(s)$. Adams and Kramer [57] found that Ag ions are adsorbed rapidly onto amorphous FeS.

526 Characteristics and Behaviour of Individual Elements

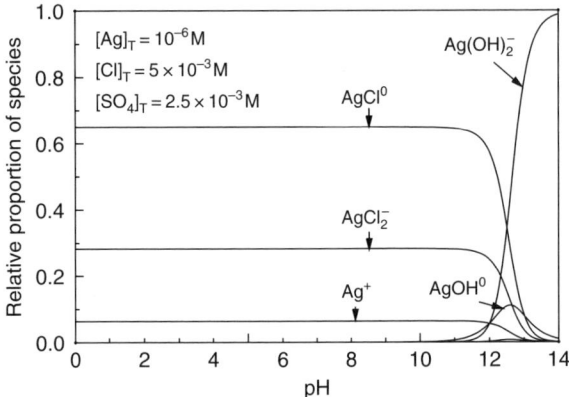

Figure 22.7 Speciation of Ag(I) in the presence of chloride and sulfate ions

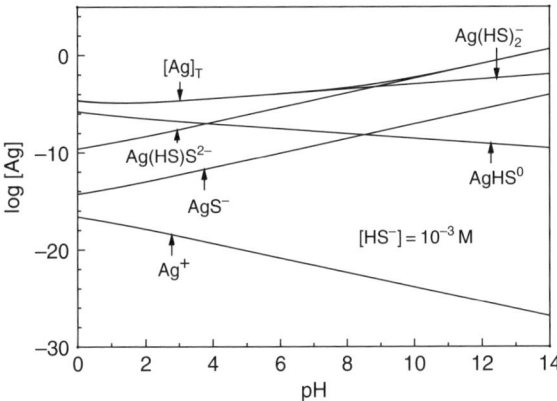

Figure 22.8 Dissolution of $Ag_2S(s)$ as a function of pH

Silver also forms a number of precipitates, with AgCl(s) and $Ag_2S(s)$ being the most common in oxidizing and anaerobic environments, respectively. The formation of the precipitate is dependant on the relative amount of chloride and sulfide in the system. A representation of $Ag_2S(S)$ dissolution is shown in Figure 22.8.

22.3.3.2 Sorption–Desorption Reactions

Silver has been shown to be relatively immobile in soils and tends to be retained in the surface layers of soils at pH values greater than 4, especially in soils with a high concentration of organic matter [8,59]. Goethite, ferrihydrite, Mn oxides, clay minerals, and humic acid have been shown to adsorb substantial amounts of Ag [60–64]. However, few binding constants have been reported for Ag in the literature and most adsorption studies are based on isotherms at relatively high concentrations of Ag.

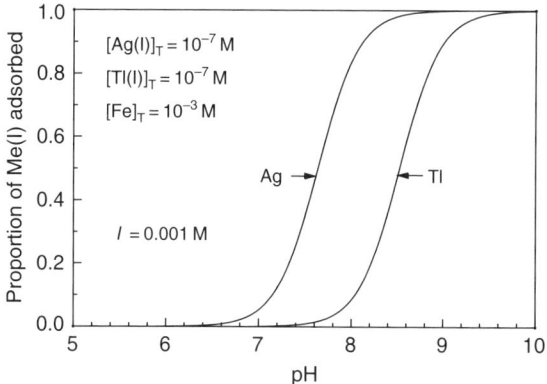

Figure 22.9 Adsorption of Ag(I) and Tl(I) onto hydrous ferric oxide modeled using the data of Dzombak and Morel [65]

An intrinsic complexation constant, $\log K^{int}$, for Ag^+ ions with hydrous ferric oxide was calculated by Dzombak and Morel [65] from batch adsorption data.

$$\equiv Fe - OH^0 + Ag^+ \rightleftharpoons \equiv Fe - OAg^0 + H^+ \quad \log K^{int} = -1.72 \quad (22.4)$$

The calculated adsorption of Ag(I) and Tl(I) onto hydrous ferric oxide as a function of pH using the diffuse layer model [65] is shown in Figure 22.9.

Silver has also been shown to adsorb onto both humic and fulvic acid fractions, with high retentive capacities of up to 30 mg kg^{-1} [8,43]. Sikora and Stevenson [62] reported conditional stability constants for Ag binding with humic acids ranging from 3.08 to 4.51 at a pH of 6.5 ($I = 0.1$ M). Varshal *et al.* [66] determined the conditional stability constant, log K, for a 1:1 silver-fulvic acid complex at pH 8.0 ($I = 0.1$ M) as 5.67.

Silver concentrations in the surface horizons of soils sampled from Ag-spiked (AgNO$_3$) samples from Japan were correlated with the organic matter contents, allophane contents, surface areas and Al concentrations of the soils [67]. A sequential extraction scheme was used to investigate the Ag fractionation of five Japanese soils [68]. Mean concentrations of Ag in the chemical fractions were; 0.8 % exchangeable, 0.5 % carbonate bound, 20.9 % organically bound, 8.8 % oxide-bound, and 60.3 % residual. Similar results were found in Welsh soils with most of the Ag enriched in the surface layer of the soils and partitioned in the organic and residual fractions [43].

22.3.3.3 Oxidation–Reduction Chemistry

Silver exists in two oxidation states, Ag(I) and Ag(0), in most natural waters. The standard electrode potential for the Ag^+/Ag^0 couple can be calculated as 0.799 V using the thermodynamic data given in Wagman *et al.* [32]. Other oxidation–reduction reactions used to construct an E_h–pH diagram for the Ag–S–O–H system are shown in Table 22.8.

The E_h–pH diagram is shown in Figure 22.10. In the presence of sulfide ions, Ag can precipitate as Ag$_2$S(s). The native Ag(s) species dominates a large portion of the diagram,

Table 22.8 Oxidation/reduction reactions in the Ag–S–O–H system

Reaction	E^0
$Ag^+ + e^- \rightleftharpoons Ag(s)$	0.799
$Ag(OH)_2^- + 2H^+ + e^- \rightleftharpoons Ag(s) + 2H_2O$	2.219
$Ag_2S(s) + H^+ + 2e^- \rightleftharpoons 2\,Ag(s) + HS^-$	−0.273
$2Ag^+ + SO_4^{2-} + 8H^+ + 8e^- \rightleftharpoons Ag_2S(s) + 4H_2O$	0.517
$2Ag(s) + SO_4^{2-} + 8H^+ + 6e^- \rightleftharpoons Ag_2S(s) + 4H_2O$	0.413

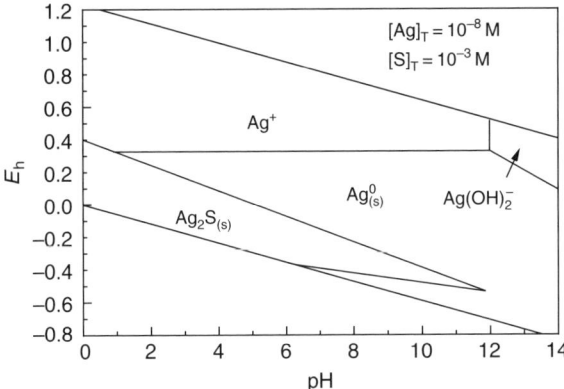

Figure 22.10 E_h–pH diagram for the system Ag–S–O–H

although under reducing conditions the dominant form of Ag under acidic conditions is the precipitate $Ag_2S(s)$.

22.4 Thallium

22.4.1 Geochemical Occurrences and Soil Concentrations

Thallium forms a small number of rare minerals, including lorandite, $TlAsS_2$, the most widespread Tl ore mineral, and crookesite, $Cu_7(Tl,Ag)Se_4$. Lorandite is often associated with the minerals sphalerite, $ZnS(s)$, and galena, $PbS(s)$. As a result, Tl is recovered as a by-product of processing of zinc and lead ores. Because the radius of the Tl^+ ion is similar to that of the K^+ ion, it is often found in K-containing minerals, such as the micas and feldspars, and its content increases from mafic to felsic rocks. Coals are often enriched in Tl, with an average reported value of 0.05 mg kg^{-1} [69], although very high average values of up to 7.5 mg kg^{-1} have been reported in coals from Guizhou Province, China [70].

Total Tl concentrations in uncontaminated soils typically range from 0.1 to 1.8 mg kg^{-1} [14] with a median value of 0.8 mg kg^{-1}. Other studies give ranges of 0.29–1.17 mg kg^{-1} and

0.08–0.91 mg kg^{-1}, respectively, for soils from China [71] and Austria [72]. A study of the content of Tl in French soils calculated a median value of 0.29 mg kg^{-1} and a range of 0.13–1.54 mg kg^{-1} [73]. Although many of the French soils were collected from sites near potential anthropogenic sources of Tl, the major factor affecting the distribution of Tl in the soils was the local bedrock geology. Thallium contents were highest in soils derived from limestones and granites. Thallium was believed to be associated with sulfide minerals in the limestone-derived soils and with micas and feldspars in the granitic soils. Very high Tl concentrations have been reported in China in soils from an area subjected to the natural processes of Tl-rich sulfide mineralization [6]. Thallium contents ranged from 1.5 to 6.9 mg kg^{-1} in undisturbed natural soils compared with <0.2 to 0.5 mg kg^{-1} in soils from an area not associated with sulfide mineralization.

22.4.2 Sources of Contamination

Thallium primarily enters the environment through the burning of coal and from the smelting of Pb and Zn ores [5]. Thallium is used in a number of electronic devices, primarily for the semiconductor industry and is also added to speciality glasses to increase their density and refractive indexes. It is used as a catalyst to create certain organic compounds and radioactive Tl compounds are used in imaging procedures for the evaluation of myocardial disease [5]. It was once widely used as a rodenticide but has been banned in the United States since the 1970s.

Thallium may enter the soil through atmospheric deposition from coal burning, the smelting of ores, or gaseous emissions from cement factories. It can also directly enter the soil from application of fly ash as a soil amendment, or application of industrial wastes and biosolids. The Tl concentration in biosolids has not been extensively studied, although one report in Canada on 16 samples found concentrations from nondetectable to 131 mg kg^{-1} [74].

22.4.3 Chemical Behavior in Soils

22.4.3.1 Soluble and Solid Phases

Thallium(III) readily hydrolyzes in water to form a series of hydroxo complexes – $TlOH^{2+}$, $Tl(OH)_2^+$, $Tl(OH)_3^0$ and $Tl(OH)_4^-$, whereas Tl(I) weakly hydrolyzes in water and forms only one hydroxo complex at high pH values (Table 22.9). Of the Tl(III) complexes, $Tl(OH)_3^0$ predominates over much of the pH range (Figure 22.11).

In addition to strong complexes with hydroxyl ions, Tl^{3+} ions also form strong complexes with Cl$^-$ ions and a much weaker complex with SO_4^{2-} ions (Table 22.9). Even at relatively low Cl$^-$ concentrations, these chloride complexes predominate in acidic conditions (Figure 22.12). The corresponding chloride and sulfate complexes with Tl^+ are much weaker than those of Tl^{3+} and higher concentrations of both Cl$^-$ and SO_4^{2-} ions need to be present for these complexes to be important constituents of soil solutions (Figure 22.13).

Both Tl(III) and Tl(I) form a number of insoluble precipitates. For Tl(III), these include a hydroxide, $Tl(OH)_3(s)$ and an oxide, $Tl_2O_3(s)$. The stability phase relationships between these two forms have not been studied, but from published solubility products

Table 22.9 Formation constants, log β, for thallium with some inorganic ligands [75]

Reaction	log β	Reaction	log β	Reaction	log β
Thallium(III)					
$Tl^{3+} + H_2O \rightleftharpoons TlOH^{2+}$	-0.6	$Tl^{3+} + Cl^- \rightleftharpoons TlCl^{2+}$	7.72	$Tl^{3+} + SO_4^{2-} \rightleftharpoons TlSO_4^+$	4.4
$Tl^{3+} + 2H_2O \rightleftharpoons Tl(OH)_2^+$	-1.4	$Tl^{3+} + 2Cl^- \rightleftharpoons TlCl_2^+$	13.48		
$Tl^{3+} + 3H_2O \rightleftharpoons Tl(OH)_3^0$	-3.3	$Tl^{3+} + 3Cl^- \rightleftharpoons TlCl_3^0$	16.48		
$Tl^{3+} + 4H_2O \rightleftharpoons Tl(OH)_4^-$	-15	$Tl^{3+} + 4Cl^- \rightleftharpoons TlCl_4^-$	18.29		
Thallium(I)					
$Tl^+ + H_2O \rightleftharpoons TlOH^0$	-13.2	$Tl^+ + 2Cl^- \rightleftharpoons TlCl_2^-$	-0.1	$Tl^+ + SO_4^{2-} \rightleftharpoons TlSO_4^-$	1.37
$Tl^+ + Cl^- \rightleftharpoons TlCl^0$	0.57				

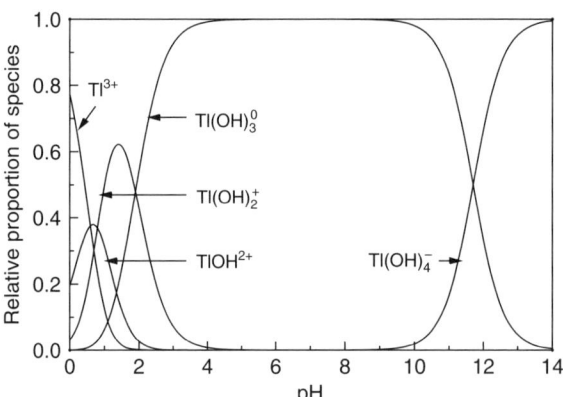

Figure 22.11 Hydrolysis of Tl(III) species in water

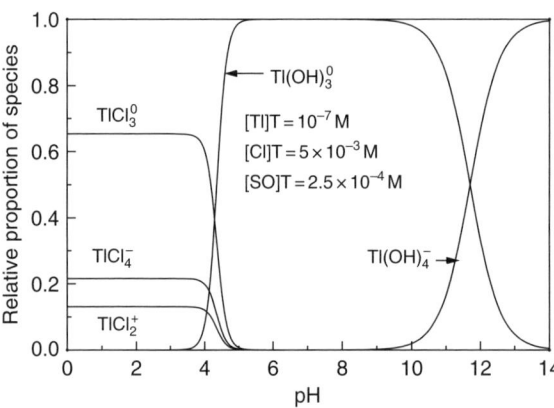

Figure 22.12 Speciation of Tl(III) species in the presence of chloride ions

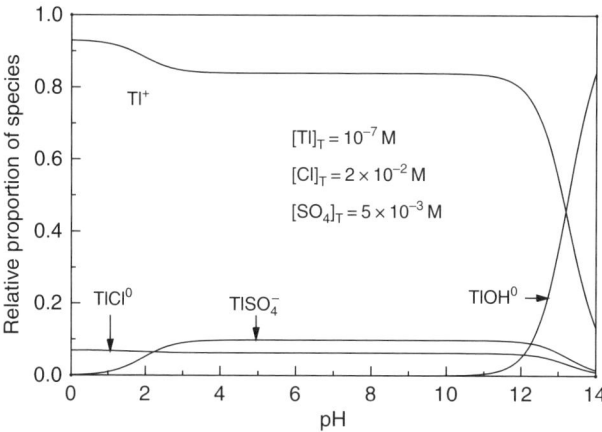

Figure 22.13 Speciation of Tl(I) species in the presence of chloride and sulfate ions

Table 22.10 Common precipitates of thallium [75]

Reaction	log K_{so}	Reaction	log K_{so}
Thallium(III)		*Thallium(I)*	
$Tl(OH)_3(s) + 3\,H^+ \rightleftharpoons Tl^{3+} + 3\,H_2O$	3.21	$Tl_2S(s) + H^+ \rightleftharpoons 2\,Tl^+ + HS^-$	−7.19
$Tl_2O_3(s) + 6\,H^+ \rightleftharpoons 2\,Tl^{3+} + 3\,H_2O$	5.90[a]	$TlCl(s) \rightleftharpoons Tl^+ + Cl^-$	−3.74

Data from [75] except [a]Biedermann and Glacer [76].

(Table 22.10) the hydroxide is the more insoluble form. Thallium oxide, $Tl_2O_3(s)$, occurs in nature as the rare mineral avicennite.

Thallium hydroxide, $Tl(OH)_3(s)$, has the lowest solubility when compared to the hydroxides of other trivalent metals, such as Al^{3+}, Cr^{3+}, Fe^{3+} and Ga^{3+} [77]. Calculations of the total concentration of free and hydroxo complexes of Tl(III) species in equilibrium with $Tl(OH)_3(s)$ using the constants given Tables 22.9 and 22.10 indicate that the upper limit of total Tl(III), $[Tl]_T$, in the absence of other complexant ligands is $10^{-6.51}$ M at pH values between about 3 and 11 (Figure 22.14). This corresponds to a Tl concentration of 0.063 mg L^{-1}.

For Tl(I), a number of precipitates are known to form [78]. These include a chloride, TlCl(s), a hydroxide, TlOH(s), a carbonate, $Tl_2CO_3(s)$ and a sulfide, $Tl_2S(s)$. Solubility products for TlCl(s), TlOH(s), and $Tl_2S(s)$ are shown in Table 22.10. The hydroxide and carbonate precipitates are too soluble to be of interest under normal environmental conditions and TlCl(s) is much more soluble than silver chloride, AgCl(s), and would only be important at very high Tl^+ and Cl^- ion concentrations. Thallium sulfide, $Tl_2S(s)$, can form under anaerobic conditions but is one of the most soluble of the metallic sulfides.

22.4.3.2 Sorption–Desorption Reactions

Thallium has been shown to be relatively mobile in soils [79] but can be retained by clays, oxide minerals, and soil organic matter. A sequential extraction scheme used to investigate

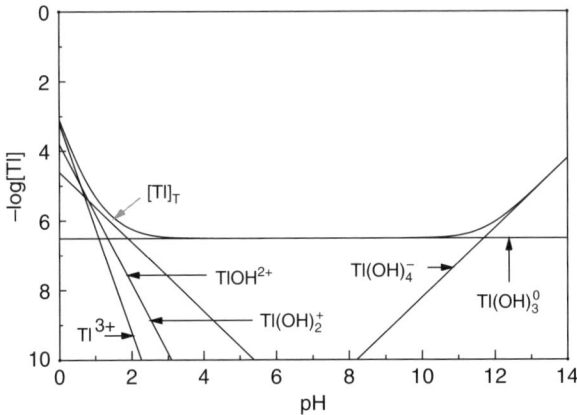

Figure 22.14 Dissolution of Tl(OH)$_3$(s) as a function of pH

Tl fractionation in polluted soils from Poland indicated that almost half of the Tl was in a nonlabile fraction and associated with the soil parent material [80]. Of the labile fraction, about 30 % was extracted with a reducing agent and presumably associated with Mn- and Fe-oxides, and about 10 % was extracted with an oxidizing agent and presumably associated with soil organic matter. A similar study on Japanese soils by Wang et al. [81] also concluded that labile Tl was mainly associated with the Mn-Fe oxide fraction. Using ammonium oxalate and EDTA as extractants Martin et al. [79] concluded that amorphous oxides of Fe and Al were the most important adsorbing surfaces for Tl-contaminated soils in Spain.

Thallium(I) has been shown to behave in a similar manner to K, Rb, and Cs in the presence of vermiculite because of its similar ionic radius and hydration energy and can enter the interlayer space of vermiculite where it is retained on air drying [82,83]. Studies on the retention of Tl(I) ions onto δ-MnO$_2$(s) and γ-Al$_2$O$_3$(s) have shown two different mechanisms of adsorption [84]. Complete adsorption of Tl(I) onto δ-MnO$_2$(s), a synthetic oxide with characteristics similar to birnessite, was complete at pH 4.7 and it was suggested that retention occurred through oxidation of Tl$^+$ ions to Tl^{3+} ions and the subsequent precipitation of Tl$_2$O$_3$(s). A similar experiment, however, conducted by Jacobson et al. [8] could not confirm by X-ray diffraction the presence of Tl$_2$O$_3$(s) precipitates. In contrast, Tl$^+$ ion adsorption onto γ-Al$_2$O$_3$(s) occurred through interaction with surface hydroxyls increasing over the pH range 8–12. An intrinsic complexation constant, log K^{int}, for Tl$^+$ ions with hydrous ferric oxide for use in the computer program MINTEQA2 has been calculated [85] using the metal adsorption data presented in Dzombak and Morel [65], the linear free energy relationship between surface complexation constants, and the first metal hydrolysis constants.

$$\equiv \text{Fe} - \text{OH}^0 + \text{Tl}^+ \rightleftharpoons \equiv \text{Fe} - \text{OTl}^0 + \text{H}^+ \quad \log K^{int} = -3.5 \quad (22.5)$$

Because Tl(I) ions very weakly hydrolyze in water (log $\beta_{-1,1,1} = -13.2$), the calculated intrinsic binding constant for Tl(I) is the weakest of all the metals and adsorption occurs at a much higher pH than for the other metals. The calculated adsorption of Tl(I) onto hydrous ferric oxide as a function of pH using the diffuse layer model [65] shown in

Figure 22.9 shows that Tl(I) retention does not occur until pH values above 7.5 and confirms the findings of Jacobson *et al.* [8] who found very little adsorption of Tl(I) ions onto ferrihydrite at a pH of 5.1.

22.4.3.3 Oxidation–Reduction Chemistry

Thallium can exist in two oxidation states, Tl(III) and Tl(I). The standard electrode potential for the Tl^{3+}/Tl^+ couple can be calculated as 1.28 V using the thermodynamic data given in Wagman *et al.* [32]. This would indicate that Tl^+ is the dominant species over most of the E_h–pH stability field for water. As discussed above, the predominant form of Tl(III) at Tl concentrations above $10^{-6.51}$ M and in the pH range 3–11 is the precipitate $Tl(OH)_3(s)$, while at concentrations below this value the soluble species, $Tl(OH)_3^0$, should predominate.

At lower redox potentials, Tl(I) predominates as the Tl^+ ion, and at extremely high pH values (< 3.4), the hydrolyzed $TlOH^0$ species forms. In anaerobic conditions, thallium sulfide, $Tl_2S(s)$, may form, although compared with other metallic sulfides it is relatively soluble. As no Gibbs free energy value is given by Wagman *et al.* [32] for $Tl(OH)_3(s)$, a value of -478.1 kJ mol^{-1} was calculated from the solubility product given by Martell *et al.* [75] and the thermodynamic data for Tl^{3+} and H_2O given in Wagman *et al.* [32]. This value was used to calculate the standard electrode potential, E^0, for the reduction of $Tl(OH)_3(s)$ to Tl^+ ions given in Table 22.11. The E_h–pH predominance diagram for Tl(III) and Tl(I) species is given in Figure 22.15.

Table 22.11 Reduction reactions involving Tl(III) and Tl(I) species

Reaction	E^0
$Tl(OH)_3(s) + 3H^+ + 2e^- \rightleftharpoons Tl^+ + 3H_2O$	1.228
$Tl(OH)_3^0 + 3H^+ + 2e^- \rightleftharpoons Tl^+ + 3H_2O$	1.357
$2Tl^+ + SO_4^{2-} + 8H^+ + 8e^- \rightleftharpoons Tl_2S(s) + 3H_2O$	0.263

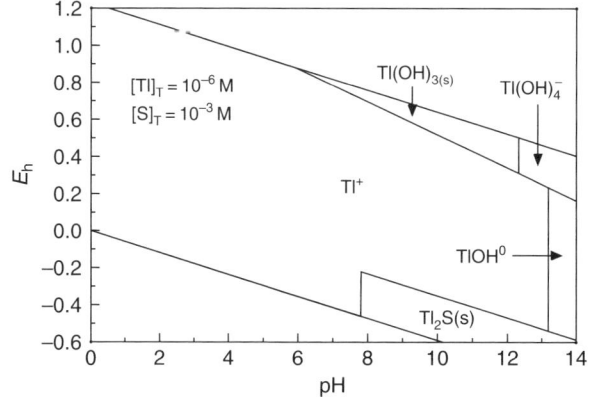

Figure 22.15 Oxidation–reduction reactions of Tl(III) and Tl(I) species

22.5 Vanadium

22.5.1 Geochemical Occurrences and Soil Concentrations

There is no single mineral ore from which vanadium is recovered. A deep-red mineral, vanadinite, $Pb_5(VO_4)_3Cl$, contains vanadium but it is uncommon. An oxyhydroxide of V, montrosite, $[(V^{3+}, V^{4+}, Fe^{2+})OOH]$, has been reported in uranium deposits with high V contents [86,87]. However, V is found as a trace element in a number of different rock materials, and because of the similarity between V^{3+} and Fe^{3+} and Al^{3+}, it is found in magnetite (iron oxide) and bauxite (aluminum ore) deposits. Similarly, because the vanadate ion behaves in a similar manner to phosphate ions, it is also found in rocks with high concentrations of phosphorus-containing minerals, and also in sandstones that have high uranium contents.

Vanadium concentrations are also high in carbon-rich deposits, such as coal, oil shale, crude oil, and tar sands, although the high concentrations in coal are believed to be associated with inorganic clay minerals [88]. Vanadium has been shown to replace Al in the octahedral sheets of clay minerals [89]. Because of the strong correlation between dissolved V and Si, the majority of V found in rivers is believed to be derived from the weathering of silicate minerals [90].

The average concentration of V in soils ranges from trace amounts up to $400\,mg\,kg^{-1}$, with an average value of $108\,mg\,kg^{-1}$ [14]. A mean value of $85\,mg\,kg^{-1}$ has been calculated for some soils of the United States [91]. The highest concentrations of V of $150-460\,mg\,kg^{-1}$ have been reported for soils derived from mafic rocks and the lowest of $5-22\,mg\,kg^{-1}$ in organic soils [2]. Aide [92] found a very strong relationship between V and Fe contents in soils and suggested that this relationship could be used to indicate anthropogenic contamination in regional soils.

In soils to which fly ash has been added, concentrations of up to $430\,mg\,kg^{-1}$ have been measured [93]. Very high concentrations of $620-1680\,mg\,kg^{-1}$ have been measured in soils in the vicinity of a vanadium mine in South Africa, with about 50% of the V occurring in the V(V) oxidation state [94].

22.5.2 Sources of Contamination

Because V is used extensively to make steel alloys for tools and construction purposes and is used in many other industrial applications, it is a common contaminant in many industrial wastes.

Vanadium contamination of soil may occur from the disposal and discharge of petroleum and coal products [2]. Other potential anthropogenic sources include phosphate fertilizers and mining activities. Vanadium concentrations in some phosphate fertilizers have been reported to range from 34 to $143\,mg\,kg^{-1}$ [18] and concentrations in some biosolids to range from 8 to $54\,mg\,kg^{-1}$ [74]. The V concentration in biosolids has not been extensively studied, although one report found concentrations from 8 to $54\,mg\,kg^{-1}$[74].

22.5.3 Chemical Behavior in Soils

22.5.3.1 Soluble and Solid Phases

Vanadium can exist in three oxidation states, V(V), V(IV) and V(III), in the presence of water at 25 °C and one atmosphere pressure. The aqueous chemistry of V(V) is extremely complex, particularly at higher concentrations. At low concentrations of V(V), five species have been reported (Table 22.12). These are the pervanadyl ion, VO_2^+, vanadic acid, $H_3VO_4^0$ and its three conjugate bases, $H_2VO_4^-$, HVO_4^{2-} and VO_4^{3-}. The species, VO_2^+, is the dominant species under acidic conditions and VO_4^{3-} under extremely alkaline conditions (Figure 22.16). Vanadic acid, $H_3VO_4^0$, is only a minor constituent within the pH range 2–5.

In addition to these mono-nuclear species, eleven soluble poly-nuclear species have been reported by Petterson *et al.* [95] and fourteen in Martell *et al.* [75] containing 2, 3, 4, 5, 6 or 10 vanadium atoms (Table 22.13). At concentrations of about 1.25 mM, the V_2, V_4,

Table 22.12 Dissociation constants, log K, for mononuclear vanadium(V) species

Reaction	log K	Reaction	log K
$VO_2^+ + 2H_2O \rightleftharpoons H_3VO_4^0 + H^+$	−3.67	$H_2VO_4^- \rightleftharpoons HVO_4^{2-} + H^+$	−8.06
$H_3VO_4^0 \rightleftharpoons H_2VO_4^- + H^+$	−3.4	$HVO_4^{2-} \rightleftharpoons VO_4^{3-} + H^+$	−13.28

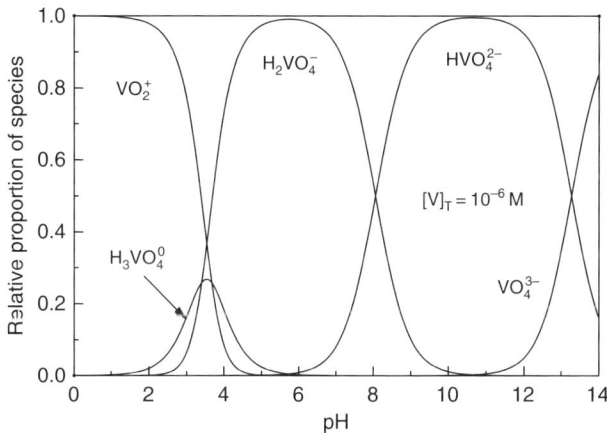

Figure 22.16 Speciation of vanadium(V) at low concentrations

Table 22.13 Polynuclear soluble vanadium(V) species

$H_2V_2O_7^{2-}$	$V_3O_{10}^{5-}$	$V_4O_{12}^{4-}$	$V_5O_{15}^{5-}$	$V_6O_{18}^{6-}$	$H_3V_{10}O_{28}^{3-}$
$HV_2O_7^{3-}$		$HV_4O_{13}^{5-}$			$H_2V_{10}O_{28}^{4-}$
$HV_2O_7^{4-}$		$V_4O_{13}^{6-}$			$HV_{10}O_{28}^{5-}$
					$V_{10}O_{28}^{6-}$

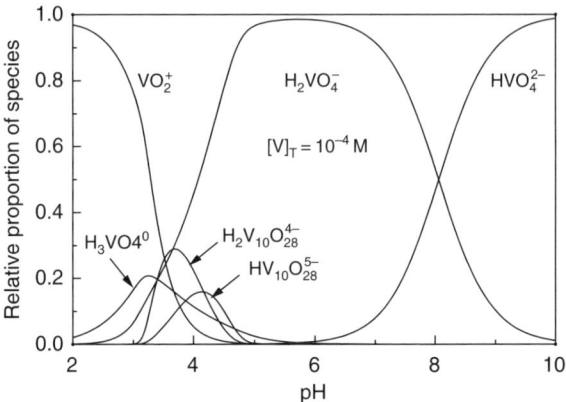

Figure 22.17 Speciation of vanadium(V) at a concentration of 0.1 mM

Table 22.14 Formation constants, log β, for vanadium(IV) with some inorganic ligands

Reaction	log β	Reaction	log β
$VO^{2+} + H_2O \rightleftharpoons VO(OH)^+ + H^+$	−5.7	$VO^{2+} + SO_4^{2-} \rightleftharpoons VOSO_4^0$	2.46
$2VO^{2+} + 2H_2O \rightleftharpoons (VO(HO))_2^{2+} + 2H^+$	−6.7	$VO^{2+} + CO_3^{2-} \rightleftharpoons VOCO_3^0$	3.47
$VO^{2+} + Cl^- \rightleftharpoons VOCl^+$	−0.1	$VO^{2+} + CO_3^{2-} + H_2O \rightleftharpoons VO(OH)CO_3^- + H^+$	1.52

V_5 and V_{10} polynuclear species are the most important [96]. The speciation of V(V) at a 0.1 mM concentration is shown in Figure 22.17 using the formation constants given in Elvington et al. [97].

The aqueous chemistry of V(IV) is dominated by the vanadyl cation, VO^{2+}, and its associated complexes with inorganic and organic cations [98] (Table 22.14). The calculated aqueous speciation of V(IV) in a simulated soil solution is shown in Figure 22.18.

Because SO_4^{2-} ions are stronger complexes for VO^{2+} than Cl^- ions, the $VOSO_4^0$ complex is much more important than the $VOCl^+$ complex even though the SO_4^{2-} ion concentration is lower than the Cl^- concentration. The aqueous chemistry of vanadium(III) is similar to that of Fe(III) in that the vanadic, V^{3+}, ion hydrolyzes in water to form the VOH^{2+} and $V(OH)_2^+$ ions (Table 22.15). By analogy with the Fe(III) system, the V^{3+} ion would be expected to form relatively weak complexes with the Cl^- ion but a strong complex with the SO_4^{2-} ion. A formation constant, log β, of 3.28 has been calculated for the VSO_4^+ ion [98]. This compares with values of 3.89, 3.81, and 2.25 for Al^{3+}, Cr^{3+}, and Fe^{3+}, respectively.

A dinuclear species, $V_2(OH)_2^{4+}$, has been shown to form but is only important at relatively high V(III) concentrations. The speciation of soluble V(III) species is shown in Figure 22.19.

A number of oxide minerals are known for V, these include vanadium pentoxide, $V_2O_5(s)$, vanadium oxyhydroxide, $VO(OH)_2(s)$, vanadium tetroxide, $V_2O_4(s)$, vanadium

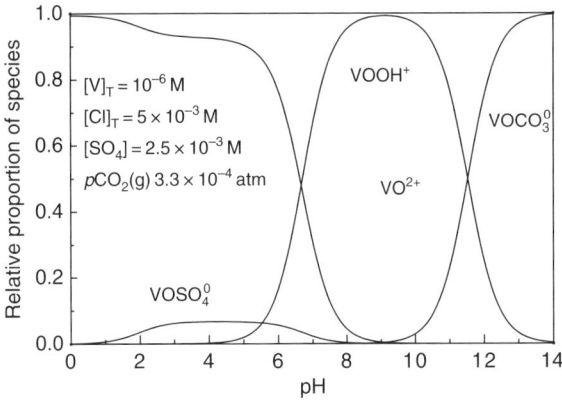

Figure 22.18 Speciation of vanadium(IV) in a simulated soil solution

Table 22.15 Formation constants, log β, for vanadium(III) with some inorganic ligands

Reaction	log β	Reaction	log β
$V^{3+} + H_2O \rightleftharpoons VOH^{2+} + H^+$	-2.3	$2V^{3+} + 2H_2O \rightleftharpoons V_2(OH)_2^{4+} + 2H^+$	-3.79
$V^{3+} + 2H_2O \rightleftharpoons V(OH)_2^+ + 2H^+$	-6.33	$V^{3+} + SO_4^{2-} \rightleftharpoons VSO_4^+$	3.28

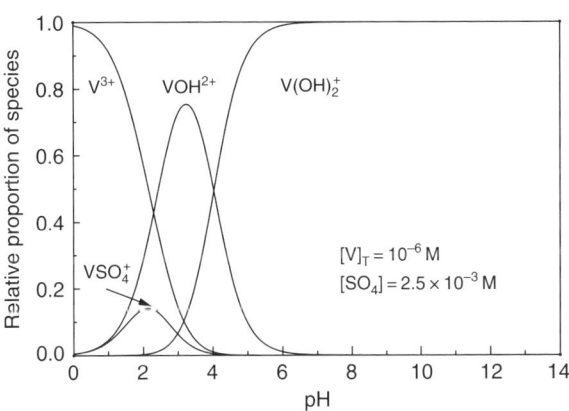

Figure 22.19 Speciation of vanadium(III) in the presence of SO_4^{2-} ions

trioxide, $V_2O_3(s)$ and vanadium hydroxide, $V(OH)_3(s)$ (Table 22.16). The oxides $V_2O_5(s)$, $V_2O_4(s)$ and $V_2O_3(s)$ are generally prepared at high temperatures (>300 °C) [99], conditions unlikely to occur in soils or sediments, and are therefore unlikely to precipitate under ambient conditions. However, $VO(OH)_2(s)$ and $V(OH)_3(s)$ can form at 25 °C [100] and so are possible vanadium oxide precipitates in soils.

Vanadium pentoxide, $V_2O_5(s)$, is the most important synthesized V compound and is used in a number of industrial processes. It may enter soils through atmospheric deposition

Table 22.16 *Some oxide, oxyhydroxide, and hydroxide precipitates of vanadium*

Reaction	log K_{so}	Reaction	log K_{so}
Vanadium(V)		Vanadium(IV)	
$V_2O_5(s) + 2H^+ \rightleftharpoons 2VO_2^+ + H_2O$	−1.36	$VO(OH)_2(s) + 2H^+ \rightleftharpoons VO^{2+} + 2H_2O$	6.11
Vanadium(III)			
$V(OH)_3(s) + 3H^+ \rightleftharpoons V^{3+} + 3H_2O$	7.59		

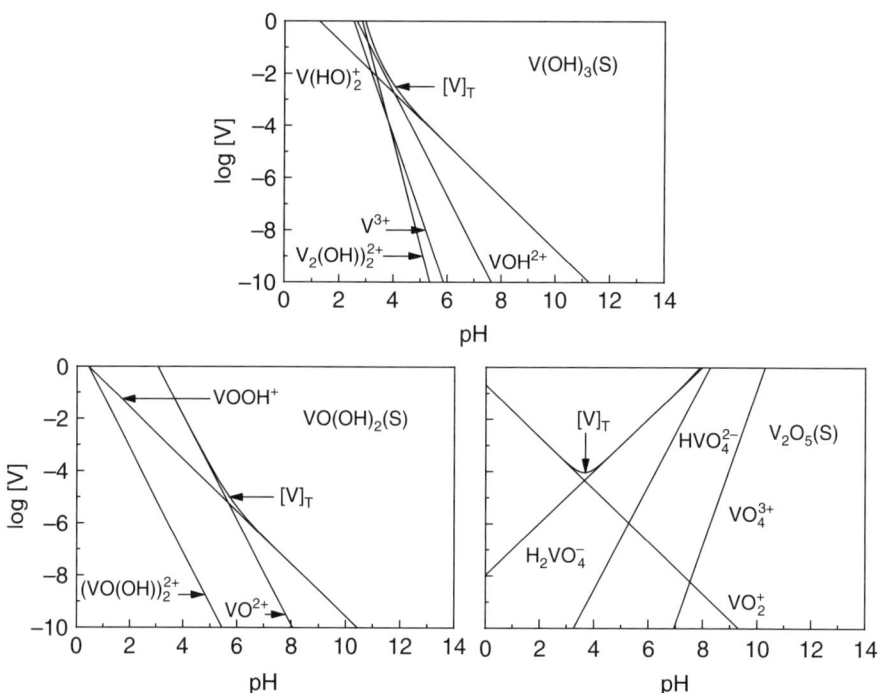

Figure 22.20 *Solubility of some vanadium oxides as a function of pH*

or disposal of wastes in the vicinity of industrial establishments. It is, however, relatively soluble in water, with a minimum solubility at about pH 4 (Figure 22.20). The solubilities of $VO(OH)_2(s)$ and $V(OH)_3(s)$, on the other hand, both decrease as the pH increases (Figure 22.20).

22.5.3.2 Sorption–Desorption Reactions

Adsorption of both V(IV) and V(V) onto oxide minerals has been interpreted in terms of the formation of inner-sphere bidentate complexes for the vanadyl ion, VO^{2+}, and monodentate complexes for the vanadate ions, $H_nVO_4^{n-1}$ [101]. The oxidation of V(IV) to V(V) is considerably enhanced on alumina and titania oxides after adsorption of the vanadyl ion [102]. More recent work on the adsorption of V(V) ions onto goethite

Table 22.17 Surface reactions and intrinsic complexation constants for V(V) onto goethite

Adsorption reaction	log K^{int}
$2 \equiv Fe-OH^0 + VO_4^{3-} + 4H^+ \rightleftharpoons \; \equiv Fe_2-OVO(OH)_4^+ + 2H_2O$	42.87
$2 \equiv Fe-OH^0 + VO_4^{3-} + 3H^+ \rightleftharpoons \; \equiv Fe_2-OVO_2(OH)^0 + 2H_2O$	41.26

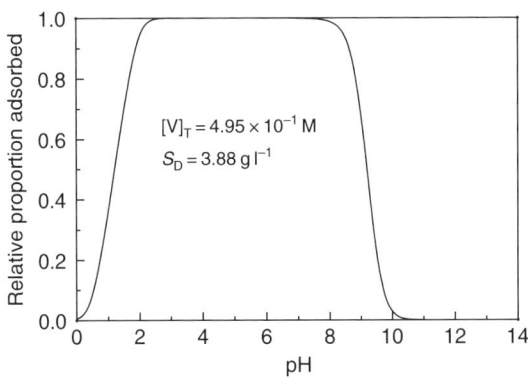

Figure 22.21 Adsorption of vanadium(V) onto goethite

[103] has demonstrated the formation of two bidentate complexes, $\equiv Fe_2OVO(OH)_2^+$ and $\equiv Fe_2OVO_2(OH)^0$ and their intrinsic formation constants were determined by batch adsorption experiments (Table 22.17).

The calculated adsorption of V(V) ions onto goethite as a function of pH using the diffuse layer model and the constants given in Peacock and Sherman [103] is shown in Figure 22.21.

22.5.3.3 Oxidation–Reduction Reactions

The vanadate ion has been shown to act as a terminal electron acceptor by a number of anaerobic organisms [104]. Microbial reduction of V(V) to V(IV) by *Shewanella oneidensis* and *Geobacter metallireduccus* results in the precipitation of a V(IV)-containing solid [105,106].

Using the thermodynamic data given in Wanty and Goldhaber [98], it is possible to construct an E_h–pH diagram for the three oxidation states of V – V(III), V(IV) and V(V). The oxidation–reduction reactions and the calculated standard electrode potential, E^0, are shown in Table 22.18 and the E_h–pH predominance diagram in Figure 22.22.

From Figure 22.22, it can be seen that V(V) species predominate in oxic environments at low pH values and in both oxic and anoxic environments at pH values greater than about 8. The vanadyl ion, VO^{2+}, and its hydrolyzed species, $VOOH^+$, do not exist at pH values above about 7 and that below this value they are stable in soil environments that are oxic to anoxic. Vanadium(III) species only exist under anoxic conditions and at pH values up to about 10, where they become unstable relative to V(V) species.

Table 22.18 Oxidation–reduction reactions for the oxidation states of vanadium

Reaction	E^0	Reaction	E^0
Vanadium(V)–vanadium(IV)		*Vanadium(IV)–vanadium(III)*	
$VO_2^+ + 2H^+ + e^- \rightleftharpoons VO^{2+} + 2H_2O$	1.00	$VO^{2+} + 2H^+ + e^- \rightleftharpoons V^{3+} + H_2O$	0.337
$H_2VO_4^- + 4H^+ + e^- \rightleftharpoons VO^{2+} + 3H_2O$	1.419	$VO^{2+} + H^+ + e^- \rightleftharpoons VOH^{2+}$	0.191
$H_2VO_4^- + 3H^+ + e^- \rightleftharpoons VOOH^+ + 2H_2O$	1.084	$VO^{2+} + H_2O + e^- \rightleftharpoons V(OH)_2^+$	−0.038
		$VOOH^+ + H^+ + e^- \rightleftharpoons V(OH)_2^+$	0.297
Vanadium(IV)–vanadium(III)			
$H_2VO_4^- + 4H^+ + 2e^- \rightleftharpoons V(OH)_2^+ + 2H_2O$	0.691		
$HVO_4^{2-} + 5H^+ + 2e^- \rightleftharpoons V(OH)_2^+ + 2H_2O$	0.029		

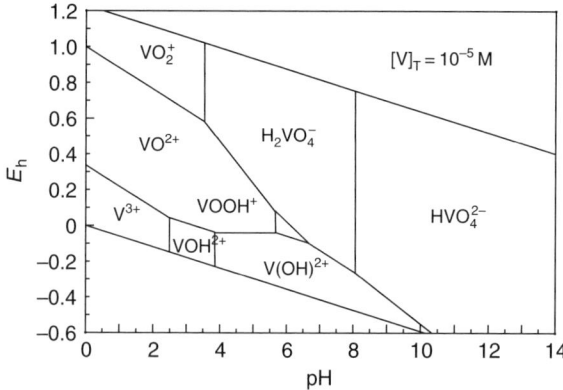

Figure 22.22 E_h–pH predominance diagram for soluble species of vanadium

22.6 Environmental and Human Health Risks

22.6.1 Molybdenum

Molybdenum is an essential element for the growth and development of both plants and animals. In plants, for example, the presence of Mo in certain enzymes allows the plant to absorb nitrogen [107]. Most plant species, except for legumes, do not exhibit symptoms of Mo-deficiency because Mo requirements are low. Deficiency symptoms in legumes are mainly exhibited as N-deficiency symptoms because of the primary role of Mo in nitrogen fixation [107]. Nitrogen metabolism, protein synthesis and sulfur metabolism are also affected by the presence of Mo.

Molybdenum has a significant influence on pollen formation, so fruit and grain formation are affected in Mo-deficient plants. Unlike the other micronutrients, Mo-deficiency symptoms are not confined mainly to the youngest leaves because Mo is mobile in plants. The characteristic Mo deficiency symptom in vegetable crops is irregular leaf blade formation known as whiptail, but interveinal mottling and marginal chlorosis of older leaves have also been observed [108].

Molybdenum deficiencies in plants are found mainly on acidic, highly weathered, sandy soils in humid regions [109]. Unlike many other micronutrients, Mo uptake by plants increases with increased soil pH. Molybdenum deficiencies in legumes may be corrected by liming acid soils rather than by increasing Mo applications. However, treatment of seeds with Mo sources may be more economical than liming in some areas.

Molybdenum is an essential micronutrient for animals because of the formation of Mo-containing enzymes, which are necessary for the health of animals [110]. In ruminants, the dietary intake of excessive Mo causes hypocuprosis – Mo-induced Cu deficiency. Domestic ruminants are much more susceptible to molybdenum toxicity than are nonruminants.

Molybdenum toxicity in ruminants has been reported in areas with organic soils; where plants grow in alkaline sloughs; where excess Mo-containing fertilizers have been applied to agricultural land; and where applications of lime appeared to increase plant molybdenum uptake [111]. Hypocuprosis in cattle has also been reported on land to which biosolids have been applied [112], on reclaimed land from mine tailings [113], and in cattle grazing on land contaminated with Mo from various industrial sources [111]. Soils underlain by black shales often contain >3 mg kg^{-1} Mo and incidences of bovine hypocuprosis have been reported on these soils [114].

22.6.2 Silver

Although Ag is widely distributed, it is not an essential element for plant growth, nor is it a normal constituent of plants [115]. In general, Ag accumulation in higher plants and fungi is expected only in areas contaminated with Ag such as tailings from Ag mines, and soils amended with Ag-containing sewage sludge [116]. In terrestrial plants, Ag concentrations are usually less than 0.1 mg kg^{-1} dry weight and are higher in trees, shrubs, and other plants near regions of Ag mining, with seeds, nuts, and fruits usually containing higher Ag concentrations than other plant parts [60].

Uptake and accumulation of Ag by terrestrial plants occur primarily at the roots, where the factors influencing the rate of uptake are plant species, soil composition, soil pH, and the form of Ag [4,43,60,116]. In a study presented by Smith and Carson [60], sprays containing 9.8 mg l^{-1} dissolved Ag were toxic to corn and sprays containing 100–1000 mg l^{-1} dissolved Ag killed tomato and bean plants. Conversely, more recent studies have shown that plants such as corn, lettuce, and soybean, planted in soils amended with Ag$_2$S(s) and sewage sludge, germinated and grew normally even at the highest concentration of Ag, with only Chinese cabbage and lettuce adversely affected at 14 mg Ag kg^{-1} dry weight soil and higher [116].

In the terrestrial environment, it has been demonstrated that denitrification is highly sensitive to Ag in soils. Concentrations ranging from 540 to 2700 mg Ag kg^{-1}, inhibit enzymes required for the P, S and N cycles of nitrifying bacteria in the soil [117,118]. Murata et al. [119] demonstrated that although the exchangeable fraction was small compared with the added amount of Ag, the effect on soil dehydrogenase activity was severe and bacterial colony growth was inhibited at levels between 0.1 and 0.5 mg Ag kg^{-1} of soil. In addition, ionic Ag can block DNA transcription and interrupt bacterial respiration and adenosine triphosphate (ATP) synthesis [112].

In a recent study, the survival, growth and bioaccumulation of Ag in earthworms (*L. terrestris*) was tested using an artificial soil containing increasing concentrations of $Ag_2S(s)$ for 28 days. The tested earthworms did not accumulate Ag but responded instead with reduced growth [120]. However, in another study conducted by Beglinger and Ruffing [121], on earthworms that were exposed to 2000 mg $Ag_2S(s)$ kg^{-1} soil for 14 days, found no effects on mortality, burrowing time, appearance or weight.

Silver has been shown to hyperaccumulate in macrofungi and several plants species. Borovicka *et al.* [122] investigated the accumulation of Ag in the macrofungi *A. solitaria* and *A. strobiliformis*. Samples had Ag concentrations 800 to 2500 times higher than soil concentrations, indicating hyperaccumulation of the element. In a recent study on the uptake of Ag by two metal-tolerant plants – *Brassica juncea* and *Medicago sativa* – Ag accumulations reached 12.4 % wt and 13.6 % wt, respectively, with Ag stored as discrete nanoparticles [123].

22.6.3 Thallium

Because of the similar ionic radii of Tl^+ (0.170 nm) and K^+ (0.164 nm), the ability of Tl^+ ions to substitute for, or interfere with, K^+ ions in biogeochemical systems is well known [124]. The toxicity of Tl to mammalian health is comparable to or greater than the toxicity of Hg, Pb, Cd, Cu, and Zn [7,125].

Thallium concentrations in plants are usually below 0.1 mg kg^{-1} dry weight [126]. Studies on the uptake of Tl by a number of agricultural crops have shown increased uptake by rape, *Brassica nupis*, [126,127] and Chinese green cabbage [6]. Plant contents up to 65 mg kg^{-1} have been reported in plants growing on Tl contaminated waste piles [12] and up to 500 mg kg^{-1} in Chinese green cabbage growing in soils associated with Tl-rich sulfide mineralization [6]. The thallium-accumulator plants kale, *Brassica oleracea acephala*, candytuff, *Iberis intermedia*, and the mountain herb, *Biscutella laevigata* (Brassicaceae) have received much attention as potential candidates for the phytoextraction of Tl from contaminated soils [129–132].

Thallium is, however, phytotoxic to many plants. Common duckweed, *Lemna minor*, growth was inhibited at Tl doses of less than 0.2 mg l^{-1} and chlorosis and death occurred at Tl doses of 10 mg l^{-1} [133]. Garden cress *(Lepidium sativa)* shoots showed reduced growth at soil concentrations of 10 mg Tl kg^{-1}[125]; photosynthesis and transpiration were reduced by 50–70 % in corn (*Zea mays* L.) and sunflower (*Helianthus annuus* L.) plants exposed to up to 10 mg l^{-1} in solution culture [134].

22.6.4 Vanadium

The high concentrations of V in many crude oils and coals have raised the suggestion that increased atmospheric deposition of V will occur through the burning of fossil fuels [135,136].

Many of the biochemical activities of V are related to the complexing ability of the vanadyl cation, VO^{2+}, and the chemical similarity of vanadate to phosphate, which allows vanadate compounds to interact with numerous enzymes in living organisms by either inhibiting or activating them [137]. Concentrations of V in plants are typically

0.27–4.2 mg kg^{-1} [138]. At low concentrations, V appears to stimulate plant growth, but at higher concentrations it is toxic. It is believed to be taken up principally as the vanadate ion, H$_n$VO$_4^{3-n}$, but is reduced to the vanadyl ion, VO^{2+}, during plant uptake [93].

Vanadium has been shown to be both essential and toxic to humans [10], with the major input source being ingestion through intake of food [139]. Studies on exposure of children to V near a V-processing operation showed no negative impacts on their health over a three-year period [140]. Accumulation of vanadate ions in bone and reduction of vanadate ions to vanadyl ions in red blood cells appear to be major detoxification mechanisms [141]. There has been increased interest in the toxic effects of V to humans because of its insulin mimetic activities and its use in the treatment of diabetes. Human trials have demonstrated a high tolerance for orally supplied V at pharmacological doses greater than 100 times that of the usual dietary intake [142].

References

[1] Rudnick, R.L.; Gao, S.; Composition of the continental crust; in Rudnick, R.L. (ed.), The Crust – Treatise on Geochemistry, Vol. 3 (Holland, H.D., Turekin, K.K., eds); Elsevier–Pergamon, Oxford, 2003, pp. 1–64.
[2] Kabata-Pendias, A., Trace Elements in Soils and Plants, 3rd edn; CRC Press, Boca Raton, FL, 2001.
[3] Scott, C.; Lyons, T.W.; Bekker, A. et al., Tracing the stepwise oxygenation of the Proterozoic ocean; Nature 2008, 452, 456–460.
[4] Ratte, H.T., Bioaccumulation and toxicity of silver compounds: a review; Environ. Toxicol. Chem. 1999, 18, 89–108.
[5] Kazantzis, G., Thallium in the environment and health effects; Environ. Geochem. Health 2000, 22, 275–280.
[6] Xiao T.; Guha J.; Boyle D. et al., Environmental concerns related to high thallium levels in soils and thallium uptake by plants in southwest Guizhou, China; Sci. Total Environ., 2004, 318; 223–244.
[7] JohnPeter, A.L.; Viraraghavan, T., Thallium: a review of public health and environmental concerns; Environ. Int. 2005, 31, 493–501.
[8] Jacobson, A.R.; McBride, M.B.; Baveye, P. et al., Environmental factors determining the trace-level sorption of silver and thallium to soils; Sci. Total Environ. 2005, 345, 191–205.
[9] Domingo, J.L., Vanadium: a review of the reproductive and development toxicology; Reprod. Toxicol. 1996, 10, 175–182.
[10] Mukherjee, B.; Patra, B.; Mahapatra, S. et al., Vanadium – an element of atypical biological significance; Toxicol. Lett. 2004, 150, 135–142.
[11] Arnon, D.I.; Wessel, G., Vanadium as an essential element for green plants; Nature 1953, 172, 1039–1040.
[12] Kaplan, D.I.; Adriano, D.C.; Carlson, C.L. et al., Vanadium: toxicity and accumulation by beans; Water Air Soil Pollut. 1990, 49, 81–91.
[13] Kurada, P.K.; Sandell, E.B., Geochemistry of molybdenum; Geochim. Cosmochim. Acta 1954, 6, 35–63.
[14] Ure, A.M.; Berrow, M.L., The elemental constituents of soils; Environmental Chemistry Vol. 2; Royal Society of Chemistry, London, 94–204, pp. 1982.
[15] Kubota, J., Areas of molybdenum toxicity to grazing animals in the western states; J. Range Manage. 1975, 28, 252–256.
[16] Neunhäyserer, C.; Berreck, M.; Insam, H., Remediation of soils contaminated with molybdenum using soil amendments and phytoremediation; Water Air Soil Pollut. 2001, 128, 85–96.
[17] Draggan, S., Molybdenum; in: Cleveland, C.J. (ed.), Encyclopedia of Earth; Environmental Information Coalition, National Council for Science and the Environment, Washington, DC, 2008

[18] McBride, M.B.; Spiers, G., Trace element content of selected fertilizers and dairy manures as determined by ICP-MS; Comm. Soil Sci. Plant Anal. 2001, 32, 139–156.

[19] Essington, M.E., Formation of calcium and magnesium molybdate complexes in dilute aqueous solution; Soil Sci. Soc. Am. J. 1993, 56, 1124–1127.

[20] Essington, M.E., Calcium molybdate solubility in spent oil shale and a preliminary evaluation of the association constants for the formation of $CaMoO_4^0$(aq), $KMoO_4^-$(aq) and $NaMoO_4^-$(aq); Environ. Sci. Technol. 1990, 24, 214–220.

[21] Öhman, L., Equilibrium and structural studies of silicon(IV) and aluminum(III) in aqueous solution. 21. A potentiometric and ^{27}Al NMR study of the system H^+–Al^{3+}–MoO^2_4; Inorg. Chem. 1986, 28, 3629–3632.

[22] Verweij, W., Equilibria and constants in CHEAQS: selection criteria, sources and assumptions; Version 5, April 2003. (www.tiscali.nl/cheaqs)

[23] Goldberg, S.; Lesch, S.M.; Suarez, D.L., Predicting molybdenum adsorption by soils using soil chemical parameters in the Constant Capacitance Model; Soil Sci. Soc. Am. J., 2002, 66, 1836–1842.

[24] Akiyoshi, S.; Hisashi, S., Adsorption characteristics of molybdenum on various soils and incorporation into soil organic matter; Jap. J. Soil Sci. Pl. Nutr. 2004, 75, 179–184.

[25] Zhang, P.C.; Sparks, D.L., Kinetics and mechanisms of molybdate adsorption/desorption at the goethite/water interface using pressure-jump relaxation; Soil Sci. Soc. Am. J. 1989, 53, 1028–1034.

[26] Goldberg, S.; Forster, H.S.: Godfrey, C.L., Molybdenum adsorption on oxides, clay minerals and soils; Soil Sci. Soc. Am. J. 1996, 60, 425–432.

[27] Gustafsson, J.P., Modeling molybdate and tungstate adsorption to ferrihydrite; Chem. Geol. 2003, 200, 105–115.

[28] Motta, M.M.; Miranda, C.F., Molybdate adsorption on kaolinite, montmorillonite and illite: Constant capacitance modelling; Soil Sci. Soc. Am. J. 1989, 53, 380–385.

[29] Bibak, A.; Borggaard, O.K., Molybdenum adsorption by aluminum and iron oxides and humic acid; Soil Sci. 1994, 158, 323–328.

[30] Cruywagen, J.J.; DeWet, H.F., Equilibrium study of the adsorption of molybdenum(VI) on activated charcoal; Polyhedron 1988, 7, 547–556.

[31] Fox, P.; Doner, H.E., Accumulation, release and solubility of arsenic, molybdenum and vanadium in wetalnd sediments; J. Environ. Qual. 2003, 2428–2435.

[32] Wagman, D.D.; Evans, W.H.; Parker, V.B. et al., The NBS tables of chemical thermodynamic properties; J. Phys. Chem. Ref. Data 1982, 11, Suppl. 2, 1–392.

[33] Tucker, M.D.; Barton, L.L.; Thompsom, B.M., Reduction and immobilizationof molybdenum by Desulfovibrio desulfuricane; J. Environ. Qual. 1997, 26, 1146–1152.

[34] Helz, G.R.; Miller, C.V.; Charnock, J.M. et al., Mechanism of molybdenum removal from the sea and its concentration in black shales: EXAFS evidence; Geochim. Cosmochim. Acta 1996, 60, 3631–3642.

[35] Shukor, M.Y.; Habib, S.H.M.; Rahman, M.F.A. et al., Hexavalent molybdenum reduction to molybdenum blue by S Marceseus strain Dr. Y6; Appl. Biochem. Biotech. 2008, 149, 33–43.

[36] Renner, H. Silver, silver compounds and silver alloys; in: Elvers, B., Hawkins, S., Russey, W., Schulz, G. (eds), Ullman's Encyclopedia of Industrial Chemistry, 5th edn; Wiley-VCH Verlag GmbH, Weinheim, 1993.

[37] Kramer, J.R.; Adams, N.W.H.; Manolopoulos, H. et al., Silver at an old mining camp, Cobalt, Ontario, Canada; Environ. Toxicol. Chem. 1999, 18, 23–29.

[38] Mukherjee, A.B.; Mukherjee, A.P., An overview of silver in India. Int. J. Surf. Min. Reclam. Environ. 1997, 11, 195–201.

[39] Boyle, R.W., The geochemistry of silver and its deposits with notes on geochemical prospecting for the element; Geol. Surv. Can., Ottawa 1969, 1–264.

[40] Purcell, T.W.; Peters, J.P., Sources of silver in the environment; Environ. Toxicol. Chem. 1998, 17, 539–546.

[41] Kramer, J.R.; Benoit, G.; Bowles, K.C. et al., Environmental chemistry of silver; in: Andren, A.W., Bober, T.W. (eds), Silver in the Environment: Transport Fate and Effects; Society of Environmental Toxicology and Chemistry, Pensacola, FL, 2002.

[42] Adriano, D.C., Trace Elements in Terrestrial Environments: Biogeochemistry, Bioavailability and Risks of Metals; 2nd edn; Springer-Verlag; New York.
[43] Jones, K.C.; Davies, B.E.; Peterson, P.J., Silver in Welsh soils: Physical and chemical distribution studies; Geoderma 1986, 37, 157–174.
[44] Figueroa, J.A.L.; Wrobel, K.; Afton, S. et al., Effect of some heavy metals and soil humic substances on the phytochelatin production in wild plants from silver mine areas of Guanajuato, Mexico; Chemosphere 2008, 70, 2084–2091.
[45] Sterckeman, T.; Douay, F.; Proix, N. et al., Assessment of the contamination of cultivated soils by eighteen trace elements around smelters in the north of France; Water Air Soil Pollut. 2002, 135, 173–194.
[46] Dumaresq, C., The occurrence of arsenic and heavy metal contamination from natural and anthropogenic sources in the Cobalt area of Ontario; MSc thesis, Carelton University, Ottawa, ON, 1993.
[47] The Silver Institute, World Silver Survey 2007 – A Summary; GFMS Limited; Washington, DC, 2007.
[48] Eckelman, M.J.; Graedel, T.E., Silver emissions and their environmental impacts: A multilevel assessment; Environ. Sci. Tech. 2007, 41, 6283–6289.
[49] World Health Organization, Silver and Silver Compounds: Environmental Aspects; WHO, Geneva, 2002.
[50] Blaser, S.A.; Scheringer, M.; MacLeod, M. et al., Estimation of cumulative aquatic exposure and risk due to silver: contribution of nano-functionalized plastics and textiles; Sci. Total Environ. 2008, 390, 396–409.
[51] Scafer, M.M.; Overdier, J.T.; Armstrong, D.E., Removal, partitioning, and fate of silver and other metals in wastewater treatment plants and effluent-receiving streams; Environ. Toxicol. Chem. 1998, 17, 630–641.
[52] US EPA, Ambient water quality criteria for silver; US Environmental Protection Agency, Washington, DC, 1980.
[53] Throbäck, I.N.; Johansson, M.; Rosenquist, M. et al., Silver (Ag^+) reduces denitrification and induces enrichment of novel nirK genotypes in soil; Microbiol. Lett. 2007, 270, 189–194.
[54] Benn, T.M.; Westerhoff, P., Nanoparticle silver released into water from commercially available sock fabrics; Environ. Sci. Technol. 2008, 42, 4133–4139.
[55] Stumm, W.; Morgan, J.J., Aquatic Chemistry: An Introduction Emphasizing Chemical Equilibria in Natural Waters; John Wiley & Sons, Inc., New York, 1981.
[56] Adams, N.W.H.; Kramer, J.R., Silver speciation in wastewater effluent, surface waters, and pore waters; Environ. Toxicol. Chem. 1999, 18, 2667–2673.
[57] Adams, N.W.H.; Kramer, J.R., Reactivity of Ag^+ ion with thiol ligands in the presence of iron sulfide; Environ. Toxicol. Chem. 1998, 17, 625–629.
[58] Bell, R.A.; Kramer, J.R., Structural chemistry and geochemistry of silver–sulfur compounds: Critical Review; Environ. Toxicol. Chem. 1999, 18, 9–22.
[59] Smith, I.C.; Carson, B.L. Trace Metals in the Environment, Vol. 2, Silver; Ann Arbor Science, Ann Arbor, MI, 1977.
[60] Dyck, W., Adsorption and co-precipitation of silver on hydrous ferric oxide; Can. J. Chem. 1968, 46, 1441–1444.
[61] Davis, J.A.; Leckie, J.O., Effect of adsorbed complexing ligands on trace metal uptake by hydrous oxides; Environ. Sci. Technol. 1978, 12, 1309–1315.
[62] Sikora, F.J.; Stevenson, F.J., Silver complexation by humic substances: conditional stability constants and nature of reactive sites; Geoderma 1988, 42, 535–363.
[63] Szabo, G.; Guczi, J.; Valyon, J. et al., Investigations of the sorption characteristics of radio-silver on some natural and artificial soil particles; Sci. Total Environ. 1995, 172, 65–78.
[64] Li, J.; Rate, A.W.; Gilkes, J., Silver ion desorption kinetics from iron oxides and soil organic matter: effect of adsorption period; Aust. J. Soil Res. 2004, 42, 59–67.
[65] Dzombak, D.A.; Morel, F.M.M., Surface Complexation Modeling; John Wiley & Sons, Inc., New York, 1990.
[66] Varshal, G.M.; Velyukhanova, T.K.; Baranova, N.N. et al., Geochemical significance of complexing silver(I) with humus acids. Geochem. Int. 1995, 32, 1–9.

[67] Hou, H.; Takamatsu, T.; Koshikawa, M.K. *et al.*, Migration of silver, indium, tin, antimony, and bismuth and variations in their chemical fractions on addition to uncontaminated soils; Soil Sci. 2005, 170, 624–639.
[68] Hou, H.; Takamatsu, T.; Koshikawa, M.K. *et al.*, Concentrations of Ag, In, Sn, Sb, and Bi, and their chemical fractionation in typical soils in Japan; Eur. J. Soil Sci. 2006, 57, 212–227.
[69] Nriagu, J.O.; History, production and uses of thallium; in: Nriagu, J.O. (ed.), Thallium in the Environment, John Wiley & Sons, Inc., New York, 1998.
[70] Dai, S.; Zeng, R.; Sun, Y., Enrichment of arsenic, antimony, mercury and thallium in a Late Permian anthracite from Xingren, Guizhou, southwest China; Int. J. Coal Geol. 2006, 66, 217–226.
[71] Wenqi, Q.; Yalei, C.; Jieshan, C.; Indium and thallium background contents in China; Int. J. Environ. Stud. 1992, 40: 311–315.
[72] Hofer, G.F.; Aichberger, K.; Hochmair, U.S.; Thalliumgehalte landwirtschaftlich genutzer Böden Oberösterreichs; Die Bodenkult. 1990, 41, 187–193.
[73] Tremel, A.; Masson, P.;Sterckeman, T. *et al.*, Thallium in French ecosystems. I. Thallium contents in arable soils; Environ. Pollut. 1997, 95, 293–302.
[74] WEAO, Fate and significance of contaminants in sewage sludge biosolids applied to agricultural landl; Final Report, Water Association of Ontario, Canada, 2001.
[75] Martell, A.F.; Smith, R.M., NIST critically selected stability constants of metal complexes; NIST Standard Reference Database 46, Version 8.0; NIST, Gaithersburg, MD, 2004.
[76] Biedermann, G.; Glacer, J., Calculation of equilibrium constants for some thallium(III) species in aqueous solutions containing different ionic media; Acta Chem. Scand. 1986, A40, 331–334.
[77] Lin, T.-S; Nriagu, J.O.; Speciation of thallium in natural waters; in: Nriagu, J.O. (ed.) Thallium in the Environment; John Wiley & Sons, Inc., New York, 1998, 31–43.
[78] Kaplan, D.I.; Mattigod, S.V., Aqueous geochemistry of thallium; in: Nriagu, J.O. (ed.) Thallium in the Environment; John Wiley & Sons, Inc., New York, 1998, 15–29.
[79] Martin, H.W.; Kaplan, D.I., Temporal changes in cadmium, thallium and vanadium mobility in soil and phytoavailability under field conditions; Water Soil Air Pollut. 1998, 101, 399–410.
[80] Jakubowska, M.; Pasieczna, A.; Zembrzuski, W. *et al.*, Thallium fractions of soil formed on flood-plain terraces.; Chemosphere 2007, 66, 611–618.
[81] Wang, L.; Kubota, M.; Higashi, T.*et al.*, Evaluation of a sequential extraction procedure for the fractionation of thallium in soil. Soil Sci. Plant Nutr. 2004, 50, 339–347.
[82] Walker, G.F.; Milne, M.A., Hydration of vermiculite saturated with various cations; Trans. Int. 4th Congr. Soil Sci. 1950, 2, 62–67.
[83] Frantz, G.; Carlson, R.M., Effects of rubidium, cesium, and thallium on interlayer potassium release from Transvaal vermiculite; Soil Sci. Soc. Am. J. 1987, 51, 305–308.
[84] Bidoglio, G.; Gibson, P.N.; O'Gorman *et al.*, X-ray absorption spectroscopy investigation of surface redox transformations of thallium and chromium on colloidal mineral surface; Geochim. Cosmochim. Acta 1993, 57, 2389–2394.
[85] HydroGeoLogic, Diffuse-layer sorption reactions for use in MINTEQA2 for HWIR metals and metalloids; Report prepared for USEPA, Athens, Georgia, 1999.
[86] Evans, H.T.; Block, S., The crystal structure of montroseite, a vanadium member of the diaspore family; Am. Miner. 1953, 38, 1242–1250.
[87] Forbes, P; Debussy, J., Characterization of fresh and altered montroseite [V,Fe]OOH. A discusssion of alteration processes; Phys. Chem. Miner. 1988, 15, 438–445.
[88] Swaine, D.J.; Trace Elements in Coal; Butterworths, London, 1990.
[89] Meunier, J.D.,The composition and origin of vanadium-rich clay minerals in Colorado Plateau Jurassic sandstones; Clays Clay Miner. 1994, 42, 391–400.
[90] Shiller, A.M.; Mao, L., Dissolved vanadium in rivers: effects of silicate weathering; Chem. Geol. 2000, 176, 13–22.
[91] Shacklette, H.T.; Boerngen, J.G., Element concentrations in soils and other surficial materials of the conterminous United States; USGeol. Prof. Pap., 1984, 1270, 105.
[92] Aide, M., Geochemical assessment of iron and vanadium relationships in oxic soil environments; Soil Sed. Contam., 2005, 14, 403–416.

[93] Morrell, B.G.; Lepp, N.W.; Phipps, D.A., Vanadium uptake by higher plants: some recent developments; Environ. Geochem. Health, 1985, 8, 14–18.
[94] Panichev, N.; Mandiwana, K.; Moema, D. *et al.*, Distribution of vanadium(V) species between soil and plants in the vicinity of vanadium mine; J. Hazard. Mater. 2006, A137, 649–653.
[95] Petterson, L.; Hedman, B.; Andersson, I. *et al.*, Multi component polyanions; Chem. Scrip. 1983, 22, 254–264.
[96] Petterson, L; Andersson, I.;Gorzsás A., Speciation in peroxovanadate systems; Coord. Chem. Rev. 2003, 237, 77–87.
[97] Elvington, K.; Baro, A.G.; Petterson, I., Speciation in vanadium bioinorganic systems. 2. An NMR, ESR, and potentiometric study of the aqueous H^+-vanadate-maltol system; Inorg. Chem. 1996, 35: 3388–3393.
[98] Wanty, R.B.; Goldhaber, M.B., Thermodynamics and kinetics of reactions involving vanadium in natural systems. Accumulation of vanadium in sedimentary rocks; Geochim. Cosmochim. Acta 1992, 56, 1471–1483.
[99] Cook, O.A., High temperature heat contents of V_2O_3, V_2O_4 and V_2O_5; J. Am. Chem. Soc. 1947, 331–333.
[100] Garrels, R.M.; Some thermodynamic relations among the vanadium oxides and their relation to the oxidation state of uranium ores of the Colorado Plateaus; Am. Min. 1953, 38, 1251–1265.
[101] Wehrli, B.; Stumm, W., Vanadyl in natural waters: adsorption and hydrolysis promotes oxygenation; Geochim. Cosmochim. Acta 1989, 53, 696–677.
[102] Wehrli, B.; Stumm, W., Oxygenation of vanadyl(IV): effect of coordinated surface hydroxyl groups and OH^-. Langmuir, 1988, 4, 753–758.
[103] Peacock, C.L.; Sherman, D.M., Vanadium(V) adsorption onto goethite (α-FeOOH) at pH 1.5 to 12: a surface complexation model based on ab initio molecular geometries and EXAFS spectrometry; Geochim. Cosmochim. Acta 2004, 68, 1723–1733.
[104] Bautista, E.M.; Alexander, M., Reduction of inorganic compounds by soil microorganisms; Soil Sci. Soc. Am. J. 1972, 36, 918–920.
[105] Carpentier, W.; Sandra, K.; De Smet, I. *et al.*, Microbial reduction and precipitation of vanadium by *Shewanella oneidensis*; Appl. Environ. Microbiol. 2003, 69, 3636–3639.
[106] Ortiz-Bernad, I.; Anderson, R.T.; Urionis, H.A. *et al.*, Vanadium respiration by *Geobacter metallireduccus*: novel strategy for in situ removal of vanadium from ground water; Appl. Environ. Microbiol. 2004, 70, 3091–3095.
[107] Rees, D.C.; Tezcan, F.A.; Haynes, C.A. *et al.*, Structural basis for biological nitrogen fixation; Phil. Trans. Roy. Soc. 2005, 363A, 971–984.
[108] Kaiser, B.N.; Gridley, K.L.; Brady, J.N. *et al.*, The role of molybdenum in agricultural plant production; Ann. Bot. 2005, 96, 745–754.
[109] Katyal, J.C.; Randhawa, N.S., Micronutrients; FAO Fertilizer and Plant Nutrition Bulletin No. 7; FAO, Rome, 1983.
[110] Boermans, H.J., Molybdenum poisoning: an introduction; The Merk Veterinary Manual; Merck & Co., Inc., Whitehouse Station, NJ, 2006.
[111] Wade, G.M., Molybdenum toxicity and hypocupris : a review; J. Anim. Sci. 1978, 46, 1078–1085.
[112] O'Connor, G.A.; Brobst, R.B.; Chaney, R.L. *et al.*, A modified risk assessment to establish molybdenum standards for land application of biosolids; J. Environ. Qual. 2001, 30, 1490–1507.
[113] Gardner, W.C.; Broersma, K.; Popp, J.D. *et al.*, Copper and health status of cattle grazing high molybdenum forage from a reclaimed mine tailings site; Can. J. Anim. Sci. 2003, 83, 479–485.
[114] Thompson, I.; Thornton, I.; Webb, J.S., Molybdenum black shales and the incidence of bovine hypocupris; J. Sci. Food Agric. 1972, 23, 879–891.
[115] Moore, J.W., Inorganic Contaminants of Surface Water: Research and Monitoring Priorities; Springer-Verlag; New York, 1991.
[116] Hirsch, M.P., Availability of sludge-borne silver to agricultural crops; Environ. Toxicol. Chem.,1998, 17, 610–616.
[117] Domsch, K.H., Effects of pesticides and heavy metals on biological processes in soil; Plant Soil 1984, 76, 367–378.

[118] Johnsson, M.; Pell, M.; Stenstöm, J., Kinetics of substrate-induced respiration (SIR) and denitrification: applications to a soil amended with silver. Ambio 1998, 27, 40–44.
[119] Murata, T.; Kanao-Koshikawa, M.; Takamatsu, T. Effects of Pb, Cu, Sb, In and Ag contamination on the proliferation of soil bacterial colonies, soil dehydrogenase activity, and phospholipid fatty acid profiles of soil microbial communities; Water Air Soil Pollut. 2005, 164, 103–118.
[120] Ewell, W.S.; Gorsuch, J.W.; Ritter, M. et al., Ecotoxicological effects of silver compounds. Proceedings, 1st Argentum International Conference on the Transport, Fate and Effects of Silver in the Environment, Madison, WI, 8–10 August, 1993, p. 9.
[121] Beglinger, J.M.; Ruffing, C.J., Effects of silver sulfide on the terrestrial earthworm. Proceedings, 5th Argentum International Conference on the Transport, Fate and Effects of Silver in the Environment, Hamilton, ON, 28 September–1 October, 1997.
[122] Borovicka, J.; Randa, Z.; Jelinek, E. et al., Hyperaccumulation of silver by *Amanita strobiliformis* and related species of the section *Lepidella*; Mycol. Res. 2007, 111, 1339–1344.
[123] Harris, A.; Bali, R., On the formation and extent of uptake of silver nanoparticles by live plants; J. Nanopart. Res. 2008, 10, 691–695.
[124] Saga, M., Thallium; Tox. Environ. Chem. 1994, 45, 11–32.
[125] Heim, M.; Wappel, O.; Markert, G., Thallium in terrestrial environments - occurrence and effects; Ecotoxicology 2002, 11, 369–377.
[126] Tremel, A.; Masson, P.;Garraud, H. et al., Thallium in French ecosystems. II. Concentration of thallium in field grown rape and some other plant species; Environ. Pollut. 1997, 97, 161–168.
[127] Pavlìková, J.; Zbìral, J.; Smatanová, M. et al., Uptake of thallium from artificially and naturally contaminated soils into rape (*Brassica napus L.*); Agric. Food Chem. 2005, 53, 2867–2871.
[128] Wierzbicka, M.; Szarck, G.; Grodzinska, K. Highly toxic thallium in plants from the vicinity of Olkusz (Poland); Ecotoxicol. Environ. Saf. 2004, 59, 84–88.
[129] Anderson, C.W.N.; Brooks, R.R.; Chiarucci, A. et al., Phytomining for nickel, thallium and gold; J. Geochem. Explor. 1999, 67, 407–415.
[130] Kurz, H.; Schultz, R.; Römheld, V., Selection of cultivars to prevent the contamination of the food chain by cadmium and thallium; J. Plant Nutr. Soil Sci. 1999, 162, 323–328.
[131] LeBlanc, M.; Petit, D.; Deran, A. et al., The phyto-mining and environmental significance of hyper-accumulation of thallium by *Iberis intermedia* from south France; Econ. Geol. 1999, 94, 109–114.
[132] Al-Najar, H.; Schultz, R.; Römheld, V., Plant availability in the rhizosphere of hyperaccumulator plants: a key factor for assessment of phytoextraction; Plant Soil 2003, 249, 97–105.
[133] Kwan, K.K.M.; Smith, S., The effect of thallium on the growth of *Lemna minor* and plant tissue concentrations in relation to both exposure and toxicity; Environ. Pollut. 1988, 52, 203–220.
[134] Carlson, R.W.; Bazzaz, F.A.; Rolfe, G.L., The effect of heavy metals on plants. II Net photosynthesis and transpiration of whole corn and sunflower plants heated with Pb, Cd, Ni and Tl; Environ. Res. 1975, 10, 113–120.
[135] Duce, R.A.; Hoffman, G.L., Atmospheric vanadium transport to the ocean; Atmos. Environ. 1976, 10, 989–996.
[136] Gélinas, Y.; Lucolte, M.; Schmit, J.P., History of the atmospheric deposition of major and trace elements in the industrialized St. Lawrence Valley, Quebec, Canada; Atmos. Environ. 2000 34, 1797–1810.
[137] Rehder, D.; Inorganic considerations on the function of vanadium in biological systems; in: Sigel, H., S. (eds), Metal Ions in Biological Systems, Vol. 31, Vanadium and Its Role in Life; Marcel Dekker, New York, 1995, pp. 1–43.
[138] Barker, A.V.; Pilbeam, D.J., Handbook of Plant Nutrition; CRS Press, Boca Raton, FL, 2007.
[139] Barceloux, D.G.; Vanadium; J. Toxicol. Clin. Toxicol. 1999, 37, 265–278.
[140] Lener, J.; Kučera, J.; Kodl, M.; Skokanová, V., Health effects of environmental exposure to vanadium; in: Nriagu, J.O. (ed.), Vanadium in the Environment, Part 2. Health Effects; John Wiley & Sons, Inc., New York, 1998, pp. 1–19.

[141] Baran, E. J., Vanadium detoxification; in: Nriagu, J.O. (ed.), Vanadium in the Environment, Part 2. Health Effects; John Wiley & Sons, Inc., New York, 1998, pp. 317–345.
[142] Thompson, K.H..; Battell, M.; McNeill, J.H., Toxicology of vanadium in mammals; in: Nriagu, J.O. (ed.), Vanadium in the Environment, Part 2. Health Effects; John Wiley & Sons, Inc., New York, 1998, pp. 21–37.

23
Gold and Uranium

Ian D. Pulford

Environmental Chemistry, Chemistry Department, University of Glasgow, Glasgow, Scotland, UK

23.1 Introduction

Gold (Au) and uranium (U) are two of the most valuable and important metals mined by humankind. They have some similar and some contrasting properties, but their uses are very different. Gold has been used for millennia for its intrinsic value in coins and jewelry, but more recently it has been used for other purposes, especially medical and sensor applications. Uranium usage, on the other hand, has developed only in the mid twentieth century for energy and weapons production.

Gold is a member of Group 11 of the transition metals. It is a malleable, ductile, soft metal that can be beaten into a very fine sheet or extruded into a thin wire. It has one of the highest electrode potentials for a metal ($E^0 = +1.691$ V), which means that it does not readily corrode. These, and other properties summarized in Table 23.1, give gold certain characteristics that have made it one of the most sought-after metals for thousands of years. Because of its colour and scarcity, along with its malleability, ductility and stability, gold has long been used for coinage and jewelry. The same properties have also made gold and its alloys a commonly used material in restorative and conservative dentistry and orthodontics [1]. Gold has a long history of medical use, and in the twentieth century compounds of gold were developed for treatment of rheumatoid arthritis and cancer [2]. A property of gold that has received considerable interest recently is its ability to form nanoparticles, typically of diameter <50 nm, which has led to the use of gold as a catalyst, for

---

Trace Elements in Soils Edited by Peter S. Hooda
© 2010 Blackwell Publishing, Ltd

Table 23.1 Physical-chemical properties of gold and uranium [15–17]

Property	Gold	Uranium
Density (g cm^{-3})	19.3	18.95
Melting point (K)	1337	1405
Boiling point (K)	3081	4018
Electrical resistivity (Ω m) at 298 K	2.35×10^{-8}	30.8×10^{-8}
Thermal conductivity (W m^{-1} K^{-1}) at 300 K	317	27.6
Common naturally occurring oxidation states	0, I, III	IV, VI
Naturally occurring isotopes (% abundance and half-life, $t_{1/2}$)	^{197}Au (100 %; stable)	^{234}U (0.0057 %: $t_{1/2}$ 2.5×10^5 y) ^{235}U (0.71 %: $t_{1/2}$ 7.04×10^8 y) ^{238}U (99.3 %: $t_{1/2}$ 4.5×10^9 y)

electrochemical sensors [3,4], and for targeted drug delivery [5]. Gold is not known to be essential to humans, plants, animals or microorganisms. Nor is it thought to be toxic, although a recent study [6] showed that growth of fungal biomass used for water clean-up treatment was completely inhibited at Au concentrations in solution >50 mg l^{-1}. The only reported effect of gold on humans is an allergic reaction in some people [7]. The main environmental concerns associated with gold are due not to its own chemical properties but more generally to problems associated with mining operations, and specifically to treatment of the ore with mercury [8–10] or cyanide [11].

Uranium is a member of the actinoid series of elements. It is a malleable, ductile metal that easily tarnishes. Other properties are summarized in Table 23.1. However, the most important property of uranium, and the main reason why it is of interest in environmental systems, is its radioactive decay [12]. Uranium-238 (^{238}U) is the most abundant isotope and, along with the second most abundant isotope, ^{235}U, is one of the primordial radionuclides that have been present on earth since its formation approximately 4.5×10^9 years ago. Much of the interest in uranium is due to the fact that ^{235}U is fissile and so forms the basis of nuclear power production. Both isotopes (^{238}U and ^{235}U) undergo radioactive decay by alpha emission, and the descendent isotopes in the decay chain decay through a series of alpha- or beta-emissions. Important isotopes in the ^{238}U decay series are ^{226}Ra, ^{222}Rn (a gas), ^{210}Pb (which can be used to date sediments over a period to about 200 years before the present), and the final product, stable ^{206}Pb (which is the basis for U/Pb dating) [12]. ^{235}U, which is the isotope used for nuclear energy production and in nuclear weapons, is obtained by enrichment of uranium ores to about 3 % ^{235}U for nuclear fuel and 93 % ^{235}U for weapons-grade uranium. A by-product of this process is depleted uranium (DU), which has a lower ^{235}U and ^{234}U content than natural uranium (atomic abundances 0.20 % and 0.0009 %, respectively; cf. values in Table 23.1). Because of its high density, DU has been used for military applications, such as armour-piercing shells and the manufacture of tank armour, and subsequently there have been concerns over its fate in the environment, and its effect on human health, following its use in the Balkans and the Middle East in the 1990s [13,14].

23.2 Geogenic Occurrence

23.2.1 Gold

The commonest gold minerals are the tellurides, $AuTe_2$ (calaverite, krennerite), Au_2Te_3 (montbrayite), and $AuAgTe_4$ (sylvanite), but these are not commercially important. Gold is usually found as the metallic form (Au^0), either in alluvial deposits, where it settles due to its high density after weathering out of gold-containing rocks (as for example in California and the Yukon, sparking the Gold Rushes in the nineteenth century, or in the Urals in Russia), or as reef- or vein-gold (for example in South Africa). The latter is formed as a result of hydrothermal metamorphosis and the gold is present as microscopic particles in quartz. Because gold is found mainly as metal particles in soils or rocks, and is not uniformly dispersed throughout, measurement of concentration in these media is affected by sampling problems [18]. Gold is a rare element in the earth's crust, with a concentration usually quoted in the range 1.1 $\mu g\ kg^{-1}$ [15,19] to 4 $\mu g\ kg^{-1}$ [20,21].

Concentrations of gold in uncontaminated soils are difficult to assess because of the metal's highly localized distribution in rocks. The problem is compounded because of two processes: (i) the tendency of gold to be held in the upper horizons, and especially organic horizons, of soils (see below), and (ii) enrichment of gold in lower horizons due to gravitational migration of the highly dense gold particles, or to the presence of gold-bearing material [22]. Many studies that quote Au concentrations in soils have been carried out because of the likelihood of contamination due to gold mining, metal smelting or some other anthropogenic input. Eisler [23] reviewed the published data for gold concentrations in abiotic materials, including soils, and highlighted the need for caution when dealing with published values due to the variety of analytical techniques used, and hence differences in the analytical limits of detection, plus the fact that some reported analyses were carried out on preconcentrated samples. The three studies of gold in soil that were reported by Eisler [23] included one that measured Au in soil around Lake Nasser in Egypt [24], which recorded a mean value of 163 $\mu g\ kg^{-1}$ in soil to a depth of 60 cm. This analysis was done using flame atomic absorption spectrophotometry (AAS), but no limit of detection was given. However, as historically gold mining is known to have taken place in areas around Lake Nasser, it may be that gold-bearing sediments contaminated these soils. In the other two studies, one was on soils around a gold mine in Nevada [25] and the other on the long-term mobility of metals in soils following application of sewage sludge [26]. The Nevada study reported a range of 5–29 (mean 13) $\mu g\ Au\ kg^{-1}$ in pre-mining deposits, which were distinguished from post-mining deposits by a zone of surface weathering.

The figures quoted by both the Egyptian and Nevada studies are high compared with values obtained recently from soils sampled along a 120 km transect in southern Norway [27], an area with no gold mining activity. This study used ICP-AES (inductively coupled plasma–atomic emission spectroscopy) analysis, with a detection limit of 0.2 $\mu g\ Au\ kg^{-1}$. The median value for soil B horizons was 0.35 $\mu g\ Au\ kg^{-1}$ (range <0.2–1.6) and for C horizons 0.2 $\mu g\ Au\ kg^{-1}$ (range <0.2–1.2). In a study of soil in the Kola Peninsula of northwest Russia, measuring Au by graphite furnace AAS with a detection limit of 0.1 $\mu g\ kg^{-1}$, the geometric means for C horizons from soils in three different catchments were 0.7, 1.0 and 1.1 $\mu g\ Au\ kg^{-1}$ [28]. The topsoils in this area are contaminated by gold due to impacts of copper and nickel smelting. In a much earlier study using neutron activation analysis [29], mean values of 1.0 μg Au kg^{-1} were measured in soils from three Arctic islands, and 0.90 μg Au kg^{-1} in weathered basalts from Tahiti. These three studies taken together suggest that a figure of about 1 μg Au kg^{-1} may be a reasonable estimate for the concentration of gold in uncontaminated soil, and is the figure generally quoted [19].

23.2.2 Uranium

Uranium is found in a number of different depositional environments, the commonest being magmatic, hydrothermal and sedimentary (especially sandstone, shale, conglomerate, calcrete and phosphate). The main ores of uranium found in deposits that are economic for mining (generally taken as a U content greater than 0.01 %, expressed as U_3O_8) are: uraninite (UO_2); pitchblende ($U_2O_5 \cdot UO_3$ or more commonly U_3O_8); carnotite ($K_2(UO_2)_2$ $(VO_4)_2 \cdot 2H_2O$); tyuyamunite ($Ca(UO_2)_2(VO_4)_2 \cdot 5-8H_2O$); coffinite ($U(SiO_4)_{1-x}OH_{4x}$); uranophane ($Ca(UO_2)_2Si_2O_7 \cdot 6H_2O$); and brannerite ($(U,Ca,Ce)(Ti,Fe)_2O_6$). Since the main interest in U in the environment is due to its radioactivity, and because of the very long half-lives of the main U isotopes, concentrations in environmental materials are often quoted in terms of specific activity (Bq kg^{-1}), as well as milligrams per kilogram (mg kg^{-1}). In order to allow direct comparison with other elements, the concentrations used here will be on a wt/wt basis. Typical concentrations of U in a variety of rock types and soil are shown in Table 23.2 (note the concentrations, which are three orders of magnitude greater than for gold). Unlike gold, uranium is not found naturally in the metallic form U^0. The U minerals are found as veins or individual grains in other rocks. The highest concentrations are found in acidic rocks such as granite. Sedimentary rocks tend to have relatively low U concentrations, except for those high in organic matter, such as shale, and phosphate deposits such as apatite, due to precipitation of insoluble U(IV) sulfide or phosphate.

As with gold, measurement of U in soils has often been performed due to concern over contamination rather than an interest in the natural levels, although some studies have tried to assess baseline concentrations against which anthropogenic inputs could be assessed. Meriwether et al. [30] measured U in soils of Louisiana, the state that has the lowest average radiation exposure rate in the United States, based on the activity of ^{214}Bi, one of the members of the ^{238}U decay series, in order to estimate U concentrations, which ranged

Table 23.2 Typical concentrations of gold and uranium in common rock types and soil. Note the difference in units [19]

Rock type	Au (μg kg^{-1})	U (mg kg^{-1})
Basalt	0.5	0.43
Granite	1.8	4.4
Limestone	2.0	2.2
Sandstone	3.0	0.45
Shale	2.5	3.7
Soil	1–20	0.7–9

from 2.35 to 3.98 mg kg^{-1}. Sheppard *et al.* [31] measured U in topsoil from seven sites across Canada by neutron activation–delayed neutron counting (DNC), with a detection limit of 0.1 mg kg^{-1}. The uranium concentrations in these soils ranged between 0.4 and 2.6 mg kg^{-1}. Uranium measured by ICP-MS (inductively coupled plasma mass spectrometry) in 77 Japanese surface soils had a median value of 2.28 mg U kg^{-1}, range 0.17–4.60 mg U kg^{-1} [32]. The general consensus is that <10 mg U kg^{-1} can be considered as a background value, although some soils in this range may be subject to low-level contamination (see Section 23.3.2).

23.3 Soil Contamination

23.3.1 Gold

Sources of gold contamination are thought to be primarily from mining and smelting activities. Van de Velde *et al.* [33], using double-focusing ICP-MS with micro concentric nebulization, measured gold concentrations in the range 0.07–0.35 ng kg^{-1} in a 140 m snow/ice core from Mont Blanc in the French-Italian Alps. This represents approximately 200 years of deposition. They concluded that the major input of gold was due to mining and smelting, based on the fact that there was no significant change in Au concentrations over that period, and that inputs from these activities would already have been important in the eighteenth century. More compelling evidence was given by the Au/Pt ratio, which was 0.6 in the Mont Blanc snow and ice over the period 1780–1900, agreeing well with the ratio found in nickel ores from Noril'sk in Russia (0.6–0.7), but contrasting with the ratios for material from natural sources (upper continental crust, 6.2; sea salt spray, 0.25; cosmic dust, 4.2). Furthermore, it was estimated that the main natural input, from rock and soil dust, accounted for only 0.26–0.65 % of the total measured concentrations.

Sewage sludge is also a source of Au input to soils. A study of sewage sludge from 16 cities in the United States, using neutron activation analysis, measured a range of 210–7000 μg Au kg^{-1} [34]. Reeves *et al.* [35], also using neutron activation analysis, reported an average of 770 μg Au kg^{-1} in sewage sludges from Werribee, Melbourne (Australia); and quoted a range of 280–56 000 μg Au kg^{-1} in sewage sludges from Germany. Gold in sewage sludge comes from specific anthropogenic activities, such as mining, smelting and jewelry making. In the soils of a transect in southern Norway [27], gold was the only element that showed a clear anthropogenic input, with elevated concentrations being measured in the surface soil (O horizon) collected in and around the city of Oslo. The range of gold concentrations in the O horizons of the transect soils was <0.2–58.2 μg Au kg^{-1}, with a median value of 1.5 μg Au kg^{-1}. Enrichment in the surface soil was clearly seen by the ratio of Au in O horizons to Au in C horizons (O/C) of 7.5, compared with a value of 1.8 for the B horizon/C horizon (B/C) ratio. The work by Boyd *et al.* [28] showed clear enrichment of surface soils due to input of gold from Cu/Ni smelters in the Kola Peninsula of Russia. The concentration of gold in 25 topsoils of a catchment 5 km south of the Monchegorsk plants ranged from 1.3 to 99.5 μg Au kg^{-1}, with a geometric mean of 9.47 μg Au kg^{-1}. In a catchment 25 km south of this plant the concentrations in 24 topsoil samples ranged from

0.2 to 5.4 μg Au kg^{-1} (mean 1.8), while in a catchment 10 km north of the Zapolyarniy plant the range was <0.1 to 5.3 μg Au kg^{-1} (mean 0.7).

Concentrations of gold in soils and deposits associated with mining are extremely variable. Care has to be taken to distinguish between soil that has been contaminated by the mining process and the actual gold-bearing deposits. In soil at a gold mine site in New South Wales, a mean gold concentration of 1632 (±76) μg Au kg^{-1} was measured and at a site in Queensland the mean value was 1216 (±144) μg Au kg^{-1} [36]. At a gold mine site in Six Mile Canyon, Nevada, Miller et al. [25] measured a range of 80–843 μg Au kg^{-1} (mean 473) in post-mining fan deposits, and 15–424 μg Au kg^{-1} (mean 166) in post mining channel deposits.

23.3.2 Uranium

Contamination of soil by uranium is caused by mining and processing operations; nuclear reactor operations; poor storage practices for wastes; accidents at nuclear sites; manufacture and testing of weapons; dispersal of fly ash from coal-powered power stations; and long-term application of phosphate fertilizers. The last two are diffuse sources, which result in generally lower U concentrations in soil than the other sources.

Contamination by uranium was measured in garden soils in Port Hope, Ontario, Canada due to waste from a uranium extracting and refining facility [37]. Concentrations of U in topsoils (0–15 cm) ranged from 7.5 to 420 mg U kg^{-1} dry soil, compared with a background value of 2.0 mg U kg^{-1}. It was thought that this contamination came from a waste dump in the town some time between 1945 and 1948, the measurements being made in 1976. Bykova et al. [38] measured U contamination of soils in Kyrgyzstan, which was a major supplier of uranium in the 1940s and 1950s, with four mines and one milling facility. This has left a legacy of 70 $\times$ 10^6 m^3 of radioactive waste in 80 waste dumps and 49 tailings sites. Uranium concentrations up to 80 mg U kg^{-1} were found in the tailings, and up to 5 mg U kg^{-1} in the waste dumps. At two sites, in central and southern Kyrgyzstan, concentrations up to 35 mg U kg^{-1} were found in topsoils in areas adjacent to tailings sites and waste dumps, compared with values up to 5 mg U kg^{-1} in nearby uncontaminated soils and up to about 3 mg U kg^{-1} in soils of northern Kyrgyzstan, which were unaffected by U mining and processing.

A recent cause for concern about U contamination in the environment has been the use of weapons and armour containing depleted uranium (DU). When a DU shell hits a hard target, such as tank armour, about 10–35 % of the shell material is dispersed as an aerosol, whereas shells that hit soft targets, or miss their target, tend to remain intact. Di Lella et al. [39] measured U in soil at an army depot in Kosovo, an area of approximately 20 000 m^2 that had been attacked by 300 DU rounds. They found total U concentrations of 0.69–31.47 mg U kg^{-1}, with the highest concentrations close to DU shell holes exceeding the local background of 1–2 mg U kg^{-1}. The spatial distribution of uranium from the DU shells was limited. Uranium concentrations were measured in soils at two firing ranges in the United Kingdom that were used for testing DU containing shells and tank armour [14]. At one site, uranium concentrations in the range 1.6–37.7 mg U kg^{-1} were measured in soil around two guns and their target areas, which compared with a value of 0.8 mg U kg^{-1} at a control site away from the firing positions. At a second firing range concentrations of 21–282 mg

U kg^{-1} were measured in soil around the target area, but in two storage areas where armour plating, some containing DU, had been stored after being used as targets for DU shells, the U concentrations were 68+/−2.5 and 18 671+/−3346 mg U kg^{-1}.

Contamination of agricultural soils by uranium has been shown as a result of high or prolonged addition of phosphate fertilizers, which can have very high U contents due to the accumulation of U in apatite deposits [40,41]. Makweba and Holm [42] measured U concentrations of 250–570 mg U kg^{-1} in fertilizers made from phosphate deposits in Tanzania, while values in the range of approximately 100–200 mg U kg^{-1} were reported in phosphate rock and fertilizers from Morocco, Tunisia, Mexico and the United States [43]. Topsoils from the long-term field trials at Rothamsted Experimental Station in the United Kingdom, which had received phosphate fertilizers since the mid nineteenth century, were enriched in U compared to soils that had received no P over that time [40]. The mean values of U in soils that received P and those which had not, were 2.42 and 2.43 mg U kg^{-1}, respectively, in the 1870s, but by 1976 this had risen to a mean of 2.88 mg U kg^{-1} in the fertilized plots compared with 2.44 mg U kg^{-1} in the unfertilized plots. Furthermore, the uranium enrichment was confined to the topsoil, suggesting little if any downward movement of U in the soil profile (see Chapter 7). Total U concentrations (expressed as UO_2^{2+}) were measured up to 50.6 mg U kg^{-1} in the top 90 cm of agricultural soils in Mexico, compared with values of 1 mg U kg^{-1} or less in nonagricultural soils [41].

23.4 Chemical Behaviour in Soils

23.4.1 Gold

Gold is found naturally in soil in three oxidation states: Au^0, Au (I), and Au (III). As previously stated, elemental gold (Au^0) is extremely stable and hence tends to persist in soil. One possible mechanism for transport in soil is as elemental colloidal gold, which undergoes gravitational settling. There has long been speculation about this mechanism, and recently the presence of naturally occurring gold nanoparticles and nanoplates of diameter <200 nm in weathering gold deposits in Australia has been shown [44]. Elemental gold may be formed by reduction of Au(I) and Au(III) species in soil.

Many of the current ideas about the forms of gold in solution have been developed from theoretical calculations using thermodynamic data and have focused especially on conditions in waters in supergene deposits, which are secondary deposits formed by weathering of Au from primary minerals. This weathering may be chemical and/or biological, and results in the mobilization, transportation and redeposition of gold. The forms of gold suggested by such studies are thought also to be found in soils, but there is much uncertainty due to differences in soil properties compared to supergene systems (pH, E_h, organic matter content, and concentrations of other ions – especially complexing ions). There is also the question of the concentration of gold in natural waters. In a survey of 132 samples of natural waters from a variety of sources, a mean of 0.101 μg Au l^{-1} was found in samples from mineralized areas, while that from unmineralized areas was only 0.002 μg Au l^{-1} [45].

Table 23.3 Common solution species of Au and U ions

Gold (I) and (III) complexes	Uranium (VI) complexes
Hydroxy species	*Noncalcareous systems*
$Au(I)OH(H_2O)^0$	UO^{2+}
$Au(III)(OH)_3(H_2O)^0$	UO_2H^+
Halide species	$(UO_2)_2(OH)_2^{2+}$
$Au(I)Cl_2^-$	$(UO_2)_3(OH)_5^+$
$Au(III)Cl_4^-$	$(UO_2)_3(OH)_7^-$
$Au(III)Br_4^-$	*Carbonate systems*
$Au(I)I_2^-$	$UO_2CO_3^0$
$Au(I)ClOH^-$	$UO_2(CO_3)_2^{2-}$
Sulfur species	$UO_2(CO_3)_3^{4-}$
$Au(I)HS(H_2O)^0$	*Sulfate species*
$Au(I)(HS)_2^0$	$UO_2SO_4^0$
$Au(I)(S_2O_3)_2^{3-}$	$UO_2(SO_4)^{2-}$
Cyanide and thiocyanate species	*Phosphate species*
$Au(I)(CN)^0$	$UO_2HPO_4^0$
$Au(I)(CNS)^0$	$UO_2PO_4^-$
Humic species	$UO_2(HPO_4)_2^{2-}$
$Au(III)(OH)_2FA^{-a}$	

[a] FA = fulvic acid.

The solution chemistry of gold is dominated by complex formation (Table 23.3) due to the instability of the Au(I) and Au(III) aquo ions, $Au(H_2O)_2^+$ and $Au(H_2O)_4^{3+}$, respectively [46]. Simple hydroxy ions for both oxidation states have been suggested as being important under most conditions, $AuOH \cdot H_2O^0$ [46,47] and $Au(OH)_3(H_2O)^0$ [48]. In environments where sulfur species are present in high concentrations, such as in the presence of oxidizing sulfide minerals, bisulfide ($Au(HS)_2^-$ and $AuHS \cdot H_2O^0$) and thiosulfate ($Au(S_2O_3)_2^{3-}$) complex ions become dominant. The thiosulfate complex is thought to be found under more alkaline, oxidizing conditions, while the sulfide complexes persist in more reducing systems. Halide complexes have also been postulated as important species for gold in solution, with chloride complexes being the most stable [18]. Early thermodynamic models to describe the solution chemistry of gold assumed that $Au(Cl)_4^-$ was the most important species in relation to gold mobility, but this assumed Au and Cl concentrations much higher than found naturally [47]. Colin et al. [48] suggested that $Au(Cl)_4^-$ was formed only if other more strongly complexing ligands are absent, and that it was stable only in very acidic, highly oxidizing, chloride-rich conditions. Cyanide (CN^-) and thiocyanate (CNS^-) ions also form strong complexes with gold. Indeed, cyanide is used in the extraction of Au from gold ores due to the formation of the stable $Au(CN)_2^-$ complex [49]. Some plants release CN^- ions, and both CN^- and CNS^- may be formed during the process of plant tissue degradation. Humic and fulvic acids can also complex gold [50,51], with N- and S-containing groups being particularly important binding sites.

In addition to control of mobility due to precipitation of elemental gold, the complex ions listed in Table 23.3, and Au^0 itself, can be adsorbed on to the surfaces of soil components, including iron oxides, manganese oxides, humic materials and sulfide [47].

Ran et al. [47] showed that adsorption depended on both the nature of the surface and the Au species in solution. They concluded that Au–chloro complexes were preferentially adsorbed onto manganese oxide (birnessite), iron oxide (goethite) and humic acid. Birnessite, with strong oxidizing and sorbing ability, was effective at adsorbing Au(I) ions such as $Au(S_2O_3)_2^{3-}$. A recent study has shown that manufactured gold nanoparticles are stabilized by interaction with humic acid [52].

The chemistry of gold in soil is therefore dominated by the presence of complex ions in solution and by the immobilizing processes of adsorption on to oxides and humic material, or precipitation as elemental gold.

23.4.2 Uranium

Under the conditions found in the natural environment, the two important oxidation states of uranium are U(IV) and U(VI). As with gold, much of our understanding of the environmental chemistry of U is based on geochemical studies, as opposed to studies specifically on soils.

U(IV) is insoluble under oxidizing conditions, and is an important species only in reduced soils ($E_h < +200$ mV), forming UO_2. In highly reduced systems, this is due to the action of iron- and sulfate-reducing bacteria [53,54]. While it has commonly been assumed that precipitation of UO_2 is an immobilizing mechanism, recent studies have shown that the product may be in the form of nanoparticles of diameter <2 nm, which may be mobile in porous soils [54]. If such particles were transported to more oxidized parts of the soil, reoxidation to U(VI) could occur. U(IV) can also bind strongly with organic matter.

The chemistry of U(VI) is dominant in most soils [55–57]. Ionic speciation (Table 23.3) is highly dependent on pH, with UO_2^{2+} being the major ion below about pH 5 and the hydroxyl complexes of U(VI) becoming more important as the pH increases above 5. Once carbonate concentration in soil solution becomes significant, carbonate complexes of U(VI) become important species, and these are dominant above a pH of about 6. U(VI) can also form complexes with sulfate and phosphate ions. These complex ions tend to increase the solubility of uranium in the environment.

A number of sorption processes have been suggested for controlling the mobility of U(VI) species in soils, and binding can occur onto all the major solid-phase components of soils: aluminosilicate clays, hydrous oxides (especially iron oxides) and humified organic matter [58]. In acidic soils, sorption of U is weak and ion exchange on negative binding sites may be the main process while cationic species such as UO_2^{2+} are dominant. As the pH increases, negatively charged binding sites increase, but the dominant solution species become the anionic carbonate complexes, and so sorption decreases. This effect of carbonate complexation producing anionic species is considered to be the most important influence of U mobility in soils [57,59]. It has also been proposed that specific sorption (chemisorption) of U(VI) species can take place onto iron oxide surfaces [55]. Sherman et al. [60] have recently proposed a surface complexation model for U(VI) on goethite (α-FeOOH) which includes all the UO_2–OH–CO_3 complexes. Values of K_d, the solid/solution partition coefficient, for U are very variable. Vandenhove et al. [57] quoted a range of 33–395 100 l kg^{-1} for clayey and organic soils, and 0.03–4500 l kg^{-1} for loamy and sandy soils. These authors suggested that pH was the best predictor of K_d for soils with a pH of 6 or greater, but that organic matter content or amorphous iron oxide were better at more acidic pH values.

Studies using selective extraction procedures have also suggested the importance of iron oxides and humic material as binding sites for U in soils. Di Lella et al. [39], using the dithionite–citrate reagent to extract U in the DU-contaminated soils in Kosovo, found 24–36 % of the total U in this fraction, which was assumed to be poorly crystalline iron oxides. Using the BCR sequential extraction scheme on the soils from the UK firing ranges, Oliver et al. [61] found that the distribution of uranium in all four fractions varied between samples, but that overall the oxidizable fraction accounted for 50 % or more of the U recovered. This was equated to uranium held by the organic fraction. In a detailed study of colloidal fractions in the same soils using ultracentrifugation, gel filtration and gel electrophoresis, evidence was found for both U-humic colloids and U-Fe/Al colloids [62].

23.5 Risks from Gold and Uranium in Soils

23.5.1 Gold

There is little, if any, evidence that gold presents a risk of toxicity in soil. An experiment to assess the use of fungal biomass for detoxifying industrial wastewater showed that growth of the fungal isolate used was inhibited completely above 50 mg Au l^{-1} [6]. This concentration, however, is many orders of magnitude higher than those found in natural waters [45]. The source of the gold used was an atomic absorption standard solution made up in 0.5 M HNO_3, which is unlike the solution species of Au found in most natural systems. More generally, there is concern about the release of nanoparticles into the environment because the behaviour of such materials is not known [52]. As the use of gold nanoparticles increases, this aspect may require more attention.

Some studies have measured the uptake of gold by natural vegetation, but again these have usually been carried out in gold-mining areas, where soil concentrations are higher than normal, and where contamination by dust blow is an issue. Typical background concentrations of gold in plant tissue have been given and range from about 1 to 10 μg Au kg^{-1} [19,63]. In the goldfields of Western Australia, a maximum value of 11 μg Au kg^{-1} was measured in tea-tree (*Melaleuca*. Myrtaceae) growing on soil with an Au content of 830 mg kg^{-1}. This compared with a background plant tissue concentration of <0.5 μg Au kg^{-1}. In general, shrub species such as *Melaleuca* had higher concentrations of Au in their tissues than trees such as *Eucalyptus* [64].

An aspect of uptake of gold by plants that has received more attention is the possibility that plants could be used to 'harvest' gold from contaminated soils or mining wastes [63]. This idea has also been applied to other metals where bioavailability, and hence plant uptake, has been increased by treating the soil with chelating agents such as EDTA, but the bioavailability of gold is more commonly increased by treatment with cyanide and thiocyanate, which both form stable complexes with Au. Anderson and co-workers have shown in both pot [63] and field experiments [65] that the phytoextraction of gold is possible using such treatments. In the pot trial the highest tissue concentration obtained was 57 mg Au kg^{-1} on a dry weight basis, growing *Brassica juncea* (Indian mustard) in a medium of finely disseminated colloidal gold particles in sand treated with 0.64 g kg^{-1} ammonium thiocyanate. However, thiocyanate treatment was less successful when used on an alkaline gold ore in Brazil because the gold–thiocyanate complex is stable only under acid conditions. Cyanide was much more effective at improving bioavailability of gold, with an average concentration of 39 mg Au kg^{-1} found in *B. juncea* growing on ore containing 0.6 mg Au kg^{-1}, which had been treated with 0.15 g NaCN kg^{-1}. These pot and field trials suggested that recovering 1 kg Au ha^{-1} by growing plants capable of taking up these amounts of gold is realistic for soil containing at least 2 g Au Mg^{-1} soil. Chapter 14 provides a comprehensive review of phytoextraction application, effectiveness and limitations.

Recent work using X-ray absorption near-edge spectroscopy (XANES) has suggested that *B. juncea* can reduce gold taken up as an Au(I)CN complex to nanoparticles of Au^0 [66]. About 50 % of the gold taken up from gold-rich soil treated with KCN had been reduced in this way. This observation may open up a way to produce gold nanoparticles in a defined size range for catalysis and other uses, which is difficult to achieve by conventional means as the particles tend to sinter together because of the low melting point of gold. The problem of producing gold nanoparticles in plants is the need to remove the plant tissue prior to extracting the gold. Enzymic digestion using 1-β-endogluconase removed 55–60 wt % of the plant tissue within 24 hours, but with a loss of 50–60 % of the gold into solution. However, the nanoparticles produced were 5–50 nm in diameter, compared with 100–1000 nm diameter for particles obtained by ashing the residual biomass from the enzymic digestion. If the amount of plant tissue destroyed enzymically can be increased to about 95 %, and loss of Au back into solution controlled, then production of small nanoparticles of gold that would have a variety of uses may be possible.

23.5.2 Uranium

Concern about risk to living organisms due to radionuclides is usually based on radiological dose rate, but for U there is evidence that there may be a greater risk due to its chemical toxicity. Sheppard *et al.* [67] have reviewed the chemical toxicity of U and have estimated predicted no-effect concentrations (PNECs) from published data. The values for soil organisms are 250 mg U kg^{-1} dry soil for terrestrial plants, and 100 mg U kg^{-1} dry soil for other soil biota; for mammals the value is 0.1 mg U kg^{-1} body weight d^{-1}.

The value of 250 mg U kg^{-1} dry soil for terrestrial plants was suggested from data reported by nine published studies, but there was considerable variation between them arising from differences in soil, plant species used and endpoint measured (survival, biomass yield, growth rate, time to emergence and so on). The study that was considered most reliable gave a range for the EC$_{25}$ value (where performance of the endpoint measured was 25 % diminished compared to the control) of 300–500 mg U kg^{-1} dry soil. Given the few studies involved, a safety factor was applied and it was concluded that the PNEC should not be less than 100 mg U kg^{-1} dry soil for terrestrial plants, and a value of 250 mg U kg^{-1} dry soil was proposed. These figures were supported by evidence from solution culture studies. But data from such studies were not used in arriving at a value of EC$_{25}$ or PNEC because of the uncertainty in extrapolating from solution U concentrations in culture solution to that in soil solution, which is affected by the solid/solution partition coefficient (K_d) for U. Similar reservations were made regarding the soil biota data, which came from studies using different organisms (for example earthworms, collembolan, N-fixing microbes) and used various endpoints (for example survival, respiration rate, enzyme activity) as criteria to assess effects.

Uptake of U by plants is highly dependent on pH, with the greatest uptake under acidic conditions where UO_2^{2+} is the dominant ionic species in solution [68]. Uptake is severely diminished above pH 6, when the carbonate complexes are the dominant U species in solution. Organic acids, such as citric acid, can increase the bioavailability of U, even in neutral soils, but it is likely that the U–citrate complex is rapidly degraded by microbial action, releasing UO_2^{2+} ions that can either be taken up by plants or sorbed onto the surfaces of soil components. Thus the prospects of using phytoremediation for the cleanup of U-contaminated soil seem limited, except possibly in acidic soils or if amendments are applied to enhance solubility of U. Vandenhove et al. [69] considered the feasibility of phytoremediation to clean up low-level U-contaminated soil, and concluded that addition of citric acid to the soil was necessary in order for this to be realistic. The problem is the low soil-to-plant transfer factor (TF), which is usually extremely low (<0.05), and only approaches a value of about 0.1 in highly contaminated soil. Use of citric acid as a soil amendment increased this value to the range 2–7 depending on plant species.

23.6 Concluding Comments

The chemistry and behaviour of gold and uranium have received relatively little attention in natural, uncontaminated soils as the concentration of both elements tends to be low, and there is little evidence of any risk to human health or ecosystem functioning. More attention has been paid to contaminated systems, especially as a result of mining and refining processes, but recent developments in the use of gold (nanoparticles for use as catalysts, sensors and medical purposes) and uranium (depleted uranium used for military purposes) have raised concerns. Both Au and U exist as complex ions in solution, and are adsorbed onto the surfaces of soil components. Their soil chemistry has been largely inferred from geochemical studies.

References

[1] Knosp, H.; Holliday, R.J.; Corti, C.W., Gold in dentistry: alloys, uses and performance; Gold Bull. 2003, 36, 93–102.
[2] Fricker, S.P., Medical uses of gold compounds: past, present and future; Gold Bull. 1996, 29, 53–60.
[3] Yáñez-Sedeño, P.; Pingarrón, J.M., Gold nanoparticle-based electrochemical biosensors; Anal. Bioanal. Chem. 2005, 382, 884–886.
[4] Pingarrón, J.M.; Yáñez-Sedeño, P.; González-Cortés, A., Gold nanoparticle-based electrochemical biosensors; Electrochim. Acta 2008, 53, 5848–5866.

[5] Paciotti, G.F.; Meyer, L.; Weinreich, D. *et al.*, Colloidal gold: A novel nanoparticle vector for tumor directed drug delivery; Drug Deliv. 2004, 11, 169–183.
[6] Moore, B.A.; Duncan, J.R.; Burgess, J.E., Fungal accumulation of copper, nickel, gold and platinum; Mater. Eng. 2008, 21, 55–60.
[7] Hostýnek, J.J., Gold: an allergen of growing significance; Food Chem. Toxicol. 1997, 35, 839–844.
[8] Taylor, H.; Appleton, J.D.; Lister, R. *et al.*, Environmental assessment of mercury contamination from the Rwamagasa artisanal gold mining centre, Geita District, Tanzania; Sci. Total Environ. 2005, 343, 111–133.
[9] Feng, X.; Dai, Q.; Qiu, G.; Li, G.; He, L.; Wang, D., Gold mining related mercury contamination in Tongguan, Shaanxi Province, PR China; Appl. Geochem. 2006, 21, 1955–1968.
[10] Pataranawat, P.; Parkpian, P.; Polprasert, C.; Delaune, R.D.; Jugsujinda, A., Mercury emission and distribution: potential environmental risks at a small-scale gold mining operation, Phichit Province, Thailand; J. Environ. Sci. Health A 2007, 42, 1081–1093.
[11] Zagury, G.J.; Oudjehani, K.; Deschênes, L., Characterization and availability of cyanide in solid mine tailings from gold extraction plants; Sci. Total Environ. 2004, 320, 211–224.
[12] MacKenzie, A.B., Environmental radioactivity: experience from the 20th century – trends and issues for the 21st century; Sci. Total Environ. 2000, 249, 313–329.
[13] Bleise, A.; Danesi, P.R.; Burkart, W., Properties, use and health effects of depleted uranium (DU): a general overview; J. Environ. Radioact. 2003, 64, 93–112.
[14] Oliver, I.W.; Graham, M.C.; MacKenzie, A.B.; Ellam, R.M.; Farmer, J.G., Assessing depleted uranium (DU) contamination of soil, plants and earthworms at UK weapons testing sites; J. Environ. Monit. 2007, 9, 740–748.
[15] Emsley, J., The Elements; Clarendon Press, Oxford, 1991, 251.
[16] Bond, G.C.; Louis, C.; Thompson, D.T., Catalysis by Gold (Catalytic Science Series, G.J. Hutchings (ed.), Vol. 6), Imperial College Press, London, 2006.
[17] Zhang, P.C.; Krumhansl, J.L.; Brady, P.V., Introduction to the properties, sources and characteristics of soil radionuclides; in: P.C. Zhang, P.V. Brady (eds), Geochemistry of Soil Radionuclides; Soil Science Society of America, Madison, WI, 2002, p. 1–20.
[18] Edwards, R.; Lepp, N.W.; Jones, K.C., Other less abundant elements of potential environmental significance; in: B.J. Alloway (ed.), Heavy Metals in Soils; Blackie Academic & Professional, Glasgow, 1995, pp. 306–352.
[19] Bowen, H.J.M., Environmental Chemistry of the Elements; Academic Press, London, 1979, p. 333.
[20] Puddephatt, R.J., The Chemistry of Gold, Monograph 16, Topics in Inorganic and General Chemistry, R. J. H. Clark (ed.); Elsevier Scientific, Oxford, p. 274, 1978.
[21] Alloway, B.J., Heavy Metals in Soils, 2nd edn; Blackie Academic & Professional, Glasgow, 1995, p. 368.
[22] Ure, A.M.; Berrow, M.L., The elemental constituents of soils; in: Environmental Chemistry; Royal Society of Chemistry, London, 1982, pp. 94–204.
[23] Eisler, R., Gold concentrations in abiotic materials, plants and animals: a synoptic review; Environ. Monit. Assess. 2004, 90, 73–88.
[24] Rashed, M.N.; Awadallah, R.M., Trace elements in Faba bean (*Vicia faba* L) plant and soil as determined by atomic absorption spectroscopy and ion selective electrode; J. Sci. Food Agric. 1998, 77, 18–24.
[25] Miller, J.R.; Rowland, J.; Lechler, P.J.; Desilets, M.; Hsu, L-C, Dispersal of mercury-contaminated sediments by geomorphic processes, Sixmile Canyon, Nevada, USA: implications to site characterization and remediation of fluvial environments; Water Air Soil Pollut. 1996, 86, 373–388.
[26] McBride, M.B.; Richards, B.K.; Steenhuis, T.; Russo, J.J.; Sauve, S., Mobility and solubility of toxic metals and nutrients in soil fifteen years after sludge application; Soil Sci. 1997, 162, 487–500.
[27] Reimann, C.; Arnoldussen, A.; Englmaier *et al.*, Element concentrations and variations along a 120-km transect in southern Norway – anthropogenic vs. geogenic vs. biogenic sources and cycles; Appl. Geochem. 2007, 22, 851–871.

[28] Boyd, R.; Niskavaara, H.; Kontas, E. et al., Anthropogenic noble-metal enrichment of topsoil in the Monchegorsk area, Kola Peninsula, northwest Russia; J. Geochem. Explor. 1997, 58, 283–289.
[29] Crocket, J.H.; MacDougall, J.D.; Harriss, R.C., Gold, palladium and iridium in marine sediments; Geochim. Cosmochim. Acta 1973, 37, 2547–2556.
[30] Meriwether, J.R.; Beck, J.N.; Keeley, D.F.; Langley, P.; Thompson, R.H.; Young, J.C., Radionuclides in Louisiana soils; J. Environ. Qual. 1988, 17, 562–568.
[31] Sheppard, S.C.; Sheppard, M.I.; Ilin, M.; Tait, J.; Sanipelli, B., Primordial radionuclides in Canadian background sites: secular equilibrium and isotopic differences; J. Environ. Radioact. 2008, 99, 933–946.
[32] Yoshida, S.; Muramatsu, Y.; Tagami, K.; Uchida, S., Concentrations of lanthanide elements, Th, and U in 77 Japanese surface soils; Environ. Int. 1998, 24, 275–286.
[33] Van de Velde, K.; Barbante, C.; Cozzi, G. et al., Changes in the occurrence of silver, gold, platinum, palladium and rhodium in Mont Blanc ice and snow since the 18th century; Atmos. Environ. 2000 34, 3117–3127.
[34] Furr, A.K.; Lawrence, A.W.; Tong, S.S.C. et al., Multielement and chlorinated hydrocarbon analysis of municipal sewage sludges of American cities; Environ. Sci. Technol. 1976, 10, 683–687.
[35] Reeves, S.J.; Plimer, I.R.; Foster D., Exploitation of gold in a historic sewage sludge stockpile, Werribee, Australia: resource evaluation, chemical extraction and subsequent utilisation of sludge. J. Geochem. Explor. 1999, 65, 141–153.
[36] Reith, F. McPhail D.C., Mobility and microbially mediated mobilization of gold and arsenic in soils from two gold mines in semi-arid and tropical Australia; Geochim. Cosmochim. Acta 2007, 71, 1183–1196.
[37] Tracy, B.L.; Prantl, F.A. Quinn J.M., Transfer of ^{226}Ra, ^{210}Pb and uranium from soil to garden produce: assessment of risk; Health Phys. 1983, 44, 469–477.
[38] Bykova, E.I.; Gorborukova, L.P.; Kostenko, E.S.; Namazbekova, S.Sh., Man-caused uranium contamination of biosphere objects in the territory of Kyrgyzstan; J. Environ. Sci. Health A 2006, 41, 2665–2682.
[39] Di Lella, L.A.; Nannoni, F.; Protano, G.; Riccobono, F., Uranium contents and ^{235}U/^{238}U atom ratios in soil and earthworms in western Kosovo after the 1999 war; Sci. Total Environ. 2005, 337, 109–118.
[40] Rothbaum, H.P.; McGaveston, D.A.; Wall, T.; Johnston, A.E.; Mattingley, G.E.G., Uranium accumulation in soils from long-continued applications of superphosphate; J. Soil Sci. 1979, 30, 147–153.
[41] Guzmán, E.T.R.; Regil, En.O.; Gutiérrez, L.R.R.; Alberich, M.V.E.; Hernández, L.R.R.; Regil, Ed.O., Contamination of corn growing areas due to intensive fertilization in the high plane of Mexico; Water Air Soil Pollut. 2006, 175, 77–98.
[42] Makweba, M.M. Holm E., The natural radioactivity of the rock phosphates, phosphatic products and their environmental implications; Sci. Total Environ. 1993, 133, 99–110.
[43] Guzmán, E.T.R.; Ríos, M.S.; García, J.L.I.; Regil, E.O., Uranium in phosphate rock and derivatives; J. Radioanal. Nucl. Chem. 1995, 189, 301–306.
[44] Hough, R.M.; Noble, R.R.P.; Hitchen, G.J. et al., Naturally occurring gold nanoparticles and nanoplates; Geol. 2008, 36, 571–574.
[45] McHugh, J.B., Concentration of gold in natural waters; J. Geochem. Explor. 1988, 30, 85–94.
[46] Vlassopoulos, D.; Wood S.A., Gold speciation in natural waters: I. Solubility and hydrolysis reactions of gold in aqueous solution; Geochim. Cosmochim. Acta 1990, 54, 3–12.
[47] Ran, Y.; Fu, J.; Rate, A.W.; Gilkes, R.J., Adsorption of Au(I, III) complexes on Fe, Mn oxides and humic acid; Chem. Geol. 2002, 185, 33–49.
[48] Colin, F.; Vieillard, P. Ambrosi J.P., Quantitative approach to physical and chemical gold mobility in equatorial rainforest lateritic environment; Earth Planet. Sci. Lett. 1993, 114, 269–285.
[49] Johnson, C.A.; Grimes, D.J.; Leinz, R.W.; Rye, R.O., Cyanide speciation at four gold leach operations undergoing remediation; Environ. Sci. Technol. 2008, 42, 1038–1044.
[50] Baker, W.E., The role of humic acid in the transport of gold; Geochim. Cosmochim. Acta 1978, 42, 645–649.

[51] Vlassopoulos, D.; Wood, S.A.; and Mucci A., Gold speciation in natural waters: II. The importance of organic complexing – experiments with some simple model ligands; Geochim. Cosmochim. Acta 1990, 54, 1575–1586.
[52] Diegoli, S.; Manciulea, A.L.; Begum, S.; Jones, I.P.; Lead, J.R.; Preece, J.A., Interaction between manufactured gold nanoparticles and naturally occurring organic macromolecules; Sci. Total Environ. 2008, 402, 51–61.
[53] Lovley, D.R.; Phillips, E.J.P.; Gorby, Y.A.; Landa, E.R., Microbial reduction of uranium; Nature, 1991, 350, 413–416.
[54] Suzuki, Y.; Kelly, S.D.; Kemner, K.M.; Banfield, J.F., Nanometre-size products of uranium bioreduction; Nature 2002, 419, 134.
[55] Duff, M.C.; Amrhein C., Uranium (VI) adsorption on goethite and soil in carbonate solutions; Soil Sci. Soc. Am. J. 1996, 60, 1393–1400.
[56] Choppin, G.R., Actinide speciation in the environment; J. Radioanal. Nucl. Chem. 2007, 273, 695–703.
[57] Vandenhove, H.; Van Hees, M.; Wouters, K.; Wannijn, J., Can we predict uranium bioavailability based on soil parameters? Part 1: Effect of soil parameters on soil solution uranium concentration; Environ. Pollut. 2007, 145, 587–595.
[58] Koch-Steindl, H.; Pröhl G., Considerations on the behaviour of long-lived radionuclides in the soil; Radiat. Environ. Biophys. 2001, 40, 93–104.
[59] Echevarria, G.; Sheppard, M.I.; Morel, J.L., Effect of pH on the sorption of uranium in soils; J. Environ. Radioact. 2001, 53, 257–264.
[60] Sherman, D.M., Peacock, C.L.; Hubbard, C.G., Surface complexation of U(VI) on goethite (–FeOOH). Geochim. Cosmochim. Acta, 2008. 72, 298–310.
[61] Oliver, I.W.; Graham, M.C.; MacKenzie, A.B.; Ellam, R.M.; Farmer, J.G., Distribution and partitioning of depleted uranium (DU) in soils at weapons test ranges – investigations combining the BCR extraction scheme and isotopic analysis; Chemosphere 2008, 72, 932–939.
[62] Graham, M.C.; Oliver, I.W.; MacKenzie, A.B.; Ellam, R.M.; Farmer, J.G., An integrated colloid fractionation approach applied to the characterisation of porewater uranium-humic interactions at a depleted uranium contaminated site; Sci. Total Environ. 2008, 404, 207–217.
[63] Anderson, C.W.N.; Brooks, R.R.; Stewart, R.B.; Simcock, R., Harvesting a crop of gold in plants; Nature 1998, 395, 553–554.
[64] Lintern, M.J.; Butt, C.R.M.; Scott, K.M., Gold in vegetation and soil - three case studies from the goldfields of southern Western Australia; J. Geochem. Explor. 1997, 58, 1–14.
[65] Anderson, C.; Moreno, F.; Meech, J., A field demonstration of gold phytoextraction technology; Mater. Eng. 2005, 18, 385–392.
[66] Marshall, A.T.; Havercamp, R.G.; Davies, C.E.; Parsons, J.G.; Gardea-Torresdey, J.L.; van Agterveld, D., Accumulation of gold nanoparticles in *Brassica juncea*; Int. J. Phytoremediation 2007, 9, 197–206.
[67] Sheppard, S.C.; Sheppard, M.I.; Gallerand, M-O.; Sanipelli, B., Derivation of ecotoxicity thresholds for uranium, J. Environ. Radioactiv. 2005, 79, 55–83.
[68] Ebbs, S.D.; Brady, D.J.; Kochian, L.C., Role of uranium speciation in the uptake and translocation of uranium by plants; J. Exp. Bot. 1998, 49, 1183–1190.
[69] Vandenhove, H.; Van Hees, M.; Van Winckel, S., Feasibility of phytoextraction to clean up low-level uranium-contaminated soil; I. J. Phytoremediation 2001, 3, 301–320.

24
Platinum Group Elements

F. Zereini[1] *and C.L.S. Wiseman*[2]

[1]*Institute for Atmospheric and Environmental Sciences, J.W. Goethe-University, Frankfurt, Germany*
[2]*Centre for Environment, University of Toronto, Toronto, Ontario, Canada*

24.1 Introduction

The metals platinum (Pt), palladium (Pd), iridium (Ir), rhodium (Rh), ruthenium (Ru), and osmium (Os) are commonly referred to as the platinum group elements (PGE). These noble metals are thermally stable and highly resistant to oxidation when exposed to atmospheric oxygen, humidity, and dilute acids and bases. They are only susceptible to the oxidizing effects of strong acids.

PGE are extremely rare and are present at concentrations of about 0.4–5 $\mu g\ kg^{-1}$ in the earth's crust [1]. Their chemical characteristics make them quite useful as catalysts in a variety of chemical and pharmaceutical processes. In particular, Pt, Pd, and Rh are widely used in a number of applications such as in hydration and dehydration reactions in the pharmaceutical industry, in polymer processing, and in the production of pesticides and dyestuffs. Since the 1980s in Europe, they have also been used as the active catalyst material in automobile catalytic converters to reduce the emissions of nitrogen oxides, hydrocarbons, and carbon monoxide.

Today, the catalytic converter industry is the most important consumer of PGE. In 2007, for instance, a total of 54.2 % of the global production of Pt was used in the production of catalytic converters [2]. Platinum was also heavily used by the jewelry-making industry, which consumed 20.4 % of available global supplies in the same year. Other primary users

Trace Elements in Soils Edited by Peter S. Hooda
© 2010 Blackwell Publishing, Ltd

of Pt include producers of chemicals, electronics, glass and mineral oil, together using 18.1 % of its total production in 2007. The catalytic converter manufacturers were also the greatest consumers of Pd and Rh in 2007, using a total of 58 % and 87 % of the global supply of these two metals, respectively. While the introduction of catalytic converters has dramatically reduced pollutant emissions from automobiles, their use has also led to elevated environmental concentrations of Pt, Pd, and Rh. PGE emissions, their concentrations in the environment, and their potential impacts on organisms are discussed in detail in three recently published books [3–5]. Despite their thermal and chemical stability, Pt, Pd, and Rh are emitted in automobile exhausts due to elevated temperatures in the converter and the mechanical abrasion of the catalyst surface. Although they are typically emitted in small amounts, observed increases in Pt, Pd, and Rh levels in the environment since the introduction of catalytic converters have been significant and are a cause of concern [3–5]. Most PGE appear to be emitted in a metallic form, which is believed to be biologically inert [6]. However, small amounts may be released in an oxidized state [7]. The toxic potential of the individual PGE seems to be highly variable. Several platinum metal salts are known to be toxic and have a significant potential to elicit hypersensitive reactions in susceptible individuals [8]. The evidence suggests that Pd may be a more reactive species in the environment due to its greater solubility and bioavailability compared with other PGE [9]. Palladium has been demonstrated to be more readily taken up by plants [10], animals, and humans than Pt and Rh (for example [11]).

This chapter reviews PGE emissions, deposition and distribution in soils, as well as the geochemical behavior of the emitted PGE in terms of their solubility, mobility, and bioavailability.

24.2 Sources of PGE in Soils

Both geogenic and anthropogenic sources can contribute to the presence of PGE in soils. These sources are described in the following two sections.

24.2.1 Geogenic Sources

PGE are normally present at extremely low concentrations in the lithosphere. Concentrations of 0.4 μg kg^{-1} for Pt and Pd, 0.1 μg kg^{-1} for Ru, 0.06 μg kg^{-1} for Rh, and 0.05 μg kg^{-1} for Os and Ir have been cited for the earth's crust [12]. These elements can occur at higher concentrations in localized deposits as a result of certain geochemical processes [13]. Such deposits become economically attractive for mining when they reach concentrations of 10 g PGE per Mg of rock. The PGE are typical products of the early crystallization of ultrabasic and basic magma. Globally, these metals occur most commonly in rock complexes with peridotite and dunite. These metals are siderophiles and have an affinity for chalcophile elements. As siderophiles, they tend to form alloys with Fe (isoferroplatinum (Pt$_3$Fe); Pt ± Ir, Os, Rh, Ru, Pd, Fe, Ni, Cu, Au; Pd ± Pt, Ir, Os, Ru, Pb; Ir ± Pt, Os, Rh, Ru, Pd, Fe, Au). Their affinity for chalcophiles causes them to commonly combine with S, As, Se, Te and Bi to form the deposits cooperite (PtS), sperrylite (PtAs$_2$), geversite (PtSb$_2$), vysotskite (Pt, Pd, Ni), stibiopalladinite (Pd$_3$Sb), and laurite (RuS$_2$).

The most economically important PGE deposits worldwide are located in the Bushveld Igneous-Complexes of South Africa, in the Stillwater Massif of Montana, USA and in the nickel- and copper-dominating gabbros and norites of Sudbury, Canada and Norilsk, Russia. Currently, the Bushveld complexes and the deposits of Norilsk yield the largest amounts of PGE in global terms. The global production of Pd for industrial purposes has almost doubled from 133 Mg in 1993 to 258 Mg in 2007 [2,14]. The production of Pt has increased at a similar rate, from 137 Mg in 1993 to 207 Mg in 2007. About 96 % of the global supply of Pt and Pd in 2007 originated from four countries: Russia, South Africa, United States, and Canada. Russia produced just over half of the global supply of Pd in 2007 (51 %), while about 34 % came from South Africa. The United States and Canada produced about 12 % of the global amount of this metal. South Africa is the primary global supplier of Pt, contributing 78 % of the total supply of this metal in 2007, followed by Russia with 12 % and the United States and Canada with 5 % [2].

PGE mining and metallurgy causes significant amounts of these metals to be emitted in the environment, particularly in the direct vicinity of these activities. In almost all investigated soils close to major PGE deposits, elevated amounts of these metals have been found. For instance, in soil samples taken in the area of the Lac des Iles deposit of Ontario, Canada, Pt concentrations of as much as 340 mg kg^{-1} have been measured [15]. Soils taken from the area of the Sudbury complex were found to contain about 100 μg Pd kg^{-1} [16]. Soils located over the copper-nickel deposit of Lac Sheen in Quebec, Canada had as much as 1600 μg Pt kg^{-1} and 33 μg Pd kg^{-1} [17]. Gunn [18] reported Pt concentrations of as much as 30 μg kg^{-1} in soils taken from over the Unst ophiolite in the Shetland Islands of Scotland.

24.2.2 Anthropogenic Sources

There are several important anthropogenic sources of PGE in the environment. For instance, the chemical industry and noble metal extraction and processing industries are known to contribute to the widespread distribution of PGE in the biosphere and troposphere [19]. According to Reimann and Niskavaara [19], 2.2 Mg of Pd are emitted with the dust from the Monschegorsk mining facility in Russia on an annual basis. Hospitals can also be a significant source of PGE in the environment. Platinum-based compounds such as cisplatin and carboplatin are commonly used in chemotherapy for the treatment of various types of cancers. The Pt tends to accumulate in the blood, urine, and liver and is primarily excreted in urine, contributing to elevated concentrations in the waste water of hospitals. In Germany, it was estimated that a total of 14.2 kg of Pt was emitted from hospitals in 1996 [20]. In addition, Pt has also been found in petrol and diesel [21]. According to Alt *et al.* [21], fertilizer can contain significant amounts of Pt (0.032–23 μg kg^{-1}). This may have contributed to the elevated levels of Pt observed for cultivated soils compared with uncultivated soils [21].

Automobile catalytic converters have received the most attention as a primary contributor of PGE to the environment. The release of PGE in automobile emissions due to the thermal and mechanical erosion of the catalyst material has led to their elevated levels in the environment, as has been documented in number of studies [3–5]. The potential of these metals to impact the environment and human health has been the focus of much debate.

Initially, automobiles with two-stroke internal combustion engines were first fitted with catalytic converters that contained Pt and Rh as the active catalysts in the 1980s in Europe. Since then, the use has shifted to three-way catalytic converters, which also contain Pd in amounts as much as 5 g per liter of engine cylinder capacity [22]. As a result, global increases in the use of Pd by the catalytic converter industry over the past 10–15 years have been slightly greater than that for Pt. The use of Pd by the motor industry increased from 22 Mg in 1993 to 136 Mg in 2007, while Pt consumption rose from 52 Mg to 131 Mg during the same period [2,23]. This greater application of Pd in catalytic converters has led to its elevated levels in soils in the vicinity of highways over the last 10 years [24]. Since the late 1990s, automobiles with diesel-fuelled engines have also been fitted with Pt-based catalytic converters. The catalytic converter industry is presently attempting to develop Pd-based catalytic converters for diesel-fuelled automobiles.

PGE emissions from automobiles fitted with catalytic converters have been investigated qualitatively and quantitatively in a variety of experimental studies, as well as in dynamometer experiments. As a result, a fair amount is known about this emission source. Dynamometer experiments have shown that Pt is emitted together with wash coat particles having a diameter of 0.1–20 μm, as well as nanoparticulates of Pt and PtO_2 clusters of <5 nm [7]. Artelt et al. [25,26] demonstrated that most of the emitted particles have a diameter >10 μm. Although most PGE appear to be emitted in metallic form (Pt^0, Pd^0), small amounts of Pt(II) and Pt(IV) may be present in exhaust fumes [7]. Automobiles with three-way catalytic converters have been found to emit Pt in the range 2–78 ng km^{-1} [26]. Emissions rates can vary, though, depending on the running condition of the motor and the age of the catalytic converter [25,27]. Older catalytic converters appear to emit considerably less Pt (up to 21 ng m^{-3}) than new catalytic converters (up to 87 ng m^{-3}) [25]. PGE emissions are also a function of speed and driving behavior. Inacker and Malessa [28], for instance, observed increases in Pt emissions with increases in speed in dynamometer tests. The highest Pt emissions were found to occur at high speeds in the range 80–140 km h^{-1}. Platinum was also observed to be emitted in greater amounts with the 'stop-and-start' characteristic of city driving compared to driving at a constant speed of 80 km h^{-1}, that is 2–3 times higher. Observed PGE concentrations in soils along roadways reflect the effects of speed and driving behavior on emissions, as discussed in the next section.

24.3 Emissions, Depositional Behavior, and Concentrations in Soils

The highest environmental concentrations of PGE have been repeatedly observed to occur along roads with the highest traffic volumes and speed limits (for example [9,29]). Similar to that for other traffic-related metal emissions, PGE-containing particles released in automotive exhaust tend to be deposited close to the source; that is, the roadside environment. The particles may be further transported with road runoff to drainage systems and may accumulate in rainwater retention ponds [30,31]. Spray or splash water from tires can also strongly influence the depositional behavior of PGE along roadways, as demonstrated by Zereini *et al.* [31]. Under dry conditions, PGE-containing particles can be transported by wind. The extent of atmospheric drift is dependent on the morphology and size of the emitted particles, as well as the meteorological conditions such as wind direction and velocity [30,32]. Local conditions such as geomorphology and the presence of vegetation can also influence how PGE will be deposited in the local environment.

Given their emission and deposition close to road networks, most investigations of the spatial distribution of PGE in soils have focused on areas close to roadways [24]. It should be noted that the measurement of environmental concentrations of PGE poses an analytical challenge. These metals are typically present at very low concentrations with interfering matrix constituents. As such, their concentrations cannot be reliably determined without the aid of pre-isolation/enrichment techniques, even when such powerful analytical instruments as ICP-MS (inductively couple plasma mass spectrometry) are used [33]. Bencs *et al.* [34] and Angelone *et al.* [35] provide overviews of the various methods available for the measurement of PGE in environmental media.

Zereini *et al.* [36] were the first to detect the existence of elevated Pt concentrations in soils along roads with higher volumes of traffic in Germany, using a proven, reliable method which combines nickel sulfide (NiS)–fire assay with GF-AAS (graphite furnace atomic absorption spectrometry [37]. Since then, a number of studies have consistently shown that PGE concentrations in soils along roads are often considerably higher than typical background levels (<1 μg kg^{-1}) and have increased over time (for example [9,10]). One recent study investigated the PGE concentrations in soils along a 23-km stretch of a four-lane motorway in Germany with a heavy traffic load of 110 000–131 000 vehicles per day [24]. The soils had mean concentrations of 83 μg Pd kg^{-1}, 132 μg Pt kg^{-1}, and 20 μg Rh kg^{-1}. The concentrations were highly variable and heterogeneously distributed, with values ranging from 20 to 191 μg kg^{-1} for Pd, from 41 to 254 μg kg^{-1} for Pt, and from 7 to 36 μg kg^{-1} for Rh. Similar patterns have been observed for other motorways in Germany [30,31]. Soils along entrance ramps to motorways have also been shown to have higher PGE concentrations than motorway exit ramps [31]. Clearly, more PGE are emitted with acceleration and associated increases in temperature, gas flow, and mechanical erosion. In addition, factors such as geomorphology, vegetation type and characteristics of the microclimate can strongly determine the variability of observed PGE concentrations in soils at different locations [24].

Environmental concentrations of PGE have increased significantly over time since the introduction of automotive catalytic converters. Zereini *et al.* [24] found that concentrations of Pt, Pd, and Rh in soils along the A5 motorway in Germany in 2004 were higher than their values measured for soils collected from the same stretch in 1994. The concentration of Pd in soils directly along the road was 15 times higher on average compared with concentrations measured in 1994, while Pt and Rh increased twice as much. This is clearly related to an increase in the use of Pd as an active catalyst in catalytic converters over the past 10–15 years. In addition, the finest soil fractions were found to contain the highest PGE concentrations. A strong relationship has been observed between Pt and Rh concentrations in soils, indicating that most PGE emissions originate from catalytic converters. A similar relationship has also been detected for Pd and Pt concentrations. A ratio of approximately 5:1 between Pt and Rh concentrations has been consistently observed in earlier studies of PGE concentrations in soils, sediments and street dust (for example [38]). This reflected the ratio that was commonly used for these elements (Pt, Rh) in the production of catalytic converters in Europe. In recent years, however, ratios between Pt and Rh concentrations and Pd and Rh concentrations in soils have shifted and are highly variable. This appears to reflect a change toward the increased use of Pd over Pt and the introduction of various types of catalytic converters with very different PGE concentrations and ratios (Pt–Pd–Rh, Pd–Rh and Pt-only catalytic converters) [24].

Concentrations of PGE in soils as a function of distance from the roadside have been investigated in a number of studies (for example [24,31]). The highest concentrations of PGE have been found to occur within 10 m of the roadside, with the greatest levels observed in soils directly along the road (Figure 24.1). Levels normally decline rapidly with increasing distance from the roadside, with geogenic background levels of <1 μg PGE kg^{-1} normally being reached beyond a distance of 10 m. Elevated levels of PGE, however, have been found to occur further away from road networks. For instance, Zereini et al. [24] found elevated levels of Pd and Pt in meadow soils collected 50 m away from a motorway in Germany. Similar findings have been observed in the United States [29] and elsewhere [30,31]. The decreasing PGE concentration gradient suggests that the size of the emitted PGE-containing particles is similar to that found by Artelt et al. [25,26] during laboratory simulations, that is mainly >10 μm.

Few studies have investigated PGE concentrations according to soil depth. One study of samples taken up to a depth of 16 cm, 0.3 m from the roadside of a motorway in Germany has shown that PGE concentrations tend to decrease rapidly with depth [24]. The highest PGE concentrations were observed within the top 4 cm depth, which contained 92 %, 82 %, and 72 % of the total Rh, Pt, and Pd measured for the sample core. Similar observations have been made in soils collected along other motorways in Germany [30]. Elevated Pd concentrations were still observed at depths of 12–16 cm, in contrast to Pt [24]. This may indicate a higher solubility and mobility of this metal compared to the other PGE.

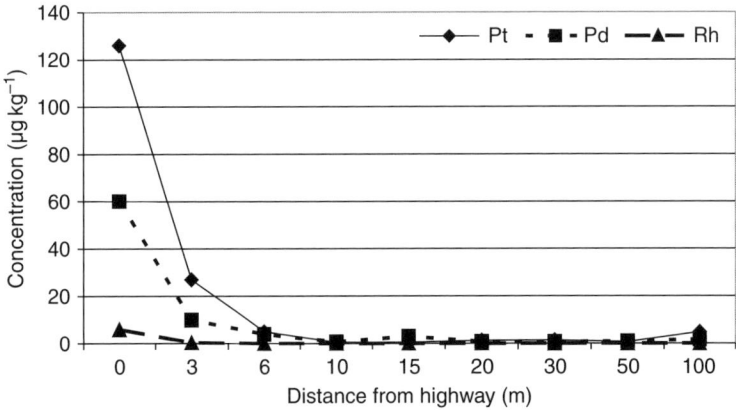

Figure 24.1 P, Pd, and Rh concentrations (μg kg^{-1}) as a function of distance from the highway A5 in 2004 (Adapted with permission from F. Zereini, C. Wiseman and W. Püttmann, Changes in palladium, platinum and rhodium concentrations and their spatial distribution in soils along a major highway in Germany from 1994 to 2004, Environ. Sci. Technol. 41, 451–456. Copyright 2007 American Chemical Society.)

24.4 Geochemical Behavior in Soils

Very little is known about the geochemical behavior of emitted PGE in the biosphere, especially in terms of their mobility and solubility in soils under natural conditions. Indirect indices such as the ratio between measured Pt and Rh concentrations, however, indicate that these two metals have a low solubility [30]. The low solubility of Pt and Rh has also been demonstrated in experimental investigations [39]. In contrast, significant amounts of Pd have been shown to be emitted in a relatively soluble form or have the potential to be rapidly transformed from a metallic state once deposited in the environment [9].

A number of factors appear to play a role in the mobilization and accumulation of PGE in the environment such as pH, E_h, and the concentration of chloride ions [40]. In soils, Pt and Rh have been found most soluble under highly acidic conditions [39]. In one study, a maximum of 0.5 % of the Pt from a three-way catalytic converter was found to be soluble in soil solution [39]. In contrast to Pt, experiments have shown that Pd is more likely to be taken up by plants [41] and terrestrial and aquatic animals [42,43]. Jarvis et al. [9] found that Pd in street dust was more soluble in water than Pt and Rh. In another study, Pd was found to occur at greater depths in soils (12–16 cm) along a highway compared to Pt [24]. These studies indicate that Pd has a higher solubility and, hence, mobility and potential bioavailability compared with other PGE.

Several studies which investigated the sorptive dynamics of PGE with soil organic material and mineral phases under laboratory conditions have shed some light on how these metals may behave in the environment. Lustig et al. [44] observed that the behavior of Pt-compounds and their capacity to bind to and complex with soils was strongly dictated by their chemical characteristics such as oxidation state. Skerstupp et al. [45] found that Pt in solution $[(Pt_4Cl_6)^{2-}]$ was 10 times more likely to be sorbed on synthetic ferrihydrite, $(Fe_{1.78}Al_{0.21})O_3 \cdot 1.8H_2O$ than the clay mineral montmorillonite. Dikikh et al. [46] investigated the sorptive behavior and mobility of water-soluble PGE emitted from automobiles on various soil minerals. They found that Pt, Pd, and Rh were more easily mobilized when bound to mineral phases with a low sorption capacity such as feldspar and quartz compared with those with a higher sorption capacity such as kaolinite.

Soil organic matter, particularly its humic compounds fraction, is capable of binding trace metals due to its high sorptive capacity. The presence of other organic materials in soil can also influence the solubility and mobility of these metals [47]. Lustig et al. [48] conducted a number of solubility experiments on Pt(0)-containing soils (Pt-black and tunnel dust) with complexing agents commonly found in the environment (amino acids, humic acids, l-methionine). They found that significant amounts of metallic Pt were solubilized in the presence of air and chelating agents with N-, P-, and/or S-donors. The evidence suggests that soil type and the presence of humic substances and clay minerals ultimately determine the extent to which soluble forms of PGE are adsorbed or immobilized [48].

24.5 Bioavailability

The solubility of PGE present in soils under natural conditions determines their bioavailability and potential to be taken up by plants and other organisms. Biomonitoring studies of

PGE emissions have shown that plants, trees, grass, and mosses in the vicinity of streets with high traffic volumes often have elevated concentrations of these metals (for example [49]). Most of the Pt found in association with plants appears to occur in particulate form on the surface. A number of studies have shown, however, that plants are quite capable of taking up water soluble Pt compounds [41,50]. Most studies have focused on the potential of $PtCl_4$ uptake by test plants grown in the nutrient solution. This compound ($PtCl_4$) is easily dissolved in water and is not representative of the automobile-emitted Pt found in soils. As such, conclusions regarding the bioavailability of PGE from catalytic converters under natural conditions cannot be made reliably. In addition to the type of metal compound present, the bioavailability of PGE is also influenced by a number of soil factors [51], as for other trace elements.

The bioavailability of Pt in the environment is also discussed in several publications (for example [41,44]). According to Ballach *et al.* [52], the root uptake of water-soluble Pt^{4+} may be due to the specific chemical characteristics of the rhizosphere; that is, interactions between roots and soil particles, and the associated release of protons and chelators. In particular, glutathione in plant cells appears to play an important role in stress reactions and the binding of trace metals [53]. Lustig and Schramel [54] found that commonly cultivated plants (onions, radishes, corn, potatoes, broadbeans) have a limited capacity to take up Pt which is emitted in automobile exhaust. For instance, potatoes only took up about 0.09 % of the total available Pt. Most of this was found in the peel (0.55 ± 0.39 $\mu g\ kg^{-1}$), while 0.29 ± 0.15 $\mu g\ kg^{-1}$ Pt was found in the inner part of the potatoes. In another study, Eckhardt and Schäfer [55] investigated the transfer coefficients of different metals from soils to plants such as spinach and garden cress and found high transfer rates for Pt, Pd, and Rh. For all observed plants, the greatest transfer was found for Pd, followed by Pt and Rh. The uptake of Pt and Rh for various plants can be compared with that for Cu and Pb, while the bioavailability of Pd may be as high as that for Zn and Cd in some instances.

Our knowledge regarding the fate of emitted PGE in soils and their impact on vegetation under natural conditions is limited. This is due, in part, to difficulties associated with the measurement of low PGE concentrations in biological and environmental materials [56] and changes regarding the stability of metal species during analytical separation procedures [29].

24.6 Conclusions

The use of catalytic converters has served to dramatically reduce pollutant emissions from the burning of fossil fuels; however, the release of PGE has led to significant increases in Pt, Pd, and Rh concentrations in soils close to road networks. This may pose a possible risk to both humans and the environment.

The analysis of PGE, especially Pd, in environmental media remains a challenge. A reliable method which could be used for the analysis of these metals in all types of environmental samples has yet to be developed. In addition, little is known about the geochemical behavior of emitted PGE-containing particles once deposited on soils in terms of their solubility and mobility. PGE emissions can be expected to increase in the future, as greater numbers of automobiles are fitted with catalytic converters and fuel cells with Pt-based catalysts. This highlights the need for continued monitoring of environmental PGE concentrations and research regarding the geochemical fate of these metals.

References

[1] Hartley, F.R., Chemistry of Platinum Group Metals; Elsevier, Amsterdam, Oxford, New York, 1991.
[2] Johnson Matthey, Platinum – Annual Report; Johnson Matthey, London, 2007, p. 104.
[3] Zereini, F.; Alt, F. (eds), Emissionen von Platinmetallen – Analytik, Umwelt- und Gesundheitsrelevanz; Springer Verlag, Berlin, Heidelberg, New York, London, 1999, p. 327.
[4] Zereini, F.; Alt, F. (eds), Anthropogenic Platinum Group Element Emissions – Their Impact on Man and Environment; Springer Verlag, Berlin, Heidelberg, New York, London, 2000, p. 308.
[5] Zereini, F.; Alt, F. (eds), Palladium Emissions in the Environment – Analytical Methods, Environmental Assessment and Health Effects; Springer Verlag, Berlin, Heidelberg, New York, London, 2006, p. 639.
[6] Wiseman, C.L.S.; Zereini, F., Airborne particulate matter, platinum group elements and human health: a review of recent evidence; Sci. Total Environ. 2009, 407 2493–2500.
[7] Schlögl R.; Indlekofer, G.; Oelhafen, P., Mikropartikelemissionen von Verbrennungsmotoren mit Abgasreinigung, Röntgen-Photoelektronenspektroskopie in der Umweltanalytik; Angewandte Chemie 1987, 99, 312–322.
[8] WHO, Environmental Health Criteria 226 – Palladium, International Programme on Chemical Safety; World Health Organization, Geneva, 2002, p. 201.
[9] Jarvis, K.; Parry, S.; Piper, M., Temporal and spatial studies of autocatalyst-derived platinum, rhodium, and palladium and selected vehicle-derived trace elements in the environment; Environ. Sci. Technol. 2001, 35, 1031–1036.
[10] Schäfer, J.; Hannker, D.; Eckhardt, J.D.; Stüben, D., Uptake of traffic-related heavy metals and platinum group elements (PGE) by plants; Sci. Total Environ. 1998, 215, 59–67.
[11] Moldovan, M.; Rauch, S.; Gómez, M.; Palacios, M.; Morrison, GM., Bioaccumulation of palladium, platinum and rhodium from urban particulates and sediments by the freshwater isopod *Asellus aquaticus*; Water Res. 2001, 35, 4175–4183.
[12] Wedepohl, K.H., The composition of the continental crust; Geochim. Cosmochim. Acta 1995, 59, 1217–1232.
[13] Mountain, B.W.; Wood, S.A., Chemical controls on the solubility, transport, and deposition of platinum and palladium in hydrothermal solutions; Econ. Geol. 1988, 83, 492–510.
[14] Johnson Matthey, Platinum – Annual Report; Johnson Matthey, London, 2003, p. 80.
[15] Fortescue, J.; Stahl, H.; Webb, J. R., Humus geochemistry in the Lac des Iles area, District of Thunder Bay, Ontario; Geology Survey, Map 808000; Canadian Geochemical Surveys, National Resources Canada, Ottawa, ON, 1988; http://edg.rncan.gc.ca/geochem/metadata_pub_e.php?id=00730 (accessed 28 November 2009).
[16] Coker, W.B.; Dunn, C.E.; Hall, G.E.M. *et al.*, The behavior of platinum group elements in the surficial environment at Ferguson Lake, N.W. T., Rottenstone Lake, Sask. and Sudbury, Ontario, Canada; J. Geochem. Explore. 1991, 40, 165–192.
[17] Wood, S.A.; Vlassopoulos, D., The dispersion of Pt, Pd and Au in surficial media about two PGE-Cu-Ni prospects in Quebec; Can. Mineral. 1990, 28, 649–663.
[18] Gunn, A.G., Drainage and overburden geochemistry in exploration for platinum group element mineralization in the Unst ophiolite, Shetland, U.K; J. Geochem. Explor. 1989, 31, 209–236.
[19] Reimann, C.; Niskavaara, H., Regional distribution of Pd, Pt, and Au-emissions from the nickel industry on the Kola Peninsula, NW-Russia, as seen in moss and humus samples; in: F. Zereini,

F. Alt (eds), Palladium Emissions in the Environment – Analytical Methods, Environmental Assessment and Health Effects; Springer Verlag, Berlin, Heidelberg, New York, 2006, pp. 53–70.

[20] Kümmerer, K.; Helmers, E.; Hubner, P. *et al.*, European hospitals as a source for Pt in the environment; Sci. Total Environ. 1999, 225, 155–165.

[21] Alt, F.; Eschnauer, H.-R.; Mergler, B.; Messerschmidt, J.; Tölg, G., A contribution to the ecology and enology of platinum; Fresenius J. Anal. Chem. 1997, 357, 1013–1019.

[22] Domesle, R., Katalysatortechnik Abschlusspräsentation; in: Edelmetallemissionen; GFS; BMBF, Hannover, 1996, pp. 8–16.

[23] Johnson Matthey, Platinum – Annual Report; Johnson Matthey, London, 1996, p. 120.

[24] Zereini, F.; Wiseman C.; Püttmann, W., Changes in palladium, platinum and rhodium concentrations and their spatial distribution in soils along a major highway in Germany from 1994 to 2004; Environ. Sci. Technol. 2007, 41, 451–456.

[25] Artelt, S.; Levsen, K.; König, H.P.; Rosner, G., Engine test bench experiments to determine platinum emissions from three-way catalytic converters; in: F. Zereini, F. Alt (eds), Anthropogenic Platinum Group Element Emissions; Springer Verlag, Berlin, New York, London, 2000, pp. 33–44.

[26] Artelt, S.; Kock, H.; Konig, H.P.; Levsen, K.; Rosner, G., Engine dynamometer experiments: platinum emissions from differently aged three-way catalytic converters; Atmos. Environ. 1999, 33, 3359–3567.

[27] Palacios, M.; Gomez, M.; Moldovan, M. *et al.*, Platinum group elements: quantification in collected exhaust fumes and studies of catalyst surfaces; Sci. Total Environ. 2000, 257, 1–15.

[28] Inacker, O.; Malessa, R., Experimentalstudie zum Austrag von Platin aus Automobilabgaskatalysatoren; in: Edelmetallemissionen, GFS; BMBF, Hannover, 1996, pp. 48–53.

[29] Ely, J.; Neal, C.; Kulpa, C.; Schneegurt, M.; Seidler, J.; Jain, J., Implications of platinum group element accumulation along U.S. roads from catalytic-converter attrition; Environ. Sci. Technol. 2001, 35, 3816–3822.

[30] Cubelic, M.; Pecoroni, R.; Schäfer, J.; Eckhardt, J.-D.; Berner, Z.; Stüben, D., Verteilung verkehrsbedingter Edelmetallimmissionen in Böden; UWSF-Z, Umweltchemie und Ökotoxikologie, 1997, pp. 249–258.

[31] Zereini, F.; Skerstupp, B.; Rankenburg, K. *et al.*, Anthropogenic emission of platinum-group elements (Pt, Pd and Rh) into the environment: concentration, distribution and geochemical behaviour in soils; J. Soil Sedim. 2001, 1, 44–49.

[32] Zereini, F.; Alt, F.; Messerschmidt, J. *et al.*, Concentration and distribution of heavy metals in urban airborne particulate matter in Frankfurt am Main, Germany; Environ. Sci. Technol. 2005, 39, 2983–2989.

[33] Gomez, M.B.; Gomez, M.M.; Palacios, M.A., Control of interferences in the determination of Pt, Pd and Rh in airborne particulate matter by inductively coupled plasma mass spectrometry; Anal. Chim. Acta 2000, 404, 285–294.

[34] Bencs, L.; Ravindra, K.; Van Grieken, R., Methods for the determination of platinum group elements originating from the abrasion of automotive catalytic converters; Spectrochim. Acta, Part B, 2003, 58, 1723–1755.

[35] Angelone, M.; Nardi, E.; Pinto V.; Cremisini, C., Analytical method to determine palladium in environmental matrices: a review; in: F. Zereini, F. Alt (eds), Palladium Emissions in the Environment – Analytical Methods, Environmental Assessment and Health Effects; Springer Verlag, Berlin, Heidelberg, New York, 2006, pp. 245–229.

[36] Zereini, F.; Zientek, C.; Urban, H., Konzentration und Verteilung von Platingruppenelementen (PGE) in Böden: Platinmetall-Emission durch Abrieb des Abgas-Katalysatormaterials; Umweltwissens. Schadstoff-Forsch. 1993, 3, 130–134.

[37] Zereini, F.; Skerstupp, B.; Urban, H., A Comparison between the use of sodium and lithium tetraborate in platinum-group elements determination by nickel sulphide fire-assay; Geostand. Newslett. 1994, 18, 105–109.

[38] Schäfer, J.; Eckhardt, J.-D.; Berner, Z.; Stüben, D., Time-dependent increase of traffic-emitted platinum-group elements (PGE) in different environmental compartments; Environ. Sci. Technol. 1999, 33, 3166–3170.

[39] Zereini, F.; Skerstupp, B.; Alt, F.; Helmers, E.; Urban, H., Geochemical behaviour of platinum-group elements (PGE) in particulate emissions by automobile exhaust catalysts: experimental results and environmental investigations; Sci. Total Environ. 1997, 206, 137–146.
[40] Fuchs, W. A.; Rose A.W., The geochemical behavior of platinum and palladium in the weathering cycle in the Stillwater Complex, Montana; Econ. Geol. 1974, 69, 332–346.
[41] Ballach, H.-J.; Wittig, R., Reciprocal effects of platinum and lead on the water household of poplar cuttings; ESPR-Environ. Sci. Pollut. Res. 1996, 3, 1–10.
[42] Zimmermann, S.; Sures, B., Uptake of palladium by the fauna; in: F. Zereini, F. Alt (eds), Palladium Emissions in the Environmental: Analytical Methods, Environmental Assessment and Health Effects; Springer Verlag, Berlin, New York, London, 2006, pp. 501–511.
[43] Ek, K.H.; Rauch, S.; Morrison, G.M.; Lindberg, P., Platinum group elements in raptor eggs, faeces, blood, liver and kidney; Sci. Total Environ. 2004, 334/335, 149–159.
[44] Lustig S.; Zang S.; Michalke B.; Schramel P., Platinum determination in nutrient plants by inductively coupled plasma mass spectrometry with special respect to the hafnium oxide interference; Fresenius J. Anal. Chem. 1996, 357, 1157–1163.
[45] Skerstupp, B.; Zereini, F.; Urban, H., Adsorption of platinum and palladium on hydrous ferric oxide (HFO) – an investigation by TXRF and XPS; Eur. J. Mineral. 1995, 7, 234.
[46] Dikikh, J.; Eckhardt, J.-D.; Berner, Z.; Stüben, D., Sorption behaviour of Pt, Pd and Rh on different soil components: results of an experimental study; in: S. Rauch (ed.), AGS Book Series, Independent volume; Springer, Berlin, 2006.
[47] Zimmermann, S.; Menzel, C.; Stüben, D.; Taraschewski, H.; Sures, B., Lipid solubility of the platinum group metals Pt, Pd, and Rh in dependence on the presence of complexing agents; Environ. Pollut. 2003, 124, 1–5.
[48] Lustig, S.; Zang, S.; Beck, W.; Schramel, P., Dissolution of metallic platinum as water soluble species by naturally occurring complexing agents; Microchim Acta 1998, 129, 189–194.
[49] Djingova, R.; Kovacheva, P.; Wagner, G.; Markert, B., Distribution of platinum group elements and other traffic related elements among different plants along some highways in Germany; Sci. Total environ. 2003, 308, 235–246.
[50] Ballach, H.-J., Accumulation of platinum in roots of poplar cuttings induces water stress; Fresenius Environ. Bull. 1995, 4, 719–724.
[51] Rosner, G.; König, H.P.; Kock, H.; Hertel, R.F.; Windt, H., Motorstandexperiment zur Untersuchung der Platin-Akkumulation durch Pflanzen; Angewante Botanik, 1991, 65, 127–132.
[52] Ballach, H.-J.; Alt, F.; Messerschmidt, J.; Wittig, R., Determinants of the phytotoxicity of platinum; in: F. Zereini, F. Alt (eds), Anthropogenic Platinum Group Emissions; Springer Verlag, Berlin, New York, London, 2000, pp. 105–114.
[53] Schützendübel, A.; Nikolova, P.; Rudolf, C.; Polle, A., Cadmium and H_2O_2-induced oxidative stress in *Populus* × *canescens* roots; Plant Physiol. Biochem. 2002, 40, 577–584.
[54] Lustig, S.; Schramel P., Platinum bioaccumulation in plants and overview of the situation for palladium and rhodium; in: F. Zereini, F., Alt (eds), Anthropogenic Platinum Group Element Emissions; Springer Verlag, Berlin, New York, London, 2000, pp. 95–104.
[55] Eckhardt, J.-D.; Schäfer J., Pflanzenverfügbarkeit, Boden-Pfalnze Transfer; in F. Zereini, F. Alt (eds); Emissionen von Platinmetallen; Springer Verlag, Berlin, Heidelberg, New York, 1999, 229–237.
[56] Hooda, P.S; Miller, A.; Edwards, A.C., The distribution of automobile catalysts-cast platinum, palladium and rhodium in soils adjacent to roads and their uptake by grass; Sci. Total Environ. 2007, 384, 384–392.

Index

Absence of gradients and Nernstian equilibrium stripping (AGNES) 97
Accelerated solvent extraction (ASE) 72–3
Acceptable daily intake (ADI) 300
Acid extraction 62, 65–7
Acids for matrix dissolution 54–5
Adsorptive stripping voltammetry (AdSV) 97
Advection 25–6
Aeolanthus biformifolius 316, 327
Akagenite 391
Alfalfa 126, 243, 334
Algaecides 384, 386
Allolobophora 159
Allophone 16
Aluminium (Al) 145, 207–8, 274, 354, 416
   As 387–8
   deficiency 183–4, 186
   Mn 485
   phytoremediation 320, 322
   Sb 394
   Sn 505
   soil amendment 354–5, 357–61, 363, 368–9
   soil chemistry 16, 18, 20–1, 23–4
   V 534
Aluminium hydroxide
   Cu 445–6
   MN 485
Aluminium oxide 6, 117, 142, 157, 357–8, 370
   As 388, 390–2
   Mn 485, 490
   Ni 468
   Se 486, 490
Aluminosilicates 6, 354, 357–61, 370
   clays 358–9, 370
   zeolites 359–61, 370
Alyssum sp. and phytoremediation 313–14, 316–20, 323, 326–8
Alyxia rubricaulis 316
Ammonium acetate 121
Ammonium phosphate 136
Andenosine triphosphate (ATP) 384, 541
Anglesite 442, 448
Anions 205, 212–13, 221–2, 242
   Cu 441, 444, 446
   exchange capacity 16
   fertilizers 140, 142–3
   immobilization 355–8
   Mo 516, 521–2
   Se 485–7, 490, 491
   soil chemistry 12–13, 15–16, 20–1, 23–4
   soil solution 81, 88–9, 91, 94, 100
   trace element deficiency 183–4, 189–90
   U 559
   Zn 428
Annelids 162
Anodic stripping voltammetry (ASV) 56, 94, 96–9
Antimonate 391, 393, 395
Antimonite 391, 393, 395
Antimony (Sb) 3, 7, 56, 87, 222, 383–400
   adsorption and desorption 391, 394–6
   Ag 386, 523
   alloy with Pb 442
   ash 361, 363
   geogenic occurrence 385–6
   risk 396, 400
   soil contamination 386
   speciation 393–4
Aporrectodea 159
Apples 507
Aqua regia 54–5, 62
Aqueous speciation modelling 199–201, 204–7, 219–21
Arabidopsis sp. 321–2, 333, 416
Arabis flagellosa 324
Arachnids 160, 164, 166

Arsenate 383, 385, 387–92, 393, 398
　fertilizers 142
　phytoremediation 332–3
　soil amendment 355, 357, 361, 365, 368
　soil chemistry 21, 24–25
Arsenic (As) 3–5, 56–7, 60, 142, 156, 331–3, 383–400
　Ag 523
　ash 361–3, 370
　bioavailability 236, 241, 243, 245, 267–8, 273, 279, 392, 396
　biosolids 115
　cancer 5, 267–8, 331, 384
　Co 474
　Cu 444
　desorption 392–3
　fertilizers 136, 139–41, 142, 147
　gypsum 369, 370
　immobilization 355–7, 358–60, 367, 370
　geogenic occurrence 385
　liming 368
　PGE 569
　phosphates 365, 370
　phytoextraction 331–3, 336
　phytoremediation 311–12, 316, 331–3
　regulatory limits 294, 299–303
　retention in soil 388–92
　risk 396–400
　Sb 7, 386
　soil chemistry 12, 20–1, 23
　soil contamination 386
　soil solution 87, 90–1, 99, 101
　speciation 387–8
　toxic element behaviour 205–7, 213
Arsenite 383–4, 385, 387–8, 390–2, 398
　fertilizers 142
　phytoremediation 332–3
　soil amendment 357
Ash 361–3, 503, 529, 534, 556
　phytoremediation 311–12, 320, 335–6
　soil amendment 354, 356, 360, 361–3, 367, 370
Astragalus sp. 316, 326
Athyrium yokoscense 324
Atomic absorption spectrometry (AAS) 56–7, 61, 89, 91
Atomic emission spectrometry (AES) 89
Atmospheric deposition 4, 9, 140, 296, 298
　Ag 523–4
　As 142, 386
　Cd and Zn 414
　Co 465
　Cr 464
　Cu 445
　F 145
　Hg 504–5
　Mn 482–3
　Mo 518
　Ni 464
　Pb 446–8, 451
　PGE 571
　Se 483–4
　Sn 500, 502–3, 505–6
　V 537–8, 542
Aulfides 68
Avicennite 531

Baghouse dust 139
Barium (Ba) 63, 361, 387
Barley (*Hordeum vulgare*) 119, 121, 186, 192, 238
　Cr 471
　Ni 473–4
Bauxite 534
Beans 541, 574
Beet 119, 473, 491
Beetles 156, 159, 162, 164
Bentonite 358–9
Berkheya coddii 319–20, 327
Bicarbonates 25
Bioaccessibility 230–3, 249–51, 254, 364
Bioaccumulation 230, 233, 251, 253–5, 396–7, 400
Bioassays 251–3, 254
Bioavailability 6–7, 11, 81–2, 91, 200, 229–55, 267–88
　As 236, 241, 243, 245, 267–8, 273, 279, 392, 396
　assessment 234–53
　Au 560–1
　B 186
　Cd 235–41, 243–5, 248–9, 252–3
　Cd and risk assessment 267, 272–3, 277, 279, 281, 288
　Cd and Zn 417, 419, 421–2, 425, 427, 574
　Co 469–70
　Cr 236, 238, 241, 243, 465, 467, 475
　Cu 187, 235–8, 241, 244, 248–9, 252–3, 279, 446, 448
　defined 230–3
　Fe 188, 235
　fractionation 64, 69, 71, 238, 239–40, 245
　Hg 243, 500
　immobilization 353–4, 364–5
　Mn 189, 235, 483–5, 489–90
　Ni 192, 237, 241, 244–5, 253, 468
　Pb 235–7, 241, 244, 249, 252–3, 448, 450
　Pb and risk assessment 273–4, 277, 279–80, 288
　PGE 568, 573–4
　phytoremediation 311, 333, 335
　regulatory limits 295, 299, 302–4
　Sb 396
　Se 485, 489–91

soil deficiency 175, 177, 179–80, 182, 186, 192
  Zn 235–8, 241, 244–5, 249, 252–3
Biodegradation 123, 366–7, 370
Biofortification 193–4, 427
BIOLOG 157, 161, 251
Biosensors 251–3, 254
Biosludge *see* Sewage sludge
Biosolids 113–29
  Cd 115–22, 125–7, 410–12, 414–15, 416–18, 419–22
  Mo 115–16, 518, 541
  phytoremediation 315, 321–2, 335
  soil amendment 354, 363–4, 367
  Tl 116, 529
  V 534
  Zn 115–16, 118–19, 122, 125–7, 410–12, 414–15, 416–18
Biotic ligand model (BLM) 162–3, 242–3
Biotite 482
Birnessite 221
  As 390–1
  Au 559
  Cu 445
  Mn 483, 485–6
Biscutella laevigata 316, 542
Bismuth (Bi) 56, 87, 523, 569
Boltzmann distribution law 15
Borate 24
Boron (B) 3–4, 24–5, 101, 482
  ash 361, 363
  deficiency 176, 183–6, 193
  fertilizers 138, 141
  phytoremediation 312, 327–8
Bottom ash 361
Brassica sp. 142, 325–6, 334, 542 *see also* Indian mustard
Broccoli 326
Buffered salt solutions extraction 62–3
Butyltin 73

Cabbage 236, 333, 473, 541–2
Caenorhabditis elegans 166–7
Cadmium (Cd) 3–7, 143–4, 409–29
  ash 361
  bioaccessibility 250
  bioavailability 235–41, 243–5, 248–9, 252–3
  bioavailability risk assessment 267, 272–3, 277, 279, 281, 288
  bioavailability with Zn 417, 419, 421–2, 425, 427, 574
  biosolids 115–22, 125–7, 410–12, 414–15, 416–18, 419–22
  crops with high content 421–2
  fertilizers 136–8, 139–41, 143–4, 147
  food chain risk 422–6
  fractionation 59–60, 69, 99
  geogenic occurrence 409–15

gypsum 368–70
Hg 497
immobilization 355, 357, 359, 360–1, 370
organic matter 366–7, 370, 417, 420
overdose 267, 272
PGE 574
phosphates 364, 370
regulatory limits 294–5, 299, 305
soil chemistry 9–11, 13, 16, 18, 20–3, 25–7, 29–30
Tl 542
toxic element behaviour 204, 211, 215, 217, 219–20
toxicity for soil dwellers 156, 159, 161, 163–6
Zn 191, 409–29
Cadmium sulfoselenide 482
Calcite 485
Calcium (Ca) 24, 60, 63, 102, 356, 360
  As 387–8
  ash 361, 363
  bioaccessibility 250
  bioavailability 255
  deficiency 5, 176, 183, 188–9
  liming 368
  Mn 485, 490
  Mo 519–20
  Pb 451
  phosphates 136, 365
  rice 425
  toxicity 164, 205–7, 214
  Zn and Cd 422, 425–6, 428
Calcium carbonate 20, 184, 186, 191
Calcium vanadate 517
Cancer and carcinogens 282, 284, 299, 303
  As 5, 267–8, 331, 384
  Cr 462, 470
  Cu 451
Canola (*Brassica rapa*) 142, 325–6, 334
Capillary electrophoresis (CE) 90–1
Capillary zone electrophoresis 241
Carambola (*Averrhoa carambola*) 324
Carbon (C) 117, 160–1, 205, 497 *see also* Dissolved organic carbon (DOC)
  deficiency 176, 183, 187, 191
Carbon dioxide 203–4, 207, 385
Carbonates 55, 63–4, 67–9, 117, 368
  Ag 527
  bioavailability 229, 238–9
  Cu 441, 446, 449
  deficiency in trace elements 182–3, 192
  Mn 484, 488
  Ni 468
  Pb 442, 448
  soil chemistry 12, 13, 18, 21, 25
  Tl 531
  toxic element behaviour 199, 203–5
  U 559, 562

Carrots (*Daucus carota*) 29, 119, 237, 507
Cassiterite 500, 505
Catalase 161
Catalytic converters 567–75
Cation exchange capacity (CEC) 16, 23–4, 31, 143, 235, 237, 298
  Ni 468, 473
Cations 62–3, 360, 362
  Ag 524
  bioavailability 241–2
  Co 474–5
  Cr 465
  Cu 441, 446, 450
  deficiency in trace elements 182–3, 186, 188, 191
  immobilization 355–6, 357–61, 365–6, 370
  Mn 481, 485, 490
  Ni 468, 475
  Pb 448
  phytoremediation 312
  soil chemistry 11–13, 15–16, 19–21, 23, 25, 30
  soil solution 87, 89, 91, 93–5, 100–2
  toxic element behaviour 201, 205, 208, 210, 212, 214–15, 218, 221–2
  trace metal toxicity 162–3
  V 536, 542
Celery 473
Cellulase 161
Centipedes 157, 162
Cerium (Ce) 274
Cerussite 442, 448
Cesium (Cs) 93, 333–4, 359, 532
Chalcophiles 568–9
Chalcopyrite 444
Charge distribution (CD) mode 213
Charge distribution multisite ion complexation model (CD-MUSIC) 209, 212–13
CHEAQS 200, 220
Chelating agents extraction 62, 64
CHESS 200
Chlorides 13, 30, 63, 199, 204, 243, 323
  Ag 524, 526
  Cu 449
  Hg 506
  PGE 573
  Se 487
  Tl 529–31
  Zn and Cd 417, 419, 426
Chlorine (Cl) 3–4, 13, 30, 60, 143, 211
  Cu 443
  deficiency 176, 185
Chlorite 392, 468, 482
Chocolate 421
Chromates 24, 361, 368, 466
Chromatographic separation 81, 85–7, 88–90

Chromatography 241–2
Chromite 461, 463
Chromium (Cr) 3–4, 7, 355–7, 370, 461–2, 474
  ash 361
  bioavailability 236, 238, 241, 243, 465, 467, 475
  biosolids 115–17, 124–7
  chemical behaviour in soil 465–7
  Cu 444
  deficiency 176, 188, 192
  fertilizers 136, 139, 140–1, 147, 464
  fractionation 55, 60, 64
  geogenic occurrence 463
  liming 368, 370
  phytoremediation 312, 315
  risks 470–2
  soil chemistry 12, 20, 23
  sources of contamination 464
  toxicity 156, 205
Chromolaena odoratum 324–5
Cinnabar (HgS) 498, 501, 506
Citrate 467
Classical model of risk assessment 268–70
Clay minerals 199, 207–8, 213–17, 220–2
Clinoptilolite 360
Clover 338
Cobalanun 186
Cobalt (Co) 3–4, 7, 60, 186–7, 210, 462–3, 474
  chemical behaviour in soil 468–70
  Cu 444
  deficiency 5, 176–7, 181–4, 186–8
  geogenic occurrence 463–4
  immobilization 360, 365, 370
  PGE 567
  phytoextraction 336
  phytoremediation 312, 316, 319, 327, 330
  risks 474
  soil chemistry 20, 23–4
  soil solution 86
  sources of contamination 465
  toxicity for soil dwellers 156, 163
Cold vapour-atomic absorption spectrometry (CV-AAS) 56–7
Common ragweed (*Ambrosia*) 330
Community-level physiological profiling (CLPP) 161
Compartment models 163–6
Complexes 11–12, 13, 24–5
  soil solution 88, 91, 93–4, 97, 101
Complexing gel integrated microelectrode (CGIME) 99–100
Composite sampling 40–1, 44, 48

Compost 176, 192, 296
　Cd 419–22
　Cd and Zn 414
　Co 465
　Cu 445
　Hg 505
　immobilization 356, 363, 366–7
Computer models for speciation 100–3
Conceptual models 271–2
Coning 41, 45
Constant capacitance model (CCM) 201, 208–9
Continuous leach inductively coupled plasma mass spectrometry (CL-ICP-MS) 64
Copper (Cu) 3–6, 187, 441–53
　Ag 444, 525
　ash 361, 370
　Au 444, 554–5
　bioavailability 187, 235–8, 241, 244, 248–9, 252–3, 279, 446, 448
　biosolids 115–16, 118, 122, 125–7
　Cd 410, 411, 414, 416
　chemical behaviour in soil 445–6
　deficiency 5, 176–8, 180–5, 187–9, 191–3, 429, 449–51, 453
　fertilizers 141, 143
　fractionation 59–60, 69
　gypsum 368, 370
　immobilization 355–60, 370
　Mn 482
　Mo 187, 444, 449–50, 541
　Ni 444, 473
　organic matter 366–7, 370, 445–6
　PGE 568, 569, 574
　phosphates 364–5, 370
　phytoremediation 312, 316, 319, 321, 327, 330, 333
　phytostabilization 334–5
　regulatory limits 302
　risks 449–52
　Sb 386
　soil chemistry 11–13, 18, 20, 23–5, 27, 29
　soil sampling 42, 48–9
　soil solution 94, 98–9
　sources and content 443–5
　Tl 542
　toxic element behaviour 210, 219
　toxicity 450–1
　toxicity for soil dwellers 156, 159, 161, 163–7
　Zn 410, 416, 428–9, 441, 444, 449
Copper sulfate 146
Corn and maize (*Zea mays*) 30, 186, 191, 316–17, 330
　Ag 541
　As 142
　biosolids 118, 122, 124, 126–9
　Cd 421
　immobilization 363, 365

Pt 574
Tl 542
Zn 427
Cress (*Lepidium sativa*) 248–9, 542, 574
Crookesite 258
Crop residues 176, 192
Crustacea 160
Cucumbers 221
Cyanide 558, 560–1

Dehydrogenase 161
Depleted uranium (DU) 552, 556–7, 560
Derivatization 86–8
Desorption 13–17
Desulfovibrio desulfuricans 355
Deterministic models 300–1
Detoxification 162, 165–7
Diamondback moths (*Plutella xylostella*) 326
Diethylenetriamine pentaacetic acid (DTPA) 64, 68
Differential pulse ASV (DPASV) 96–7
Diffuse layer model (DLM) 201, 208–9, 212–13
Diffusive gradients in thin films (DGT) 244–5, 246–9, 254–5
　soil solution 91–4
Dilute acid solutions 63–4
Dimethylarsinic acid (DMAA) 385
Direct current plasma (DCP) 57
Dissolution 18–19
Dissolved organic carbon (DOC) 183, 187, 191, 242
　toxic element behaviour 205, 210, 220
Donnan membrane technique (DMT) 94–6, 241
Dose–response assessment 268–9, 280–4, 285, 288
　regulatory limits 297, 300
Drought 180
Diethylenetriamine pentaacetic acid (DTPA) 117, 121, 234–8
Duckweed (*Lemna minor*) 542
Dunite 568
Durum wheat 417, 421
Dynamic energy budget (DEB) 158

Earthworms 233, 251–2, 364, 472, 542
　trace metal toxicity 156, 158–62, 164–5, 166–7
ECOSAT 200, 220, 242
Ecotoxicity 141, 367–8 *see also* Phytotoxicity
　bioavailability 251–2, 254–5
　regulatory limits 296, 297–8
　soil dwellers 156, 158, 165–7
Electrospray ionization mass spectrometry (ESI-MS) 90
Elsholtzia splendens 244, 249

Enchytraeids 156, 166
Energy dispersive X-Ray fluorescence
    (EDXRF) 61, 245
EQ3/6 200
Escherichia coli 253
Ethylenediamine tetraacetic acid (EDTA)
    62, 64, 68, 86
    bioavailability 234–8, 243–4, 249
    phytoremediation 320, 330–1, 334
Eucalyptus 560
Europium (Eu) 60, 359
Exchangeable fraction 65–7, 69
Excretion 272, 275, 277, 302
    Cu 441, 450–1
    Hg 508
    Pb 450
    PGE 569
Exposure assessment 268–9, 272–80,
    284–5, 287
Exposure pathways 270, 271–2, 274,
    288, 300
Extended X-ray absorption fine structure
    (EXFAFS) spectroscopy 73–4, 245,
    363, 365, 416

Ferrihydrite 354–6, 357–8
    Ag 526
    As 390–1
    Mo 521
    PGE 573
    Tl 533
Ferrites 468
Ferritin 166
Fertilizers 4–5, 7, 9, 135–48, 245, 354
    see also Nitrogen fertilizers; Phosphate
    fertilizers
    ash 361
    biosolids 114, 118
    Cd and Zn 411–13, 414–17, 419, 427
    Co 465
    Cr 136, 139–41, 147, 464
    Hg 136, 139, 141, 505
    long-term accumulation 139–41
    micronutrients 135, 139, 147
    Mn 482–3
    Mo 518
    Ni 140–1, 464
    organic matter 366
    phytoremediation 311–12, 317–21, 323–5,
        328, 330, 332, 334, 336
    Se 136, 482–4, 487
    trace element deficiency 181, 187, 193
    U 136, 140, 147, 556–7
    V 136, 534
FIAM 242–4, 253
Field capacity (FC) 247
Field-flow fractionation (FFF) 71, 83–5
Field portable X-ray fluorescence (FPXRF) 61

Fish 9, 12
FITEQL 201–3
Flame atomic absorption spectrometry
    (F-AAS) 55–6, 58
Flax 419, 421, 426
Flooding 21–2, 424–5
    Mn 488, 490
    Se 490–1
Flow field-flow fractionation (FlFFF) 85
Fluorine (F) 3–4, 145, 176, 443
    fertilizers 136, 140–1, 145, 147–8
Fluorite 145
Flux Donnan membrane (FDM) technique 96
Fly ash 361–3, 529, 534, 556
Food chain 159, 229, 321–5
    Cd 409, 422–6
    Cr 462, 471, 475
    fertilizer contamination 141–2, 144,
        146–7
    Hg 500, 509
    Se 483, 491
    Zn 409, 427–9
Four layer model 208
Fractionation 53–4, 61–74, 81–103
    analysis 54–5, 59
    bioavailability 64, 69, 71, 238, 239–40, 245
    Se 489, 491
    soil sampling 45, 54
Free ionic species 229–33, 241–4
Freundlich model 16–17
Fungi 191–3, 328, 507, 541–2
Fungicides 411, 413, 482, 498, 507
    As 384, 386

Galena (PbS) 442, 447–8, 528
Garden crops 273–4, 422, 426, 429
    bioavailability risk 267, 272, 273–4
Garnierite 468
Gas chromatography (GC) 87–9
Gaussian model 201, 203, 220
Gel filtration chromatography 85
Gel integrated microelectrode (GIME) 99–100
Gel permeation chromatography (GPC) 85
Geobacter metalliredussus 539
GEOCHEM 242
Geochemists 200
Germanium (Ge) 87, 497
Gibbsite 207, 221, 391
Glucose oxidase 161
Goethite 354–5, 360
    Ag 526
    As 390–3
    Au 559
    Cu 445
    Mo 521
    Sb 391, 395–6
    toxic element behaviour 209–10, 211–13,
        220–1

U 559
V 538–9
Gold (Au) 3, 7, 99, 551–2, 562
  chemical behaviour in soil 557–9
  Cu 444, 554–5
  geogenic occurrence 553–4
  Hg 502, 504–5
  PGE 568
  phytoextraction 336
  phytoremediation 329
  risks 560–1
  Sb 384–5
  soil contamination 555–6
Graphite furnace atomic absorption spectrometry
    (GF-AAS) 55–6, 89
Grasses 29, 273, 314, 318,
    335, 359
  As 400
  Co 186–7, 474
  Sb 400
  Zn 427
Grinding 41, 45–6, 69
Gynura pseudochina 324
Gypsum 319, 368–70

Hafnium (Hf) 60
Halloysite 392, 449
Hanging mercury drop electrode (HMDE) 96
Harvestmen 162, 164
Haumaniastrum robertii 316, 327
Hazard identification 268–9, 270–2
Hematite 354–5, 389–91, 445, 521
Hemp dogbane (*Apocynum*) 330
Herbicides 313–15, 318, 384, 386
High-performance liquid chromatography
    (HPLC) 88–91
Holmium (Ho) 60
Hornblende 482
Hot spots 40, 42, 44, 49
Humic acid 68, 242
  Ag 526–7
  Au 558–9
  Cu 446
  Mn 484, 490
  Mo 522
  Se 484, 490
  Sb 395
  soil amendment 356, 359, 366
  soil chemistry 24–5
  soil solution 84, 93, 97–8, 101
  toxic element behaviour 199–200, 204–5,
      217–21
Hydroxyapatite (HA) 364–5
Hydride generation (HG) 87
Hydride generation atomic absorption
    spectrometry (HG-AAS) 56
Hydrides 56, 87, 89
Hydrocarbons 161

Hydrochloric acid 55, 63
Hydrofluoric acid 54–5
Hydrogen (H) 145, 176, 203–4,
    218
Hydrous ferric oxide (HFO) 209, 210–13,
    220–1
  Ag 527
  Mo 521
  Tl 527, 532
Hydroxides 64, 68, 118, 207
  As 388, 390
  Cr 465
  Cu 445–6
  Mn 482–3
  phytoremediation 320
  soil amendment 354, 356
  soil chemistry 18–21, 24
  Tl 529, 531
  trace element deficiency 182–4,
      186–9
  V 537–8
Hydroxyls 203–5, 208
Hyperaccumulators 6, 316–17
  Ag 542
  As 331–3
  Cd 321–4
  Co 327
  Cs 334
  Hg 328
  Ni 317–20, 327, 472
  Pb 329–30
  phytoremediation 312–15,
      316–17, 338
  Se 325–7, 337
  Tl 333
  wildlife 337
  Zn 416

Iberis intermedia 333, 542
Illite 23, 214–15, 217, 222
  As 389–90, 392
  Mo 521
  Pb 448
Imogolite 445, 449
Impatien violaeflora 324
Indian mustard 6, 243, 542, 561
  phytoremediation 315, 320, 327, 329,
      330–1, 334
Inductively coupled plasma (ICP)
    techniques 57–9
Inductively coupled plasma–atomic emission
    spectrometry (ICP-AES) 56–8
Inductively coupled plasma–mass spectrometry
    (ICP-MS) 83, 85, 87–91, 95, 241
  analysis and fractionation 56, 57–9, 60–1,
      64, 71
Inductively coupled plasma–optical emission
    spectrometry (ICP-OES) 57, 61, 91

Ingestion 274, 334–5, 450
  bioavailability 267–70, 272, 274, 275,
    277–80, 288
  regulatory limits 294, 299, 301–2
Inhalation 275–6, 507
  bioavailability 267–70, 272, 275–6,
    278–80
Insecticides 384, 386
Insects 160
Instrumentation neutron activation analysis
    (INAA) 59–60
Iodine (I) 3, 4, 267
  deficiency 5, 176–7, 181, 183, 267
Ion activity product 101
Ion-exchange chromatography 88–9
Ion-exchange resins 86–7
Ion-selective electrodes (ISEs) 93–4, 241
Ionic strength 182, 201–2, 219
Iridium (Ir) 567–8
Iron (Fe) 3–4, 5, 60, 188
  As 387–8
  bioaccessibility 250
  bioavailability 188, 235
  biosolids 122
  Cd and Zn 416, 421–2, 424–6
  Cu 441, 444, 449
  deficiency 5, 176–8, 180, 182–9, 191, 193,
    425, 427–9
  Mn 481, 485
  PGE 567, 568
  phytoremediation 319–20, 321, 333
  rice 425
  Sb 394
  Se 493
  shellfish 421
  Sn 505
  soil amendment 354–6, 360, 363, 368
  soil chemistry 13, 16, 18, 20–1,
    22–4, 29
  soil solution 86
  toxic element behaviour 207–8
  V 534
iron hydroxide 64, 118, 186, 188,
    354, 356
  Co 469–70
  Cu 445–6
  Mn 483, 485, 487
  Sb 391, 395–6
Iron oxide 6, 67, 142, 157
  As 388, 391, 392–3
  Au 558–9
  bioavailability 229, 230, 238–9
  biosolids 117–18
  Cd 420, 429
  Cr 465, 470, 475
  Cu 445
  deficiency 186, 188–9, 191
  fractionation 63, 65–9

  Hg 506
  Mn 483, 485, 487, 490
  Mo 516
  Ni 468, 475
  Pb 448, 449
  phytoremediation 312, 320, 335
  Se 482, 486, 490
  soil amendment 354–8,
    360, 370
  soil chemistry 11–13, 16, 20–1,
    23–5
  soil solution 83
  Tl 532
  U 559–60
  V 534
  Zn 416–17, 429
Iron oxyhydroxide 71
Irrigation 142, 424, 482
Isotopic dilution (ID) techniques 245–6, 254
Itai-itai disease 9, 423

Justicia procumbens 324

Kaolinite 16, 23, 207, 214–17, 222, 358–9
  As 389, 390, 392
  Mo 521
  Pb 448–9

Langmuir model 16–17
Lanthanum (La) 60, 274
Layered double hydroxide (LDH) 416, 468
Leaching 5, 71–3, 156
  analysis and fractionation 55, 62, 64,
    67–9, 71–3
  bioavailability 230, 234, 238, 247
  biosolids 114
  Cd and Zn 415, 428
  Co 470
  Cr 467
  Cu 445
  fertilizers 140–1, 147
  immobilization 353, 359, 363, 367
  Mg 505
  Mn 490
  phytoremediation 311–12, 319, 327, 329,
    331, 334
  Se 487, 490–1, 493
  Sn 503
  soil chemistry 10–11, 26–7, 31
  soil sampling 39
  soil solution 82, 87
  trace element deficiency 182–4, 186,
    188–91
Lead (Pb) 4–7, 145–6, 329–31, 441–53
  Ag 523, 525
  analysis and fractionation 60, 64, 69
  ash 361, 370
  bioaccessibility 250

bioavailability 235–7, 241, 244, 249, 252–3, 448, 450
bioavailability and risk assessment 273–4, 277, 279–80, 288
biosolids 115–17, 124–7
  Cd 410–11
  chemical behaviour in soil 448–9
  Cu 444
  fertilizers 136, 139–41, 145–7
  gypsum 368, 370
  Hg 502
  immobilization 355–61, 370
  Mo 517, 519
  organic matter 367, 370, 448
  PGE 568, 574
  phosphates 364–5, 370
  phytoextraction 329–31, 334, 336
  phytoremediation 311–12, 315, 321, 329–31, 333, 338
  phytostabilization 334–5, 338
  regulatory limits 301
  risks 449–52
  Sb 386
  Sn 497, 503
  soil chemistry 11–13, 20, 23–5, 27, 29
  soil sampling 42
  soil solution 87–8, 94, 99
  sources and content 446–8
  Tl 528–9, 542
  toxic element behaviour 210–11, 215, 217, 219
  toxicity 451–2, 453
  toxicity for soil dwellers 156, 159, 161, 164–6
  Zn 410–11, 428
Lead phosphate 315
Lepidocrocite 355, 391
Lettuce (*Lactuca sativa*) 30, 119, 121, 236, 360
  Ag 541
  Cd 411, 417–20, 422–3, 425–6
  Cr 472
  Hg 507
  Ni 221, 473
  Zn 338, 417–20, 422, 429
Ligands 65, 68, 72–3, 162–3, 164
  As 392
  Au 558
  bioavailability 243
  biosolids 120
  Cd 142
  Co 469
  Cr 465, 467
  immobilization 355, 358–9
  Mn 484
  Mo 521
  Ni 467
  Pb 448
  phytoremediation 323

  Sb 396
  Se 491
  soil chemistry 13, 23–4, 25, 30
  soil solution 81–2, 87, 97, 101–3
  Tl 530
  toxic element behaviour 199, 204, 211, 217
  trace element deficiency 182, 188
  U 147
  V 536–7
Lignin 181
Lignite 486
Lime and liming 9, 29, 193, 367–8
  Cd and Zn 418
  Cr 475
  fertilizers 143
  Hg 505
  Mn 482–3
  Mo 541
  Ni 468, 473, 475
  Se 483
  soil amendment 354, 356, 364, 367–8, 370
Limonite 358
Liquid chromatography (LC) 88–9
Liquid chromatography–mass spectrometry (LC-MS) 241
Liquid-liquid extraction 86–8
Lithium (Li) 63
Litter decomposition 160–1
Livestock 312, 335, 337
  As 333, 384, 386
  bioavailability 287–8
  Cd and Zn 413–14, 419, 421, 425, 429
  Co 186–7, 463, 470, 474
  Cr 462, 470
  Cu 187, 449
  Fe 188
  Hg 508–9
  Mo 515, 518, 541
  Ni 462
  Pb 442, 450, 451–2
  regulatory limits 296
  Se 190, 325–7, 482, 483–4
  trace element deficiency 176–7, 178–82, 192
  Zn 191
Lorandite 528
Lowest observed adverse effect level (LOAEL) 281, 299
Low-molecular weight organic acids (LMWOA) 64, 122–3
Lumbricus 159

Macro-arthropods 159
Maize *see* Corn and maize (*Zea mays*)

Magnesium (Mg)   63, 102, 205, 208, 356, 360–1
  Cu   444
  deficiency   176, 188–9
  Mn   490
  Mo   519–20
  phytoremediation   319, 322
  soil chemistry   23–4
  Zn toxicity   428
Magnetite   534
Manganese (Mn)   3–4, 7, 60, 188–9, 481–93
  As   388, 391
  bioavailability   189, 235, 483–5, 489–90
  biosolids   122
  Cd and Zn   416, 424
  chemical behaviour in soil   484–90
  concentrations and sources   482–3
  deficiency   5, 176–8, 180, 182–6, 188–9, 191, 193, 490–1
  immobilization   357–60, 364–5, 370
  phytoremediation   312, 316, 319, 322
  risks   490–3
  soil chemistry   11–13, 20–1, 23–4
  toxic element behaviour   207–8, 210
Manganese hydroxide   64, 118, 482–3, 485
Manganese oxide   6, 67, 83, 357–8, 482–5, 487–8, 490
  Ag   526
  As   391
  Au   558–9
  bioavailability   229–30, 238–9
  biosolids   117–18
  Cd   420
  Co   469–70, 475
  Cr   466, 470, 475
  Cu   445
  fractionation   63, 65–9
  Ni   468, 475
  Pb   448–9
  phytoremediation   312, 320, 335
  Se   486
  soil amendment   355, 357–8, 370
  soil chemistry   13, 16, 20, 23
  Tl   532
  Zn   416–17
Manganese oxyhydroxide   71
Manganese sulfate   146
Manganite   483, 485
Manure   9, 114, 140, 176, 192, 366
  Cd   413–15, 419–22
  Cr   464, 472
  Cu   414, 445
  Hg   505
  Mn   482–3
  Mo   518
  N   414
  Ni   464, 468
  Se   483
  Zn   413–15
Marine bacterium (*Vibrio fischeri*)   252
Matrix dissolution   54–5
Maximum concentration level (MCL)   384, 396
Maximum permissible concentrations (MPCs)   253
Maximum water holding capacity (MWHC)   247–8
Medicago sativa   542
Mehlich-I   121
Mercurial lyase (MerB)   328–9
Mercuric reductase (MerA)   328–9
Mercury (Hg)   3–5, 7, 497–500
  Ag   525
  bioavailability   243, 500
  biosolids   115–16
  chemical behaviour in soil   506
  electrodes   96–7, 99
  fertilizers   136, 139, 141, 505
  fractionation   53–4, 60, 88, 93
  geogenic occurrence   501–2
  global demand and supply   499
  isotopes   498
  phytoremediation   311, 328–9
  risks   498, 500, 507–9
  soil chemistry   13, 20, 23
  soil contamination   503–5
  Tl   542
  toxic element behaviour   203–6, 211
  toxicity for soil dwellers   156, 166
Metabolomics   167
Metallothionein   166, 422
Methanearsonic acid (MAA)   385
Methylmercury   5, 9, 12, 86, 87
Methyltin   12
Microarthropods   156, 162–3
Microelectrodes   93, 95, 98–100
Micronutrients   4–5, 139, 177–8, 235, 367, 441
  deficiency   176–84, 188, 190–1, 193
  fertilizers   135, 139, 147
MICROQL   200–2, 221
Microsensors   98–100
Microtox   252
Microwave-assisted continuous-flow sequential extraction   70
Microwave-assisted leaching   72
Microwave-induced plasma (MIP)   57
Millipedes   157, 161
MINEQL   200
MINTEQA2   200–1, 220, 242
Mites   156, 159, 164, 166
Mobility and mobilization   25–7, 199, 353–71
  Ag   526
  As   302, 396
  Au   553, 558
  bioavailability   230, 233–5

biosolids 117
Bo 184
Co 469, 474
Cr 465, 467, 471, 475
Cu 187
Fe 188
fractionation 62–3, 65, 69, 74
Hg 506–7
Mn 490
Mo 189, 540
Ni 468, 472
PGE 568, 573
regulatory limits 293, 302–3
Sb 396
Se 490
Sn 506
soil chemistry 10–11, 19–27, 28–9, 32
Tl 531
trace element deficiency 180, 182, 183
U 559
Molluscs 159, 160, 166
Molybdates 24, 517, 519, 520–2
Molybdenite 516–17
Molybdic acid 518–19
Molybdenosis 187
Molybdenum (Mo) 3–4, 7, 101, 515–16
ash 361, 363
biosolids 115, 116, 518, 541
chemical behaviour in soil 518–22
Cu 187, 444, 449–50, 541
deficiency 176–7, 181, 183–5, 187, 189–90, 191, 515, 540–1
geochemical occurrences 517–18
Mn 482
phytoremediation 334, 338
risks 540–1
soil chemistry 12, 20–1
soil contamination 518
toxic element behaviour 205, 213
Monobutyltin 503
Montmorillonite 23, 214–15, 217, 222, 359
As 389, 390, 392
Cu 445
Mo 521
Pb 448–9
PGE 573
Se 486
Mushrooms *see* Fungi
Myriapods 159–60

Nematodes 156, 158, 162, 166–7
Neodymium (Nd) 274
Neutron activation analysis (NAA) 56, 59–60
NICA–Donnan model 205, 217–19, 220, 242

NICA model 217–18
Nickel (Ni) 3–4, 6–7, 192, 221, 317–21, 462, 474
ash 361, 370
Au 554, 555
bioavailability 192, 237, 241, 244–5, 253, 468
biosolids 115–16, 121–2, 125–7
Cd 410–11
chemical behaviour in soil 467–8
Co 474
Cu 444, 473
deficiency 176, 183, 185, 188, 192
fertilizers 140–1, 464
fractionation 60, 86
geogenic occurrences 463
immobilization 355–6, 357, 359, 370
organic matter 367, 370, 474
PGE 567–8, 569
phytoextraction 336–7
phytoremediation 311–12, 316–17, 327, 330, 338
phytomining 317–21, 329, 336, 338
phytostabilization 334–5
regulatory limits 302
risks 472–4
soil chemistry 11–12, 16, 20, 23, 24
sources of contamination 464
toxic element behaviour 204, 210, 215–16, 219, 221
toxicity for soil dwellers 156, 161, 163
wildlife 337
Zn 413
Nitrate 13, 21, 55, 63, 189, 360
Cu 449
Hg 505
Pb 450
Nitric acid 55, 63, 67
Nitric-hydrochloric acid (aqua regia) 54–5, 62
Nitrification 161
Nitrogen (N) 160–1, 367, 414
Ag 541
biosolids 113–14, 115–16
Cd 417, 419
deficiency 176, 187, 189, 192, 193
fixation 4, 6, 186–7, 189, 251
Mo 515, 540
phytoremediation 319, 323, 335
Zn 417, 419, 429
Nitrogen fertilizers 140, 143–4, 146–7, 419, 505
phytoremediation 319, 323
No observed adverse effects level (NOAEL) 280–2, 299
No observed effect concentrations (NOECs) 295, 297–8
Nyssa sylvatica 327

Oats (*Avena sativa*) 29, 126, 412, 472, 473, 491
One-compartment model 163–4
Onions 574
ORCHESTRA 200–1, 219–20, 242
Organic acids in biosolids 121–5, 127–9
Organic matter 68, 217–19, 365–7
   Ag 527
   analysis and fractionation 55, 63, 65–9
   Au 557
   bioavailability 229–30, 237, 239, 242, 273
   biosolids 113–14, 116–18, 120
   Cd 366–7, 370, 417, 420
   Co 470, 475
   Cr 470–2, 475
   Cu 366–7, 370, 445–6
   deficiency in trace elements 183–4, 186–9, 191–2
   fertilizers 140, 142–3
   Hg 506
   immobilization 354–5, 358–9, 363, 365–7, 368, 370
   Mn 483, 488, 490
   Mo 522
   Ni 367, 370, 475
   Pb 367, 370, 448
   PGE 573
   phytoremediation 315, 335
   Se 486, 490
   soil chemistry 24–5, 27, 29, 30
   soil sampling 43, 45, 46–8
   soil solution 81, 85, 94, 100–2
   Tl 531
   toxic element behaviour 199, 217–19, 220, 222
   trace metal toxicity 161, 163
   U 559–60
   Zn 366–7, 370, 417, 429
Osmium (Os) 567, 568
Oxalate 69
Oxides 67, 207–13, 229, 354–8 *see also* Aluminium oxide; Iron oxide; Manganese oxide
   Ag 527
   As 385, 389, 390
   analysis and fractionation 55, 63, 65–7
   biosolids 117, 118
   Cd and Zn 416–17, 420, 429
   Cr 462
   Cu 445, 449
   deficiency in trace elements 183–4, 186, 188–9, 191
   Pb 442, 448–50
   phytoremediation 312, 320, 335
   soil amendment 354–8, 360, 370
   soil chemistry 18, 21, 23–4
   Tl 529, 531–2
   trace element behaviour 199–200, 207–13, 214, 220, 222

V 537–8 *see also* Aluminium oxide; Iron oxide; Manganese oxide
Oxyanions 11–12, 21, 23, 393, 471, 518
   soil amendment 356, 360–2, 365, 368, 370
Oxygen (O) 176
Oxyhydroxides 18, 23, 71, 207
   As 389
   Co 469
   Sb 391, 394, 395–6
   V 534, 536, 538

Palladium (Pd) 3, 7, 567–74
Paper mill sludge 354, 362, 367
Particle-induced X-ray emission (PIXE) 60, 245
Particle sizes 71, 83–6, 116, 120
   fractionation 54, 57–8, 61, 65, 71
Peanuts 421
Peas 491
Peat 362, 366, 370
Pecan 320, 413
Perchlorate 140
Perchloric acid 55
Peridotite 568
Peroxidise 161
Pesticides 9, 156, 159, 167
   As 385
   Hg 498
   Pb 145, 442
   PGE 567
pH 19–23, 142, 163–4, 298
   Ag 524, 526–7
   As 385, 387–93, 396
   ash 361–3
   Au 557
   bioavailability 230, 232, 235, 237, 239, 242–3, 245, 250
   bioavailability and risk assessment 273, 277
   biosolids 116, 119–20, 122
   Cd 143–4, 410–18, 420, 422, 424, 427–8
   Co 468–70, 474
   Cr 462, 465–6, 470–2, 474
   Cu 445–6
   deficiency in trace elements 182–4, 186–91, 193
   F 145
   fractionation 62, 63, 66–7, 72
   Hg 506
   immobilization 355–6, 357–64, 370
   liming 367–8
   Mn 484, 485–7, 490
   Mo 519, 521, 541
   Ni 468, 473–4
   organic matter 366–7
   PGE 573
   phosphates 364–5

phytoremediation 311–12, 318–25, 327, 332,
    335, 338
  Sb 393–6
  Se 486–7, 489, 490
  Sn 503, 505
  soil chemistry 10, 17–25, 29, 31, 32
  soil solution 82–3, 89, 95, 100–1
  Tl 527–8, 530–3
  toxic element behaviour 199–203, 205, 207,
    210–12, 214, 219, 221
  U 559–60, 562
  V 535–40
  Zn 410–18, 420, 422, 424, 427–8
Phosphatase 161
Phosphate fertilizers 136–8, 146, 414, 419,
    464–5
  Hg 505
  Mn 483
  Mo 518
  phytoremediation 319, 321, 323, 330, 332
  Se 483, 487
  soil contaminants 135, 136–8, 139–40, 142–7
  U 556–7
  V 534
Phosphates 55, 164, 183, 191, 364–5 see also
    Phosphate fertilizers
  As 384, 392
  bioavailablity 229, 236
  biosolids 117
  Cd and Zn 410–11, 414, 416, 419
  Cu 446
  inorganic 136
  Mn 484, 488
  Ni 468
  Pb 448
  phytoremediation 315, 323, 330, 332,
    334–5
  Se 486–7
  soil amendment 354–6, 364–5, 370
  soil chemistry 12–13, 18
  U 554, 559
  V 542
Phosphogypsum 411
Phosphorite 410
Phosphorus (P) 56, 245, 364–5
  Ag 541
  ash 363
  biosolids 113–15
  Cd and Zn 414, 417, 419
  deficiency 176, 187, 189–90, 191–3
  Pb 451
  soil amendment 364–5, 367, 369
  V 534
PHREEQC 200–1, 220
Phyllanthus serpentinus 316
Phyllosilicates 389–90, 394
Physicochemical properties 53–4, 65,
    114, 298

As 388, 400
Hg 502
immobilization 366, 370
Ni 467
Sb 400
soil chemistry 10, 13, 28–9
soil solution 81, 99
Physiologically based pharmacokinetic (PBPK)
    models 284
Physiologically based extraction test
    (PBET) 250–1
Phytate 427
Phytoaccumulation 383, 400 see also
    Hyperaccumulators
Phytoavailability 29, 64, 122–5
  biosolids 116–18, 120, 122–5, 126, 129
  Cd and Zn 415–17, 419–20, 425,
    427–8
  Fe 427
  Pb 330
  phytoremediation 311, 315, 317, 320, 330,
    333, 335
Phytochelatins 317, 328, 332, 422–3
Phytoextraction 6, 315–17, 333–4, 336
  As 331–3, 336
  Au 329, 560
  B 327–8
  Cd and Zn 423
  Co 327
  Hg 328
  Ni 336, 337
  Pb 329–31, 334, 336
  phytormediation 311–17, 320–5, 333–4,
    337–8
  Se 325–7, 331
  Tl 542
  wildlife 336–7
Phytomining 311, 314, 317–21, 327, 329,
    336–8
  Ni 317–21, 329, 336, 338
Phytorcmcdiation 6, 311–39, 462, 493, 562
Phytosiderophores 187, 188, 323, 427
Phytostabilization 334–5, 354
  phytoremediation 311, 314–15, 329,
    333–5, 338
Phytotoxicity 141, 143, 427–9
  As 384
  Cd 409, 418, 422–3
  Co 474
  Cr 472
  Ni 472
  phytoremediation 317, 322, 327, 331,
    334–5, 338
  soil amendment 356, 363–4, 370
  Tl 542
  Zn 409–10, 413, 418, 422–3, 427–9
Phytovolatization 311–12, 325–7, 337
  Hg 328–9

Plant uptake 28–31, 141, 162, 177–9, 248–9, 335
  Ag 541
  As 142
  Au 560–1
  B 184, 186
  bioavailability 230–1, 233, 235–6, 238–40, 243–5, 247–9, 253–5
  bioavailability and risk assessment 268, 272, 273, 288
  biosolids 114, 116–21, 122–5, 126–9
  Cd 143–4, 409, 411, 413–20, 422–6, 427–9
  Co 470, 474
  Cr 467, 470–2, 474
  Cu 187, 446, 450, 574
  Fe 188
  Hg 500, 507
  immobilization 354, 356, 359–60, 367
  Mn 189, 489, 490
  Mo 541
  Ni 413, 474
  Pb 146, 574
  PGE 568, 573–4
  regulatory limits 296, 300
  Se 190, 490–1, 493
  soil chemistry 10, 13, 15, 28–31
  Tl 542
  U 147, 562
  V 543
  Zn 191, 409, 411, 413–20, 422–6, 427–9
Platinum (Pt) 3, 7, 444, 555, 567–75
Platinum group elements (PGE) 444, 567–75
Poppy seed 421, 426
Pollution-induced community tolerance (PICT) 161–2
Pollution swapping 305
Polydimethylsiloxane (PDMS) 87
Potassium (K) 60, 63, 114, 245, 356, 360
  ash 361, 363
  deficiency 176, 186, 193
  fertilizers 140, 146–7
  Pb 447
  phytoremediation 323, 333
  Tl 532
Potatoes 13, 221, 574
Powellite 520
Prairie dogs 326
Preanalysis treatment 40–1, 45–7, 48–9, 69
Precipitation 14, 18–19
Predicted no-effect concentrations (PNECs) 561
Prediction models 100–2
Probabilistic models 300–1
Protection level 294–5
Proteomics 167
Protons 207–17, 218–19, 243
Provisional tolerable weekly intake (PTWI) 143
Pseudomonas fluorescens 253

Pteris vitatta 314, 316, 331–3
Pyrite 21
Pyrolusite 483
Pyrolysis 336
Pyromorphites 356, 364–5, 442
Pyruvate 384

Quantitative risk assessment (QRA) 287–8, 299
Quartering 41, 45

Radioactivity 551–2, 554–5
Radiochemical neutron activation analysis (RNAA) 59
Radioisotopes 461–3, 481
Radionuclides 59–60, 135–6, 333, 336, 359, 561
Radishes 121, 244, 473, 574
Rape (*Brassica nupis*) 542
Raphanus sativus 118
Red fescue (*Festuca rubra*) 335
Red root pigweed (*Amaranthus retroflexus*) 334
Redox potential 19–23, 120, 201, 219
  Ag 527–8
  As 385, 387–8, 396
  Au 557
  bioavailability 230, 255
  Co 468–70, 475
  Cr 466, 475
  Cu 441
  deficiency in trace elements 182–3, 186, 188–90
  Hg 506
  Ni 462, 468
  PGE 573
  Sb 395–6
  Se 489
  Sn 505
  soil amendment 355, 358, 367
  soil solution 81–3, 100–1
  Tl 533, 539–40
Reference dose (RfD) 282
Regulations 5–6, 156–7, 287
  biodegradable waste 366–7
  biosolids 114–5
  derivation 293–4, 296–301
  Hg 498, 504
  limits 293–305
Residual fraction 65, 68
Reversed-phase high-performance liquid chromatography (RP-HPLC) 88–9
Reversed-phase ion-pair chromatography (RP-IPC) 88–9
Rhizobium 6
Rhizofiltration 334
Rhizosphere 28–9, 63–4, 122–5, 142
  bioavailability 232, 238, 243, 254–5
  biosolids 119, 121, 122–5, 127–9

deficiency 175, 180, 187–8, 191
PGE 574
phytoremediation 320, 325, 328, 332
soil solution 81–3
Rhodium (Rh) 3, 7, 567–8, 570–4
Rice 5, 250
    Cd 10, 409–10, 415, 417, 419, 421–6
    Mn 484, 491
    phytoremediation 321, 323–5, 329, 331, 333, 336
    Zn 423–5, 427
Ricinus communis 474
Riffling 41
Risk characterization 268–9, 284–7
    errors 284–5
Rodenticides 516, 529
Root exudates 122–7, 129
Root zone *see* Rhizosphere
Rubber 411
Rubidium (Rb) 532
Ruthenium (Ru) 567–8
Ryegrass (*Lolium multiflorum*) 29, 273

Samarium (Sm) 60
Sample preparation 82–3, 103
Seafood 9, 12, 142, 160
    shell fish 9, 421, 425–6
SEC-ICP-MS 85–6
Sedimentation field-flow fractionation (SdFFF) 84–5
Sedum alfredii 324–5
Selenates 325–6, 484–5, 486–7, 490–1, 493
Selenides 484–5, 488, 491, 493
Selenite 484–5, 486–7, 490–1, 493
Selenium (Se) 3–5, 7, 190, 213, 365, 482–93
    Ag 523
    analysis and fractionation 53–4
    bioavailability 485, 489–90, 491
    biosolids 115
    chemical behaviour in soil 484–90
    concentrations and sources 483–4
    Cu 444
    deficiency 5, 176–7, 181, 190, 193, 482–3, 491–3
    fertilizers 136, 482–4, 487
    PGE 569
    phytoremediation 311–12, 315–16, 325–7, 328, 331
    risks 490–3
    soil chemistry 12, 20, 23
    soil solution 87–8
    toxicity for soil dwellers 156
    wildlife 336–8, 482, 493
Selenium dioxide 482
Selenosulfides 493
Sensitivity of soil use 294–5, 301

Separation techniques 81
Sequential extraction procedures (SEPs) 238–40, 245
Sequential extraction 65–71
sesquioxides 183, 192
Sewage sludge 5–7, 9, 64, 113–29
    Ag 516, 524, 541
    Au 553, 555
    bioavailability 237–40, 253, 287
    Co 465
    Cr 464
    Cu 445
    Hg 505
    Mn 482–3
    Mo 518
    Ni 464, 468
    regulatory limits 294–6, 298, 301, 304–5
    Se 483–4
    Sn 503
    soil amendment 354, 366–7, 371
    Zn 429
Shell fish 9, 421, 425–6
Shewanella sp. 355, 539
Siderophiles 568
Silene vulgaris 244, 249
Silica (Si) 20, 55, 208, 359–61, 534
Silicates 54–5, 65, 68, 117, 355
    Cu 443, 444, 446
    Mn 485, 490
    Ni 468
    Pb 447, 448
    phytoremediation 312, 320
    Se 486
    soil chemistry 18, 23
    V 534
Silicon (Si) 20, 55, 208, 274, 359–61, 497, 534
Silver (Ag) 3, 7, 13, 20, 222, 515–16
    Cd and Zn 410
    chemical behaviour in soil 524–8
    Cu 444, 525
    geochemical occurrences 523
    Hg 502
    Mo 519
    risks 541–2
    Sb 386, 523
    sources of soil contamination 523–4
    Tl 531
Single chemical extraction procedures 234–8
Single extraction 61–6, 68
Size-exclusion chromatography (SEC) 71, 83, 85–6, 88–9
Smectites 214–15
Snails 158, 160
Sodium (Na) 59, 60, 63, 360–1, 485
    deficiency 176, 184, 186
Sodium borohydride 87

Soil constituents 23–5
Soil sampling 39–49, 54
 errors 41–3, 45, 48–9
 scale 48–9
Soil solution 10–13, 16, 20–1, 28–30,
  81–103, 240–4
 bioavailability 229–30, 231–3, 235, 240–4,
  246–9, 255
 biosolids 118, 120, 122–3
 fractionation 61–2, 64
 immobilization 353
 Mn 484, 487
 Mo 519
 Se 488
Soil stabilization 6, 353–4, 369, 371
Solid-phase extraction (SPE) 86–7
Solid-phase microextraction (SPME) 87
Solid phase soil 428
 bioavailability 229–30, 231–3, 238, 240,
  242–3, 245–9
 biosolids 120, 122
 chemistry 10–13, 16, 23, 25, 28, 30
 fractionation 54–5, 59, 61, 63, 74, 81,
  92–3
Sorption 13–17
Soybean 118, 324, 421, 428
Speciation 6, 7, 11–12, 27, 163, 183, 240–4
 Ag 524–6
 As 385, 387–8, 389, 398–9, 400
 bioavailability 230–3, 234–5, 238,
  240–4, 247
 Cd and Zn 422
 chemical modelling 199–222
 Cr 467
 equipment 73–4
 fractionation 53–4, 56, 58–9, 65, 69–70,
  72, 74
 immobilization 363
 Mn 484, 490
 Mo 520
 model names 200
 phytoremediation 333
 Sb 393–5, 398–9, 400
 Se 484, 490
 soil solution 81–103
 Tl 530–1
 V 535–7
Species-component matrix 203–4
Species sensitivity distributions (SSDs) 157,
  163, 295, 297
Sphalerite 415, 416, 528
Spiders 159, 160, 162
Spinach (*Spinacea oleracea*) 29–30, 422,
  425–6, 574
 bioavailability 236–7
Springtails 156, 158–9, 164, 166–7
Square wave anodic stripping voltammetry
  (SWASV) 96, 99

Stanleya pinnata 326
Stern model 201
Stibnite 384, 386, 395
Stinging nettle (*Urtica dioica*) 30–1
Stockholm humic acid model 220
Stratification of soil 43, 48
Stripping chronopotentiometry (SCP) 97
Strontium (Sr) 23, 60, 93, 359
 fertilizers 140, 147
Succinate 384
Sudden infant death syndrome 384–5
Sugar cane 491
Sulfates 13, 20–1, 189, 325–6, 355–6
 Ag 526
 As 392
 Cu 441, 449
 Pb 442, 448
 Se 487
 Tl 259, 531
 toxic element behaviour 199, 204–5
 U 559
Sulfides 55, 65, 68, 83
 Ag 525–7, 531
 As 387
 Au 558
 Cd 415, 424
 Cu 441, 444
 Hg 506
 Mo 516–17
 Ni 462
 Pb 442, 447
 Sb 393
 Se 482
 Sn 500
 soil chemistry 12–13, 18, 20–1
 Tl 529, 531, 533, 542
 U 554
 Zn 415, 424
Sulfur (S) 21, 69, 164, 323, 361
 Ag 523–6, 528, 541
 As 387
 Cd and Zn 415–16
 Cu 444, 449
 deficiency 176, 187, 190
 Pb 442, 447–8
 PGE 569
 Sb 385, 386
 Se 482
Sunflowers (*Helianthus annuus*) 142, 250, 542
 Cd and Zn 417, 419, 421, 425–6
Supercritical fluid extraction (SFE) 72–3, 86
Superphosphate 137, 140, 146–7
Surface complexation models 207–19, 362
 toxic element behaviour 199–201, 203,
  207–19, 221
Swiss chard (*Beta vulgaris*) 412
Synchrotron X-ray microfluorescence
  spectroscopy 48

Tantalum (Ta) 60
Tea-tree (*Melaleuca*) 560
Tellurium (Te) 87, 482, 523, 569
Tension lysimeters 82
Thallium (Tl) 3, 7, 222, 316, 515–17, 527
  biosolids 116, 529
  chemical behaviour in soil 529–33
  geochemical occurrences 528–9
  immobilization 360, 369–70
  phtyoextraction 333, 336
  risks 516–17, 542
  sources of contamination 529
Thermoelectric atomic absorption spectrometry (ET-AAS) 55–6
Thin film mercury electrode (TFME) 96
Thiocyanate 558, 560–1
Thiomolybdate 450
Thiobacillus ferrooxidans 355
Thlaspi caerulescens 6, 313–14, 316, 321–5, 337, 423
Thorium (Th) 60, 140, 147, 359
Threshold toxicity values 294–5
Tin (Sn) 3, 7, 56, 243, 497–500
  alloy with Pb 442
  ash 361, 363
  chemical behaviour in soil 505
  geogenic occurrence 500–1
  isotopes 498
  phytoremediation 315, 333
  risks 499, 506–7
  soil solution 87–8
  sources of contamination 502–3
Titanium (Ti) 60, 274, 485, 490
Tobacco (*Nicotiana* sp.) 122, 321, 426
  Cd 419, 421–3, 425–6
Tolerable daily intake (TDI) 299–300
Tolerance index (TI) 28
Tomatoes 221, 472–4, 541
Toxicogenomics 167
Toxicokinetics 163–6
Tracer studies 274–5, 278, 280
Transcriptomics 167
Transfer factor (TF) 28
Triple layer model (TLM) 202, 208
Triple superphosphate (TSP) 364–5
Tungsten (W) 334, 365
Two-compartment model 164

Unbuffered neutral salt solutions 234–8, 244, 255
Unbuffered salt solutions 62–3, 65
Ultrafiltration (UF) 71, 83–4
Uncertainty factors 282

Uranium (U) 3, 7, 147, 334, 534, 551–2, 562
  chemical behaviour in soil 558–60
  fertilizers 136, 140, 147, 556–7
  geogenic occurrence 554–5
  risks 561–2
  soil chemistry 13, 20
  soil contamination 556–7
Urease 161

Vacuum filtration 82–3
Vanadinite 534
Vanadium (V) 3, 7, 20, 60, 222, 515–17
  ash 361
  chemical behaviour in soil 535–40
  Cu 444
  fertilizers 136, 534
  geochemical occurrence 534
  regulatory limits 297
  risks 517, 542–3
  sources of contamination 534
Vermiculite 334, 445, 468, 486, 532
  toxic element behaviour 207, 214–15
Vineyard 48–9
Vitamin B_{12} 186
Voltammetry 96–8, 241
Volatilization 506

Wageningen Donnan membrane technique (WDMT) 95–6
Water-soluble fraction 65–7
WHAM 200, 205, 217, 219–20, 242
Wheat 6, 142, 357, 473
  bioavailability 236–8, 249
  biosolids 117–18, 121
  Cd and Zn 419, 421, 426
  trace element deficiency 186, 191, 193–4
Wildlife 177, 179, 334–5, 336–7, 338
  Hg 508
  Se 336–8, 482, 493
  Zn 337, 429
Willow 312–13, 336
Woodlice 157, 161, 166
Wulfenite 517

X-ray absorption fine structure (XAFS) spectrometry 73–4
X-ray absorption near-edge structure (XANES) spectroscopy 73–4, 561
X-ray absorption spectroscopy (XAS) 73, 245, 389, 415–16
X-ray based spectroscopic techniques 245
X-ray fluorescence spectroscopy (XRF) 60–1, 245, 468

Yttrium (Y) 55, 274

Zeolites 359–61, 370
Zinc (Zn) 3–7, 26–7, 42, 45–6, 191–2,
    409–29
  Ag 523, 525
  analysis and fractionation 59–60, 64, 65
  As 398
  ash 361, 363, 370
  bioacessibility 250
  bioavailability 235–8, 241, 244–5, 249,
    252–3
  bioavailability and risk 273, 287
  biosolids 115–16, 118–19, 122, 125–7, 410,
    414–15, 416–18
  Cu 410, 416, 428–9, 441, 444, 449
  deficiency 5, 176–8, 180–9, 191–4, 425, 427
  fertilizers 139–41, 143–4, 147
  food chain 427–9
  geogenic occurrence 409–15
  Hg 497, 502, 504
  immobilization 355–7, 359–61, 370
  Mn 482
  Ni 413
  organic matter 366–7, 370, 417, 429
  PGE 574
  phosphates 364–5, 370
  phytoremediation 312–14, 316–17, 319,
    321–5, 330, 338
  phytostabilization 334–5
  regulatory limits 302
  rice 423–5
  Sb 398
  shellfish 421
  Sn 503
  soil chemistry 11–13, 16, 18–20, 23–7,
    29–31
  soil solution 102
  Tl 528–9, 542
  toxic element behaviour 210, 215–16, 219
  toxicity for soil dwellers 156–7, 159,
    162–6
  wildlife 337, 429
Zinc oxide 146
Zinc sulfate 146
Zirconium (Zr) 55, 274